# Perspectives in Neural Computing

W0043861

Springer-Verlag London Ltd.

*Also in this series:*

J.G. Taylor
The Promise of Neural Networks
3-540-19773-7

Maria Marinaro and Roberto Tagliaferri (Eds)
Neural Nets - WIRN VIETRI-96
3-540-76099-7

Adrian Shepherd
Second-Order Methods for Neural Networks: Fast and Reliable Training Methods for Multi-Layer Perceptrons
3-540-76100-4

Jason Kingdon
Intelligent Systems and Financial Forecasting
3-540-76098-9

Dimitris C. Dracopoulos
Evolutionary Learning Algorithms for Neural Adaptive Control
3-540-76161-6

M. Kárný, K. Warwick and V. Kůrková (Eds)
Dealing with Complexity: A Neural Networks Approach
3-540-76160-8

John A. Bullinaria, David W. Glasspool and George Houghton (Eds)
4th Neural Computation and Psychology Workshop, London,
9-11 April 1997: Connectionist Representations
3-540-76208-6

Maria Marinaro and Roberto Tagliaferri (Eds)
Neural Nets - WIRN VIETRI-97
3-540-76157-8

L.J. Landau and J.G. Taylor (Eds)
Concepts for Neural Networks: A Survey
3-540-76163-2

Thomas Lindblad and Jason M. Kinser
Image Processing using Pulse-Coupled Neural Networks
3-540-76264-7

L. Niklasson, M. Bodén and T. Ziemke (Eds)

# ICANN 98

**Proceedings of the 8th International Conference on Artificial Neural Networks, Skövde, Sweden, 2-4 September 1998**

## Volume 2

 Springer

Lars Niklasson, PhD
Mikael Bodén, PhD
Tom Ziemke, MSc
Department of Computer Science, University of Skövde, PO Box 408,
54128 Skövde, Sweden

*Series Editor*
J.G. Taylor, BA, BSc, MA, PhD, FInstP
Centre for Neural Networks, Department of Mathematics, King's College,
Strand, London WC2R 2LS, UK

ISBN 978-3-540-76263-8    ISBN 978-1-4471-1599-1 (eBook)
DOI 10.1007/978-1-4471-1599-1

British Library Cataloguing in Publication Data
A catalogue record for this book is available from the British Library

Library of Congress Cataloging-in-Publication Data
International Conference on Artificial Neural Networks (European
  Neural Network Society)   (8th : 1998 : Skövde, Sweden)
     ICANN 98 : proceedings of the 8th International Conference on
  Artificial Neural Networks, Skövde, Sweden, 2-4 September 1998 / L.
  Niklasson, M. Bodén, and T. Ziemke, eds.
        p.      cm. -- (Perspectives in neural computing)
     Includes bibliographical references (p. ).
     1. Neural networks (Computer science)--Congresses.    I. Niklasson,
  Lars F.   II. Bodén, Mikael B.   III. Ziemke, T. (Tom), 1969-
  IV. Series.
  QA76.87.I56    1998
  006.3'2--dc21                                        98-28433

Typesetting: Camera-ready by contributors
34/3830-543210 Printed on acid-free paper

# Contents, Volume 2: Poster Presentations

## Poster Presentations: Theory

### Spotlight Presentations: Theory I: Algorithms

### Spotlight Presentations: Theory II: Dynamical Systems, Time Series

### Spotlight Presentation: Theory III: Signal Decomposition Methods

### Poster Session I: Theory

**Poster Session II: Theory**

## Poster Presentations: Applications

### Spotlight Presentations: Applications I: Process Control, Diagnosis

### Spotlight Presentations: Applications II: Image Processing

## Poster Presentations: Computational Neuroscience and Brain Theory

### Spotlight Presentations: Computational Neuroscience and Brain Theory (1)

### Spotlight Presentations: Computational Neuroscience and Brain Theory (2)

# Poster Presentations: Connectionist Cognitive Science and Artificial Intelligence

### Spotlight Presentation: Connectionist Cognitive Science and Artificial Intelligence

### Poster Session I: Connectionist Cognitive Science and Artificial Intelligence

# Poster Presentations: Autonomous Robotics and Adaptive Behavior

### Spotlight Presentations: Autonomous Robotics and Adaptive Behavior (1)

### Poster Session II: Autonomous Robotics and Adaptive Behavior

## Poster Presentations: Hardware and Implementations

### Spotlight Presentations: Hardware and Implementations

# Poster Presentations:
# Theory

# A Study on Functional–Link Neural Units with Maximum Entropy Response

Simone Fiori and Francesco Piazza *

Dept. Electronics and Automatics – University of Ancona, Italy
E-mail: simone@eealab.unian.it

**Abstract.** In this paper a preliminary study on an unsupervised learning theory for functional-link neural units is presented. A set of learning equations for this neural topology based on an information-theoretic approach is given. Then, learning and approximation capabilities are investigated through computer simulations concerning density shaping.

## 1  Introduction

Over the recent years, information-theoretic optimization by neural networks has become an important research field. Since the pioneering work of Linsker and Plumbley (see for instance [4,7] and references therein), several authors were involved in the study of the unsupervised neural learning driven by entropy-based criteria, with current applications to blind separation of sources [1,2,11], linear estimation and time-series prediction [9], probability density shaping [1,6,11–13], and unsupervised classification [10]. Moreover, the same techniques have been proven to be effective in other contexts, as in computer applications, like uniform hash map searching [3,5].

Particularly, the aim of the different techniques for neural density shaping is to find a non-linear transformation of an input random process (with unknown statistics) that maximizes the entropy of the transformed process. Then, the first derivative of the found function approximates the probability density distribution of the input random process [9,10], with a degree of accuracy depending on the structure of the neural network. Usually, the neural topologies used in the literature involve semi-linear neural units, that is, linear combiners followed by static sigmoidal non-linearities.

In this paper, a preliminary study concerning the use of more complex and flexible neural units endowed with functional links [8,14] is presented. In order to test for the learning and approximation capabilities of the proposed structure, cases of density shaping are tackled and discussed through computer simulations.

## 2  Learning of maximum entropy connection strengths

In this paper the following input-output description for a functional-link neuron is assumed:

$$y = \Psi(x; a) = \text{sgm}(z) \ , \quad z = \varphi(x; a) \ , \tag{1}$$

---

* This research was partially supported by the Italian MURST.

where sgm$(\cdot)$ is a sigmoidal function, bounded above and below, continuous and strictly increasing; $\varphi(\cdot)$ is a strictly monotonic polynomial depending upon parameters in $a = (a_0, a_1, \ldots, a_n)$. If $x$ is supposed to be a stationary continuous-time random process $x = x(t) \in \mathcal{X}$ with probability density function (pdf) $P_x(x)$, then $y$ will be a random process $y = y(t) \in \mathcal{Y}$ too, with a pdf denoted here with $P_y(y; a)$. The differential entropy of the random process $y(t)$ defined as:

$$H_y(a) \overset{\text{def}}{=} - \int_{\mathcal{Y}} P_y(\eta; a) \log P_y(\eta; a) d\eta \qquad (2)$$

can be related to the differential entropy of the random process $x(t)$ by means of the fundamental formula: $P_y = P_x/|\psi|$ , $\psi(x; a) \overset{\text{def}}{=} \Psi'(x; a)$. Using that substitution in the formula (2), yields:

$$H_y = - \int_{\mathcal{X}} \frac{P_x}{|\psi|} \log\left(\frac{P_x}{|\psi|}\right) |\psi| d\xi = H_x + \int_{\mathcal{X}} P_x \log |\psi| d\xi , \qquad (3)$$

where by definition, $|\psi|$ assumes the expression:

$$|\psi(x; a)| = \text{sgm}'[\varphi(x; a)] \left| \frac{d\varphi(x; a)}{dx} \right| . \qquad (4)$$

The entropy $H_y$ depends upon coefficients $a_k$ $(k = 0, 1, 2, \ldots, n)$ by means of $\psi(x; a)$, thus:

$$\frac{\partial H_y(a)}{\partial a_k} = \int_{\mathcal{X}} \frac{P_x(\xi)}{|\psi(\xi; a)|} \frac{\partial |\psi(\xi; a)|}{\partial a_k} d\xi . \qquad (5)$$

By definition of $\psi$, it is easy to see that the partial derivatives involved in the integral (5) read:

$$\frac{\partial |\psi|}{\partial a_k} = \text{sgm}''[\varphi] \frac{\partial \varphi}{\partial a_k} \left| \frac{d\varphi}{dx} \right| + \text{sgm}'[\varphi] \frac{\partial}{\partial a_k} \left| \frac{d\varphi}{dx} \right| ,$$

whereby direct calculations lead to:

$$\frac{1}{|\psi|} \frac{\partial |\psi|}{\partial a_k} = \frac{\text{sgm}''[\varphi]}{\text{sgm}'[\varphi]} \frac{\partial \varphi}{\partial a_k} + \frac{\partial}{\partial a_k} \left| \frac{d\varphi}{dx} \right| \cdot \left| \frac{d\varphi}{dx} \right|^{-1} .$$

To find a vector of parameters $a$ maximizing the entropy of the neuron response $y(t; a)$, a set of continuous-time learning equations derived by the gradient steepest ascent method is used here. Following this way, such equations have to be written in the form $da/dt = \partial H_y/\partial a$, that is:

$$\frac{da}{dt} = \int_{\mathcal{X}} P_x(\xi) \left[ \frac{\text{sgm}''[\varphi(\xi; a)]}{\text{sgm}'[\varphi(\xi; a)]} \frac{\partial \varphi(\xi; a)}{\partial a} + \frac{1}{\varphi'(\xi; a)} \frac{\partial \varphi'(\xi; a)}{\partial a} \right] d\xi , \qquad (6)$$

for $k = 0, 1, \ldots, n$. Besides, from equations (3) and (4) the exact expression of the quantity $G \overset{\text{def}}{=} H_y - H_x$, hereafter referred to as "entropy gap", is obtained:

$$G(a) = \int_{\mathcal{X}} P_x(\xi) \log\{\text{sgm}'[\varphi(\xi; a)]|\varphi'(\xi; a)|\} d\xi . \qquad (7)$$

It should be noted that $G(a)$ is not guaranteed to be positive, since it actually represents a difference among entropies. Anyway, as $H_y(a)$ maximizes, $G(a)$ maximizes too, thus *maximizing the response entropy may be conceived as the maximization of the entropy gap between the original and the squashed processes.*

## 3  A simple case-study

In this paragraph the simple case-study concerning the following neural structure is discussed:

$$\text{sgm}(z) = \frac{1}{2} + \frac{1}{2}\text{erf}(z) = \frac{1}{\sqrt{\pi}} \int_0^z \exp(-u^2)du \ , \ \varphi(x;a) = a_0 + a_1 x \ , \quad (8)$$

together with an excitation endowed with a Laplacian distribution $P_x(x) = \frac{\rho}{2}e^{-\rho|x-\mu|}$, where $\rho > 0$. Equation (8) represents the input–output relation of a sigmoidal neuron with one weight and one bias. This is an interesting case in the theory since it is possible to find solutions of the differential system (6) in a closed form. In fact, the relevant quantities involved in the integrals are found to be $\text{sgm}''(z)/\text{sgm}'(z) = -2z$ and $\varphi'(x;a) = a_1$, whereby the others follow. Thus system (6) rewrites, in this case:

$$\frac{da_0}{dt} = -\rho \int_{-\infty}^{+\infty} e^{-\rho|\xi-\mu|}(a_0+a_1\xi)d\xi \ , \ \frac{da_1}{dt} = \frac{1}{a_1} - \rho \int_{-\infty}^{+\infty} e^{-\rho|\xi-\mu|}\xi(a_0+a_1\xi)d\xi \ ,$$

or, after direct calculations:

$$\frac{da_0}{dt} = -2(a_0 + a_1\mu) \ , \ \frac{da_1}{dt} = \frac{1}{a_1} - 2(a_0 + a_1\mu)\mu - 4a_1\rho^{-2} \ . \quad (9)$$

Vanishing the time-derivatives is the fastest way to determine the equilibrium points $(a_0, a_1)^e$ for the above differential system. They are found to be:

$$(a_0, a_1)^e = (+\mu\rho/2, -\rho/2) \text{ and } (-\mu\rho/2, +\rho/2) \ . \quad (10)$$

It would be interesting to give in this simple case the exact expression of the entropy gap (7), that is:

$$G(a_0, a_1) = \log\left(\frac{1}{\sqrt{\pi}}|a_1|\right) - [a_0^2 + 2a_0a_1\mu - (2\rho^{-2} + \mu^2)a_1^2] \ .$$

Note that the entropy gap is invariant with respect to a sign exchange among $a_0$ and $a_1$, coherently with results (10).

## 4  Three more complex examples

It is worth to consider the more complex neural structure described by functions:

$$\text{sgm}(z) = \tanh(z) \ , \ \varphi(x;a) = a_0 + e^{a_1}x + e^{a_3}x^3 + \cdots + e^{a_n}x^n \ , \quad (11)$$

with $n$ being an *odd* integer. Note that, due to the exponential structure of the coefficients and the odd value of the degree of the polynomial, the property $\varphi'(x; a) > 0$ holds true for any value of $x$ (providing that $a_1, a_3, \ldots, a_n > -\infty$). With the structural functions as above, the relevant learning quantities are found to be:

$$\frac{\text{sgm}''(z)}{\text{sgm}'(z)} = -2\tanh(z) = -2y \ , \ \varphi'(x; a) = e^{a_1} + 3e^{a_3}x^2 + \cdots + ne^{a_n}x^{n-1} \ ,$$

and the others follow.

In order to simulate learning equations (6) particularized with functions (11), their instantaneous, stochastic, discrete-time approximations can be used. They are expressed by:

$$a_0(t + 1) = a_0(t) - 2 \cdot \eta_0 \cdot y \ ,$$

$$a_k(t + 1) = a_k(t) + \eta_k \cdot e^{a_k} \cdot \left(\frac{k}{\varphi'} - 2xy\right) x^{k-1} \ ,$$

for $k = 1, 2, \ldots, n$ with $\eta_0, \eta_1, \eta_2, \ldots, \eta_n$ being sufficiently small positive learning rates. Moreover, as learning performance index, an estimation of the entropy gap $H_y - H_x$ may be obtained by averaging over the learning epochs the instantaneous, stochastic approximation of the right hand of the expression (7):

$$\text{entropy gap } G \sim \ \log[(1 - y^2)\varphi'] \ .$$

It is known that the quantity $\psi(x; a)$ plays a central role in density shaping by neural units [9,11], since it approximates the pdf of the excitation $x(t)$ (with a degree of accuracy related to sgm$(\cdot)$ and $\varphi(\cdot, \cdot)$ both in the stationary and in the non-stationary case). To test for the probability density shaping capability of the proposed neural structure, three experiments are considered in what follows.

First, an excitation with probability distribution function:

$$P_x(x) = \frac{1}{2\sqrt{2\pi}\sigma} \left\{ \exp\left[-\frac{(x - \mu_1)^2}{2\sigma^2}\right] + \exp\left[-\frac{(x - \mu_2)^2}{2\sigma^2}\right] \right\}$$

was presented to the functional link neuron (11) with $n = 3$. Simulation results are shown in Figure 1. Each epoch counts 60 input samples; a total of 50 epochs (corresponding to 3000 samples) was used. Two more experiments were performed with a polynomial $\varphi(x; a)$ with degree $n = 7$. The first one was performed with a real-world (musical) excitation. Results with batches counting 50 samples and 50 epochs are shown in Figure 2. The Figure 3 shows simulation results for a sinusoidal excitation with the pdf $P_x(x) = 1/(\pi\sqrt{1 - x^2})$. There each epoch counts 100 samples for a total of 40 epochs. In all cases the neural unit endowed with functional links, with a relatively small number of parameters to be learnt, seems to possess interesting approximation capability. However, in the third case presented here, the neuron seems to show the worst performance. Likely, this is caused by the rapid growing of the true pdf near the edges $x = \pm 1$, since $\psi(x; a)$ hard-limits near those points because of the very small value of $\tanh'(z)$. This shows that the architecture exhibits an approximation capability well-delimited over the input range.

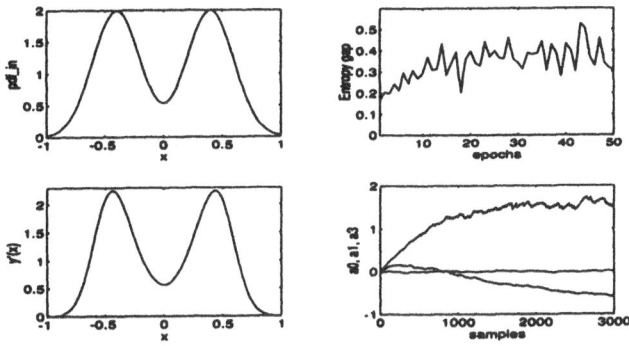

**Fig. 1.** Entropy maximization for a two-overlapping-Gaussian excitation.

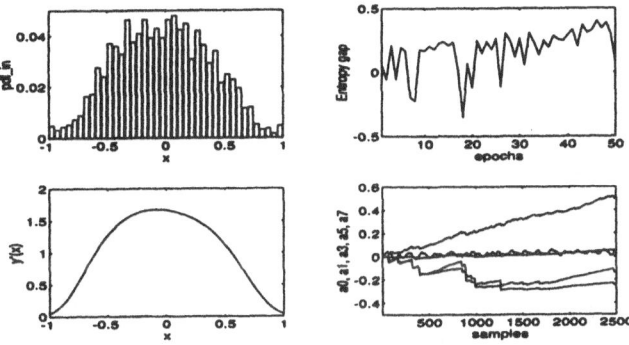

**Fig. 2.** Entropy maximization for a real-world excitation.

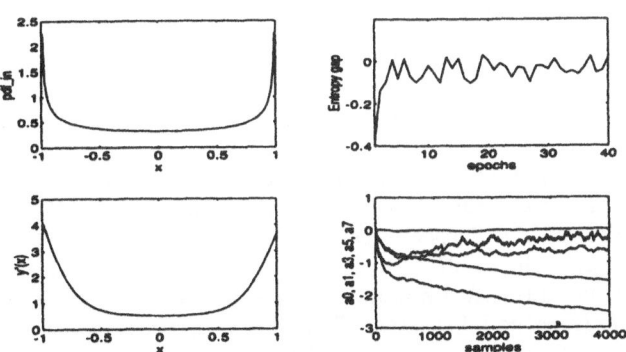

**Fig. 3.** Entropy maximization for a sinusoidal excitation.

# 5 Conclusion

In this work, a new learning theory for functional-link neural units, based on an information-theoretic approach, is presented. Then, learning and approximation capabilities shown by different units were investigated through density shaping problem. Simulation results confirm the effectiveness of the proposed learning theory and the good flexibility exhibited by the non-semi-linear topologies. About the approximation capability of the functional-link neural units, analytical results in connection with regularization and RBF theories are currently under investigation. Applications of the proposed neural architecture for density shaping to estimation, identification and time-series analysis are expected.

# References

1. S.-I. AMARI, T.-P. CHEN AND A. CHICOCKI, *Stability Analysis of Learning Algorithms for Blind Source Separation*, Neural Networks, Vol. 10, No. 8, pp, 1345 – 1351, 1997
2. A.J. BELL AND T.J. SEJNOWSKI, *An Information Maximisation Approach to Blind Separation and Blind Deconvolution*, Neural Computation, Vol. 7, No. 6, pp. 1129 – 1159, 1996
3. S. FIORI, P. BUCCIARELLI AND F. PIAZZA, *Blind Signal Flatting Using Warping Neural Modules*, Proc. Int. Joint Conf. on Neural Networks (IJCNN), Vol. 2, pp. 2312 – 2317, 1998
4. R. LINSKER, *Local Synaptic Rules Suffice to Maximize Mutual Information in a Linear Network*, Neural Computation, Vol. 4, pp. 691 – 702, 1992
5. B.S. MAJEWSKI, N.C. WORMALD, G. HAVAS AND Z.J. CZECH, *A Family of Perfect Hashing Methods*, Computer Journal, Vol. 39, pp. 547 – 554, 1996
6. P. MORELAND, *Mixture of Experts Estimate A-Posteriori Probabilities*, Artificial Neural Networks, pp. 499 – 505, Springer-Verlag, 1997
7. M.D. PLUMBLEY, *Efficient Information Transfer and Anti-Hebbian Neural Networks*, Neural Networks, Vol. 6, pp. 823 – 833, 1993
8. Y.-H. PAO, *Adaptive Pattern Recognition and Neural Networks*, Addison-Wesley Publishing Company, 1989 (Chpt. 8)
9. Z. ROTH AND Y. BARAM, *Multidimensional Density Shaping by Sigmoids*, IEEE Trans. on Neural Networks, Vol. 7, No. 5, pp. 1291 – 1298, Sept. 1996
10. A. SUDJIANTO AND M.H. HASSOUN, *Nonlinear Hebbian Rule: A Statistical Interpretation*, Proc. Int. Conf. on Neural Networks (ICNN), Vol. 2, pp. 1247 – 1252, 1994
11. A. TALEB AND C. JUTTEN, *Entropy Optimization - Application to Source Separation*, Artificial Neural Networks, pp. 529 – 534, Springer-Verlag, 1997
12. V. VAPNIK, *The Support Vector Method*, Artificial Neural Networks, pp. 263 – 271, Springer-Verlag, 1997
13. Y. YANG AND A.R. BARRON, *An Asymptotic Property of Model Selection Criteria*, IEEE Trans. on Information Theory, Vol. 44, No. 1, pp. 95 – 116, Jan. 1998
14. S.M. ZURADA, *Introduction to Neural Artificial Systems*, West Publishing Company, 1992

# Mean Field Theory based on Belief Networks for Approximate Inference

Wim Wiegerinck*†and David Barber‡
Stichting Neurale Netwerken, University of Nijmegen
Nijmegen, The Netherlands

## Abstract

Exact inference in large, densely connected belief networks is computationally intractable, and approximate schemes are therefore of great
importance. In the context of approximate inference in sigmoid belief
networks, mean field theory has received much interest. In this method
the exact log-likelihood is bounded from below using a mean field approximating distribution. In the standard mean field theory, the approximating distribution is assumed to be factorial. In this paper we propose
to use a (tractable) belief network as an approximating distribution. We
show that belief networks fit very well into mean field theory, and no
additional bounds are required. We derive mean field equations which
provide an efficient iterative algorithm to optimize the parameters of the
approximating distribution. Simulation results on an inference problem
indicates a considerable improvement over existing mean field methods.

## 1  Introduction

Belief networks provide a rich framework for probabilistic modeling and reasoning [1]. Due to the directed graphical structure in these models, exact inference
requires only local computations, in contrast to undirected approximations. In
practice, this means that networks of reasonable size are tractable, as long as
the 'neighborhoods' are small. However, the complexity of inference scales exponentially with the size of the neighborhoods and, as a result, large densely
connected networks can only be handled with approximate methods. As the
size of conditional probability tables also scales with the size of the neighborhoods, it is convenient to parametrize large models in a compact way, e.g. as
noisy-OR networks[1] or sigmoid belief networks[2].

In mean field approximations of large networks this compact parametrization is exploited[3]. However, we will show that the graphical structure of
the model can be pushed much further to provide a more accurate, bounded
approximation for inference, without incurring much more computational overhead.

The paper is organized as follows. In section 2 we review the standard mean
field theory using factorized models to approximate sigmoid belief networks, as

---

*http://www.mbfys.kun.nl/~wimw

†This research is supported by the Technology Foundation STW, applied science division
of NWO and the technology programme of the Ministry of Economic Affairs

‡http://www.mbfys.kun.nl/~davidb Supported by the Real World Computing Project.

proposed by [3]. In section 3 we show how this theory can be extended in a natural way using belief networks. In section 4 we apply the method on a toy problem from [3, 4].

## 2 Mean Field Theory

Given a probability model $P(S)$, we wish to compute the likelihood $P(S_V)$ that the set of visible variables $V \equiv v_1, \ldots, v_N$ is in state $S_V \equiv S_{v_1}, \ldots S_{v_N}$. This involves the summation over exponentially many states of the remaining (hidden) variables $H$, $P(S_V) = \sum_{\{S_H\}} P(S_V, S_H)$. In a sigmoid belief network [2] with binary units ($S_i = 0/1$) the probability that the variable $S_i = 1$ is

$$P(S_i = 1 | \mathrm{pa}(S_i)) = \sigma \left( \sum_j J_{ij} S_j + h_i \right) \qquad (1)$$

where $\sigma(z) \equiv (1 + e^{-z})^{-1}$. The parents of $S_i$ are denoted $\mathrm{pa}(S_i)$; the biases are $h_j$ and the weights are $J_{ij}$ such that $J_{ij} = 0$ for $S_j \notin \mathrm{pa}(S_i)$. Mean field theory for such networks is based on the following lower bound[1] on the log likelihood [3] for any approximating distribution $Q(S_H)$ and parameters, $\xi_i \in \mathbb{R}$,

$$\ln P(S_V) \geq \mathcal{F}_V[Q, \xi] = \sum_{ij} J_{ij} \langle S_i S_j \rangle_Q + \sum_i (h_i - \sum_j \xi_j J_{ji}) \langle S_i \rangle_Q - \sum_i h_i \xi_i$$
$$- \sum_i \ln \left\langle e^{-\xi_i z_i} + e^{(1-\xi_i) z_i} \right\rangle_Q - \sum_{\{S_H\}} Q(S_H) \ln Q(S_H) \qquad (2)$$

where $z_i = \sum_j J_{ij} S_j + h_i$ and $\langle \cdot \rangle_Q$ is the average with respect to the mean field distribution $Q(S_H)$. Since this inequality holds for any $Q$ and $\xi$, one can make the bound as tight as possible by optimizing $\mathcal{F}_V[Q, \xi]$ with respect to $Q$ and $\xi$.

### 2.1 Factorized models

To make the bound $\mathcal{F}_V[Q, \xi]$ tractable, standard mean field theory restricts itself to factorized models

$$Q(S_H) = \prod_{i \in H} Q(S_i) \qquad (3)$$

Using the parametrization, $q_i \equiv Q(S_i = 1)$, all terms of (2) are easy to compute, e.g. $\langle S_i \rangle_Q = q_i$, $\langle S_i S_j \rangle_Q = q_i q_j$, and $\left\langle e^{-\xi_i z_i} \right\rangle_Q = e^{-\xi_i h_i} \prod_j (1 - q_j + q_j e^{-\xi_i J_{ij}})$. The entropy factorizes nicely into a tractable sum of entropies per site,

$$\sum_{\{S_H\}} Q(S_H) \ln Q(S_H) = \sum_i q_i \ln q_i + (1 - q_i) \ln(1 - q_i) \qquad (4)$$

---

[1] This bound results from the Kullback-Leibler divergence between the true distribution and an approximating distribution on the hidden units.

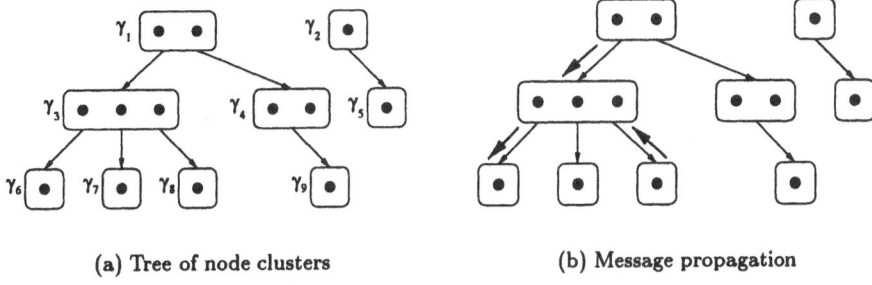

| (a) Tree of node clusters | (b) Message propagation |

Figure 1: Graphical tree structure of the approximating distributions $Q$.

A fast, two step iterative procedure to optimize $\mathcal{F}_V[Q, \xi]$ with respect to $Q$ and $\xi$, is the following [3]. First the gradient with respect to the $q_i$'s is set equal to zero, which yields the mean field equations

$$q_i = \sigma \left( h_i + \sum_j [J_{ij} q_j + J_{ji}(q_j - \xi_j) + K_{ji}] \right) \tag{5}$$

where $J_{ij}^s = J_{ij} + J_{ji}$, and

$$K_{ij} = -\frac{\partial}{\partial q_j} \ln \left\langle e^{-\xi_i z_i} + e^{(1-\xi_i) z_i} \right\rangle_Q . \tag{6}$$

The $q_i$'s are optimized by iteration of the mean field equations (5) while the $\xi_i$'s remain fixed. In the second step, the $\xi_i$'s are optimized using (2) while the $q_i$'s remain fixed. Note that the optimization of each $\xi_i$ can be performed by one-dimensional optimizations.

Recently [5, 4] have proposed to improve the bound (2) using a mixture model. Unfortunately, the entropy term $\sum_{\{S_H\}} Q(S_H) \ln Q(S_H)$ is not tractable for mixtures, and an additional bound is needed. In the following section we propose an alternative class of models which generalize the standard mean field theory straightforwardly, without requiring any additional bound.

## 3 Mean Field Theory Using Belief Networks

We show here how mean field theory can be extended in a natural way by using tractable belief networks for the approximating distribution $Q$. For convenience we restrict ourselves to trees of clusters of nodes[2],

$$Q(S_H) = \prod_\gamma Q(S_\gamma | S_{\mathrm{pa}\gamma}) \tag{7}$$

---

[2]This does not preclude disconnected branches - see fig. 1(a).

where disjoint clusters $\gamma \subset H$ form a partition of $H$. The clusters $\{\gamma\}$ are ordered $\gamma_1 < \gamma_2 < \ldots \gamma_k$ and pa$\gamma$ is either empty, or contained in *one* of the predecessors of $\gamma$(see fig. 1(a)). The tractability of these models is determined by the size of the $\gamma$'s, which we assume to be small. For each cluster $\gamma$, the conditional distributions in (7) contain $(2^{|\gamma|} - 1) \times 2^{|pa\gamma|}$ parameters.

## 3.1 Computing the mean field bound

We now show that $\mathcal{F}_V[Q, \xi]$ in (2) is tractable and computable by *local* computations and simple message propagations (we refer the reader to standard texts, such as [1]). First of all, the marginal probability distributions on $\gamma$, can be computed using the recursion

$$Q(S_\gamma) = \sum_{S_{pa\gamma}} Q(S_\gamma | S_{pa\gamma}) Q(S_{pa\gamma}) \tag{8}$$

The terms $\langle S_i \rangle_Q$ with $i \in \gamma$, and $\langle S_i S_j \rangle_Q$ with $i, j \in \gamma$ can be computed by summation of $Q(S_\gamma)$ over all states $S_\gamma$ with $S_j = 1$ and $S_i = 1, S_j = 1$ respectively. To compute terms of the form $\langle S_i S_j \rangle_Q$ for which $i \in \gamma_i$ and $j \in \gamma_j$, with $\gamma_i \neq \gamma_j$ having a common ancestor $\gamma_0$, standard message passing algorithms can be used[1], see fig. 1(b).

We write the terms $\langle e^{-\xi_i z_i} \rangle_Q = e^{-\xi_i h_i} \sum_{\{S_H\}} R(S_H)$, where $R(S_H) = \prod_\gamma R(S_\gamma | S_{pa\gamma})$ and $R(S_\gamma | S_{pa\gamma}) \equiv Q(S_\gamma | S_{pa\gamma}) \exp(\sum_{j \in \gamma} -\xi_i J_{ij} S_j)$. Note that $R$ and $Q$ have similar graphical structures, and we can therefore use message propagation techniques again to compute $\langle e^{-\xi_i z_i} \rangle_Q$. The last term to consider is the entropy term, which decouples into a sum of averaged entropies per $\gamma$,

$$\sum_{\{S_H\}} Q(S_H) \ln Q(S_H) = \sum_\gamma \sum_{S_{pa\gamma}} Q(S_{pa\gamma}) \sum_{S_\gamma} Q(S_\gamma | S_{pa\gamma}) \ln Q(S_\gamma | S_{pa\gamma}) \tag{9}$$

We conclude that all terms in $\mathcal{F}_V[Q, \xi]$ are tractable without the need of additional approximations.

## 3.2 Mean field equations

To derive mean field equations, we differentiate (2) with respect to the parameters $Q(S_\gamma^i | S_{pa\gamma})$, *i.e.*, the $i$-th state of the conditional probability distribution for cluster $\gamma$ (of which there are $n_\gamma = 2^{|\gamma|} - 1$ states), analogous to section 2.1,

$$\begin{aligned} &(Q(S_\gamma^1 | S_{pa\gamma}) \ldots Q(S_\gamma^{n_\gamma} | S_{pa\gamma})) \\ &= \vec{\sigma} \left( \frac{\sum_{ij} J_{ij}^i \nabla_\gamma \langle S_i S_j \rangle_Q + \sum_i (h_i - \sum_j \xi_j J_{ji}) \nabla_\gamma \langle S_i \rangle_Q + K + L}{Q(S_{pa\gamma})} \right) \end{aligned} \tag{10}$$

where the gradient $\nabla_\gamma$ is with respect to $Q(S_\gamma^1 | S_{pa\gamma}) \ldots Q(S_\gamma^{n_\gamma} | S_{pa\gamma})$. Furthermore,

$$K = -\nabla_\gamma \sum_i \ln \left\langle e^{-\xi_i z_i} + e^{(1-\xi_i)z_i} \right\rangle_Q \tag{11}$$

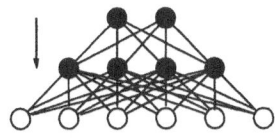

Figure 2: Graphical structure of the 2-4-6 nodes sigmoid belief network. Open circles: visible units $V$. Filled circles: hidden units $H$.

and

$$L = -\sum_{\gamma'} \sum_{S_{\text{pa}\gamma'}} [\nabla_\gamma Q(S_{\text{pa}\gamma'})] \sum_{S_{\gamma'}} Q(S_{\gamma'}|S_{\text{pa}\gamma'}) \ln Q(S_{\gamma'}|S_{\text{pa}\gamma'}) \quad (12)$$

$\vec{\sigma}(x_1, \ldots, x_{n_\gamma}) \equiv (1 + \sum_j e^{x_j})^{-1}(e^{x_1}, \ldots, e^{x_{n_\gamma}})$ is the generalized sigmoid function. Finally, $Q(S_{\text{pa}\gamma}) \equiv 1$ if $\text{pa}\gamma = \phi$. The explicit evaluation of the gradients can be performed efficiently, again using standard message propagation.

To optimize the bound, we again use a two step iterative procedure as described in section 2.1. Note that the optimization with respect to the $\xi_i$'s in the second step remains decoupled.

# 4 Simulations

To compare the method with existing mean field approximations [3, 4], we examined a toy benchmark problem in a three layer (2-4-6 nodes) sigmoid belief network in which the last 6 nodes are visible, fig. 2. We generated 500 networks with parameters $\{J_{ij}, h_j\}$ drawn randomly from the uniform distribution over $[-1, 1]$. The lower bound values $\mathcal{F}_V$ for several approximating structures (including 'standard mean field') are compared with the true log likelihood, using the relative error $\mathcal{E} = \mathcal{F}_V / \ln P(S_V) - 1$, fig. 3. These show that considerable improvements can be obtained when belief networks are used. Note that a 5 component mixture model ($\approx 80$ variational parameters) yields $\mathcal{E} = 0.01139$ on this problem [4]. The results also suggest that one should exploit knowledge about the graphical structure of the model. For instance, the chain (fig. 3(b)) with no graphical overlap with the original graph shows hardly any improvement over the standard mean field approximation. On the other hand, the trees model (fig. 3(c)), which has about the same number of parameters, but a larger overlap with the original graph, does improve considerably over the mean field approximation (and even over the 5 component mixture model). By increasing the overlap, as in fig. 3(d), the improvement gained is even greater.

# 5 Discussion

The use of tractable belief networks fits well in mean field theory. Their ability to exploit the graphical structure of the model seems very powerful. Note that our approach (see also [6]) is very different from suggested methods [7] of using mean field theory to strip away intractable parts of the graph and compute

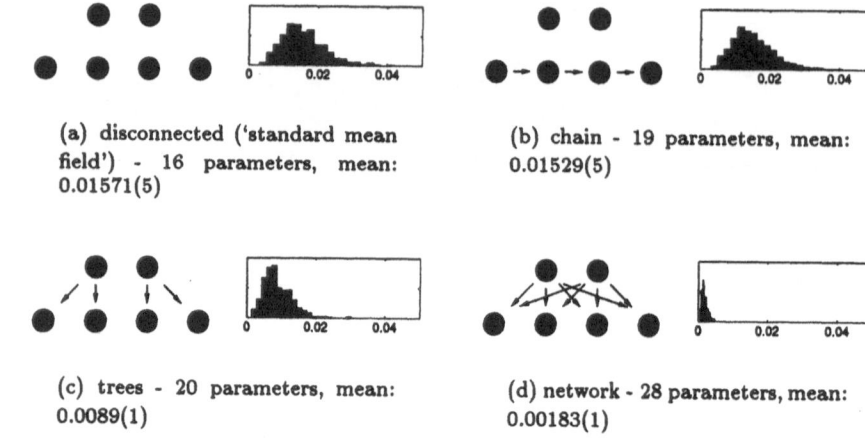

(a) disconnected ('standard mean field') - 16 parameters, mean: 0.01571(5)

(b) chain - 19 parameters, mean: 0.01529(5)

(c) trees - 20 parameters, mean: 0.0089(1)

(d) network - 28 parameters, mean: 0.00183(1)

Figure 3: Graphical structures of the approximating distributions on $H$ (cf. fig. 2 ). For each structure, histograms of the relative error between true log likelihood and the lower bound is plotted. Horizontal scale have been fixed to [0,0.05] in all plots.

with the remaining *fixed* tractable substructure. In contrast, we use *variational* tractable structures together with mean field theory which leads, in general, to a more powerful approximation. We believe that such approaches will prove beneficial for learning large belief networks.

We have also developed a similar approach based on *undirected* graphical approximations, which have roughly the same accuracy, although the optimization procedure is implemented in a different way [6].

[1] J. Pearl. *Probabilistic Reasoning in Intelligent systems: Networks of Plausible Inference*. Morgan Kaufmann Publishers, Inc., 1988.

[2] R. Neal. Connectionist learning of belief networks. *Artificial Intelligence*, 56:71–113, 1992.

[3] L.K. Saul, T. Jaakkola, and M.I. Jordan. Mean field theory for sigmoid belief networks. *Journal of Artificial Intelligence Research*, 4:61-76, 1996.

[4] C.M. Bishop, N. Lawrence, T. Jaakkola, and M. I. Jordan. Approximating Posterior Distributions in Belief Networks using Mixtures. In *Advances in Neural Information Processing Systems*, volume 10. MIT Press, 1998. In press.

[5] T.S. Jaakkola and M.I. Jordan. Approximating posteriors via mixture models. In M.I. Jordan, editor, Proceedings NATO ASI *Learning in Graphical Models*. Kluwer, 1997.

[6] D. Barber and W. Wiegerinck. Tractable undirected approximations for graphical models. In *ICANN'98: International Conference on Artificial Neural Networks, Skövde*, 1998.

[7] L. K. Saul and M. I. Jordan. Exploiting Tractable Substructures in Intractable Networks. In D. S. Touretzky, M. C. Mozer, and M. E. Hasselmo, editors, *Advances in Neural Information Processing Systems*, volume 8. MIT Press, 1996.

# On the Use of Local RBF Networks to Approximate Multivalued Functions and Relations

Klaus Hahn

Institute of Biomathematics and Biometrics of the GSF - National Research Center
for Environment and Health, Postfach 1129, D-85758 Oberschleißheim, Germany
E-mail: hahn@gsf.de

Thomas Waschulzik

Technology Center Informatics, Intelligent Systems, FB 3,
University of Bremen, Bibliothekstr. 1, D-28359 Bremen, Germany

### Abstract

A connectionist model made up of a combination of RBF networks
is proposed; the model decomposes multivalued dependencies into local
single valued functions; theory and applications are presented.

## 1  Introduction

We present a new network structure which is modelled according to the "implicit function theorem" [1]. It roughly states that multivalued functions and relations, which can be described by the zeros of an implicit function, can locally be represented by single valued functions. In this network, the local functions are realized by feedforward networks (RBF). They are incorporated into a global network by a symmetric topological encoding of the in- and output spaces and by a product of error functions. The latter represent separated classes of local functions. Via a least square training of the global network it is decided which one of the local networks generalizes best in a special region. This optimal network is then used for the local generalization of the multivalued function.

This construction performs interpolation and classification tasks on the same automatic control level as standard RBF networks. Moreover, it uses the generalization quality and the transparent parametrization of feedforward networks in the treatment of multivalued functions and relations. A regularization (smoothing) term is included in the model. Such networks are relevant in image smoothing, where discontinuities like edges pose a severe problem to numerous filters [2]. In addition, the use of least square training allows a precise treatment of the regression problem. Moreover in the field of non unique inverse problems, like in spectral analysis - where parameters of the unknown spectrum are to be determined -, or like in medical image analysis - where e.g. from the noisy image of a tumour its sharp contour (the regression curve) shall be reconstructed -, one frequently has to deal with relations.

## 2  The global network and its local RBF constituents

In this section, the essentials of the construction will be developed. The (local) single valued mappings used, are for notational simplicity one dimensional:

$\mathcal{R} \to \mathcal{R}$. In contrast to such single valued functions, which can be described e.g. as mappings $x \mapsto y$, multivalued functions and relations map $x$ not only to a single value $y$, but also to several discrete $y$ values or even to intervals. Thus for relations, we cannot expect to find an expression like $y = \text{network}(x, p)$ which gives all y-values depending on $x$, after some training process determining the parameters $p$ of the generalizing network. To find the y-values given $x$ for a relation, we therefore use the least square error function which was the basis of the training process and calculate from this function, $error(x, y)$, the zeros along the $y$ axes for a fixed $x$ (the case "given $y$" is treated in a symmetrical way).

Though relations globally differ from functions, they can locally be described by functions $x \mapsto y$ or $y \mapsto x$, at least if they are not too pathological. Since no coordinate axis is discriminated the input and output spaces will be encoded in a symmetric way; i.e. we introduce a x-axis-RBF set $\{\Phi_i(x) | i = 1 \text{ to necessary}$ number of radial basis functions to cover that region of the x-axis which is occupied by the orthogonal projection of the relation on $x\}$, and a y-axis-RBF set $\{O_j(y) | \ldots\}$ in an analogous way. This encoding of the in- and output spaces can be called topological, since adjacent points are described (encoded) by RBF's with adjacent centres. A local function $x \mapsto y$ may then be described by the term:

$$error(x, y)_{x \mapsto y, j} = \left[ O_j(y) - sig\left( \sum_i w_{ij} \Phi_i(x) \right) \right]^2 \tag{1}$$

where the sigmoid, $sig(x) = 1/(1 + \exp[-4x])$, has scaling properties. The center of $O_j(y)$ localizes the neighborhood in the $y$ space, where $x \mapsto y$ can be trained to the data via $error(x, y)_{x \mapsto y, j}$; where $O_j(y) \approx 0$ this neighborhood has reached its limits. Looking for the zeros of $error(x, y)_{x \mapsto y, j}$ after training we find two problems:

a) there are two generalizing solutions, symmetric to the center of $O_j(y)$,

b) to reduce the neighborhood not too much, we must use a broad function $O_j(y)$, with the consequence, that the valley of zeros is unpleasantly flat.

These problems can be circumvented by summation over (1)

$$error(x, y)_{x \mapsto y} = \sum_j \left[ O_j(y) - sig\left( \sum_i w_{ij} \Phi_i(x) \right) \right]^2 \tag{2}$$

This represents all local functions in the direction $x \mapsto y$, which are regarded in some multivalued interpolation problem. Equation (2) can be interpreted as an implicit function in the sense mentioned in the Introduction. To illustrate this Ansatz, we show in Fig. 1 an interpolation, where via (2) a multivalued function $x \mapsto y$ with crossing and discontinuities is approximated by the training points indicated.

The mappings $y \mapsto x$ are summarized by the symmetric expression

$$error(x, y)_{y \mapsto x} = \sum_k \left[ \Phi_k(x) - sig\left( \sum_l w_{kl} O_l(y) \right) \right]^2 \tag{3}$$

The advantages of (2) and (3), however, have a price: if the training to the data cannot be performed perfectly as is usually the case, the error surfaces (2)

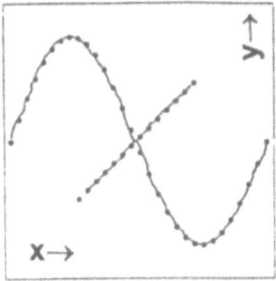

Fig. 1  A multivalued function $x \mapsto y$ is generalized by (2), (44 training points).

or (3) show no more perfect zeros as solution curves, but a course of minima somewhat above 0. The precise detection of this course is not trivial and will be discussed in section 3.

Up to now only multivalued functions can be treated. In order to generalize a relation (e.g. a spiral), a product of (2) and (3) is used:

$$error(x, y) = error_{x \mapsto y}(x, y) \cdot error_{y \mapsto x}(x, y) \tag{4}$$

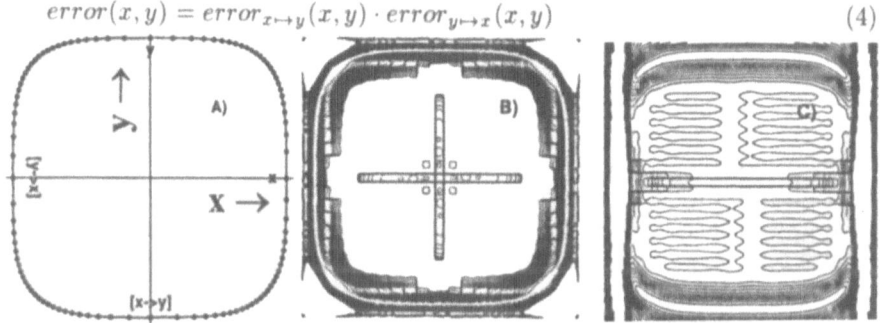

Fig. 2  The errror surfaces of a relation described by four local functions

An example of a generalizing error function (4) - after training - is given in Fig. 2, panel B); the valley of minima follows the pattern "suggested" by the training points given in panel A). The error functions of the active local functions in the two factors of (4), in (2) and (3), are shown in panel C). As (2) generalizes the maps $x \mapsto y$ only, clear minima are present only at the top and bottom of panel C) (in panel A) the coordinate axes are given). To find the minima of (3) in C) just rotate panel C) by 90°. Panel A) additionally shows the continuous course of minima of panel B).

*Summarizing:* the error functions ["implicit functions"] (2), (3) and (4) define global networks by the course of minima ["zeros"]. These courses generalize multivalued functions ((2), (3)) and relations (4). The global networks are composed of RBF's ["local functions"] which generalize the data locally. As will be exemplified in Fig. 4, for some interpolation tasks, not only the networks are finally relevant, but also their extensions to the error surfaces.

## 3  Training process and ridge detection

In this section, the calculational procedure will be explained. The training points are the input, the course of minima in the error surface finally is the output, which must be analyzed by some algorithm. The number of basis functions is given initially by the recipe: Choose the centres and widths of the Gaussians $\Phi_i(x)$ and $O_j(y)$ heuristically (in case of exact data: width of RBF's $\approx$center of RBF's distance $\approx$mean training point distance, in case of noisy data: width $\approx$center distance $\approx$standard deviation of training point density). For a relation the weights $w_{ij}$ are obtained by a minimization of (4) plus a regularization term:

$$\min_w \Big[\sum_{p=1}^{ntraining} error(x_p, y_p) + \alpha \sum_{ij} (const + w_{ij})^2\Big] \tag{5}$$

where $(x_p, y_p)$ are training points and $\alpha$ is the regularization strength. For Fig. 1 - Fig. 4 $\alpha \approx 0.05$ was used, for noisy data, $\alpha \approx 0.5$; const = 0.5 to achieve a flat error surface. Via the fit (5), the error function "decides" where (2) or (3) is generalizing best (builds a course close to zero), the w-matrix in the other term causes a bounded factor only. Therefore it is sufficient to use the same w-matrix in the factors (2) and (3) of (4). In contrast to a Spline interpolation, the training points need not to be ordered, the network generalizes to the nearest neighbours, sequential learning is possible. In this work a standard procedure of the NAG numerical library was used to perform the minimization. With respect to smoothness of the local RBF-networks as penalty term the weight-decay regularization was used [3]. Finally a fine tuning of the parameters can be performed via cross validation.

According to this training process, the error surface is minimized in the training points, the generalization extends these minima to smooth courses, which constitute the generalizing continua. A strict mathematical definition of courses or ridges is still an object of controverse debates [4]. However, there exists a numerical algorithm, which solves the problem of ridge calculation on a finite grid. To apply a variant of this watershed algorithm, $-error(x, y)$ is computed and discretized. In order to identify the pairs $(x, y)$ which belong to the modelled multivalued function, the crest lines are extracted [5]. Although the original watershed algorithm detects only watershed and not crest lines in general, it can be modified by inserting artificial borders into the original image, before the watershed is computed. With this modified algorithm, it is possible to detect crest lines or ridges without adding artefacts. Fig. 3 and Fig. 4 demonstrate further applications of the described method.

## 4 Noisy data

In case of a quadratic error function, one can show for a continuous set of noisy training points that training of the weigth matrix $w$ for a standard RBF network (e.g. for the $x \mapsto y$ map) leads to a separation of the error function into a bias and a variance term. The minimum of the bias defines the network, which approximates the regression (e.g. $\langle y|x \rangle$) [3]. Similar reasoning leads for (2) and (3) to the same separation, the bias term of (2) is:

$$error_{x \mapsto y}(x, \langle y|x \rangle) = \sum_j \Big[\langle O_j|x \rangle - sig\Big(\sum_i w_{ij}\Phi_i(x)\Big)\Big]^2 \tag{6}$$

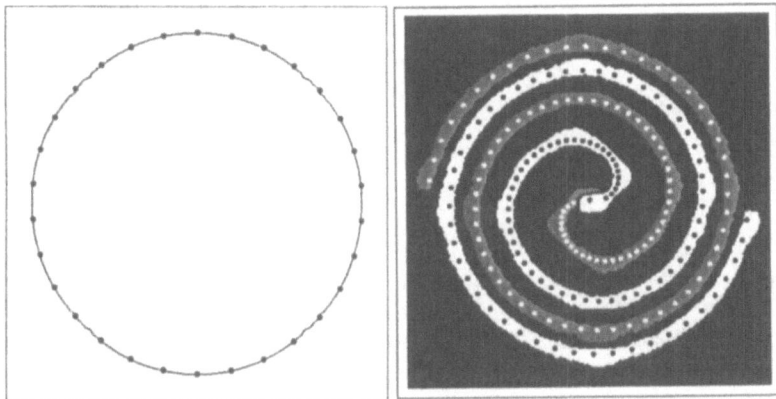

Fig. 3  A circle is defined by 30 points, the generalizing course is shown.
Fig. 4  The two spiral problem is treated by (4). Not only the courses are shown,
but a three level colouring of the error surfaces.

where $p(y|x)$ denotes the conditional distribution of training points $y$ for fixed
$x$ and

$$\langle f|x\rangle = \int f(y)p(y|x)dy \tag{7}$$

is a conditional mean of some function $f$.

In case of a Gaussian distribution $p(y|x) = e^{-\left(y-\langle y|x\rangle\right)^2/\left(2\sigma(x)^2\right)}/\sqrt{2\pi\sigma(x)}$, we
find for $O_j(y) = e^{-(y-m_j)^2/(2s_j^2)}$:

$$\langle O_j|x\rangle = e^{-\left(\langle y|x\rangle-m_j\right)^2/\left(2(s_j^2+\sigma(x)^2)\right)}s_j/\sqrt{s_j^2+\sigma(x)^2} \tag{8}$$

For (3) the bias term is completely symmetric to (6). Since the course of (4) is
defined by the minima of (2) or (3), in case of noisy data, (4) is replaced by the
product of (6) and its symmetric counterpart. To exploit this modification, we
must assume an approximate analytical form for the conditional point densities
$p(y|x)$ and $p(x|y)$ (both densities include the unknown regression parameters
$\langle y|x\rangle$, $\langle x|y\rangle$ and the standard deviations $\sigma(x), \sigma(y)$ as known). In regions of
the plane, where (6) is active, the network approximates the regression $\langle y|x\rangle$,
where the counterpart is active the regression $\langle x|y\rangle$. The training process is
still performed by (5), which is free of distribution assumptions. In Fig. 5 and
Fig. 6 applications of the described procedure are shown.

## 5  Summary and outlook

A new variant of a RBF network is proposed. It is applied to the approximation of multivalued functions and relations and seems to be a reasonable
tool for generalization tasks, for non unique inverse problems and for special
classification problems. The model generalizes exact and noisy data. In case of
noisy data in the calculational procedure (not in training), assumptions about

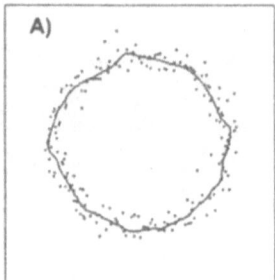 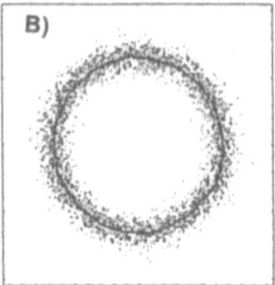

Fig. 5   The training points in panel A) are derived by a radial Gaussian distribution for 200 equidistant angles. The calculated curve is an estimator for the generating circular regression. In panel B) 5000 angles are used to demonstrate consistency.

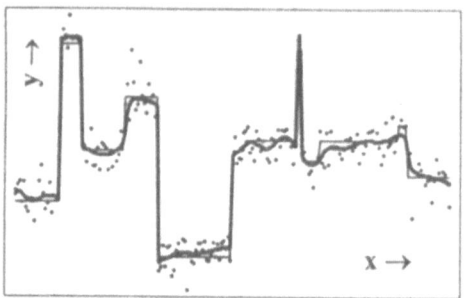

Fig. 6   The ability of (6) to approximate noisy discontinuous functions is demonstrated; thick line: (6), thin line:   true regression, 200 training points are used [2].

the underlying point distributions enter. The network is defined for relations up to now by a symmetric Ansatz which originates in the use of two coordinate systems. In case of noisy data, however, the introduction of more coordinate systems might reduce the bias of this model. The model is presented in the plane only. Extensions to more dimensions are straightforward at least for the case of discontinuous image smoothing, Fig. 6. In case of true relations the construction of the model indicates how extensions to higher dimensions should be formulated.

# References

[1] R. Courant and F. Ford, Introduction to Calculus and Analysis, Springer-Verlag, Berlin Heidelberg New York, 1989.

[2] C. K. Chu, I. Glad, F. Godtliebsen, J. S. Marron, Edge preserving smoothers for image processing, preprint, 1996.

[3] C. Bishop, Neural Networks for Pattern Recognition, Clarendon Press, Oxford, 1995.

[4] J. J. Koenderink and A. J. van Doorn, Image Structure, Invited Lecture, Mustererkennung 1997, 19. DAGM-Symposium, Braunschweig, Springer

[5] L. Najan, R. Vaillant, Topological and geometric corners by watershed, in Hlavac, Sara (Eds.): CAIP 95 Proceedings, LNCS 970, Springer, 1995.

# Learning Higher Order Boltzmann Machines using Linear Response

M.A.R. Leisink* and H.J. Kappen[†]

Department of Biophysics
University of Nijmegen, Geert Grooteplein 21,
NL 6525 EZ Nijmegen, The Netherlands[‡]

March 23th, 1998

### Abstract

Boltzmann machines are able to represent some probability distribution but the exact learning algorithm needs a time that is exponential in the number of neurons. The approximation method called Linear Response is not only applicable to machines with only second order interactions, but can be extended to Boltzmann machine with third and higher order interactions. It is shown that this can be used to estimate probability distributions which have strong third or higher order correlations.

## 1 Introduction

A Boltzmann machine is a network of stochastic variables (neurons), which value is either plus or minus one. All neurons are linked to each other with symmetric weights $w_{ij} = w_{ji}$. Due to this symmetry the stationary probability distribution is given by the Boltzmann-Gibbs distribution $P(\vec{s})$, which is a known function of the weights and thresholds of the network [1].

Since the computation of this distribution requires an amount of time proportional to $2^N$, where $N$ is the number of neurons, an approximation is needed. Kappen and Rodríguez [2] have shown an approximation method called Linear Response which is an order $N^3$ algorithm.

We extend this approximation to higher order networks. These higher order networks have not only first and second order interactions $\theta_i$ and $w_{ij}$ but also third and higher order interactions like $w_{ijk}$ or $w_{ijklm}$.

We show that it is possible to derive do the Linear Response approximation similar to the second order case. Moreover in the absence of hidden units it is possible to obtain an immediate approximation of the weights instead of learning the machine using gradient descent.

---

*http://www.mbfys.kun.nl/~martijn

[†]http://www.mbfys.kun.nl/~bert

[‡]This research is supported by the Technology Foundation STW, applied science division of NWO and the technology programme of the Ministry of Economic Affairs.

## 2 Approximation of the Free Energy

A higher order Boltzmann machine has a probability distribution

$$P(\vec{s}) = \frac{1}{Z} \exp(-E(\vec{s})) \tag{1}$$

where

$$Z = \sum_{\text{all } \vec{s}} \exp(-E(\vec{s})) \tag{2}$$

and

$$E(\vec{s}) = -\sum_i \theta_i s_i + \alpha E_{int}(\vec{s}) \tag{3}$$

The interaction energy in the last equation is given by all interactions in the system

$$E_{int}(\vec{s}) = -\frac{1}{2} \sum_{ij} w_{ij} s_i s_j - \frac{1}{6} \sum_{ijk} w_{ijk} s_i s_j s_k - \ldots \tag{4}$$

For $\alpha = 1$ we have the fully connected Boltzmann machine; for $\alpha = 0$ the model is decoupled.

Calculation of the free energy $F = -\log Z$ requires a time that is exponential in the number of neurons. Therefore we need an approximation. We will follow the work of Plefka [3] and expand this free energy in $\alpha$, since the model is calculable for $\alpha = 0$.

First we introduce new variables

$$m_i = \langle s_i \rangle = -\frac{\partial F}{\partial \theta_i} \tag{5}$$

and perform a standard Legendre transformation to make these $m_i$ the new independent variables.

$$G(\vec{m}, \vec{w}, \alpha) = F(\vec{\theta}, \vec{w}, \alpha) + \sum_i \theta_i m_i \tag{6}$$

where $G$ is known as the Gibbs free energy which has $m_i$ and the weights as independent variables.

We approximate this $G$ by a Taylor expansion around $\alpha = 0$

$$G(\vec{m}, \vec{w}, \alpha) \approx G(\vec{m}, \vec{w}, 0) + \alpha \left. \frac{\partial G}{\partial \alpha} \right|_{\alpha=0} \tag{7}$$

Since we know that $G(\vec{m}, \vec{w}, 0)$ represents the Gibbs potential of the decoupled spins, we obtain

$$G(\vec{m}, \vec{w}, \alpha) \approx \sum_i \left\{ \frac{1+m_i}{2} \log \frac{1+m_i}{2} + \frac{1-m_i}{2} \log \frac{1-m_i}{2} \right\}$$
$$+ \alpha \langle E_{int} \rangle_{\alpha=0} \tag{8}$$

and the mean field variables $m_i$ are calculated by using the property of the Legendre transformation that $\theta_i = \partial G/\partial m_i$, where we put in the approximated Gibbs free energy.

# 3   Estimating the Correlations

Notice that in the exact case the correlations can be expressed as higher order derivatives of the free energy

$$\langle s_i \rangle = -\frac{\partial F}{\partial \theta_i}$$

$$\langle s_i s_j \rangle = -\frac{\partial^2 F}{\partial \theta_i \partial \theta_j} + \langle s_i \rangle \langle s_j \rangle \tag{9}$$

$$\langle s_i s_j s_k \rangle = -\frac{\partial^3 F}{\partial \theta_i \partial \theta_j \partial \theta_k} + \langle s_i \rangle \langle s_j s_k \rangle + \langle s_j \rangle \langle s_k s_i \rangle$$
$$+ \langle s_k \rangle \langle s_i s_j \rangle - 2 \langle s_i \rangle \langle s_j \rangle \langle s_k \rangle$$

From the Legendre transformation we know the relationship between the higher order derivatives of $F$ and those of $G$

$$\frac{\partial^2 F}{\partial \theta_i \partial \theta_j} = -\left( \frac{\partial^2 G}{\partial m_k \partial m_l} \right)^{-1}_{ij}$$

$$\frac{\partial^3 F}{\partial \theta_i \partial \theta_j \partial \theta_k} = -\sum_{\alpha\beta\gamma} \frac{\partial^2 F}{\partial \theta_i \partial \theta_\alpha} \frac{\partial^2 F}{\partial \theta_j \partial \theta_\beta} \frac{\partial^2 F}{\partial \theta_k \partial \theta_\gamma} \frac{\partial^3 G}{\partial m_\alpha \partial m_\beta \partial m_\gamma} \tag{10}$$

If we make use of equation 8, with $\alpha$ set to one, we can estimate the higher order derivatives of $G$

$$\frac{\partial G}{\partial m_i} \approx \tanh^{-1} m_i + \frac{\partial \langle E_{int} \rangle}{\partial m_i} \tag{11}$$

$$\frac{\partial^2 G}{\partial m_i \partial m_j} \approx \frac{\delta_{ij}}{1 - m_i^2} + \frac{\partial^2 \langle E_{int} \rangle}{\partial m_i \partial m_j} \tag{12}$$

$$\frac{\partial^3 G}{\partial m_i \partial m_j \partial m_k} \approx \frac{2\delta_{ijk} m_i}{\left(1 - m_i^2\right)^2} + \frac{\partial^3 \langle E_{int} \rangle}{\partial m_i \partial m_j \partial m_k} \tag{13}$$

where the derivatives of $\langle E_{int} \rangle$ are simple expressions in terms of $m_i$ and the weights.

If we want to approximate the correlations given the thresholds and the weights, we first calculate the mean field variables $m_i$ using equation 11 and then we use equation 12 to estimate the derivatives of $G$. After that we use the Legendre transformation to find the derivatives of $F$. Using equations 9 we can find the estimated correlations.

If we, however, know all the wanted correlations (as is the case if there are no hidden neurons) we can use equation 9 to calculate the derivatives of $F$ and the Legendre transformation to obtain the derivatives of $G$. Since our approximated Gibbs free energy is linear in the weights, we can use equation 12 to make a direct estimate of the weights.

# 4 Results

We demonstrate the Linear Response approximation on a Boltzmann machine which has thresholds, second and third order weights. Our target distribution will be a Boltzmann distribution itself with only second and third order interactions

$$Q(\vec{s}) = \frac{1}{Z} \exp(\sum_{i<j} w_{ij} s_i s_j + \sum_{i<j<k} w_{ijk} s_i s_j s_k) \tag{14}$$

where $Z$ is the normalization constant. The weights are random from a Gaussian with zero mean and a standard deviation $\sigma/\sqrt{N}$ for the second order and $\sigma/N$ for the third order weights. $N$ is the number of neurons. We have used a network with 10 neurons, since in that case it is possible to compare the results with the exact calculations.

First we have trained our Boltzmann machine using the exact learning algorithm and using the Linear Response approximation. We have learned the target distribution $Q(\vec{s})$ with a third order Boltzmann machine as well as with a second order Boltzmann machine (which we expect to give worse results). We use the Kullback divergence [4] to measure the distance between the target and the learned probability distribution.

In figure 1 the Kullback divergence is plotted versus the standard deviation $\sigma$. In the upper graph we used a target with third order interactions only; in the lower graph the target has both second and third order interactions. Since we have expanded $G$ in the weights, we expect worse results as $\sigma$ increases. Notice that the Kullback divergence of the exact third order machine is always zero since in that case the task is realizable.

From figure 1 we conclude that for targets in which third order correlations play a significant role the third order Linear Response approximation is useful as long as $\sigma$ is not too large. (For a second order models it is known that the mean field approximation used here breaks down at $\sigma = 1/2$ for large $N$ [5].)

Secondly we assess the quality of the approximation to compute the correlations of a given Boltzmann distribution. We have initialized the second and third order weights of a network of 10 neurons with zero mean and standard deviation $\sigma/\sqrt{N}$ for the second order interactions and $\sigma/N$ for the third order interactions. We have plotted the exact correlations versus the estimated ones for a network with $\sigma = 0.25$. Figure 2 shows two graphs for $\langle s_i s_j \rangle$ and $\langle s_i s_j s_k \rangle$. We see that the approximated second and third order correlations are almost equal to the exact ones.

# 5 Discussion

In this paper we extended the Linear Response approximation to Boltzmann machines with higher order interactions. We showed that good approximations can be obtained as long as the weights in the target distribution are not too large.

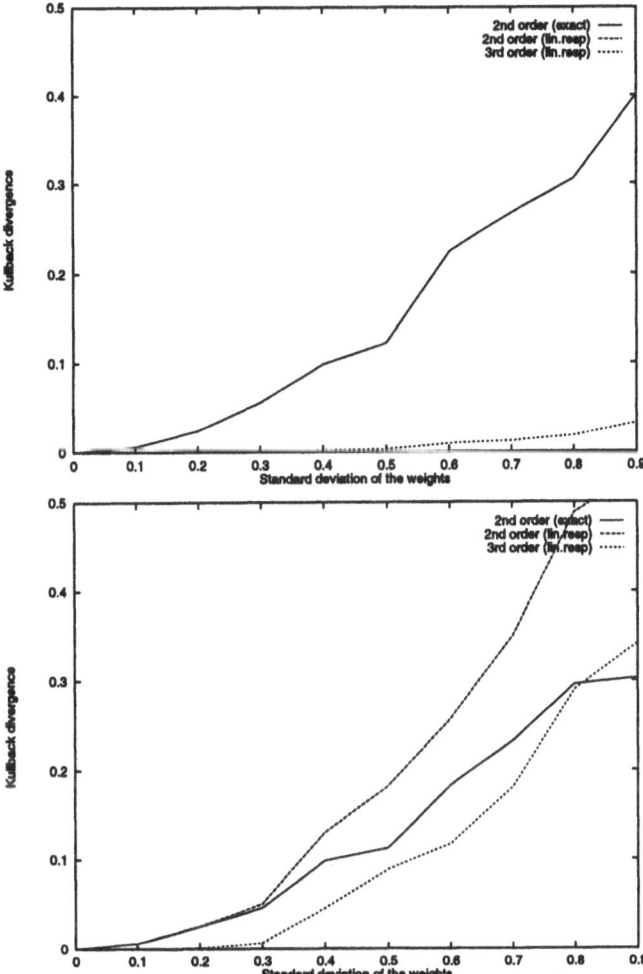

Figure 1: A network of 10 neurons learning a distribution a) with only third order interactions and b) with both second and third order interactions. Each point is an average over 10 random problems. Notice that in the upper graph the two second order lines are on top of each other.

It is possible to expand the Gibbs free energy $G$ upto the second order of $\alpha$. For a second order Boltzmann machine this brings in the TAP-term as was shown by Plefka [3] and Kappen et al. [2]. The same can be done for higher order Boltzmann machines which increases the accuracy of the estimation. The inversion of equation 12 however might be no longer possible since the derivatives of $G$ are not linear in the weights in that case.

Although the theory presented is in principle valid for any order of the Boltzmann machine, only third and maybe fourth order will be useful in practice. The calculation time is polynomial, but it can be fairly large, since the

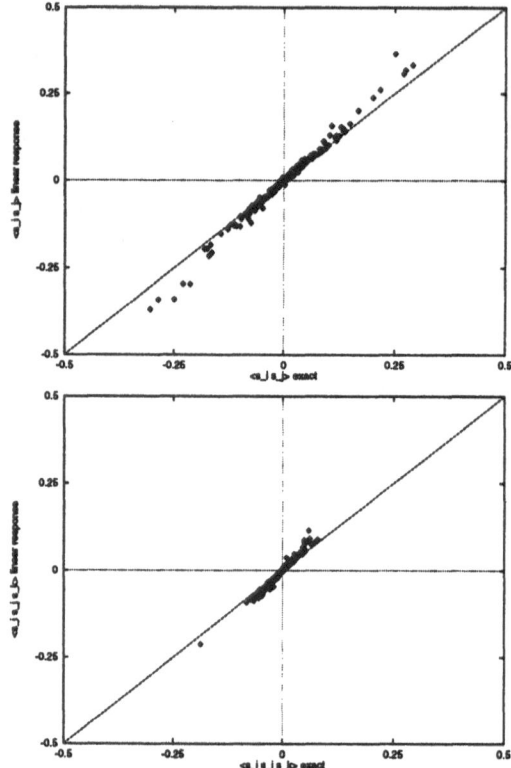

Figure 2: The estimation error of the second and third order correlations in a network of 10 neurons. a) $\langle s_i s_j \rangle_{ex}$ vs. $\langle s_i s_j \rangle_{lr}$ b) $\langle s_i s_j s_k \rangle_{ex}$ vs. $\langle s_i s_j s_k \rangle_{lr}$

exponent is proportional to the order of the Boltzmann machine. Furthermore the storage of all the correlations might be a problem.

# References

[1] J. Hertz, A. Krogh, R.G. Palmer, Introduction to the Theory of Neural Computation, Chapter 7, Addison-Wesley (1991), ISBN 0-201-50395-6.

[2] H.J. Kappen and F.B. Rodríguez, Boltzmann Machine learning using Mean Field theory and Linear Response correction, Proceedings NIPS, MIT Press (1998). In press.

[3] T. Plefka, Convergence condition of the TAP equation for the infinite-ranged Ising spin glass model, J. Phys. A: Math. Gen. 15 (1982) 1971–1978.

[4] S. Kullback, Information Theory and Statistics, Wiley, New York (1959).

[5] H. Takayama and K. Nemoto, Spin glass properties of a class of mean-field models, J. Phys.: Condens. Matter 2 (1990) 1997–2007.

# Order Parameters
# for Self-Organizing Maps

Angelika Spitzner
Daniel Polani

Institut für Informatik, Johannes Gutenberg-Universität
D-55099 Mainz, Germany
polani@informatik.uni-mainz.de

## Abstract

We introduce and discuss different approaches to construct order parameters for Kohonen's Self-Organizing Maps. As one approach the notion
of an order parameter in the sense of Haken's synergetics is studied and
contrasted with organization measures using SOM structure information.

## 1 Introduction

Self-organizing neural models have been first introduced by [1]. Among them
Kohonen's *Self-Organizing Map* (SOM) has found particular attention [2], since
it provides a model for organizational processes in the brain as well as for adaptive nonlinear principal curve and manifold analysis and feature mapping [3].
The SOM training process is paradigmatic for many self-organizing processes
studied in neural systems. It displays an important difference to supervised
learning concepts: In the latter, the progress of learning can be quantified
by comparison to the teacher signal. The SOM training process, however,
is defined only by the dynamics of the learning rule and does not possess a
canonical measure quantifying its training progress and the *organization* of the
self-organizing map. Such a measure would be relevant for theoretical considerations of SOMs as well as for practical purposes, like e.g. finding termination
criteria for the training.

The lack of a canonical organization measure leads to different definitions
sensitive to different incongruent properties of the SOM training process. This
is reflected in the wide selection of organization measures used in literature.
Among them are the well-known quantization error, then a generalized inversion
measure [4], the *topographic product* [5], the relative connection length variation
from [6], a "geodesic" organization measure [7], and measures [8, 9] based on a
Hebbian mechanism from [10]. While all those measures use knowledge about
the geometrical interpretation of a SOM to determine its degree of organization,
more general approaches restrict themselves to the consideration of the SOM
dynamics. In [11] the SOM is reduced to a *deterministic dynamical system*
(DDS). This idea is used in related models to derive quantities measuring the
formation of neural maps in close relationship to the Liapunov functions for the
corresponding system [12]. For the SOM DDS, however, an energy function,
which would be a natural choice as organization measure, does not exist [11]. It

is not even known whether this system possesses a Liapunov function in general [13, 14].

To determine a quantity parametrizing the SOM training process we therefore consider it as a special case of a general *self-organizing process*. Although there is no generally agreed upon precise mathematical definition, the notion "self-organizing" is used for a wide class of processes with certain intuitively evident properties [15]. For characterization of such processes, the notion of *order parameters* is introduced. In the literature there are different views of order parameters: In statistical physics an order parameter is a quantity signaling that a system (or an ensemble of systems) is in a certain *phase* in the sense of (non-equilibrium) thermodynamics. To construct an order parameter for a given system, it is necessary to have an interpretation for the system, like in [16], where the order parameter is introduced ad hoc for the network model considered. An order parameter for SOMs thus requires a geometrical interpretation, e.g. as discrete topological description of probability distributions, like the measures mentioned in the last paragraph.

A second notion of an order parameter has been introduced by Haken in context of his synergetics theory [17]. He considers the dynamics of a high-dimensional dynamical system near a fixed point, i.e. an equilibrium. The current system state is specified by the deviations from the equilibrium (the so-called *modes*). Mode space is partitioned into short-lived and long-lived modes. The relevant long-lived modes dominate the long-term behaviour of the system and are called *master modes*, while the short-lived are called *slave modes* in the terminology of synergetics. To reduce complexity, the system is restricted to the master modes space by adiabatic elimination. The master modes, i.e. the degrees of freedom of the restricted system, are interpreted as order parameters of the system. Above approach is closely related to the restriction of a system's dynamics near the equilibrium to the dynamics on its *center manifold* [15], a notion known from the theory of dynamical systems. In this picture, the order parameters can be viewed as a parametrization for the center manifold.

Our paper casts a view on both notions of order parameters. In Sec. 2 we will give definitions and conventions. In Sec. 3 we will make Haken's notion for our calculations more precise and present some results. Simulation studies concerning the notion of order parameters from statistical physics then will be based on the measure from [4] and will be the topic of Sec. 4.

# 2 The Self-Organizing Map: Remarks

## 2.1 Definitions and Conventions

We consider the SOM as a map $w : A \to V$ from a discrete finite set $A$ of (formal) *neurons* to a convex subset $V$ of $\mathbb{R}^d$, the *input space*, mapping each neuron $j$ to its *weight* $w_j$. On $A$ the metric $d_A$ is induced by the adjacency structure of an undirected graph without weights, the *Kohonen graph*, $A$ being its vertex set, on $V$ we choose the Euclidean metric. During the training we obtain a sequence $\left(w(t)\right)_{t=1,2,\ldots}$ using the standard learning rule

$$\Delta w_j(t) := \epsilon(t) \cdot h(i^*_{w(t)}(x(t)), j)(x(t) - w_i(t)) \tag{1}$$

for all neurons $j$, $x(t)$ being the training input at time $t$, $\epsilon(t)$ the learning rate, $h$ the activation profile and $i^*_{w(t)}(x(t))$ a neuron $i$ minimizing the distance $d_V(x(t), w_i(t))$. In the following we will write $i^*$ instead of $i^*_{w(t)}$ for notational convenience. In our calculations and simulations the Kohonen graph is linear or (in Sec. 4) square, the learning rate is constant over time as well as the activation profile and the latter is further chosen as Gaussian on $A$ considered as embedded into $\mathbb{R}^1$ or $\mathbb{R}^2$, resp. and given as $h(i,j) = e^{-\|i-j\|^2/\sigma^2}$ for $i, j \in A$.

## 2.2 The SOM as Dynamical System

To construct order parameters in the sense of synergetics we transform the SOM into a DDS. For every SOM state represented by $w(t)$ at a given time step $t$ we set $\dot{w}(t) := F(w(t))$ with $F(w(t)) := E_x(w(t+1) - w(t))$, where $E_x$ denotes taking the expectation value of its argument over the distribution of training signals $x$ and the argument of $E_x$ is calculated from the learning rule Eq. (1). It follows that this definition can be easily extended to a dynamical system over continuous time (note there is no explicit time dependency in Eq. (1)) on the set of maps $w$. For the uniform distribution on input space $[0, 1]$ considered here $F$ can be explicitly given for the linear map with open boundaries [11] and – with a few modifications – for the linear map with periodic boundary conditions [18]. The view as dynamical system is closely related to learning with a very small constant learning rate $\epsilon$ [3, 13].

# 3 Order Parameters: The Synergetics View

For the SOM DDS, one can apply adiabatic elimination, if necessary, after some transformation [17, 18]. We can not give here the full procedure; the main idea, however, is to calculate the Jacobian $J$ of $F$ at the equilibrium state. The eigenvectors of $J$ at the equilibrium determine the different canonical modes, i.e. "eigendeviations" of the system from the equilibrium. If the eigenvalues are strongly negative, the corresponding deviation is strongly damped by the dynamics and only short-lived. It is called *slave mode* in the terminology of synergetics. If they are only slightly negative (close to 0) or even positive, they are long-lived and called *master modes*, because they control the long-time behaviour of the system. The corresponding slave modes are determined to a large degree by the value of the master modes; the calculation of this functional dependence is precisely the purpose of aforementioned adiabatic elimination. In this view the values assumed by the master modes function as order parameters. As alternative interpretation one could view those master modes also as the SOM's *relevant degrees of freedom*, namely those degrees of freedom that persist after factoring out the short-lived transient "noise". Synergetics interprets the existence of a small number of dominant master modes in a high-dimensional system as self-organization.

In its purest fashion, this approach assumes a clear separation between master and slave modes, i.e. clearly separated clusters of eigenvalues. But we can also turn this view around: The degree of clustering among the eigenvalues of the system can be considered as a measure for the degree of self-organization. Note that the method as discussed is restricted to an analysis of the local neighbourhood of the equilibrium. On the other hand it uses only the dynamical systems properties and does in no way require the interpretation of the SOM as a topographic and/or spatial mapping.

Now we show to what extent the synergetics concept of order parameters can be applied to SOMs. With a 50-neuron linear grid SOM on $[0, 1]$ at the equilibrium, Fig. 1 shows the eigenvalue distributions for different widths $\sigma$ of the activation profile. In Fig. 1(a) for every $\sigma$ the 50th eigenvalue (i.e. the largest one, since the eigenvalues are sorted in ascending order) is 0 (line parallel to $x$-axis). This reflects the fact that with periodic boundary conditions the SOM can be freely translated without changing the dynamics. With increasing $\sigma$ the distance between the second largest eigenvalue (number 49) and 0 also increases until $\sigma \approx 15$, indicating that with the activation profile encompassing larger neighbourhoods the ordering forces take over control of the collective SOM behaviour. In the same region also a few of the modes with strongest dampening (eigenvalues 1-4) split off. With further increase of $\sigma$, however, the collective dampening (and thus the tendency to return to equilibrium, i.e. to order) decreases again, since the activation profile begins to "wrap around" the unit interval boundaries for $\sigma > 15$ and the ordering effects are leveled off. Note that there is no dominant master mode except for the translation mode.

Fig. 1(b) shows the behaviour for the open boundary case. Because of the boundaries, the eigenvalue 0 translational mode cannot prevail and when $\sigma$ increases, this leads to collective freezing of the system, i.e. the SOM becomes increasingly rigid. However, in the region of $\sigma \approx 18$, where the activation profile begins to "probe" the grid boundaries the system produces a dominant mode and also the difference between maximal and minimal eigenvalue and between maximal eigenvalue and the cluster of similar eigenvalues attains a maximum. The dominant mode could then be used to parametrize a state near the equilibrium. However its dominance is not very prominent, indicating that the SOM is a mainly "democratic" system close to the equilibrium.

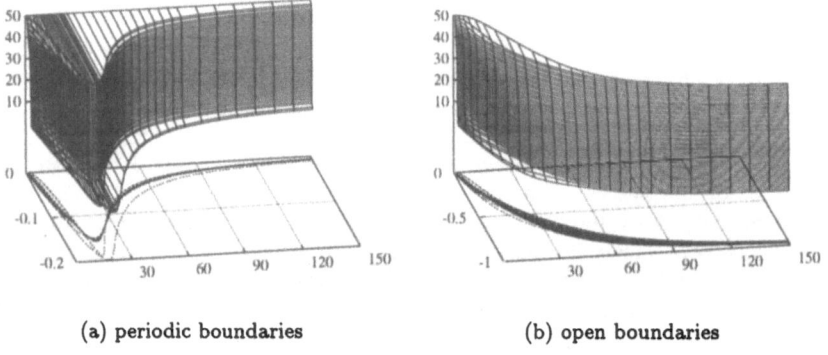

(a) periodic boundaries       (b) open boundaries

Figure 1: Eigenvalue distributions for different widths of the activation profile. The $x$-axis denotes the activation width $\sigma \in (0, 150]$, the $y$-axis the eigenvalues and the $z$-axis the number of the eigenvalue, the eigenvalues being sorted and numbered in ascending order.

# 4    A "Classical" Order Parameter

To consider order parameters in the sense of statistical physics, one has to make use of further knowledge about the system, i.e. in our case the geometrical interpretation of the SOM. As example for an order parameter, consider the inversion measure from [4]: $\mu_I := \frac{1}{|C_K| \cdot (|A| - \epsilon)} \sum_{(j,k) \in C_K} D(j, k)$. $C_K$ is the set of edges $(j, k)$ of the Kohonen graph, $D(j, k)$ is the number of neurons $l \neq j, k$, whose receptive fields $i^{*-1}(l)$ intersect with the line between $j$ and $k$. This sum is then normalized according to the number of possible connections and intersections. It measures how well the Kohonen graph matches the neighbourhood of the neuron weights in input space. A small value signifies good topology preservation, a high one distortions. We performed 50 SOM runs of 20000 training steps to obtain approximations for an equilibrium ensemble of SOMs for constant learning rates $\epsilon$ and activation widths $\sigma$, evaluating and averaging $\mu_I$ at the end of the runs. Fig. 2(a) shows the resulting landscape for a linear grid trained with the uniform distribution on $[0, 1]$ (training of the linear grid with the uniform distribution on $[0, 1]^2$ yields a very similar picture), Fig. 2(b) the same for a square grid SOM trained on $[0, 1]^2$.

$\mu_I$ separates the $\sigma/\epsilon$ plane (which correspond to state variables of statistical physics) into different, clearly distinct regions, acting as detector for the different state classes, i.e. "phases". The two phases correspond to ordered ($\mu_I \approx 0$) vs. disordered ($\mu_I \approx 0.3$ or $0.25$) SOMs. For the linear SOM larger $\epsilon$ and larger $\sigma$ improves ordering. Larger $\sigma$ corresponds to a stronger collective dynamics, which is known to be able to enhance ordering [3]. For the square SOM, increasing $\epsilon$ improves ordering. But $\sigma$ attains an optimal value for ordering at a certain value. Increasing $\sigma$ beyond that value impairs ordering. This reflects the higher complexity of the ordering process in the two-dimensional map.

# 5    Conclusions and Outlook

We have contrasted two different approaches to define order parameters for SOMs. The first one from the synergetics, has the advantage of viewing SOMs

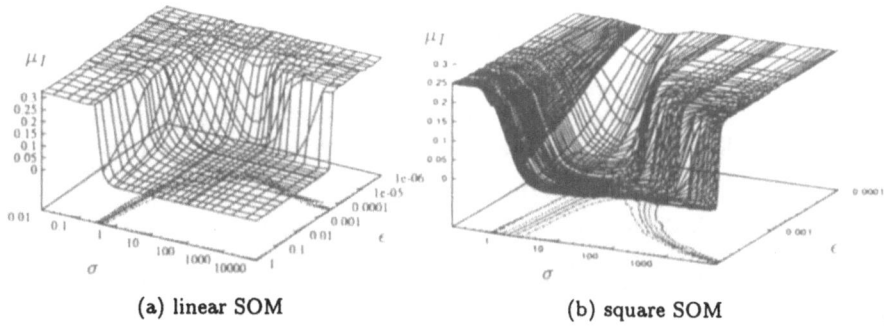

(a) linear SOM          (b) square SOM

Figure 2: Inversion measure landscape for ensembles of linear and square grid SOMs trained with different fixed learning rates and activation widths.

just as dynamical systems and not requiring any geometric interpretation. Its disadvantage is that in the current form it is only defined close to the equilibrium and that the SOMs display no prominent "master" mode. The second is an order parameter in the "classical" sense. These order parameters act as organization measures for SOMs and thereby serve the double purpose of enabling a closer analysis of the SOM dynamics and to be able to evaluate the performance of a SOM in a given task. In the future we plan to adapt the "synergetic" parameter class to include states far from the equilibrium and with configurations different from it, to achieve an order parameter defined purely by SOM dynamics. The classical approach to order parameters using geometrical interpretations, promises to be useful already in this stage. In future we would like to study the phase diagrams using other measures, determine more precisely to which degree Figs. 2(a) and 2(b) still contain transient behaviour and whether the boundaries in fact realize sharp phase boundaries.

# References

[1] C. von der Malsburg. Self-organization of orientation sensitive cells in the striate cortex. *Kybernetik*, 14:85–100, 1973.

[2] Teuvo Kohonen. *Self-Organization and Associative Memory*, volume 8 of *Springer Series in Information Sciences*. Springer, 3rd edition, May 1989.

[3] Helge Ritter, Thomas Martinetz, and Klaus Schulten. *Neuronale Netze*, 1994

[4] Stéphane Zrehen and François Blayo. A geometric organization measure for Kohonen's map. In *Proc. of Neuro-Nîmes*, pages 603–610, 1992.

[5] Hans-Ulrich Bauer and Klaus R. Pawelzik. Quantifying the neighbourhood preservation of self-organizing feature maps. *IEEE Trans. Neural Networks*, 3(4):570–579, 1992.

[6] P. Demartines and F. Blayo. Kohonen's self-organizing maps: Is the normalization necessary? *Complex Systems*, 6(2):105–123, April 1992.

[7] S. Kaski and K. Lagus. Comparing self-organizing maps. In C. von der Malsburg, W. von Seelen, J. C. Vorbrüggen, and B. Sendhoff, editors, *Proc. ICANN96*, 1996.

[8] Thomas Villmann. *Topologieerhaltung in selbstorganisierenden neuronalen Merkmalskarten*. PhD thesis, Universität Leipzig, 1996.

[9] Daniel Polani. Organization measures for self-organizing maps. In Teuvo Kohonen, editor, *Proceedings of the Workshop on Self-Organizing Maps (WSOM '97)*, pages 280–285. Helsinki University of Technology, June 1997.

[10] Thomas Martinetz and Klaus Schulten. Topology representing networks. *Neural Networks*, 7(2), 1994.

[11] E. Erwin, K. Obermayer, and K. Schulten. Self-organizing maps: ordering, convergence properties and energy functions. *Biol. Cybern.*, 67:47–55, 1992.

[12] Laurenz Wiskott and Terrence Sejnowski. Objective functions for neural map formation. TR INC-9701, The Salk Institute for Biological Studies, San Diego, CA, 1997.

[13] M. Cottrell, J.C. Fort, and G. Pagès. Two or three things that we know about the Kohonen algorithm. In Michel Verleysen, editor, *Proceedings of the European Symposium on Artificial Neural Networks (ESANN)*, pages 235–244, Brussels, 1994.

[14] M. Cottrell, J. C. Fort, and G. Pagès. Theoretical aspects of the som algorithm. In *Proc. WSOM '97*, pages 246–267, Espoo, Finland, June 1997.

[15] G. Jetschke. *Mathematik der Selbstorganisation*. Vieweg, Braunschweig, 1989.

[16] David Saad and Sara A. Solla. Exact solution for on-line learning in multilayer neural networks. *Phys. Rev. Lett*, 74(21):4337–4340, 1995.

[17] Hermann Haken. *Synergetic Computers and Cognition*. Springer, 1991.

[18] Angelika Spitzner. Parameter der Dynamik von Kohonen-Karten. Diplomarbeit, Universität Mainz, Institut für Informatik, 1997.

# Slope Centering: Making Shortcut Weights Effective

Nicol N. Schraudolph

nic@idsia.ch

IDSIA, Corso Elvezia 36

6900 Lugano, Switzerland

http://www.idsia.ch/

### Abstract

Shortcut connections are a popular architectural feature of multi-layer perceptrons. It is generally assumed that by implementing a linear sub-mapping, shortcuts assist the learning process in the remainder of the network. Here we find that this is not always the case: shortcut weights may also act as distractors that slow down convergence and can lead to inferior solutions. This problem can be addressed with *slope centering*, a particular form of *gradient factor centering* [1]. By removing the linear component of the error signal at a hidden node, slope centering effectively decouples that node from the shortcuts that bypass it. This eliminates the possibility of destructive interference from shortcut weights, and thus ensures that the benefits of shortcut connections are fully realized.

## 1 Shortcuts

Shortcut weights bypass a given hidden node by connecting its inputs directly to the node(s) it projects to. They are a popular architectural feature of multi-layer perceptrons, in particular those with more than one hidden layer. They are generally thought to be beneficial to the learning process by providing a linear sub-network that a) backpropagates error gradients to preceding layers without the blurring and attenuation associated with the passage through a layer of hidden nodes, and b) frees the bypassed hidden node(s) from responsibility for the linear component (now implemented by the shortcuts) of the mapping that it is to learn.

Here we take a closer look at the second argument. It is true that with shortcuts added, a nonlinear network generally attains larger capacity and may therefore be able to better approximate a given mapping. What about the dynamics of gradient descent in such a network though — how do shortcuts affect learning in the bypassed hidden node? Our experiments with single hidden layer networks — where backpropagation through shortcuts does not play a role — suggest that shortcuts actually slow down convergence, and may lead to inferior solutions (see Section 3).

This should not come as a surprise — after all, the simultaneous adaptation of additional parameters (the shortcut weights) at non-infinitesimal step sizes must necessarily add noise to the error signal. The shortcuts add degrees of freedom to which the bypassed hidden nodes are also coupled, and such redundant parametrization impedes optimization by gradient descent. In essence, while bypassed hidden nodes need no longer concern themselves with the linear component of the mapping to be learned, how are they to know this? In order to reap the full benefit of shortcut weights, the linear component must be explicitly subtracted from the error signal for bypassed nodes. With *slope centering* we introduce an efficient and effective way of doing this.

## 2  Slope Centering

Consider a hidden node with net input $y$ and activation given by $z = f(y)$, where $f$ is a nonlinear (typically sigmoid) function. According to the chain rule, the error gradient with respect to some objective $E$ acquires the factor $f'(y)$ — the node's current *slope* — as it is backpropagated through $f$:

$$\frac{\partial E}{\partial y} = f'(y) \frac{\partial E}{\partial z} \tag{1}$$

We can split this gradient into two orthogonal components:

$$\frac{\partial E}{\partial y} = \left[ f'(y) - \langle f'(y) \rangle \right] \frac{\partial E}{\partial z} + \langle f'(y) \rangle \frac{\partial E}{\partial z} \tag{2}$$

where $\langle \cdot \rangle$ denotes averaging over input patterns. This average evolves gradually on the slow time scale of the network's weight dynamics; for the pattern-dependent computation of the instantaneous gradient in (2) we can therefore assume $\langle f'(y) \rangle \approx$ const. Note that this means that the second term in (2) is linear: it is the error that would be backpropagated by a linear node in place of the present (nonlinear) hidden node. Conversely, by subtracting this linear error from it, we have made the other (first) term purely nonlinear. We have thus split the hidden node's error signal into its linear and nonlinear components.

As we have argued above, the linear component of the error should be removed for hidden nodes that are bypassed by shortcuts. This leads us to propose backpropagating error signals in that case via

$$\frac{\partial E}{\partial y} := \left[ f'(y) - \langle f'(y) \rangle \right] \frac{\partial E}{\partial z} \tag{3}$$

Since this differs from (1) in that the slope $f'(y)$ has been centered about zero by subtracting its average, we refer to this technique as *slope centering*. Note that (3) does not describe the error gradient proper; it rather expresses that gradient projected into the null space of the shortcut weights.[1] We have thus effectively decoupled the hidden node from its shortcuts; our empirical results (see Section 3) show that such an orthogonalization of parameters can be highly beneficial to the network's learning process.

---

[1] Rather than introduce additional notation to that effect, we have abused the partial derivative notation here as a placeholder for the slope-centered backpropagated error signal.

In online learning, the average slope $\langle f'(y) \rangle$ in (3) must be approximated, typically by an exponential running average. When batch learning is used, the average slope over the current batch may be approximated by the value computed for the previous batch. Alternatively, one may calculate the slope-centered error (3) for the current batch exactly by accumulating $\sum f'(y)$, $\sum \partial E/\partial z$, and $\sum f'(y) \partial E/\partial z$, then using the fact that

$$\left\langle \left[ f'(y) - \langle f'(y) \rangle \right] \frac{\partial E}{\partial z} \right\rangle = \left\langle f'(y) \frac{\partial E}{\partial z} \right\rangle - \langle f'(y) \rangle \left\langle \frac{\partial E}{\partial z} \right\rangle \qquad (4)$$

## 3  Empirical Results

We now demonstrate the effect of slope centering on speed and reliability of convergence as well as generalization performance in feedforward networks trained by accelerated gradient descent. After describing the general setup of the experiments, we present our respective results for two benchmarks: the toy problem of symmetry detection in binary patterns, and a difficult vowel recognition task.

### 3.1  Experimental Setup

For each benchmark we ran four experiments examining the separate and combined effect of shortcuts and slope centering. Each experiment consisted of a number of runs starting from different initial weights drawn from a zero-mean Gaussian distribution with standard deviation 0.3. Training was done in batch mode; the hidden-to-output weights of the network were updated *before* backpropagating error through them [2]. In order to make the optimization as efficient as possible, we tested a number of acceleration methods, then chose the combination of two (*vario-η* and *bold driver*) that yielded the fastest reliable convergence overall.[2] Vario-$\eta$ [3, 4, page 48] sets the local learning rate for each weight inversely to the standard deviation of its stochastic gradient:

$$\Delta w_{ij} = \frac{-\eta\, g_{ij}}{\varrho + \sigma(g_{ij})}, \quad g_{ij} \equiv \frac{\partial E}{\partial w_{ij}}, \quad \sigma(u) \equiv \sqrt{\langle u^2 \rangle - \langle u \rangle^2}, \qquad (5)$$

with the small positive constant $\varrho = 0.1$ preventing division by near-zero values. The global learning rate $\eta$ was adjusted by the bold driver technique [5, 6, 7, 8]: after each batch in which the error did not increase by more than $\varepsilon = 10^{-10}$ (for numerical stability), $\eta$ is increased by 2%. Otherwise, the last weight change is undone, and $\eta$ decreased by 50%. Due to the amount of recomputation they require, we did count those "failed" epochs in our performance figures.

### 3.2  Symmetry Detection Problem

A fully connected feedforward network with 8 inputs, 8 hidden units (tanh nonlinearity) and a single logistic output is to learn the symmetry detection

---

[2]Note that any performance advantage for slope centering reported thereafter has thus been realized *on top of* a state-of-the-art accelerated gradient method as control.

| slopes: topology: | conventional | | | | centered |
|---|---|---|---|---|---|
| | **mean ± st.d.** quartiles | direct comparison: # of faster runs | | | **mean ± st.d.** quartiles |
| short- cuts? **no** | **65.4 ± 15.9** 57/62/70 | | 52 – 48 | * | **51.6 ± 16.2** 43/64.5/∞ |
| | | 81 | 0 ⎪ 61 | 4 | |
| | | 17 | 39 ×⎪ 99 | 95 | |
| **yes** | **90.4 ± 31.1** 69.5/80/102 | | 0 – 100 | | **33.1 ± 8.6** 28/31/35 |

\* Mean and standard deviation exclude 34 runs which did not converge.

Table 1: The number of epochs required to converge to criterion on the symmetry detection task, with *vs.* without slope centering and/or shortcuts. Runs with identical random seeds are also compared directly (may sum to less than 100 due to ties).

task: given a binary pattern (composed of ±1s) at the input, it is to signal whether the pattern is symmetric about its middle axis (target = 1) or not (target = 0). The network was trained on all 256 patterns, using a cross-entropy loss function and *error centering* [1, 9], until the root-mean-square error of its output fell below 0.01. Since the complement of a symmetric bit pattern is also symmetric, the symmetry detection task has *no* linear component at all — we therefore expected shortcuts to be of minimal benefit here.

Table 1 shows that indeed adding shortcuts alone was not beneficial — it slowed down convergence in over 80 of the 100 runs performed, and significantly increased the c.v. (coefficient of variation). Subsequent addition of slope centering, however, brought about an almost 3-fold increase in learning speed, and restored the original c.v. of about 1/4. When used together, slope centering and shortcuts never increased convergence time, and on average cut it in half. By contrast, slope centering *without* shortcuts failed to converge about 1/3 of the time. This is no surprise: since slope centering projects the backpropagated gradient into the null space of (linear) shortcut weights, the hidden nodes can no longer reduce the linear component of the error signal on their own.

## 3.3   Vowel Recognition Problem

We also tested our approach on Deterding's speaker-independent vowel recognition data [10], which has been adopted [11] as a popular [12, 13, 14, 15] neural network benchmark. The task is to recognize the eleven steady-state vowels of British English in a speaker-independent fashion, given 10 spectral features of the speech signal. The data consists of 990 patterns: 6 instances for each of the 11 vowels spoken by each of 15 speakers. We follow the conventional split into training (first 8 speakers) and test set (remaining 7).

We trained fully connected feedforward networks with 10 inputs, 22 logistic hidden units, and 11 logistic output units by minimization of cross-entropy loss. The target was 1 for the output corresponding to the correct vowel, 0 for all others. After each epoch, the network's generalization ability was measured in

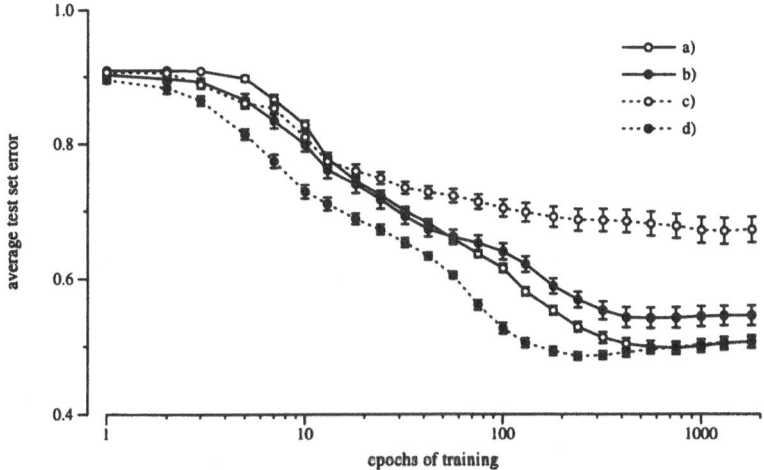

Figure 1: Evolution of the average test set error while learning the vowel recognition task with shortcut weights (filled marks) and/or slope centering (dashed lines). The two techniques worsen performance individually, while their conjunction improves it.

terms of its misclassification rate on the test set. For the purpose of testing, a maximum likelihood approach was adopted: the network's highest output for a given test pattern was taken to indicate its classification of that pattern.

Figure 1 shows how the test set error (averaged over 25 runs) evolved during training. While the addition of shortcut weights alone (b) worsened generalization performance (*cf.* a), in conjunction with slope centering (d) it resulted in faster convergence (by a factor of five), and to lower test test errors. The use of slope centering *without* shortcuts again proved ill-advised (c).

# 4   Conclusion

We have introduced slope centering, a technique that modifies a hidden node's error signal so as to eliminate interference from shortcut weights that bypass it. Using an already accelerated gradient method as a baseline, we found in two benchmarks that while shortcuts alone worsened performance, their combination with slope centering further sped up learning significantly. Slope centering is one example of the more general *gradient factor centering* approach [1]. It has long been known that centering input and hidden unit activities is beneficial [16, 17], and recently we have extended this notion to the centering of error signals [9]. In future work, we will investigate the centering of additional gradient factors, such as those that occur in networks with multiplicative nodes.

## Acknowledgment

This work was supported by the Swiss National Science Foundation under grant numbers 2100–045700.95/1 and 2000–052678.97/1.

# References

[1] N. N. Schraudolph, "On centering neural network weight updates", In *Tricks of the Trade* [18], ◁ ftp://ftp.idsia.ch/pub/nic/center.ps.gz ▷.

[2] S. Shah, F. Palmieri, and M. Datum, "Optimal filtering algorithms for fast learning in feedforward neural networks", *Neural Networks*, 5:779–787, 1992.

[3] R. Neuneier and H. G. Zimmermann, "How to train neural networks", In *Tricks of the Trade* [18].

[4] H. G. Zimmermann, "Neuronale Netze als Entscheidungskalkül", in *Neuronale Netze in der Ökonomie: Grundlagen und finanzwirtschaftliche Anwendungen*, H. Rehkugler and H. G. Zimmermann, Eds., pp. 1–87. Vahlen, Munich, 1994.

[5] A. Lapedes and R. Farber, "A self-optimizing, nonsymmetrical neural net for content addressable memory and pattern recognition", *Physica*, **D** 22:247–259, 1986.

[6] T. P. Vogl, J. K. Mangis, A. K. Rigler, W. T. Zink, and D. L. Alkon, "Accelerating the convergence of the back-propagation method", *Biological Cybernetics*, **59**:257–263, 1988.

[7] R. Battiti, "Accelerated back-propagation learning: Two optimization methods", *Complex Systems*, **3**:331–342, 1989.

[8] R. Battiti, "First- and second-order methods for learning: Between steepest descent and Newton's method", *Neural Computation*, 4(2):141–166, 1992.

[9] N. N. Schraudolph and T. J. Sejnowski, "Tempering backpropagation networks: Not all weights are created equal", in *Advances in Neural Information Processing Systems*, D. S. Touretzky, M. C. Mozer, and M. E. Hasselmo, Eds. 1996, vol. 8, pp. 563–569, The MIT Press, Cambridge, MA.

[10] D. H. Deterding, *Speaker Normalisation for Automatic Speech Recognition*, PhD thesis, University of Cambridge, 1989.

[11] A. J. Robinson, *Dynamic Error Propagation Networks*, PhD thesis, University of Cambridge, 1989.

[12] M. Finke and K.-R. Müller, "Estimating a-posteriori probabilities using stochastic network models", in *Proceedings of the 1993 Connectionist Models Summer School,*, M. C. Mozer, P. Smolensky, D. S. Touretzky, J. L. Elman, and A. S. Weigend, Eds., Boulder, CO, 1994, Lawrence Erlbaum Associates, Hillsdale, NJ.

[13] M. Herrmann, "On the merits of topography in neural maps", in *Proceedings of the Workshop on Self-Organizing Maps*, T. Kohonen, Ed. Helsinki University of Technology, 1997, pp. 112–117.

[14] S. Hochreiter and J. Schmidhuber, "Unsupervised coding with LOCOCODE", in *Proceedings of the 7th International Conference on Artificial Neural Networks,*, Lausanne, Switzerland, 1997, pp. 655–660, Springer Verlag, Berlin.

[15] G. W. Flake, "Square unit augmented, radially extended, multilayer perceptrons", In *Tricks of the Trade* [18].

[16] B. Widrow, J. M. McCool, M. G. Larimore, and C. R. Johnson, Jr., "Stationary and nonstationary learning characteristics of the LMS adaptive filter", *Proceedings of the IEEE*, 64(8):1151–1162, 1976.

[17] Y. LeCun, I. Kanter, and S. A. Solla, "Eigenvalues of covariance matrices: Application to neural-network learning", *Phys. Rev. Letters*, 66(18):2396–2399, 1991.

[18] G. B. Orr and K.-R. Müller, Eds., *Tricks of the Trade: How to Make Neural Networks Really Work* (working title), To appear in *Lecture Notes in Computer Science*. Springer Verlag, Berlin, 1998.

# Constrained Second-Order Recurrent Networks for Finite-State Automata Induction

Stefan C. Kremer

Computing and Information Science Department,
University of Guelph, Guelph, Ontario, N1G 2W1, Canada
E-mail: skremer@snowhite.cis.uoguelph.ca

Ramón P. Ñeco, Mikel L. Forcada
Departament de Llenguatges i Sistemes Informàtics,
Universitat d'Alacant, E-03071 Alacant, Spain.
E-mail: {neco,mlf}@dlsi.ua.es

## Abstract

This paper presents an improved training algorithm for second-order dynamical recurrent networks applied to the problem of finite-state automata (FSA) induction. Second-order networks allow for a natural encoding of finite-state automata in which each second-order connection weight corresponds to one transition in a finite-state automaton. In practice, however, when trained using gradient descent, these networks almost never assume this type of encoding and sophisticated algorithms must be used to extract the encoded automata. This paper suggests a simple modification to the standard error function for second-order dynamical recurrent networks which encourages these networks to assume natural FSA encodings when trained using gradient descent. This obviates the need for cluster-based extraction techniques and provides a simple method for guaranteeing the stability of the network for arbitrarily long sequences. Initial results also suggest that fewer training strings must be presented to achieve convergence using the modified error.

## 1 Introduction

A number of researchers have trained discrete-time recurrent neural networks (DTRNN) to behave like deterministic finite-state automata (DFA) (see, e.g., [1, 2]. The internal representation of learned DFA states can deteriorate for long strings, due to the use of real-valued sigmoid functions which never reach the bounds [3]. In this case, we say that the network is *unstable*. For this reason, it is difficult to make predictions about the generalization performance of trained DTRNN. As a partial solution to this problem, some authors force learning (or modify the architecture) to occur such that state values form clusters [3, 4]. In this paper we show a slightly modified learning algorithm with respect to that used in [1] to learn DFA using second-order DTRNN, which obviates the need for cluster-based extraction of the DFA, and obtains a stable behavior for long strings.

# 2 Encoding DFA in DTRNN

## 2.1 Definitions

**Deterministic finite-state automata (DFA):** Formally a DFA $M$ is a 5-tuple, $M = (Q, \Sigma, \delta, q_I, F)$, where $Q$ is a set of $|Q|$ states labelled $q_1, ..., q_{|Q|}$; $\Sigma$ is a set of input symbols labelled $\sigma_1, ..., \sigma_{|\Sigma|}$; $\delta$ is the state transition function, $\delta : Q \times \Sigma \to Q$; $q_I$ is the initial state, and $F$ is the set of accepting states. A string is accepted by the DFA $M$ if an accepting state is reached; otherwise, the string is rejected.

**Discrete-time recurrent neural networks (DTRNN):** A DTRNN [5] can be defined in a way that is parallel to the above definition of DFA [2]. A DTRNN is a 6-tuple $N = (X, U, Y, \mathbf{f}, \mathbf{h}, \mathbf{x}_0)$, where $X = [S_0, S_1]^{n_X}$ is the state space, with $S_0$ and $S_1$ the values defining the range of outputs of the transfer functions, and $n_X$ the number of state units; $U = \mathcal{R}^{n_U}$ is the set of possible input vectors, with $n_U$ the number of input lines; $Y = [S_0, S_1]$ is the output space of the network (one output unit); $\mathbf{f} : X \times U \to X$ is the *next-state function*, which computes a new state $\mathbf{x}[t]$ from the previous state $\mathbf{x}[t-1]$ and the input just read $\mathbf{u}[t]$; $\mathbf{h} : X \times U \to Y$ is the output function; and finally $\mathbf{x}_0$ is the initial state. In this paper we use a second-order DTRNN (2ODTRNN) [1, 2] such that the $i$-th coordinate of the next-state function is : $\mathbf{f}_i(\mathbf{x}[t-1], \mathbf{u}[t]) = g\left(\sum_{j=1}^{n_X} \sum_{k=1}^{n_U} W_{ijk} x_j[t-1] u_k[t]\right)$, and the output is: $y = \mathbf{h}(\mathbf{x}[t-1], \mathbf{u}[t]) = g\left(\sum_{j=1}^{n_X} \sum_{k=1}^{n_U} W_{0jk} x_j[t-1] u_k[t]\right) \in Y$ where $g(x) = 1/(1 + e^{-x})$. All weights in the above equations are real numbers.

## 2.2 Encoding

An advantage of 2ODTRNN over their first-order counterparts is that the former allow a one-to-one mapping between the states of an automaton and the processing units in the network (see [6]) whereas the latter do not (see [7]). To encode a given DFA, a 2ODTRNN is created which takes input vectors equal in dimensionality to the number of input symbols. In particular, we assume that component $u_k[t]$ of each input vector is equal to 1 if the input symbol at time $t$ is $\sigma_k$ and 0 for all other input symbols (*one-hot* encoding). We will use a similar encoding to map automaton states to DTRNN unit activations. We say that, if after presentation of a string, the activation value of the output unit $y$ is close to 1, then the string is accepted, while if it is close to 0, the string is rejected. We use the *state units*, to represent the state of the automaton using a one-hot encoding; that is, $\mathbf{x}_i[t]$ is high if the state of the automaton at time $t$ is $q_i$ and low otherwise. Thus, the DTRNN will have $n_X = |Q|$ state units. The initial state is represented by unit 1. The objective of the encoding scheme used in this paper is to ensure that when the current input symbol and state are encoded in the network as defined above, the operation of the network guarantees that the next activation vector corresponds to the next state of the automaton. To ensure this operation by the network, we require that

$$W_{ijk} = \begin{cases} +H & \text{if } \delta(q_j, \sigma_k) = q_i \\ -H & \text{otherwise} \end{cases} \qquad W_{0jk} = \begin{cases} +H & \text{if } \delta(q_j, \sigma_k) \in F \\ -H & \text{otherwise} \end{cases}$$

where $H > 0$. It is easy to prove that a second-order DTRNN of size $n_X = |Q|$ may easily emulate a DFA of size $|Q|$ using the encoding defined above for sufficiently large values of $H$. The proof presented here is a special case of a more general scheme described in [8] and [9]. To our knowledge, this construction using no biases and weights that are either $-H$ or $H$ has never been considered before. We define that a state vector $\mathbf{x}$ of the network is an $\epsilon$-valid representation of state $q_i$ for some $\epsilon$ in $(0, 0.5)$ if: (1) the state $x_i$ of the unit representing state $q_i$ is *high*, that is, within the interval $[1 - \epsilon, 1]$; (2) the state of all the other units representing state is *low*, that is, within the interval $[0, \epsilon]$; and (3) the state of the output unit is high if $q_i \in F$ and low otherwise. As can be seen, there is a symmetric forbidden interval $(\epsilon, 1 - \epsilon)$ which is inaccessible to $\epsilon$-valid representations.

Then, given the following conditions: (1) the initial state of the network is an $\epsilon$-*valid representation* of the initial state of the DFA; (2) the activation function is $g(x) = 1/(1 + \exp(-x))$ and thus $S_0 = 0$, $S_1 = 1$; (4) the weights are chosen as defined above; then, for every automaton of size $|Q|$ there is always a set of pairs $(H, \epsilon)$ such that the DTRNN always assumes valid representations of state regardless of the length of the input string, and it is always possible to find the minimum value of $H$ by varying $\epsilon$.

The proof proceeds by induction on the length of strings. The base case (zero length) is guaranteed (independently of $\epsilon$) by the choice of initial state made above. And for the induction step, we want to find conditions on the pair $(H, \epsilon)$ such that the induction step holds, that is, that a network in an $\epsilon$-valid state moves to the corresponding $\epsilon$-valid state after reading any input symbol. Let us assume, without loss of generality, that $q_j$ is the state before reading symbol $\sigma_k$ and that $\delta(q_j, \sigma_k) = q_i$. That is, the state before reading the symbol is an $\epsilon$-valid representation of $q_j$, in particular the following worst (weakest) case, when $\mathbf{x}_j[t - 1]$ is $1 - \epsilon$ and all other $\mathbf{x}_l[t - 1]$ with $l \neq j$ are all $\epsilon$. After reading symbol $\sigma_k$ we have to make sure that: (1) $\mathbf{x}_i[t]$ is high; (2) all $\mathbf{x}_n[t]$ with $n \neq i$ are low; (3) the state of the acceptance neuron is high if $q_i \in F$ and low otherwise.

Ensuring a high value for $\mathbf{x}_i[t]$ in the worst case leads to the following inequality: $1 - \epsilon \leq g(H - |Q|H\epsilon)$. Ensuring a low value for $\mathbf{x}_n[t]$ $(i \neq n)$ is more difficult in the following special case: when, for all, $l \neq j$ $\delta(q_l, \sigma_k) = q_n$. In that case, the following inequality has to hold: $\epsilon \geq g(-H + |Q|H\epsilon)$. Both inequalities are equivalent, since $g(x) = 1 - g(-x)$ and $g(x)$ is a strictly growing function. Identical inequalities are obtained by considering the behavior of the state of the acceptance unit. The corresponding equation relating $H$ and $\epsilon$, $\epsilon = g(-H + H|Q|\epsilon)$ ensures a minimum $H$ for a given $\epsilon$, which can be found by taking $dH/d\epsilon = 0$ and solving iteratively the resulting equation. The resulting pair $(H, \epsilon)$ ensures the validity of the induction step, and so, the correct behavior (validly represented states) of the DTRNN for strings of any length. Example values of $H$ are 2, 3.113, 3.589, 3.922, 6.224 for $|Q| = 2, 3, 4, 5, 30$.

# 3  Training Algorithm

We use a modification to the error used in the gradient descent algorithm which encourages the DTRNN to adopt the above encoding. More precisely, we augment the standard performance error measure with an encoding error measure that evaluates the difference between the network's current weights and the above encoding. This causes the gradient descent algorithm to favor these desirable (stable and easily interpretable) encodings over less favorable encodings while simultaneously attempting to match the desired outputs.

We identify two desired criteria for our encodings: (1) all weights in the network should approach $+H$ or $-H$ and, since we are interested only in DFA, (2) only one state unit should be high for a given input and current state. The second criterion can also be specified by saying that for each state unit $i$, input unit pair, $(j, k)$, for each value of $i$, only one of the weights $W_{ijk}$ should be positive. We can measure the degree to which a network satisfies these criteria by two simple metrics that measure an encoding error.

We define the error of the network as $E = \alpha E_{\text{perf}} + \beta E_H + \gamma E_1$, where $E_{\text{perf}}$ represents the standard RMS performance error measuring the difference between the actual and desired output. $E_H \equiv \frac{1}{4} \sum_{ik} \sum_{j\neq 0} (W_{ijk}^2 - H^2)^2$ is minimized when all connection weights assume a value $-H$ or $+H$. $E_{1+} \equiv \frac{1}{2} \sum_{j\neq 0} \sum_k (\sum_{i\neq 0} W_{ijk} - H(3 - N_{\text{SODRN}}))^2$ is minimized when only one of the connection weights leading away from a given state and input unit is $+H$ and the others are all $-H$ and is zero when exactly one of the $|Q|$ weights leading out of each state-input pair, $(j, k)$ is $H$, and all of the other $|Q| - 1$ weights are $-H$. The parameters $\alpha$, $\beta$ and $\gamma$ are used to control the effect of each error term[1].

We use an incremental technique which presents strings to the network in order of increasing length. First we present to the network the strings in the learning set up to length $L$. We say that the network has learned all these strings when two conditions are satisfied: (1) the output is correct, and (2) for each $j$ and $k$, only one weight $W_{ijk}$ is positive. When both conditions are satisfied, we add to the learning set strings of length $L + 1$. We repeat the process until all the strings in the training set are correctly classified (we used all strings up to length 9 for these experiments). Training the net using this method, we find that the network incrementally learns the grammar rules or state transitions. This incremental learning technique can be viewed as an alternative to inserting symbolic knowledge covering these rules as in [10].

---

[1] In all the experiments reported here we used $\alpha = 1.1$, $\beta = 0.1$ and $\gamma = 0.1$.

In our simulations, we use a value of $H$ that is different for each set of weights $C_{jk} = \{W_{ijk}|i = 1, 2, ..., |Q|\}$, so we have $|\Sigma||Q|$ distinct values of $H$, one value for each pair $(j, k)$, $H_{jk}$, the mean value of the set $C_{jk}$ for each training epoch[2].

# 4 Experiments

We examined the performance of the learning algorithm on a class of benchmark problems for grammar induction: The Tomita grammars [1] which were also used to evaluate the automaton extraction techniques described earlier [1]. These grammars are regular grammars over a binary alphabet ($\Sigma = \{0, 1\}$).

In all the experiments, the presentation of training samples is performed as follows. For each given string length, $L$, we use the same training algorithm used in [1] and [10]. The network only gets to see some small randomly-selected fraction of the learning set (we have used 2 strings). When the network classifies the current learning set correctly or when it reaches a maximum number of epochs (50 in our experiments), 2 more strings are added to the current learning set. In Table 1 we show some preliminary results using the technique described. In this table we show, for each Tomita grammar tested, the maximum length of the strings presented, the number of strings presented during learning, epochs required and the value of the tolerance used ($\tau$). If we compare these results with those reported in [1], it reveals that while our method offers a simpler extraction technique, we require many more epochs to achieve convergence. This, however, is an unfair comparison, since we present significantly fewer training strings per epoch (in particular, the largest numbers of epochs are often devoted to the shortest string lengths). This makes our approach more suitable for problems in which training data is sparse. It should further be noted that our average string length (especially for early epochs) is much less than that in [1].

| Tomita grammar | Max. length | Strings presented | Epochs required | $\tau$ | Epochs in [1] | Strings in [1] |
|---|---|---|---|---|---|---|
| 2 | 4 | 30 | 1521 | 0.2 | 46 | 1024 |
| 4 | 6 | 126 | 1332 | 0.3 | 133 | 1024 |
| 6 | 4 | 30 | 733 | 0.1 | 733 | 1024 |
| 7 | 7 | 254 | 2916 | 0.1 | 186 | 1024 |

Table 1: Results obtained learning Tomita grammars.

---

[2]This means that we are in fact not computing a true gradient since, in our equations, we assumed that the value of $H$ did not depend on the current weights (future work may consider using a true gradient).

# 5 Concluding Remarks

In this paper, we have suggested a simple modification to the standard error definition for second-order DTRNN which encourages these networks to assume natural encodings of DFA when trained using gradient descent. The automata learned using the modified error are strongly biased towards having $|Q|$ states if the network has $|Q|$ state units. Future work will examine what happens when there is a mismatch between the number of states and processing units. It is hypothesized that these networks might be able to infer simple automata that are reasonable approximations to the desired automata even when exact automata compatible with the training sample would be much bigger, and that our new algorithm is good for sparse training data.

**Acknowledgements:** We acknowledge funding by the Spanish Comisión Interministerial de Ciencia y Tecnología (CICyT) under project TIC97-0941. R.P. Ñeco has been supported by the Spanish Generalitat Valenciana. The authors would like to thank Dr. Lee Giles and NEC Research Institute (Princeton, NJ) for providing software to simulate 2ODTRNN.

# References

[1] Giles, C.L., Miller, C.B., Chen, D. *et al.* (1992) Learning and Extracting Finite State Automata with Second-Order Recurrent Neural Networks, *Neural Computation* 4(3):393–405.

[2] Pollack, J.B. (1991) The Induction of Dynamical Recognizers, *Machine Learning* 7:227–252.

[3] Zeng, Z., Goodman, R., Smyth, P. (1993) Learning Finite State Machines With Self-Clustering Recurrent Networks, *Neural Computation* 5(6):976–990.

[4] Das, S., Mozer, M. (1998) Dynamic On-Line Clustering and State Extraction: An Approach to Symbolic Learning, *Neural Networks* 11(1):53–64.

[5] Haykin, S. (1994) *Neural Networks, A Comprehensive Foundation*, New York, NY: Macmillan.

[6] Goudreau, M.W., Giles, C.L., Chakradhar, S.T. *et al.* (1994) First-Order Versus Second-Order Single-Layer Recurrent Neural Networks, *IEEE Transactions on Neural Networks*, 5(3):511–513.

[7] Kremer, S.C. (1995) On the Computational Power of Elman-Style Recurrent Networks, *IEEE Transactions on Neural Networks* 6(4):1000–1004.

[8] Carrasco, R.C., Forcada, M.L., Valdés, M.A., Ñeco, R.P. (1998) A stable encoding of finite-state machines in discrete-time recurrent neural nets with sigmoid units, to be published.

[9] Ñeco, R.P., Forcada, M.L., Carrasco, R.C. *et al.* (1998) Encoding of Sequential Translators in Discrete-Time Recurrent Neural Nets, *Dep. de Llenguatges i Sistemes Informàtics, Universitat d'Alacant*, Technical Report DLSI-TR-01-98.

[10] Omlin, C.W., Giles, C.L. (1996) Stable Encoding of Large Finite-State Automata in Recurrent Neural Networks with Sigmoid Discriminants, *Neural Computation* 8:675–696.

# Exact Learning Curves for EKF Training

B Schottky and D Saad

Neural Computing Research Group, Aston University
Birmingham B4 7ET, Great Britain
{schottba,saadd}@aston.ac.uk

**Abstract**

We formulate a learning algorithm for online learning in neural networks using the principles of the Extended Kalman Filter approach. This gives rise to a Bayesian learning scheme where the learning rate parameter is adapted automatically. The approach is applied to regression problems whereby the rule to be learned is non-stationary, examining both linear and nonlinear cases.

## 1  Introduction

Online learning is an important learning paradigm in the context of neural networks, especially for large networks and non-stationary tasks. A continuous stream of training examples is used for adapting sequentially a set of parameters, gradually improving the approximation to the underlying rule realised by the system. This rule is often referred to as 'teacher', whereas the approximating system is termed 'student'.

A widely used approach for regression problems is to define an objective measure for the discrepancy between the desired and the actual output produced by the current estimate of the rule. The parameters of the student are then updated by gradient descent on this error measure controlled by a learning rate parameter $\eta$. One problem of this 'ad hoc' approach is the choice of the learning rate. The optimal schedule, depending on the characteristics of the system [1, 2], is generally not known, so heuristic estimates have to be used. For instance in a noisy but learnable stationary scenario, one has to decay the learning rate to zero asymptotically as $1/t$ ($t$ counts the training examples) with a specific prefactor [1]. If, however, the underlying rule is drifting, the learning rate has to stay finite, keeping track of the changing rule.

A principled alternative is to use Bayesian methods; this allows to incorporate one's knowledge (or belief) in a statistical manner, from which the optimal learning process may be inferred. Moreover, if the assumptions are precise one gets optimal performance (whereas one has to pay a price if they are not). In the Bayesian approach current knowledge about the rule is kept in the student's posterior distribution rather than in a single estimate. This distribution is then updated due to a new example using the Bayes rule. In principle the Bayes rule *is* already an online algorithm: the complete knowledge is represented in a posterior distribution of some parameters; this distribution is updated due to a new example, the example itself can then be discarded. However in general one has to approximate the true posterior since the exact form requires storing all examples. A general framework for Bayesian online learning using a Gaussian approximation to the posterior has been proposed in [3]. Whereas this algorithm is efficient for classification problems [3, 4] it is computationally very expensive for regression and is applicable to very simple architectures only.

An approximation rendering the algorithm feasible is to use the Extended Kalman Filter (EKF) approach [5]. We have shown how this method can be applied to neural networks solving regression problems in [6], independently this was derived in [7].

In this approximation, the posterior is represented by a multidimensional Gaussian distribution and the update of it's two parameters, the mean and the variance, is carried out by linearising the update dynamics and the teacher's response (the 'update' and 'measurement' in signal processing terms) which are in general nonlinear. It is known that depending on the type and severeness of these nonlinearities the EKF approach might give poor results.

The algorithm is applicable to any smooth network. In this paper however we present the update rules only for a nonlinear perceptron. The main contribution of this paper is then to provide an analysis of the algorithm using methods from statistical physics. We derive exact learning curves in the limit of large system size $N$ and can therefor asses the behaviour for several parameter settings like noise variance or drift speed. We use these tools to address briefly two questions: How is the performance compared to a non-Bayesian approach and how does the nonlinearity in particular influence the performance of the EKF approach.

Moreover one should notice, that the models and results presented here are also interesting from the signal processing point of view where EKF is commonly used, providing insight to the performance of Bayesian online learning in this context.

## 2   The model

We consider an unknown teacher rule $f_T$ which maps an input pattern $\boldsymbol{\xi} \in \mathbf{R}^N$ to an output value $z \in \mathbf{R}$,

$$z = f_T(\mathbf{w}_0, \boldsymbol{\xi}) + \zeta, \tag{1}$$

where $\zeta$ is an additional noise term representing the corruption process considered here to have a normal distribution of zero mean and variance $\sigma_T^2$. The parameter vector $\mathbf{w}_0$ is unknown and specifies the rule within a given general architecture.

In a non-stationary scenario the couplings $\mathbf{w}_0$ are time dependent, so we have

$$\mathbf{w}_0(t+1) = v(\mathbf{w}_0(t), \boldsymbol{\rho}_t), \tag{2}$$

where $t$ is the discrete time variable and $\boldsymbol{\rho}_t$ represents a set of random variables driving the nondeterministic part of the evolution.

The information about the underlying rule is provided by a stream of random examples $(\boldsymbol{\xi}^t, z^t)$ where $z^t$ is the corrupted teacher response at time $t-1$ due to Eq.(1) and is used for updating the student parameter vector $\mathbf{w}$ which specifies the student rule $f_S(\mathbf{w}, \boldsymbol{\xi})$. The EKF based learning algorithm can be applied to any smooth network architecture. This analysis however will be restricted to the case of a nonlinear perceptron in a realisable scenario, so we choose $f_T = f_S = f$ with

$$f(\mathbf{w}, \boldsymbol{\xi}) = \phi(\mathbf{w} \cdot \boldsymbol{\xi}), \tag{3}$$

where the transfer function $\phi$ will be specified later.

The teacher vector $\mathbf{w}_0$ is restricted to be normalised to one, $|\mathbf{w}_0| = 1$ and the non-stationarity is chosen to be a random drift

$$\mathbf{w}_0(t+1) \cdot \mathbf{w}_0(t) = 1 - \frac{\delta_T}{N}, \tag{4}$$

where the coefficient $\delta_T$ controls the drift 'speed'.

The main quantity of interest is the Bayesian generalisation error given by

$$\epsilon_g(t) = \left\langle \left( \phi(\mathbf{w}_0(t) \cdot \boldsymbol{\xi}) - \langle \phi(\mathbf{w} \cdot \boldsymbol{\xi}) \rangle_{\rho_t(\mathbf{w})} \right)^2 \right\rangle_{\boldsymbol{\xi}}, \tag{5}$$

with $\rho_t(\mathbf{w})$ being the current estimate of the posterior.

# 3   The update equations for EKF Learning

In a Bayesian approach the current knowledge about the teacher couplings is contained in the posterior distribution approximated in the EKF framework by a Gaussian (for a derivation of the EKF approach see [5, 6, 7]

We represent the posterior $\rho(\mathbf{w})$ for the couplings at time $t$ by the mean $\hat{\mathbf{w}}(t)$ and the variance matrix $C(t)$, so $\rho(\mathbf{w}) = \mathcal{N}(\hat{\mathbf{w}}(t), C(t))$. Given the case (3) one gets the update equations for a given new example $(\boldsymbol{\xi}^{t+1}, z^{t+1})$

$$\hat{\mathbf{w}}(t+1) = \hat{\mathbf{w}}(t) + \frac{\phi'(z^{t+1} - \phi)}{\sigma_S^2 + \phi'^2 \boldsymbol{\xi}^{t+1'} C(t) \boldsymbol{\xi}^{t+1}} C(t) \boldsymbol{\xi}^{t+1} \tag{6}$$

$$C(t+1) = C(t) - \frac{\phi'^2 C(t) \boldsymbol{\xi}^{t+1'} \boldsymbol{\xi}^{t+1} C(t)}{\sigma_S^2 + \phi'^2 \boldsymbol{\xi}^{t+1'} C(t) \boldsymbol{\xi}^{t+1}} + \frac{2\delta_S}{N^2} I, \tag{7}$$

where $\phi$ and $\phi'$ replace $\phi(\hat{\mathbf{w}}(t) \cdot \boldsymbol{\xi}^{t+1})$ and $\phi'(\hat{\mathbf{w}}(t) \cdot \boldsymbol{\xi}^{t+1})$. We have introduced two new variables: $\sigma_S$ and $\delta_S$. Ideally these should be chosen to correspond to the true noise variances, i.e. $\sigma_S = \sigma_T$ and $\delta_S = \delta_T$. However, if the true noise variances are not known (as is generally the case) one has to use estimated values and use them to update the posterior.

The scaling in this scenario is $O(\hat{\mathbf{w}}_i) = 1/\sqrt{N}$, $O(C_{ij}) = 1/N$ and $O(\xi_i) = 1$ keeping in mind that we later take the limit $N \to \infty$.

# 4   Isotropic assumption and equations for the order parameter

In the statistical mechanics approach we are interested in the system's behaviour for large system size $N \to \infty$. In this so called 'thermodynamic limit' one can capture the system behaviour by a small set of order parameters [8]. In general one has to consider the whole covariance matrix $C$, even in this limit. This is, however, quite expensive in time and memory. A cheaper version of the algorithm is to use an isotropic approximation

$$C(t) = \frac{\eta(t)}{N} I \tag{8}$$

where $I$ is the $N \times N$-identity matrix. Then, employing the thermodynamic limit one can describe the system by the order parameters $Q = \mathbf{w}_0 \cdot \mathbf{w}_0$, $R = \hat{\mathbf{w}} \cdot \hat{\mathbf{w}}$ and $\eta$. By introducing the continuous time $\alpha = t/N$ the evolution of these parameters is given by

$$\frac{dQ}{d\alpha} = 2\eta \left\langle \hat{\mathbf{w}} \cdot \boldsymbol{\xi} \frac{\Delta\phi'}{\sigma_S^2 + \eta\phi'^2} \right\rangle + \eta^2 \left\langle \frac{\Delta^2\phi'^2}{(\sigma_S^2 + \eta\phi'^2)^2} \right\rangle \tag{9}$$

$$\frac{dR}{d\alpha} = \eta \left\langle \mathbf{w}_0 \cdot \boldsymbol{\xi} \frac{\Delta\phi'}{\sigma_S^2 + \eta\phi'^2} \right\rangle - \delta_T R \tag{10}$$

$$\frac{d\eta}{d\alpha} = -\eta^2 \left\langle \frac{\phi'^2}{\sigma_S^2 + \eta\phi'^2} \right\rangle + 2\delta_S \tag{11}$$

where $\Delta$ is the difference between the student's (with weight vector $\hat{\mathbf{w}}$) and the noise corrupted teacher's responses, $\Delta = \phi(\mathbf{w}_0 \cdot \boldsymbol{\xi}) + \zeta - \phi(\hat{\mathbf{w}} \cdot \boldsymbol{\xi})$; $\phi'$ is to be taken at $\hat{\mathbf{w}} \cdot \boldsymbol{\xi}$ and the average $\langle \ldots \rangle$ is over the random patterns $\boldsymbol{\xi}$ of length $|\boldsymbol{\xi}| = \sqrt{N}$ and noise $\zeta$. The joint statistics of the random variables $\mathbf{w}_0 \cdot \boldsymbol{\xi}$ and $\hat{\mathbf{w}} \cdot \boldsymbol{\xi}$ is entirely determined by $Q$ and $R$ (being a two dimensional Gaussian), so that the above equations are a closed system which can be solved numerically given some initial conditions.

## 5 Results for the nonlinear perceptron

We address now briefly two questions: How good is the approach compared to gradient descent learning and how does the performance depend on specific system parameters, i.e. on the nonlinearity of the transfer function and the drift speed of the rule.

We focus on the case were $\sigma_T$ and $\delta_T$ are known (setting $\sigma_S = \sigma_T$ and $\delta_S = \delta_T$) and we fix the noise rates to $\sigma_S = \sigma_T = 0.3$. As the transfer function we introduce $\phi(x) = \text{erf}(ax/\sqrt{2})$, where the parameter $a$ controls the nonlinearity in the system.

In fig.1 we compare EKF- and gradient descent learning where we have fixed $a = 1$. For the latter we show the results for three fixed learning rates $\eta$. For a stationary task (left figure) the learning rate has to be small for good asymptotic results (it actually has to be annealed to zero for $\alpha \to \infty$), but this causes the performance at the beginning of the learning process to deteriorate. To get better results one would have to impose an explicit learning rate schedule. For learning the drifting rule there is asymptotically an optimal nonzero learning rate which is however not known. EKF learning on contrast yields superior results and the choice of the 'effective' learning rate is done automatically.

Turning to the second question one should notice that for nonlinear transfer function the update equations use a linearization around the actual mean. This approximation is getting worse as the nonlinearity increases and as the posterior distribution tails become more significant. Therefore, cases with high nonlinearity and large drifting speed would lead to a bad performance of the algorithm. The control parameter $a$ and a variation of $\delta_T$ allows one to investigate these effects.

Fig.2 shows the learning curves for several values of the non linearity parameter $a$ for a stationary (left) and non-stationary task. The theoretical results show, in the region investigated, that for drifting rules the asymptotic performance worsens with

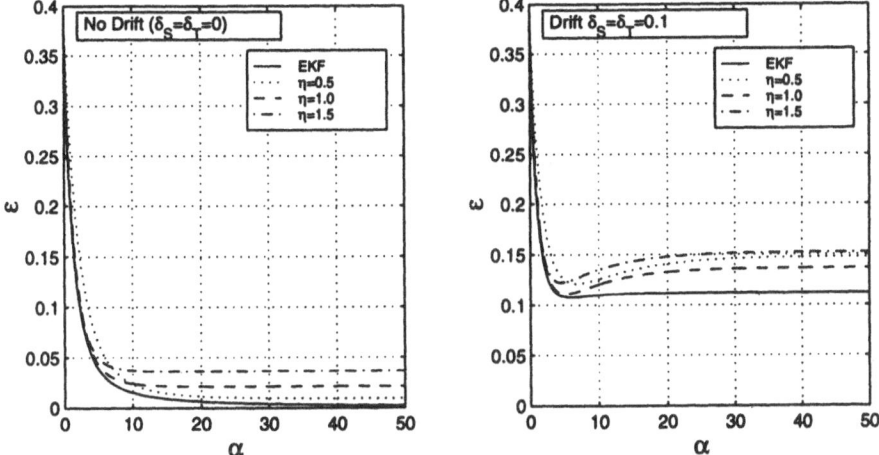

Figure 1: The generalisation error for EKF- and gradient descent learning (the latter for several values of the learning rate $\eta$) is compared for a stationary (left) and non stationary task. The Bayesian approach shows there superior performance while determining the learning rate schedule automatically.

increasing nonlinearity; for the case $a = 3$ the behaviour even diverges. This means that there is a transition to a non-converging phase depending on the specific system parameters: for this cases the EKF algorithm fails completely.

## 6  Conclusions

We have presented an EKF based Bayesian learning scheme for neural networks solving regression problems. Beside being based on a principled approach this algorithm avoids the problem of choosing training parameters like the learning rate, which has to be done heuristically in other learning schemes. Using methods from statistical mechanics we obtain exact results for the learning curves.

We showed the efficiency of the algorithm, in several cases outperforming the simple gradient descent. However we showed as well, that due to the approximations involved the EKF based learning scheme fails for highly nonlinear mappings when the rule to be learned is drifting. Our description gives exact results for this generally known phenomenon and opens the field for further analysis.

More extensive investigations of the system behaviour depending on the various parameters is under way and the analysis can be extended to more complicated network architectures (the algorithm itself is directly applicable to any smooth network). Another open question is the possible improvement by allowing for a non-isotropic Gaussian posterior distribution. This will be particularly beneficial for low noise rates and at the beginning of the learning process and might well prove crucial for learning in more complicated networks. Finally there is the question of model evaluation in an online manner [7] in analogy to model evaluation for batch learning [9].

**Acknowledgements:** We would like to thank the Leverhulme Trust for their support (F/250/K)

540

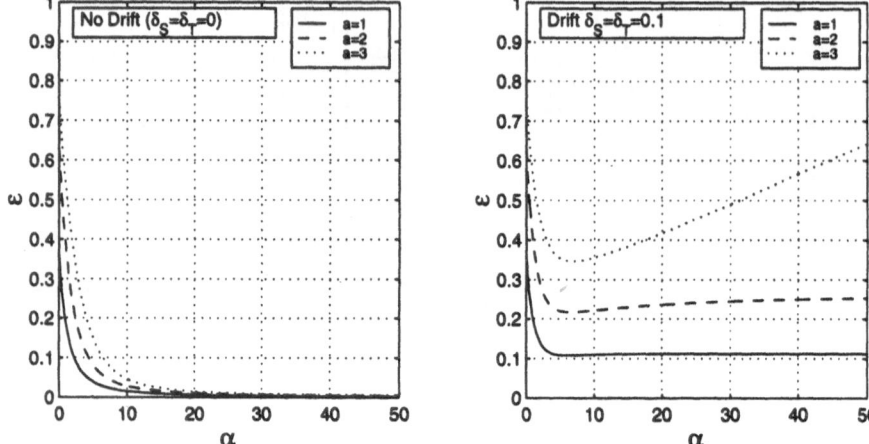

Figure 2: The graphs show the Bayesian generalisatoin error for a stationary (left) and a drifting task (right). Whereas there is always convergence for $\delta_S = \delta_T = 0$ this is not the case for a drifting concept: when the nonlinearity is "sufficiently large" (here $a = 3$) the system diverges. Simulations confirm the theoretical curves.

# References

[1] Leen TK, Schottky B and Saad D, Two approaches to optimal annealing, Proceedings of NIPS*97, Eds. M.I. Jordan, M.J. Kearns and S.A. Solla, MIT press, Cambridge MA, 1998.

[2] Saad D and Rattray M, Globally Optimal Parameters for On-line Learning in Multilayer Networks, *Phys. Rev. Lett.*, **79**, 2578-2581, 1997.

[3] Opper M, Online versus offline learning from random examples: General Results, *Phys.Rev.Lett.* **77**, 4671-4674, 1996

[4] Winther O and Solla S, Bayesian online learning in the perceptron, *preprint*, 1997 and Optimal Bayesian online, *preprint*, 1997

[5] Welch G and Bishop G, An introduction to the Kalman filter, *UNC-CH Computer Science Technical Report 95-041*, 1995

[6] Schottky B and Saad D, An online algorithm for neural networks using Extended Kalman Filter, *preprint*

[7] deFreitas JFG, Niranjan M and Gee AH, Hierarchical Bayesian-Kalman Models for regularisation and ARD in sequential learning, *preprint*, 1997

[8] Saad D and Solla S, On-line learning in soft committee machines, *Phys. Rev. E* **52** 4225 (1995)

[9] MacKay DJC, Bayesian interpolation, *Neural Computation* **4**, 415-447, 1992 and The evidence framework applied to classification networks, *Neural Computation* **4**, 720-736, 1992

# Sparse Regression: Utilizing the Higher-order Structure of Data for Prediction

Aapo Hyvärinen

Helsinki University of Technology
Laboratory of Computer and Information Science
P.O. Box 2200, FIN-02015 HUT, Finland
Email: aapo.hyvarinen@hut.fi

### Abstract

Independent component analysis and the closely related method of sparse coding model multidimensional data as linear combinations of independent components that have nongaussian, usually sparse, distributions. Such a modelling approach is especially suitable in large dimensions, as it avoids the curse of dimensionality. It also seems to represent important properties of sensory data. In this paper we show how to use these models for regression. If the joint density of two random vectors is modelled by independent component analysis, it is possible to obtain simple algorithms to compute the maximum likelihood predictor of one of the vectors when the other vector is observed. The obtained predictors are nonlinear, but in contrast to such nonparametric methods as MLP, the nonlinearities are not chosen ad hoc: They are directly determined by the density approximation.

## 1 Introduction

Independent component analysis (ICA) [3, 8] is a recently developed statistical technique whose goal is to express observed random variables $x_1, x_2, ..., x_m$ as linear combinations of unknown component variables, denoted by $s_1, s_2, ..., s_n$. The components $s_i$ are, by definition, mutually statistically independent, and zero-mean. Let us arrange the observed variables $x_i$ into a vector $\mathbf{x} = (x_1, x_2, ..., x_m)^T$ and the independent components $s_i$ into a vector $\mathbf{s}$, respectively; then the linear relationship is given by

$$\mathbf{x} = \mathbf{As} \tag{1}$$

Here, $\mathbf{A}$ is an unknown $m \times n$ matrix, called the mixing matrix. The basic problem of ICA estimation is then to estimate the mixing matrix $\mathbf{A}$ and the realizations of the independent components $s_i$ using only observations of the mixtures $x_j$. This means that we try to approximate the joint density of $\mathbf{x}$ as precisely as possible by the densities of sums of independent random variables. We assume here that $n \geq m$, in order to have a nonsingular joint density.

ICA is closely related to the method of sparse coding. Sparse coding [1, 9] is a method for finding a neural network representation of multidimensional data in which only a small number of neurons is significantly activated at the same time. Equivalently, this means that a given neuron is activated only rarely. In this paper, we assume that the representation is linear. Denoting by $\mathbf{x} = (x_1, x_2, ..., x_m)^T$ the observed $m$-dimensional random vector that is input to a neural network, and by $\mathbf{s} = (s_1, s_2, ..., s_n)^T$ the vector of the sparse component variables, we obtain a representation that is of the same form as (1). The idea in sparse coding is to find the matrix $\mathbf{A}$ and a method of determining the $\mathbf{s}$ as a function of $\mathbf{x}$ and $\mathbf{A}$ so that the components $s_i$ are as 'sparse' as possible. A random variable $s_i$ is called sparse when it has a probability density function with a peak at zero, and heavy tails; for all practical purposes, sparsity is equivalent to supergaussianity or leptokurtosis (positive kurtosis) [7]. A fundamental result is that the estimation of the ICA model for sparse data is roughly equivalent to sparse coding, see e.g. [7].

ICA has been applied to blind source separation [8] and feature extraction [2]. Sparse coding has been applied to denoising [6], and especially to neurophysiological modelling [9]. The purpose of this paper is to show how the ICA data model (and thus sparse coding) can be applied to *regression* (prediction). The motivation for this application is that the ICA model seems to capture certain essential properties of the multidimensional densities of natural sensory data, as has been argued in [1, 9]. On the other hand, specifying the joint density of random variables is sufficient to be able to predict a subset of them using observations of the other variables.

We show in this paper how the maximum likelihood principle can be used to derive predictions of a subset of the $x_i$ based on observations of the other variables. This gives a method of regression that is parametric, yet nonlinear, and which can be expected to be especially suitable for sensory data. The method is a direct generalization of ordinary linear regression; indeed, if the independent components $s_i$ were gaussian, Eq.(1) would simply give multivariate gaussian distributions, and maximum likelihood regression would be linear. This approach of *sparse regression* is especially suitable in large dimensions, as it avoids the curse of dimensionality.

## 2    Using the ICA model for regression

Assume that we have approximated the joint density of the $x_i$ with the ICA model in (1). This means that we have estimated the matrix $\mathbf{A}$, and approximated the densities of the $s_i$ by some suitable parametric method. The estimation of the matrix $\mathbf{A}$ can be accomplished by any method of ICA estimation, see e.g. [2, 7]. Suitable approximations for the density are discussed in Section 3; in many cases, the densities of the $s_i$ may also be approximately known a priori. After performing the ICA estimation, we thus have a relatively simple approximation of the joint density of the vector $\mathbf{x}$.

Now, assume that we observe next a realization of just a subset of the variables $x_i$. For notational simplicity, this can be taken to be the $k$ first components of $\mathbf{x}$. Denote the vector of observed variables by $\mathbf{x}_o$ and the vector of the remaining variables $x_{k+1}, ..., x_m$ by $\mathbf{x}_-$. (We make no difference between the notations for the random variables and their realizations.) Denote the corresponding upper and lower parts of $\mathbf{A}$ by $\mathbf{A}_o$ and $\mathbf{A}_-$, which means that (1) can be expressed as

$$\begin{pmatrix} \mathbf{x}_o \\ \mathbf{x}_- \end{pmatrix} = \begin{pmatrix} \mathbf{A}_o \\ \mathbf{A}_- \end{pmatrix} \mathbf{s}. \tag{2}$$

The goal is now to predict $\mathbf{x}_-$ when $\mathbf{x}_o$ is given.

To predict $\mathbf{x}_-$, we could simply choose the value that maximizes the conditional probability, given the observed $\mathbf{x}_o$. In other words, define

$$\hat{\mathbf{x}}_- = \arg \max_{\mathbf{x}_-} p(\mathbf{x}_-|\mathbf{x}_o) \tag{3}$$

Denote by $\mathbf{W}$ the inverse of $\mathbf{A}$, split to two parts as above:

$$\mathbf{s} = (\mathbf{W}_o \quad \mathbf{W}_-) \begin{pmatrix} \mathbf{x}_o \\ \mathbf{x}_- \end{pmatrix} \tag{4}$$

Denote further by $\mathbf{w}_o^i$ and $\mathbf{w}_-^i$ the $i$-th row of $\mathbf{W}_o$ and $\mathbf{W}_-$, respectively, and by $p_i$ and $f_i(s_i) = -\log p_i(s_i)$ the probability density of $s_i$ and the corresponding (negative) log-density, respectively. Then (3) can be expressed as

$$\hat{\mathbf{x}}_- = \arg \min_{\mathbf{x}_-} \sum_{i=1}^n f_i((\mathbf{w}_o^i)^T \mathbf{x}_o + (\mathbf{w}_-^i)^T \mathbf{x}_-). \tag{5}$$

This maximum likelihood estimator gives a parametric prediction (regression) of $\mathbf{x}_-$. If the data is nongaussian, the $f_i$ are nonquadratic, and so the regression defined by (5) is not linear.

## 3 Approximating the densities of the independent components

To obtain the maximum likelihood estimator in (5) in practice, we need approximations of the (log-)densities of the independent components to be able to maximize (5). In this section, we discuss a method for accomplishing this in the case where the number of independent components equal the dimension of the data space, i.e. $n = m$. In the case $n > m$, it may be necessary to use prior knowledge to approximate the distributions of the $s_i$; for example, sparse distributions might be approximated by the Laplace distribution [5, 9].

In fact, it will be seen below (Section 4) that we need in practice only an approximation of the *derivative* of the log-density, i.e. $-f_i'$, which is called the score function. This can be accomplished by the method originally given in [10]. Approximate $f_i'$ by a sum of basis functions:

$$f_i'(s) = \sum_{j=1}^J b_j(i)\phi_j(s) \tag{6}$$

where $\phi_1(s) = s$ always, and the other functions might include $\tanh(s)$, $\mathrm{sign}(s)$, or $s^3$. It was then shown in [10] that the vectors $\mathbf{b}(i)$ of the coefficients $b_j(i)$ can be obtained as

$$\mathbf{b}(i) = [\mathbf{M}(i)]^{-1}\mathbf{h}(i) \tag{7}$$

where the elements $m_{pq}(i)$ of the matrix $\mathbf{M}(i)$ are defined as

$$m_{pq}(i) = E\{\phi_p(s_i)\phi_q(s_i)\}, \text{ for } p,q = 1, ..., J \tag{8}$$

and the elements $h_j(i)$ of the vector $\mathbf{h}(i)$ are defined as

$$h_j(i) = E\{\phi_j'(s_i)\}, \text{ for } j = 1, ..., J. \tag{9}$$

Thus the coefficients $b_j(i)$ are quite easy to estimate. Estimating $\mathbf{h}(i)$ and $\mathbf{M}(i)$ is simple from (8) and (9) by replacing the expectations with respect to $s_i$ by sample averages of estimates of $\mathbf{s}$. Such estimates can be obtained by a simple inverse: $\hat{\mathbf{s}} = \mathbf{A}^{-1}\mathbf{x}$. Plugging these estimates in (7) then gives the coefficients to be used in the expansion (6).

## 4  Computing the maximum likelihood estimate

The main step in the ML estimation procedure is to perform the minimization in (5). If the dimension of $\mathbf{x}_-$ is only one, we have a simple 1-D optimization problem, i.e. line search that can be solved by any classical method. Otherwise, we can develop a gradient descent method, which gives simply:

$$\Delta\hat{\mathbf{x}}_- \propto -\sum_i f_i'((\mathbf{w}_o^i)^T\mathbf{x}_o + (\mathbf{w}_-^i)^T\hat{\mathbf{x}}_-)\mathbf{w}_-^i. \tag{10}$$

The gradient descent can be started, e.g. from the point given by linear regression. The $f_i'$ are approximated as in Section 3.

## 5  Minimum mean-square estimator

It is also possible to use the minimum mean-square (MMS) estimator. This is obtained by computing the mean of the posterior distribution:

$$\hat{\mathbf{x}}_- = \int \mathbf{x}_- \prod_{i=1}^{n} p_i((\mathbf{w}_o^i)^T\mathbf{x}_o + (\mathbf{w}_-^i)^T\mathbf{x}_-)|\det\mathbf{W}|d\mathbf{x}_-. \tag{11}$$

In the classical gaussian case, the ML and MMS estimators are equivalent, but this is not at all true in general. The integration in (11) can only be done in practice using numerical methods, which essentially restricts the estimator to the case where $\mathbf{x}_-$ is one-dimensional. Here we need an approximation of the probability densities $p_i$ themselves; for methods to approximate them, see, e.g. [5].

# 6  Summary of the method

Thus we have derived a method of maximum likelihood regression for data distributed according to (1). We call this method *sparse regression*, because in most cases in practice, the densities of the independent components are sparse [9]. To summarize, the basic method is as follows:

1. Using a sample of $\mathbf{x}$, estimate the mixing matrix $\mathbf{A}$ in (1), by any suitable ICA estimation method.

2. Either fix the densities of the independent components by using prior knowledge, or (if $n = m$) estimate the corresponding score functions $f_i'$ by (6).

3. Given an observation $\mathbf{x}_o$ of some of the variables $x_i$, obtain a nonlinear estimate $\hat{\mathbf{x}}_-$ of the missing variables by minimizing (5) either by line search, if $\mathbf{x}_-$ is 1-D, or otherwise by (10).

The obtained predictors are nonlinear, but in contrast to some nonparametric methods, the nonlinearities used in the method are not chosen ad hoc: they are directly determined by the density approximation. The approach of sparse regression is especially suitable in large dimensions, as it avoids the curse of dimensionality. Indeed, it is closely related to the method of projection pursuit regression [4]. The method contains ordinary linear regression as a special case, because any multivariate gaussian distribution can be expressed by Eq. (1) (simply by taking the $s_i$ i.i.d. gaussian, and using a square root of the covariance matrix as the mixing matrix), and linear regression is ML regression for gaussian variables. Thus the method should always be better than linear regression, which is not always the case with nonparametric methods, which often need specific forms of regularization to prevent overfitting. The ICA model, in contrast, has very few parameters (usually of the same order $O(n^2)$ as linear regression), and thus overfitting is not a problem.

# 7  Simulations

We illustrate the nonlinear character of sparse regression by a simple 3-D example. In this example, a random $3 \times 3$ orthogonal matrix was taken as the mixing matrix, and the distributions of the independent components were assumed Laplacian [5, 9]. We then predicted $x_3$ using $x_1$ and $x_2$. Because the mixing matrix was orthogonal, the variables $x_i$ were uncorrelated, and therefore linear regression of $x_3$ would have given zero for all values of $x_1$ and $x_2$ (all the variables had zero mean).

The sparse regression of $x_3$ is depicted in Fig. 1. Taking into account the higher-order structure of $\mathbf{x}$ makes the regression clearly nonlinear. In fact, in this special case of Laplace densities, the regression is piecewise inear. This is because the ML estimator of $s$ is a piecewise linear function of $\mathbf{x}_o = (x_1, x_2)$. In fact, this form of nonlinearity can be interpreted as a competition: the

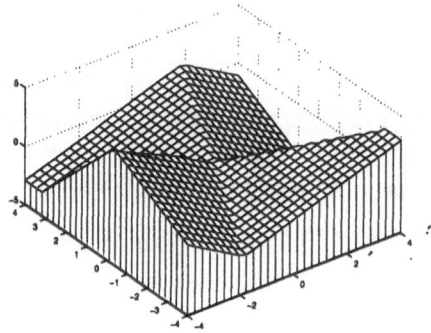

Figure 1: An illustration of sparse regression. The joint density of 3 random variables was represented as the densities of linear combinations of 3 laplacian variables, and one of the random variables was predicted by the two others. The prediction curve is nonlinear, in fact, it is piecewise linear.

independent components (or the neurons corresponding to them) compete for representation. The two independent components which can best represent the given data point are activated, and give a linear regression whose coefficients depend on which neurons win the competition [9].

# References

[1] H.B. Barlow. Unsupervised learning. *Neural Computation*, 1:295–311, 1989.

[2] A.J. Bell and T.J. Sejnowski. The 'independent components' of natural scenes are edge filters. *Vision Research*, 37:3327–3338, 1997.

[3] P. Comon. Independent component analysis – a new concept? *Signal Processing*, 36:287–314, 1994.

[4] P.J. Huber. Projection pursuit. *The Annals of Statistics*, 13(2):435–475, 1985.

[5] A. Hyvärinen. Sparse code shrinkage: Denoising of nongaussian data by maximum likelihood estimation. Technical report, Helsinki University of Technology, Laboratory of Computer and Information Science, 1998.

[6] A. Hyvärinen, P. Hoyer, and E. Oja. Sparse code shrinkage for image denoising. In *Proc. IEEE Int. Joint Conf. on Neural Networks*, pages 859–864, Anchorage, Alaska, 1998.

[7] A. Hyvärinen and E. Oja. A fast fixed-point algorithm for independent component analysis. *Neural Computation*, 9(7):1483–1492, 1997.

[8] C. Jutten and J. Herault. Blind separation of sources, part I: An adaptive algorithm based on neuromimetic architecture. *Signal Processing*, 24:1–10, 1991.

[9] B. A. Olshausen and D. J. Field. Sparse coding with an overcomplete basis set: A strategy employed by V1? *Vision Research*, 37:3311–3325, 1997.

[10] D.-T. Pham, P. Garrat, and C. Jutten. Separation of a mixture of independent sources through a maximum likelihood approach. In *Proc. EUSIPCO*, pages 771–774, 1992.

# A Linear Programming Neural Circuit Model

József Bíró

High Speed Networks Laboratory

Dept. of Telecommunications and Telematics

Technical University of Budapest

Budapest, Hungary

e-mail: biro@ttt-atm.ttt.bme.hu

Miklós Boda

Ericsson, Traffic Analysis and Network Performance Laboratory

Budapest, Hungary

e-mail: Miklos.Boda@era-t.ericsson.se

### Abstract

In this paper we present a neural circuit model which can solve linear programming problems. The main feature of the model is that it takes into account saturating behaviour of circuit elements in a possible realizations. The neural model can be viewed as a gradient system and its operation is based on the penalty function approach of solving linear programming tasks. In the paper the properties of the model are discussed including stability and that how to utilize the saturation in the neurons for obtaining better performance.

## Introduction

One of the largest application areas of artificial neural networks is to solve optimization problems. In the last decade considerable attention has been paid for optimization neural networks. These systems are considered as a potentially efficient hardware solutions for solving large-scale or complex optimization tasks. Although many problematic, and therefore challenging questions arise in connection with hardware realizations, in principle optimization neural networks could work very fast as a parallel computational structure in truly distributed and parallel implementation.

Optimization neural networks are feed-back systems, after initialization they are iterating or relaxing until reaching a (hopefully) stable equilibrium state. This equilibrium should represent an exact or approximate solution of the optimization task. Stability questions play central role in the analysis of optimization neural networks.

After a brief overview of linear programming neural nets the model of $LPNN_S$ is introduced. After that stability analysis is performed. In Section 4 we show how to utilize the saturation function in the neurons to produce "better" solution than those of any other neural net based purely on the penalty function approach. Finally, a numerical example is presented.

# 1 Linear Programming Neural Network Models

Optimization neural nets are usually designed for nonlinear programming and based on penalty function approach [3], [5], [2], [1]. It means that if we have the following type of constrained optimization task

$$\text{Minimize } f(x), \quad \text{subject to } g_i(x) \leq 0, \ i = 1, \dots, p \qquad (1)$$

where $x \in \mathbf{R}^n$, $f$ and $g_i : \mathbf{R}^n \to \mathbf{R}$ are scalar valued functions of $n$ variables, then it can be transformed to an unconstrained objective function $f(x) + P(x)$, where $P(x)$ comprises the penalty functions which are responsible for fulfilling the constraints. The neural nets are seeking for the optimum of $f(x) + P(x)$. These networks can also be used for solving linear programming tasks, however, there are some neural model which are developed especially for linear programs.

In [6] a neural network is presented which can solve linear programming problems with equality constraints. In this case, inequalities should be transformed to equalities by introducing slack variables. The disanvantegous property of the network is that its system equation contains nondifferentiable functions, probably this is the reason why no stability analysis is performed. However, it is demonstrated that the discrete-time implementation of the system can converge to the right solution of an example problem.

In [8] a recurrent neural network for linear programming is presented. The main feature of this network that it can recover the exact solution of the linear programming task. This is obtained so that the derivatives of the linear objective function is multiplied by an exponentially decaying parameter which makes the network much slower than any other network based on the simple penalty function approach.

In [7] a continuous-time linear programming network is presented which is based on the logarithmic barrier function method differing significantly from the penalty function approach. The system called LP-Net is governed by coupled differential equations. Although it is shown by simulation that the network may be stable and robust against limited numerical precision of analog devices, nor rigorous stability analysis is performed neither any implementation is reported.

A different approach can be found in [4] for solving linear programs. The neural network presented are based on the modified Lagrangean multiplier approach which is a mix of penalty method and Lagrangean multiplier method. The resulted system equations conform coupled differential equations which makes the network more complex than those based on simple penalty method and gradient search.

# 2  The Neural Model

We consider the following linear programming problem

$$\text{Minimize } \sum_{k=1}^{n} c_k x_k \text{ s.t. } \vec{D}_i x - B_i \leq 0 \;,\; i = 1, \ldots, p \;,\; x_k \geq 0 \qquad (2)$$

The system of differential equation governing the operation is as follows

$$\frac{dz_k}{dt} = -c_k - \alpha_k \sigma\left(\frac{\partial P(x)}{\partial x_k}, T\right) \;,\; x_k = \Theta(z_k) \; k = 1, \ldots, n \qquad (3)$$

$\Theta(.)$ can be any of monotone increasing functions with finite saturation values at $\infty$, $-\infty$, $\alpha_k$ is a constant with the property of $\alpha_k > |c_k|$, $T$ is a positive constant controlling the shape of $\sigma()$ and for which $T < 1$, $\sigma()$ is a sigmoid-like function *saturating* the derivatives of the penalty function, and $P(x)$ is any kind of penalty function operating on the constraints. Without any restriction we can also assume that $\sigma(\infty, T) = 1$ and $\sigma(-\infty, T) = -1$ for any positive $T$. If $\sigma(y_k, T_k) = \tanh(y_k/T_k)$ is chosen then the steepness of $\sigma$ is well-controllable by $T$.

The role of the saturation function $\sigma()$ in the model is very important from realization point of view, because it provides a more realistic approach in which the saturation behaviour of circuit elements can be taken into account.

According to this $CON$ and $DEC$ type neurons can be defined. $DEC$ neurons contain integrators and $\Theta$ as nonlinearity. The inputs of the $k^{\text{th}}$ $DEC$ neuron are $\sigma(y_k, T)$ with $-\alpha_k$ weight and $-c_k$. The output is $x_k$. The $k^{\text{th}}$ $CON$ neuron consists of two subsequent nonlinear operation $\partial P(x)/\partial x_k$ and $\sigma$. Its inputs are $x_1, \ldots, x_n$. The architecture of these neurons are shown in Fig 1.

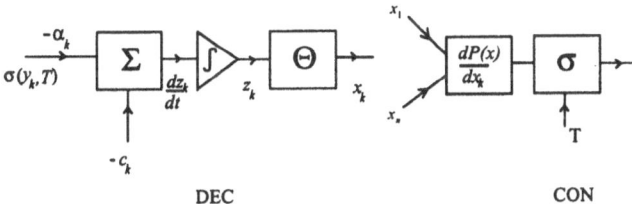

DEC                                              CON

Figure 1: The $DEC$ and $CON$ neurons in $LPNN_S$

# 3  Stability Analysis

We can have observed from the dynamics that $LPNN_S$ is not a usual gradient system. The derivatives of the penalty function does not appear in the network directly, but they influence the operation through a nonlinear transformation $\sigma$. The question is: Is there any function which decreases along any trajectory of $x$

produced by the network? For finding the answer let us consider the following function:

$$L(x) = \vec{c}^T x + \sum_{k=1}^{n} \int_{\xi_k=0}^{x_k} \alpha_k \sigma(\frac{\partial P(\xi)}{\partial \xi_k}, T) d\xi_k \tag{4}$$

It can easily be shown that this is a Lyapunov function of the system. Taking the time derivative of $L(x)$ we obtain the following formula

$$\frac{dL(x)}{dt} = \sum_k \frac{dx_k}{dt} \left( c_k + \alpha_k \sigma(\frac{\partial P(x)}{\partial x_k}, T) \right) \tag{5}$$

Identifying $dz_k/dt$ based on the dynamics of the network we obtain $dL(x)/dt = \sum_k dx_k/dt * dz_k/dt$ which is apparently less than or equal to zero due to the strictly monotone transformation between $x_k$ and $z_k$. $L(t)$ is also lower bounded because $x$ is in a finite bounded region and every continuously differentiable function like $L(t)$ is bounded over a finite set of state variables $x$. Consequently, the proposed neural network is stable in Lyapunov sense.

# 4    Solution Feasibility and Optimality

An important and non-trivial property of $LPNN_S$ is that the state variables $x$ converge towards the feasibility region if the network operates in *saturation mode*. The system is said to be in saturation mode if the following inequalities hold: $|\partial P(x)/\partial x_k| > \tilde{y}_k$ , $\forall k$ where $\tilde{y}_k$ is defined such that

$$\alpha_k \sigma(\tilde{y}_k, T) = |c_k| \tag{6}$$

Note, that it is always possible to find such $\tilde{y}_k$ because $|c_k|/\alpha_k < 1$ due to the definition of $\alpha_k$. The definitions above imply that $\tilde{y}_k$ is positive. In Fig 2 the saturation region is also depicted.

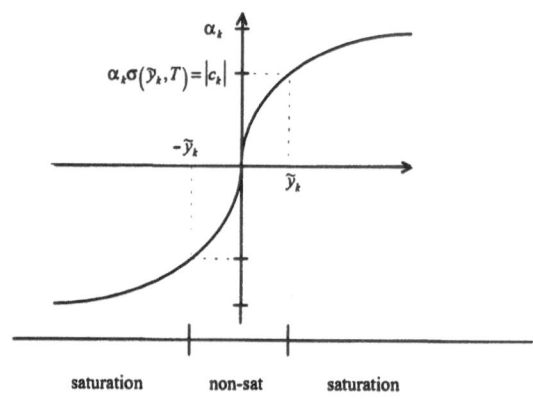

Figure 2: The saturation and non-saturation regions of $LPNN_S$

Now, we show that the time derivative of the penalty function is decreasing if $x$ is in the saturation region. In saturation mode $\partial P(x)/\partial x_k$ and $dz_k/dt$

have opposite sign. It follows from the fact that if $y_k = \partial P(x)/\partial x_k > \tilde{y}_k$ then $\alpha_k \sigma(y_k, T) > |c_k|$ and from the dynamics $dz_k/dt = -c_k - \alpha_k \sigma(y_k, T) < -c_k - |c_k| \leq 0$ Similarly, if $y_k = \partial P(x)/\partial x_k < -\tilde{y}_k$ then $\alpha_k \sigma(y_k, T) < -|c_k|$ and from the dynamics $dz_k/dt = -c_k - \alpha_k \sigma(y_k, T) > -c_k + |c_k| \geq 0$ Due to the strictly monotone increase of $\Theta$ it is also given that $dx_k/dt$ and $dz_k/dt$ have the same sign. Thus, $\partial P(x)/\partial x_k$ and $dx_k/dt$ have the opposite sign, too.

If the time derivative is written as a complete differential

$$\frac{\partial P(x)}{\partial t} = \sum_k \frac{\partial P(x)}{\partial x_k} \frac{dx_k}{dt} \tag{7}$$

it can be observed that every element of the sum is negative, consequently, the time derivative of $P(x)$ is also negative. It means that the network always converges towards the feasibility region.

Further issue is that where the equilibrium point is located. For the final state of the network the following equalities hold $c_k = -\alpha_k \sigma(y_k^*, T)$ Because $\sigma$ is an odd function $c_k = \alpha_k \sigma(-y_k^*, T)$ also hold. Taking the absolute value of both side of the equalities above and the definition of $\tilde{y}_k$ (6) we obtain that $|y_k^*| = \tilde{y}_k$. The meaning of this result is that the network equilibrium is associated with the boundary of the saturation region. Summarizing the behaviour of the neural system: $LPNN_S$ converges towards the feasibility region provided it operates in saturation mode then finally gets stuck on the boundary of the saturation region.

One can ask why so interesting this property is. It is because the "size" of the non-saturation region can be controlled by parameter $T$. If we use the simple $\tanh(y_k/T)$ function then the non-saturation intervals $(-\tilde{y}_k, \tilde{y}_k)$ can be arbitrarily small with $T \to 0$. In the limiting sense $T = 0$ the solution becomes feasible and optimal, hence, $LPNN_S$ is 'asymptotically exact dynamic solver'.

It is also clear that the border of the saturation region, and thus the solution produced by the network lies outside the feasible region for finite steepness of $\sigma$ which is the realistic case. A natural question to emerge is how close this solution to the feasible region is and how to measure the infeasibility of the equilibrium point in order to compare the quality of the solution with that produced by other neural networks.

In connection with this we present results on a subset of the family of linear programming tasks. We define this subset with the following properties.

- $c_k \leq 0$, $\sum_k D_{ik} > 0$, and $x_k > 0$, $\forall k, i$

- Parameter $T$ is adjusted such that the equation $y_k = \alpha_k \sigma(y_k, T)$ should have such a solution also which differs from 0 and the positive solution should be larger than $-c_k$. It can easily be realized because $\alpha_k > |c_k|$ by definition. More concisely, this condition can be described as (see Fig. 3)

$$T < \frac{-c_k}{\tanh^{-1}\left(\frac{-c_k}{\alpha_k}\right)}, \quad \forall k \tag{8}$$

Now, we introduce a *solution infeasibility measure* which depends on the underlying linear programming task and penalty function used in the neural network. This measure can be expressed as $SIF(\beta, x) = \sum_i \beta_i \Omega_i(x)$ where

$\beta_i = \sum_k D_{ik}$ and $\Omega_i(x) = \partial P(x)/\partial g_i(x)$ Since $\beta_i > 0, \forall i$ due to the assumption $SIF(\beta, x)$ is like a penalty function, therefore, it is a reasonable solution infeasibility measure.

One of the main features of $LPNN_S$ in solving the family of linear programs defined above is that it always produces a *better* solution according to the measure defined than those of linear programming neural networks which do not contain $\sigma$ nonlinearity, like Kennedy & Chua's neural circuit [5]. In what follows we argue for this statement by rigorous analysis. First, let us consider a "conventional" linear programming neural network following strictly the penalty function approach. The dynamics of this network can be described by $dz_k/dt = -c_k - \partial P(x)/\partial x_k$ Let $\hat{x}^*$ designate the equilibrium state produced by this network while $x^*$ is the final state of $LPNN_S$. The derivatives $\partial P(x)/\partial x_k$ of the penalty function at these equilibriums are marked by $\hat{y}_k^*$ and $y_k^*$, respectively. From the dynamics we can obtain that $\alpha_k \sigma(y_k^*, T) = \hat{y}_k^*$ It also holds that $-c_k = \hat{y}_k^*$ and therefore $-c_k = \alpha_k \sigma(y_k^*, T)$. It immediately yields the inequality $y_k^* \le \hat{y}_k^*$ due to the appropriate steepness of $\sigma()$ adjusted by $T$. Fig 3 illustrates the relation between $y_k^*$ and $\hat{y}_k^*$ Equality holds if $-c_k = 0$. Since $\partial P(x)/\partial x_k = \sum_i \partial P(x)/\partial g_i(x) * \partial g_i(x)/\partial x_k$ $\partial P(x)/\partial g_i(x) = \Omega_i(x)$ and $\partial g_i(x)/\partial x_k = D_{ik}$ the inequality $y_k^* \le \hat{y}_k^*$ can be formulated as $\sum_i D_{ik}\Omega_i(x^*) \le \sum_i D_{ik}\Omega_i(\hat{x}^*)$ , $k = 1, \ldots, n$ If we sum up all these inequalities for $k$ we obtain $\sum_i \beta_i \Omega_i(x^*) < \sum_i \beta_i \Omega_i(\hat{x}^*)$ which accords to the solution infeasibility measure defined above. Along this result we can also conclude that the nonlinear function $\sigma()$ saturating the penalty derivatives plays central role not only in realistic modeling of neural network, but also in obtaining "better" performance.

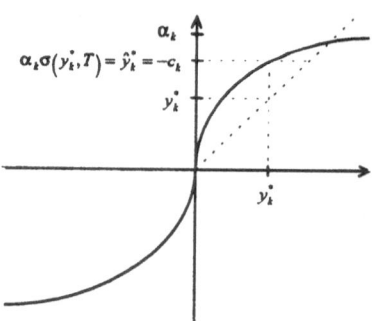

Figure 3: The relation between $\hat{y}_k^*$, $y_k^*$ in $LPNN_S$

# 5 A Numerical Example

Let us consider the following very simple one-dimensional task: Minimize $-x$ subject to $0 \le x \le B$, where $B$ is positive. Clearly the solution is $x = B$. We apply quadratic penalty functions with controlling parameter 0.5. In this case a linear programming neural network without $\sigma$ minimizes the function

$$\hat{\Phi}(x) = -x \text{ if } x \le B \text{ , otherwise } -x + 0.5(x - B)^2 \tag{9}$$

while $LPNN_S$ seeks for the minimum of

$$\Phi(x) = -x \text{ if } x \le B \text{ , otherwise } -x + \int_0^x \alpha \tanh((\xi - B)/T)d\xi \tag{10}$$

where $\alpha$ is 3. The minima of these functions are $\hat{x}^* = 1 + B$ and $x^* = T \tanh^{-1}(1/3) + B$ , respectively. In this case, the solution infeasibility measure of these solutions become $SIF(1, \hat{x}^*) = 1$ , $SIF(1, x^*) = T \tanh^{-1}(1/3)$ We can see that the solution $x^*$ is better than $\hat{x}^*$ according to $SIF()$ if $T < 1/\tanh^{-1}(1/3)$ which is in accordance with the condition formulated in (8).

# References

[1] J. Bíró, E. Halász, T. Trón, M. Boda, and G. Privitzky. Neural networks for exact constrained optimization. In *Lecture Notes in Computer Science (Proceedings of the International Conference on Artificial Neural Networks -ICANN'96)*, Bochum (Germany), June 1996.

[2] J. Bíró, Z. Koronkai, L. Ast, T. Trón, and M. Boda. Analyses of extended and generalized optimization neural networks. *Journal of Artificial Neural Networks*, 2(4):401–409, August 1995.

[3] L.O. Chua and G. Lin. Nonlinear programming without computation. *IEEE Trans. on Circuits and Systems*, 52(2), February 1984.

[4] A. Cichocki and R. Unbehauen. Neural networks for solving systems of linear equations and related problems. *IEEE Trans. on Circuits and Systems*, 39(2):124–138, February 1992.

[5] M.P. Kennedy and L.O. Chua. Neural networks for nonlinear programming. *IEEE Trans. on Circuits and Systems*, 35(5):554–562, May 1988.

[6] S. Hui S. H. Zak, V. Upatising. Solving linear programming problems with neural networks: A comparative study. *IEEE Trans. on Neural Networks*, 6(1):94–104, 1995.

[7] K. Zikan T.P. Caudell. A neural network architecture for linear programming. In *Proc. of IEEE ICNN'92*, pages 91–96, October 1992.

[8] J. Wang. Analysis and design of a recurrent neural network for linear programming. *IEEE Trans. on Circuits and Systems*, 40(9):613–618, September 1993.

# Learning Invariance Manifolds

Laurenz Wiskott

Computational Neurobiology Laboratory
The Salk Institute for Biological Studies
San Diego, CA 92186-5800
wiskott@salk.edu, http://www.cnl.salk.edu/CNL/

### Abstract

A new algorithm for learning invariance manifolds is introduced that allows a neuron to learn a non-linear transfer function to extract invariant or rather slowly varying features from a vectorial input sequence. This is generalized to a group of neurons, referred to as a Gibson-clique, to learn slowly varying features that are uncorrelated. Since the transfer functions are non-linear, this technique can be applied iteratively to learn more and more complex and invariant features in a hierarchical architecture. Two simple examples demonstrate the general properties of the learning algorithm.

## 1  Introduction

*Third [...], the process of perception must be described. This is not the processing of sensory inputs, however, but the extracting of invariants from the stimulus flux.* (GIBSON, 1986, p. 2)

Learning invariant representations is one of the major problems in neural systems. The approach described in this paper is conceptually most closely related to work by BECKER & HINTON (1995) and STONE (1996). The idea is that while an input signal may change quickly due to changes in the sensing conditions, e.g. scale, location, and pose of the object, certain aspects of the input signal change slowly or rarely only, e.g. the presence of a feature or object. The task of a neural system in learning invariances is therefore the extracting of slow aspects from the input signal.

On an abstract level, the input $\mathbf{x} = \mathbf{x}(t)$ of a sensor array can be viewed as a trajectory in a high-dimensional input space. Many points in this space can represent the same feature if they only differ in their sensing conditions. One can imagine that these points lie on a manifold (e.g. LU ET AL., 1996), which may be called *invariance manifold*. Looking at an object under varying sensing conditions means that the trajectory lies within the invariance manifold. Saccading to a new object, for instance, will cause a jump in the trajectory with a component perpendicular to the manifold. Here, a single manifold is defined by an equipotential surface of a scalar transfer function $g(\mathbf{x})$ in the N-dimensional space. The set of all equipotential surfaces defines a (continuous) family of manifolds. This can be extended to a set of transfer functions $g_i(\mathbf{x})$ providing a set of manifold families.

The proposed algorithm differs from the work by BECKER & HINTON (1995) and STONE (1996) in the mathematical formulation, one distinct feature being that input signals are individually combined in a non-linear fashion, which follows the idea that complex non-linear computation can be performed by the dendritic tree (MEL, 1994). Furthermore, the system is formulated as a learning algorithm rather than an online learning rule, and it is naturally generalized to a group of output neurons, here referred to as a *Gibson-clique*.

## 2 The Learning Algorithm

Consider a neuron that receives an $N$-dimensional input signal $\mathbf{x} = \mathbf{x}(t)$ where $t$ indicates time and $\mathbf{x} = [x_1, ..., x_N]^T$ is a vector. The neuron is able to perform a non-linear transformation on this input defined as a weighted sum over a set $\mathbf{h} = [h_1, ..., h_M]^T$ of $M$ non-linear functions $h_m = h_m(\mathbf{x})$ (usually $M > N$). Here polynomials of degree two are used, but other sets of non-linear functions could be used as well. Applying $\mathbf{h}$ to the input signal yields the non-linearly expanded signal $\mathbf{h}(t) \equiv \mathbf{h}(\mathbf{x}(t))$. The set of weights $\mathbf{w} = [w_1, ..., w_M]^T$ is subject to learning and the final output of the neuron is given by $y(t) \equiv g(\mathbf{x}(t)) \equiv \mathbf{w}^T \mathbf{h}(\mathbf{x}(t))$. $g$ is called the transfer function of the neuron. Notice that it is much more complex than the common sigmoidal functions employed in conventional model neurons, since it combines individual components of the input signal in a non-linear fashion.

The objective is to optimize the weights such that the output is as invariant as possible, i.e. it has a minimal mean square of the time derivative

$$\Delta(y) \equiv \langle \dot{y}^2 \rangle = \mathbf{w}^T \langle \dot{\mathbf{h}} \dot{\mathbf{h}}^T \rangle \mathbf{w}, \tag{1}$$

a quantity that will be referred to as the $\Delta$-value. $\langle \cdot \rangle$ indicates the temporal mean.

This objective alone would lead to a system that learns constant features that do not change at all over the input signal or, more likely, a system that learns no features at all by setting $\mathbf{w} = 0$. Thus a constraint needs to be imposed such that the output signal conveys some information. This also means that a feature needs to change at least slowly or rarely, so that it is actually not invariant features that are learned but slowly varying features or, for short, slow features. However, the manifolds themselves defined by $g = $ const can still be considered invariance manifolds. The constraint that features have to convey some information is here formalized by requiring that the variance of the output signal be unity,

$$\langle y^2 \rangle - \langle y \rangle^2 = \mathbf{w}^T \underbrace{\left( \langle \mathbf{h} \mathbf{h}^T \rangle - \langle \mathbf{h} \rangle \langle \mathbf{h}^T \rangle \right)}_{=\mathbf{I}} \mathbf{w} = \mathbf{w}^T \mathbf{w} = 1. \tag{2}$$

It is assumed here that the signals produced by the non-linear functions $h_m$ have zero mean and a unit covariance matrix. A sphering stage has to be applied to an arbitrary set of non-linear functions $\mathbf{h}'$ to derive the set $\mathbf{h}$ with these properties.

Minimizing the $\Delta$-value under this constraint is equivalent to finding the normalized eigenvector with minimal eigenvalue for matrix $\langle \mathbf{h}\mathbf{h}^T \rangle$. The minimal eigenvalue is equal to $\langle \dot{y}^2 \rangle$. The eigenvectors of the next higher eigenvalues produce uncorrelated neurons with the next higher $\Delta$-values. These can be useful if several uncorrelated features need to be extracted. They can also be used to propagate enough information through a cascade of transfer functions (cf. the second example in Section 3).

It is useful to measure the invariance of signals not by the $\Delta$-value directly but by a measure that has a more intuitive interpretation. A good measure may be an index $\eta$ defined by $\eta(y) \equiv \sqrt{\Delta(y)\,T/(4\pi^2)}$ if $t \in [0,T]$. For a pure sine wave $\sin(n\,2\pi\,t/T)$ with an integer number of oscillations $n$ the index $\eta$ is just the number of oscillations, i.e. $\eta = n$. Thus the index $\eta$ of an arbitrary signal indicates what the number of oscillations would be for a pure sine wave of same $\Delta$-value, at least for integer values of $\eta$.

# 3 Examples

The properties of the learning algorithm are now illustrated by two examples. The first example is about learning complex cell behavior based on simple cell outputs. Since the required transfer function is a second degree polynomial, one Gibson-clique of degree two is sufficient for this task. The second example is abstract and requires a more complicated transfer function, which can be approximated by three Gibson-cliques in succession.

The example for learning complex cell behavior based on simple cell outputs follows the view that simple cells can be modeled by Gabor wavelets (JONES & PALMER, 1987), while a complex cell combines the responses of several simple cells of same orientation and location but different phase, e.g. taking the square sum of a cosine and a sine Gabor-wavelet. The response of a Gabor-type simple cell to a visual stimulus continuously moving across the visual field can be modeled by a combination of time varying amplitude $a(t)$ and phase $\phi(t)$, which have the form of low-pass filtered Gaussian noise here. Two simple cells of same orientation and location have same amplitude and phase modulation but a constant phase difference depending on the phase shift of their receptive fields (see Fig. 1). In this example the signals of three simple cells, two at same location and one at a different location, are modeled by $x_1(t) = a_1(t)\sin(4\pi t + 2\phi_1(t))$, $x_2(t) = a_1(t)\sin(4\pi t + 2\phi_1(t) + \pi/4)$, and $x_3(t) = a_2(t)\sin(4\pi t + 2\phi_2(t))$, $t \in [0,1]$. Notice that the phase difference between the first two simple cells is $45°$ and not $90°$. The latter would be more convenient, but is not necessary. The complex-cell response that can be extracted from these three simple cells is the amplitude signal $a_1(t)$.

Fig. 1 left top shows a sketch of the receptive fields of the three hypothetical simple cells. The generating amplitude and phase signals $a_1(t)$ and $\phi_1(t)$ are shown below. On the right is shown the input signal $\mathbf{x}(t)$, cross-sections through the learned transfer function $\mathbf{g}(\mathbf{x})$ (arguments not varied are set to zero, e.g. $g_1(x_2, x_3)$ means $g_1(0, x_2, x_3)$), and the extracted output signal $\mathbf{y}(t)$. All signals have unit variance and all graphs range from $-4$ to $+4$, including the grey value scale of the contour plots. Time axes range from 0 to 1.

The amplitude comodulation and 45° phase relationship of the first two simple cells is reflected in the elliptic form of trajectory plot $x_2(t)$ vs. $x_1(t)$. The slow component of this signal is its distance from the center weighted according to the elliptic shape. The third simple cell has no relationship to the first two (see, for example, trajectory plot $x_3(t)$ vs. $x_2(t)$). The first component of the learned transfer function, $g_1(\mathbf{x})$, correctly represents the elliptic shape of the $x_2(t)$ vs. $x_1(t)$ trajectories and completely ignores signal $x_3(t)$. It therefore extracts the amplitude signal $a_1(t)$ as desired (compare signals $a_1(t)$ and $y_1(t)$). The correlation coefficient between $a_1(t)$ and $y_1(t)$ is 0.97. The $\eta$-index of $y_2(t)$ is almost as high as the one of $x_2(t)$ which indicates that it does not represent another invariance (as one would also expect from the way the input signal was generated). Thus, $g_2(\mathbf{x})$ could be discarded.

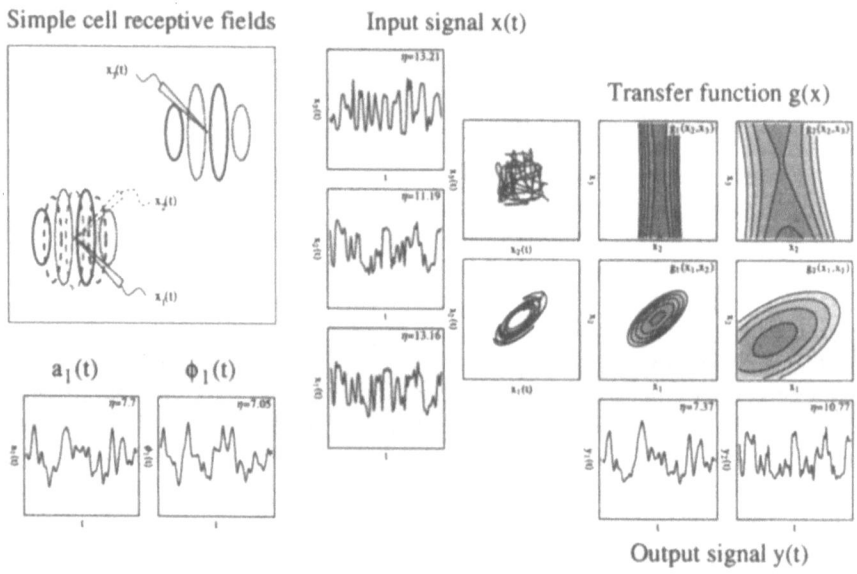

Figure 1: Learning complex cell response with one Gibson-clique.

The first example was particularly easy because a second degree polynomial was sufficient to recover the slow feature well. The second example is more complex. First generate a slowly and a fast varying random time series $x_s(t)$ and $x_f(t)$, respectively. Both signals are normalized to unit variance and then mixed to provide the input signal $\mathbf{x}(t) \equiv [x_f, \sin(2x_f) + 0.5x_s]^T$. The task is to extract the slowly varying random time series $x_s(t)$

The transfer function required to extract the slow feature $x_s$ cannot be well approximated by a second degree polynomial. One might therefore use third or higher degree polynomials. However, one can also iterate the learning algorithm, applying it with second degree polynomials repeatedly, leading to transfer functions of second, fourth, eighth, sixteenth degree etc. To avoid an explosion of the signal dimensionality only the first components of the output signal of one Gibson-clique should be used as an input for the next clique. In this example only three components of an output signal are transfered to the next

Gibson-clique. This cuts down the computational cost of this iterative scheme significantly compared to the direct scheme of using higher degree polynomials.

Figure 2 shows slow and fast random time series, input signal, and transfer function as well as output signal for three Gibson-cliques in succession. The plotted transfer functions always include the transformations done by previous Gibson-cliques, too. The correlations between selected output signals and $x_s(t)$ are illustrated at the bottom left. These are the output signals with the highest correlation with $x_s(t)$, namely $y_3(t)$, $y_2(t)$, and $y_1(t)$ for the the first, second, and third Gibson-clique, respectively. The corresponding correlation coefficients are 0.48, 0.76, and 0.94. Several Gibson-cliques are necessary to approximate the wave-like invariance manifolds that can be seen in the $x_2(t)$ vs. $x_1(t)$ trajectory plot. Notice that each Gibson-clique is trained in an unsupervised manner and that no back-propagation of any kind of error-signal is required. It is interesting that the correlation between $y_2(t)$ of the third Gibson-clique and the fast random time series $x_f(t)$ is 0.98, which means that also $x_f(t)$ is represented.

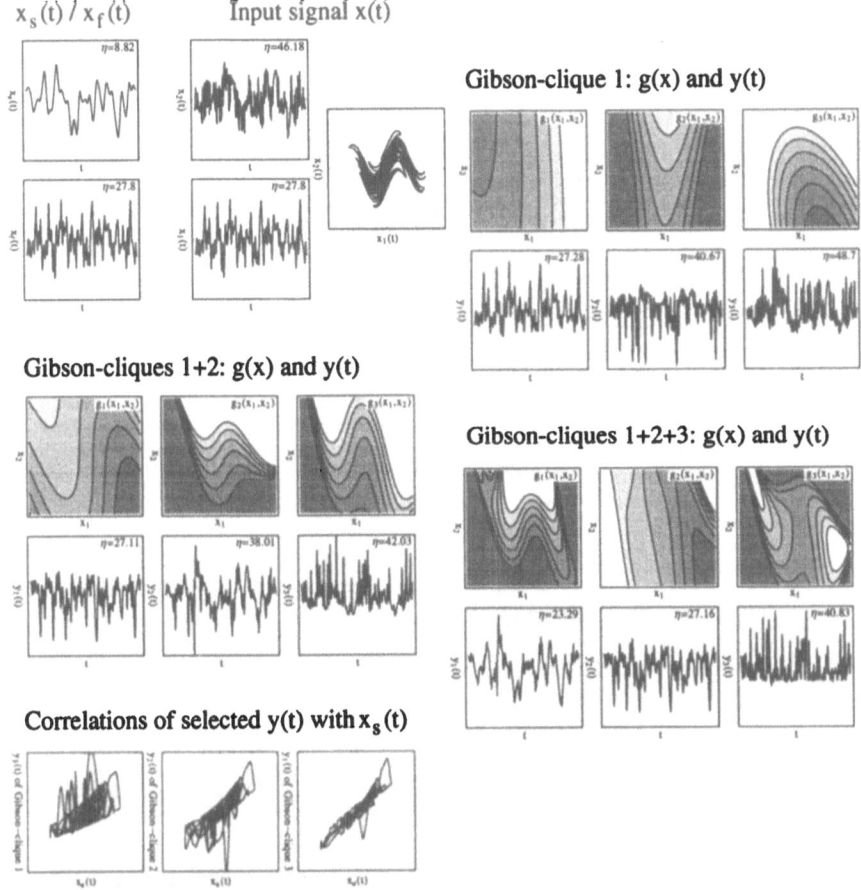

Figure 2: Hidden slow signal discovered by three Gibson-cliques in succession.

Limited space does not permit showing more complex examples but it should be mentioned that more complex multi-layered hierarchical networks of Gibson-cliques have been trained. Such a network, connected to a one-dimensional input array, was trained with moving objects and it developed neural responses that were more and more translation invariant and specific to more and more complex objects as one proceeds in the hierarchy to the top, much akin to the mammalian visual system.

# 4 Conclusion

A new unsupervised learning algorithm has been presented and tested on two simple examples. With the algorithm a group of neurons, referred to as a Gibson-clique, can be trained to learn a high-dimensional non-linear transfer function to extract slow components from a vectorial input signal. Since the learned transfer functions are non-linear, the algorithm can be applied iteratively, so that complex transfer functions can be learned in a multi-layer network of Gibson-cliques with limited computational effort.

**Acknowledgment**

I am grateful to Terrence Sejnowski for his support and valuable feedback. The author has been partially supported by a Feodor-Lynen fellowship by the Alexander von Humboldt-Foundation, Bonn, Germany.

# References

BECKER, S. AND HINTON, G. E. (1995). Spatial coherence as an internal teacher for a neural network. In CHAUVIN, Y. AND RUMELHART, D. E., editors, *Backpropagation: Theory, Architecture and Applications.*, pages 313–349. Hillsdale, N.J. : Lawrence Erlbaum Associates.

GIBSON, J. J. (1986). *The Ecological Approach to Visual Perception.* Lawrence Erlbaum Associates, London. Originally published in 1979.

JONES, J. P. AND PALMER, L. A. (1987). An evaluation of the two dimensional Gabor filter model of simple receptive fields in cat striate cortex. *J. of Neurophysiology*, 58:1233–1258.

LU, H.-M., HECHT-NIELSEN, R., AND FAINMAN, S. (1996). Geometric properties of image manifolds. In *Proc. of the 3rd Joint Symp. on Neural Comp.*, volume 6, pages 53–60, San Diego, CA. Univ. of California.

MEL, B. W. (1994). Information processing in dendritic trees. *Neural Computation*, 6:1031–1085.

STONE, J. V. (1996). Learning perceptually salient visual parameters using spatiotemporal smoothness constraints. *Neural Computation*, 8(7):1463–1492.

# Developmental Evolution of an Edge Detecting Retina

Alistair G Rust, Rod Adams and Stella George

Department of Computer Science, University of Hertfordshire, UK

Hamid Bolouri

Engineering Research & Development Centre, Univ. of Hertfordshire, UK

Biology 216-76, California Institute of Technology, USA

{*a.g.rust, r.g.adams, s.j.george, h.bolouri*}@herts.ac.uk

## Abstract

The task addressed in this paper is the evolution of an artificial retina with an on-centre/off-surround response which performs edge detection. Evolutionary optimisation is performed on the parameters of a developmental model. The model is capable of creating three dimensional, multi-layer neural networks by modelling the outgrowth of neuron-to-neuron connectivity. A genetic algorithm is used to optimise the developmental parameters, measured against a target retina structure.

The first stage of evolution adapts the parameters of outgrowth rules and the developmental environment. We show that this type of development can be sensitive to noisy conditions (perturbations in neuron positions). This limitation can be overcome by incorporating overgrowth and pruning. Staged evolution of these processes is shown to result in robust development.

## 1 Evolutionary Artificial Neural Networks

Evolutionary Artificial Neural Networks (EANNs) have been widely studied as a means of optimising ANNs for specific applications [1]. Commonly the weights and biases of networks are evolved, with the network architecture either remaining fixed [2] or also subject to evolution [3, 4, 5]. Architectures are adapted in terms of the number of neurons, neuron connectivity and neuron transfer functions. Models with architecture evolution, normally incorporate forms of developmental mechanisms which encode rules controlling the creation of networks. The developmental mechanisms modelled are abstracted from biological neural development but with various degrees of biological plausibility. Evolution acts on the developmental rules.

This paper addresses two issues concerning the evolution of ANN architectures. Firstly, the most biologically defensible models tend to incorporate the greatest number of parameters. This creates large search spaces, which can make evolving the desired functionality intractable [4]. Secondly, current models are sensitive to noise, such as the initial positions of neurons [3, 5].

The long term aim of our work is to be able to automatically create ANNs of arbitrary complexity, such that they are not limited to stereotyped architectures. We present a method of robust ANN evolution, where the evolutionary process becomes steadily more complex to meet the demands of the application.

## 2 Previous Work

We have been implementing a 3D model of biological development, in which neuron-to-neuron connectivity is created through interactive self-organisation [6, 7]. Development occurs as a number of overlapping stages, which govern how neurons extend axons and dendrites, collectively termed neurites. Neurons grow within an artificial, embryonic environment, into which neurons and their neurites emit local chemical gradients. The growth of neurites is influenced by the local gradients and the following sets of interacting, developmental rules:

- *Intrinsic growth rules* control the times at which neurites branch and the directions of growth post-branching.

- *Interactive overgrowth rules* enable growing neurites to branch in response to gradient conditions within their local developmental environment.

- Interspersed with growth, spontaneous neural activity processes regulate the growth rate of neurites [7]. The cumulative effects of activity are used by *pruning rules* to remove individual neurons and synapses once growth is completed.

The developmental rules are controlled by parameters, much in the same way as genes can be thought of as parameters for biological development. Evolution then becomes the identification of optimal sets of the developmental parameters. We have previously carried out some preliminary investigations on evolving the growth parameters using a genetic algorithm [6]. The remainder of the paper addresses evolving the developmental processes in stages.

## 3 The Retina Model

The modelling of the mammalian retina has been chosen as the testbed application. The retina was chosen as it has been extensively studied and does not require learning through synapse modification, which simplifies initial modelling. The aim has been to model the on-centre/off-surround response in the retina in order to perform edge detection [8].

We are specifically modelling the formation of triad junctions, which are thought to be responsible for the edge detection response. In the current implementation triad junctions are formed using two phases of outgrowth. Initially a layer of cones and a layer of bipolar cells grow together. The junctions formed by this outgrowth phase become the targets for horizontal cell outgrowth. A valid triad junction is one innervated by two different horizontal cells. Invalid triad junctions are automatically pruned.

Implicit in our modelling procedure is that the required functionality of the retina can be produced by determining the underlying structure. Cones

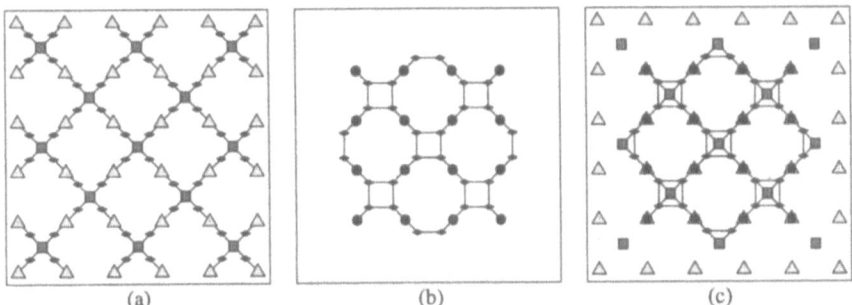

Figure 1: Plan views of the 6x6 cone target retina. Cones are shown as triangles, bipolar neurons as squares, triads junctions as diamonds and horizontal neurons as circles. (a) Cone-to-bipolar connections. (b) Triad junctions formed by horizontal neurons with the synapses in (a). (c) Complete retina combining (a) and (b).

transmit input signals to their triad junctions. Horizontal cells average the signal value they receive through their connections to the triad junctions. The response of bipolar cells is determined by the input signal levels from the cones (centre) and horizontal cells (surround), namely:

$$bipolar\_output = \sum c_{i\pm1,j\pm1} - k \sum h_{i\pm1,j\pm1}$$

where $c$ is the response of a cone, $h$ the averaged response of a horizontal neuron, $k$ is a constant of proportionality, and $i$ and $j$ are vertical and lateral indexes respectively of the bipolar neuron.

## 3.1  Fitness Function Design

In this implementation, a retina's functionality is directly related to its geometry. Hence, we are investigating whether evolution can be driven by a fitness function derived entirely from a target retinal architecture and connectivity. We therefore use fitness functions based on the connectivity between neighbouring neurons in the different layers. This significantly reduces the computation required since test patterns do not need to be applied to developed networks and evaluated. Although the desired retina functionality is not used as a measure of fitness, the functional performances of resulting structures are considered. Ultimately we intend to compare this approach with evolution based on functionality alone or a combination of both approaches.

To reduce the time taken to evolve solutions, a retina consisting of 36 cones, 13 bipolars and 16 horizontal cells is used. The target architecture is illustrated in Figure 1. Edge-effects are negated by calculating the fitness value using only the central bipolar neuron and 4 central horizontal neurons. Each layer of neurons possesses its own set of parameters, hence neurons in the same layer grow under the same developmental controls. Therefore, the parameters found on the small retina, can be directly applied to larger retinas.

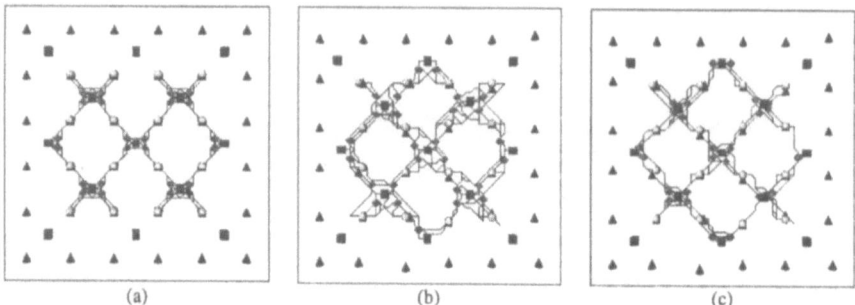

(a)             (b)             (c)

Figure 2: Retina architectures grown using the parameters of the best evolved individuals. Connections are centralised since for edge triad junctions to form, further horizontal neurons would be required to give the correct innervation. The numbers in brackets summarise GA information: number of parameters evolved, genome length in bits and number of generations respectively. (a) Intrinsic growth rules, symmetrical retina [20,35,2]. (b) Intrinsic & interactive overgrowth rules, perturbed retina [25,60,48]. (c) Intrinsic, interactive overgrowth & pruning rules, perturbed retina [33,62,55]

## 4 Results

Evolution was performed using the GENESIS genetic algorithm (GA) package [9]. In all cases the population size was 50 with a mutation rate of 0.001. Selection was rank based and the elitist strategy was used. Once a stage of evolution was complete a 32x32 cone retina was grown (1024 cones, 481 bipolar neurons and 900 horizontal neurons) and functionality was investigated.

Evolution was initially carried out on a retina with symmetrically placed neurons. Figure 2(a) shows that in the absence of any variability, intrinsic developmental rules are sufficient to find the target retina structure whereby the actual functional response (Figure 3(c)) matches the desired response (Figure 3(b)). The set of evolved intrinsic growth parameters was then used to grow a 32x32 retina, where the positions of neurons were perturbed. (Neurons could vary in all 3 directions by 1 unit with a 25% probability, where each unit represents a 10% displacement in position.) Under such noisy conditions the intrinsic rules fail to produce adequate functionality as illustrated in Figure 3(d).

To improve the performance of the developmental model, overgrowth rules were added into the evolutionary process. These parameters permit growing neurons to produce extra branches, determined by the local developmental environment. The new population was seeded using the best parameters from the symmetrical outgrowth case. The best parameters were allowed to vary but within a tighter genetic range than previously permitted.

Each set of evolved parameters was used to create 3 retinas having different initial neuron positions. This was to prevent the evolutionary process adapting to the characteristics of a single perturbed retina. The average fitness of the grown networks was then compared to the target structure. The fitness function

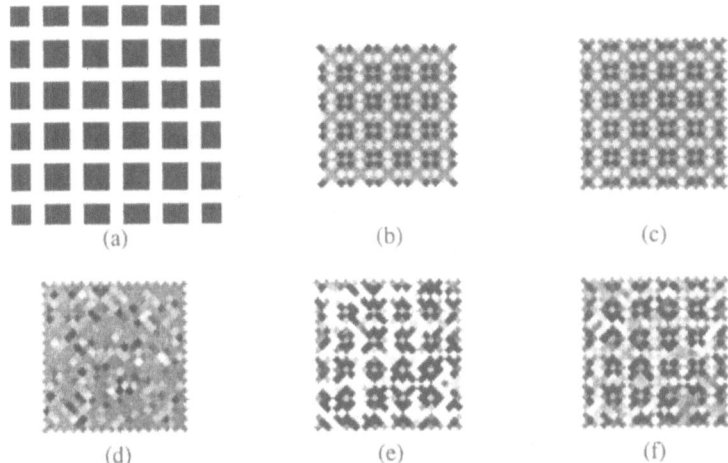

Figure 3: Desired and evolved functionality for a 32x32 cone retina, where images are shown three times their original size. (a) Input image. (b) Desired output. (c)-(f) are outputs of retinas grown using the parameters from the best individuals from the smaller retina. (c) Intrinsic growth rules, symmetrical retina. (d) Intrinsic growth rules, perturbed retina. (e) Intrinsic & interactive overgrowth rules, perturbed retina. (f) Intrinsic, interactive overgrowth & pruning rules, perturbed retina.

encouraged connectivity without harshly penalising multiple connections.

Figure 3(e) shows that the addition of the interactive overgrowth rules, results in a significant improvement in structure and hence functionality over intrinsic rules only. Compared to Figure 3(d) edges are more distinguishable. However, black and white pixels in Figure 3(e) indicate that those particular bipolar neurons are saturated due to having too many connections (see Figure 2(b)).

To reduce the level of multiple connections whilst retaining connectivity, parameters which control pruning mechanisms were added into the evolutionary process. The best parameter set from the previous stage was again used to seed the new genome, also with restricted genetic variability. Multiple retinas were again grown and evaluated using the original fitness function for the outgrowth parameters. With the addition of the pruning parameters, the effects of noise on the functionality of the retina are again reduced, as seen in the architecture of Figure 2(c) and functionally in Figure 3(f).

## 5  Discussion

The results have shown that it is possible to evolve a correctly functioning retina from symmetrically placed neurons using a target structure and intrinsic growth rules only. For perturbed neurons i.e. more complex situations, the complexity of the developmental programme needed to be increased through the addition of overgrowth and pruning, implemented using interactive self-organisation.

The evolution was also carried out in stages incorporating parameters from previous results and evolving them alongside new parameters. In this way the search space is increased in an orderly manner where evolution is channelled through developmental constraints. This contrasts with other models where all the developmental rules are co-evolved. Presenting such large, global search spaces can cause evolution to stall [4].

The addition of overgrowth and pruning shows the developmental model to be robust under noisy conditions. This stage of ANN design is hence made to be less susceptible to errors caused by previous stages of artifical development, such as cell migration [3, 5].

# 6 Conclusion

A strategy of evolving small scale models, whose parameters could be successfully scaled up, and staged evolution was used to reduce the search space and computational complexity. For an artificial retina, where functionality is directly related to structure, it is possible to use a target structure to evolve an edge detection capability even under noisy conditions, where neurons are subject to perturbations. Future work will focus on evolving ANNs where evolutionary progress is evaluated on functionality rather than structure.

# References

[1] Yao X. A review of evolutionary artificial neural networks. International Journal of Intelligent Systems 1993; vol 8: 4: 539-567

[2] Nolfi S. Evolving non-trivial behaviors on real robots: A garbage collecting robot. To appear in: Robotics and Autonomous Systems 1998

[3] Cangelosi A, Parisi D, Nolfi S. Cell division and migration in a 'genotype' for neural networks. Network 1994; 5:497-515

[4] Dellaert F, Beer R. A developmental model for the evolution of complete autonomous agents. In: Proceedings of SAB'96. MIT Press, Cambridge,MA, 1996

[5] Kodjabachian J, Meyer J-A. Evolution and development of modular control architectures for 1-D locomotion in six-legged animats. Submitted for publication to Evolutionary Computation; 1998

[6] Rust AG, Adams R, George S, Bolouri H. Designing development rules for artificial evolution. In: Proceedings of ICANNGA'97. Springer-Verlag, 1997, 508-511

[7] Rust AG, Adams R, George S, Bolouri H. Activity-based pruning in developmental artificial neural networks. In: Proceedings of ECAL'97 MIT Press, Cambridge, MA, 1997, 224-233

[8] Dowling JE. The Retina: An Approachable Part of the Brain. 1st ed, Harvard University Press, London, 1987

[9] Grefenstette JJ, GENESIS 5, ftp.aic.nrl.navy.mil:/pub/galist/src/ 1990

# Activity Driven Update
# in the Neural Abstraction Pyramid

Sven Behnke and Raúl Rojas

Institute of Computer Science, Free University of Berlin

14195 Berlin, Germany

{ behnke | rojas }@inf.fu-berlin.de

## Abstract

The *Neural Abstraction Pyramid* is a hierarchical neural architecture for image interpretation based on image pyramids and cellular neural networks and inspired by the principles of information processing found in the visual cortex. In this paper we extend the model by describing a parallel mechanism of bottom-up attention control, the *Activity Driven Update* of processing elements. We apply this mechanism to the binarization of handwritten ZIP-codes in a real-world application. The experimental results indicate that updating only a fraction of the processing elements is sufficient for good binarization. Both speed and performance of the application were improved with the new method.

## 1 Introduction

The human brain manages to focus its limited resources on the relevant visual stimuli of complex scenes. This is done by a mechanism called *attention control* that is driven by salient visual stimuli and the interpretation goal. Psychophysical evidence [3] suggests that attention comprises two components: a bottom-up, fast, primitive mechanism that selects stimuli based on their saliency and a second, slower, top-down mechanism, the spotlight of attention, that is under cognitive control. Both processes compete to select visual stimuli for detailed investigation. As a result the focus of attention is moved from one location to the next either using the covert spotlight or by overt saccadic eye movements.

In this paper we focus on bottom-up attentional processes that work on a finer time scale, namely in the first few milliseconds of a fixation. In [5] it has been shown that latencies can improve image segmentation. The idea is to delay a stimulus based on its relative value. We propose a method named *Activity Driven Update* of processing nodes in the *Neural Abstraction Pyramid*. The update sequence depends on the saliency of a stimulus. This leads to short delays that can be used to improve image interpretation. Salient parts are interpreted first and provide via horizontal and vertical feedback links a larger context for the interpretation of the more ambiguous image parts.

The paper is organized as follows: In the next section we give a brief summary of the Neural Abstraction Pyramid [2] architecture and algorithms. The proposed Activity Driven Update is presented in section 3. Section 4 describes a first application, the binarization of handwriting. The paper concludes with a discussion of the experimental results and gives an outlook of future work.

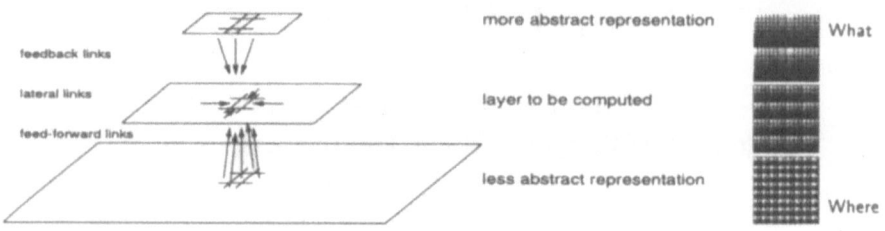

Figure 1: Sketch of the Neural Abstraction Pyramid.

## 2 Neural Abstraction Pyramid

The Neural Abstraction Pyramid [2] is a hierarchical neural architecture for image interpretation that is based on the ideas of image pyramids and cellular neural networks. It is inspired by the principles of information processing found in the visual cortex. Algorithms for this architecture are defined in terms of local interactions of processing elements that utilize horizontal as well as vertical feedback loops. The goal is to transform a given image into a sequence of more and more abstract representations while the level of detail decreases.

The main features of the Neural Abstraction Pyramid architecture are:

- *Pyramidal shape:* Layers of neural processing elements (*nodes*) are arranged vertically to form a pyramid (see Fig. 1).

- *Analog representation:* The nodes of each layer describe the image in a two dimensional representation. The level of abstraction of these representations increases with height, while the level of detail (spatial resolution) decreases. The bottom layer stores the given image (a signal). Subsymbolic representations are present in intermediate layers, while the highest layers contain almost symbolic descriptions. The representation consists of some *quantities* that can have *values* from a finite interval.

- *Local interaction:* Each node is connected to some nodes from its neighborhood via directed *weighted links*. The shared weights of all nodes in a layer are described by a common *template*. The types of links are:
    - *Feed-forward links:* perform feature extraction,
    - *Lateral links:* facilitate consistent image interpretation,
    - *Feedback links:* propagate interpretation hypotheses downwards.

- *Discrete time computation:* The update of a node's values for time step $t$ depends only on the input values at $(t-1)$. The update can be done:
    - *Layer by layer:* All nodes of a layer are updated simultaneously. The layers are processed in a predetermined sequence, e.g. bottom-up.
    - *Priority driven:* The update sequence depends on a priority that can be defined e.g. in terms of the presence of reliable inputs.

- *Multiscale representation:* Quantities can be stored as image pyramid.

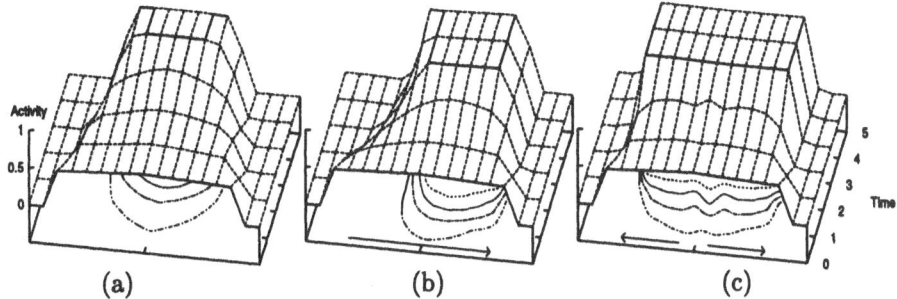

Figure 2: Update methods: (a) buffered, (b) unbuffered, (c) activity driven.

The described architecture has been designed to facilitate the development of image interpretation algorithms that utilize both horizontal and vertical feedback loops. They have to be implemented in a way that honors the principles of *Gestalt psychology*, e.g. proximity, continuity, closure, and simplicity. To make this possible, it is necessary to specify for each layer simple consistent representations that model the objects potentially present in the image. The link weights and update rules have to be selected such that they favor simple and consistent representations instead of complicated or inconsistent ones.

Image interpretation works *iteratively*. First the given image is fed into the bottom layer. In the course of computation the interpretations spread upwards via feed-forward links at locations where little ambiguities exist. These partial results provide via lateral and feedback links a larger context for the interpretation of the more ambiguous stimuli. The quality of the interpretation increases and after a few iterations the interpretation is stable.

# 3 Activity Driven Update

The interpretation performance of the Neural Abstraction Pyramid depends on the update sequence of the nodes as Fig. 2 illustrates using a simple example. It is shown how a one dimensional stimulus develops in three different update modes. We used a plateau stimulus (shown in the front) that increases slightly from the edges (0.5) towards the middle (0.56). The successive cuts show how the stimulus develops over time under a dynamic that is described by: $q_x^{t+1} = \max(0, \min[1, q_x + (q_{x-1}^{t(+1)} + q_{x+1}^{t(+1)})/2 - 0.5])$. Neighboring nodes have excitatory links and the activity is mapped to $[0, 1]$ using a negative bias and saturation. The update modes differ in the handling of the neighboring activities.

(a) *Buffered update* is conservative. All nodes have to be computed in time step $t$ until the resulting activities can be used in step $(t + 1)$. This makes the result independent of the update sequence within a time step. All nodes can be computed in parallel since no dependencies exist. On the other hand, the interpretation is slow, because information travels horizontally with a speed of only one node per time step.

(b) *Unbuffered update* computes the nodes in a predetermined order, here from left to right. The resulting activity $q_{x-1}^{t+1}$ of a node is used immediately to compute the activity $q_x^{t+1}$ of its right neighbor. The interpretation converges much faster, since information travels the full length from left to right within the same time step. However, the information flow from right to left is still slow which results in an undesired asymmetric response of the system.

(c) *Activity Driven Update* uses the same unbuffered strategy to speed up convergence. It prevents undesired asymmetric responses by making the update sequence dependent on the input. The nodes are computed in the sequence of their activity with the most active node computed first. Fast communication occurs now from the more active to the less active image parts. Since the activities represent confidence of interpretation, the image parts that are easy to interpret are computed first which in turn bias and speed up the interpretation of the more ambiguous image parts. If multiple interpretations compete, the one that first receives support from the context is likely to win. Activity Driven Update also speeds up computation, because the nodes with activity zero will never get active and so they don't need to be updated. In most applications the vast majority of nodes will be inactive.

Ordered update does not require global communication. If integrate-and-fire nodes are used, those that receive a stimulus that fits their receptive field will fire earlier than nodes that get a suboptimal stimulus. The firing nodes trigger their neighbors via excitatory links, if they are close enough to the firing threshold. This leads to an avalanche effect that causes a fast traveling activity wave. The wave stops if it crashes with a wave from the opposite direction or enters locations that have an activity that is too low to be triggered. If the nodes need about the same refractory time, all nodes will become activated synchronously again. This synchronization could be a basis for feature binding.

## 4    Binarization of Handwriting

We illustrate Activity Driven Update in the Neural Abstraction Pyramid architecture using a real-world application. The task is to separate handwriting in the foreground from the background in gray level images that have been provided by Siemens AG and show scanned ZIP-codes from German letters. This task is nontrivial, since the envelopes are mostly made of dark paper.

A histogram based thresholding technique, similar to [4] has been used for binarization in the original ZIP-code recognition system. Its limitations become visible when a gray level gradient, broken lines or noise are present in the image.

A more powerful binarization method that is based on the Neural Abstraction Pyramid architecture was proposed in [2] (see there for more details). The idea is to detect the lines and to assign their corresponding pixels to the foreground. The fact that lines in handwriting usually exhibit good continuity is exploited. Three levels of abstraction (see Fig. 3) are used in the pyramid:

- Level 0: gray values (in), foreground (out) / background separation
- Level 1: edges in four orientations (–), (|), (/), and (\)
- Level 2: lines in four orientations

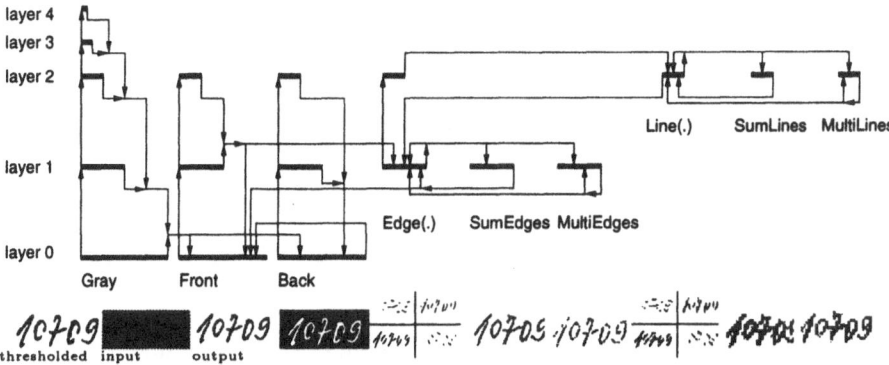

Figure 3: Sketch of the binarization application with quantities and links.

The values of the quantities are interpreted as confidences about the presence of a certain feature at the corresponding image position. The values are computed as saturated weighted sum of the input values. The interaction of the quantities is described by the following templates (see Fig. 3):

- *Gray value:* The original image is stored as a pyramid in quantity Gray.

- *Foreground and background:* Front computes the image details that are darker than their surround while Back contains the brighter background. Both templates inspect the immediate neighborhood of a node more closely and have links that facilitate a region-growing process by excitatory feedback from upper levels. In addition, Front receives inhibitory input from Back and excitatory input from SumEdges to strengthen responses that are supported by detected edges and to remove the noise.

- *Edges:* The distributed edge representation uses four quantities Edge(.) of different orientations. In addition, a quantity SumEdges represents the sum of all and MultiEdges signals the existence of more than one orientation at the same location. The Edge(.)-template functions are:
  - *Edge detection:* two levels of the Front values are inspected,
  - *Cooperation with edges of similar orientation:* via excitatory links from edge elements that would form a good continuation,
  - *Cooperation with lines of the same orientation:* via an excitatory link from the corresponding line element,
  - *Competition with edges at the same position:* via inhibitory input from MultiEdges and SumEdges.

- *Lines:* The line representation is done similarly. The edge representation is inspected to detect lines. Line elements cooperate and compete as well.

In [2] the update of the nodes was done layer by layer with buffered inputs. There line completion and noise removal could be demonstrated. It was investigated, how the performance of a ZIP-code recognition system [1] that is used in a large scale application can be improved by the pyramid binarization.

# 5 Results and Discussion

A set of 503 hard images that were rejected by the original recognition system was selected from 4134 handwritten five-digit ZIP-codes. These images were binarized using the Neural Abstraction Pyramid, again presented to the recognition system, and the results of both runs were combined. The original system had an acceptance rate of 84.56% with 1.17% of the accepted ZIP-codes being substitutions. Using the described modification 169 images were accepted additionally, while only one of these was substituted. Thus, the overall acceptance rate improved to 88.65% without decreasing the systems reliability.

In the new experiments we augmented the binarization pyramid by the proposed Activity Driven Update mechanism in the Edge(.) and in the Line(.) quantities. This improved both speed and performance of the system. Now 209 of the 503 selected hard images were accepted and none of these was substituted. This improved the overall acceptance rate to 89.62% and lowered the substitution rate to 1.11% of the accepted digits. The average running time for the entire recognition was 0.8 seconds per ZIP-code on a Pentium-II-266.

In this paper the extension of the Neural Abstraction Pyramid by an Activity Driven Update mechanism has been proposed. This mechanism facilitates bottom-up attention control. It accelerates convergence by recomputing only a small fraction of the nodes. Since the update sequence depends on the image content, an improved interpretation performance could be achieved. The Activity Driven Update was tested in a real-world application, the binarization of handwritten ZIP-codes. The new technique improved both speed and performance of the recognition. In the future we plan to investigate, how the Activity Driven Update can be combined with a top-down attentional process and how spontaneous synchronization can be used for segmentation and binding.

# References

[1] Sven Behnke, Marcus Pfister, and Raúl Rojas. Recognition of handwritten digits using structural information. In *Proceedings of International Conference on Neural Networks, Houston*, volume 3, pages 1391–1396, 1997.

[2] Sven Behnke and Raúl Rojas. Neural abstraction pyramid: A hierarchical image understanding architecture. In *Proceedings IJCNN'98–Anchorage*, volume 2, pages 820–825, May 1998.

[3] J. Braun and D. Sagi. Vision outside the focus of attention. *Perception and Psychophysics*, 48:45–58, 1990.

[4] J. Kittler and J. Illingworth. Minimum error thresholding. *Pattern Recognition*, 19(1):41–47, 1986.

[5] R. Opara and F. Wörgötter. Using visual latencies to improve image segmentation. *Neural Computation*, 8(7):1493–1520, 1996.

# Jacobian Neural Network Learning Algorithms

André Elisseeff and Hélène Paugam-Moisy

Laboratoire de l'Informatique du Parallélisme - URA CNRS 1398

Ecole Normale Supérieure de Lyon, F-69364 LYON cedex 07, FRANCE

aelissee , hpaugam @ens-lyon.fr

### Abstract

Starting from an analytic description of a multilayer network, as a function of its weights, we first define a learning algorithm which both determines the network architecture and learns all the examples with any given precision. Second, we define a regularization technique for improving the generalization ability of the network. In case of several output units, a modular approach is proposed for reducing the number of weights. We analyse the properties of the algorithm and prove that the network can be learned in polynomial time. The complexity of the task of learning neural networks is discussed in conclusion.

## 1 Previous Results

Consider a multilayer network with $N_I$ input units, $N_H$ hidden units (one hidden layer) and $N_O$ output units. Both inputs and outputs are real-valued. The hidden units compute a non-linear function $f$ which will be specified later on. The output units are assumed to be linear. A learning set of $N_P$ input examples is given and fixed. For all $p \in \{1..N_P\}$, the $p^{th}$ example is defined by its input $d_p \in \mathbb{R}^{N_I}$ and the corresponding desired output $t_p \in \mathbb{R}^{N_O}$.

The learning set can be represented by an *input matrix* $\mathcal{D} = [d_1, \ldots, d_{N_P}]^T$ and a *target matrix* $\mathcal{T} = [t_1, \ldots, t_{N_P}]^T$. For all $h \in \{1..N_H\}$, $w_h^1 \in \mathbb{R}^{N_I}$ is the vector of weights between all the input units and the $h^{th}$ hidden unit, and $W^1 = [w_1^1, \ldots, w_{N_H}^1]$ is the *input weight matrix*. Similarly, the *output weight matrix* $W^2$ is defined by $W^2 = [w_1^2, \ldots, w_{N_O}^2]$.

The input matrix $\mathcal{D}$ is assumed to be fixed. Thus the network output matrix can be expressed as a function $g$ which depends on the weights only

$$g : \mathbb{R}^w \longrightarrow \mathbb{R}^{N_P N_O}$$
$$(W^1, W^2) \mapsto g(W^1, W^2) = F(\mathcal{D}W^1)W^2$$

where $w = N_I N_H + N_H N_O$ is the number of weights and $F$ is a matrix operator which transforms a $n$ by $m$ matrix $A$ into a $n$ by $m$ matrix $F(A)$ according to the relation $[F(A)]_{ij} = f([A]_{ij})$, $i \in \{1..n\}$ and $j \in \{1..m\}$. The jacobian matrix of $g$ at point $(W^1, W^2)$ is denoted by $J(W^1, W^2)$. Assume that all the activation functions $f$ are analytic, hence $g$ and the jacobian are analytic.

An analytical approach has been developed in a previous work [1] for calculating the required number of hidden units to have an onto jacobian matrix. A notion of $\mathcal{H}$-function has been defined for the activation function $f$ of the hidden units.

**Definition 1** *A function $f : \mathbb{R} \to \mathbb{R}$, in $C^1(\mathbb{R})$, is a $\mathcal{H}$-function iff*
$$\exists x_0, \ \forall |x| > x_0, \ f'(x) \neq 0 \quad \text{and} \quad (\forall a \in \mathbb{R} / |a| > 1) \ \lim_{x \to \pm\infty} | \tfrac{f'(ax)}{f'(x)} | = 0$$

Definition 1 includes Gaussian, tanh and logsig functions, hence the family of networks to which our approach can be applied is wide. The main theorem of [1] can be reformulated as follows (a detailed proof is developed in [2])

**Theorem 1** *For having an onto jacobian matrix $J$ for an input matrix $\mathcal{D}$ whose rows are in general position, a network with one hidden layer of $N_H = \lceil N_P N_O/(N_I + N_O) \rceil$ units is necessary and $N_H = \lceil N_P/(N_I + N_O) \rceil N_O$ hidden units are sufficient, with linear output units and an activation function which is a $\mathcal{H}$-function for hidden units.*

Furthermore, if we assume that the jacobian is analytic, then a random choice of $(W^1, W^2)$, satisfying the conditions of theorem 1, provides an onto jacobian $J(W^1, W^2)$ with probability one, almost surely. Next section shows how this approach gives way to defining new learning algorithms.

# 2 Learning Algorithms

**Definition 2** *A network $\eta$-learns a problem $(\mathcal{D}, \mathcal{T})$ iff ($\|.\|$ for euclidian norm) $\exists (W_1, W_2) \in \mathbb{R}^w$ s.t. $\|g(W_1, W_2) - \mathcal{T}\| \leq \eta$*

From theorem 1, we can deduce that $\eta$-learning is achievable $\forall \eta > 0$ as soon as $N_H \geq 2 \lceil N_P/(N_I + N_O) \rceil N_O$. In this section, we set $N_O = 1$, for clarity in the formula. Hence output and target matrices $g$ and $\mathcal{T}$ become output vector $g$ and target vector $t$. However, all the algorithms (figures 2.1,2.2) will be summarized with several output units $N_0$.

## 2.1 Learning examples with precision $\eta$

Let $X = (W^1, W^2)$ and $Y = (V^1, V^2)$ be two sets of $w$ weights each, for two one hidden layer networks with $N_H \geq \lceil N_P/(N_I + 1) \rceil$. From theorem 1, we can assume that the jacobian $J = J(W^1, W^2)$ defined at $X$ is onto. Then a new network, with $2w$ weights, can be defined by combining these two networks with a coefficient $\lambda \in \mathbb{R}$. The resulting network is still a one hidden layer network and its output $z$ is

$$(1) \quad z = \frac{1}{\lambda}(g(Y) - g(X)) = \frac{1}{\lambda}(g(X + \epsilon) - g(X)) = \frac{1}{\lambda}(J\epsilon + R(\epsilon))$$

where $\epsilon = Y - X$ and $R(\epsilon)$ is the rest of the Taylor expansion. Since the norm of this term is proportional to $\|\epsilon\|^2$, we have $R(\epsilon) = \mathcal{O}(\|\epsilon\|^2)$ for small values of $\epsilon > 0$. Hence

$$\|z - \frac{1}{\lambda}J\epsilon\| = \mathcal{O}(\frac{\|\epsilon\|^2}{\lambda})$$

In order to control the network error $\|z - t\|$, we can set $\epsilon = \lambda J^{-1}t$, since the jacobian matrix $J$ is assumed to be onto. Hence

$$(2) \quad \|z - \frac{1}{\lambda}J\epsilon\| = \mathcal{O}(\frac{\|\epsilon\|^2}{\lambda}) \quad \Leftrightarrow \quad \|z - t\| = \mathcal{O}(\lambda\|J^{-1}t\|^2)$$

Let $Y = X + \epsilon$ and choose a very small value for $\lambda$, then the network output approximates $\frac{1}{\lambda} J\epsilon$ which is equal to the target $t$. This approximation is controlled by $\mathcal{O}(\lambda \|J^{-1}t\|^2)$ which is linked to the norm of the Hessian matrix around the point $Y$. Assume that the activation functions have a first and a second derivatives which are bounded, then the term $R(\epsilon)$ is bounded by :

$$\|R(\epsilon)\| \le \frac{1}{2}max(\|w^2\|\|f''\|_\infty, \|f'\|_\infty)(\|\mathcal{D}\|^2 + 2\sqrt{N_P N_H}\|\mathcal{D}\|)$$

From this inequality, a bound on the error of the network can be derived

$$(3)\ \|z - t\| \le \frac{\lambda}{2}\|J^{-1}t\|^2 max(\|w^2\|\|f''\|_\infty, \|f'\|_\infty)(\|\mathcal{D}\|^2 + 2\sqrt{N_P N_H}\|\mathcal{D}\|)$$

and a bound on $\lambda$ can be inferred such that $\|z - t\| \le \eta$. Finally, the quantity $z = \frac{1}{\lambda}(g(Y) - g(X))$ corresponds to the output of a multilayer network with $2w$ weights which $\eta$-learns the problem $(\mathcal{D}, t)$. We have proved that $\eta$-learning is possible as soon as a network can be composed of two subnetworks which verify $N_H \ge \lceil N_P/(N_I + 1)\rceil$. Hence $\eta$-learning can be achieved, for any value of $\eta > 0$, by a network with at least $2N_P$ weights.

| | |
|---|---|
| 1. | Take an initial network with $N_H = \lceil \frac{N_P N_O}{N_I + N_O} \rceil$ hidden units |
| 2. | Choose random weights $X$ for that network |
| 3. | Calculate the Jacobian matrix $J$ and inverse it |
| 4. | Calculate the scalar $\lambda$ according to $\eta$ (formula (3)) |
| 5. | Calculate $\epsilon = \lambda J^{-1}\mathcal{T}$ |
| 6. | Calculate the weights $Y$ for the second network |
| 7. | Compose both networks in a new one : $\frac{1}{\lambda}(g(Y) - g(X))$ |

Figure 1: Jacobian learning algorithm, with precision $\eta$.

## 2.2 Learning with regularization

According to a common observation formalized in [3], high sensitivity to input changes induces bad generalization ability. The output of the network is scaled by $\frac{1}{\lambda}$ which can be very high if a good accuracy is required for learning (equation (3)). Hence the network should generalize poorly, which can be avoided by a regularization technique. Starting from the previous algorithm, we add, as a constraint in the calculation of $\epsilon$, the minimization of the norm of the output variations, by evaluating a gradient around each input example $d_p$ :

$$z(d_p) = \frac{1}{\lambda}(J_p\epsilon + R_p(\epsilon))$$

where $J_p$ is the $p^{th}$ row of $J$ and $R_p$ is the rest of the Taylor expansion of $z(d_p)$. Then

$$\nabla_{d_p}(z(d_p)) = \frac{1}{\lambda}\left(\nabla_{d_p}(J_p\epsilon) + \nabla_{d_p}(R_p(\epsilon))\right)$$

$\epsilon$ being supposed to be small, $R_p(\epsilon)$ is negligible in front of $J_p\epsilon$. Since all the functions are analytic, the Taylor expansion and the gradient can be swapped and $\|\nabla_{d_p}(R_p(\epsilon))\|$ is proportional to $\|\epsilon\|^2$ whereas $\|\nabla_{d_p}(J_p\epsilon)\|$ is proportional to $\|\epsilon\|$. This last term can be written as $\nabla_{d_p}(J_p\epsilon) = B_p\epsilon$. Hence we have $\|\nabla_{d_p}(z(d_p))\| \approx \|\frac{1}{\lambda}(B_p\epsilon)\|$. The inversion of $J$ can be considered as the minimization, for $\epsilon$, of $\|\frac{1}{\lambda}J\epsilon - t\|^2$. Then we can introduce the sum of the norms of the gradients and minimize the error $\mathcal{E} = \|\frac{1}{\lambda}J\epsilon - t\|^2 + \gamma\sum_{p=1}^{N_P}\|\frac{1}{\lambda}B_p\epsilon\|^2$. The regularization coefficient $\gamma$ modulates the influence between both components. We differentiate this expression to find $\epsilon$

$$(4)\quad \epsilon = \lambda\left(J^TJ + \gamma\sum_{p=1}^{N_P}B_p^TB_p\right)^{-1}J^Tt$$

By introducing this regularization term, we do not control the error and the $\eta$-learning is no longer guaranteed. But this drawback concerns all the learning methods which are based on a regularization technique.

1. Take an initial network with $N_H = \left\lceil\frac{N_PN_O}{N_I+N_O}\right\rceil$ hidden units
2. Choose random weights $X$ for that network
3. Calculate and inverse the jacobian matrix $J$
4. Calculate the scalar $\lambda$ according to $\eta$ (formula (3))
5. Calculate the $N_PN_O$ gradients $B_p$ and $A = \lambda\left(J^TJ + \gamma\sum_{p=1}^{N_PN_O}B_p^TB_p\right)^{-1}J^T$
6. Calculate $\epsilon = AT$
7. Calculate the weights $Y$ for the $2^{nd}$ network
8. Compose both networks in a new one : $\frac{1}{\lambda}(g(Y) - g(X))$

(a) Global approach

1. Take an initial network with $N_H = \left\lceil\frac{N_P}{N_I+1}\right\rceil$ hidden units
2. Choose random weights $X$ for that network
3. Calculate and inverse the Jacobian matrix $J$
4. Calculate the scalar $\lambda$ according to $\eta$ (formula (3))
5. Calculate the $N_P$ gradients $B_p$ and $A = \lambda\left(J^TJ + \gamma\sum_{p=1}^{N_P}B_p^TB_p\right)^{-1}J^T$
6. $\forall s \in \{1,..,N_O\}$, calculate $\epsilon_s = At_s$, with $\mathcal{T} = [t_1,\ldots,t_{N_O}]$
7. Calculate the weights $Y_s$ for the $s^{th}$ subnet
8. Compose initial network $X$ with all the subnetworks $Y_s$ to form the final network.

(b) Modular approach

Figure 2: Regularized jacobian learning algorithms: two variants.

## 2.3 Modular approach

Assume now $N_O > 1$. Two approaches can be considered for networks with several output units. Either the learning algorithm of figure 2(a) is applied, composing two networks of roughly $N_PN_O$ weights each, which is the *global approach*, or one network per output unit can be built. The idea is to choose only one weight matrix $X$ for a one output network such that the Jacobian at this point is non zero, and then to choose $Y_1,\ldots,Y_{N_O}$ for the subnetworks corresponding to each of the $N_0$ outputs. The *modular approach* consists in learning $N_O$ subnetworks with one output unit and $N_P$ weights each (figure 2(b)). This variant reduces the total number of weights, e.g. for $N_O = 2$ the global network has $4N_P$ weights, whereas the modular one has $3N_P$ weights.

# 3  Analysis and experiments

## 3.1  Complexity analysis

The values of the most time-consuming steps are related to the complexity of matrix computations which is bounded by $\mathcal{O}((N_PN_O)^3)$. The complexity of the regularized algorithm (figure 2(a)) is $\mathcal{O}\left(N_I(N_P N_O)^3\right)$ as soon as $N_P \geq N_I$ (which is realistic in applications). The complexity of the modular algorithm (figure 2(b)) is bounded by $\mathcal{O}\left(N_I N_P^3 + N_P^2 N_O\right)$ as soon as $N_P \geq N_I$. This makes the modular approach very attractive since less parameters and also less computational time are required.

The polynomial complexity bounds of the jacobian learning algorithms must be compared to usual learning algorithms for multilayer networks. The Perceptron learning rule has been proved to converge, but its computation can require an exponential time. For backpropagation, the complexity bounds are even not defined since convergence is not ensured.

## 3.2 Numerical computation

The main drawback of the jacobian learning algorithms is the possibility of a numerical overflow in the calculation of $\epsilon$. Since the bound on $\lambda$ is very small, the value of $\epsilon$ is low, whereas the value of $w^2$ is very high. For exact computation, the number coding must be implemented such that both low and high numbers have a precise memory representation, which is not verified on usual computers, but could be reached with the help of a multi-precision library (e.g. GMP). In all the classification experiments presented in next section, this problem has never been met but it has occured with some input patterns in regression. We propose, as an heuristic, to add a bias input unit to the network. Intuitively, the bias decreases the singularity of the jacobian matrix $J$ and hence decreases the norm of $J^{-1}$ and increases the value of $\lambda$.

## 3.3 Experimental results

| Classification | Learning | | | Generalization | | |
|---|---|---|---|---|---|---|
| Problem | original | JNN | reg.JNN | original | JNN | reg.JNN |
| Breiman wave-forms | – | 100% | 92,3% | 81% | 74% | 82,8% |
| Sonar discrimination | 99,8% | 100% | 99% | 90,4% | 78% | 92,3% |

Table 1: Performance for initial (JNN) and regularized (reg.JNN) algorithms, compared with original results found in literature

| | Simple Int. | Radial | Harmonic | Additive | Comp. Int. |
|---|---|---|---|---|---|
| SMART | 0.018 | 0.016 | 0.16 | 0.0086 | 0.049 |
| BPL | 0.017 | 0.026 | 0.21 | 0.019 | 0.070 |
| reg.JNN | 0.011 | 0.0082 | 0.024 | 0.053 | 0.061 |

Table 2: Generalization performance of the regularized algorithm (reg.JNN), compared to the SMART algorithm (other model) and to the BPL method of [4] (same model, other learning rule), on five regression problems.

Several experiments have been performed on benchmark problems, (a) in classification: Breiman wave-forms [5] and sonar signals [6], (b) in regression: five types of functions defined in [4]. The algorithms have been programmed with Matlab. The activation function is *arctan* and each experiment takes less than five minutes, on a Sparc station, including tuning the hyperparameter $\gamma$ (optimal values are usually around $10^{-2}$). Tables 1 and 2 shows a comparison of performance between our jacobian algorithms and results found in literature for other methods (with the same experiment protocol). In regression (table 2), error is measured by the Fraction of Variance Unexplained which is directly linked to the quadratic error.

## 4 Conclusion

This article has presented a polynomial time learning algorithm which determines the network architecture and learns all the examples with any given precision $\eta$. A regularized variant of this algorithm has shown very good performance on several benchmark problems, both in classification and in regression.

In literature, polynomial algorithms for learning multilayer networks have been defined only for very large networks [7] or specific problems [8]. Furthermore, the task of learning had been proved to be NP-complete for small networks [9, 10]. Our algorithms play a great part in filling the gap between these two families of results, since they compute, in polynomial time, networks which are general purpose, with a small number of parameters (especially for the modular approach) compared to other polynomial learning algorithms.

In further issues, we would like to control the growth of the network architecture w.r.t. the number of examples $N_P$ and to find a threshold on $N_P$ from which it would be no longer necessary to increase the network size for ensuring good performance in generalization. Another purpose would be to search a theoretical solution for avoiding numerical inconsistency (e.g. a better control of initial weights).

## References

[1] Elisseeff and Paugam-Moisy. Size of multilayer networks for exact learning : analytic approach. In *Proc. of NIPS*96*, pages 162–168. MIT Press, 1997.

[2] Elisseeff and Paugam-Moisy. Size of multilayer networks for exact learning: analytic approach. Research Report 96-16, LIP, 1996.

[3] Dimopoulos, Bourret, and Lek. Use of some sensitivity criteria for choosing networks with good generalization ability. *Neural Proc. Letters*, 2(6):1–4, 1995.

[4] Hwang, Lay, Maechler, Martin, and Schimert. Regression modeling in backpropagation and projection pursuit learning. *IEEE NN*, 5(3):342–353, 1994.

[5] Breiman, Friedman, Olshen, and Stone. *Classification and regression trees*. Wadsworth Inc., Belmont California, 1984.

[6] Gorman and Sejnowski. Analysis of hidden units in a layered network trained to classify sonar targets. *Neural Networks*, 1:75–89, 1988.

[7] Roy, Kim, and Mukhopadhyay. A polynomial time algorithm for the construction and training of a class of multilayer perceptrons. *Neural Networks*, 6:535–545, 1993.

[8] Cohen. Learning noisy perceptrons by a perceptron in polynomial time. In *38th Symposium on the Foundations of Computer Science*, pages 514–523, 1997.

[9] Judd. *Neural Network Design and the Complexity of Learning*. MIT Press, 1990.

[10] Vu. On the infeasibility of training neural networks with small squared error. In *Proc. of NIPS*97*, 1998. (to appear).

# Quadratic Concepts

Jörg C. Lemm

Institut für Theoretische Physik I, Universität Münster

D–48149 Münster, Germany

E-mail: lemm@uni-muenster.de

### Abstract

Gaussian processes have recently become popular in the empirical learning community. They encompass many classical methods of statistics, e.g., radial basis functions or various splines, and are technically convenient due to the fact that Gaussian integrals can be performed analytically and the corresponding saddle point equations (to be solved for a maximum a–posteriori approximation or empirical risk minimization) are linear. At the same time, however, this technical advantage implies a severe practical limitation. Linear equations, i.e., quadratic and therefore convex error surfaces, forbid the implementation of genuine non–convex prior knowledge. For example, one may want to implement the belief that individual earthquakes or electrocardiograms tend to be similar to either a prototype $A$ OR a prototype $B$. Quadratic concepts (or non–zero mean Gaussian processes in Bayesian interpretation) are used as building blocks for implementation of non–convex prior knowledge exploring possibilities to go beyond Gaussian processes. Continuous data functions are introduced based on the observation that learning is, implicitly or explicitly, based on an infinite number of data. It is shown how empirical measurement of an infinite amount of data is possible by a–posteriori control.

## 1   Introduction

In the setting of empirical learning available training data are used to obtain information about new test situations. Clearly, this generalization from training to test data requires knowledge about their dependencies. In this paper, such knowledge concerning dependencies between training and test data, in some contexts also known as rules or axioms, will be called *prior knowledge*.

To be specific, we consider a typical function approximation problem. Assume a given set of training data $D_T = \{(x_i, y_i | 1 \leq i \leq n\}$ sampled i.i.d. from an unknown but fixed "true state of Nature". The aim is to obtain an approximation function $h(x)$ to predict unknown outcomes $y$ for test situations $x \in \mathcal{X}$ by $y = h(x)$.

Relying on the fact that the generalization ability of any learning system is crucially based on the dependencies it implements, our goal has to be a strict empirical measurement and control of the prior or "dependency" data $D_P$ which represent our prior knowledge. It is interesting to note that for the common situation with an infinite number of potential test situations $x \in \mathcal{X}$ also the number of dependencies to be controlled empirically becomes infinite.

*Empirical measurement of an infinite number of data, however, seems at first glance impossible.* On the other hand, for an infinite set $\mathcal{X}$ of test situations any learning system has to use an infinite number of data, either explicitly or implicitly. To discuss this empirical measurement problem let us have a closer look at two examples:

1. A simple bound on $h(x)$ for every $x$ like $h(x) \leq a$, $\forall x \in \mathcal{X}$ corresponds for infinite $\mathcal{X}$ to an infinite number of data.

2. Similarly, deviations from exact symmetries may be bounded. Let $\sigma$ denote a one–to–one transformation on $\mathcal{X}$. Then, $\|h(x) - h(\sigma(x))\| < a$ describes a bound on the deviation from an exact symmetry under $\sigma$, also corresponding to an infinite number of data. The prototypical example is smoothness i.e., approximate symmetry under infinitesimal translations.

Like the number of training data can only be finite for practical reasons also the number of actually appearing test situations in the future can only be finite. The key point is now, that from all possible dependencies only those related to actual test situations have to be controlled empirically. Measurement devices, for example, only have to be active at the, always finite, number of actually appearing test situations. Thus, there is an easy way to enforce bounds without actually measuring an infinite number of times. To be specific, bounds are often the consequence of using realistic, non–ideal measurement devices:

1. A simple bound $h(x) < a$ is implemented by using a measurement device with cut–off at $a$.

2. Assume that, in addition to an upper and lower bound on $h$, input noise or input averaging with respect to $\sigma$ is present in the measurement device we use. That means that we do not have perfect control over the value of $x$. In that situation fixing $x$ at the measurement device still allows that $\sigma(x)$ produces the observable result $y$. The resulting effective function is necessarily smooth with respect to the transformations $\sigma$ which generate the input noise. (Compare [1, 4, 5].)

Thus, one can say: *Infinite a–priori information can be empirically measured by a–posteriori control at the time of testing.* From this point of view, related to that of constructivism, (also infinite) a–priori information can (and should) be explicitly related to empirical control of the application situation.

# 2 Quadratic concepts

As a first step empirical dependency control requires an explicit and readable formulation of prior knowledge. By explicit formulation we mean in the following the expression of prior knowledge directly in terms of the functions values $h(x)$, like it is done in regularization theory or for stochastic processes. In an implicit implementation of dependencies, on the other hand, single function values $h(x)$ are not parameterized independently. Examples include neural networks, linear, additive or tensor models. Also the realization of learning algorithms can induce dependencies, e.g., due to restricted initial conditions and stopping rules.

In the regularization framework an approximation $h(x)$ is chosen to minimize a regularization or error functional $E(h)$. Prior knowledge can be represented by a regularization term added to the training error. A typical example of a smoothness related regularization functional for $d$–dimensional $x = \{x_{(1)}, \cdots, x_{(d)}\}$ is

$$E(h) = \sum_i^n \left( y_i - h(x_i) \right)^2 + \int_{-\infty}^{\infty} d^d x \sum_{l=1}^{d} \sum_k \gamma_k \left( \frac{\partial^k h(x)}{\partial x_{(l)}^k} \right)^2 . \tag{1}$$

It is well known that (for approximation problems with log–loss [5]) a regularization or error functional can be interpreted as up to a constant proportional to the negative logarithm of the posterior probability $p(h|D) \propto p(D|h) \propto e^{-\beta E(h)}$ with $D = D_T \cup D_P$. For differentiable functionals $E$ the optimal solution $h^*$ is found by setting the functional derivative of $E$ with respect to $h$ to zero. Clearly, the functional $E$ must depend on each single function value $h(x)$ to determine a complete function $h \in \mathcal{H}$. Thus, for an infinite set $\mathcal{X}$ the $x$–integral (or a sum over $x$, respectively) corresponds to an infinite number of (prior) data terms. For quadratic terms like in Eq.(1) the common structure of the training data and prior terms can best be seen by writing them as scalar products. Let angular brackets denote scalar products and matrix elements of symmetric operators, e.g. $\langle t|h \rangle = \int dx\, t(x) h(x)$ and $\langle t\,|K\,|\,h \rangle = \int dx\, dx' t(x) K(x, x') h(x') = \langle t|h \rangle_K$ where $x$ can be $d$–dimensional. We obtain for a mean–square training error term

$$(y_i - h(x_i))^2 = \int_{-\infty}^{\infty} d^d x\, \delta(x - x_i) \Big( h(x) - t_i(x) \Big)^2 = \langle h - t_i\,|P_i|\,h - t_i \rangle,$$

with $t_i \equiv y_i$ and $P_i$ the projector into the space of functions on $x_i$. Similarly, a typical prior term becomes by partial integration for vanishing boundary terms

$$\int_{-\infty}^{\infty} d^d x \sum_{l=1}^{d} \left( \frac{\partial^k h(x)}{\partial x_l^k} \right)^2 = -\Big\langle h - t_0\,\Big|\Delta^k\Big|\,h - t_0 \Big\rangle,$$

with $t_0 \equiv 0$ and negative (semi) definite $d$–dimensional Laplacian $\Delta$ with kernel $\Delta(x, x') = \delta(x - x') \sum_{l=1}^{d} \frac{\partial^2}{\partial x_l^2}$. *Notice that here, analogously to finite training data $y_i$, a "continuous data" or template function function $t_0 \equiv 0$ corresponding to infinite data has been introduced explicitly.* Indeed, there is no reason except technical convenience why the special function $t_0 \equiv 0$ should be preferred over other functions $t(x)$ which are not identically equal to zero. As long as only one continuous data or template function $t(x)$ enters the functional, $t_0 \equiv 0$ can always be achieved by solving for $h(x) - t(x)$ instead for $h(x)$. This is possible within classical regularization functionals of the form (1). If, however, more than one template function appears in a functional, then only one of them can be set to zero. Such cases will be studied below. We summarize: *Regularization* *functionals have to contain continuous data (template functions) to determine a function h completely. In the special case of classical functionals of form (1) the template function can be chosen identically zero, so it does not appear explicitly in the formalism. In the general case there is no reason not to consider non–zero template functions.* Having motivated the use of continuous data or template functions we now define a quadratic concept. Assume $h$ to be in some Hilbert space $\mathcal{H}$.

**Definition.** (Quadratic concept.) A quadratic *concept* is a pair $(t, K)$ consisting of a *template function* $t(x) \in \mathcal{H}$ and a real symmetric, positive semi–definite operator on $\mathcal{H}$, the *concept operator* $K$. The operator $K$ defines a *concept distance* $d^2(h) = \langle h - t | K | h - t \rangle = \|h - t\|_K^2$ on subspaces where it is positive definite. The maximal subspace in which the positive semi–definite $K$ is positive definite is the *concept space* $H_K$ of $K$. The corresponding hermitian projector $P_K$ in this subspace $H_K$ is the *concept projector*.

**Remark 1** (Approximate symmetries): Typical concept operators $K$ are related to symmetries. Let for example $Sh(x) = h(\sigma(x))$ with $\sigma$ a permutation within $\mathcal{X}$, i.e., one–to–one. Then $K = (I - S)^T (I - S)$, with identity $I$ and $^T$ denoting the transpose, defines a symmetry concept $d_S^2(h) = \langle h - Sh | h - Sh \rangle$. Similarly, assume $S(\theta) = e^{\theta s}$ to be a continuous symmetry (Lie) group, parameterized by $\theta$ and with infinitesimal generators $s$. Then an infinitesimal symmetry concept for $S \approx 1 + \theta s$ is defined by $d_s^2(h) = \langle sh | sh \rangle$. For infinitesimal translations (smoothness) $s = \partial/\partial x$.

**Remark 2** (Gaussian processes): A quadratic concept defines a Gaussian process according to $p(D|h) \propto e^{-d^2(h)/2}$ with covariance operator $K^{-1}$. We remark however, that while Gaussian processes can also be defined for continuous $\mathcal{X}$ [2, 7] we do not discuss continuum limits for the non–Gaussian extensions below. In these cases we refer to a lattice approximation.

**Remark 3** (Support vector machine): Expanding $h = \sum_m n_m \Phi_m(x)$ in a basis of eigenfunctions of $K = \sum_m \lambda_m \Phi_m \Phi_m^T$ one obtains $E(h) = \sum_i (y_i - h(x_i))^2 + \langle h | K | h \rangle = \sum_i (\sum_m n_m \Phi_m(x) - y_i)^2 + \lambda_m |n_m|^2$. Replacing the mean–square training error by Vapnik's $\epsilon$–insensitive error yields a support vector machine with kernel $K$ [6]. In general, flat regions of $E(h)$ as they appear in the $\epsilon$–insensitive error and other robust error functions have the technical advantage that they do not contribute to the gradient and can be ignored within a saddle point approximation.

**Remark 4** (Templates): Templates $t(x)$ can be constructed directly by experts. They also can represent a structural hypothesis realized by a parameterized learning system like a neural network. Templates can be used for transfer by choosing them as the output of a learning system trained for a similar situation. For finite spaces templates can in principle be estimated by sampling.

**Remark 5** (Covariances): Covariances $K^{-1}$ can be given directly by experts. They do not necessarily have to be local but can include non–local correlations. As already remarked they are often constructed from symmetry considerations. For finite spaces covariances can in principle also be estimated by sampling.

# 3 Combination of concepts

Classical regularization functionals consist of a sum of quadratic concepts. In a probabilistic interpretation this corresponds to a combination by AND. Typically, for example, a training data term AND a prior term is approximated by $h$. The sum of quadratic concepts, however, is again a quadratic concept. Analogously, a product of Gaussians is Gaussian. Straightforward calculation shows that a sum of squared distances $d_i^2$ with concept operators $K_i$ can be written $E(h) = \sum_i^N E_i(h) = \sum_i^N d_i^2(h)/2 = d^2(h)/2 + V$, with squared distance $d^2(h) = \langle h - \bar{t} | K | h - \bar{t} \rangle$, template average $\bar{t} = K^{-1}\tilde{t}$, $\tilde{t} = \sum_i^N K_i t_i$, $K = \sum_i K_i$, and $h$–independent minimal component energy $V = \left( \sum_i^N \langle t_i | K_i | t_i \rangle - \langle \bar{t} | K | \bar{t} \rangle \right)/2$, which has the structure of a variance up to a factor $2N$. The linear stationarity equation for a functional $E = d^2(h)/2 + V$ reads $0 = K(\bar{t} - h) = \tilde{t} - Kh$. For positive definite, i.e., invertible $K$, this has solution $h = \bar{t} = K^{-1}\tilde{t}$ which can be solved in a space with dimension smaller or equal to $n$ [7, 3] . Now let us present types of non–convex error functionals.

**Example:** Consider an image reconstruction task where we expect the image of a face. Thus, we may choose concepts with partial template functions for eyes, nose and mouth and require the reconstructed image to approximate the given pixel data AND the eye, nose and mouth templates. Typically, however, the constituents of a face can appear in many different variations. Eyes may be open OR closed, blue OR brown but also translated, scaled or otherwise deformed. Such OR–like combinations of alternative concepts are examples of non–convex prior knowledge.

In a probabilistic interpretation of alternative concepts representing disjunct events indexed by $i$, this yields the mixture model

$$E_M(h) = -\ln \sum_i^N p(D, i|h) - c = -\ln \left( \sum_i^N p(i) e^{-\beta(E_i(h) - F_i) + c} \right), \qquad (2)$$

with component energies $E_i(h) = d_i^2(h)/2 = \langle h - t_i | K_i | h - t_i \rangle / 2$, arbitrary constant $c$ and $p(D, i|h) = p(i) e^{-\beta(E_i(h) - F_i)}$. If $i$–dependent, the normalization integrals $\beta F_i = -\ln \left( \int dt_i e^{-\beta E_i} \right)$ have to be calculated so they do not interfere with the mixture probabilities $p(i)$. In that case the model has the structure of a disordered ("spin–glass–like") system. The model has the stationarity condition $0 = t_M(h) - K_M(h)h$, with $K_M = \sum_i p(D, i|h) K_i$ and $t_M = \sum_i p(D, i|h) K_i t_i$. The parameter $\beta$ is known as inverse temperature and interpolates between a convex AND at high temperature and a non–convex OR at low temperature.

Products are another possibility to implement OR–like structures leading to technically convenient polynomial models

$$E_{LG}(h) = \frac{1}{2} \prod_{i=1}^N d_i^2(h). \qquad (3)$$

The stationarity equation is $K_{LG}(h)h = t_{LG}(h)$ where $K_{LG}(h) = \sum_i M_i(h)K_i$ and $t_{LG}(h) = \sum_i M_i(h)K_it_i$ with $M_i(h) = \prod_{k \neq i} d_k^2(h)$. The model resembles the Landau–Ginzburg treatment of phase transitions in statistical mechanics. Numerical studies of solvable mixture and polynomial models have been performed and will be reported elsewhere.

# 4 Conclusions

The a–priori information required for function approximation corresponds to an infinite number of data. The paper discusses template functions as example of an explicit implementation of infinite data. An infinite number of data can be empirically measured by a–posteriori control. While in classical convex regularization functionals template functions representing infinite number of data can be chosen identically zero, general non–convex a–priori information requires explicitly non–zero template functions.

**Acknowledgements** The author was supported by a Postdoctoral Fellowship (Le 1014/1–1) from the Deutsche Forschungsgemeinschaft and a NSF/CISE Postdoctoral Fellowship at the Massachusetts Institute of Technology. He also wants to thank Federico Girosi, Tomaso Poggio, Jörg Uhlig, Achim Weiguny, and Chris Williams for discussions.

# References

[1] Bishop, C.M.: Training with noise is equivalent to Tikhonov regularization. *Neural Computation* **7** (1), 108–116, 1995.

[2] Doob, J.L.: *Stochastic Processes*. Wiley, New York, 1953.

[3] Girosi, F., Jones, M., and Poggio, T.: Regularization Theory and Neural Networks Architectures. *Neural Computation* **7** (2), 219–269, 1995.

[4] Leen, T.K.: From Data Distributions to Regularization in Invariant Learning. *Neural Computation* **7**, 974–981, 1995.

[5] Lemm, J.C. Prior Information and Generalized Questions. A.I.Memo No. 1598, C.B.C.L. Paper No. 141, Massachusetts Institute of Technology, 1996, (available at http://planck.uni–muenster.de:8080/lemm).

[6] Vapnik, V.N.: The Nature of Statistical Learning Theory. Springer, 1995.

[7] Wahba, G.: Spline Models for Observational Data. SIAM, 1990.

# Complexity of Boolean Computations for a Spiking Neuron*

Michael Schmitt[†]

Lehrstuhl Mathematik und Informatik

Fakultät für Mathematik

Ruhr-Universität Bochum

D-44780 Bochum, Germany

mschmitt@lmi.ruhr-uni-bochum.de

### Abstract

We investigate the computational power of a model for a spiking neuron in the Boolean domain by comparing it with traditional neuron models such as threshold gates (or McCulloch-Pitts neurons) and sigma-pi units (or polynomial threshold gates). In particular, we estimate the number of gates required to simulate a spiking neuron by a disjunction of threshold gates and we establish tight bounds for this threshold number. Furthermore, we analyze the degree of the polynomials that a sigma-pi unit must use for the simulation of a spiking neuron. We show that this degree cannot be bounded by any fixed value. Our results give evidence that the use of continuous time as a computational resource endows single-cell models with substantially larger computational capabilities.

## 1   Introduction

Models of single neurons like the threshold gate or the sigmoidal gate are theoretically well investigated and widely used in applications. These and related types of models, however, do not capture the phenomenon of timing which is believed to be essential for computations in biological neural systems (see e.g. Rieke et al. [1]). Spiking neurons, or leaky-integrate-and-fire neurons as they are called in biophysics and theoretical neurobiology, are an attempt to explore the computational significance of this phenomenon.

We investigate in this paper a simple version of a spiking neuron, the so-called "spiking neuron of type A" according to the terminology of Maass [2]. Despite its simplicity, the model has sufficient qualities so that one can observe the impact that the use of time as a resource has on the computational capabilites of a single neuron.

We focus on the computation of Boolean functions. In Section 2 the definition of the model and references to related work are given. We compare the computational capabilities of a spiking neuron with those of traditional neuron

---

*Work supported in part by the ESPRIT Working Group NeuroCOLT No. 8556

[†]This research was done while the author was with the Institute for Theoretical Computer Science at the Technische Universität Graz, A-8010 Graz, Austria.

models: threshold gates and sigma-pi units. Disjunctions of threshold gates are universal computational devices for the class of Boolean functions provided that the number of gates is not restricted. Therefore, the question arises how many threshold gates are required to simulate a spiking neuron by a disjunction of threshold gates. We analyze this threshold number of a spiking neuron in Section 3. A single sigma-pi unit is also capable to compute any Boolean function provided that there is no bound on the degree of the polynomial that the unit may use. In Section 4 we study the degree that a sigma-pi unit must have to be able to simulate a spiking neuron.

Due to space limitations, all proofs are omitted in this extended abstract. They can be found in the complete version [3].

## 2 Basic concepts and related work

A *Boolean function* on $n$ variables is a function $f : \{0,1\}^n \to \{0,1\}$. A *threshold gate* (or *McCulloch-Pitts neuron*) with $n$ inputs is characterized by $n + 1$ real numbers: the *weights* $w_1, \ldots, w_n$ and the *threshold* $\theta$. It computes a Boolean function $f$ where $f(x_1, \ldots, x_n) = 1$ if and only if $\sum_{i=1}^{n} w_i x_i \geq \theta$. A Boolean function computable by a threshold gate is also called *threshold function*.

In this paper we focus on a simple model for a spiking neuron, the so-called "spiking neuron of type A" [2]. For the sake of brevity we will refer to it as *a spiking neuron* henceforth throughout this paper.

A *spiking neuron* $v$ receives inputs in the form of short pulses, also known as *spikes*, from $n$ input neurons $a_1, \ldots, a_n$. For $i = 1, \ldots, n$ there is a connection from $a_i$ to $v$ characterized by two numbers: the weight $w_i \in \mathbb{R}$ and the delay $d_i \in \mathbb{R}^+$ (where $\mathbb{R}^+ = \{x \in \mathbb{R} : x \geq 0\}$). Time is treated as a continuous variable denoted by $t$. We assume that if input neuron $a_i$ *fires*, that is, emits a spike, at time $t_i$, this generates a rectangular pulse in $v$ of the form $h_i(t - t_i)$ with

$$h_i(t) = \begin{cases} 0 & \text{for } t < d_i \text{ or } t \geq d_i + 1, \\ w_i & \text{for } d_i \leq t < d_i + 1. \end{cases}$$

Neuron $v$ is assumed to fire as soon as the sum

$$P_v(t) = \sum_{i=1}^{n} h_i(t - t_i)$$

of these pulses reaches a certain threshold $\theta_v$. The functions $h_i$, known as *postsynaptic potentials*, model the effect that a firing of neuron $a_i$ has on the *membrane potential* $P_v$ at the so-called trigger zone of $v$, which is the location where the decision is made whether a spike is to be emitted by $v$.

The firing threshold $\theta$ of a biological neuron is a function depending on the time which has passed since its last firing. For Boolean computations we assume that the neuron has not fired for a sufficiently large period of time, so that its firing threshold has returned to its so-called resting value. Therefore we treat the threshold $\theta$ as a constant throughout this paper.

The neuron model that we consider here is a simple version of a *leaky integrate-and-fire neuron*. Further discussions of this and other neuron models can be found in the brief article by Softky and Koch [4] or in the surveys by Gerstner [5] and Maass [2]. The latter also contain comprehensive lists of references to relevant literature from biophysics and neurobiology. The main simplification of our model is the embodiment of the postsynaptic potential as a rectangular pulse rather than a continuous function of a similar shape. Rectangular pulses are widely used in silicon implementations of networks of spiking neurons such as described by Murray and Tarassenko [6].

A spiking neuron may be viewed as a digital or analog computational element, depending on the type of temporal coding that is used. In the following we restrict our analysis to the coding of binary values. For this *binary coding* we assume that input neuron $a_i$ fires at time 0 if it encodes a 1, and that it does not fire at all if it encodes a 0. Correspondingly, we assume that $v$ outputs a 1 if it fires as a result of this input from $a_1, \ldots, a_n$, and that $v$ outputs a 0 if it does not fire.

We view the delays $d_i$ as "programmable parameters" of the spiking neuron, in addition to the weights $w_i$ of its synapses. Hence a learning algorithm may change the function computed by the spiking neuron not only by modifying its synaptic efficacies but also by tuning the transmission delays between neurons. For a discussion of the biological relevance of such mechanisms see Maass and Schmitt [7] and the references in there.

When considering the computation of Boolean functions, it is not hard to see (and was already stated in [2]) that the type of a spiking neuron studied here is at least as powerful as a threshold gate. The latter can be simulated by using the same weights and assigning equal values to all delays. Moreover, in [2] a concrete Boolean function has been exhibited that can be computed by a single spiking neuron, but requires at least $n/(2\log((n/2)+1))$ threshold gates when computed by a circuit. Further results in [7] indicate that the expressive power of a spiking neuron with $n$ variable delays is much higher than that of a threshold gate: When the weights are fixed and the delays are programmable, the VC-dimension of a spiking neuron is $\Theta(n \log n)$, whereas for a threshold gate with variable weights it is $\Theta(n)$. (The VC-dimension can be considered a measure for the variety of a class of functions; see [7] for details.) The significant increase of the VC-dimension from a threshold gate to a spiking neuron demonstrates the considerably large impact of a delay as a programmable parameter on the computational power of a single-cell model.

In the following sections we provide further evidence for the large computational capabilities of this neuron model. First we establish bounds for its threshold number. Then we study its relation to sigma-pi units.

# 3   On the threshold number

The *threshold number* of a Boolean function $f$ is the smallest number $k$ such that $f$ can be computed by a disjunction of $k$ threshold gates. The threshold number of a computational unit, such as a threshold gate or a spiking neuron, is defined as the maximum threshold number of any Boolean function that can be computed by the unit. Obviously, a threshold gate has threshold number 1. Calculating the threshold number of a computational unit is a natural way to measure how much more powerful the unit is than a threshold gate.

Investigations of the threshold number of Boolean functions can be traced back to Jeroslow [8]. He has shown that every Boolean function on $n$ variables has threshold number at most $2^{n-1}$, and further that for each $k$, where $1 \leq k \leq 2^{n-1}$, there is a Boolean function having threshold number $k$, with the parity function attaining the maximum value.

Maass and Schmitt [7] have shown that every Boolean function computable by a spiking neuron can be computed by a disjunction of at most $2n-1$ threshold gates. This implies a threshold number of at most $2n-1$ for the spiking neuron. The following result shows that even less than $n$ gates are sufficient.

**Theorem 1** *The threshold number of a spiking neuron with $n \geq 2$ inputs is at most $n - 1$.*

By an explicit construction Maass [2] has shown that there is a Boolean function that can be computed by a spiking neuron, whereas any threshold circuit that computes this function has at least $n/(2\log(n/2 + 1))$ gates. This immediately implies that $n/(2\log(n/2+1))-1$ is a lower bound for the threshold number of a spiking neuron. We will now considerably improve this bound.

**Theorem 2** *The threshold number of a spiking neuron with $n$ inputs is at least $\lfloor n/2 \rfloor$.*

Theorems 1 and 2 together provide asymptotically tight bounds for the threshold number of a spiking neuron.

**Corollary 3** *The threshold number of a spiking neuron with $n$ inputs is $\Theta(n)$.*

The linear upper bound from Theorem 1 raises the question if having a small threshold number is sufficient for a Boolean function to be computable by a spiking neuron. This conjecture is refuted by the following result.

**Proposition 4** *For each $n \geq 2$ there is a Boolean function on $n$ variables that has threshold number 2 and cannot be computed by a spiking neuron.*

We remark that the threshold number 2 in Proposition 4 is minimal with this property since every Boolean function with threshold number 1 is a threshold function and hence computable by a spiking neuron.

## 4   On the threshold order

A threshold gate can be considered as a linear separator, that is, a hyperplane in the $n$-dimensional space which classifies the set of Boolean vectors depending on which side they lie with respect to the hyperplane. A natural generalization of this notion is to allow for higher order terms in the weighted sum such that one arrives at polynomial surfaces as classifiers. Neuron models employing polynomials as activation functions are also known as sigma-pi units (see, e.g., Softky and Koch [4]). Such models are computationally more powerful than threshold gates even in the Boolean domain and even when the degree of the polynomial is bounded by some small constant as in the case of, for instance, quadratic surfaces.

In the following we compare the computational capabilities of the spiking neuron with those of polynomial neurons. We present a lower bound for the spiking neuron in terms of the so-called threshold order. The *threshold order* of a Boolean function $f$ is the smallest number $m$ such that $f$ has a separator of degree $m$, but no separator of degree $m - 1$. A *separator of degree $m$* for a Boolean function $f$ on $n$ variables is given by a weight vector $w_1, \ldots, w_d$ of dimension $d = \sum_{i=1}^{m} \binom{n}{i}$ and a threshold $\theta$, such that

$$\sum_{S \subseteq \{1, \ldots, n\}, 0 < |S| \leq m} w_S \prod_{i \in S} x_i \geq \theta \iff f(x) = 1.$$

We define the threshold order of a computational unit as the largest threshold order of any Boolean function the unit can compute.

The notion of a threshold order has been introduced by Minsky and Papert [9]. (They simply called it order.) In particular they showed that each Boolean function on $n$ variables has threshold order at most $n$ with the parity function being of order exactly $n$. Obviously, a threshold gate has threshold order 1. We show that the threshold order of a spiking neuron is significantly larger.

**Theorem 5** *The threshold order of a spiking neuron with $n$ inputs is $\Omega(n^{1/3})$.*

This result shows that the threshold order of a spiking neuron grows arbitrarily large with the number of inputs. Thus, a polynomial threshold gate, or sigma-pi unit, with a degree that is bounded by a constant is computationally less powerful than a spiking neuron.

**Corollary 6** *There is no sigma-pi unit with fixed degree that can simulate a spiking neuron.*

The currently best upper bound is established in the following result.

**Proposition 7** *The threshold order of a spiking neuron with $n \geq 2$ inputs is at most $n - 1$.*

Finally, by analogy with the proof of Proposition 4 we obtain the following statement as a counterpart to the result on the threshold number.

**Corollary 8** *For each $n \geq 2$ there is a Boolean function on $n$ variables that has threshold order 2 and cannot be computed by a spiking neuron.*

# 5 Conclusions

Time is undoubtedly an essential parameter in neural computation. We have analyzed a neuron model that uses the timing of single firing events to represent input and output values and has transmission delays as programmable parameters. We have shown that the computational capabilities of this model are significantly larger than those of a threshold gate or a sigma-pi unit with fixed degree. In particular, we have established a tight lower bound of $\Omega(n)$ for the threshold number of a spiking neuron, which is the number of gates required to simulate a spiking neuron by a disjunction of threshold gates. Moreover, we have shown that the threshold order, which is the degree of the polynomial used by a sigma-pi unit for simulating a spiking neuron, must grow as $\Omega(n^{1/3})$.

The lower bounds presented here bear far-reaching consequences for the following reason: It is not hard to see that these lower bounds remain valid for models of spiking neurons that employ more realistic functions as postsynaptic potentials. In fact, there is a very mild condition that such a model must satisfy in order for these bounds to hold: It is sufficient that the postsynaptic potential is non-zero for a definite period of time. If this requirement is met the ideas provided here show that the model becomes more powerful by using time as a computational resource.

# References

[1] F. Rieke, D. Warland, W. Bialek and R. de Ruyter van Steveninck, *SPIKES: Exploring the Neural Code*, The MIT Press, Cambridge, Mass., 1996.

[2] W. Maass, Networks of spiking neurons: The third generation of neural network models, *Neural Networks* **10** (1997) 1659–1671.

[3] M. Schmitt, On computing Boolean functions by a spiking neuron, TU Graz, 1998, available via http://www.lmi.ruhr-uni-bochum.de/mschmitt/.

[4] W. Softky and C. Koch, Single-cell models, in: *The Handbook of Brain Theory and Neural Networks*, ed. M. A. Arbib, The MIT Press, Cambridge, Mass., 1995, 879–884.

[5] W. Gerstner, Time structure of the activity in neural network models, *Physical Review E* **51** (1995) 738–758.

[6] A. Murray and L. Tarassenko, *Analogue Neural VLSI: A Pulse Stream Approach*, Chapman & Hall, London, 1994.

[7] W. Maass and M. Schmitt, On the complexity of learning for a spiking neuron, in: *Proceedings of the Tenth Annual Conference on Computational Learning Theory COLT'97*, ACM Press, New York, 1997, pp. 54–61.

[8] R. G. Jeroslow, On defining sets of vertices of the hypercube by linear inequalities, *Discrete Mathematics* **11** (1975) 119–124.

[9] M. L. Minsky and S. A. Papert, *Perceptrons: An Introduction to Computational Geometry*, expanded edition, The MIT Press, Cambridge, Mass., 1988.

# Unsupervised Time Series Segmentation by Predictive Modular Neural Networks

Vas. Petridis and Ath. Kehagias

Dept. of Electrical and Computer Eng.,
Aristotle Univ. of Thessaloniki, Greece

**Abstract**

Consider a switching time series, produced by several randomly activated sources. The separation of incoming data into distinct classes may be effected using predictive modular neural networks, where each module is trained on data from a particular source. We present a mathematical analysis regarding the convergence of a quite general class of *competitive, winner-take-all* schemes which allocate data into classes, one class corresponding to each active source.

## 1   Introduction

Consider a time series generated by several randomly activated sources. *Time series segmentation* involves finding the active source at every time step. This has been examined in [3] (using local experts [2]) and in [1], where the following method is used. The observed time series is used as input to a bank of neural network neural predictive modules; at every time step the new observation $y_t$ is *allocated* to the neural predictive module which yields minimum prediction error; then each module is retrained on the data so far allocated to it. In this manner each neural neural predictive module may be *associated* with a particular source, exhibiting minimum prediction error when this source is activated; hence, at every time step the active source is identified by the neural predictive module which has minimum error. In order to *train* each module, labeled data from each source must be available. If training must take place concurrently with segmentation, using the unlabeled measurements of the time series, then accurate segmentation requires *on-line, unsupervised data allocation* to the neural predictive *modules*. In this paper we examine the properties of a general class of *competitive data allocation schemes* which utilize *predictive modular neural networks*.

## 2   Parallel Data Allocation

The *source* time series $Z(t), t = 1, 2, ...$, takes values in a finite *source* set $\Theta = \{1, 2, ..., K\}$; the *observation* time series $Y(t), t = 1, 2, ...$, takes values in $\Re$ (the set of real numbers). $Y(t)$ is generated by a function $Y(t) = F_{Z(t)}(Y(t-1), Y(t-2), ..., Y(t-L))$; where $F_1(.), F_2(.), ..., F_K(.)$ are functions from $\Re^L$ to $\Re$. Hence, $Y(t)$ is determined by past observations and the current source. The *time series segmentation* consists in producing $\widehat{Z}(t)$, an estimate of $Z(t)$, for times $t = 1, 2, \ldots$ , which is equivalent to finding the source which is active for $t = 1, 2, \ldots$ . Using $K$ neural predictive modules of the form $\widehat{Y}_k(t) = \widehat{F}_k(Y(t-1), Y(t-2), \ldots, Y(t-L))$ we can compute the prediction $\widehat{Y}_k(t)$ (for $k = 1, \ldots, K$) and set $\widehat{Z}(t) = \underset{k=1,2,\ldots,K}{\arg\min} \left| Y(t) - \widehat{Y}_k(t) \right|$; i.e. $Y(t)$ is allocated to the neural network of minimum prediction error. The problem can be decomposed into two subproblems: *data allocation* and *predictor training*; here we will deal with the former since, with accurately allocated data, the neural predictive module training subproblem can be solved using a variety of training algorithms. The following classification / training algorithm (a variation of which appears in [1]) is used.

At $t = 0$    $K$ predictors are randomly initialized
For $t =$    $1,2,\ldots$
      Observe $Y(t)$.
      For $k = 1, 2, \ldots, K$ compute $\widehat{Y}_k(t)$ and $|Y(t) - \widehat{Y}_k(t)|$.
      Assign $Y(t)$ to predictor nr. $\widehat{Z}(t)$
      (where $\widehat{Z}(t) = \underset{k=1,2,\ldots,K}{\arg\min} \left| Y(t) - \widehat{Y}_k(t) \right|$).
      Retrain each predictive module on all data assigned to it.
Next $t$

Data allocation is performed in a competitive, winner-take-all manner, to the predictor of minimum error. The question discussed in this paper is whether (a) each neural predictive module will specialize in one source, accepting all or most data generated by this source and rejecting data from other sources, or (b) a neural predictive module will obtain data from more than one sources.

# 3 Convergence

**Two sources.** If two sources are active, the source process $Z(t)$ takes values in $\{1, 2\}$; at time $t$ we have $\Pr\left(Z(t) = i\right) = \pi_i$, $i = 1, 2$. Obviously $\pi_1 + \pi_2 = 1$; it is also assumed that: for $i = 1, 2$ we have $0 < \pi_i < 1$. The observation $Y(t)$ is given by $Y(t) = F_{Z(t)}(Y(t-1), Y(t-2), \ldots, Y(t-L))$. For two neural predictive modules ($i = 1, 2$) we have $\widehat{Y}_i(t) = \widehat{F}_i(Y(t-1), Y(t-2), \ldots, Y(t-L))$. The *allocation* process $W(t)$ takes values in $\{1, 2\}$; $W(t) = i$ means that $Y(t)$ is allocated to the $i$-th neural predictor. If, at time $t$ , $y_t$ is generated by source $i$ and allocated to neural predictor $j$, then $Z(t) = i$ and $W(t) = j$. The processes $M_{ij}(t)$ ($t = 1, 2, \ldots$ , $i, j = 1, 2$) are defined by

$$M_{ij}(t) = \begin{cases} 1 & \text{if} \quad Z(t) = i, W(t) = j \\ 0 & \text{else;} \end{cases}$$

and the processes $N_{ij}(t)$ (where $t = 1, 2, \ldots$ and $i, j = 1, 2$) are defined by $N_{ij}(t) = \sum_{s=1}^{t} M_{ij}(s)$. Hence $N_{ij}(t)$ indicates the *total* number of source $i$

samples assigned to neural predictive module $j$, up to time $t$. The variable $X(t)$, denotes the total *specialization* of the system:

$$X(t) = [N_{11}(t) - N_{21}(t)] + [N_{22}(t) - N_{12}(t)].$$

The data assignment probabilities (for neural predictive module 1) depend on $X(t)$. In case $X(t)$ is large and positive, at least one of $[N_{11}(t) - N_{21}(t)]$ and /or $[N_{22}(t) - N_{12}(t)]$ must be large and positive, which means that either neural predictive module nr.1 has received a large surplus of source nr.1-generated data, or neural predictive module nr.2 has received a large surplus of source nr.2-generated data, or both. Similar remarks hold in case $X(t)$ is large and negative. Hence, it is reasonable to assume that the data assignment probabilities (for neural predictive module 1) depend on $X(t)$:

$$f(n) = \Pr\left(W(t) = 1 \,|\, Z(t) = 1, X(t-1) = n\right),$$
$$g(n) = \Pr\left(W(t) = 1 \,|\, Z(t) = 2, X(t-1) = n\right).$$

In other words, $f(n)$ is the probability that neural predictive module 1 accepts a datum from source 1, given that so far it has accepted $n$ more data from source 1 than from source 2, while $g(n)$ is the probability that neural predictive module 1 accepts a datum from source 2, given that so far it has accepted $n$ more data from source 1 than from source 2. Regarding these probabilities, the following assumptions are made.

A1    For $n = ..., -1, 0, 1, ...$    $f(n) > 0$, $\lim\limits_{n \to -\infty} f(n) = 0$, $\lim\limits_{n \to +\infty} f(n) = 1$,

A2    For $n = ..., -1, 0, 1, ...$    $g(n) > 0$, $\lim\limits_{n \to -\infty} g(n) = 1$, $\lim\limits_{n \to +\infty} g(n) = 0$.

By assumption **A1**, if neural predictive module 1 has accumulated many more data from source 1 than from source 2, then it will be very likely to accept an additional datum generated from this source and will be very unlikely to accept an additional datum from source 2. This is reasonable: if the neural predictive module has been trained on data mostly originating from source nr. 1, rather than from nr. 2, than it will exhibit improved performance on source nr.1 data and deteriorated performance on source 2 data. Similar remarks can be made regarding assumption **A2**. It must be stressed that A1 and A2 refer to the combination of time series, network architecture, training law and data allocation algorithm. *It is not necessary to take into account particular characteristics of any of the above components*; it is only required that A1 and A2 hold true, which may be the case for various combinations of time series, network architecture, training law and data allocation algorithm.

The data allocation procedure described above, implies that $X(t)$ is Markovian. The transition probabilities (for $m, n = 0, \pm 1, \pm 2, ...$) can be obtained from the data allocation method and are $p_{n,m} = 0$ if $|n - m| \neq 1$, $p_{n,n-1} = \pi_2 \cdot g(n) + \pi_1 \cdot (1 - f(n))$, $p_{n,n+1} = \pi_1 \cdot f(n) + \pi_2 \cdot (1 - g(n))$. Hence, convergence can be studied using methods from the theory of Markov chains. We have established two convergence theorems; in this paper we omit the proofs because of space limitations. The first theorem ensures convergence of $X(t)$.

**Theorem 1** *If conditions* **A1, A2** *hold, then*

$(i) \quad \Pr \left( \lim_{t \to \infty} |X(t)| = +\infty \right) = 1,$

$(ii) \quad \Pr \left( \lim_{t \to \infty} X(t) = +\infty \right) + \Pr \left( \lim_{t \to \infty} X(t) = -\infty \right) = 1.$

From (i) it is seen that total specialization goes to infinity; from (ii) it is seen that at least one neural predictive module will (in the long run) accumulate either a lot more source nr.1 samples than source nr.2 samples $(X(t) \to +\infty)$ or a lot more source 2 samples than source 1 samples $(X(t) \to -\infty)$. The total probability that one of these two events will take place is one, i.e. one neural predictive module will certainly specialize in one of the two sources.

**Theorem 2** *If conditions* **A1, A2** *hold, then*

$$\Pr \left( \lim_{t \to \infty} \frac{N_{21}(t)}{N_{11}(t)} = 0 \,\middle|\, \lim_{t \to \infty} X(t) = +\infty \right) = 1,$$

$$\Pr \left( \lim_{t \to \infty} \frac{N_{12}(t)}{N_{22}(t)} = 0 \,\middle|\, \lim_{t \to \infty} X(t) = +\infty \right) = 1,$$

$$\Pr \left( \lim_{t \to \infty} \frac{N_{11}(t)}{N_{21}(t)} = 0 \,\middle|\, \lim_{t \to \infty} X(t) = -\infty \right) = 1,$$

$$\Pr \left( \lim_{t \to \infty} \frac{N_{22}(t)}{N_{12}(t)} = 0 \,\middle|\, \lim_{t \to \infty} X(t) = -\infty \right) = 1.$$

Theorem 2 states that, with probability one, *both* neural predictive modules will specialize, one in each source and in a "strong" sense . For instance, if $X(t) \to +\infty$, then the *proportions* $\frac{N_{21}(t)}{N_{11}(t)}$ (nr. of source 2 samples divided by nr. of source 1 samples *assigned to neural predictive module* 1) and $\frac{N_{12}(t)}{N_{22}(t)}$ (nr. of. source 1 samples divided by nr. of source 2 samples *assigned to neural predictive module* 2) both go to zero ; this means that "most" of the samples on which neural predictive module 1 was trained come from source 1 and, also, that "most" of the time a sample of source 1 is assigned (classified) to the neural predictive module which is specialized in this source; similar remarks hold for neural predictive module 2. Hence we can identify source $i$ with neural predictive module $i$, for $i = 1, 2$. A completely symmetric situation holds when $X(t) \to -\infty$. By Theorem 1, $X(t)$ goes either to $+\infty$ or to $-\infty$, so specialization of both neural predictive modules (one in each source) is guaranteed.

Only a very brief sketch will be of the proofs of the above theorems is given. Regarding Theorem 1, it is proved that the Markovian process $X(t)$ is *transient*; i.e. that w.p.1 (with probability one) $X(t)$ will spend only a finite amount of time in any particular state. Then it follows that $|X(t)|$ must go to infinity w.p.1, which is (i); (ii) follows easily. Regarding Theorem 2, we exploit the fact that, if $X(t)$ goes to $\infty$, then transitions to lower states are highly improbable. Such transitions are "counted" by the process $N_{21}(t)$. This process is dependent, but it can be compared to an auxiliary process $\overline{N}_{21}(t)$, which has a larger probability of transitions to lower states and is independent;

in fact it is a sequence of Bernoulli trials, and its properties are easily obtained. By appropriate construction of $\overline{N}_{21}(t)$ it can be proved that $\overline{N}_{21}(t)/t$ goes to zero with probability one; then relating $\overline{N}_{21}(t)$ and $N_{21}(t)$ it can be shown that also $\overline{N}_{21}(t)/t$ goes to zero with probability one. A similar argument, depending on an uxiliary process $\overline{N}_{11}(t)$ is used to show that $N_{11}(t)/t$ goes to $\pi_1$. Then the first conclusion of the theorem follows easily. The remaining conclusions are proved similarly.

**Many Sources.** The case of more than two sources is treated here by an informal argument; a more formal presentation in terms of convergence theorems will be reported in the future. Consider the case of $K$ sources ($K > 2$) and a data allocation scheme starting with two neural predictors, and adding more predictors "as needed" (for instance whenever the prediction error exceeds a certain threshold). *Consider two sources*: source 1 and *composite* source $[2, 3, ..., K]$. By Theorems 1 and 2, in the long run one neural predictive module will mostly receive data from one source. In the long run the second neural predictive module will mostly receive data from the other source. Hence the data are separated into two sets: those generated by source 1 and those generated by all other sources. Reapplying the data allocation scheme on the composite data set, we will obtain two new neural predictive modules, with one specializing in source 2 data and the other specializing in sources 3, 4, ... , $K$. After sufficient time has elapsed, neural predictive module 2 will specialize in source 2, while neural predictive module 3 will mostly receive data from sources 3,4, ..., $K$. The same argument can be repeated for sources 3, 4, ..., $K$, adding neural predictive modules as needed, resulting in one neural predictive module specializing in each source.

# 4 Experiments

**Exp. Group A.** Four sources have been used: (a) for $Z(t) = 1$, a logistic time series of the form $y(t) = f_1(y(t-1))$, where $f_1(x) = 4x(1-x)$; (b) for $Z(t) = 2$, a tent-map time series of the form $y(t) = f_2(y(t - 1))$, where $f_2(x) = 2x$ if $x \in [0, 0.5)$ and $f_2(x) = 2(1-x)$ if $x \in [0.5, 1]$; (c) for $Z(t) = 3$, a double logistic time series of the form $y(t) = f_3(y(t-1)) = f_1(f_1(y(t-1)))$ and (d) for $Z(t) = 4$, a double tent-map time series of the form $y(t) = f_4(y(t - 1)) = f_2(f_2 y(t - 1)))$. The four sources are activated consecutively, each for 100 time steps, giving an overall period of 400 time steps. Ten such periods are used, resulting in a 4000-steps time series. The task is to discover the four sources and the switching schedule by which they are activated. At every step $y_t$ is mixed with additive white noise uniformly distributed in the interval $[-A/2, A/2]$. The neural predictive modules used are 1-5-1 sigmoid neural networks. In every experiment performed, all four sources are eventually identified. This takes place at some time $T_c$, which is different for every experiment. After time $T_c$, a classification figure of merit is computed. It is denoted by $c = T_2/T_1$, where $T_1$ is the *total* number of time steps after $T_c$, and $T_2$ is the number of *correctly classified* time steps after $T_c$. Table 1 shows the results of the experiments.

**Exp. Group B**. Here we consider a time series obtained from three sources of the Mackey-Glass type. The time series evolves in continuous time and satisfies the differential equation : $\frac{dy}{dt} = -0.1y(t) + \frac{0.2y(t-t_d)}{1+y(t-t_d)^{10}}$. For each source a different value of the delay parameter $t_d$ was used, namely $t_d$= 17, 23 and 30. The time series is sampled in discrete time, at a sampling rate $\tau = 6$, with the three sources being activated alternately, for 100 time steps each. The final result is a time series with a switching period of 300 and a total length of 4000 time steps. The time series is observed at various levels of additive observation noise; results expressed in terms of the parameters $c$ and $T_c$ appear in Table 1. Segmentation is again quite accurate for fairly high noise levels.

Table 1: Segmentation Results for Experiment Groups A and B

| | | | | | | |
|---|---|---|---|---|---|---|
| | $A$ | 0.00 | 0.05 | 0.10 | 0.15 | 0.20 |
| Exp. Group A | $T_c$ | 500 | 1800 | 1800 | 800 | 2200 |
| | $c$ | 0.982 | 0.969 | 0.947 | 0.529 | 0.529 |
| | $A$ | 0.00 | 0.05 | 0.10 | 0.15 | 0.20 |
| Exp. Group B | $T_c$ | 1700 | 1100 | 1300 | 3500 | 1200 |
| | $c$ | 0.978 | 0.977 | 0.853 | 0.935 | 0.664 |

# 5 Conclusion

In this paper we have presented two theorems (regarding the convergence of competitive data allocation) which state that, if the *general* conditions **A1** and **A2** are satisfied, data allocation will result in succesful predictor specialization. The competitive data allocation method may also be called "parallel", in contradistinction to a "serial" method, where a threshold is fixed and the first predictive module with prediction error below the threshold receives the new incoming datum (at every time step the predictive modules are considered in a specified order).We have performed a convergence analysis of the serial data allocation case, which yields results similar to the parallel case; this analysis is presented elsewhere. Hence we now have general conditions which ensure convergence of the data allocation scheme for both the serial and parallel case.

# References

[1] A. Kehagias and V. Petridis, "Time Series Segmentation using Predictive Modular Neural Networks", *Neural Computation*, 1997, vol.9, pp.1691-1710.

[2] M.I. Jordan and R.A. Jacobs. 1994. "Hierarchical Mixtures of Experts and the EM Algorithm", *Neural Computation*, vol.6, pp. 181-214.

[3] K. Pawelzik, J. Kohlmorgen and K.R. Muller. 1996. "Annealed Competition of Experts for a Segmentation and Classification of Switching Dynamics", *Neural Computation*, vol.8, pp.340-356.

# ICE - an Incremental Hybrid System for Continuous Learning

Michael Tagscherer

FORWISS - Bavarian Research Center for Knowledge-Based Systems

Am Weichselgarten 7, 91058 Erlangen, Germany

tagscherer@forwiss.de

## Abstract

ICE is a new incremental construction algorithm of a hybrid system for continuous learning tasks. The basis of the hybrid system is a radial basis function (RBF) network layer. The second layer consists of local models. The two layers are closely combined with a strong interaction. The number of RBF-neurons and the number of local models have not to be determined in advance, which is one of the main advantages of ICE. Another advantage over existing methods are useful network outputs already during the initial learning phase. The development of the described approach is embedded in a larger project[1] that is primarily concerned with system identification tasks for industrial control such as steel processing.

## 1  Introduction

The approximation of nonlinear *and* time-variant functions are the main requirements for system identification tasks for industrial control such as steel processing [1]. While nonlinear function approximation is a well-known application for neural networks, the approximation of nonlinear functions that change over time poses many additional problems which are the focus of our current research.

Because of the dynamics of industrial processes, system identification especially with neural networks has to be done continuously throughout the life-time of the process. We refer to this as *continuous learning*; other common, but depending on the context sometimes misleading expressions are on-line learning, adaptation, sequential, incremental or life-long learning.

### 1.1  Continuous Learning and Neural Networks

The *stability-plasticity dilemma* is one of the main problems of continuous learning tasks. While the system should be able to follow changes of the time-varying function as quickly and accurately as possible (i.e. optimal plasticity), previously acquired information should not be "forgotten" (i.e. stability) [2]. In

---

[1]This research was sponsored by the Federal Ministry of Education, Science, Research and Technology under grant number 01 IN 505 B.

this case, the popular multi-layer perceptrons (MLPs) are not suited. Because of their global activation functions, global information can be forgotten while presenting local information. When local information (i.e. data in a particular area of the input-space) is presented during continuous learning, the shape of the nonlinear function represented by the MLP can change also in other regions of the input-space, thus "forgetting" what was learned during previous training in those regions.

## 1.2 Localized Units and Localized Models

One approach to cope with this problem is to use *localized units*, e.g. radial basis function (RBF) networks [3]. Not only because they are much faster trainable, RBF-networks are in many cases a better choice than MLPs. Nevertheless some difficulties have to be resolved when using RBF-networks, e.g. the optimal topology (i.e. how many neurons are needed), the right placement of each neuron and the extension of each receptive field. Too few neurons usually result in a poor approximation of the target function, while too many neurons can cause overfitting. Another problem is the size of the receptive fields. If for example the receptive fields are too small, a large number of neurons is needed to cover the input-space.

Instead of using localized units, another approach is to use a certain number of *localized models*. Each model is valid for a different region of the input-space. The main problem of these approaches is to find a good representation of the target function by different models, i.e. it is necessary to find a useful clustering of the input-space. Thus, the quality of the approach depends on the quality of the respective clustering method. Counterpropagation networks which use Kohonen networks for clustering are an example for these kind of approaches [4] [5]. One disadvantage of those systems is the weak combination of the two different layers. First, the clustering must be nearly completed before at the second layer useful models can be adapted.

In the following section a new algorithm, called Incremental Clustering and Evaluation (ICE) algorithm will be described. The ICE-algorithm combines advantages of RBF networks and localized models to a new hybrid system. Noteworthy is that the number of RBF-neurons, the number of models and the size of their receptive fields are determined incrementally during the presentation of the training patterns.

# 2 Hybrid System and ICE-Algorithm

ICE is a new incremental construction algorithm of a hybrid system for continuous learning tasks. In the following the base system for the generation of multible models will be described. The foundation of the hybrid system is a radial basis function (RBF) network layer. The second layer consists of local models.

In contrast to methods like counterpropagation-networks, the ICE-model layer and the RBF cluster layer described below are closely combined. For example the RBF-layer uses information from the model-layer to decide if new RBF-neurons are needed and the model layer uses the RBF-neurons as the base for its models. With this strong interaction a useful network output is already available during the initial learning phase. This is very important for industrial applications, e.g. plants can be put into operation much faster.

The main idea of the ICE-algorithm is that the individual models *compete* with each other. In this way they shrink their receptive fields in a mutual interaction.

## 2.1 Network and Model Design

The base of each model are $n$ special RBF-neurons. In the following we call those RBF-neurons *prototypes*. The usage of more than one prototype is a major difference to common RBF approaches where one local model is typically based on one RBF-neuron. In contrast, each ICE-model can have *more* than one prototype, e.g. in the case of linear models the goal of the ICE-algorithm is to create $i + 1$ prototypes per model where $i$ is the input-space dimension. The reason is that a linear model of dimension $i$ can be determined by $i + 1$ input points.

In combination with the strong interaction between the different ICE-layers the usage of more than one prototype per model results in a better representation of the past training patterns within each model. This makes a clustering with less models possible. A clustering with less models is easier to survey and to handle. With regard to industrial control tasks, such as steel processing, a better acceptance of such systems can be expected.

## 2.2 Initial Configuration and the First Model

The initial ICE-network configuration starts from *scratch*. No prototypes and no models are existing. The first prototype will be created after the presentation of the first input-output pattern. After that the first model will be adapted (cf. Fig. 1a). In case of linear models, the parameters can be calculated directly.

The first prototype will be generated at the position of the first input pattern. Because of no competing models or competing prototypes respectively, the receptive field can be *infinite*. The prototype offset (output value at the prototype center) is set to the value of the first target value. In the next step, this prototype, especially the combination of center and offset, is the base for the first model.

Until enough prototypes are created, the output of the model is set to the mean of the offsets of the current available prototypes within the model area. As long as not enough prototypes are available a model will give a constant output with the mean value of the available prototype offsets.

While presenting new training patterns, new prototypes are created. The center and offset location are determined by the same procedure as for the first pattern. If enough prototypes are generated the model can be adapted (cf. Fig. 1b).

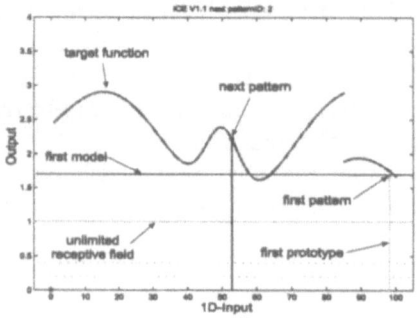 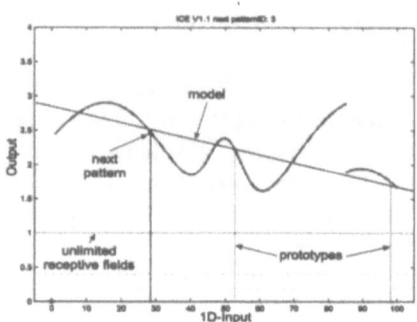

(a) After presentation of the first pattern.

(b) After presentation of the second pattern.

Figure 1: (a) First prototype (with infinite receptive field) and the first model. (b) First model after presentation of the second training pattern.

## 2.3 Adapting and Creating Models

As long as the model quality is sufficient, the active prototype (prototype with highest activation) moves into the direction of the current training pattern, i.e. the center moves into the direction of the target input (in a similar way like Kohonen networks [4]) and the offset moves into the direction of the target output value (cf. Fig. 2a). Different approaches are possible to determine the distance of the prototype movement, e.g. depending on the *age* of the prototype (*age* = number of prototype adaptations).

The first step is the extension of the network with a new *competing* prototype. In the same way as described in section 2.2, the prototype center and its offset are determined. The only difference is that the prototype receptive fields of competing models have to be shrinked mutually. The shrinking process is comparable with the Dynamic Decay Adjustment algorithm (DDA) [6]. At the center of competing ICE-prototypes (prototype of a competing model) a upper limit activation is permitted. Using this criterion, all receptive fields can be determined in a straight forward way. The shrinking of the receptive field is followed by the adaptation of the new model. The borders of the models are determined by the activations of their prototypes (cf. Fig. 2b and Fig. 2c).

## 3 Experimental Results

The first experiments are done with nonlinear noncontinuous functions with 10% noise. In all tests useful network outputs are produced already during the initial learning phase (cf. Fig. 2d). Remarkable is that the training patterns are "presented only once". Even with few training patterns a comprehensible coverage of the target function with the linear models are found.

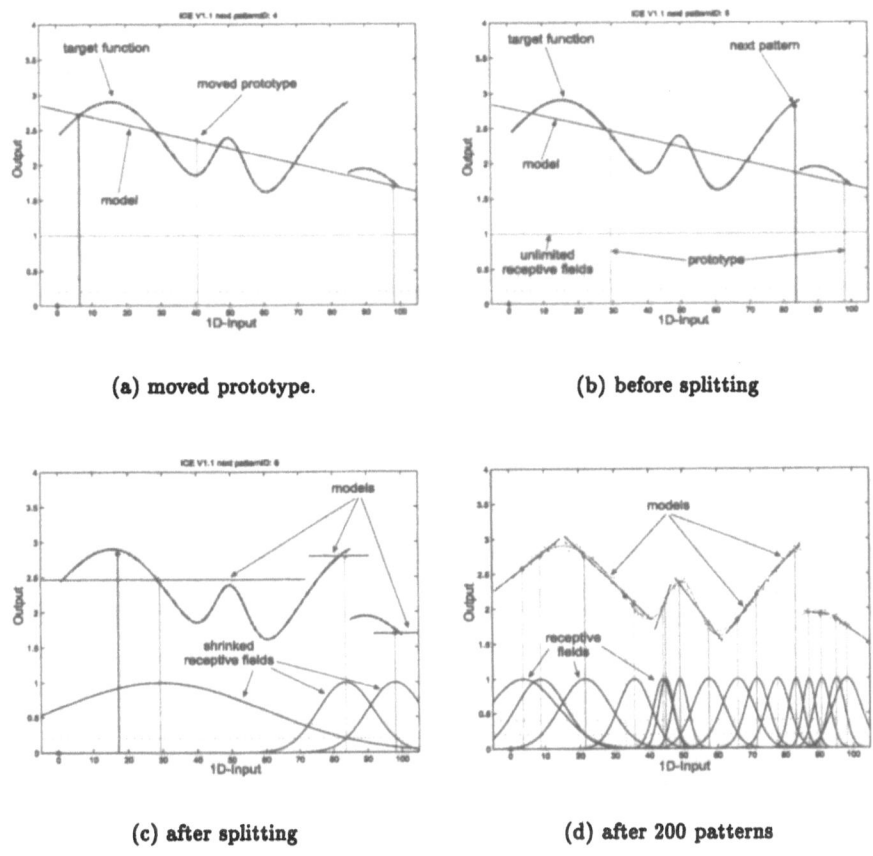

(a) moved prototype.

(b) before splitting

(c) after splitting

(d) after 200 patterns

Figure 2: (a) Prototype moves into the direction of the last training pattern. The generation of new models is shown in (b) and (c). The old model (b) is splitted and the receptive fields of competing models are shrinked (c). (d) shows the result after 200 training patterns (10% noise)

In experiments with up to 11-dimensional input-space patterns, comparable results to the 1D-examples presented above are found. Comparable means already after the first training pattern useful outputs were available. In comparison to the method with the smallest RMS error the RMS error of ICE was slightly higher. But the method with the smallest RMS error needs the first 20000 of 100000 patterns for training. In contrast to that the ICE-algorithm is able to predict useful output after the first pattern is presented. Another disadvantage of approaches like MLPs is that many experiments were necessary to find the best network-topology. In contrast to that an advantage of the ICE-algorithm is that the number of RBF-neurons and the number of local models have not to be determined in advance. According to the problem the ICE-algorithm generates an appropriate topology automatically. The number of models and prototypes respectively and the extension of the receptive field are automatically determined by the ICE-algorithm.

# 4  Conclusion

The ICE-algorithm is a new incremental algorithm for classification and function approximation problems and can be used as a base for continuous learning tasks. One advantage of the ICE-algorithm is that the network starts from *scratch*, i.e. no network topology parameter has to be determined in advance. All parameters, e.g. the number of RBF-neurons in the first layer of the hybrid system or the number of localized models in the second layer, are adjusted autonomously based on the training data presented.

Even if linear models were used in order to get a better understanding of the algorithm and its behaviour, the ICE-algorithm can be easily extended to nonlinear models. Furthermore the prototype activation can be used as a *confidence value* of the network output, which is increasingly important in real world applications that strongly rely on the performance gained by neural networks. Beyond that useful network outputs are already available when the second pattern is presented and during the whole initial learning phase. This is very important for industrial applications, e.g. plants can be put into operation much faster.

# References

[1] Thomas Martinetz, Peter Protzel, Otto Gramckow, and G. Sörgel. Neural network control for steel rolling mills. In B. Kappen and S. Giele, editors, *Neural Networks: artificial intelligence and industrial application*, Berlin, 1995. Springer-Verlag.

[2] G. Carpenter and S. Grossberg. A massively parallel architecture for a self-organizing neural pattern recognition machine. *Computer Vision, Graphics, and Image Processing*, 37:54–115, 1987.

[3] T. Poggio and F. Girosi. Regularization algorithms for learning that are equivalent to multilayer networks. *Science*, 247:978–982, 1990.

[4] T. Kohonen. Self–oganized formation of topologically correct feature maps. *Biological Cybernetics*, 43:59–69, 1982.

[5] R. Hecht-Nielsen. Counterpropagation networks. In M. Caudill and C. Butler, editors, *Proceedings of the First IEEE International Conference on Neural Networks*, pages 19–32, Volume 2, San Diego, CA, 1987. SOS Printing.

[6] Michael R. Berthold and Jay Diamond. Boosting the performance of rbf networks with dynamic decay adjustment. In D. Touretzky G. Tesauro and J.Alspector, editors, *Advances in Neural Information Processing Systems 7*, volume 7. Morgan Kaufmann Publishers, San Mateo, California, 1994.

# Synthesis of Probabilistic Automata in $p$RAM Neural Networks

Marcílio C. P. de Souto*, Neural Systems, Imperial College, London, UK
Teresa B. Ludermir and Wilson R. de Oliveira
Depart. de Informática, Univ. Fed. de Pernambuco, Recife, Brazil

**Abstract.** This paper extends previous results on the computability of RAM-based neural networks to single layer sequential $p$RAM neural networks. We present an algorithm to map any probabilistic automaton into a single layer sequential $p$RAM network. We also propose a recognition algorithm, which is based on the way probabilistic automata work, to be used with the resulting architectures. With our algorithms, we extend the computational power of these networks from finite automata to that of probabilistic automata.

## 1 Introduction

We study the computational capabilities of single layer sequential RAM-based neural network with *probabilistic* RAMs ($p$RAMs). The $p$RAM is an extension of the Probabilistic Logig Node (PLN) of Aleksander [9] in which continuous probabilities can be stored at the locations of the RAM memories [4]. This kind of node can input and output both binary and continuous values [6]. Networks built with $p$RAMs have often been trained by means of reinforcement procedures [5]. An important aspect about $p$RAMs and their reinforcement training algorithm is the fact that they are hardware implementable, and chips are commercially available [2].

Single layer sequential RAM-based networks have been used to handle problems with temporal patterns [1, 7, 10]. For instance, in [1, 7] this kind of topology was used with $p$RAMs for sequential pattern verification of signatures and regular languages, respectively. All these systems simulate the behaviour of finite state automata. Thus, they can learn only regular languages, which are the simplest languages in the Chomsky hierarchy [8]. The languages approached in this paper, on the other hand, belong to the class of weighted regular languages, which contains all regular languages [12]. We present an algorithm to map any probabilistic automaton, which are weighted regular language acceptors, to a single layer sequential $p$RAM network.

## 2 Probabilistic Automata

Probabilistic automata recognise the class of weighted regular languages [3]. Such a class of languages includes properly all regular languages. Also, this class relates

---

* Supported by the Brazilian Federal Agency for Postgraduate Studies (CAPES) grant No. 0704/95-8. E-mail: m.desouto@ic.ac.uk

to all other languages in the Chomsky hierarchy. For instance, there are context-free, context-sensitive and recursive languages that can be generated by weighted regular grammars. A probabilistic automaton $\mathbf{A}$ is a 5-tuple $\mathbf{A} = (\Sigma, Q, \delta, q_0, F)$ where $\Sigma$ and $Q$ are sets of inputs symbols and internal states, respectively; $\delta$ is a mapping $\Sigma \times Q$ into the set of $n \times n$ stochastic state transition matrices (where $n$ is the number of states in $Q$); $q_0 \in Q$ is the initial state; and $F \subseteq Q$ is a finite set of final states. In the Markovian terminology, a probabilistic automaton is a controlled process where the input symbol determines which of several transition matrices will be applied at each step. This system will accept all input sequences (sentences) which obey the following: after the last symbol has been fed, the system is at least in one of its final states and the probability of such states is above some threshold.

An example of a probabilistic automaton, which recognises the weighted regular language that first appeared in [12], is the following: $\mathbf{A} = (\Sigma, Q, \delta, q_0, F)$, where $\Sigma = \{0, 1\}$, $Q = \{000, 001, 010, 011, 100, 101\}$, $q_0 = 000$, $F = \{011\}$ and $\delta$ is given by the look-up table in Figure 1(a). The language recognised by $\mathbf{A}$ with threshold $\lambda = 0.5$ is $T(\mathbf{A}, \lambda) = \{1^m 01^n, p(x) \mid 0 < m \leq n\}$ which is a context-free language. In order to simplify our synthesis algorithm, we will assume, without loss of generality, that the probabilistic automata in this paper are generated from grammars in the normal form. This means that there will exist at most two transitions under a certain symbol leaving a given state.

| (input,state) | next state(s) | probability(ies) | | (input,state) | next state(s) | probability |
|---|---|---|---|---|---|---|
| (0,000) | {101} | {1.0} | | (0,000) | $p_2, p_1, p_2$ | 1.0 |
| (0,001) | {011 , 100} | {0.5 ; 0.5} | | (0,001) | $p_1, p_1, p_1$ | 0.5 |
| (0,010) | {011 , 101} | {0.5 ; 0.5} | | (0,010) | $p_1, p_1, p_1$ | 0.5 |
| (0,011) | {101} | {1.0} | | (0,011) | $p_2, p_1, p_2$ | 1.0 |
| (0,100) | {101} | {1.0} | | (0,100) | $p_2, p_1, p_2$ | 1.0 |
| (0,101) | {101} | {1.0} | | (0,101) | $p_2, p_1, p_2$ | 1.0 |
| (1,000) | {001} | {1.0} | | (1,000) | $p_1, p_1, p_2$ | 1.0 |
| (1,001) | {001 , 010} | {0.5 ; 0.5} | | (1,001) | $p_1, p_1, p_1$ | 0.5 |
| (1,010) | {010} | {1.0} | | (1,010) | $p_1, p_2, p_1$ | 1.0 |
| (1,011) | {011 , 100} | {0.5 ; 0.5} | | (1,011) | $p_1, p_1, p_1$ | 0.5 |
| (1,100) | {100} | {1.0} | | (1,100) | $p_2, p_1, p_1$ | 1.0 |
| (1,101) | {101} | {1.0} | | (1,101) | $p_2, p_1, p_2$ | 1.0 |

**Fig. 1.** (a) Transition table. (b) Transition table codified

## 3 Model

Our model will consists of a single layer sequential network with pRAM nodes operating synchronously, called SpRAMN (Figure 2). Such a structure, without loss of generality, will be regarded as a language acceptor. Also, as we will define later, such a class of networks has to adhere to a rigid encoding of its output. In this paper, we use pRAMs with binary inputs and continuous outputs. The output for this kind of node will be the value stored at the currently addressed location. By analogy to a probabilistic automaton, we will assume that the network has: (1) an input alphabet, where each symbol is represented by a vector $\mathbf{I}$ of $l$ bits; and (2) a set $S$ of internal states, represented by a vector of $m$ bits, where a vector $s_0 \in S$ is chosen to be the network initial state. There is also a set $R \subseteq S$ standing for the network accepting states (i.e., the final states in probabilistic automata). The network transition function is defined as follows.

The set $S$ of internal states is represented by a set $X$ of $m$ pRAMs. The output of such nodes are fed back to the network input. However, since we are interested in networks simulating the behaviour of the probabilistic automata previously defined, our network will be constrained to have, at each time step, at most two possible next states. Also, as we deal only with binary input pRAMs, we must define a binary coding for such states. In order to do so, we restrict the set

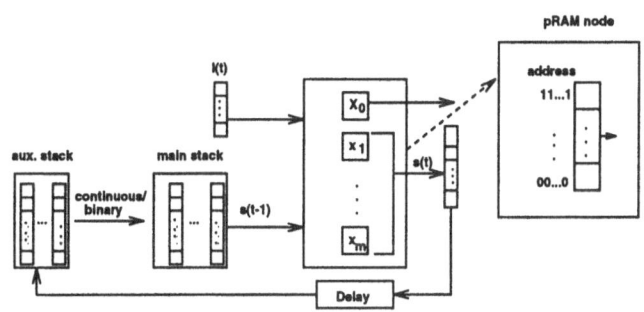

**Fig. 2.** A single layer sequential pRAM network

of possible storable values for the nodes in $X$ to six different values $\{p_1, ..., p_6\}$ ($p_i \in [0, 1]$), which will have a one-to-one relationship to the following set of binary vectors $\{(0), (1), (0, 0), (0, 1), (1, 0), (1, 1)\}$. Thus, we can transform the continuous output into binary vectors, as well as map any two binary vectors into a single vector of $p_i$ values. For instance, a vector $(p_4, p_5, p_5)$ is mapped into the two vectors $\{(0, 1, 1), (1, 0, 0)\}$ (compare the **next state(s)** entries for the index $(0, 001)$ in Figures 1(a) and 1(b)). In other words, we associate a set of at most two binary vectors with the values stored in the nodes of $X$, which will represent the actual network next state(s). So, even storing continuous values, the network will have a finite number of possible states.

Furthermore, there is a node $X_0 \notin X$, which is a feedforward node, responsible for the network transition probabilities. That is, such a node stores values in the range $[0, 1]$. In brief, our model will consist of $n = (m + 1)$ pRAMs with $(n + l)$ input lines. The outputs of these nodes depend both on the previous network state (i.e., $s(t - 1)$) and on the external input $\mathbf{I}(t)$. At each instant, the state of this system is a vector in $\{p_1, ..., p_6\}^m$ which is mapped into a set of at most two binary vectors. Additionally, the output $y_0 \in [0, 1]$ of the first node is the probability associated with the current transitions. Thus, we model a pRAM network as a probabilistic automaton. We also state that one can simulate all probabilistic automata by these networks.

## 4 Implementation of probabilistic automata

Here, we present an extension of the results in [11] to SpRAMNs. In that work, Ludermir provided an algorithm to implement any weighted regular language in neural networks with $m$-state PLN nodes (MPLN); that is, she showed that such networks can simulate any probabilistic automaton. The MPLN node is a pRAM node in which both input and output are binary. Since we map the network continuous-valued states (finite) into binary numbers, the kind of *p*RAM node used in this paper is equivalent to the MPLN node.

A drawback with Ludermir's approach is that the architectures generated out of the weighted regular grammars are irregular in that they could have several layers with different number of nodes, and each node could have a different fan-in. Thus, these network do not fit in any specific class of architectures used in practice. For instance, based on such an approach, one can claim that all problems solved by SpRAMNs can be described as probabilistic automata. However, Ludermir's approach does not allow one to claim that SpRAMNs can implement an arbitrary probabilistic automaton. Thus,in the next section, we will present an extension of that analysis to SpRAMNs.

### 4.1 Synthesis algorithm

A probabilistic automaton $\mathbf{A} = (\Sigma, Q, \delta, q_0, F)$ is reduced to an SpRAMN $\mathbf{P}$ as follows:

1. A mapping $\Sigma \to \{0, 1\}^l$, where $l = \lceil \log_2 |\Sigma| \rceil$, is defined. Such a mapping transforms each element in $\Sigma$ to a vector of binary values. Thus, the network external input $\mathbf{I}$ will be a vector with $l$ elements.
2. A mapping $Q \to \{0, 1\}^m$, where $m = \lceil \log_2 |Q| \rceil$ is defined. So, the automaton states are also coded as binary vectors. Thus, in order to codify the states in $Q$, a vector of $m$ bits is needed. These binary vectors will represent the set $S$ of network valid states. Then, we can define a set $X$ of $m$ pRAMs to represent these states. In this case, we assume, without loss of generality, that there is one-to-one relationship between the nodes in $X$ and the bits in the coding vector.
3. A node $X_0$ is defined to store the probabilities assigned to the transitions.
4. The network initial state $s_0$ is defined as the vector standing for $q_0 \in Q$.
5. The set $R$ of network accepting states will consist of vectors representing the states in $F \subseteq Q$.

From **Steps 1, 2**, and **3**, it follows that $\mathbf{P}$ has to have $n = (m + 1)$ pRAMs with $(l + n)$ input lines. Next, in order to implement $\delta$, a function has to be assigned to each pRAM. This can be done by setting the contents of the nodes in $X$ so that they can output the next states for all pairs of input symbol and state. Likewise, we set the contents of node $X_0$ so that this node outputs the probabilities assigned to transitions which lead the network to its current states. This can be done as follows. First, we represent $\delta$ by using a look-up table, **((input,state),next state(s),probability(ies))**, like the one in Figure 1(a). Since a pRAM could be also seen as a look-up table, it is straightforward to regard the pairs **(input,state)** as forming addresses to the pRAMs in $\mathbf{P}$. In contrast, the entries **next state(s)** and **probabilit(ies)**, respectively, define what the nodes in $X$ and node $X_0$ have to output for each pair. In the entries **next state(s)**, we assume that the set of next states in each one is in lexical order. In addition, by using the coding defined in Section 3, we can transform each of these entries into a single vector of continuous values (Figure 1(b)). So, after this step, these values could be directly stored in the contents of the nodes in $X$. Now, recall that for a given state and input symbol, the possible next states in the look-up table representing the automata transitions were coded in lexical order. Also, there will exist at most two transitions leaving a given state under a given symbol. So, with regard to the probabilities assigned to the these transitions, it is only necessary to store the one concerning the first next state. The omitted one is the complement. Finally, the memory positions not employed to represent the states in $Q$ could be labelled as rejected states. Thus, the network will not be trapped in undefined configurations.

In summary, our algorithm generates an SpRAMN which has its number of node logarithmic on the number of automaton states, and each node has a number of memory locations linear in the number of such states. Once the network is generated, it could be used with the new recognition algorithm presented in the next section, which makes such a network behave like a probabilistic automaton.

## 4.2 A probabilistic recognition method

In general, a sequential RAM-based network with probabilistic nodes, when fed with an input pattern, will randomly follow only a single path in its space state. Thus, in terms of probabilistic automata, the computation of the probabilities of the final states is not possible. Nevertheless, since we have defined that the values stored in the nodes of $X$ represent vectors of bits tied to the possible next states of the SpRAMN, we can transform these values again in binary vectors. Hence, we can recover the set of network current states. These states could be stored in a stack and then used together with the next input symbol read. The result of accessing the network with this symbol and the state in the top of the stack is stored in an auxiliary stack (Figure 2). In other words, the current symbol will be kept in the input line until the main stack is empty. This means that, in our model, for each current symbol, there might be a number of intermediate steps before the next symbol is fed. When the main stack is empty, the continuous vectors in the auxiliary stack are decoded and transfered to the main stack. This procedure is repeated until the end of the sentence has been reached. This way, the network can go on all distinctive paths that a string $\omega$ can follow and keep track of the probability of the states.

There are different ways to calculate and store these probabilities. For instance, we could have a variable associated with each valid network state. The purpose of this variable is to store the probability for each state at each time step. The computation of such a probability could be done by multiplying the transition probability of each state along the distinctive paths followed by the sentence. If, at a given time step, the same state appears in different paths, the probabilities associated with this state could be summed up and the paths merged. Then, in the next time step, only one copy of such a state needs to be used. At the end of this process, when the last symbol is fed, if both (1) the states that network ultimately arrives contains at least one of the network accepting (final) states, and (2) the probabilities for these states are above the threshold, the sentence is accepted by the network. In order to process a sentence, in the worst case, such an algorithm will make $L * N$ accesses to the memory locations in the network, where $L$ is the length of the sentence and $N$ the number of network valid states.

## 5  Discussion and conclusion

We extend the results in [11], which regards the implementation of weighted regular languages in arbitrary MPLN networks to SpRAMNs. Our method not only allows the construction of any probabilistic automaton in SpRAMNs, but also increases the functions that can be computed by such networks. Although our method currently involves no training, this method could be used to create the structure of the network and its initial probabilities from a partially known probabilistic automaton. Then, such a network could be submitted to some kind of learning algorithm. For instance, its training could involve only the change of the transition probabilities stored in $X_0$ or changes in the contents of the nodes in $X$, representing the network states. If changes are allowed in the contents of such nodes, we could end up with a complete different structure. However, practical shortcomings of these results, such as the feasibility of defining a learning rule for our system, need to be investigated further.

## References

1. P. J. L. Adeoadato and J. G. Taylor. Recurrent neural networks with pRAMs. In *Proc. of ICANN95*, vol. 1, pp. 607–612. Springer-Verlag, 1995.
2. T. G. Clarkson, C. K. Ng, D. Gorse, and J. G. Taylor. Learning probabilistic RAM nets using VLSI structures. *IEEE Trans. on Computers*, 41(12):1552–1561, 1992.
3. K. S. Fu. *Syntactic Pattern Recognition and Applications*. Prentice-Hall, 1982.
4. D. Gorse and J. G. Taylor. On the equivalence properties of noisy neural and probabilistic RAM nets. *Physics Letters A*, 131(6):326–332, 1988.
5. D. Gorse and J. G. Taylor. Reinforcement training strategies for probabilistic RAMs. In *Proc. of NEURONET90*, pp. 180–184, 1990.
6. D. Gorse and J. G. Taylor. A continuous input RAM-based stochastic neural model. *Neural Networks*, 4:657–665, 1991.
7. D. Gorse and J. G. Taylor. Enconding temporal structure in probabilistic RAM networks. In *Proc. of IEE Int. Conf. on ANN*, pp. 369–372, UK, 1991.
8. J. E. Hopcroft and J. D. Ullman. *Introduction to Automata Theory, Languages, and Computation*. Addison-Wesley Publishing Company, 1979.
9. W. K Kan and I. Aleksander. A probabilistic logic neuron network for associative learning. In *Proc. of the IEEE Int. Conf. on AANs*, vol. II, pp. 541–548, 1987.
10. T. B. Ludermir. A feedback network for temporal pattern recognition. In *Parallel Proc. in Neural Syst. and Computers*, pp. 395–398, Amsterdam, 1990.
11. T. B. Ludermir. Logical neural nets and distributed implementations of weighted regular languages. In *Proc. of IEE Int. Conf. on ANNs*, pp. 158–162, UK, 1991.
12. A. Salomaa. Probabilistic and weighted grammars. *Inf. and Control*, 15:529–544, 1969.

# Neural Modeling of Nonlinear Differential Equations with Discrete Measurements A Lagrangian Approach

J. Wagenhuber

Siemens AG, Corporate Technology, Information & Communications,
Otto-Hahn-Ring 6, D-81730 Munich, Germany.

### Abstract

We present a new algorithm for the adaptation of parameters or weights of a continous dynamical system, given by differential equations, to a discrete set of measurement data. This method works about a factor equal to the dimension of the system's state space faster than conventional sensitivity analysis and does not assume a continuous measurement signal. Representing a process by a system of differential equations allows the incorporation of a-priori knowledge, which especially in technical applications is given by differential relationships between certain process variables, whereas a system representation using discrete, recurrent equations renders the inclusion of a-priori knowledge more difficult. This algorithm can also be generalized to the modeling of spatio-temporal systems. It computes the gradient of the model-data mismatch directly using an abstract formalism of Lagrange multipliers in function spaces.

## 1 Introduction

The modeling of complex systems with unknown dynamics is an important and in most cases an ambitious task with relevance to a wide area of applications such as chemical process design and control and other problems. With data driven approaches and the use of nonlinear parametric representations for dynamical processes, such as artificial neural networks (see Refs. [1] especially for chemical processes), methods for nonlinear dynamic modeling gain importance in theory and applications. In the following we present an algorithm for the adaptation of continuous dynamical systems to a given set of discrete data which works for arbitrary objective functions and is faster than conventional adaptation procedures based on sensitivity analysis of the system (about a factor equal to the dimension of the state space). The complex dynamic system can either be quantified by a chemically, physically or otherwise motivated model or by the use of *artificial neural networks*. Between these extremes a description with *hybrid models*, i.e. a combination of both model types mentioned above, exists. Hybrid system descriptions resemble the most realistic situation in modeling problems arising in technical applications: a-priori knowledge of scaleable, defined parts in the system which can be modeled physically (e.g. reactors) together with completely unknown system components quantified by artificial neural networks (e.g. chemical reactions). A-priori knowledge for a given process is often available only in the form of differential equations excluding the possibility of incorporating this knowledge into a discrete, i.e. recurrent iterative system or into a static relationship between process variables. The most natural form of physically motivated models with neural components is therefore given by a continuous dynamical system in state space representation

$$\frac{d\mathbf{x}}{dt}(t) = \mathbf{f}(\mathbf{x}(t), \mathbf{u}(t), \mathbf{w}), \qquad \mathbf{y}(t) = \mathbf{m}(\mathbf{x}(t), \mathbf{w}), \tag{1}$$

where $\mathbf{x}$ denotes state variables which evolve according to a system $\mathbf{f}$ of nonlinear differential equations. The function $\mathbf{m}$ describes the instantaneous mapping from the state variables $\mathbf{x}$ to a set $\mathbf{y}$ of measurement values. Eventually existing control inputs are represented by $\mathbf{u}$. The differential equation as well as the static function can be represented by neural nets or in a hybrid formulation and both neural weights and physical parameters are collected in a vector $\mathbf{w}$. Note, that the representation (1) covers also spatio-temporal systems, i.e. partial differential equations of parabolic or hyperbolic type, where the discretized forms of the partial differential operators must be used, giving a very high dimensionality in state space. In data driven modeling the weights are adapted using a given set of measurements $\boldsymbol{\eta}_n$ of $\mathbf{y}$ at discrete times $t_n$. The mismatch between model and measurements is quantified by an objective function $E$, e.g. the usual sum of quadratic errors

$$E = \frac{1}{2} \sum_n \|\mathbf{y}(t_n) - \boldsymbol{\eta}_n\|^2, \tag{2}$$

and adaptation of the model to measurement data is done by minimizing $E$.

Optimizing of $E$ with the usual gradient-based algorithms requires the determination of the gradients $dE/d\mathbf{w}$ and $dE/d\mathbf{x}_0$ of the cost function with respect to the weights and to the initial state of the system. A well-known conventional technique to calculate this derivatives is sensitivity analysis [2]. In this approach sensitivity matrices $d\mathbf{x}(t)/d\mathbf{w}$ and $d\mathbf{x}(t)/d\mathbf{x}_0$ are determined, which quantify how sensitive the time evolution of the model reacts upon a change in the weights $\mathbf{w}$ or the initial states $\mathbf{x}_0$. An additional calculation then gives the desired gradients.

# 2 Weight Adaption in Dynamical Systems — Lagrangian Formalism

In the following we will construct an economic algorithm for the direct calculation of the gradients of the modeling error function with respect to the model parameters and to the initial states of the system. In this approach we avoid the calculation of whole matrices (sensitivities) in order to collapse them into single gradient vectors $dE/d\mathbf{w}$ and $dE/d\mathbf{x}_0$ of the cost function. For convenience reasons the sum of square errors (2) is assumed, since the transfer of this algorithm to general forms of $E$ is easily done. Without restriction of generality we assume further, that the dynamic system of Eq. (1) has to be modeled within a time interval $[t_0, T]$. Minimization of the model error $E$ has to be performed within the functional space $\mathscr{C}[t_0, T]$ of at least piecewise continuously differentiable functions under the additional constraint, that the dynamic equations of the model (1) must be fulfilled. We rewrite the dynamical equation of (1) as an integral equation

$$\mathbf{x}(t) = \mathbf{x}_0 + \int_{t_0}^{t} \mathbf{f}(\mathbf{x}(t'), \mathbf{u}(t'), \mathbf{w})\, dt', \tag{3}$$

where the vectorial function $\mathbf{x}(t)$ is an element of the space $\mathscr{C} = \mathscr{C}[t_0, T]$. The difference between right and left hand sides of this equation defines a function $\boldsymbol{\Phi}$ mapping the function space $\mathscr{C}$ into itself

$$\boldsymbol{\Phi}(\mathbf{x}, \mathbf{w}, \mathbf{x}_0)(t) := \mathbf{x}_0 + \int_{t_0}^{t} \mathbf{f}(\mathbf{x}(t'), \mathbf{u}(t'), \mathbf{w})\, dt' - \dot{\mathbf{x}}(t). \tag{4}$$

The relation $\boldsymbol{\Phi}(\mathbf{x}) = \mathbf{0}$ is valid exactly when the state space function $\mathbf{x}(t)$ fulfills the model equations (1) with initial state $\mathbf{x}(t_0) = \mathbf{x}_0$. Therefore the gradients $dE/d\mathbf{w}$ and $dE/d\mathbf{x}_0$ are to be calculated with respect to the above constraint. We handle this constraint with an extension of the Lagrange method onto functional spaces (Appendix A). From the rightmost equation in (13) after a little algebra we get a defining relation for the Lagrange multiplier, the left hand side given by

$$\frac{\partial E}{\partial \mathbf{x}} \cdot \Delta \mathbf{x} = \int_{t_0}^{T} \sum_n \delta(t - t_n) \left[ \left( \frac{\partial \mathbf{m}(t_n)}{\partial \mathbf{x}} \right)^{\mathbf{T}} \cdot \frac{\partial E}{\partial \mathbf{y}(t_n)} \right]^{\mathbf{T}} \cdot \Delta \mathbf{x}(t)\, dt,$$

where $\Delta \mathbf{x}(t)$ denotes a test function from $\mathscr{C}$ and the Dirac delta function reflects the fact, that only a discrete set of measurements exists. In a similar way the functional derivative $\partial \boldsymbol{\Phi}/\partial \mathbf{x}$ of (4) builds the left hand side of the rightmost equation in (13) giving the adjoint differential equation

$$\frac{d\boldsymbol{\lambda}(t)}{dt} + \left( \frac{\partial \mathbf{f}(t)}{\partial \mathbf{x}} \right)^{\mathbf{T}} \cdot \boldsymbol{\lambda}(t) = \sum_n \delta(t - t_n) \left( \frac{\partial \mathbf{m}(t_n)}{\partial \mathbf{x}} \right)^{\mathbf{T}} \cdot (\boldsymbol{\eta}_n - \mathbf{y}(t_n)) \tag{5}$$

where we have used an obvious shorthand notation for the arguments of $\mathbf{f}$. The differential equation system (5) must be solved in *reversed time* with the end condition $\boldsymbol{\lambda}(T) = \mathbf{0}$, its solution $\boldsymbol{\lambda}(t)$ giving an integral representation of the linear functional $\mathbf{L}$ (the Lagrange multiplier).

To get finally defining relationships for the desired gradients $dE/d\mathbf{w}$ and $dE/d\mathbf{x}_0$ the leftmost equation in (13) must be applied to the constraint (4) with the settings $\mathbf{p} = \mathbf{w}$ and $\mathbf{p} = \mathbf{x}_0$. After a lengthy but straightforward calculation this results in

$$\frac{dE}{d\mathbf{x}_0} = \boldsymbol{\lambda}(t_0), \quad \frac{dE}{d\mathbf{w}} = \int_{t_0}^{T} \left( \frac{\partial \mathbf{f}(\mathbf{x}(t'), \mathbf{u}(t'), \mathbf{w})}{\partial \mathbf{w}} \right)^{\mathbf{T}} \cdot \boldsymbol{\lambda}(t')\, dt'. \tag{6}$$

Therefore the integral representation $\boldsymbol{\lambda}$ of the Lagrange multiplier corresponds exactly to the sensitivity of the objective function with respect to initial state of the system (1), whereas determination of a weight gradient requires an additional integration.

# 3  Algorithm for Calculating the Gradients

The adjoint system (5) together with the relations in (6) outlines the calculation of the desired gradients $dE/d\mathbf{w}$ and $dE/d\mathbf{x}_0$ of the cost function. In contrast to conventional approaches as described in Refs. [3], where the measurement is assumed as a continuous signal, we have to deal with $\delta$-like singularities at the inhomogeneous part of the adjoint differential equations. This resembles the determination of Green's functions analogous to the physical examples of evolving electrical fields out of point-like charges or the response of a mechanical system to a sudden external impact. Each mismatch between model and data then serves as such an impact which propagates (in reversed time) according to the adjoint system and contributes to the desired gradients of the cost function. It can easily be shown, however, that a $\delta$-contribution to the inhomogeneous part of a differential equation can be handled exactly by a solution of the *homogeneous part* with a corrected initial condition. Because of the linearity of the adjoint equations (5) the contributions of all measurements $\boldsymbol{\eta}_n$ add up. Treating further the desired gradients as time dependent variables we therefore finally formulate the following

**Algorithm.** Solve the following initial value problem in *reversed time*

$$\frac{d}{dt}\frac{dE}{d\mathbf{x}_0}(t) = -\left(\frac{\partial \mathbf{f}(\mathbf{x}(t), \mathbf{u}(t), \mathbf{w})}{\partial \mathbf{x}}\right)^{\mathbf{T}} \cdot \frac{dE}{d\mathbf{x}_0}(t),$$

$$\frac{d}{dt}\frac{dE}{d\mathbf{w}}(t) = -\left(\frac{\partial \mathbf{f}(\mathbf{x}(t), \mathbf{u}(t), \mathbf{w})}{\partial \mathbf{w}}\right)^{\mathbf{T}} \cdot \frac{dE}{d\mathbf{x}_0}(t),$$

$$\frac{d\mathbf{x}}{dt}(t) = \mathbf{f}(\mathbf{x}(t), \mathbf{u}(t), \mathbf{w}). \tag{7}$$

The end conditions are given by

$$\frac{dE}{d\mathbf{x}_0}(T) = \mathbf{0}, \quad \frac{dE}{d\mathbf{w}}(T) = \mathbf{0}, \quad \mathbf{x}(T). \tag{8}$$

Every time $t_n$ a measurement value $\boldsymbol{\eta}_n$ is reached, including at the end $T$ and beginning $t_0$ of the interval $[t_0, T]$, the integration process must be stopped and the actual value of $dE/d\mathbf{x}_0$ corrected according to the following mapping procedure

$$\frac{dE}{d\mathbf{x}_0}(t_n) \longmapsto \frac{dE}{d\mathbf{x}_0}(t_n) + \left(\frac{\partial \mathbf{m}(\mathbf{x}(t_n), \mathbf{w})}{\partial \mathbf{x}}\right)^{\mathbf{T}} \cdot (\mathbf{y}(t_n) - \boldsymbol{\eta}_n). \tag{9}$$

After this correction the integration process is reinitialized and continued until a new measurement value or the beginning $t_0$ is reached. If the end condition $\mathbf{x}(T)$ for the state space variables is not known otherwise, it must be determined by a conventional forward propagation of the dynamical system (1).

The desired gradients of the objective function are given by the solution values at the beginning of the considered interval $[t_0, T]$

$$\frac{dE}{d\mathbf{x}_0} = \frac{dE}{d\mathbf{x}_0}(t_0), \quad \frac{dE}{d\mathbf{w}} = \frac{dE}{d\mathbf{w}}(t_0) + \sum_n \left(\frac{\partial \mathbf{m}(\mathbf{x}(t_n), \mathbf{w})}{\partial \mathbf{w}}\right)^{\mathbf{T}} \cdot (\mathbf{y}(t_n) - \boldsymbol{\eta}_n). \tag{10}$$

The additional expression in $dE/dw$ results from the direct dependence of $E$ upon $\mathbf{w}$ via the function $\mathbf{m}$ which maps the internal states of the system into the space of measurable quantities. It vanishes, when $\mathbf{m}$ is not dependent on any parameters or weight respectively.

# 4  Performance

In this section we estimate the performance of the Lagrangian approach for gradient calculation and compare it with the conventional sensitivity analysis. The most performant way for this is the calculation of columns of the sensitivity matrices $d\mathbf{x}(t)/d\mathbf{w}$ and $d\mathbf{x}(t)/d\mathbf{x}_0$ respectively, adding up to a total of $n + w$ columns, where $n$ denotes the number of state space variables $\mathbf{x}$ and $w$ the number of weights $\mathbf{w}$. Each column is the solution of a differential equation of dimension $2n$, because $n$ is the length of each column and also the dimension of the dynamic system (1). Sensitivity analysis of a dynamical system therefore corresponds in essence to the solution of $n + w$ differential equations each with dimension $2n$. In contrast Eq. (7) shows, that the Lagrange method for calculating the gradients $dE/d\mathbf{w}$ and $dE/d\mathbf{x}_0$ requires essentially the integration of a single differential equation with dimension $2n + w$. Under normal conditions the adjoint system (7) can be solved with explicit numerical algorithms. In this case, which does not require the solution of matrix equations, the CPU time scales about linearly with dimension. Therefore a comparison between both alternative algorithms shows

$$\frac{2n \times (n + w)}{2n + w} > \frac{2n \times (n + w)}{2n + 2w} = n.$$

Thus the Lagrangian method should be about a factor of $n$, the dimension of the state space, faster than the conventional sensitivity analysis.

# 5  Acknowledgements

We acknowledge financial support of this work by the German Bundesministerium für Bildung, Wissenschaft, Forschung und Technologie (BMBF) under grant No 03 D 0022 A.

# Appendix A: Abstract Lagrangian Formalism

Suppose we have given a functional space $\mathscr{C}$ with an objective function $E$, defined as a mapping from this space onto the real numbers

$$\begin{aligned} E: \mathscr{C} &\longrightarrow \mathbb{R} \\ \mathbf{x} &\longmapsto E(\mathbf{x}) \end{aligned} \tag{11}$$

with an additional constraint $\boldsymbol{\Phi}(\mathbf{x}, \mathbf{p}) = \mathbf{0}$ by a mapping

$$\begin{aligned} \boldsymbol{\Phi}: \mathscr{C} \times \mathscr{P} &\longrightarrow \mathscr{C} \\ (\mathbf{x}, \mathbf{p}) &\longmapsto \boldsymbol{\Phi}(\mathbf{x}, \mathbf{p}) \end{aligned} \tag{12}$$

of $\mathscr{C}$ into itself, which depends also on a parameter vector $\mathbf{p} \in \mathscr{P}$, where the parameter space $\mathscr{P}$ normally is a finite dimensional real space $\mathbb{R}^p$.

Then the gradient $dE/d\mathbf{p}$ of the cost function with respect to the parameters, assuming the additional constraint is fulfilled, is given by

$$\frac{dE}{d\mathbf{p}} = \mathbf{L} \cdot \frac{\partial \mathbf{\Phi}}{\partial \mathbf{p}}, \quad \text{where} \quad \frac{\partial E}{\partial \mathbf{x}} = -\mathbf{L} \cdot \frac{\partial \mathbf{\Phi}}{\partial \mathbf{x}}. \tag{13}$$

$\mathbf{L}$, defined by the rightmost equation above, is a linear functional mapping the functional space $\mathscr{C}$ into the real domain. In the usual case of finite dimensions the linear functional $\mathbf{L}$ can be represented by a real vector, i.e. by a set of real numbers which in this case are known as the Lagrangian multipliers.

**Proof.** The constraint can be resolved explicitly in the neighbourhood of a given parameter setting (implicit function theorem). To get the derivative of this explicit relation we differentiate the constraint getting

$$\frac{\partial \mathbf{\Phi}}{\partial \mathbf{x}} \cdot \frac{d\mathbf{x}}{d\mathbf{p}} + \frac{\partial \mathbf{\Phi}}{\partial \mathbf{p}} = 0 \implies \mathbf{x} = \mathbf{x}(\mathbf{p}) \quad \text{with} \quad \frac{d\mathbf{x}}{d\mathbf{p}} = -\left(\frac{\partial \mathbf{\Phi}}{\partial \mathbf{x}}\right)^{-1} \cdot \frac{\partial \mathbf{\Phi}}{\partial \mathbf{p}}.$$

Using the chain rule

$$\frac{dE}{d\mathbf{p}} = \frac{\partial E}{\partial \mathbf{x}} \cdot \frac{d\mathbf{x}}{d\mathbf{p}} \implies \frac{dE}{d\mathbf{p}} = \underbrace{-\frac{\partial E}{\partial \mathbf{x}} \cdot \left(\frac{\partial \mathbf{\Phi}}{\partial \mathbf{x}}\right)^{-1}}_{:=\mathbf{L}} \cdot \frac{\partial \mathbf{\Phi}}{\partial \mathbf{p}}.$$

With the definition of $\mathbf{L}$ inserted above, Eqs. (13) follow immediately.

# References

[1] S. P. Chitra. Neural Net Applications in Chemical Engineering. AI Expert, pp. 20–25, November 1992

I. M. Galván et. al. The Use of Neural Networks for fitting Complex Kinetic Data. Computers chem. Engng. 1996, 20:1451-1465

J. Wagenhuber, H.-J. Zander, R. Dittmeyer. Dynamic Modeling of Chemical Reaction Systems with Neural Networks and Hybrid Models. In: Selected Papers of the First European Congress on Chemical Engineering (ECCE-1), AIDIC Conference Series Vol. 2, 1997, pp. 311–318

[2] E. Baake, M. Baake, H. G. Bock, and K. M. Briggs. Fitting ordinary differential equations to chaotic data. Phys. Rev. A 1992, 45:5524–5529

M. Guay, and D. D. McLean. Optimization and Sensitivity Analysis for Multiresponse Parameter Estimation in Systems of Ordinary Differential Equations. Computers chem. Engng. 1995, 19:1271–1285

[3] B. A. Pearlmutter. Learning state space trajectories in recurrent neural networks. Neural Comp. 1989, 1:263–269

V. López, R. Huerta, and J. R. Dorronsoro. Recurrent and Feedforward Modeling of Coupled Time Series. Neural Comp. 1993, 5:795–811

# Simple Synchrony Networks : Learning to Parse Natural Language with Temporal Synchrony Variable Binding

Peter C.R. Lane* and James B. Henderson

Department of Computer Science, University of Exeter

Prince of Wales Road, EXETER EX4 4PT, UK

{pclane,jamie}@dcs.exeter.ac.uk

## Abstract

The Simple Synchrony Network (SSN) is a new connectionist architecture, incorporating the insights of Temporal Synchrony Variable Binding (TSVB) into Simple Recurrent Networks. The use of TSVB means SSNs can output representations of structures, and can learn generalisations over the constituents of these structures (as required by systematicity). This paper describes the SSN and an associated training algorithm, and demonstrates SSNs' generalisation abilities through results from training SSNs to parse real natural language sentences.

## 1 Introduction

Temporal Synchrony Variable Binding (TSVB) [1] extends the representational ability of a connectionist network to include entities. The original motivation behind TSVB was for a network to represent variables and so carry out symbolic reasoning [1]. Henderson [2] argues that this extension further gives connectionist networks an inherent ability to learn generalisations across entities. This ability allows TSVB networks to learn the kinds of regularities that arise from a compositional generative grammar, which [3] uses to describe the property of systematicity. In particular, with tasks involving language, TSVB networks will generalise information learned about one syntactic constituent to other syntactic constituents.

This paper begins by describing the basic idea behind TSVB, which is that with pulsing units entities can be represented using the timing of pulses. The pulsing binary-threshold units of [1] provide a connectionist model of structures and symbolic reasoning, but are not suitable for training with standard algorithms such as backpropagation [4]. To develop an architecture for TSVB networks that can use backpropagation, we start with the Simple Recurrent Network (SRN) architecture [5] and extend it with units that pulse. The resulting Simple Synchrony Network (SSN) architecture has two SRN components, one standard SRN that represents overall context, and one TSVB SRN that represents a set of entities. This information is combined to compute the output for each entity. As for SRNs, SSNs can be trained using Backpropagation Through Time [4]. Finally, we demonstrate the ability of SSNs to

*Supported by the Engineering and Physical Sciences Research Council, UK.

represent structure and learn generalisations over structural constituents with results from experiments training SSNs to parse a corpus of natural language sentences.

# 2   Temporal Synchrony Variable Binding

Temporal Synchrony Variable Binding [1] is a connectionist technique for representing entities. We will adopt the following central characteristics of TSVB:
- the division of each time period into discrete phases,
- pulsing units, to compute within each phase independently of other phases,
- non-pulsing units, to compute across all phases, combining information about several entities.

Thus in each time step the network cycles through the set of entities, using pulsing units to compute about each entity independently. To communicate information between entities there are also non-pulsing units, which compute across all phases and thereby represent information about the overall context. Being able to compute in terms of entities, overall context, and their interactions is a crucial feature of the architecture proposed below.

The use of TSVB in a connectionist network has two important consequences. The first is that the output can represent structure. The activation of output units in a particular phase can represent information about a particular constituent in the structure, including its structural relationships to other constituents. The second consequence is that TSVB networks inherently generalise information learned about one entity to other entities. Because different times (i.e. phases) are used to represent different entities, and the same link weights are used at every time, the same learned information is applied to every entity. This argument is used in [2] to demonstrate that TSVB networks possess inherent systematicity [3].

# 3   Simple Synchrony Networks

Rather than starting with an existing TSVB architecture and making it compatible with backpropagation, we choose to start with an existing backpropagation architecture and add TSVB. A natural choice is the Simple Recurrent Network (SRN) architecture [5]. This architecture already handles temporal sequences. An SRN accepts a sequence of input patterns and produces a sequence of internal and output patterns. TSVB requires such sequences both for each entity and for the overall context. The known effectiveness of learning in SRNs implies that each of these sequences can be learned effectively, but this is not sufficient, since the sequences are not all independent. The Simple Synchrony Network (SSN) architecture minimizes the amount of interaction between these sequences while maintaining the generality of the computations that can be performed. This approach preserves the effectiveness of learning in these networks.

## 3.1   The architecture and training algorithm

Figure 1 illustrates a SSN architecture. The two recurrent components at the bottom are each SRNs (minus their output layers). The lefthand SRN

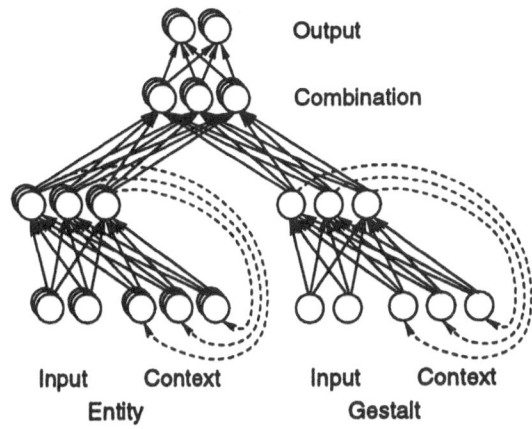

Figure 1: A Simple Synchrony Network. The pulsing units are depicted as several units stacked on top of each other, because they store activations for several entities.

(the entity component) consists of pulsing units, and computes a distributed representation of each entity independently. The righthand SRN (the gestalt component) uses nonpulsing (i.e. standard) units, and computes a distributed representation of the overall context. Thus as a SSN processes a sequence of inputs, it computes one sequence of representations for each entity plus one sequence of representations for the overall context. After each step, the upper hidden layer combines the overall context representation with each entity's representation to produce the output pattern for each entity.

What allows a SSN to learn effectively is the way it combines information between the separate sequences of representations computed by its SRN components. One type of interaction is when information about entities is needed to compute the overall context. In [1] this is done with logical AND and OR combinations across entities, but these cannot be used with backpropagation. Continuous combination operations, such as summation, do not in practice generalise to greater or fewer numbers of entities, and thus cannot be used. The solution adopted in SSNs is to push these dependencies down into the input layer. Any information that is input about an entity and might be relevant to the overall context must also be represented in the input to the gestalt component. As long as information about a fixed number of entities is *input* at any one time (in our application this is one), there will be no need for the input representation to generalise over different numbers of entities.

Another type of interaction between sequences is when information about the overall context is needed to compute information about entities. One approach would be to allow this interaction to take place at every time step. However this would lead to many different times at which any given piece of information could be communicated. In practice these alternatives compete, leading to ineffective learning. The remaining possibilities are communication at the time when the information is input and/or communication at the time when the information is needed in the entity's output. All these options are somewhat effective, but the latter appears to work the best, so that is the one used in this paper. This is done with the combination layer shown above the two SRN components.

The only remaining type of interaction is between the sequences of representations for different entities. Because we do not presuppose any organisation to the set of entities (such as a stack or a tape), it is sufficient for all such interaction to go through the representation of the overall context, as covered above.

SSNs are trained using the same algorithm as can be used for SRNs, Back-propagation Through Time (BPTT) [4]. BPTT works by unfolding the network into one copy per time period, and then applying standard backpropagation to the resulting feed-forward network, using weight sharing to keep the different copies of the network the same. For SSNs, within each time period's copy BPTT must also make a copy of the pulsing units for each entity. Therefore, each link to or from a pulsing unit will be duplicated once per time period and once per entity. During backpropagation training the weight for each such copy of the link must be updated in an identical fashion.

## 3.2   Representing structure

One of the strengths of the SSN architecture is its ability to identify entities in its output, and so output a representation of structure. To illustrate this we consider a SSN computing the syntactic structure for the sentence "John loves a woman", as shown in figure 2. Each constituent in the syntactic structure is represented as an entity, and the structure is output as a set of relationships between these entities. The pattern of input-output discussed here is used in the experiments reported in the next section.

Firstly, we must define what is to be input to the entity and gestalt components of the SSN. In language processing, every word is relevant to the overall context. Therefore, every word is input to the gestalt component of the SSN, one word per time period, as is done with a standard SRN. Further, each word may introduce a new syntactic constituent. Therefore each word is also input to the entity component of the SSN in a previously unused phase.

The syntactic structure for a sentence can be specified incrementally using three output units, *Grandparent*, *Parent* and *Sibling*. This output format is illustrated in figure 2 as a sequence of pieces of structure. When accumulated together these pieces completely specify the entire syntactic structure, as shown at the bottom of figure 2.

There are two cases to this output format. If the current word's constituent has already been introduced by another word, then the *Parent* output identifies that constituent from those that have been previously introduced, as shown for "woman" in figure 2. Otherwise the *Parent* output identifies the constituent that was introduced with the current word. In this case the structural position of this parent constituent must also be identified. If the parent constituent is part of a constituent that has already been introduced, then the *Grandparent* output unit identifies that constituent during the current time period, as shown for "a". If the parent constituent is part of a constituent that is introduced later, then the *Sibling* output unit will identify the current parent constituent during the time period when the later constituent is introduced. This is shown in the period for "loves", which identifies the parent of "John" as the *Sibling*.

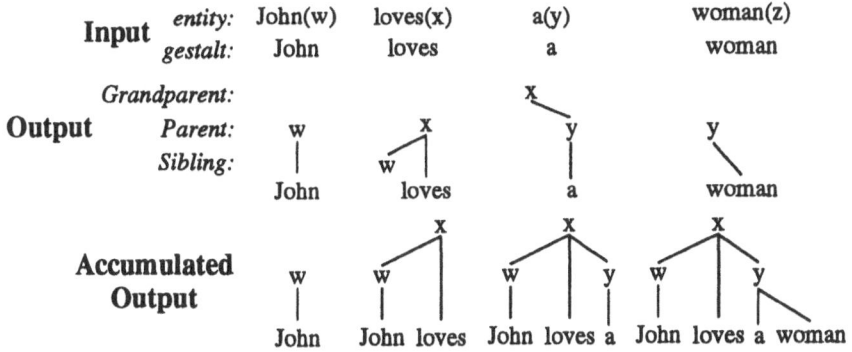

Figure 2: The input and output information for "John loves a woman". Phases are shown as variables (w,x,y, and z).

# 4   Learning to parse natural language

Although connectionist networks have been applied to tasks involving language learning in the past, there has been no convincing application to learning to parse naturally occurring sentences. We have conducted experiments in training SSNs to parse natural language using the Susanne[1] corpus as a source of preparsed sentences taken from newspaper reports. This section describes these experiments and our current results.[2]

We used the input format described above, but with part-of-speech tags as the input instead of words. For example, the sentence "John loves a woman" would be input as "NP VVZ AT NN". This reduces the overhead of training because less data is required, since there are many fewer part-of-speech tags than words. Both the entity and gestalt inputs use a localist representation of each letter within a part-of-speech tag. Since the part-of-speech tags are at most three letters long, we have three banks of input units per component. The target output is an unlabelled parse tree, represented with the output format also described above. To convert from the continuous outputs of the network to a discrete structure, we first take maximums across competing outputs, then convert the resulting structural relationships to a set of constituents.

We compare our results to the current state-of-the-art for mapping word tags to labelled parse trees, which are Probabilistic Context Free Grammars [7]. The standard evaluation of performance compares the constituents output by the model (the SSN in this case) to the constituents in the corpus, to determine the percentage of the output constituents that are correct (precision), and percentage of the correct constituents that are output (recall). Reported figures for both precision and recall are around 75% [7].

We trained a range of SSNs on a training set formed from sentences of length less than thirty words (13,523 words in total). Each network was trained until the sum-squared error reached a minimum, and results obtained on a cross-validation set. The cross-validation set consisted of 4,700 words with an average

---

[1]The Susanne corpus is sponsored by the Economic and Social Research Council (UK) with the University of Sussex as grantholder.

[2]These and some related experiments are discussed in more detail in [6].

(unrestricted) sentence length of 21.6 words. The best two of these networks were then selected for testing. They each had 20 units in their gestalt and entity hidden layers, but one had 10 units in its combination layer, and the other had 20. The test set consisted of 4,602 words with an average sentence length of 26.2 words. The average performance of these two networks on the training set was 71.6% precision, 75.8% recall, on the cross-validation set, 68.2% precision and 73.8% recall, and on the test set, 62.6% precision and 69.4% recall.

# 5 Conclusion

This paper has presented a new connectionist architecture, Simple Synchrony Networks (SSNs), which is trained using an extension of Backpropagation Through Time. SSNs combine the characteristics of Simple Recurrent Networks to learn about patterns across time with the characteristics of Temporal Synchrony Variable Binding to represent entities. We demonstrate the ability of SSNs to represent structure and generalise learned information across entities with experiments training SSNs to parse natural language sentences. We conclude that the SSN is a simple but significant new architecture extending the impressive generalisation abilities of connectionist networks in pattern matching tasks to the more complex domains typical of higher level cognition.

# References

[1] Shastri L and Ajjanagadde V. From simple associations to systematic reasoning: A connectionist representation of rules, variables, and dynamic bindings using temporal synchrony. *Behavioral and Brain Sciences*, 16:417–494, 1993.

[2] Henderson J. A connectionist architecture with inherent systematicity. *Proceedings of the Eighteenth Conference of the Cognitive Science Society, La Jolla, CA*, 1996.

[3] Fodor J A and Pylyshyn Z W. Connectionism and cognitive architecture: a critical analysis. *Cognition*, 28:3–71, 1988.

[4] Rumelhart D E, Hinton G E, and Williams R J. Learning internal representations by error propagation. In D.E. Rumelhart and J.L. McClelland, (eds.), *Parallel Distributed Processing, Vol 1*. MIT Press, Cambridge, MA., 1986.

[5] Elman J L. Finding structure in time. *Cognitive Science*, 14:179–211, 1990.

[6] Henderson J and Lane P. A connectionist architecture for learning to parse. *Proceedings of the Association of Computational Linguistics*, 1998.

[7] Charniak E. Statistical Techniques for Natural Language Parsing. *AI Magazine*, forthcoming.

# Gaussian Processes for Switching Regimes

Amos Storkey

Neural Systems Group, Imperial College, London amoss@ic.ac.uk

**Abstract.** It has been shown that Gaussian processes are a competitive tool for nonparametric regression and classification. Furthermore they are equivalent to neural networks in the limit of an infinite number of neurons. Here we show that the versatility of Gaussian processes at defining different textural characteristics can be used to recognise different regimes in a signal switching between different sources.

## 1 Introduction

The use of Gaussian processes [3] to tackle many of the standard neural network problems was reintroduced by Williams [6] and prompted by recent work showing that neural networks and Gaussian processes were closely related. Neal [2] showed that in the limit of an infinite number of neurons, the two were equivalent. It was also noted the linear models and radial basis functions were special cases of Gaussian processes [1]. Rasmussen showed that Gaussian processes were competitive on a number of benchmark problems [4]. Here we look at the problems of non stationary signals, specifically the case of switching signals. This situation has received attention in the past [5]. Here we show that Gaussian processes are a useful tool for tackling this problem.

## 2 Gaussian Processes for Regression

Consider a set of points $\{x_i\}$, which consist of the points in input space at which we will later receive data $\{x_i\}$ $i = 1, 2, \ldots, n$ and the points $\{x_i\}$ $i = n+1, 2, \ldots, m$ at which we would like to make predictions. We use a superscript $D$ (for DATA) to denote an m-vector truncated to the elements $i = 1, 2, \ldots, n$, and a superscript $P$ (for PREDICTION) to denote an m-vector truncated to the elements $i = n+1, \ldots, m$.

We suppose for now that there is a true unknown function $f(x)$ which generates datum $f_i$ at point $x_i$. This datum is corrupted by measurement noise $\eta_i$, assumed for now to be Gaussian, mean zero, variance $\sigma^2$. We define the random variable $y_i$ by

$$y_i = \begin{cases} f_i + \eta_i \ i = 1, 2, \ldots, n \\ f_i \qquad i = n+1, \ldots, m \end{cases}$$

So $y_i$ combines the possible values of the data to be received (including measurement noise) with the possible values of the predictions (without measurement noise). Now $y = (y_1, y_2, \ldots, y_m)$ contains all the values of interest, and so we wish to find some prior distribution over $y$.

We define this distribution in two stages. First of all we have assumed that $\eta_i$ is $Gauss(0, \sigma^2)$. We now assume that the prior function over $\boldsymbol{f} = (f_1, f_2, \ldots, f_m)$ can be expressed as a multivariate Gaussian

$$P(\boldsymbol{f}|H) = \frac{1}{Z'} \exp\left(-\frac{1}{2}(\boldsymbol{f} - \boldsymbol{\mu})^T C^{-1}(\boldsymbol{f} - \boldsymbol{\mu})\right)$$

where $\boldsymbol{\mu} = \boldsymbol{\mu}(H)$ is some mean vector, $C = C(H)$ is some covariance matrix and $Z'$ is the relevant normalisation constant. $H$ stands for any set of hyperparameters, which, for now, are assumed to be known.

Then $\boldsymbol{f}$ and $\boldsymbol{\eta}$ are independent Gaussian random variables, and so $\boldsymbol{y}$ is a sum of independent Gaussian distributed random variables, and therefore has a prior distribution of

$$P(\boldsymbol{y}|H) = \frac{1}{Z} \exp\left(-\frac{1}{2}(\boldsymbol{y} - \boldsymbol{\mu})^T Q^{-1}(\boldsymbol{y} - \boldsymbol{\mu})\right)$$

where

$$Q_{ij} = \begin{cases} C_{ij} + \delta_{ij}\sigma^2 & i, j \leq n \\ C_{ij} & \text{otherwise} \end{cases}$$

For future use, we partition Q into the form

$$\begin{pmatrix} Q^{DD} & Q^{DP} \\ Q^{PD} & Q^{PP} \end{pmatrix}$$

where $Q^{DD}$ is $n \times n$ and $Q^{PP}$ is $(m-n) \times (m-n)$. Note that $Q^{PD} = (Q^{DP})^T$.

Suppose we have now received data at points $\boldsymbol{x}^D$ given by $\boldsymbol{y}^D = \boldsymbol{y}^*$. Then we obtain the posterior distribution

$$P(\boldsymbol{y}^P|\boldsymbol{y}^D = \boldsymbol{y}^*, H) = \frac{P(\boldsymbol{y}, \boldsymbol{y}^D = \boldsymbol{y}^*|H)}{P(\boldsymbol{y}^D = \boldsymbol{y}^*|H)}$$

$$= \frac{Z^D}{Z} \exp\left(-\frac{1}{2}(\boldsymbol{y} - \boldsymbol{\mu})^T Q^{-1}(\boldsymbol{y} - \boldsymbol{\mu}) - \frac{1}{2}(\boldsymbol{y}^D - \boldsymbol{\mu}^D)^T(Q^{DD})^{-1}(\boldsymbol{y}^D - \boldsymbol{\mu}^D)\right)$$

This simplifies to the posterior distribution we want

$$P(\boldsymbol{y}^P|\boldsymbol{y}^D = \boldsymbol{y}^*, H) = \frac{1}{Z^P} \exp\left(-\frac{1}{2}(\boldsymbol{y}^P - \hat{\boldsymbol{y}})^T S^{-1}(\boldsymbol{y}^P - \hat{\boldsymbol{y}})\right)$$

where $S = (Q^{PP} - Q^{PD}(Q^{DD})^{-1}Q^{DP})$ and $\hat{\boldsymbol{y}} = Q^{PD}(Q^{DD})^{-1}\boldsymbol{y}^D + \boldsymbol{\mu}$ by the partitioned inverse equations. Note that we only need to invert matrices $Q^{DD}$ and $S$, which are $n \times n$ and $(m-n) \times (m-n)$ respectively. There is no need to invert any $m \times m$ matrices such as $Q$. Here the formulation in [6, 1, 4] is extended to the multivariate predictor case.

This Gaussian process approach has a number of advantages. These are

- The posterior distribution can be calculated analytically.
- The prior form is very flexible: many different forms of covariance matrix can be used, each giving a different type of textural structure to the signal.

– Prior knowledge about functional forms can meaningfully be represented by a Gaussian process: the hyperparameters relate directly to length scales.

There is a computational disadvantage to this method: It involves calculating the inverse of an $n \times n$ matrix, involving $o(n^3)$ computations. Hence the computational power needed increases significantly with the size of the dataset, making it less suitable for cases where many data are available.

## 3 Types of Covariance Functions

We have said nothing yet of the form of the covariance function $C$. In fact for the Gaussian distributions above to be meaningful for all points in input space, the distributions need to satisfy the Chapman-Kolmogorov equations. This is done if the covariance function is that of a Gaussian process. For this to be the case, the covariance function must be positive semidefinite symmetric, and $C_{ij}$ must depend on variables $x_i$ and $x_j$, and no other $x_k$. Furthermore the mean $\mu_i = \mu(x_i)$.

Given a set of scaling hyperparameters $\theta_1, \theta_2, r_l$, a common choice for $C$ is

$$C(x_i, x_j; H) = \theta_1 \exp\left(-\frac{1}{2}\sum_{l=1}^{L} \frac{(x_i^{(l)} - x_j^{(l)})^2}{r_l^2}\right) + \theta_2$$

where $L$ is the dimension of the input space, and $l$ counts through each dimension. This corresponds to saying that the closer points are in input space, the more correlated their function values will be, and that the function is smooth.

## 4 Determining Different Signal Regimes: Gaussian Process Mixtures

Very often the data under study has not been generated from a stationary process. A common example of this is where a number of different signal sources are present, and the observable signal is created by switching between these different regimes.

Here this situation is modelled with a mixture of Gaussian processes. Latent variables represent which of the current regimes generated a sample datum. Then different hyperparameters or covariance structures can be used to represent the characteristics of the different regimes.

The great benefit of Gaussian processes is that the covariance matrix structure can represent many different signal structures and textures, from smooth curves to random fractal textures.

As it is not known which regime is generating the signal at any point, and the structure of the signals is unknown, these variables/parameters are given prior distributions which should be integrated over.

Let $s_k$ denote the regime which generated datum $k$. For now let us assume there are two possible regimes. Then we can form the Gaussian process prior

$$P(y|H) = \sum_{s \in B(m)} p_s \Phi_s(y; H)$$

where $B(m)$ is the set of binary vectors of length $m$. $\Phi$ is a Gaussian kernel of the form

$$\Phi_s = \frac{1}{Z_s} \exp\left(-\frac{1}{2}(y - \mu(H))^T Q_s(H)^{-1}(y - \mu(H))\right)$$

with $\mu(H) = (\mu_{s_1}, \mu_{s_2}, \ldots, \mu_{s_m})^T$ where $\mu_1, \mu_2 \in H$. The covariance function $Q_s$ is given by

$$(Q_s)_{ij} = \begin{cases} \theta_{1s} \exp\left(-\frac{1}{2}\sum_{l=1}^{L} \frac{(x_i^{(l)} - x_j^{(l)})^2}{r_{l_s}^2}\right) + \theta_{2s} + \delta_{ij}\sigma^2 & \text{for } s = s_i = s_j \\ 0 & \text{otherwise} \end{cases}$$

which says that within a given regime we have the usual smooth functions, but there is no correlation between the points in different regimes.

Now let us assume that the switching regime is a Poisson process, rate $\lambda$. Therefore the probability of $y(t_k) = y(t_{k-1})$ is given by the probability of an even number of switches between the two time points:

$$\cosh(\lambda(t_k - t_{k-1})) \exp(-\lambda(t_k - t_{k-1}))$$

Hence $P(y(t_k) \neq y(t_{k-1})) = \sinh(\lambda(t_k - t_{k-1})) \exp(-\lambda(t_k - t_{k-1}))$ where $\lambda$ is a hyperparameter. Other switching priors could equally well be used, and this formalism could easily be extended to multiple regimes.

All the priors are now defined, and the problem can be passed through the usual Gaussian process machinery. The first level of inference involves a tractable Gaussian marginalisation. The second level of inference involves an intractable integration over the hyperparameters and latent variables.

## 5  Integrating-Out or Maximisation?

Calculating the inverse covariance for a Gaussian process is computationally intensive. Therefore integrating out hyperparameters using Monte-Carlo Markov chain approaches can be very slow for large data sizes. The approach we take here is to sample from the posterior over the latent variables and the use a maximum posterior value for the hyperparameters. This is a form of GEM algorithm, where a sample distribution over the latent indicator variables is used instead of the true distribution.

The great benefit of this approach is that the latent variables can be Gibbs sampled, and each Gibbs sample step involves changing only one row/column of the covariance matrix. Hence the partitioned inverse equations can be used to calculate the inverse of the matrix in $o(n^2)$ flops, reducing the computational load significantly. Furthermore, because the covariance matrix has a block structure, the cost of inverting matrices is reduced. The steps of the algorithm are:

- Choose suitable values of the hyperparameters, $H$.
- Choose suitable starting values for the latent variables $s$.
- Gibbs sample the latent variables to get an E-step expression for $P(s|data)$:

$$P(s|data) \leftarrow P(s|H, data) = \frac{1}{Z} \int ds P(data|H, s) P(s) \simeq \sum_k \delta(s - s_k)$$

for a sample $\{s_k\}$

- Move towards the maximum posterior value of the hyperparameters, assuming that this distribution is the true distribution for $s$. This involves maximising

$$\int dP(s|data) \ln P(H|s, data) \simeq \frac{1}{G} \sum_{g=1}^{G} \ln P(H|s_g, data)$$

where $G$ is the chosen sample size (the GM step).
- Repeat the steps until suitably near convergence.

## 6   Example

These methods were tested on a number of toy problems. We introduce one of them. Here a signal is generated from two smooth functions of different regularity and size. In this example, the signal in figure 1 was used. It was generated by the function illustrated, made up of two separate sin waves. The only prior information given was that mentioned above. Hence no knowledge was presumed about the functional form, or periodicity of the data.

When the Gaussian process was tested on this problem it was generally able to distinguish the different regimes. The graphs in figure 1 give the predictive mean, and error bars for the two signals. The true signals are given as solid lines.

The methods were also tested on other similar problems, and problems where the signal mean differed between the signals. The model distinguished between the different regimes. Problems are sometimes encountered when the Poisson prior is such that switching is infrequent. This means that local maxima in the posterior of the latent variables are surrounded by regions of very low probability, and so the Gibbs sampler can get stuck, and not properly sample the whole data space. Occasionally resetting the Gibbs sampler with different starting positions appears to help solve this problem.

## 7   Conclusions

Gaussian processes can represent many types of functions because of the versatility of the covariance structure. This enables regimes with different second order statistical properties to be recognised, while at the same time allowing prior information about the signal form to be properly represented. These methods could be extended to higher dimensions, for example to recognise different textures in two dimensional data.

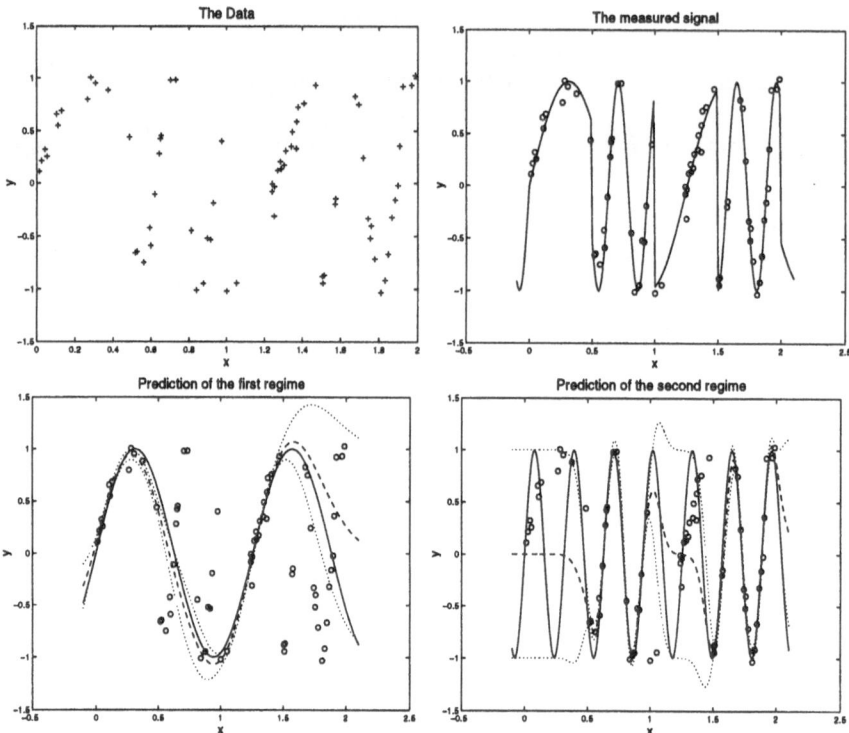

**Fig. 1.** A test problem. Predictors (dashed) and error bars (dotted) for the two generating functions (solid)

## References

1. M. N. Gibbs and D. J. C. MacKay. Efficient implementation of Gaussian processes. Preprint, 1997.

2. R. Neal. *Lecture Notes in Statistics 118: Bayesian Learning for Neural Networks.* Springer Verlag, 1996.

3. A. O'Hagan. On curve fitting and optimal design for regression. *Journal of the Royal Statistical Society B*, 40:1–42, 1978.

4. C. E. Rasmussen. *Evaluation of Gaussian Processes and other Methods for Non-Linear Regression.* PhD thesis, University of Toronto., 1996.

5. A. S. Weigend, M. Mangeas, and A. N. Srivastava. Nonlinear gated experts for time series: Discovering regimes and avoiding overfitting. *International Journal of Neural Systems*, 6:373–399, 1995.

6. C. Williams. Regression with Gaussian processes. In S. W. Ellacott, J. C. Mason, and I. J.Anderson, editors, *Mathematics of Neural Networks: Models, Algorithms and Applications.* Kluwer, 1995. Published 1997.

# Artificial Neural Networks as Approximators of Stochastic Processes

M. R. Belli, M. Conti, P. Crippa, S. Orcioni, C. Turchetti

Dept. of Electronics, University of Ancona, Ancona, Italy

e-mail: turchetti@eealab.unian.it

**Abstract**

Artificial (or biological) Neural Networks must be able to form by learning internal memory of the environment to determine decisions and subsequent actions to stimuli. By assuming that environment is essentially stochastic it follows that the mathematical framework for learning information from environment is the theory of stochastic processes approximation. The aim of this paper is to show that classes of neural networks capable of approximating stochastic processes exist.

## 1 Introduction

Learning by experience is an essential activity in biological brain for achieving information from environment. The experience so collected gives to the brain the background needed to determine decisions and subsequent actions to stimuli coming from the environment. This information to be effectively used must be maintained, in principle, indefinitely. Therefore, accordingly to Kohonen [1], the brain must be able to form *internal memory* of the sensory environment and its history. As recently pointed out by Poggio et al. [2] the ability in learning of a neural network is closely related to the approximating capabilities.

The capability of neural networks in approximating arbitrary non-random input-output mappings has been demonstrated in the earlier works of Cybenko [3], Funahashi [4], Hornik et al. [5] with reference to multilayer feed forward networks, in the work of Park et al. [6] with reference to Radial Basis Neural Networks and in the work of Conti et al. [7] with reference to Approximate Identity Neural Networks (AINNs). However it should be noted that neural networks (biological or artificial) operates in an environment in which all the signals are undoubtedly stochastic. This occurs either because a noise could be superimposed to a signal (non-stochastic) or, more generally, because the signal could be inherently stochastic in that a predictable part of the signal cannot be identified. Thus neural networks should have the ability in learning stochastic processes or, more generally, input-output transformations of stochastic processes.

The aim of this work is to show that some classes of Artificial Neural Networks (ANNs) exist such that they are capable of providing arbitrarily approximation, in the mean square sense, to prescribed stationary stochastic processes.

The networks so defined constitute a new model for neural processing in which the network acts as a universal approximator of stochastic processes. The model is consistent with the need of learning from a stochastic environment and extends previous results concerning approximating capabilities of ANNs.

## 2 Representation of a class of nonstationary processes by Brownian Motion

Let us consider a stochastic process (s. p.) $\varphi(t) = \varphi(t, \omega)$ or $\{\varphi(t), t \in T\}$ such that for every $t \in T$ the random variable (r. v.) $\varphi(\omega)$, defined on a fixed probability space $\{\Omega, B, P\}$, satisfies the conditions

$$E\{\varphi(t)\} = 0, \ E\{|\varphi(t)|^2\} < \infty \tag{1}$$

where $E\{\}$ represents the expectation of a r.v. The set of all such r.v.'s forms a real Hilbert space $L^2(\Omega)$. We consider a class of real-valued random processes $\{\varphi(t), t \in T\}$ whose covariance function $B(t, s) = R_{\varphi\varphi}(t, s) = E\{\varphi(t)\varphi(s)\}$ admits the representation

$$B(t, s) = \int_0^{+\infty} \Phi^T(t, \lambda) \cdot \Phi(s, \lambda) dF(\lambda) \tag{2}$$

where $\Phi(t, \lambda) = (g_1(t, \lambda), g_2(t, \lambda))^T$ is a family of real-valued functions of the variable $\lambda$ which depend on the parameter $t \in T$, and $F(\lambda)$ is a monotone non-decreasing function, which defines a Stieltjes measure on the measurable sets $\Delta\lambda = [\lambda, \lambda + \Delta\lambda)$ of the real line $[0, +\infty)$. Processes of such kind constitute a wide class in that it includes all the stationary processes other than many non-stationary processes. It is well known from the theory of s.p. [10-11] that $\varphi(t)$ admits the following canonical representation:

$$\varphi(t) = \int_0^{+\infty} \Phi^T(t, \lambda) \cdot dy(\lambda) \tag{3}$$

where $\{y_k(\lambda), \lambda \in \Lambda\}$, $k = 1, 2$ are orthogonal increments (o. i.) processes such that

$$E\{|dy_k(\lambda)|^2\} = dF(\lambda), \quad k = 1, 2 \tag{4}$$

and the integral in (3) is a stochastic integral. However even if eq. (3) is useful from a theoretical point of view it is not easy to implement by neural networks because an o. i. process satisfying eq. (4) is required. Instead, as the Brownian Motion (BM) process is straightforward to generate, it is more convenient to derive a representation in terms of such a process. To this end it can be shown that the following canonical representation holds:

$$\varphi(t)=\int_0^{+\infty}\Psi^T(t,\lambda)\cdot dv(\lambda)=\int_0^{+\infty}\left[c(\lambda)\Phi^T(t,\lambda)\right]\cdot dv(d\lambda) \tag{5}$$

where $\{v_k(\lambda),\lambda\in\Lambda\}$, $k=1,2$, are independent BM processes for which

$$E\left\{|dv_k(\lambda)|^2\right\}=dG(\lambda)=\sigma^2 d\lambda \tag{6}$$

and $c(\lambda)$ is a non-negative function such that

$$F(\lambda)=\int_0^{\lambda}[c(\xi)]^2 dG(\xi) \tag{7}$$

# 3 Approximation of stochastic processes by neural networks

We refer to the class of networks named Approximate Identity Neural Networks (AINNs) recently proposed in [7] and defined by the input-output relationship

$$S_n(x)=\sum_{i=1}^{n}a_i u_i(x_1)u_i(x_2).. \tag{8}$$

where $u_i(x_1)u_i(x_2)..$ are approximate identity functions. Networks of such kind have the property that they are able to approximate well a wide class of functions. Thus, with these considerations in mind, we consider the function

$$\Psi_n^T(t,\lambda)=\sum_{k=1}^{n}a_k^T u_k(t)u_k(\lambda) \tag{9}$$

so that the process

$$\varphi_n(t)=\int_0^{+\infty}\Psi_n^T(t,\lambda)\cdot dv(d\lambda) \tag{10}$$

can be defined. By combining (9) and (10) we have

$$\varphi_n(t)=\sum_{k=1}^{n}b_k^T u_k(t) \tag{11}$$

where $b_k^T=a_k^T\cdot\int_0^{+\infty}u_k(\lambda)dv(\lambda)$ are random variables. Eq. (11) defines a neural network with random parameters, which will be called Stochastic AINN (SAINN). It can be shown that as $\Psi_n^T(t,\lambda)$ approximates $\Psi^T(t,\lambda)$, that is

$$\int_0^{+\infty} \left| \Psi^T(t,\lambda) - \Psi_n^T(t,\lambda) \right|^2 d\lambda < \varepsilon \ , \tag{12}$$

then from the properties of stochastic integral, it results

$$E\left\{ \left| \varphi(t) - \varphi_n(t) \right|^2 \right\} < \varepsilon \tag{13}$$

that is the SAINN defines a s. p. $\varphi_n(t)$, which approximates in mean square the given process $\varphi(t)$.

# 4  An application example

As an application example of the theory developed we refer to the signal

$$\varphi(t) = x(t)\cos(\lambda_0 t) \tag{14}$$

where $\lambda_0$ is a constant and $x(t)$ is a stationary process (wide sense) with covariance function given by

$$E\{x(t)x(s)\} = R_{xx}(\tau) = \exp(-\alpha|\tau|) \ \text{ with } \ \tau = t - s \ . \tag{15}$$

The process $\varphi(t)$ is non-stationary being its covariance function given by

$$R_{\varphi\varphi}(t,s) = E\{x(t)x(s)\}\cos(\lambda_0 t)\cos(\lambda_0 s) \ . \tag{16}$$

By using the spectral theory of stationary processes it is straightforward to show that it results

$$B(t,s) = R_{\varphi\varphi}(t,s) = \int_0^{+\infty} \left[ g_1(t,\lambda)g_1(s,\lambda) + g_2(t,\lambda)g_2(s,\lambda) \right] F(d\lambda) =$$

$$= \int_0^{+\infty} \Phi^T(t,\lambda) \cdot \Phi(s,\lambda) F(d\lambda) \tag{17}$$

where

$$g_1(t,\lambda) = \cos(\lambda t)\cos(\lambda_0 t), \ g_2(t,\lambda) = \text{sen}(\lambda t)\cos(\lambda_0 t) \text{ and } F(d\lambda) = \frac{2\alpha}{\alpha^2 + (2\pi\lambda)^2} d\lambda \ .$$

Thus the functions $\Psi(t,\lambda) = (h_1(t,\lambda), h_2(t,\lambda))^T$ to be approximated are given by

$$h_1(t,\lambda) = \frac{1}{\sigma}\sqrt{\frac{2\alpha}{\alpha^2 + (2\pi\lambda)^2}} \cos(\lambda t)\cos(\lambda_0 t) \tag{18}$$

$$h_2(t,\lambda) = \frac{1}{\sigma}\sqrt{\frac{2\alpha}{\alpha^2 + (2\pi\lambda)^2}} \text{sen}(\lambda t)\cos(\lambda_0 t) \ . \tag{19}$$

The training of the neural network has been performed by minimizing the error

$$E = \int_T \int_\Lambda \left| \Psi^T(t,\lambda) - \Psi_n^T(t,\lambda) \right|^2 \sigma^2 \, d\lambda dt$$

thus ensuring the convergence of the two processes in the time interval $T$. Fig.1 shows the covariance function $R_{\varphi\varphi}(t,s)$ of the process $\varphi(t)$ as given by (17).

For comparison the behavior of the covariance function $R_{\varphi_n\varphi_n}(t,s)$ for the approximating process $\varphi_n(t)$ as obtained from (10) is reported in Fig.2. As you can see this behavior is very close to $B(t,s)$.

It is remarkable to note that the approximation of the covariance function $R_{\varphi\varphi}(t,s)$, is a necessary (not in general sufficient) condition for approximating a given process. However the behavior of the covariance function encompasses many of the statistical properties of a process. Thus even if the minimization of the error E guarantees the convergence of the process $\varphi_n(t)$ to $\varphi(t)$, comparison of their covariance functions is useful to confirm the validity of the approach.

# References

[1]  T. Kohonen, "Self-Organisation and Associative Memory," S. Verlag, Berlin, 1988.

[2]  T. Poggio, F. Girosi, "Networks for Approximation and Learning," IEEE Proc., vol. 78, No. 9, pp.1481-1497, Sept.1990.

[3]  G. Cybenko, "Approximation by superposition of sigmoidal function," Math. Control, Systems, Signal, vol. 2, pp. 303-314, 1989.

[4]  K. Funahashi, "On the approximate realisation of continuous mappings by neural networks," Neural Networks, vol. 2 , pp. 183-192, 1989.

[5]  K. Hornik, M. Stinchcombe, H. White "Multilayer feed forward networks are universal approximators," Neural Networks, vol. 2, pp. 395-403.

[6]  J. Park, I. W. Sandberg, "Universal Approximation using Radial-Basis-Function Networks," Neural Computation, vol. 3, pp. 246-257,1991.

[7]  M. Conti, C. Turchetti, "Approximation of dynamical systems by continuous-time recurrent approximate identity neural networks," Neural, Parallel & Scientific Computations vol. 2, pp. 299-322, 1994.

[8]  M. Conti, S. Orcioni, C. Turchetti, "A class of Neural Networks based on Approximate Identity for Analog IC's Hardware Implementation," IEICE Trans. Fundamentals, Vol.E77-A, No.6, pp.1069-1079, June 1994.

[9]  J. L. Doob "Stochastic Processes," J. Wiley & Sons, 1990.

[10] I. I. Gihman, A. V. Skorohod, "The Theory of Stochastic Processes," Springer-Verlag, Berlin 1974.

[11] Y. V. Prohorov, Y. A. Rozanov, "Probability Theory," Springer-Verlag, Berlin 1969.

632

[12] D. C. Champeney, "Fourier Theorems," Cambridge Univ. Press, Great Britain 1987.

[13] R. L. Wheeden, A. Zygmund, "Measure and Integral," M. Dekker Inc., N. Y. 1977.

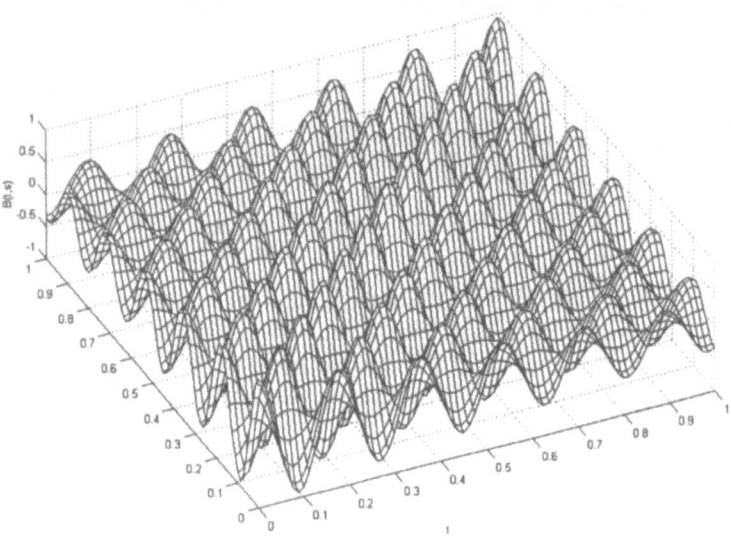

Fig.1 $R_{\varphi\varphi}(t,s)$, $t,s \in [0,1]$ and $\lambda_0 = 35$ as computed from (17)

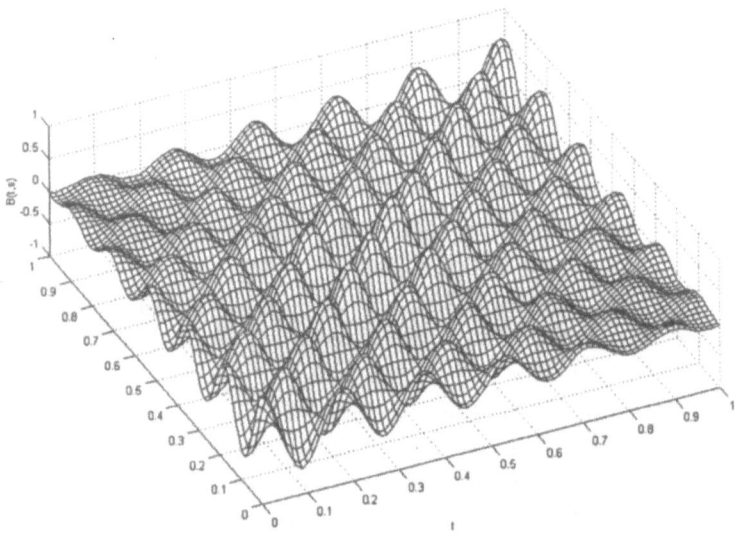

Fig.2 $R_{\varphi_n\varphi_n}(t,s)$, $t,s \in [0,1]$ and $\lambda_0 = 35$, as computed from stochastic process realizations

# Piecewise Affine Neural Networks and Nonlinear Control

Charles-Albert Lehalle*

CMLA (DIAM), Ecole Normale Supérieure de Cachan,
Paris, France

Robert Azencott

CMLA (DIAM), Ecole Normale Supérieure de Cachan,
Paris, France

## Introduction

Linear control of systems governed by an equation such as $x_{n+1} = Ax_n + Bu_n$ (where $x$ and $u$ belong to finite dimensional vector spaces) is widely known [4]. Control design is more complex when dealing with nonlinear systems (for instance [1] or [3]). A well-known method consists in linearizing the system at positions that occur the most often, then solving this serie of linear control problems, and finally patching such local controls together [8].

The piecewise affine neural networks are identical to feed-forward neural networks except that their activation function is continuous piecewise affine rather than sigmoidal. These neural networks can implement quite generic continuous piecewise affine functions on polyhedral cells. Therefore they stand between a collection of local linear controls and a nonlinear control deduced straight from the nonlinear system (this is not always possible, especially when the exact dynamic of the model is unknown [6] and [10]).

On the other hand, since the learning phase of a neural network by gradient retropropagation in a closed loop is difficult [7], a neural network accurate initialization based on linearizations of the system to be controlled is a real benefit. Besides, piecewise affine neural networks can be considered as approximations of neural networks with sigmoidal activation ; this allow some results to be extended to more standard perceptrons.

The aim of this paper is to exhibit the main properties of this new kind of neural networks and to show how the training of these networks can be initialized from linear functions, especially in a control environment

---

*In the framework of a thesis directed by Pr. R. Azencott at the CMLA (DIAM research group) Ecole Normale Supérieure de Cachan, the thesis is carried out at the Research Department of Renault.

First, the action of a piecewise affine neural network is specified by a partition of its input space such that on each polyhedral cell generated by this partition, the network action is affine. Then the slopes and the constants of these affine pieces are characterized as functions of the network weights. The number of cells are determined as a function of the number of hidden neurons. Then a basic result is that in a given hypercube any continuous piecewise affine function can be emulated by a piecewise affine neural network.

The next section presents a methodology to initialize piecewise affine neural networks for controlling nonlinear systems and shows an illustration.

# 1 Properties of piecewise affine perceptrons

**Definition 1 (Piecewise affine perceptron)** *a piecewise affine perceptron (PAP) from $\mathbb{R}^d$ to $\mathbb{R}^a$ with* **one hidden layer** *of $N$ neurons is a function such as :*

$$\forall X \in \mathbb{R}^d, \ \Psi(X) = \Phi\left(W^{(2)} \cdot \Phi\left(W^{(1)} \cdot X + b^{(1)}\right) + b^{(2)}\right) \tag{1}$$

*which is totally specified by $W = [W^{(1)}, b^{(1)}, W^{(2)}, b^{(2)}]$ where : $W^{(1)}$ is a matrix in $\mathcal{M}_{(d,N)}(\mathbb{R})$, $W^{(2)}$ in $\mathcal{M}_{(N,a)}(\mathbb{R})$, $b^{(1)}$ a vector in $\mathbb{R}^N$, and $b^{(2)}$ in $\mathbb{R}^a$. $\Phi$ is a function which applies the response function $g$ to each coordinate of a given vector. For piecewise affine neural networks : $g(x) = x$ for $-1 \leq x \leq 1$, $g(x) = 1$ for $x > 1$, and $g(x) = -1$ for $x < -1$.*

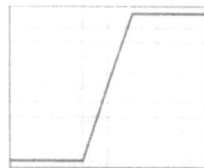 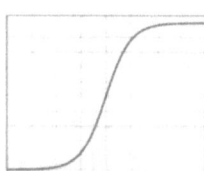

Figure 1: The activation function of a piecewise affine neural network and a sigmoidal one.

Such neural networks can be considered as approximations of perceptrons with sigmoidal activation function rather than piecewise affine. This allow some results on PAP to be extended to those standard perceptrons.

## 1.1 Partition of space into polyhedral cells

Let $h(X) = \Phi\left(W^{(1)} X + b^{(1)}\right)$ be the vector of hidden units of a given PAP (the $j$th coordinate of $h$ is the activation the $j$th hidden neuron) and $\mathcal{H}_-(i)$ and $\mathcal{H}_+(i)$ the hyperplanes normal to $W_i^{(1)}$ (the $i$th row vector of $W^{(1)}$) and containing respectively the points $\tilde{x}_i^+$ and $\tilde{x}_i^-$ defined by (where $b_i^{(1)}$ is the $i$th coordinate of $b^{(1)}$) :

$$\left\{ \begin{array}{rcl} \left\langle W_i^{(1)}, \tilde{x}_i^+ \right\rangle & = & 1 - b_i^{(1)} \\ \left\langle W_i^{(1)}, \tilde{x}_i^- \right\rangle & = & -1 - b_i^{(1)} \end{array} \right. \tag{2}$$

**Definition 2 (Initial partition)** *The initial partition associated to a PAP is the set $C$ of polyhedral cells generated by the intersections of the $N$ parallel hyperplanes pairs $(\mathcal{H}_-(i), \mathcal{H}_+(i))$ defined by (2).*

**Proposition 1 (First layer action)** *On each polyhedral cell in the initial partition $C$ of a PAP, $h(X)$ is affine or constant with respect to $X$. $h(X)$ is continuous on the whole input space.*

**Terminal partition :** the terminal partition of a PAP is the partition generated by the intersections of parallel hyperplanes pairs $(\mathcal{H}'_-(c,j), \mathcal{H}'_+(c,j))$ depending on the second layer weights and on the value of $h(X)$ on each cell in the initial partition. The adjacent cells of this partition where the PAP has the same behavior are joined.

**Proposition 2 (Second layer action)** *On each polyhedral cell in the terminal partition $C'$ of a PAP, $\Psi(X)$ is affine or constant with respect to $X$. $\Psi(X)$ is continuous on the whole input space.*

## 1.2 PAP action on each cell

The PAP has an affine action on each polyhedral cell in the terminal partition ; the slope and the constant of this affine function depend on the non-saturated coordinates of the first layer and the second layer weights.

**Proposition 3 (PAP specification)** *A PAP with $N$ hidden neurons has a continuous piecewise affine behavior on each polyhedral cell of its terminal partition. Besides, the number of cells in its terminal partition is lower or equal to the number of cells in its initial partition.*

The second part of this proposition comes from the monotony of the activation functions of a PAP.

## 1.3 PAP and continuous piecewise affine functions

Proposition 3 asserts that a PAP is a continuous piecewise affine function, reciprocally :

**Theorem 1 (Continuous piecewise affine function representation)**
*Given an hypercube $\mathcal{K}$ of $\mathbb{R}^d$ and a continuous piecewise affine function $f$ on a polyhedral partition of $\mathcal{K}$, there is at least one PAP coinciding with $f$ on the whole hypercube.*

To prove this, first consider the set $\mathcal{E}$ of hyperplanes generating the polyhedral cells on which $f$ is affine ; hyperplanes have to be added to $\mathcal{E}$ outside the hypercube until the obtained set $\hat{\mathcal{E}}$ contains only pairs of parallel hyperplanes. Then other parallel hyperplanes pairs have to be added to have enough parameters to generate the slope and constant of $f$ on each resulting cell.

There is in fact an infinity of PAPs coinciding with $f$ on the given hypercube.

## 1.4 Number of cells

**Definition 3 (Generic set of hyperplanes)** *A generic set $\mathcal{E}$ of $n$ hyperplanes is a set where any intersection of $k$ of its elements is $(d-k)$ dimensional if $d < k$, otherwise empty.*

If the parameters of PAP are randomly chosen, the Lebesgue measure of the set of PAPs which hyperplanes $(\mathcal{H}_+(i))$ and $(\mathcal{H}_-(i))$ are not generic is zero.

**Proposition 4 (Number of cells on the input space)** *The number of $d$-dimensional polyhedral cells generated by the partition of space by a generic set of $N$ pairs of parallel hyperplanes pair $(\mathcal{H}_+(i), \mathcal{H}_-(i))$ is :*

$$B_d(N) = \sum_{k=0}^{d} \binom{N}{k} 2^k \tag{3}$$

$B_d(N)$ is the number of cells generated by the first layer ; because of the action of the second one, the number of cells at the PAP output is lower or equal to the results given by (3). This result come from [2] completed by [5].

**Behavior of the number of cells with respect to the number of hidden neurons.** When the number $N$ of hidden units is large, the number of polyhedral cells has is equivalent to $(2N)^d/d!$ ; so in a situation with numerous hidden neurons, adding a neuron multiplies the number of cells by $1 + d/(2N)$.

# 2 Application to control problems

## 2.1 Approximation of linear controls

Given two vector spaces $\mathcal{X}$ (the state space) and $\mathcal{U}$ (the control space), and $I(x, u)$ a norm on $\mathcal{X} \times \mathcal{U}$, the purpose is to find a function $u = g(x, x^*)$ (from $\mathcal{X}^2$ into $\mathcal{U}$) such that the trajectories of a state variable $x$ with dynamics :

$$\begin{cases} x_{n+1} &= f(x_n, u_n) \\ u_n &= g(x_n, x^*) \end{cases} \tag{4}$$

(where $f$ is differentiable in $\mathcal{X} \times \mathcal{U}$) verify $\lim_{n \to \infty} I(x_n - x^*, u_n) = 0$ for any initial state $x_0$ and any given state $x^*$ in $\Omega$ a fixed set.

The function $g$ here is a PAP, the automatic learning will be a gradient retropropagation of the cost function. To control (4), one can first select some linear approximations ($\hat{f}_i$) of $f$ at chosen points ($x_i$), then design linear controls ($K_i$) each of them being optimal around $x_i$ for (4) where $f$ is replaced by its linear version $\hat{f}_i$, and finally initialize a PAP with a continuous piecewise affine function $K$ constructed to be equal to each $K_i$ around $x_i$. The main PAP property (that one can construct at least one PAP coinciding with a given continuous piecewise affine function) is used here.

The methodology consists in constructing a PAP such that it could emulate a lot of different nonlinear functions around $K$ when its weights are submitted to small changes, this is done by choosing carefully among all PAPs approximately equal to $K$. Thus the automatic learning will cause the PAP **to emulate a nonlinear function that will fix the deficiencies of** $K$. This will be an efficient initialization for a PAP before running it through automatic learning. Automatic learning is achieved by gradient retropropagation on a PAP.

## 2.2   A simple example : control of a standard spring

To illustrate the PAP ability to solve control problems, let's consider the following school case : control design of a standard spring with position and speed to be controlled by applicating a force on it ([9], $\ddot{l} = -kl + u - fl^2$). The chosen cost function $I$ is quadratic in $u$ and $x$ ($I(u,x) = u^2 + x^2$). A linear controller has been designed by linearizing the spring dynamics, it has been used to initialize a PAP, and a sigmoidal version of this network has been used. FIGURE 2 shows the cost function evolution for a linear control (a) and for a neural network during the learning phase (b). The initial positions are randomized in a fixed range. The effect of the learning on the control is clear : one can see that the PAPquickly emulate a nonlinear controller.

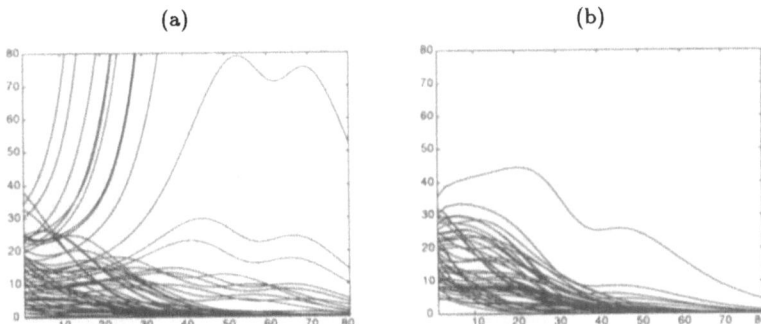

Figure 2: Evolution of the quadratic cost function through time for standard spring trajectories : for a linear control (a) and a PAP (b) initialized with it during its learning phase.

# Conclusion — future applications

One of the main purposes of this study is to use piecewise affine perceptrons (PAP) to tune a neural network designed for adaptive control of automotive engine combustion. In this context, the engine torque has to be rapidly driven to a desired value varying with time. A PAP will determine the command to apply to actuators (dealing with the fuel and air quantity entering the combustion chamber) state variables being given; the methodology is to initialize it from linear controls deduced from linearizations of a strongly nonlinear engine model, and then to let it evolve by automatic learning to a nonlinear control that will fix the linear control deficiencies, this second part is an ongoing collaboration with the Research Center of Renault.

As we have shown, a PAP can implement quite generic continuous piecewise affine functions on polyhedral cells. Besides, their relative simplicity allows their use to easily synthetize sets of local affine functions. Linear control deduced from linearizations of the set of dynamics seems to be an efficient way to initialize the PAP.

# References

[1] B. Bonnard. Contrôlabilité des systèmes non linéaires. *CR Acad des Sciences Paris*, Ser I(292):535–537, 1981.

[2] R. Buck. Partition of space. *American Mathematician Monthly*, 50:541–544, Nov 1943.

[3] C. Byrnes. Control theory, inverse spectral problem and real algebraic geometry. *Proceedings of the conference held at Michigan Technology*, 1983.

[4] d'Azzo. *Linear Control System Analysis and Design, Conventional and Modern*. McGraw Hill, 1988.

[5] H. Edelsbrunner. *Algorithms in Combinatorial Geometry*, chapter Fundamental Concepts in Combinatorial Geometry. Springer-Verlag, 1987.

[6] L. Hunt and R. Su. Design for multi-input nonlinear systems. *Proceedings of the conference held at Michigan Technology*, pages 268–297, 1983.

[7] S. Jagannarthan. Multilayer discrete-time neural-net controller with guaranteed performance. *IEEE trans. on Neural Networks*, 7(1):107–130, jan 1996.

[8] B. Jakubczyk. On linearization of control systems. *Bulletin de l'Académie Polonaise des Sciences, Ser Sci Math*, XXVIII(9-10):517–522, 1980.

[9] E. Sontag. *Mathematical Control Theory*. Springer-Verlag, 1990.

[10] H. Sussmann and R. Brockett. *Differential Geometric Control Theory*. Birkhauser, 1982.

# Some Complexity Results for Perceptron Networks

Barbara Hammer

Department of Mathematics/Computer Science, University of Osnabrück
Osnabrück, Germany
hammer@informatik.uni-osnabrueck.de

## Abstract

The loading problem is the problem to decide if a neural architecture can map a training set correctly with an appropriate choice of the weights. The following results will be shown:

The loading problem is NP-complete for any feedforward perceptron architecture with at least two neurons in the first hidden layer and varying input dimension. Further, it is NP-complete if the input dimension is fixed, but if the number of neurons can vary in an architecture with at least two hidden layers. Finally, for a recurrent perceptron network with fixed architecture and fixed input dimension, but arbitrary input length the loading problem is solvable in polynomial time.

## 1 Introduction

Neural networks are successfully used for speech recognition, image processing, controlling, time series prediction, etc. Very different learning algorithms and architectures have proved to work well in these tasks of machine learning. But before a network is trained with some algorithm in a practical application it is interesting to consider the principle complexity of the learning task.

Here, we will deal with an even simpler correlated problem, just to decide if a pattern set can be stored correctly with a neural network. This is NP-complete for a 3 node network with perceptron activation with varying input dimension [1] and some results for the sigmoidal activation exist as well [2, 3]. But the results are restricted to the 3 node resp. other special architectures. Here, the completeness result in the perceptron case is generalized to an arbitrary multilayer feedforward network with at least 2 nodes in the first hidden layer.

Another interesting question is the complexity to decide if a training set can be stored into a network with fixed input dimension, but varying number of hidden neurons as stated in [4]. This question arises e.g. if a minimum network which fits certain patterns shall be found. For multilayer architectures with at least two hidden layers this question turns out to be NP-complete as well.

Whereas these results indicate that for high dimensional inputs resp. large networks training may take a prohibitive amount of time, for each single fixed architecture a polynomial algorithm exists.

When dealing with data where the input dimension is not fixed, i.e. sequences or lists, one can use recurrent networks for their classification. Of course, the completeness results of the feedforward case transfer to this case. But here, for a fixed architecture and input dimension, another variable, the input length, occurs. Fortunately, the deciding problem is polynomial with respect to the input length for any fixed architecture, too.

## 2 The feedforward case

**Definition 1** *One neuron $o$ with input dimension $n$, weights $\mathbf{w} \in \mathbb{R}^n$ and bias $\theta \in \mathbb{R}$ computes the function $\mathbb{R}^n \to \{0,1\}$,*

$$o(\mathbf{x}) = \begin{cases} 1 & \text{if } \langle \mathbf{w}, \mathbf{x} \rangle + \theta \geq 0, \\ 0 & \text{otherwise.} \end{cases}$$

*A **multilayer feedforward perceptron** (MLP) with input dimension $n$ consists of $h$ hidden layers with $n_1, \ldots, n_h$ neurons $o_j^i$, where the neurons of layer $i$ have input dimension $n_{i-1}$ ($n_0 := n$), and one output neuron $o$ with input dimension $n_h$. The network computes the function $N : \mathbb{R}^n \to \{0,1\}$, $N(\mathbf{x}) = o(y_1^h, \ldots, y_{n_h}^h)$ where the output of the neurons is defined recursively*

$$y_j^i = \begin{cases} o_j^i(y_1^{i-1}, \ldots, y_{n_{i-1}}^{i-1}) & \text{if } i > 0, \\ x_j & \text{if } i = 0 \text{ and } \mathbf{x} = (x_1, \ldots, x_n). \end{cases}$$

If we speak of an architecture we refer to the set of networks with the same number of hidden neurons but arbitrarily chosen weights and biases. An architecture will be specified by the parameters $(n, h, n_1, \ldots, n_h)$.

**Definition 2** *The **loading problem** is the following problem: Given a neural architecture $(n, h, n_1, \ldots, n_h)$ and a training set, i.e. a set $\{(\mathbf{x}_i, y_i) \in \mathbb{R}^n \times \{0,1\} \mid i = 1, \ldots, m\}$ of patterns. Does there exist weights and biases such that the corresponding network $N$ maps the patterns correctly, i.e. $N(\mathbf{x}_i) = y_i \; \forall i$?*

If the architecture is fixed, i.e. $(n, h, n_1, \ldots, n_h)$ equals a fixed tuple for any instance, this is solvable in polynomial time. In [5] an algorithm of [6] is modified for the case $h = 1$ which can be generalized as follows: In the network each neuron defines a hyperplane in some $\mathbb{R}^t$ if $t$ is the input dimension of the neuron. To obtain an upper bound for the number of mappings which differ on a finite set of points, it is sufficient to consider only a finite number of hyperplanes: The output of the neuron remains the same on a finite input set if the hyperplane is substituted by the hyperplane with maximum distance to the set of points. But this hyperplane is already determined by at most $t + 1$ points, the points with minimum distance on both sides of the hyperplane. If these points are fixed, it remains to solve a system of linear equations to find appropriate weights. Now, we can start at the first layer and proceed as follows:

```
l := 1,  j := 1,  p_1 := x_1,  ...,  p_m := x_m,  call(rec(l,j,{p_1,...,p_m})).
rec(l,j,{p_1,...,p_m}):
  Consider neuron number j in layer l.
  For all choices of at most n_{l-1}+1 points from p_1,...,p_m:
    The choice defines a hyperplane.
    Compute the corresponding weights for the neuron.
    If j < n_l:  j := j+1, call(rec(l,j,{p_1,...,p_m})).
    If j = n_l, l < h:  p_i := (o_1^l(p_i),...,o_{n_l}^l(p_i))  (i = 1...m),
                        j := 1, l := l+1, call(rec(l,j,{p_1,...,p_m})).
    If j = n_l, l = h:  p_i := (o_1^l(p_i),...,o_{n_l}^l(p_i))  (i = 1...m), solve the
                        linear programming problem (o(p_i) = y_i)_{i=1...m}.
```

For any unit $o_j^l$ there are at most $(2 \cdot m^{n_{l-1}+1})^2$ subsets of points which define the hyperplane. Therefore we have no more than $4^h m^{2(n+n_1+\cdots+n_{h-1}+h)} \cdot n_1 \cdot \ldots \cdot n_h$ recursive steps in the algorithm. This is polynomial in $m$. Even in the real number model the algorithm is polynomial in the size of the training set if $h \geq 1$. Further, it is polynomial for $h \geq 1$ even if $n_h$ is not fixed.

For varying architecture, i.e. varying $(n, h, n_1, \ldots, n_h)$ the problem is still in NP, because we can guess appropriate weights – they can be chosen polynomial in the input set – and test if the network maps the patterns correctly. But the problem becomes NP-complete even if $h = 1$, $n_1 = 2$ and only $n$ varies [1].

Of course this could go back to the very restricted function class with only two hidden neurons. Indeed, if we take $n_1 \geq m$ the problem becomes tractable because in this case any consistent pattern set can be loaded. Therefore the question arises whether loading is intractable for an architecture with a realistic number of neurons but which is not correlated to the size of the pattern set (for binary inputs a question in [5]). In [7] Judd addresses the problem of loading with larger architectures but with a special restricted structure and growing output dimension. A more realistic setting is addressed in the following:

**Theorem 3** *For fixed $h \geq 1$, $n_1 \geq 2$, $n_2$, $\ldots$, $n_h$ and varying $n$ the loading problem is NP-complete.*

**Proof:** In [8] it is shown that for fixed $h = 1$ and $n_1 \geq 2$, but varying input dimension $n$ the loading problem is NP-complete if the output computes $o(x_1, \ldots, x_{n_1}) = x_1 \wedge \ldots \wedge x_{n_1}$ for $x_i \in \{0, 1\}$. We will reduce the loading problem with an architecture given by $(n, 1, n_1)$, $n_1$ fixed and $o = \text{AND}$ to a loading problem for a general feedforward architecture $(\tilde{n}, h, n_1, \ldots, n_h)$, $h, n_2, \ldots, n_h$ fixed, $\tilde{n} = n + n_1 + 1$. Assume there is given a pattern set $S = \{(\mathbf{x}_i, y_i) \in \mathbb{R}^n \times \{0, 1\} \mid i = 1, \ldots, m\}$ for an architecture of the first type. We enlarge $S$ to guarantee, that a bigger architecture necessarily computes an AND in the layers following the first hidden layer. Define

$$
\begin{aligned}
\tilde{S} = \ & \{((\mathbf{x}_i, 0, \ldots, 0), y_i) \in \mathbb{R}^{\tilde{n}} \times \{0, 1\} \mid i = 1, \ldots, m\} \\
& \cup \{((0, \ldots, 0, \tilde{\mathbf{z}}_i, 1), 0), ((0, \ldots, 0, \bar{\mathbf{z}}_i, 1), 1)) \mid i = 1, \ldots, n_1(n_1 + 1)\} \\
& \cup \{((0, \ldots, 0, \mathbf{p}_i, 1), q_i) \mid i = 1, \ldots, 2^{n_1}\}
\end{aligned}
$$

where $\tilde{\mathbf{z}}_i, \bar{\mathbf{z}}_i, \mathbf{p}_i \in \mathbb{R}^{n_1}$, $q_i \in \{0, 1\}$ are constructed as follows:

Choose $n_1 + 1$ points on each hyperplane $H_i = \{\mathbf{x} \in \mathbb{R}^{n_1} \mid$ the $i$-th component of $\mathbf{x}$ is $0\}$, such that $n_1 + 1$ of these points lie on one hyperplane exactly if they lie on one $H_i$. Denote the points $z_1, z_2, \ldots$. Define $\tilde{\mathbf{z}}_i \in \mathbb{R}^{n_1}$ such that $\tilde{\mathbf{z}}_i$ equals $\mathbf{z}_i$ except for the $i$th component which equals a small value $\epsilon$. Define $\bar{\mathbf{z}}_i$ in the same way with the $i$th component $-\epsilon$. $\epsilon$ can be chosen such that if one hyperplane in $\mathbb{R}^{n_1}$ separates at least $n_1 + 1$ pairs $(\tilde{\mathbf{z}}_i, \bar{\mathbf{z}}_i)$ these are exactly the $n_1 + 1$ pairs corresponding to the $n_1 + 1$ points on one hyperplane $H_i$ and the separating hyperplane nearly coincides with $H_i$. (See Fig.1.) This is due to the fact that $n_1 + 1$ points $\mathbf{z}_i$ do not lie on one hyperplane exactly if $\det \begin{pmatrix} 1 & \cdots & 1 \\ z_1 & \cdots & z_{n_1+1} \end{pmatrix} \neq 0$. $\mathbf{p}_i$ are all points in $\{-1, 1\}^{n_1}$, $q_i = 1 \Leftrightarrow \mathbf{p}_i = (1 \ldots 1)$.

*Assume, $S$ can be loaded* with a network $N$ of the first architecture. We construct a solution of $\tilde{S}$ with a network $\tilde{N}$ of the second architecture: For neuron $\tilde{o}_i^1$ in the first layer of $\tilde{N}$ choose the first $n$ weights and the bias like the weights resp. bias of neuron $o_i^1$ in $N$. The next $n_1$ weights are 0 except for the

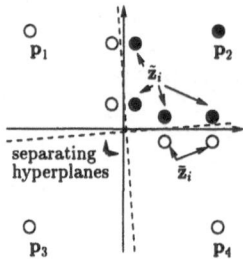

Figure 1: Additional points in $\tilde{S}$

$i$th one which is 1, the $n + n_1 + 1$st weight is $-\theta_i$ if $\theta_i$ is the bias of neuron $o_i^1$ in $N$. The neurons in the other layers compute an AND. This maps $\tilde{S}$ correctly.

*Assume, $\tilde{S}$ can be loaded* with a network $\tilde{N}$ of the second architecture. The points $(\ldots, \tilde{z}_i, 1)$ and $(\ldots, \bar{z}_i, 1)$ are mapped differently, therefore each pair is separated by at least one of the hyperplanes defined by the neurons in the first hidden layer. Because $n_1(n_1 + 1)$ points have to be separated, the hyperplanes defined by these neurons nearly coincide in the last $n_1$ dimensions with the hyperplanes $H_i$ we have used in the construction of $\tilde{z}_i, \bar{z}_i$. We can assume that the point $(\ldots, \tilde{z}_i, 1)$ is mapped to 1 by the neuron corresponding to $H_i$, maybe we have to change the sign of the weights and the biases before. Then, the values of $\{(0, \ldots, 0, \mathbf{p}_i, 1) \mid i = 1, \ldots, 2^{n_1}\}$ are mapped to the entire set $\{0, 1\}^{n_1}$ by the neurons of the first hidden layer and the rest of the network necessarily computes an AND of these values. Consequently, the network $N$ maps $S$ correctly if the weights in the first hidden layer equal the weights of $\tilde{N}$ restricted to dimension $n$ and if the output unit computes AND. $\qquad\square$

**Theorem 4** *For fixed $h \geq 2$, $n \geq 2$ and varying $n_1, \ldots, n_h$ the loading problem is NP-complete.*

**Proof:** In [6] it is shown that the problem to decide if two sets of points $P$ and $Q$ in $\mathbb{R}^2$ can be separated by $k$ lines, i.e. each pair $p \in P$ and $q \in Q$ lie on different sides of at least one line, is NP-complete. Consider the training set

$$S = \{(\mathbf{x}_i, y_i) \in \mathbb{R}^2 \times \{0, 1\} \mid (\mathbf{x}_i \in P \wedge y_i = 1) \vee (\mathbf{x}_i \in Q \wedge y_i = 0)\} .$$

for two sets of points $P$ and $Q$. $S$ can be loaded by a network with two inputs, 2 hidden layers, $n_1 = k$, and $n_2 = |P|$ exactly if $P$ and $Q$ are separable:

*Assume, $S$ can be loaded.* The hidden nodes in the first layer define $k$ lines in $\mathbb{R}^2$. Each point $p \in P$ is separated from each point $q \in Q$ by at least one line because, otherwise, the corresponding patterns are mapped to the same value.

*Assume, $P$ and $Q$ are separable by $k$ lines.* Define the weights of the neurons in the first hidden layer according to these lines. Let $P = \{p_1, \ldots, p_m\}$. Let the hidden unit $o_j^2$ in the second hidden layer compute $o_j^2(x_1, \ldots, x_k) = (\neg)x_1 \wedge \ldots \wedge (\neg)x_k$ where the $\neg$ takes place at $x_i$ if $p_j$ lies on the negative side of the $i$th hyperplane. Especially, $o_j^2(p_j) = 1$ and $o_j^2(q) = 0 \, \forall q \in Q$. Consequently, if the output unit computes an OR, the pattern set $S$ is mapped correctly. Obviously, the argumentation can be transferred to $h \geq 2$ and $n \geq 2$. $\qquad\square$

The latter theorem addresses problem (12.13) of [4] in the perceptron case. Note that for $h = 1$ and varying $n_1$ the loading problem is polynomial as already mentioned. As a consequence of these results, there exist severe computational limitations to train networks with large input dimension resp. more than two hidden layers and a large number of neurons.

# 3 Recurrent networks

It is a remarkable property of neural networks that they are capable of handling a simple kind of structured data, lists, in a natural way. Recurrent networks are a tool to deal with lists or time series, i.e. inputs with an a priori unlimited length. Let $(\mathbb{R}^n)^*$ define the set of lists with elements in $\mathbb{R}^n$.

**Definition 5** *Any function* $f : \mathbb{R}^{n+r} \to \mathbb{R}^r$ *induces a function* $\tilde{f} : (\mathbb{R}^n)^* \to \mathbb{R}^r$,

$$\tilde{f}([\mathbf{x}_1, \ldots, \mathbf{x}_l]) = \left\{ \begin{array}{ll} (0, \ldots, 0) & \text{if } l = 0, \\ f(\mathbf{x}_l, \tilde{f}([\mathbf{x}_1, \ldots, \mathbf{x}_{l-1}])) & \text{otherwise.} \end{array} \right.$$

*A* **recurrent network** *consists of two MLPs. The first network is given by an architecture* $(n + r, h, n_1, \ldots, n_h)$. *It has* $r$ *instead of one output neuron, and computes* $f_1 : \mathbb{R}^{n+r} \to \{0, 1\}^r$. *The second network is given by an architecture* $(r, \bar{h}, \bar{n}_1, \ldots, \bar{n}_{\bar{h}})$ *and computes* $f_2 : \mathbb{R}^r \to \{0, 1\}$. *The recurrent network computes the function* $f_2 \circ \tilde{f}_1 : (\mathbb{R}^n)^* \to \{0, 1\}$.

Since this is a generalization of MLPs the NP-completeness results transfer to recurrent networks, too. But in practical applications, training recurrent networks seems very difficult even for a fixed architecture for long input sequences [9]. Therefore the question arises if the loading problem for a fixed recurrent architecture, but arbitrary input length is polynomially solvable.

**Theorem 6** *Assume there is given a fixed recurrent architecture and an arbitrary training set* $S = \{(\mathbf{x}_i, y_i) \in (\mathbb{R}^n)^* \times \{0, 1\} \mid i = 1, \ldots, m\}$. *There exists a polynomial algorithm to decide if* $S$ *can be mapped by the architecture correctly.*

**Proof:** Assume, the architecture is given by $((n + r, h, n_1, \ldots, n_h), (r, \bar{h}, \bar{n}_1, \ldots, \bar{n}_{\bar{h}}))$ and the lengths of the input sequences in $S$ sum to $l$. If the values of the points in $S$ are computed by a recurrent network the following steps occur:

For one unit in the first hidden layer of the network at one recursive computation step the actual input is considered, i.e. one of $l$ possible values, the value of the context units is considered, i.e. one of $2^r$ values in $\{0, 1\}^r$, this vector is multiplied with the weights, and the sign determines the value of the unit.

For any other unit the value of the neurons in the previous layer is considered, i.e. an element of $\{0, 1\}^t$ for some $t$, the vector is multiplied with the weights, and the sign of this computation determines the value of the unit.

Consequently, we get $P(l) = l\,n_1 2^r + n_2 2^{n_1} + \cdots + n_h 2^{n_{h-1}} + r 2^{n_h} + \tilde{n}_1 2^r + \tilde{n}_2 2^{\tilde{n}_1} + \ldots + \tilde{n}_{\bar{h}} 2^{\tilde{n}_{\bar{h}-1}} + 2^{\tilde{n}_{\bar{h}}}$ polynomials of degree one for the weights that are computed at the recursive computation of the values of $S$. The output of the network is entirely determined by the sign of each of these polynomials. On the other hand, if we know the sign of each polynomial, we can test in time polynomial in $l$ whether each list $\mathbf{x}_i$ in $S$ is mapped to $y_i$.

Consequently, to solve the loading problem for $S$ it is sufficient to consider $P(l)$ polynomials of degree one for the weights, find all possible sign vectors that occur, if the polynomials are computed for some concrete weights, and test, if at least one of these sign vectors describes a solution. There exists an algorithm which finds all possible sign vectors and is polynomial in the number of the polynomials and the coefficients, i.e. polynomial in $S$ [10]. □

As a consequence for a fixed recurrent architecture the complexity to train sequences is growing only polynomially with the length of the sequences.

# 4   Discussion

It has been shown that for MLPs the complexity of training is growing extremely with growing input dimension resp. number of neurons whereas for a fixed architecture in the recursive case the additional input length of the recursive data only causes a polynomial increase of complexity.

Although the architectures considered here – multilayer networks with an arbitrary number of hidden layers and units – are realistic, the activation function, a hard limiter, is not. It would be interesting to get comparable results for the sigmoidal function or at least another continuous activation which is used in practical applications.

Further in the reductions showing the NP-completeness we have used real inputs which are in special positions. Is it possible to get similar results for points which are nearly orthogonal? Of course, for orthogonal points the loading problem is trivial, the answer is always 'yes'. Maybe a restriction of the input set such that the correlation of the (appropriately scaled) inputs is limited still ensures polynomial learnability even for a growing architecture. In [8] this question is considered for binary inputs.

# References

[1] A. Blum and R. Rivest. Training a 3-node neural network is NP-complete. *Neural Networks*, 5:117–127, 1992.

[2] J. Šíma. Back-propagation is not efficient. *Neural Networks*, 9:1017–1023, 1996.

[3] B. Hammer. Training a sigmoidal network is difficult. In M.Verleysen (ed.), *European Symposium on Artificial Neural Networks*, D-facto publications, Brussels, pp.255–260, 1998.

[4] M.. Vidyasagar. *A Theory of Learning and Generalization*. Springer, 1997.

[5] B. DasGupta, H. T. Siegelmann, and E. D. Sontag. On the complexity of training neural networks with continuous activation. *IEEE Transactions on Neural Networks*, 6:1490–1504, 1995.

[6] N. Megiddo. On the complexity of polyhedral separability. *Discrete and Computational Geometry*, 3:325–337, 1988.

[7] J.S. Judd. *Neural Networks Design and the Complexity of Learning*. MIT, 1990.

[8] M. Schmitt. *Komplexität neuronaler Lernprobleme*. Peter Lang, 1996.

[9] Y. Bengio, P. Simard, and P. Frasconi. Learning long-term dependencies with gradient descent is difficult. *IEEE Trans. on Neural Networks*, 5:157–166, 1994.

[10] S. Basu, R. Pollack, and M.-F. Roy. A new algorithm to find a point in every cell defined by a family of polynomials. *Journal of the ACM*, 43:1002–1045, 1996.

# Multilayer Neural Networks for Classification: A Pedagogical Theorem

T. M. Ellerbrock

Department of Mathematical Physics, Bielefeld University

Bielefeld, Germany

thomas@physik.uni-bielefeld.de

### Abstract

A multilayer perceptron is not likely to learn an unknown classification, if it is not given a "representative" training set. Proposing a definition of "representativity", we have proved that multilayer perceptrons are able to realize "adequate solutions" to classification problems whenever the training set is representative. The notion of "adequate solutions" is a rather natural way to give the notion of "generalization" a precise mathematical meaning from the viewpoint of topology and combinatorial geometry. This way, connectivity properties of the classes to be learned are taken into account. In contrast to the known results concerning approximation capabilities of networks, our consideration expresses generalization performance explicitly by means of the training set and without reference to the true, but unknown classification. Furthermore, these results can be used as a basis for network learning and growing algorithms.

**Keywords:** Generalization, Approximation capabilities, Representative training sets, Modularity, Triangulation

## 1 Introduction

With each classification is associated a classification mapping $f_C$. For classifications with exactly two classes $f_C$ may be zero on one class and unity on the other one. There are many publications which report the approximation capabilities of multilayer perceptrons [3, 5, 1, 8, 4, 6]. Hornik [5] and Cybenko [1] especially had derived results for classifications. For example, Cybenko [1] had proved that for any $\epsilon > 0$ there is a subset $M$ of the input space having Lesbegue measure at least $1 - \epsilon$ and that there exists a single hidden layer perceptron realizing a function $f_N$ with $|f_C(\mathbf{x}) - f_N(\mathbf{x})| < \epsilon$, $\forall \mathbf{x} \in M$. For a review and some new results see [7].

In contrast to the publications cited above and in contrast to all research known by the author, here

- approximation capabilities of multilayer perceptrons are not expressed in terms of some unknown target function or classification, but with respect to the training or a test set, which is much more closer to practice.

- the representivity of the training set is taken into account.

- the capability of nets to approximate sets (classes) instead of functions is described.

- topological aspects are integrated. The classes of a classification may not be connected, but may consist of several separated parts. Even if $f_C$ would be known, that did not help. There is no universal subset $M$ of the input space and there is no general value for $\epsilon$ such that the condition " $|f_C(\mathbf{x}) - f_N(\mathbf{x})| < \epsilon, \forall \mathbf{x} \in M$ " guarantees $f_N$ to reflect the connectedness of all conceivable classes.

- approximation is directly related to generalization. A good generalization should reflect the connectivity properties of the classification to learn. We show when it does so.

## 2  Notations and facts

$\mathcal{N}$ denotes the natural numbers without zero. $\mathcal{R}^d$ is the d-dimensional, real *Euclidean space*. Suppose $A \subset \mathcal{R}^d$. $A^\circ$ denotes the *interior*, $\overline{A}$ the *closure* and $\partial A := \overline{A} \setminus A^\circ$ the *boundary* of $A$ in the natural topology of $\mathcal{R}^d$. We use aff $A$ to refer to the *affine hull* of $A$. The boundary of $A$ with respect to the topological space (aff $A, \mathcal{T}_A$), with $\mathcal{T}_A$ being the relative topology of aff $A$, is called *relative boundary* and is denoted by rbd $A$. Let $\mathbf{p}_1, \ldots, \mathbf{p}_{k+1}$ be affinely independent and aff $\{\mathbf{p}_1, \ldots, \mathbf{p}_{k+1}\} = $ aff $A$. Then $\mathbf{x} \in$ aff $A$ can uniquely be expressed as an affine combination of the vectors $\mathbf{p}_1, \ldots, \mathbf{p}_{k+1}$ with coefficients $\lambda_1, \ldots, \lambda_{k+1}$, which are called the *barycentric coordinates* of $\mathbf{x}$. conv $A$ denotes the *convex hull* of A. A *k-simplex* $S$ in $\mathcal{R}^d$ is the convex hull of $k + 1$ affinely independent points in $\mathcal{R}^d$ called its vertices and referred to as vert $S$. If $S$ and $S'$ are both simplices and vert $S \subseteq$ vert $S'$, then $S$ is called a *face* or a *subsimplex* of $S'$, written $S \prec S'$. *dimA* denotes the dimension of the set $A \subset \mathcal{R}^d$. If $\dim S = \dim S' - 1$, then $S$ is called a *facet* of $S'$.

## 3  Topologically representative sets for classifications

**Definition 3.1 (Classification (mapping), Class boundary)**
*Let $C_+$ and $C_-$ be two non-empty, disjoint subsets of $\mathcal{R}^d$, one of them closed, and $C_+ \cup C_- = \mathcal{R}^d$. Then we call $C := (C_+, C_-)$ a classification of $\mathcal{R}^d$. Define $f_C : \mathcal{R}^d \rightarrow \{-1, +1\}$ such that $f_C^{-1}(-1) := C_-$ and $f_C^{-1}(+1) := C_+$. Then we call $f_C$ a classification mapping for $\mathcal{R}^d$. The class boundary $\partial C$ of $C$ is defined to be $\partial C := \partial C_+ = \partial C_-$.*

Solving a *classification problem* means, roughly speaking, finding a mapping $f_N : \mathcal{R}^d \rightarrow \{-1, +1\}$ which is a sufficiently good approximation of the unknown classification mapping $f_C$. Alternatively, we can express a classification problem as finding a sufficiently good approximation of the unknown class boundary.

**Definition 3.2 (Comprehensive training set)** *We call a finite subset $T$ of $\mathcal{R}^d$ a comprehensive training set for a classification $C$, if $T$ satisfies:*

$$\overline{C_+} \subseteq (\operatorname{conv} T)^{\circ} \quad or \quad \overline{C_-} \subseteq (\operatorname{conv} T)^{\circ}$$

In contrast to the usual definition, we now introduce triangulations of point sets in $\mathcal{R}^d$ not merely to be simplicial complexes but graphs of neighborly simplices.

**Definition 3.3 (Triangulation)** *Let $T$ be a finite subset of $\mathcal{R}^d$, $d \geq 2$. We define a triangulation of $T$, triang $T$, to be an undirected graph $\left( V_\triangle(T), E_\triangle(T) \right)$ with vertex set $V_\triangle(T)$ and edge set $E_\triangle(T)$ such that the following conditions hold*

   *i) If $S \in V_\triangle(T)$, then $S$ is a $d$-simplex and vert $S \subseteq T$.*

   *ii) If $\mathbf{x} \in T$, then a simplex $S \in V_\triangle(T)$ exists so that $\mathbf{x} \in$ vert $S$.*

   *iii) Let $S$ and $S'$ be in $V_\triangle(T)$ and suppose $S^\star \prec S$ and $S'^\star \prec S'$. Then $S^\star \cap S'^\star = \emptyset$ or $S^\star \cap S'^\star = \operatorname{rbd} S^\star \cap \operatorname{rbd} S'^\star$, and this intersection is a subsimplex of $S^\star$ and $S'^\star$.*

   *iv)* $\displaystyle\bigcup_{S \in V_\triangle(T)} S = \operatorname{conv} T$

   *v) $E_\triangle(T) := \{(S,S') \mid S,S' \in V_\triangle(T)$ and $S \cap S'$ is a $(d-1)$-simplex$\}$*

**Definition 3.4 (Chromatic attributes)** *Suppose a classification $C$, a training set $T$, a triangulation triang $T$ and a set $A$ in $\mathcal{R}^d$ are given. If $A \subseteq C_+$ or $A \subseteq C_-$, then we call $A$ monochrome, otherwise we call $A$ polychrome, respectively. Let $S \in V_\triangle(T)$. If vert $S$ is monochrome, we call $S$ chromatically closed, otherwise we call it chromatically open, respectively.*

A triangulation of the training set enables us to characterize the structure of the training set given. The underlying structure will become more apparent, if we select the polychrome simplices from the triangulation.

**Definition 3.5 (The cluster graph $\left( V_\triangle^{cl}(T,C), E_\triangle^{cl}(T,C) \right)$ and cluster)**

   *i) $V_\triangle^{cl}(T,C) := \{S \in V_\triangle(T) \mid S$ chromatically open$\}$*

   *ii) $E_\triangle^{cl}(T,C) := \{(S,S') \in E_\triangle(T) \mid S \cap S'$ chromatically open$\}$*

We call $\left( V_\triangle^{cl}(T,C), E_\triangle^{cl}(T,C) \right)$ cluster graph and its components clusters.

It is obvious that an adequate solution to a classification problem will be unlikely to be found, if the given training set is not representative. We propose a definition of representativity from a topological point of view.

**Definition 3.6 (Topological representativity)** *If $T$ satisfies the conditions i) and ii), then we call $T$ a training set roughly topologically representative of $C$ with respect to $\operatorname{triang} T$. If $T$ satisfies conditions i) - iii), we call it a training set fine topologically representative of $C$ with respect to $\operatorname{triang} T$.*

*i) $T$ is comprehensive for $C$.*

*ii)*
$$\left( \begin{array}{c} S \in V_\Delta(T) \text{ is chromatically} \\ \text{open with respect to } C \end{array} \right) \iff \left( \begin{array}{c} S \in V_\Delta(T) \text{ is polychrome} \\ \text{with respect to } C \end{array} \right)$$

*iii)*
$$\left( \begin{array}{c} S \in V_\Delta^{d-1}(T) \text{ is chromatically} \\ \text{open with respect to } C \end{array} \right) \Leftrightarrow \left( \begin{array}{c} S \in V_\Delta^{d-1}(T) \text{ is polychrome} \\ \text{with respect to } C \end{array} \right)$$

*with* $\quad V_\Delta^{d-1}(T) := \{ S^\star \mid \exists S \in V_\Delta(T) \,:\, S^\star \prec S \text{ and } \dim S^\star = d-1 \}$

# 4 Topologically adequate solutions to classification problems

**Definition 4.1 (Topologically adequate solution for classifications)**
*If $C$ is a classification and $T$ is a finite training set in $\mathcal{R}^d$, then we call $(C,T)$ a classification problem. Suppose $L$ is a classification in $\mathcal{R}^d$. Let $\partial C_\mu$ and $\partial L_\nu$ denote components of $\partial C$ and $\partial L$, respectively. We define the classification $L$ in $\mathcal{R}^d$ to be a roughly topologically adequate solution of $(C,T)$ with respect to $\operatorname{triang} T$, if $L$ satisfies the conditions i) to iii). If $L$ satisfies the conditions i) to iv), we call it a fine topologically adequate solution.*

*i)* $T \cap L_+ = T \cap C_+ \neq \emptyset \neq T \cap L_- = T \cap C_-$

*ii)* $\forall \mathbf{x} \in \partial C \;\exists\, S_x \in V_\Delta^{cl}(T,C)$ *so that* $\mathbf{x} \in S_x$ *and* $S_x \cap \partial L \neq \emptyset$

*iii)* $\forall \mathbf{y} \in \partial L \;\exists\, S_y \in V_\Delta^{cl}(T,C)$ *so that* $\mathbf{y} \in S_y$ *and* $S_y \cap \partial C \neq \emptyset$

*iv) For all $\partial L_\nu$ there exists one and only one cluster $(V,E)$ of the cluster graph and it exists a $\partial C_\mu$ so that:*

$$\partial L_\nu \subseteq \bigcup_{S \in V} \quad \text{and} \quad \partial C_\mu \subseteq \bigcup_{S \in V}$$

A roughly topologically adequate solution $L$ to $(C,T)$ is an approximation of $\partial C$ within the area of chromatically open simplices of the triangulation. It classifies the set $T$ the same way as $C$ does. Usually, when a classification problem is encountered, $C$ is unknown. Solving a classification problem $(C,T)$ means concluding the unknown classification from the known examples $T$, i.e. generating a classification $L$ which approximately fits $C$. The intention of the definition above is to make more precise what "good" approximation should mean.

In the general case $\partial C$ is not connected. A roughly topologically adequate solution which is not fine topologically adequate does not take into account the components of $\partial C$, whereas a fine topologically adequate solution does.

**Theorem 4.1** *Suppose $C$ and $L$ are two classifications in $\mathcal{R}^d$. If $C_+ \cap T = L_+ \cap T$ and $C_- \cap T = L_- \cap T$, and if $T$ is roughly topologically representative of $C$ and $L$ with respect to $\operatorname{triang} T$, then $L$ is a roughly topologically adequate solution to the classification problem $(C, T)$, and $C$ is a roughly topologically adequate solution to the classification problem $(L, T)$.*

**Definition 4.2 (General position of a triangulation)**
*Suppose a triangulation $\operatorname{triang} T$ of a finite training set $T \in \mathcal{R}^d$ and a classification $C$ in $\mathcal{R}^d$ are given. Again $\hat{V}_\triangle^{d-1}(T) := \{S^\star \mid \exists S \in V_\triangle(T) : S^\star \prec S \text{ and } \dim S^\star \leq d - 1\}$. We call $\operatorname{triang} T$ to be in a general position with respect to $C$, if for all $\mathbf{x} \in S^\star \cap \partial C$, $S^\star \in \hat{V}_\triangle^{d-1}(T)$, there exists a neighborhood $U$ of $\mathbf{x}$ so that $(S^\star \cap U) \cap C_+ \neq \emptyset \neq (S^\star \cap U) \cap C_-$ holds.*

**Theorem 4.2** *Let $C$ and $L$ be two classifications in $\mathcal{R}^d$ and $T$ be a finite training set in $\mathcal{R}^d$. Suppose $C_+ \cap T = L_+ \cap T$ and $C_- \cap T = L_- \cap T$. If $\operatorname{triang} T$ is in a general position with respect to $C$ and $L$, and if $T$ is fine topologically representative of $C$ and $L$ with respect to $\operatorname{triang} T$, then $L$ is a fine topologically adequate solution to the classification problem $(C, T)$, and $C$ is a fine topologically adequate solution to the classification problem $(L, T)$.*

# 5 Neural solutions to classification problems

In this section we introduce multilayer perceptron solutions to classification problems.

**Definition 5.1 (Modular multilayer perceptron of tanh-sgn type)**
*Suppose $\mathbf{N}_1, \ldots, \mathbf{N}_n$, $n \in \mathcal{N}$ are multilayer perceptrons with $d$ input neurons and one output neuron each. Let the hidden neurons have $\tanh()$ as activation function and the output neurons have $\operatorname{sgn}()$ as activation function. Suppose $\mathbf{N}_0$ is a multilayer perceptron with $n$ input neurons and one output neuron. Let the hidden neurons and the output neuron of $\mathbf{N}_0$ have $\operatorname{sgn}()$ as their activation function. We call the networks $\mathbf{N}_1, \ldots, \mathbf{N}_n$ hidden modules and the network $\mathbf{N}_0$ output module. We now connect the hidden modules with the output module so that the input neurons of the output module become identical with the output neurons of the hidden modules. Let us call the resulting network $\mathbf{N}$ a modular multilayer perceptron of tanh-sgn type.*

**Theorem 5.1 (A pedagogical theorem)**
*Suppose a training set $T$, $\operatorname{triang} T$, a classification $C$ and the associated classification mapping $f_C$ in $\mathcal{R}^d$, $d \geq 2$, are given. Let us denote by $f_N : \mathcal{R}^d \to \{-1, +1\}$ a mapping which is realized by a multilayer perceptron $\mathbf{N}$ with $d$ input neurons and one output neuron which has $\operatorname{sgn}()$ as its activation function. Let us define $N := (N_+, N_-)$ with $N_\pm := f_N^{-1}(\pm 1)$. Then:*

> *i) If $T$ is a training set roughly topologically representative of $C$ with respect to $\operatorname{triang} T$, then there exists a modular perceptron $\mathbf{N}$ of tanh-sgn type so that $N$ is a roughly topologically adequate solution to the classification problem $(C, T)$.*

*ii)* *If* $T$ *is a training set fine topologically representative of* $C$ *with respect to* triang $T$, *and if* triang $T$ *is in a general position with respect to* $C$, *then there exists a modular perceptron* N *of tanh-sgn type so that* $N$ *is a fine topologically adequate solution to the classification problem* $(C, T)$.

The concepts introduced here can also be used to construct learning and growing algorithms for multilayer perceptrons. This is done by the author in [2]. To this end, we triangulate the training set and remove all training inputs not pertaining to chromatically open simplices. When this is done, for each cluster a growing simplicial complex is generated by adding successively adjacent simplices to some initial one. For each growing simplicial complex, a neural module (multilayer perceptron) is grown in parallel and trained with the associated training inputs. Finally, these modules are connected by an output neural module resulting in a modular multilayer perceptron of tanh-sgn type.

# References

[1] G. Cybenko. Approximation by superposition of a sigmoidal function. *Mathematics of Control, Signals and Systems*, 2:303–314, 1989.

[2] T. M. Ellerbrock. "title yet not known". Ph.D. Thesis, University of Bielefeld, Germany, Department of Mathematical Physics, 1998. A short version will also be available from my homepage:
http://www. physik. uni-bielefeld.de/ ~thomas/welcome. html.

[3] K-I. Funahashi. On the approximate realisation of continuous mapping by neural networks. *Neural Network*, 2:183–192, 1989.

[4] K. Hornik. Some new results on network approximation. *Neural Network*, 6:1069–1072, 1993.

[5] K. Hornik, M. Stinchcombe, and H. White. Multilayer feedforward networks are universal approximators. *Neural Network*, 2:359–366, 1989.

[6] M. Leshno, V. Y. Lin, A. Pinkus, and S. Schocken. Multilayered feedforward networks with a non-polynomial activation function can approximate any function. *Neural Network*, 6:861–867, 1993.

[7] F. Scarselli and A. C. Tsoi. Universal approximation using feedforward neural networks: A Survey of some existing methods, and some new results. *Neural Network*, 11(1):15–37, 1998.

[8] H. White. *Artificial Neural Networks: Approximation and Learning Theory.* Blackwell, Oxford UK, Cambridge USA, 1992. (With A. R. Gallant, K. Hornik, M. Stinchcombe and J. Wooldridge).

# An Experimental Comparison of Neural ICA Algorithms

Xavier Giannakopoulos, Juha Karhunen, and Erkki Oja

Lab. of Computer and Information Science, Helsinki Univ. of Technology

P.O. Box 2200, 02015 HUT, Espoo, Finland

### Abstract

Several neural algorithms for Independent Component Analysis (ICA) have been introduced lately, but their computational properties have not yet been systematically studied. In this paper, we compare the accuracy, convergence speed, computational load, and other properties of five prominent neural or semi-neural ICA algorithms. The comparison reveals some interesting differences between the algorithms.

## 1 Introduction

Independent Component Analysis (ICA) [1, 2] is an unsupervised technique which tries to represent the data in terms of statistically independent variables. Recently, efficient new neural learning algorithms [3, 4, 5, 6, 7, 8] have been developed for ICA and applied to the closely related blind source separation (BSS) and other problems [3]. However, a serious experimental comparison of ICA algorithms is still lacking. In this paper, we present first results on such a comparison, reported in detail in [9].

We consider the standard linear data model used in ICA and BSS [2, 7, 10]:

$$\mathbf{x}(t) = \mathbf{A}\mathbf{s}(t) = \sum_{i=1}^{m} s_i(t)\mathbf{a}_i. \tag{1}$$

Here the components $s_i(t)$, $i = 1, \ldots, m$, of the column vector $\mathbf{s}(t)$ are the $m$ unknown, mutually statistically independent components (or source signals) at time or index value $t$. For simplicity, they are assumed to be zero mean and stationary. The components of the $m$-dimensional data vector $\mathbf{x}(t)$ are some linear mixtures of these independent components or sources. The $m \times m$ mixing matrix $\mathbf{A}$ is an unknown full rank constant matrix. Its columns $\mathbf{a}_i$ are the basis vectors of ICA. At most one of the independent components $s_i(t)$ is allowed to be Gaussian.

For learning the ICA expansion (1), an $m \times m$ inverse or separating matrix $\mathbf{B}(t)$ is updated so that the $m$-vector

$$\mathbf{y}(t) = \mathbf{B}(t)\mathbf{x}(t) \tag{2}$$

becomes an estimate $\mathbf{y}(t) = \hat{\mathbf{s}}(t)$ of the independent components. The estimate $\hat{s}_i(t)$ of the $i$:th independent component may appear in any component $y_j(t)$ of $\mathbf{y}(t)$. The amplitudes $y_j(t)$ are scaled to have a unit variance.

# 2 Neural ICA or BSS algorithms

In several neural ICA or BSS algorithms, the data vectors $\mathbf{x}(t)$ are preprocessed by whitening (sphering) them: $\mathbf{v}(t) = \mathbf{V}(t)\mathbf{x}(t)$. Here $\mathbf{v}(t)$ denotes the $t$:th whitened vector satisfying $E[\mathbf{v}(t)\mathbf{v}(t)^T] = \mathbf{I}$, where $\mathbf{I}$ is the unit matrix, and $\mathbf{V}(t)$ is an $m \times m$ whitening matrix. Whitening can be done in many ways [7]. After prewhitening the subsequent separating matrix $\mathbf{W}(t)$ can be taken orthogonal, which often improves the convergence. Thus in whitening approaches the total separating matrix is $\mathbf{B}(t) = \mathbf{W}(t)\mathbf{V}(t)$.

Because of limited space, we describe the algorithms included in our study only briefly. For more details, see the references and [9].

**Fixed-point (FP) algorithms.** One iteration of the generalized fixed-point algorithm for finding a row vector $\mathbf{w}_i^T$ of $\mathbf{W}$ is [3, 11]

$$
\begin{aligned}
\mathbf{w}_i^* &= E\{\mathbf{v}g(\mathbf{w}_i^T\mathbf{v})\} - E\{g'(\mathbf{w}_i^T\mathbf{v})\}\mathbf{w}_i \\
\mathbf{w}_i &= \mathbf{w}_i^*/\|\mathbf{w}_i^*\|.
\end{aligned}
\tag{3}
$$

Here $g(t)$ is a suitable nonlinearity, typically $g(t) = t^3$ or $g(t) = \tanh(t)$, and $g'(t)$ is its derivative. The expectations are in practice replaced by their sample means. Hence the fixed-point algorithm is not a truly neural adaptive algorithm. The algorithm requires prewhitening of the data. The vectors $\mathbf{w}_i$ must be orthogonalized against each other; this can be done either sequentially or symmetrically [11, 9]. Usually the algorithm (3) converges after 5-20 iterations.

**Natural gradient algorithm (ACY).** Originally proposed in [6] on heuristic grounds, this popular and simple neural gradient algorithm was later on derived from information-theoretic criteria [4]. The algorithm does not require prewhitening. The update rule for the separating matrix $\mathbf{B}$ is

$$
\Delta\mathbf{B} = \mu_k[\mathbf{I} - \mathbf{g}(\mathbf{y})\mathbf{y}^T]\mathbf{B}.
\tag{4}
$$

The notation $\mathbf{g}(\mathbf{y})$ means that the nonlinearity $g(t)$ is applied to each component of the vector $\mathbf{y} = \mathbf{B}\mathbf{x}$. The learning parameter $\mu_k$ is usually a small constant. In practice, $\mathbf{B}$ is often updated using small batches of data [5].

**Extended Bell-Sejnowski algorithm (ExtBS).** The update rule is otherwise the same as in (4) but whitening is used to improve the convergence properties:

$$
\Delta\mathbf{W} = \mu_k[\mathbf{I} - \mathbf{g}(\mathbf{y})\mathbf{y}^T]\mathbf{W}
\tag{5}
$$

where now $\mathbf{y} = \mathbf{W}\mathbf{v}$. In the extended form, kurtosis is estimated on-line for handling both super-Gaussian and sub-Gaussian sources [12], and the learning parameter $\mu_k$ is optimized using a momentum term and simulated annealing [12, 13].

**EASI algorithm.** Introduced as an adaptive signal processing algorithm in [10], EASI can be applied as a neural learning algorithm as well. The general update formula for the $\mathbf{B}$ contains two extra terms compared to (4):

$$
\Delta\mathbf{B} = \mu_k \left[ \frac{\mathbf{I} - \mathbf{y}\mathbf{y}^T}{1 + \mu_k\mathbf{y}^T\mathbf{y}} - \frac{\mathbf{g}(\mathbf{y})\mathbf{y}^T + \mathbf{y}\mathbf{g}(\mathbf{y}^T)}{1 + \mu_k|\mathbf{y}^T\mathbf{g}(\mathbf{y})|} \right] \mathbf{B}
\tag{6}
$$

In this comparison, we used the normalized version of EASI given above. In the unnormalized version, the terms in the denominator of (6) are left out, which may cause stability problems especially for nonlinearities growing faster than linearly [10].

**RLS algorithm for a nonlinear PCA criterion (NPCA-RLS).** The basic symmetric version, adapted for the BSS problem using prewhitened data vectors $\mathbf{v}(t)$, is [8]

$$
\begin{aligned}
\mathbf{z}(t) &= \mathbf{g}(\mathbf{W}(t-1)\mathbf{v}(t)) = \mathbf{g}(\mathbf{y}(t)), \\
\mathbf{h}(t) &= \mathbf{P}(t-1)\mathbf{z}(t), \\
\mathbf{m}(t) &= \mathbf{h}(t)/(\beta + \mathbf{z}^T(t)\mathbf{h}(t)), \\
\mathbf{P}(t) &= \frac{1}{\beta}\mathrm{Tri}\left[\mathbf{P}(t-1) - \mathbf{m}(t)\mathbf{h}^T(t)\right], \\
\mathbf{e}(t) &= \mathbf{v}(t) - \mathbf{W}^T(t-1)\mathbf{z}(t), \\
\mathbf{W}(t) &= \mathbf{W}(t-1) + \mathbf{m}(t)\mathbf{e}(t)^T.
\end{aligned}
\tag{7}
$$

The forgetting constant $0 < \beta \leq 1$ should be close to unity. The notation Tri means that only the upper triangular part of the argument is computed and its transpose is copied to the lower triangular part. The NPCA-RLS algorithm (7) is a recursive least-squares version of the nonlinear PCA algorithm [7]. The learning parameter is determined so that it becomes roughly optimal [8].

# 3 Experimental results

## 3.1 Artificially generated data

We have thus far made simulations using mainly artificially generated data, because only then the accuracy and convergence speed of the algorithms can be measured reliably. For real-world data, the true independent components (or their best approximations) are unknown.

The experimental setup was the same for each algorithm in order to make the comparison fair; see [9]. We used the original MATLAB codes provided by the authors whenever possible, such as [14] for the fixed-point algorithms. For the experiments, both sub-Gaussian and super-Gaussian sources were generated [9], and the mixing matrix $\mathbf{A}$ consisted of uniformly distributed random numbers.

The accuracy was measured using two performance indexes. The first one, $E_1$, is defined by [4]

$$
E_1 = \sum_{i=1}^{n}\left(\sum_{j=1}^{n}\frac{|p_{ij}|}{max_k|p_{ik}|} - 1\right) + \sum_{j=1}^{n}\left(\sum_{i=1}^{n}\frac{|p_{ij}|}{max_k|p_{kj}|} - 1\right)
\tag{8}
$$

where $\mathbf{P} = (p_{ij}) = \mathbf{BA}$ should be a permutation matrix if the sources have been separated perfectly. The second performance index $E_2$ is otherwise the same but the absolute values are replaced by squares in (8). For both indices, the greater the value, the poorer the performance. The minimum is zero.

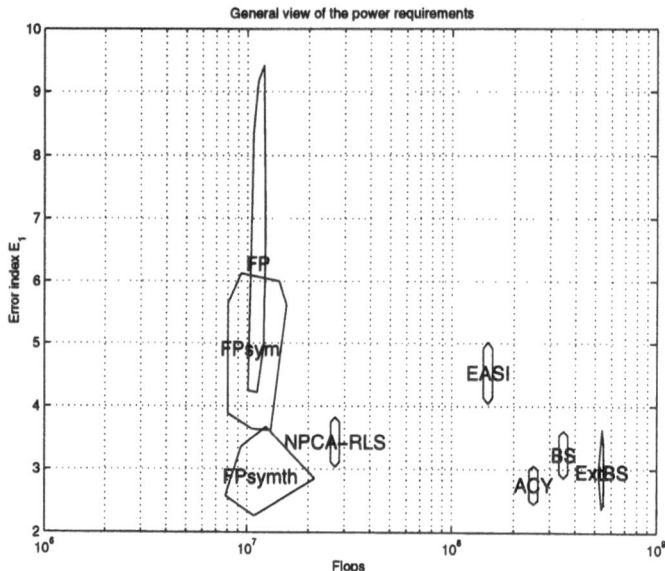

Figure 1: Power requirements in flops vs. error index $E_1$. The boxes typically contain 80% of the 100 trials.

Figure 1 is a schematic diagram of the results of basic experiments measuring both the accuracy and the computational load (in floating point operations) of the tested algorithms. The number of sources (independent components) was 10. Clearly, fixed-point algorithms require the smallest amount of computation. The symmetric version with a tanh nonlinearity (FPsymth) is the most accurate of them, while the accuracy of the basic FP algorithm using sequential orthogonalization is poorer than for the other algorithms. Of the adaptive algorithms, NPCA-RLS converges fastest. The natural gradient algorithm (ACY) and extended Bell-Sejnowski algorithm achieve a good final accuracy, but their computational load is much higher.

In figure 2, the error (square root of the index $E_2$) is plotted as a function of the number of super-Gaussian sources (for which all the algorithms worked). Generally, the natural gradient algorithm (ACY) and various modifications of it (BS, ExtBS, WACY) have the best accuracy, behaving very similarly as expected. Fixed-point algorithm (FP) has the poorest accuracy, but its error increases only slightly after 7 sources. For an unknown reason, the error of the EASI and NPCA-RLS algorithm has a peak around 5-6 sources. However, the error of all the algorithms is tolerable for most practical purposes.

In experiments where the number of sources was increased, it was necessary to replace the cubic nonlinearity $g(t) = t^3$ with the more stable $g(t) = \tanh(t)$ in the EASI algorithm and with $g(t) = \tanh(t) - t$ in the ACY algorithm to make them converge with more than 10 sources.

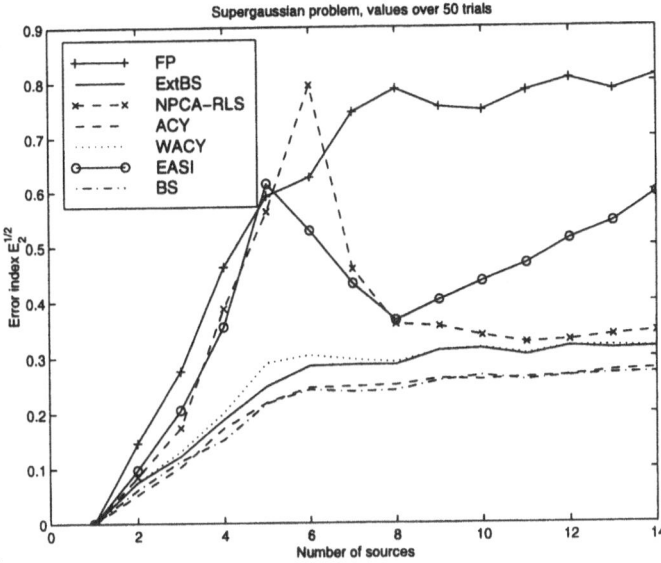

Figure 2: Error as a function of the number of sources.

When Gaussian noise was added to the data, the first conclusion was that degradation of the results is smooth at least until the noise power increases up ʹto -20dB of the signal power. Another observation was that once there is even a little of noise present in the data, the error strongly depends on the condition number of the mixing matrix. This holds both for equivariant algorithms which compute the separating matrix **B** directly and for algorithms employing prewhitening, except for NPCA-RLS which behaves differently.

A general conclusion on the experiments with artificial data is that for designing an efficient ICA algorithm, one should split the problem into different parts. These include at least the following choices: the algorithm; the nonlinearity; and the control structure. Of course, the dependencies between these constituent parts must be taken into account. Such a design allows one to make a good practical compromise between efficiency, robustness, precision, and other relevant requirements of the problem at hand.

## 3.2 Real-world data

We have recently extended our comparisons to real-world data [9], trying to find good projection pursuit directions using ICA algorithms. In projection pursuit [15], the goal is to find for visualization purposes one-dimensional projections of multidimensional data containing as much "interesting" structural information as possible. ICA seems to be a good tool for projection pursuit, because it provides non-Gaussian projections containing meaningful structural information as suggested already in [7].

As an example, we considered 5-dimensional data set consisting of measurements made on two species of crabs. All the ICA algorithms except for (4) found two good ICA basis vectors separating the species and also to a certain extent males and females of both species; see [9]. The results were clearly better than those given by standard PCA.

# 4 Conclusions

The main conclusions of this experimental comparison of ICA algorithms are as follows. The semi-neural fixed-point algorithm converges fastest, and can deal with both sub-Gaussian and super-Gaussian sources without on-line estimation of the kurtosis required by the other algorithms. Of the adaptive neural algorithms, the recursive least-squares algorithm usually has the smallest computational load. The accuracy of the Bell-Sejnowski algorithm and its natural gradient extensions is very good. The main factor affecting the final accuracy of the algorithms is the choice of the nonlinearity.

# References

[1] P. Comon, *Signal Processing*, vol. 36, pp. 287-314, 1994.

[2] C. Jutten and J. Herault, *Signal Processing*, vol. 24, no. 1, pp. 1-10, July 1991.

[3] E. Oja et al., in S.-I. Amari and N. Kasabov (Eds.), *Brain-Like Computing and Intelligent Information Systems*, Springer, Singapore, 1997, pp. 167-188.

[4] H. Yang and S.-I. Amari, *Neural Computation*, vol. 9, pp. 1457-1482, 1997.

[5] A. Bell and T. Sejnowski, *Neural Computation*, vol. 7, pp. 1129-1159, 1995.

[6] A. Cichocki and R. Unbehauen, *IEEE Trans. on Circuits and Systems-1*, vol. 43, pp. 894-906, Nov. 1996.

[7] J. Karhunen, E. Oja, L. Wang, R. Vigario, and J. Joutsensalo, *IEEE Trans. on Neural Networks*, vol. 8, pp. 486-504, May 1997.

[8] J. Karhunen and P. Pajunen, in *Proc. 1997 Int. Conf. on Neural Networks*, Houston, Texas, June 1997, pp. 2147-2152.

[9] X. Giannakopoulos, "Comparison of adaptive independent component analysis algorithms," Dipl.Eng thesis made for EPFL, Switzerland, at Helsinki Univ. of Technology, Finland, 58 p. Available at http://www.cis.hut.fi/~xgiannak/.

[10] J.-F. Cardoso and B. Hvam Laheld, *IEEE Trans. on Signal Processing*, vol. 44, pp. 3017-3030, Dec. 1996.

[11] A. Hyvärinen, in *Proc. IEEE Int. Conf. on Acoustics, Speech, and Signal Processing*, Munich, Germany, April 1997, pp. 3917-3920.

[12] M. Girolami and C. Fyfe, in *Proc. 1997 Int. Conf. on Neural Networks*, Houston, Texas, June 1997, pp. 1788-1791.

[13] M. McKeown et al., *Proc. Natl. Acad. Sci. USA*, vol. 95, pp. 803-810, 1998.

[14] *Fast ICA MATLAB package*. Available at http://www.cis.hut.fi/projects/ica/fastica.

[15] J. Friedman, *J. of Amer. Stat. Assoc.*, vol. 82, no. 397, pp. 249-266, March 1987.

# The Principal Independent Components of Images

Björn Arlt, Rüdiger Brause

FB Informatik, J.W.Goethe-Universität Frankfurt/Main, Germany

E-mail: {arlt,brause}@informatik.uni-frankfurt.de

### Abstract

This paper proposes a new approach for the encoding of images by only a few important components. Classically, this is done by the Principal Component Analysis (PCA). Recently, the Independent Component Analysis (ICA) has found strong interest in the neural network community. Applied to images, we aim for the most important source patterns with the highest occurrence probability or highest information called *principal independent components* (PIC).

For the example of a synthetic image composed by characters this idea selects the salient ones. For natural images it does not lead to an acceptable reproduction error since no a-priori probabilities can be computed. Combining the traditional principal component criteria of PCA with the independence property of ICA we obtain a better encoding. It turns out that this definition of PIC implements the classical demand of Shannon's rate distortion theory.

## 1 Introduction

Classically, the encoding of images by only a few important components is done by Principal Component Analysis (PCA). One common solution is to cut the image into smaller patches or "subimages" which are transformed linearly by projecting them on the eigenvectors of their associated covariance matrix. It is well known that the transformed components with the highest variance (the *principal components*) yield an optimal reconstruction of the original subimages in the mean square error sense. However, for the criterion of minimal redundancy encoding, the PCA is suboptimal.

Recently, the Independent Component Analysis (ICA) has become subject to many research activities and several algorithms have been proposed by different authors, e.g. [1, 2, 3]. Here, the goal is to obtain linearly transformed components which are as independent as possible (the *independent components*). This corresponds to the minimisation of the mutual information between the transformed components and therefore reduces the overall encoding amount [1, 2].

Applied to image encoding, the ICA approach assumes that each observed signal vector $x = (x_1,...,x_n)^T$ (an image containing $n$ pixels) is a linear mixture $x = Ms$ of $n$ unknown independent source signals $s = (s_1,...,s_n)^T$. The unknown mixing matrix $M$ must be non-singular; its columns can be viewed as "image primitives". To recover the sources signals, one has to determine a demixing matrix $B$ with $s = Bx$.

There are several conditions involved in the demixing process [1]: in general, the recovered source signals (denoted by $y = (y_1,...,y_n)^T$ for clarity) are scaled and permuted versions of the original sources. Furthermore, at most one of the source signals s should have a Gaussian probability distribution or else the separation will become ambiguous. This is why the recovered sources y are conventionally assumed to be non-Gaussian random variables having unit variance.

As proposed in [1, 3] the determination of **B** reduces to the computation of an orthogonal matrix $W_{ICA}$ if the observed signals x are prewhitened. This can be done by a simple PCA transform of the image vectors and scaling the obtained PCA components to unit variance. The corresponding prewhitening (or *sphering*) transform is denoted by the matrix $W_{PCA}$.

Together with the convenient assumption that the recovered source signals are centered, i.e. $\langle y \rangle \equiv 0$, we have the following ICA relation

$$y = W_{ICA} W_{PCA} (x - \langle x \rangle) = B (x - \langle x \rangle) = BM (s - \langle s \rangle) = DP (s - \langle s \rangle) \tag{1}$$

where **D** is an unknown diagonal matrix and **P** an also unknown permutation matrix.

In this model the number of independent sources is assumed to be equal to the number of image pixels. Nevertheless, we expect that for a good representation covering most of the input data some of the sources are less important than others. Thus we aim for an ordering criterion which prefers the essential source signals called *principal independent components* (PIC).

## 2 An event-oriented image model

Due to the intuitive notion of "importance" we propose that principal independent components should have a high occurrence probability. Therefore, we consider images to be composed of the superposition of many small, independent image primitives, just like a single neuron of the retina sees the world by a limited focus, which appear with a certain probability. As a further restriction, we assume that only one of two possible states is assigned to each primitive: present in the superposition or not. This leads to the formulation of *image events* $\omega_i$ (denoting the presence of primitive $i$) and $\neg\omega_i$ (denoting its absence). The task consists now of determining the most important events, i.e. those with highest probability $P(\omega_i)$.

Applied to eq. (1), the image primitives are represented by the columns of the mixing matrix **M**, and the source signals $s_i$ encode the associated image events by

$$s_i = \begin{cases} 1 & \text{for } \omega_i \quad \text{(primitive } i \text{ is present)} \\ 0 & \text{for } \neg\omega_i \quad \text{(primitive } i \text{ is not present)} \end{cases}$$

Thus, the average $\langle s_i \rangle \equiv \bar{s}_i$ of a source signal $s_i$ and its variance $\sigma_{is}^2$ are given by

$$\bar{s}_i \equiv \langle s_i \rangle = P(s_i=1) \cdot 1 + P(s_i=0) \cdot 0 = P(s_i=1) = P(\omega_i) \tag{2}$$

$$\sigma_{is}^2 = \langle s_i^2 \rangle - \bar{s}_i^2 = P(s_i=1) \cdot 1^2 + P(s_i=0) \cdot 0^2 - \bar{s}_i^2 = \bar{s}_i - \bar{s}_i^2 = \bar{s}_i (1 - \bar{s}_i) \tag{3}$$

Suppose that we have already computed the demixing matrix **B** in eq. (1). The recovered source signals $y_i$ are scaled and permuted versions of the centered original sources $s_i$. Because the permutation **P** is unknown (and, in fact, of no interest) we assume $P \equiv I$ and concentrate on the non-zero scaling factors $a_i$ satisfying

$$y_i = a_i (s_i - \bar{s}_i)$$ (4)

Since the recovered sources have zero mean and unit variance $\sigma_{iy}^2$ the following relation holds:

$$1 = \sigma_{iy}^2 = \langle y_i^2 \rangle = \langle (a_i (s_i - \bar{s}_i))^2 \rangle = a_i^2 (\langle s_i^2 \rangle - \bar{s}_i^2) = a_i^2 \sigma_{is}^2 = a_i^2 \bar{s}_i (1 - \bar{s}_i)$$ (5)

Now, if we ignore the centering terms in eq. (1), we can express the transformation of the source average $\langle s \rangle$ to the observed average $\langle x \rangle$ and to the recovered source average $\langle y \rangle$ by

$$\langle x \rangle = M \langle s \rangle \qquad \text{and} \qquad \langle y \rangle = B \langle x \rangle = BM \langle s \rangle$$ (6)

Note that here $\langle y \rangle$ is obviously non-zero unless for all $i$ the probabilities $P(\omega_i)$ are zero. With eqs. (4), (6) we have

$$\langle y_i \rangle = a_i \bar{s}_i$$ (7)

Combining eqs. (5), (7) gives the desired relation for the occurrence probabilities

$$1 = (\langle y_i \rangle / \bar{s}_i)^2 \bar{s}_i (1 - \bar{s}_i) \qquad \text{or} \qquad P(\omega_i) = \bar{s}_i = \langle y_i \rangle^2 / (1 + \langle y_i \rangle^2)$$ (8)

By this we obtained a measure to order the observed ICA components according to their decreasing occurrence probabilities, i.e. $i \geq j \Leftrightarrow P(\omega_i) \geq P(\omega_j)$.

Furthermore, if $P(\omega_i) \leq 0.5$ holds for all $i$, the components $y_i$ are ordered by their decreasing marginal entropy $H(y_i)$, because $H(y_i)$ is a convex function of the probability $P(\omega_i)$ and monotonically increasing up to its local maximum (located at $P(\omega_i) = 0.5$) [4].

## 3 Recovering the occurrence probabilities of events

To validate the theoretical results of the previous section, we computed a synthetic image according to the model in eq. (1). As image primitives we chose 16 pictures of 8×8 pixels visualising the letters 'A'...'P'. From these, 4096 different random linear mixtures were calculated and used as training samples. After prewhitening with the transform $W_{PCA}$ we presented the samples to a hierarchical ICA network similar to the one proposed in [3] with *tanh* non-linearities. The image primitives along with the eigenimages and the recovered primitives are shown in Figure 1a-c.

For the whitened PCA components we observed near-Gaussian distributions (Figure 1d) while the distributions of the ICA components are slightly "blurred" versions of the original occurrence probabilities, see Figure 1e.

The initial and the estimated occurrence probabilities of the first four sources are listed in Table 1 (the error is due to the imperfectly learned demixing matrix B). Also shown are their observed and their original marginal entropy (computed on 8 bit coefficients) compared to the marginal entropy of the first four whitened PCA components. Obviously, the single source information is reduced dramatically. Because of the "blurred" probability distributions, the marginal entropy of the recovered sources is still higher than the original entropy. However, by applying a rigorous quantization strategy we should be able to achieve further reduction [4].

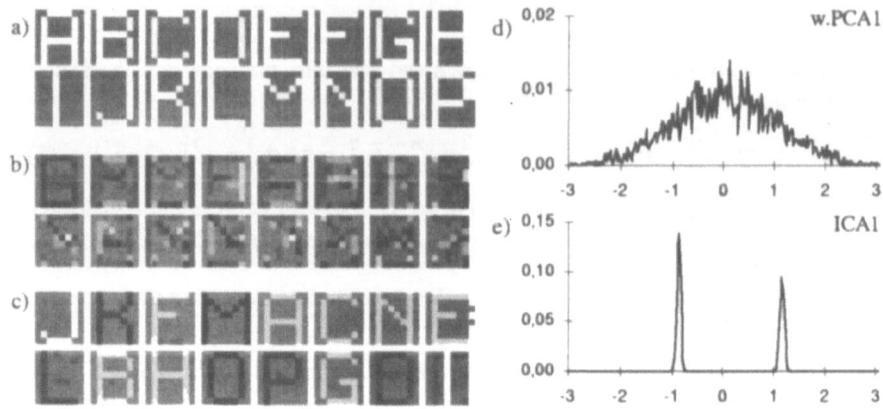

Figure 1: a) The image primitives, b) the eigenimages, and c) the recovered image primitives of the synthetic image. The probability distributions of the first whitened PCA component and of the first ICA component are shown in d) and e) respectively. To obtain the histograms the 4096 samples were quantified into 256 intervals on the horizontal axis.

| source | probability initial | probability estim. | error | compo-nent | observed entropy | compo-nent | observed entropy | original entropy |
|--------|---------|--------|--------|---------|----------|---------|----------|----------|
| 'J' | 0.444 | 0.463 | -0.019 | w.PCA1 | 7.398 | ICA1 'J' | 3.800 | 0.991 |
| 'K' | 0.415 | 0.322 | 0.092 | w.PCA2 | 7.408 | ICA2 'K' | 4.555 | 0.980 |
| 'F' | 0.696 | 0.732 | -0.036 | w.PCA3 | 7.322 | ICA3 'F' | 4.745 | 0.886 |
| 'M' | 0.624 | 0.618 | 0.006 | w.PCA4 | 7.405 | ICA4 'M' | 4.164 | 0.955 |

Table 1: Four of the source letters, their associated initial and estimated occurrence probabilities. Also shown are the observed and original marginal entropy of the four recovered sources and the first four whitened PCA components (in *bits*).

## 4   Independent components of natural images

Since the initial goal of our examinations is the efficient encoding of images with only a few important components we searched for the PIC of natural images. In our simulations a picture called *Cactus* was divided into 4543 subimages (size: 8×8=64 pixels) which were randomly chosen as training samples [4]. After centering and prewhitening of the samples we determined the matrix **B**. The corresponding image primitives were very similar to those already known in the literature, see e.g. [5].

Here, the measured probability distributions of the sources were not bimodal. This excluded the event model of section 2 for calculating the occurrence probabilities and therefore prevented an order of the sources by most probable image events. Instead we calculated the marginal entropy of the recovered sources as a new ordering criterion which is closely related to the probability ordering (see section 2).

We found that especially all the ICA components had nearly the same information; there were no components which differed much from the others. Furthermore, the marginal entropy of the ICA components was just slightly smaller than the one of the whitened PCA components.

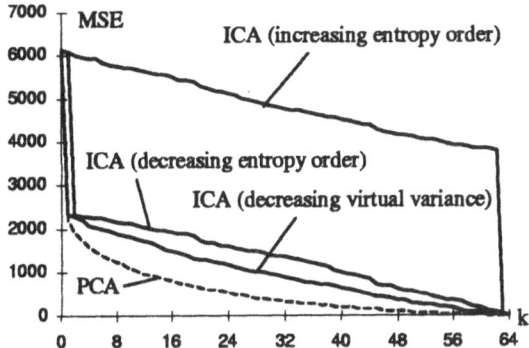

Figure 2: Decreasing the MSE by adding components.

Another measure for "importance" is the quality of the image restoration. Reconstructing the image by its first $k$ components and comparing it with the original one gives the average error for neglecting the $n-k$ components. Therefore we compared the optimal MSE (mean square error) contribution of the PCA components (ordered by decreasing variance) to those of the ICA components (ordered by increasing and decreasing entropy). For the latter we defined a third ordering criterion called the *virtual variance*

$$\text{var}^*(y_i) \equiv \text{var}\left(\frac{b_i}{\|b_i\|}\left(x - \langle x \rangle\right)\right) = \frac{\text{var}(y_i)}{\|b_i\|^2} = \frac{1}{\|b_i\|^2} \tag{9}$$

which considers the fact that the norm of a row $b_i$ of the matrix $B$ is in general not equal to unity. Consequently, an ICA component with higher virtual variance is assumed to be more important. Figure 2 shows the obtained error functions. In case of the ICA, ordering the components by their decreasing virtual variance gives the best results. However, our simulations showed that the subjective quality of image restoration by a few ICA components is not acceptable.

## 5 PIC and rate distortion theory

When the number of components in the transform approach for encoding images is reduced, the full space of image components (dimensions) is reduced to a subspace. The subspace of the ICA components is characterised by its information content whereas the subspace of the PCA components is characterised by its low MSE reconstruction error. Since the principal components of PCA cannot be replaced for obtaining a small MSE, their encoding information should be reduced by ICA. This idea can be performed in two ways:

1. Get the first $k$ PCA components with an acceptable MSE. Then, by an ICA transform, we will get the same number of encoding coefficients but with less information, i.e. less encoding bits.

2. For the same amount of encoding information as the $k$ PCA components take, we can also get $p$ more ICA transformed PCA components. Since these $k+p$ base vectors of the ICA transform span the same space as the $k+p$ PCA components, the resulting image quality will be enhanced as if $p$ more PCA components were added.

Thus the approach starting with the search for principal independent components leads to the error-bounded maximal information for each channel. This is classically known as the *rate distortion theory* [6] and has a broad range of applications in the telecommunication area.

The first one of the ideas above can be implemented if we order the $k$ ICA components according to their decreasing virtual variance and encode only the first $k' < k$ components with low additional reconstruction error. This results in a further reduction of the number of encoding bits. To validate the latter idea we computed the ICA components of the first $k$ PCA components for $k = 16,...,21$. We found that for the same information rate about one additional ICA component can be encoded with an error reduction of 5%.

Finally, we examined the influence of quantization on the MSE and the overall information rate. For $k=16$ and $k=20$ the resolution of the PCA components and their associated ICA components was set to 5, 6, 7 and 8 bit. Lowering the resolution down to 6 bit did not increase the MSE significantly whereas the information rate decreased by about 43% (!). Again, the information gain of the ICA over the PCA was about one additional component.

As a remarkable result we observed that for $k=20$ and a resolution of 6 bit both the resulting MSE and the encoding amount of the ICA components were superior to the corresponding representation with $k=16$ components quantified to 8 bit (MSE $\approx$ 13%, information rate $\approx$ 54%). A systematic investigation of this behaviour is subject to future research.

# References

[1] Comon P. Independent component analysis – a new concept?. Signal Processing 1994; 36: 287–314

[2] Amari S, Cichocki A, Yang HH. A new learning algorithm for blind signal separation. In: Touretzky D, Mozer M, Hasselmo M (ed) Advances in Neural Information Processing Systems 8, MIT Press, Cambridge MA, 1996, pp 757–763

[3] Hyvärinen A, Oja E. Independent component analysis by general non-linear Hebbian-like learning rules. Technical Report A41, Helsinki University of Technology, Laboratory of Computer and Information Science, 1996

[4] Arlt B, Brause R. The principal independent components of images. Internal Report 1/98, J.W.Goethe-Universität Frankfurt, Germany, 1998

[5] Olshausen BA, Field DJ. Natural image statistics and efficient coding. Network: Computation in Neural Systems 1996; 7: 333-339

[6] Shannon CE, Weaver W. The mathematical theory of information. University of Illinois Press, Urbana, 1949

# Pattern Formation in Locally Connected Oscillatory Networks

Margarita Kuzmina

Keldysh Institute of Applied Mathematics, RAS
Moscow, Russia
kuzmina@spp.keldysh.ru

Irina Surina

RRC Kurchatov Institute
Moscow, Russia
surina@isssph.kiae.ru

### Abstract

The subject of our study is a class of networks consisting of locally connected nonlinear oscillators. In spatially continual limit these oscillatory networks can be considered as oscillatory media governed by a system of reaction-diffusion equations. Formation of spatio-temporal patterns in nonlinear active media (wave trains, standing waves, targets and shock structures, spiral waves, stripe patterns, cluster states ) is the subject of interest in physical, chemical, biological problems.

Here the results of analytical study of 1D oscillatory media corresponding to closed and unclosed chains of limit-cycle oscillators are presented. Diffusion instability (caused by coupling) has been analysed. The analysis is reduced to the problem of existence of growing solutions of the second order ODE system for arbitrary spatial harmonics. The conditions of standing waves existence has been clarified as well.

## 1 Introduction

Our previous studies on oscillatory systems were devoted to the networks of limit-cycle oscillators coupled via arbitrary Hermitian matrix of connections. Associative memory networks of Hopfield type were designed and their main characteristics were analysed [1].

In the present paper we study similar oscillatory networks from another viewpoint: as a model of active oscillatory medium. Depending on a local coupling template (defining the neighbors of each network unit) these networks can be considered as 1D, 2D or nD spatially distributed arrays of large number of processing units. Similarly to locally connected neural networks known as cellular neural networks (CNN) [2, 3], locally connected oscillatory networks may be naturally regarded as cellular oscillatory networks.

Cellular oscillatory networks were already used and can be used further for modelling of a variety of phenomena in physics, chemistry, biology, neurophysiology. In particular, 1D and 2D networks of locally connected Wilson-Cowan oscillators were successfully used for modelling of brain cortical oscillations. A single oscillator of the model network is formed by a couple of connected excitatory and inhibitory neurons [10].

Pattern formation in discrete cellular networks is closely related to formation of dissipative structures in the corresponding nonlinear media. Various models of nonlinear media governed by the systems of reaction-diffusion equations were studied since 70s. [4 - 8]. Belousov-Zhabotinskii oscillating chemical reaction in a thin layer of fluid and oscillatory media of Ginzburg-Landau oscillators belong to the most familiar examples of active media. A considerable scope of oscillatory media studies exists. The strict mathematical results [5 - 7], physical level results [8, 9] and computer modelling [3, 8] could be mentioned as examples.

Here we present qualitative mathematical analysis of 1D media representing spatially continual limit for closed and unclosed chains of limit-cycle oscillators of Ginzburg-Landau type. The significant feature of the governing reaction-diffusion system is that the diffusion operator is not diagonal and in some parametrical range is not positively defined.

## 2 Homogeneously Connected Oscillatory Chains and Related 1D Oscillatory Media

We consider oscillatory network model consisting of limit-cycle oscillators possessing two degrees of freedom. Limit cycle of a single oscillator is the circle of unit radius in the plane. Dynamical equations governing the dynamics of the network of $N$ coupled oscillators are:

$$\dot{u}_j = (1 + i\omega_j - |u_j|^2)u_j + \sum_{k=1}^{N} W_{jk}(u_k - u_j), \quad j = 1, ..., N. \tag{1}$$

Here the variable $u_j(t) = R_j(t)\exp(i\theta_j(t))$ defines the state of $j$-th oscillator ($R_j$ and $\theta_j$ are the amplitude and the phase of oscillations, respectively), $\omega_j$ is its cycle frequency. The first term in the right-hand side of (1) defines the intrinsic dynamics of a free isolated oscillator, while the second one, responsible for interaction, is specified by the matrix of connections $W = [W_{jk}]$.

In the case of homogeneously locally connected oscillatory chains the matrix of interaction in (1) can be written as

$$W_{jk} = \begin{cases} d = \kappa e^{i\chi} = d_1 + id_2 & \text{if } k = j - 1, j + 1 \\ 0 & \text{if } k \neq j - 1, j + 1 \end{cases} \tag{2}$$

where $d = \kappa e^{i\chi}$ is the coupling strength in the chain. To transfer from dynamical system (1) to spatially continuous description, one should introduce a spatial variable $x \in [0, l] \subset R^1$ and a complex-valued function $u(x, t)$ instead of $u_j(t)$. Then the reaction-diffusion equation, representing spatially continual limit of dynamical description (1), can be easily derived:

$$u_t = (1 + i\omega(x) - |u|^2)u + d \cdot u_{xx}, \qquad (3)$$

where $u(x,t) = u_1(x,t) + iu_2(x,t)$ and $u_{xx} \equiv \partial^2 u/\partial x^2$ is 1D Laplacian $\Delta u$ in spatially 1D case. Below we consider oscillatory media with $\omega(x) = \omega = const, x \in [0, l]$.

The equation (3) can be rewritten in terms of real-valued two-component vector-function $\mathbf{u} = (u_1, u_2)^\top$:

$$\mathbf{u} = \hat{F}(\mathbf{u})\mathbf{u} + \hat{D}\mathbf{u}_{xx}, \qquad (4)$$

where

$$\hat{F}(\mathbf{u}) = \begin{bmatrix} 1 - u_1^2 - u_2^2 & -\omega \\ \omega & 1 - u_1^2 - u_2^2 \end{bmatrix} \qquad \hat{D} = \begin{bmatrix} d_1 & -d_2 \\ d_2 & d_1 \end{bmatrix}. \qquad (5)$$

Oscillatory media governed by reaction-diffusion equation (RDE) (3) represent a special case of Ginzburg-Landau oscillatory media [7, 8]. However, the diffusion operator is of more general type for RDE (3).

## 3 Diffusion Instability. Types of Spatio-Temporal Patterns

As one can easily obtain, RDE (3) possesses the following properties.

1. In the case $\omega(x) = \omega = const, x \in [0, l]$, the RDE (3) can be reduced to that one with $\omega = 0$ for the function $w(x,t) = u(x,t)e^{-i\omega t}$. So, if $\omega(x) = const$, it is sufficiently to analyse only the RDE with $\omega = 0$.

2. The function $u_0(x,t) = e^{i\theta_0}$ is the spatially homogeneous solution to RDE (3) at $\omega = 0$.

3. To analyse the properties of nonlinear RDE it is often quite helpful to use an expansion of its solutions into the series on orthonormalized system of eigenfunctions $\{X_m(x)\}$ of the corresponding linear scalar diffusion operator. For RDE (4) at $\omega = 0$ we put:

$$u_1(x,t) = \sum_{m=1}^{\infty} X_m(x)P_m(t), \quad u_2(x,t) = \sum_{m=1}^{\infty} X_m(x)Q_m(t). \qquad (6)$$

For medium corresponding to unclosed chain the boundary conditions for RDE are: $u_{1t}(0,t) = u_{2t}(l,t) = 0$. It gives $X_m(x) = \cos(\sigma_m x)$, $\sigma_m = \pi m/l$. In the case the following system of coupled ODE for $\{P_m(t), Q_m(t)\}$ can be derived:

$$\dot{P}_0 = P_0 - 1/2P_0R_0 + \sum_{m=1}^{\infty} P_m R_m \qquad (7)$$

$$\dot{Q}_0 = Q_0 - 1/2Q_0R_0 + \sum_{m=1}^{\infty} Q_m R_m \qquad (8)$$

$$\dot{P}_k = P_k - \sigma_k^2(d_1 P_k - d_2 Q_k) - 1/2 \sum_{k=1}^{m} P_{k-m} R_m - 1/2 \sum_{m=1}^{\infty} (P_{k+m} R_m + P_m R_{k+m})$$

$$(9)$$

$$\dot{Q}_k = Q_k - \sigma_k^2(d_2 P_k + d_1 Q_k) - 1/2 \sum_{k=1}^{m} Q_{k-m} R_m - 1/2 \sum_{m=1}^{\infty} (Q_{k+m} R_m + Q_m R_{k+m}),$$

$$(10)$$

where

$$R_0 = 1/2[P_0^2 + Q_0^2 + \sum_{m=1}^{\infty} (P_m^2 + Q_m^2)], \qquad (11)$$

$$R_m = 1/2 \sum_{l=1}^{m} (P_m P_{m-l} + Q_m Q_{m-l}) + \sum_{l=1}^{\infty} (P_m P_{m+l} + Q_m Q_{m+l}). \qquad (12)$$

The "moment" system (7)-(12) is in complete agreement with the analogous system derived in [7] for the case of oscillatory medium of Ginzburg - Landau oscillators with real-valued interaction.

Now the behavior of some types of RDE solutions can be discussed.

## 3.1 Diffusion instability of spatially homogeneous solution

Spatially homogeneous solution $u_0(x,t) = e^{i\theta_0}$ can lose the stability for some parameters of diffusion operator under some types of spatial structure of perturbations. This kind of instability inherent to nonlinear media is known as diffusion instability (because it is caused by the presence of diffusion). Elucidation of diffusion instability parametrical domain can be reduced to the analysis of RDE linearized around $u_0(x,t)$. Let us consider oscillatory medium corresponding to unclosed chain, put

$$\mathbf{u} = \mathbf{u}_0 + \tilde{\mathbf{u}}, \quad \mathbf{u}_0 = (1,0)^\top,$$

and use the expansion (6) for the solution $\tilde{u}$ of linearized RDE. Then we obtain the following second order ODE for $T_k(t) = (P_k(t), Q_k(t))^\top$, defining time behavior of $k$-th spatial harmonics:

$$\dot{T}_k = \hat{B}(\sigma_k)T_k, \quad \hat{B}(\sigma_k) = \begin{bmatrix} -(2 + d_1\sigma_k^2) & d_2\sigma_k^2 \\ -d_2\sigma_k^2 & d_1\sigma_k^2 \end{bmatrix} \qquad (13)$$

The eigenvalues of $\hat{B}(\sigma_k)$, that can be easily calculated in the explicit form, provide the information on diffusion instability with respect to perturbation of the spatially homogeneous state by $k$-th spatial harmonics. In particular, the following result can be obtained: the diffusion instability with respect to perturbation of arbitrary spatial structure occurs in parametrical range $\chi \in [3\pi/4, \pi]$ of angles $\chi$, defining oscillatory interaction accordingly to (2).

### 3.2 Wave trains

Plane wave trains are RDE solutions of the form $u(x, t) = U(z)$, where $z = \omega t - kx$. Strict results on small amplitude wave train solution existence were obtained in [5]. These wave trains arise as a result of bifurcation from a uniform spatially homogeneous state. One-parametrical family of wave trains was shown to exist in the case of a special class of RDE systems — so called $(\lambda - \omega)$-systems. The RDE (4) belongs to the class of $(\lambda - \omega)$-systems in the case of diagonal diffusion operator, i.e., at real-valued interaction.

### 3.3 Target patterns, spiral waves, shock structures

In the case of 2D oscillatory media the well known target patterns and rotating spiral waves exist. Strict analysis of these structures is based on the theory of "slowly varying waves" [6], which — locally in space and time — are close to plane wave trains. This study demands the deriving and analysis of dispersion relations. Impringing wave trains (analogous to converging target patterns) and shock structures that accompany target patterns were also studied in detail [6].

### 3.4 Standing Waves. Cluster States

Modulated standing waves are special RDE solutions with separated variables $x$ and $t$. In the case of oscillatory media related to unclosed chains these are the solutions of the form

$$u(x, t) = T_0 e^{-i\omega t} + T_k e^{-i(\omega t + \gamma)} \cos(kx) \tag{14}$$

The existence of standing waves for RDE (4) can be established either with the help of moment system (7)-(12) or by direct substitution of (14) into the RDE. In this way one can obtain four equations: two equations for $T_0^2$, $T_k^2$, the dispersion equation reflecting the relation between $\omega$ and $k$ and the algebraic equation for $\tan(\gamma)$. Analysis of the algebraic equation shows the existence of real-valued solutions for $\tan(\gamma)$. Therefore, standing wave solutions to RDE (4) exist. The parametrical domain of their existence still remains to be revealed.

Cluster states are RDE solutions with separated variables of another type: they correspond to medium decomposition into synchronously oscillating subdomains (clusters). The own amplitude, phase shift and frequency of oscillations are inherent to each cluster. Irregular oscillations of clusters are possible as well.

All the listed types of spatio-temporal patterns were confirmed experimentally in CO oxidation oscillating reaction on platinum crystal surface [9].

## Conclusive Remarks

The results of qualitative mathematical analysis of RDE governing the formation of spatio-temporal structures in 1D active oscillatory media are presented. The study of 1D media should be considered as an initial step of study of dissipative structures in 2D media. The ability of 2D nonlinear media to form a rich variety of spatio-temporal patterns seems to be promising from the viewpoint of modelling of 2D locally connected networks of visual cortex. To attain this objective the model of oscillatory network consisting of limit-cycle oscillators with modifiable cycle radius and center location, governed by natural generalization of dynamical system (1), can be proposed.

## Acknowledgment

This study was partially funded by Russian FFR, grant n. 96-01-00084.

## References

[1] Kuzmina M. G., Manykin E. A., Surina I. I.: Recurrent associative memory network of nonlinear coupled oscillators. In: Artificial Neural Networks (ICANN'97), Springer-Verlag, Berlin, 1997, pp.433-238 (Lecture notes in computer science, no.1327)

[2] Chua L.O., Roska T. Stability of class of nonreciprocal cellular neural networks. IEEE Trans. CS, 1990; 37:1520-1527

[3] Thiran P., Crounse K.R., Chua L.O., Hasler M. Pattern formation properties of autonomous cellular neural networks. IEEE Trans. CS, 1995; 42:757-774.

[4] Kuramoto Y., Tsuzuki T. On the formation of dissipative structures in reaction-diffusion systems. Progr. of Theor Phys., 1975; 54, no.3

[5] Howard L.N., Kopell N. Wave trains, shock fronts and transition layers in reaction-diffusion equations. SIAM-AMS Proc., 1974; 8:1-12.

[6] Howard L.N., Kopell N. Slowly varying waves and shock structures in reaction-diffusion equations. Studies in Appl.Math., 1977; 56:95-145.

[7] Ahromeeva T.S., Malinetskii G.G. New properties of nonlinear dissipative systems. KIAM Preprint, 1983, no.118 (in Russian)

[8] Falcke M., Engel H., Neufeld M. Cluster formation, standing waves and stripe patterns in oscillatory active media with local and global coupling. Phys. Rev. E, 1995; 52:763-771.

[9] Jakubith S., Rotermund H.H., Engel W., von Oertzen A., Ertl G. Spatiotemporal concentration patterns in a surface reaction: propagating and standing waves, rotating spirals, and turbulence. Phys. Rev. Lett., 1990; 65:3013-3016.

[10] Wang D.L., Terman D. Locally Excitatory Globally Inhibitory Oscillator Networks. Transactions on Neural Networks, 1995; 6:283-286

# LOCOCODE versus PCA and ICA

Sepp Hochreiter
Technische Universität München
80290 München, Germany

Jürgen Schmidhuber
IDSIA, Corso Elvezia 36
CH-6900-Lugano, Switzerland

## Abstract

We compare the performance of three unsupervised learning algorithms on visual patterns that are mixtures of few underlying sources: "Independent Component Analysis" (ICA), "Principal Component Analysis" (PCA), and our new method "*Low-complexity coding and decoding*" (LO-COCODE). ICA and PCA fail to separate the sources no matter whether their number is known or not. LOCOCODE, however, always separates them. It also codes with fewer bits per pixel than ICA and PCA.

## 1 Introduction

Recently several methods have been proposed for separating and extracting independent sources of given data: "Independent Component Analysis" (ICA, e.g. [3, 1, 2, 11]), methods enforcing sparse codes [4, 6, 12, 10], and "*low-complexity coding and decoding*" (LOCOCODE) [8, 9] based on *Flat Minimum Search* (FMS) [7]. Previous research already highlighted some of LOCOCODE's advantages [8]. Here we experimentally compare ICA, "Principal Component Analysis" (PCA), and LOCOCODE on visual data. Our criteria are: (1) Are the underlying statistical causes of the data discovered and separated? (2) What is the input reconstruction error? (3) How many bits per pixel are needed to code the input?

## 2 The compared methods

For PCA a standard MATLAB routine is used. ICA is realized by the JADE algorithm (Joint Approximate Diagonalization of Eigen-matrices, see [3]). JADE is based on whitening and subsequent joint diagonalization of 4th-order cumulant matrices. We used the MATLAB JADE version obtained via FTP from sig.enst.fr.

LOCOCODE is realized by training a 3-layer autoassociator (AA) by *Flat Minimum Search* (FMS) [7]. Each layer is fully connected to the next. The hidden layer represents the code. FMS is a general, gradient-based regularization method for finding low-complexity networks (that can be described with few bits of information and require low weight precision) with low, tolerable training error. Such nets tend to exhibit high generalization capability. During learning FMS automatically prunes weights and units, and minimizes output sensitivity with respect to remaining weights and units. See [7] for details. It has been shown that FMS-based LOCOCODE will result in *sparse* codes if inputs are describable by relatively few features (such as edges in images) [9].

# 3 Experiments

To measure the information conveyed by the various codes of the input data we train a standard backprop net on the training set used for code generation. Its inputs are the code components; its task is to reconstruct the original input. The average MSE on a test set is used to determine the reconstruction error.

Coding efficiency is measured by the average number of bits needed to code a test set input pixel. The code components are scaled to the interval $[0, 1]$ partitioned into $I$ discrete intervals — this results in $I$ possible discrete values reflecting an input noise assumption (large $I \rightarrow$ little noise). Assuming independence of the code components we estimate the probability of each discrete code value by Monte Carlo sampling on the training set. To obtain the bits per pixels (Shannon's optimal value) on the test set we divide the sum of the negative logarithms of all code component probabilities (averaged over the test set) by the number of input components.

## 3.1 Experiment 1: noisy independent bars

We use a standard benchmark task: the input is a $5 \times 5$ pixel grid with horizontal and vertical bars at random, independent positions (10 possible bar locations). Each bar is activated with probability $\frac{1}{5}$. The inputs are noisy: pixels of activated bars randomly vary in $[0.1, 0.5]$. Input units not affected by currently active bars adopt activation $-0.5$. Then Gaussian zero mean noise with variance 0.05 is added to each input. The task is to extract the statistically independent features (the bars), and is adapted from [5, 6] but even more difficult because vertical and horizontal bars may be mixed in the same input.

**Experimental conditions.** The LOCOCODE-trained AA has 25 input, 25 output, and 25 hidden units (HUs), although just 10 HUs are needed for optimal coding. Biased sigmoid output units are active in $[-1, 1]$, HUs are active in $[0, 1]$. Normal weights are initialized in $[-0.1, 0.1]$, bias weights with -1.0, the learning rate is 1.0. The net is trained on 500 randomly generated patterns for 5,000 epochs. $E_{tol} = 2.5$ (see [7]). The test set consists of 500 off-training set exemplars. For PCA and ICA, 1,000 training exemplars are used.

LOCOCODE **results:** see Figure 1 and Table 1. 15 of the 25 HUs are pruned away. LOCOCODE extracts an optimal (factorial) code which exactly mirrors the pattern generation process. It automatically finds the correct number of sources.

**PCA and ICA results:** see Figure 2 and Table 1. PCA codes and ICA-15 codes are unstructured and dense. For ICA-10 codes some sources are recognizable. They are not separated though: ICA and PCA fail to extract the true input causes and the optimal features. But at least PCA/ICA codes with 10

**input -> hidden**             **hidden -> output**

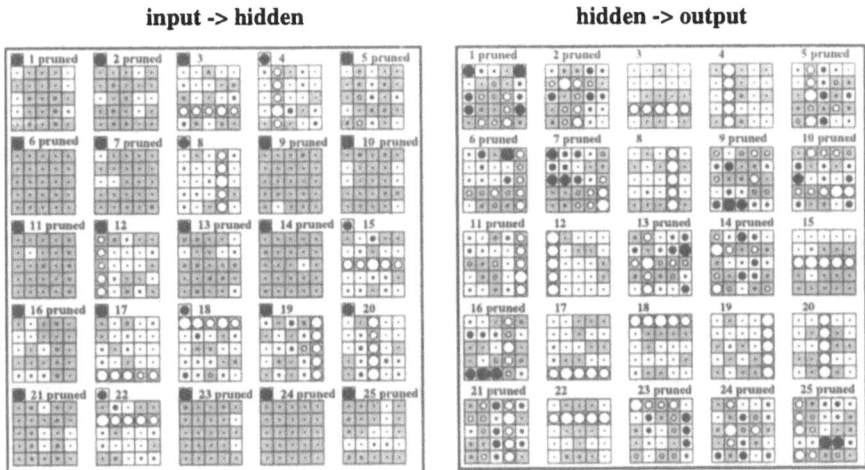

Figure 1: *Independent noisy bars. Left:* LOCOCODE *'s input-to-hidden weights. Right: hidden-to-output weights.*

components do convey as much information as 10-component codes found by LOCOCODE.

**PCA**            **ICA 10**

**ICA 15**

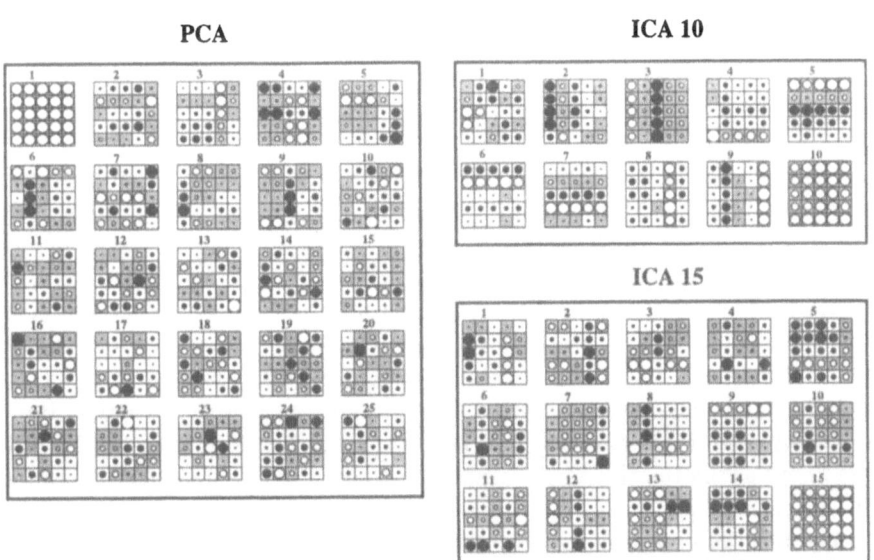

Figure 2: *Independent noisy bars. PCA and ICA: weights to code components (ICA with 10 and 15 components). Only ICA-10 codes reflect a few sources, but they do not achieve the quality of codes obtained through* LOCOCODE.

## 3.2 Experiment 2: village image

As in Experiment 1 the goal is to extract features from visual data, this time the aerial shot of a village. Figure 3 shows two images with $150 \times 150$ pixels, each taking on one of 256 gray levels. They are mostly dark except for certain white regions. $7 \times 7$ pixels subsections, corresponding to 49 inputs/outputs, from the left (right) image are randomly chosen as training (test) inputs, where gray levels are scaled to input activations in $[-0.5, 0.5]$. Targets are scaled to $[-0.7, 0.7]$.

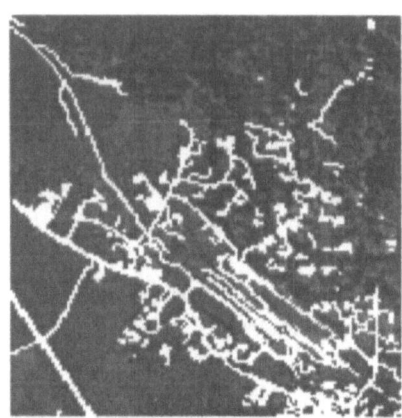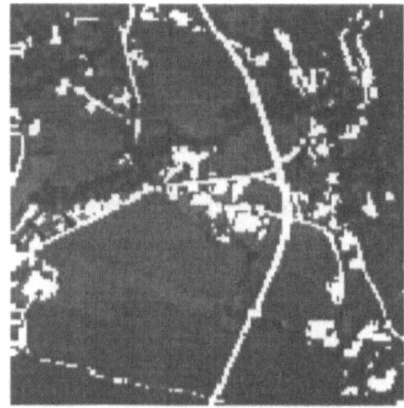

Figure 3: *Village image. Image sections used for training (left) and testing (right).*

**Experimental conditions.** Like in Experiment 1, except that training is stopped after 150,000 training examples, $E_{tol} = 3.0$. For PCA and ICA, 3,000 training exemplars are used.

LOCOCODE **results:** see Figure 4 and Table 1. 9 to 11 HUs survive the 6 trials. The entire input is covered by white on-centers of surviving units that exhibit on-center-off-surround weight structures. This allows for detecting all white regions in the input field. Since most bright spots are connected, output/input units near an active output/input unit tend to be active, too.

**PCA and ICA results:** see Table 1. PCA-10 codes and ICA-10 codes are about as informative as 10-component codes found by LOCOCODE. In fact, PCA's eigenvalues indicate that there are about 10 significant code components. LOCOCODE automatically discovers this.

# 4   Conclusion

LOCOCODE achieves success solely by reducing information-theoretic (de)coding costs. Unlike previous approaches it does not depend on explicit terms

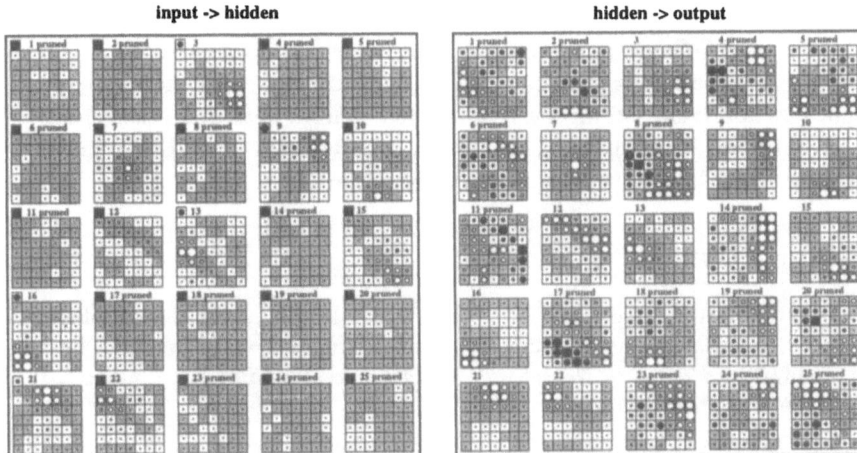

Figure 4: *Village. Left:* LOCOCODE*'s input-to-hidden weights. Right: hidden-to-output weights. Most units are essentially pruned away.*

| Exp. | input field | meth. | num. comp. | rec. error | code type | code efficency – reconst. 20 | 100 |
|---|---|---|---|---|---|---|---|
| bars | 5 × 5 | LOC | 10 | 1.05 | sparse | 0.84 - 1.15 | 1.37 - 1.06 |
| bars | 5 × 5 | ICA | 10 | 1.02 | sparse | 1.09 - 1.22 | 1.68 - 1.03 |
| bars | 5 × 5 | PCA | 10 | 1.03 | dense | 1.06 - 1.13 | 1.66 - 1.04 |
| bars· | 5 × 5 | ICA | 15 | 0.71 | dense | 1.60 - 1.11 | 2.50 - 0.73 |
| bars | 5 × 5 | PCA | 15 | 0.72 | dense | 1.58 - 0.82 | 2.47 - 0.72 |
| village | 7 × 7 | LOC | 10 | 8.29 | sparse | 0.37 - 8.52 | 0.69 - 8.29 |
| village | 7 × 7 | ICA | 10 | 7.90 | dense | 0.46 - 8.44 | 0.80 - 7.91 |
| village | 7 × 7 | PCA | 10 | 9.21 | dense | 0.46 - 9.60 | 0.80 - 9.22 |
| village | 7 × 7 | ICA | 15 | 6.57 | dense | 0.70 - 7.40 | 1.20 - 6.58 |
| village | 7 × 7 | PCA | 15 | 8.03 | dense | 0.69 - 8.43 | 1.19 - 8.04 |

Table 1: *Overview over experiments: name of experiment, input field size, coding method, code size, reconstruction error, nature of code observed on the test set. PCA's and ICA's code sizes are prewired.* LOCOCODE*'s, however, are found automatically. The final 2 columns show the coding efficiency measured in bits per pixels and the reconstruction error, for code components mapped to 20 and 100 discrete intervals.* LOCOCODE *exhibits superior coding efficiency.*

enforcing independence or zero mutual information among code components, or sparseness.

Codes obtained by ICA, PCA and LOCOCODE convey about the same information, as indicated by the reconstruction error. But LOCOCODE's coding efficiency is much higher: it needs fewer bits per input pixel.

PCA does not separate data sources in the noisy bars experiment. ICA sometimes does, to a limited extent. LOCOCODE always does. Unlike ICA it does not need to know in advance the number of independent sources — it simply prunes superfluous code components: LOCOCODE seems more appropriate than ICA for visual coding tasks where few sources determine the input.

**Acknowledgements.** This work was supported by *DFG grant SCHM 942/3-1* from "Deutsche Forschungsgemeinschaft".

# References

[1] S. Amari, A. Cichocki, and H.H. Yang. A new learning algorithm for blind signal separation. In David S. Touretzky, Michael C. Mozer, and Michael E. Hasselmo, editors, *Advances in Neural Information Processing Systems 8*, pages 757–763. The MIT Press, Cambridge, MA, 1996.

[2] A. J. Bell and T. J. Sejnowski. An information-maximization approach to blind separation and blind deconvolution. *Neural Computation*, 7(6):1129–1159, 1995.

[3] J.-F. Cardoso and A. Souloumiac. Blind beamforming for non Gaussian signals. *IEE Proceedings-F*, 140(6):362–370, 1993.

[4] P. Dayan and R. Zemel. Competition and multiple cause models. *Neural Computation*, 7:565–579, 1995.

[5] G. E. Hinton, P. Dayan, B. J. Frey, and R. M. Neal. The wake-sleep algorithm for unsupervised neural networks. *Science*, 268:1158–1161, 1995.

[6] G. E. Hinton and Z. Ghahramani. Generative models for discovering sparse distributed representations. Technical report, University of Toronto, Department of Computer Science, Toronto, Ontario, M5S 1A4, Canada, 1997. A modified version to appear in *Philosophical Transactions of the Royal Society* **B**.

[7] S. Hochreiter and J. Schmidhuber. Flat minima. *Neural Computation*, 9(1):1–42, 1997.

[8] S. Hochreiter and J. Schmidhuber. Unsupervised coding with Lococode. In W. Gerstner, A. Germond, M. Hasler, and J.-D. Nicoud, editors, *Proceedings of the International Conference on Artificial Neural Networks, Lausanne, Switzerland*, pages 655–660. Springer, 1997.

[9] S. Hochreiter and J. Schmidhuber. Feature extraction through LOCOCODE. Technical Report FKI-222-97 (revised version), Fakultät für Informatik, Technische Universität München, 1998. Submitted to *Neural Computation*.

[10] M. S. Lewicki and B. A. Olshausen. Inferring sparse, overcomplete image codes using an efficient coding framework. In M. I. Jordan, M. J. Kearns, and S. A. Solla, editors, *Advances in Neural Information Processing Systems 10*, 1998. To appear.

[11] L. Molgedey and H. G. Schuster. Separation of independent signals using time-delayed correlations. *Phys. Reviews Letters*, 72(23):3634–3637, 1994.

[12] B. A. Olshausen and D. J. Field. Emergence of simple-cell receptive field properties by learning a sparse code for natural images. *Nature*, 381(6583):607–609, 1996.

# TDSEP – an efficient algorithm for blind separation using time structure

Andreas Ziehe and Klaus-Robert Müller

GMD FIRST Rudower Chaussee 5, 12489 Berlin, Germany

e-mail:{ziehe, klaus}@first.gmd.de

### Abstract

An algorithm for blind source separation based on *several* time-delayed second order correlation matrices is proposed. The technique to construct the unmixing matrix employs first a whitening step and then an approximate simultaneous diagonalisation of several time-delayed second order correlation matrices. Its efficiency and stability are demonstrated for linear artificial mixtures with 17 sources.

## 1  Introduction

Blind source separation is an increasingly popular data analysis technique. It has been applied successfully to the so called cocktail party problem (e.g. [9, 3, 2, 5, 7, 12, 1]) and to various problems in biomedical data processing (e.g. [10, 13, 14]).

Usually it is assumed that the observed signals x are constituted of linearly mixed sources s, which are unknown, but mutually statistically independent.

$$x_i(t) = \sum_{j=1}^{n} a_{ij} s_j(t) \qquad 1 \le i, j \le n \quad \text{i.e.} \quad \mathbf{x} = \mathbf{A}\mathbf{s}. \tag{1}$$

Since neither s nor the mixing process $\mathbf{A}$ are known and we have to estimate the inverse $\mathbf{C}$ of the mixing matrix blindly $\mathbf{y}(t) = \mathbf{C}\mathbf{x}(t) = \mathbf{C}\mathbf{A}\mathbf{s}(t) = \mathbf{\Lambda}\mathbf{P}\mathbf{s}(t) \propto \mathbf{s}$, only driven by the known (mixed) measurements x. The unmixing is only possible up to scaling $\mathbf{\Lambda}$ and permutation $\mathbf{P}$. Ways to achieve this unmixing usually rely on minimizing a cost function that enforces statistical independence[1]. Mostly this involves an explicit or implicit calculation of higher order moments, which can be difficult and in the case of scarce data or outliers error prone and furthermore computationally expensive.

Since many natural signals, like speech signals or biomedical signals (EEG, MEG) have a significant time structure, it is obvious to use the time-delayed second order correlations for source separation [11, 3]. So in the following we will always assume a time structure (a pronounced autocorrelation) of the sources s. In this work we propose to use several (> 2) time delayed correlation

---

[1]In a way the statistical independence assumption is the price we have to pay for *blind* separation, i.e if knowledge about the sensor array would be taken into account we could relax this assumption.

matrices for an exploitation of the temporal signal structure and discuss the implications of the choice of the delay parameters (section 2). Furthermore we illustrate the efficiency of our approach for a large example of mixed acoustic and synthetic signals (section 3).

## 2 Methods

Along the lines of Molgedey and Schuster [11] we define the following cost function

$$l_1(C_{ij}) = \sum_{i \neq j} \langle y_i(t)y_j(t) \rangle^2 + \sum_{i \neq j} \langle y_i(t)y_j(t+\tau) \rangle^2, \tag{2}$$

where $\tau$ is a certain time lag and $\langle \ \rangle$ denotes a time average. Inspecting Eq.(2) closer, we see that a minimum is reached, if the equal time and time lagged cross-correlations vanish simultaneously, i.e. Eq.(2) enforces *decorrelation over time*. The parameter $\tau$ must be chosen very carefully, so that the two correlation matrices carry maximally different information. To circumvey this selection problem we propose to use *several* time lagged correlation matrices (TDSEP - Temporal Decorrelation source SEParation), i.e. to minimize the generalized cost function

$$l_2(C_{ij}) = \sum_{i \neq j} \langle y_i(t)y_j(t) \rangle^2 + \sum_{k=1}^{N} \sum_{i \neq j} \langle y_i(t)y_j(t+\tau_k) \rangle^2, \tag{3}$$

Depending on the character of the problem it could become important to have both: delayed second order correlations and higher order moments. We can once more generalize Eq.(3) to obtain a cost function which makes use of the time structure *and* the information carried by the higher order moments

$$l_3(C_{ij}) = \sum_{k=0}^{N} \sum_{i \neq j} \langle y_i(t)y_j(t+\tau_k) \rangle^2 + \sum_{k=0}^{N} \sum_{i \neq j} \langle f(y_i(t))g(y_j(t+\tau_k)) \rangle^2, \tag{4}$$

where $\tau_0 = 0$. Using only the second term and setting $N = 0$ we can e.g. retrieve the ICA learning rule of Jutten and Herault [9] with some appropriate non-linear functions $f, g$. On the other side, we could also fully depend on the second order time structure by using the first term in equation 4. So the amount of prior knowledge available for the very problem addressed would allow us to choose whether to rely more on higher order moments or whether to use temporal information since the signals carry temporal structure. Since we analyse speech, music or biomedical signals, time structure alone is sufficient for unmixing, so – due to space limitations in this contribution – we will study only the use of Eq.(3) in this work.

**Gradient descent** Eq.(2)-(4) need to be minimized with respect to $C$, for example with simple gradient descent, $\Delta C \propto -\eta \partial l / \partial C$ ($\eta$ is the learning rate), or other more sophisticated, e.g. 2nd order, minimization methods. The solution of (2)-(4) starts to become a quite tricky optimization problem, if the parameter space increases, i.e. if the number of sources is large. Clearly in (4), so far, we have to resort to gradient decent optimization, but for Eq.(2)-(3) it is possible to find a clever method to keep the computational burden manageable.

**Simultaneous Diagonalisation** We define a (time lagged) sample estimate of the correlation matrix as $\Sigma_{\tau(\mathbf{x})} = \langle \mathbf{x}(t)\mathbf{x}^T(t+\tau) \rangle$. Now we can write the optimization of the cost function as a linear algebra problem and to solve it via an eigenvalue decomposition of the correlation matrices. For *two* lagged correlation matrices $\Sigma_{(\mathbf{x})} = \langle \mathbf{x}\mathbf{x}^T \rangle$ and $\Sigma_{\tau(\mathbf{x})}$ the optimization problem can be solved via simultaneous diagonalisation [11] $(\Sigma_{\tau(\mathbf{x})}\Sigma_{(\mathbf{x})}^{-1})\mathbf{A} = \mathbf{A}\Lambda$. The quality of the solution, however, depends strongly on the very choice of $\tau$ (cf.Fig.1).

**Whitening and Jacobi Rotations** An alternative technique that can be applied for the joint diagonalisation of two (or more) matrices proceeds in two steps: (1) whitening and (2) several Jacobi rotations to achieve an approximate simultaneous diagonalisation of the matrix set. This method has the advantage to be numerically more stable than the explicit solution of the simultaneous diagonalisation problem above. In step 1 we find a linear transform $\mathbf{W}$, such that the first term in Eq.(3) is set to zero explicitly. This means we use prior problem knowledge to restrict the solution of the optimization problem and by that to speed convergence. The whitening transform $\mathbf{W}$ can be determined by a principal component analysis or by taking the inverse square root of the covariance matrix via an eigenvalue decomposition as follows [6]:

$$\mathbf{W} = \Sigma_{(\mathbf{x})}^{-\frac{1}{2}} = (\mathbf{V}\,\Lambda\,\mathbf{V}^T)^{-\frac{1}{2}} = \mathbf{V}\,\Lambda^{-\frac{1}{2}}\,\mathbf{V}^T.$$

This transform $\mathbf{W}$ gives a representation of the sensor signals $\mathbf{x}$ in a new basis. We call these transformed signals $\mathbf{z}$. After the pre-whitening step, any time delayed correlation matrix of the transformed signals $\mathbf{z}$ should be (approximately) a diagonal matrix up to a transformation $\mathbf{Q}$ [3]. By construction, matrix $\mathbf{Q}$ is an orthogonal matrix, i.e. $\mathbf{Q}\mathbf{Q}^T = \mathbf{I}$. For the special case of two matrices, the rotation matrix $\mathbf{Q}$ can be obtained by the eigenvalue decomposition [6] of the time delayed correlation matrix

$$\Sigma_{\tau(\mathbf{z})} \quad = \quad \langle \mathbf{z}(t)\,\mathbf{z}^T(t+\tau) \rangle = \mathbf{Q}^T\,\Sigma_{\tau(\mathbf{s})}\,\mathbf{Q} = \mathbf{Q}^T\Lambda_\tau\,\mathbf{Q}. \tag{5}$$

For more than two matrices a trick proposed by Cardoso, which is based on the method, that Jacobi [8] published in 1846, can be used. The basic idea is, that one can approximate the rotation matrix $\mathbf{Q}$ by a sequence of elementary rotations $T_k(\phi_k)$ each trying to minimize the off diagonal elements of the respective $\Sigma_{\tau(\mathbf{x})}$ matrices, where the rotation angle $\phi_k$ can be calculated in closed form (see Cardoso [4]). The final rotation is then obtained by $\mathbf{Q} = \prod_k T_k(\phi_k)$.

Concatenation of both transforms (whitening $\mathbf{W}$ and rotation $\mathbf{Q}$) yields an estimate of the mixing matrix:

$$\hat{\mathbf{A}} = \mathbf{W}^{-1}\mathbf{Q}.$$

Summarizing: (1) one can choose one delay $\tau$ with heuristics or prior knowledge or (2) one can resort to determine $\mathbf{Q}$ such that several time delayed correlation matrices are simultaneously approximately diagonalised.

**Averaging** As a further simplification we suggest to average over the set of delay matrices $\tau_k (k = 1, \ldots, m)$ :   $\Sigma_{\text{avg}} = \frac{1}{m} \sum_{k=1}^{m} (\Sigma_{\tau_k(\mathbf{z})})$ and use this averaged correlation matrix to compute one rotation $\mathbf{Q}$, instead of taking several Jacobi rotations $T_k(\phi_k)$ for every lagged correlation matrix. This idea will of course only give a crude approximation to the true minimization of Eq.(3), but it is very efficient in particular in high dimensional problems. This approximation becomes less crude if the assumption holds that the variance of the set of $\Sigma_{\tau_k(\mathbf{z})}$ matrices around the mean $\Sigma_{\text{avg}}$ is small.

# 3   Experimental results

To illustrate the theoretical reasoning, numerous experiments were made. In order to evaluate the separation quality of the unmixed signals $\hat{\mathbf{y}}(t) = \hat{\mathbf{A}}^{-1}\mathbf{x}(t) = (\hat{\mathbf{A}}^{-1}\mathbf{A})\mathbf{s}(t)$, first a suitable measure is needed. This shows, that the closer the matrix $P_{ij} = (\hat{\mathbf{A}}^{-1}\mathbf{A})_{ij}$ is to a permutation matrix, the better the source signals are reconstructed. The interference of source $i$ on channel $j$ is defined as $\frac{|p_{ij}|}{\max_k(|p_{ik}|)}$. If the true mixing matrix is known, one can define the performance index as averaged interference [3, 1]

$$SIR(P) \;=\; \frac{1}{n}\sum_{i=1}^{n}\left(\frac{1}{n-1}\sum_{j=1}^{n-1}\frac{|p_{ij}|}{\max_k(|p_{ik}|)} - 1\right). \tag{6}$$

We compiled three different test datasets, each mixed on the computer by a randomly choosen, quadratic mixing matrix. (I) filtered gaussian sources

$$
\begin{aligned}
s_1(t) &= s_1(t-1) + 0.8\,s_1(t-2) + 0.6\,s_1(t-3) + \nu_1(t)\\
s_2(t) &= s_2(t-1) + 0.7\,s_2(t-2) + 0.3\,s_2(t-3) + \nu_2(t)\\
s_3(t) &= s_3(t-1) + 0.9\,s_2(t-2) + 0.15\,s_3(t-3) + \nu_3(t)\\
s_4(t) &= \sin(2\pi\,3\,t)
\end{aligned}
$$

where   $\nu_i(t)$   is white gaussian noise $\sim N(0,1)$ ,

(II) real acoustic signals (speech, music)

$s_1(t)$   = speech: "Bon giorno signora!",   $s_2(t)$ = music: "piccolo flute",

(III) a hybrid mixture of 17 real acoustic signals and synthetic sources (super-, subgaussian and gaussian).

Table 1 shows the performance indices (large SIR corresponds to bad performance) for various settings of $\tau$. Clearly the performance can decrease drastically if the wrong delay is chosen, whereas a large number of delays always gives a stable (but not necessarily optimal) solution. So, in a practical experiment, where we have no knowledge, the use of many delays will always bring us to the safe side. It is interesting to note that the performance index can

| SIR | $\eta_{best}$ | $\tau_{worst}$ | $\tau_{1..10}$ | $\tau_{1..50}$ | $\tau_{avg10}$ | JADE v1.5 |
|---|---|---|---|---|---|---|
| gaussians + sinus | 0.0032 | 973.6634 | 0.0054 | 0.0045 | 0.0221 | 0.2508 |
| speech + music | 0.0049 | 1.7815 | 0.0064 | 0.0051 | 0.0050 | 0.0049 |
| 17 sources | 0.6284 | 11.4242 | 0.2799 | 0.2668 | 0.3035 | 1.1977 |
| M FLOPS | $\eta_{best}$ | $\tau_{worst}$ | $\tau_{1..10}$ | $\tau_{1..50}$ | $\tau_{avg10}$ | JADE v1.5 |
| gaussians + sinus | 0.96 | 0.96 | 6.10 | 25.37 | 5.52 | 5.32 |
| speech + music | 0.32 | 0.32 | 2.00 | 8.39 | 1.80 | 0.69 |
| 17 sources | 13.23 | 13.23 | 85.96 | 351.72 | 76.62 | 999.2 |

Table 1: Separation performance (SIR) and number of M FLOPS (in matlab implementation) using different strategies.

vary by more than two orders of magnitude and has a rather wiggly structure (cf. Fig.1)[2]. To explain this let us remember that as an assumption of the TDSEP algorithm we needed a pronounced time structure in the signals. In cases with bad performance indices the autocorrelation functions of the sources were too similar. This leads to ill-conditioned eigenvalue problems. As a baseline comparison we show the results for the JADE algorithm [3], which uses a simultaneous diagonalisation of 4th order cumulant martrices. TDSEP with several delays is working similarly good and in some cases even better than JADE (here the generic drawbacks of JADE with respect to gaussian signals are crucial). Comparing the number of floating point operations we observe that JADE performs well for small problems, but for large number of sources clearly TDSEP is favoured. The simple averaging approximation also yields astonishingly good results, if we remember its simplistic assumptions.

## 4 Conclusion

We proposed the TDSEP algorithm (and some simplifications) for blind source separation based on only time lagged second order correlations. TDSEP takes two steps: (1) whitening and (2) a rotation (or several elementary (Jacobi) rotations) to achieve an approximate simultaneous diagonalisation of the set of time lagged correlation matrices. For larger artificially mixed examples it was illustrated that the algorithm works well and computationally very efficient. This fact predisposes TDSEP for the analysis of biomedical signals, where we have as much as 49 channels and long recording times [14].

Future research will further study the simultaneous use of higher order and second order methods for source separation and the application of TDSEP to biomedical signals.[3]

**Acknowledgements** A.Z. was partly funded by DFG under contracts JA 379/51 and JA 379/71. We thank N.Murata for valuable discussions.

---

[2]Other data sets yield similar results.
[3]Further information can be obtained by http://candy.first.gmd.de.

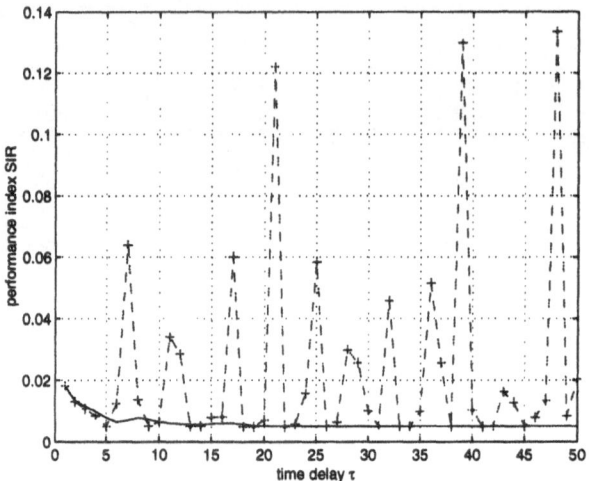

Figure 1: Performance index using one time delay $(0, \tau,$ dashed) vs. several time delays $(0, \dots, \tau,$ solid) for speech and music.

# References

[1] S. Amari, A. Cichocki, H. H. Yang. In *NIPS 95*, p. 882–893. 1996.

[2] A. J. Bell and T. J. Sejnowski. *Neural Computation*, 7(6):1129–1159, 1995.

[3] J.-F. Cardoso, A. Souloumiac. *IEEE Proceedings-F*, 140(6):362–370, 1994.

[4] J.-F. Cardoso, A. Souloumiac. *SIAM J.Mat.Anal.Appl.*, 17(1):161 ff, 1996.

[5] G. Deco and D. Obradovic. *Neural Computation*, 7(2):338–348, 1995.

[6] G.H. Golub and C.F. van Loan. *Matrix Computation*. The Johns Hopkins University Press, London, 1989.

[7] A. Hyvärinen and E. Oja. *Neural Computation*, 9(7):1483–1492, 1997.

[8] C.G.J. Jacobi. *Crelle J. reine angew. Mathematik*,vol. 30, p. 51–94, 1846.

[9] Ch. Jutten and J. Herault. *Signal Processing*, p. 1–10, 1991.

[10] S. Makeig, T-P. Jung, D. Ghahremani, A.J. Bell, T.J. Sejnowski. *PNAS*, (94):10979–10984, 1997.

[11] L. Molgedey and H.G. Schuster. *Phys. Rev. Letters*, p. 3634–3637, 1994.

[12] N. Murata, K.-R. Müller, A. Ziehe, S. Amari. In *NIPS 96*, p. 599–607. The MIT Press, 1997.

[13] R.N. Vigario. *EEG and clinical Neurophysiology*, (103):395–404, 1997.

[14] A. Ziehe, K.R. Müller, G. Nolte, B.-M. Mackert, G. Curio. Artifact removal in magneto-neurography with time delayed second order correlations. submitted. 1998.

# Optimal Cross-Validation Split Ratio: Experimental Investigation

Cyril Goutte and Jan Larsen

Department of Mathematical Modelling
Technical University of Denmark, Lyngby, Denmark
E-mail: `cg,jl@imm.dtu.dk`

### Abstract

Cross-validation is a widespread method for assessing the generalisation ability of a model in order to tune a regularisation parameter or other hyper-parameters of a learning process. The use of cross-validation requires to set yet an additional parameter, the *split ratio*. Few texts have investigated theoretically the asymptotic setting of this ratio, and no consensus has emerged. In this contribution, we investigate the sensitivity and optimal setting of the split ratio on a particular model, a non-parametric kernel estimator with adaptive metric.

## 1   Cross-validation

Most efficient learning procedures require the setting of an extra learning parameter, or "hyper-parameter". Neural networks typically use a regularisation parameter weighting a weight decay [1], or the extent of pruning [2]. Estimating the "optimal" hyper-parameter is the topic of active current research in the statistical learning community [3]. Let us consider a typical learning problem: modelling an input-output relationship based on some empirical data $\mathcal{D} = \left(x^{(i)}, y^{(i)}\right)_{i=1,\ldots N}$ sampled from an unknown, but fixed joint probability $p(x, y)$. The performance of a candidate model $f$ is measured by the generalisation error or expected risk:

$$G(f) = \int \ell(x, y, f) \, p(x, y) \, \mathrm{d}x \qquad (1)$$

where the risk $\ell(x, y, f)$ measures how "close" the model is to the data. For regression, the usual choice is the quadratic difference: $\ell(x, y, f) = (y - f(x))^2$. A principled approach to setting the hyper-parameter is to minimise the generalisation error, or an estimate thereof, as (1) can usually not be calculated due to the unknown input-output distribution. Apart from the algebraic estimators in line with Akaike's FPE [4], several data-intensive alternatives are commonly used. The *split sample* method sets aside $V$ of our $N$ data points to provide an empirical estimator of (1):

$$\widehat{G}_{SS} = \frac{1}{V} \sum_{i=1}^{V} \ell\left(x^{(i)}, y^{(i)}, f^{-V}\right) \qquad (2)$$

where for notational purpose (and without loss of generality) the first $V$ data are left out for validation. $f^{-V}$ is the model estimation based on the remaining data, $i = V + 1, \dots N$. Large values of the split ratio $\gamma = V/N$ leave few data for model estimation (eg neural network training) while small values lead to an unreliable generalisation estimator. *Cross-validation*[1] [5] attempts to increase this reliability by averaging over several splits. In *L-fold cross-validation*, $\mathcal{D}$ is split in $L$ disjoint subsets $(S_j)_{j=1,\dots L}$ of roughly equivalent size, $\bigcup_{j=1}^{N} S_j = \mathcal{D}$. The validation error (2) calculated on each $S_j$, using the remaining data for model estimation, is averaged over the subsets:

$$\widehat{G}_{CV} = \frac{1}{N} \sum_{j=1}^{L} \sum_{i \in S_j} \ell\left(x^{(i)}, y^{(i)}, f^{-j}\right) \tag{3}$$

where $f^{-j}$ is the model using $\bigcup_{i \neq j} S_i$. Equation (3) estimates the *average* generalisation error. Cross-validation supposedly enhances the reliability of the generalisation estimator at a cost of increased computational requirements. The split ratio becomes $\gamma = 1/L$, yet another parameter that potentially influences the performance of the resulting model. Little theoretical analysis has been devoted to its setting: [6] shows that for linear model selection, $\gamma$ should asymptotically tend to 1, a surprising result that indicates that the relative amount of data left for model estimation should tend to 0. A similar result was obtained by [7] in a different setting, while a recent communication [8] suggests that small split ratios should be favoured for the split sample method. Accordingly, we expect the optimal $\gamma$ to either approach 1 or 0 when $N$ increases. The experiments presented in section 3 address this important issue.

## 2   Adaptive metric kernel regression

Adaptive metric kernels are a natural extension to the standard non-parametric kernel estimator [9]. Estimation is performed on the basis of the available data:

$$\widehat{y} = f_{\mathbf{H}}(x) = \frac{\sum_{k=1}^{N} K\left(d_{\mathbf{H}}\left(x, x^{(k)}\right)\right) y^{(k)}}{\sum_{k=1}^{N} K\left(d_{\mathbf{H}}\left(x, x^{(k)}\right)\right)} \tag{4}$$

where $d_{\mathbf{H}}^2(u, v) = (u - v)^{\top} \mathbf{H}(u - v)$ defines the metric in the input space. For $P$-dimensional data, $\mathbf{H}$ is a $P \times P$ positive definite *smoothing matrix*. $\mathbf{H}$ can be parameterised by its Cholesky decomposition, or the more parsimonious diagonal parameterisation presented in [10] and adopted here: $\mathbf{H} = \mathrm{diag}(h_1^2, h_2^2, \dots h_P^2)$. The smoothing parameters $h_j$ are adapted by minimising the cross-validation error:

$$\widehat{G}_{CV} = \frac{1}{N} \sum_{j=1}^{L} \left[ \sum_{i \in S_j} \left( y^{(i)} - f_{\mathbf{H}}^{-j}\left(x^{(i)}\right) \right)^2 \right] \tag{5}$$

---

[1]The split sample method is sometimes referred to as "cross-validation" in the neural networks literature. This is inconsistent with the definition of [5] and we will here reserve the term exclusively to the averaging method (3).

where $\mathcal{D}$ is split into $L$ subsets $S_j$ and $f_{\mathbf{H}}^{-j}$ is the estimation excluding $S_j$:

$$f_{\mathbf{H}}^{-j}(x) = \frac{\sum_{k \notin S_j} K\left(d_{\mathbf{H}}\left(x, x^{(k)}\right)\right) y^{(k)}}{\sum_{k \notin S_j} K\left(d_{\mathbf{H}}\left(x, x^{(k)}\right)\right)} \tag{6}$$

and the derivatives of (5) with respect to smoothing parameter $h_p$ is [10]:

$$\frac{\partial \widehat{G}_{CV}}{\partial h_p} = -\frac{2}{N} \sum_{j=1}^{L} \sum_{i \in S_j} \varepsilon_{ii} \left( \frac{\sum_{k \notin S_j} \varepsilon_{ki} \frac{\partial K_{ik}}{\partial h_p}}{\sum_{k \notin S_j} K_{ik}} \right) \tag{7}$$

with $\varepsilon_{ki} = (y^{(k)} - f_{\mathbf{H}}^{-j}(x^{(i)}))$ and $K_{ik} = K(d_{\mathbf{H}}(\mathbf{x}^{(i)}, \mathbf{x}^{(k)}))$. The smoothing matrix is adapted by first order multi-dimensional minimisation of $\widehat{G}_{CV}$, using a conjugate gradient algorithm with approximate line search [11, appendix B].

# 3 Experiments

Let us first consider a system inspired from [12] and described by:

$$y = 10 \sin\left(\pi x_1 x_2\right) + 20 \left(x_3 - \frac{1}{2}\right)^2 + 10 x_4 + 5 x_5 + \epsilon \tag{8}$$

with iid noise $\epsilon \sim \mathcal{N}(0, 1)$. The input vector $x$ contains 10 uniformly distributed values, $x_1 \dots x_{10}$, $x_i \sim \mathcal{U}([0, 1])$, so that $x_6$ to $x_{10}$ are irrelevant. 10 dimensions is rather large in our context: in order to uniformly sample the input space in 5 evenly spaced locations on each dimensions (eg 0, 0.25, 0.5, 0.75 and 1), nearly $10^7$ data points are required. We have shown in previous experiments that the adaptive metric handles the irrelevant inputs efficiently [10].

We consider 4 increasing sample sizes: 100, 200, 500 and 1000 points. A large, $2^{13}$ points test set is generated independently from (8) in order to assess the generalisation ability. For each sample size $N$, we generate a number of datasets and investigate the following split ratios: $1/N$ (aka *leave-one-out* (LOO) cross-validation), 0.1 (the popular 10-fold cross-validation), 0.2, 0.3, 0.4 and 0.5. For each sample and each split ratio, the smoothing parameters are adapted by minimising (5). We estimate the resulting generalisation on the test set. Though only one value of $\gamma$ yields minimum generalisation, the differences might not be significant: this effect is assessed using a paired t-test on the squared residuals obtained from the test set. Accordingly, for each experiment, we generate 6 p-values assessing the significance of the difference in generalisation between each split ratio and the optimum.

Figure 1 summarises the results: for each sample size, we show the empirical distribution of the optimal split ratio. This suggests that as the number of available data grows, the optimal split ratio tends to decrease: the average optimal split ratio decreases from 27% for $N = 100$ to 11% for $N = 1000$. A $\chi^2$-test shows that the empirical distribution departs from uniform (at a $10^{-3}$ level) for $N > 100$. Another way of comparing the different split ratios is to

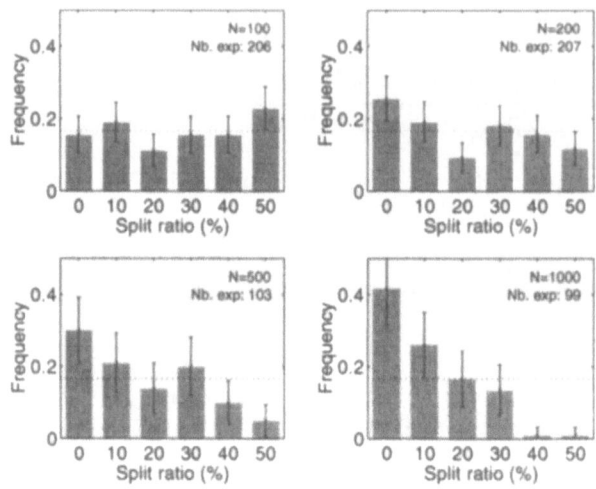

Figure 1: Empirical distribution of the generalisation-optimal split ratio for four sample sizes. Error bars indicate 2 standard deviations, the x-axis label '0' is the LOO case ($\gamma = 1/N$) and the dotted line corresponds to the uniform distribution. Read: for $N = 1000$, LOO was optimal in more than 40% of the experiments.

consider the *average* generalisation error (AGE). Figure 2 shows the percentage by which the AGE for each $\gamma$ exceeds the minimum AGE for this sample size. Eg for $N = 500$, $\gamma = 0.4$ yields on average 1% higher AGE than $\gamma = 1/N$. While figure 1 shows that optimal generalisation is achieved for a fair number of $\gamma > 0.3$ for law sample sizes, figure 2 suggests that small split ratios minimise the AGE in all cases.

Figure 4 sheds some light over this intriguing effect. In each sub-plot we consider the experiments for which the corresponding split ratio is either optimal, or non significantly worse at a 5% level according to the p-values mentioned above. The curve shows the excess generalisation for each split ratio compared to the optimal one (which obviously has a 0 excess). This plot was obtained for $N = 200$ but the other sample sizes display similar effects. This shows that when small split ratios are optimal (top row), choosing a large $\gamma$ is heavily penalised, while choosing a small ratio when a large one is optimal (bottom row) yields a limited excess in generalisation error. While large split ratios might lead to the smallest generalisation error in a number of cases, the will perform badly on average.

These experiments were carried out for a specific model and a specific (artificial) problem, raising the question of how general the results are. We could argue that as theoretical work suggests a clear asymptotic behaviour (towards 1 or 0), one experiment where $\gamma$ tends towards 0 is sufficient evidence against the alternative hypothesis. However, in order to show that this behaviour is not dependent on this particular artificial problem or experimental conditions (noise level, variable interaction), we have carried out the same set of experiments on 5 problems studied in [13]. Contrary to (8), which combines modelling and input selection, we will now tackle pure modelling problems (we refer the reader

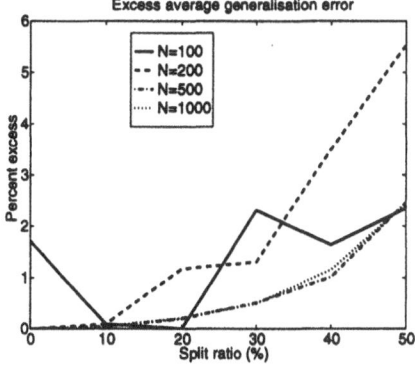
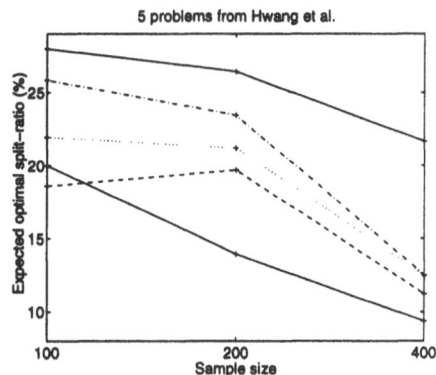

Figure 2: Excess (in percents) average generalisation error, for the 4 sample sizes. Label '0' is the LOO ($\gamma = 1/N$).

Figure 3: Evolution of the average optimal split-ratio with increasing sample size for the 5 problems of [13].

to [13] and [14] for a description). One aspect of the results is presented in figure 3, where we plot the expected optimal $\gamma$, ie the average of the empirical distribution of the generalisation optimal split ratio (cf figure 1) for our five problems (hence 5 curves). Figure 3 shows that on these problems again the optimal $\gamma$ tends to decrease with increasing sample sizes. This suggests that the pattern exhibited with the systematic study of our first problem is highly reproducible. Due to the limits imposed on this paper, we will reserve more detailed results and comments to the conference.

## 4   Conclusion

This contribution addresses the use of cross-validation for optimising various hyper-parameters in non-linear modelling. Optimal setting of hyper-parameters, eg regularisation parameters for neural networks or smoothing parameters in kernel regression, is crucial for obtaining high generalisation ability. Cross-validation involves an additional parameter, the split ratio or fraction of data used for validation. As theoretical studies indicate that the optimal split ratio is very dependent on the modelling scenario and other assumptions, we pursue an experimental investigation. We model a number of artificial problems of various difficulties using adaptive metric kernel regression. The results of these experiments suggest the following conclusions: 1) The optimal split ratio tends to zero as the amount of data grows; 2) Small split ratios seem to be a safe bet, even when the optimal ratio is fairly large; 3) The results depend on the measure of optimality, eg generalisation error or average thereof.

**Acknowledgements**   This work was supported by BIOMED II grant number BMH4-CT97-2775 and by the Danish Research Councils through the Computational Neural Network Center and Center for Neuroinformatics. JL furthermore acknowledges the Radio Parts Foundation for financial support.

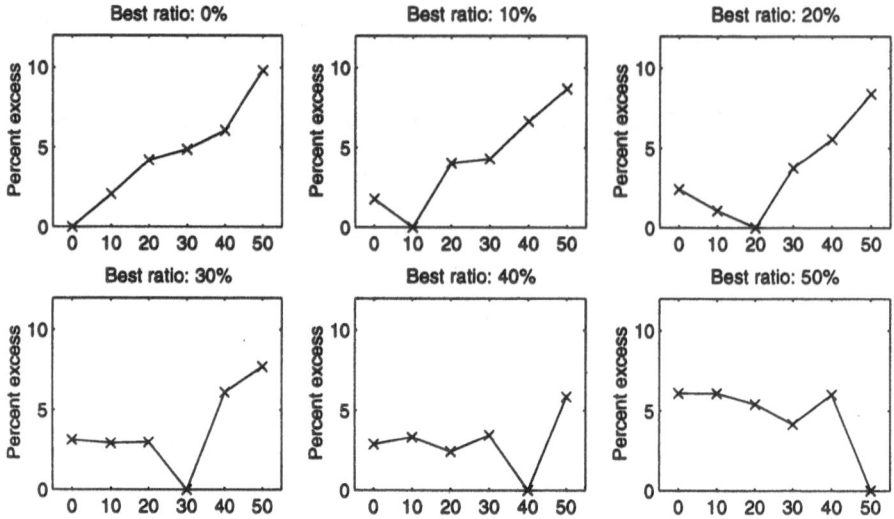

Figure 4: Average excess generalisation errors associated with each $\gamma$ compared to the optimal choice ($N = 200$). '0' stands for LOO. Read: when $\gamma = 30\%$ is optimal (bottom left), LOO yields 3% excess generalisation, compared to 8% for $\gamma = 50\%$.

# References

[1] Krogh A, Hertz JA, A simple weight decay can improve generalization. In: Moody et al. (eds), Advances in Neural Inf Proc Systems 4, 1992, pp 950–957

[2] Le Cun Y, Denker JS, Solla SA, Optimal brain damage. In: Touretzky (ed), Advances in Neural Information Processing Systems 2, 1990, pp 598–605

[3] Larsen J, Svarer C, Andersen LN, Hansen LK, Adaptive regularization in neural network modeling. Orr et al. (eds), The Book of Tricks, Springer-Verlag, 1998

[4] Akaike H, Fitting autoregressive models for prediction. Annals of the Institute of Statistical Mathematics, 1969, 21:243–247

[5] Stone M, Cross-validatory choice and assessment of statistical predictions. Journal of the Royal Statistical Society B, 1974, 36:111–147, with discussion

[6] Shao J, Linear model selection by cross-validation. Journal of the American Statistical Association, 1993, 88:486–494

[7] Larsen J, Hansen LK, Empirical generalization assessment of neural network models. In: Girosi et al. (eds), NNSP V – Proc IEEE Workshop, 1995, pp 42–51

[8] Kearns M, A bound on the error of cross validation using the approximation and estimation rates, with consequences for the training-test split. Neural Computation, 1997, 9:1143–1161

[9] Härdle W, Applied nonparametric regression. Cambridge University Press, 1990

[10] Goutte C, Larsen J, Adaptive metric kernel regression. In: NNSP VIII – Proceedings of the IEEE Workshop, 1998, in press

[11] Rasmussen CE, Evaluation of Gaussian processes and other methods for nonlinear regression. PhD thesis, Dept Computer Science, Univ of Toronto, 1996

[12] Friedman J, Multivariate adaptive regression splines. Ann Stat 1991, 19:1–141

[13] Hwang JN, Lay SR, Maechler M et al., Regression modelling in back-propagation and projection pursuit learning. IEEE Trans Neural Networks, 1994, 5:342–353

[14] Neal R, http://www.cs.utoronto.ca/~delve/data/hwang/hwangDetail.html

# On the Convergence Properties of the Temporal Kohonen Map and the Recurrent Self-Organizing Map

Markus Varsta, Jukka Heikkonen, and Jouko Lampinen
Helsinki University of Technology, Finland

Jose del R. Millán
European Commission, Joint Research Centre,
Institute for Systems Informatics and Safety, Italy

**Abstract**

This paper compares two Self-Organizing Map (SOM) based approaches for temporal sequence processing: The Recurrent Self-Organizing Map (RSOM) and Temporal Kohonen Map (TKM). The convergence properties of these algorithms are studied, and their difference in learning is emphasized both theoretically and with simulations. The results show that RSOM is superior over TKM.

## 1 Introduction

The Recurrent Self-Organizing Map (RSOM) [1, 2] is an unsupervised method for learning temporal dependencies in data presented to the map in sequences. The method was inspired by the Temporal Kohonen Map (TKM) [3]. In this paper the convergence properties of these two methods are compared and the results are then demonstrated with two simulations.

In the next section is brief overview of the SOM [4]. In section 3 two variants of the SOM, TKM and RSOM for temporal sequence processing are described. In the last section their learning rules are analyzed and the differences discovered are then related to practise with two simulations. The analysis shows how the optimal weights for both approaches are the same but in practise only RSOM can converge at the optimal weights.

## 2 The Self-Organizing Map

The Self-Organizing Map (SOM) [4] is lattice of competitive units that tries to follow the local orientation of the input manifold.

The training algorithm of the SOM is based on unsupervised learning, which can be either iterative or batch based. In the iterative approach a sample, input vector $x(n)$ at step $n$, from the input space $V_I$, is picked and compared against the weight vector $w_i$ of the unit with index $i$ in the map $V_M$. The best matching unit $bmu$ for the input pattern $x(n)$ is selected using some metric based criterion, such as

$$\|x(n) - w_b\| = \min_{i \in V_M} \|x(n) - w_i\|, \tag{1}$$

where the parallel vertical bars denote the Euclidean vector norm. The weights of the best matching and the units in its topologic neighborhood are updated toward $x(n)$ with rule

$$w_i(n+1) = w_i(n) + \gamma(n)h_{ib}(n)(x(n) - w_i(n)) , \qquad (2)$$

where $i \in V_M$ and $0 \leq \gamma(n) \leq 1$ is a scalar valued adaptation gain. The neighborhood function $h_{ib}(n)$ gives the excitation of unit $i$ when the best matching unit is $b$. A typical choice for $h_{ib}$ is a Gaussian function.

In batch training the gradient is computed for the entire input set and the map is updated toward the estimated optimum for the set. Unlike with the iterative training scheme the map can reach an equilibrium state where all units are exactly at the centroids of their regions of activity [4]. In practise batch training can be realized with a two step process. First, each input sample is assigned $bmu$ using Eq. 1. Second, the weights are updated with:

$$w_i = \sum_{j \in N_i} h_{ij}\Omega_j c_j / \sum_{j \in N_i} h_{ij}\Omega_j , \qquad (3)$$

where $c_j$ is the centroid of the set of input samples for which $j$ is the $bmu$ and $\Omega_j$ is the cardinality of the set, $N_i$ is the neighborhood of the unit $i$ and $h_{ij}$ is the value of the neighborhood function for $j \in N_i$. When using batch training usually few iterations over the training set are sufficient for convergence.

After a number of input vector presentations the mapping has formed, i.e., the weight vectors will specify the centroids of clusters covering the input space. The point density function of these centroids is related to the actual density in [5].

## 3   SOM:s for Temporal Sequence Processing

For temporal sequence processing Chappell and Taylor [3] proposed a modification to the original SOM. This modification, Temporal Kohonen Map (TKM), is not only capable of separating different input patterns but is also capable of giving context to patterns appearing in sequences.

Since with the normal SOM the outputs are computed separately for each input pattern the map is sensitive only to the last input pattern. In the TKM the outputs are replaced with leaky integrators (Eq. 4) which, once activated, gradually lose their activity when input moves away.

$$V_i(n) = dV_i(n-1) - (1/2)\|x(n) - w_i(n)\|^2 , \qquad (4)$$

where $0 \leq d < 1$ can be viewed as a time constant, $V_i(n)$ is the activation of the unit $i$ at time $n$, $w_i(n)$ is the weight vector of the unit $i$ and $x(n)$ is the input pattern. The general solution of Eq. 4 is

$$V_i(n) = -(1/2)\sum_{k=0}^{n-1} d^{(n-k)}\|x(k) - w_i(k)\|^2 + d^n V_i(0) , \qquad (5)$$

where the involvement of the earlier inputs is explicit. The map is trained in normal manner with Eq. 2, which updates the weights of the $bmu$ and its neighbors toward the last input $x(n)$.

### 3.1   Recurrent Self-Organizing Map

Moving the leaky integrators from the unit outputs into the inputs gives rise to the modified TKM called Recurrent Self-Organizing Map (RSOM) [1, 2].

With this modification the input can be modeled as

$$y_i(n) = (1 - \alpha)y_i(n - 1) + \alpha(x(n) - w_i(n)) , \qquad (6)$$

for the temporally leaked difference vector at each map unit. In Eq. 6, $0 < \alpha \leq 1$ is the leaking coefficient analogous to $d$ in the TKM, $y_i(n)$ is the leaked difference vector while $x(n)$ and $w_i(n)$ have their previous meanings. Since the feedback quantity in RSOM is a vector instead of a scalar it also captures the direction of the error and can be exploited in weight update when training the map. As a consequence the training rule can now take the form:

$$w_i(n + 1) = w_i(n) + \gamma(n)h_{ib}(n)y_i(n) . \qquad (7)$$

The $bmu$ is simply the unit with the shortest $y$

$$y_{bmu} = \min_{i \in V_M} ||y_i||. \qquad (8)$$

## 4 Comparison of TKM and RSOM

Brief mathematical analysis is sufficient to show that when seeking to maximize activity in Eq. 4, TKM should learn the same weights as the corresponding RSOM when we seek to minimize $||y||$. First consider TKM and Eq. 5. Taking derivative of Eq. 5 with respect to $w$ yields $\partial V/\partial w = -\sum_{k=0}^{n-1} d^{(n-k)}(x(k)-w)$. At the optimum $\partial V/\partial w = 0$, as it maximizes $V$, hence

$$w = \sum_{k=0}^{n-1} d^{(n-k)} x(k) / \sum_{k=0}^{n-1} d^{(n-k)}. \qquad (9)$$

Now we can repeat essentially the same analysis for RSOM. Consider Eq. 6, which for a sequence can be written $y(n) = \alpha \sum_{k=0}^{n-1}(1 - \alpha)^{(n-k)}(x(k) - w)$. The square of the norm of $y(n)$ is

$$||y(n)||^2 = \alpha^2 (\sum_{k=0}^{n-1}(1 - \alpha)^{(n-k)}(x(k) - w))^T (\sum_{k=0}^{n-1}(1 - \alpha)^{(n-k)}(x(k) - w)) . \qquad (10)$$

The update rule seeks to minimize $||y(n)||$, thus the derivative $\partial ||y(n)||^2/\partial w$ is 0 when $w$ is optimal, hence

$$w = \sum_{k=0}^{n-1}(1 - \alpha)^{(n-k)} x(k) / \sum_{k=0}^{n-1}(1 - \alpha)^{(n-k)}. \qquad (11)$$

From Eq. 11 and Eq. 9 one observes how the optimal weights are linear combinations of the inputs in the sequences and the results obtained for both models are essentially the same. As a consequence TKM and RSOM can only distinguish sequences with different responses at the leaky integrators of the map. This restriction cannot, however, be overcome with a single layer map like TKM or RSOM.

### 4.1 Learning algorithms

The difference between TKM and RSOM is the learning algorithm. The learning rule of the RSOM is gradient descent to minimize $||y||$, thus the map explicitly seeks to learn the weights suggested by Eq. 11. With TKM the situation, however, is more complicated. To simplify the following analysis, the neighborhood is non zero for the $bmu$ only. This compromise does not have major impact on the results.

Consider a sequence $s$ of discrete samples and denote the last sample of $s$ $x_s$. In the TKM, when we seek to to maximize the activity of the $bmu$ for $s$, the optimal weights can be computed with Eq. 9. Examining Eq. 5 reveals that the equation defines a parabola, thus the $bmu$ for $s$ is the unit closest to $w_s$, $w_b = \min_{i \in V_M} ||w_i - w_s||$, where $V_M$ is the map. After choosing the $bmu$, it is updated toward the last sample of $s$, $x_s$.

In steady state further training causes no changes in weights. Though in the iterative training scheme reaching steady state is not possible, criteria for a steady state can be defined and their impact considered. Consider a set of discrete sequences $S$ and a map $V_M$. The update rule seeks to minimize the sum of squares distance from the last samples $x_s$:s of the sequences $s \in S$ to the corresponding $bmus$. As a consequence in steady state

$$\sum_{s \in S_i} ||w_i - x_s||^2 = \min_{l \in V_M} \sum_{s \in S_i} ||w_l - x_s||^2, \tag{12}$$

where $w_i$ are the weights of the unit $i$, $S_i \subset S$ is the subset of sequences for which $i$ is the $bmu$ and $x_s$:s are the last samples of the sequences $s \in S_i$. The optimal weights with respect to Eq. 12 are the centroids of the last samples of subsets $S_i \subset S, \forall i \in V_M$, formally:

$$w_i = (1/\Omega_{S_i}) \sum_{s \in S_i} x_s , \tag{13}$$

which essentially is Eq. 3 without neighborhood. These weights are a necessary condition for a steady state because the gradient of Eq. 12 is zero only when Eq. 13 is satisfied.

On the other hand the weights $w_i$ also have to satisfy a maximum activity criterion to be steady state weights, otherwise some unit $l \neq i$ would be updated toward Eq. 13 and thus the map would not be steady. This criterion, in the case of the TKM, can be formalized:

$$\sum_{s \in S_i} ||w_i - w_s|| = \min_{l \in V_M} \sum_{s \in S_i} ||w_l - w_s||. \tag{14}$$

The weights optimally satisfying Eq. 14 are

$$w_i = (1/\Omega_{S_i}) \sum_{s \in S_i} w_s , \tag{15}$$

in manner analogous to Eq. 13.

The necessary conditions for steady state are stated by Eq. 13, which fixes the weights, and Eq. 14, which fixes the $bmus$. In practise these conditions are very difficult—if not impossible—to satisfy, which is demonstrated in the following example.

Let us consider a 49 unit map initialized with near optimal weights with respect to Eq. 14 for a discrete 1D input space of seven inputs: $4, 8, 12, 16, 20, 24$ and $28$ with minute additive Gaussian distributed noise $N(0,0.065)$. We set $d = 0.1429$, which was chosen for even the distribution of the weights over the input space. All random sequences of the seven inputs were considered equally likely by training the map by picking one of the inputs and updating the weights with the iterative training scheme at each step. Fig. 1 shows progress of a sample run for the TKM. Notice, how soon after the training starts the units are drawn toward the extremes of the map leaving only a few units to cover the entire input space. This phenomena is a direct consequence

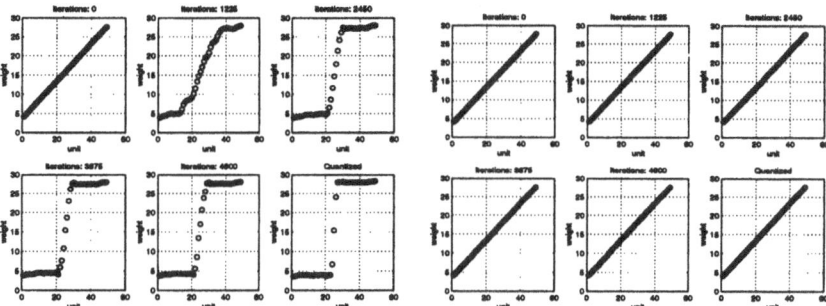

Figure 1: A map initialized with near optimal weights and trained with the TKM approach. Notice how most of the units are drawn into the edges.

Figure 2: A map initialized with near optimal weights and trained with the RSOM approach. Hardly any change can be observed in the distribution of the weights.

of the discrepancy in the derivative of the energy function (Eq. 4) which guides the *bmu* selection, and the update rule in Eq. 1. In Fig. 2 the RSOM yields unchanged result as the update rule explicitly seeks to satisfy and to improve Eq. 8.

The reason why the TKM draws the units toward the edges of the input space is intuitively explained in Fig. 3, where 49 2D inputs are in a regular grid. Random sequences of these inputs are considered. The arrows drawn at each input depict the mean difference between weights suggested by Eq. 13 and Eq. 15 for sequences ending with the input. Notice how the arrows form a gradient field of a smooth bump with the peak in the middle, hence when training the map with the TKM rule the units tend to slide on this surface toward the edges of the input space, and are eventually captured by corner inputs. This 2D case is demonstrated in Figs. 4 and 5, with the results of training maps with the TKM and the RSOM approaches using the same input vectors shown in Fig. 6. The leaking coefficients were 0.85 for $\alpha$ and 0.15 for $d$.

## 5 Conclusions

In this paper the RSOM approach was compared against the TKM which inspired RSOM. The reasons to different learning results were explored and demonstrated with examples. In [3] the question of existence of a learning algorithm to learn optimal weights for the TKM was raised. The simple solution in the form of the RSOM partially answers this question but possibly at the cost of biological plausibility which motivated the original TKM. Another aspect of both TKM and RSOM raised in the paper is that they are at best only capable of differentiating sequences with different responses to the leaking operators in their units. In fact ,in general, their ability to distinguish inputs is no better than that of a normal map of equal size. In some cases, however, the compromise in resolution by leaking activity is beneficial.

692

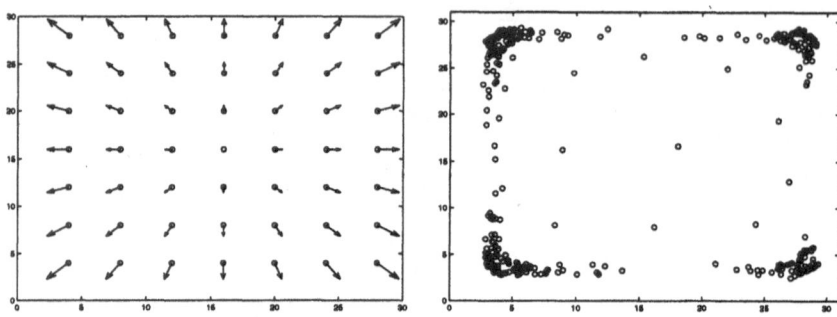

Figure 3: The mean differences of the optimal weights suggested by Eq. 13 and Eq. 15.

Figure 4: A 20 × 20 map trained with the TKM approach in 2D input space. Weight vectors in the input space are drawn with small circles (o).

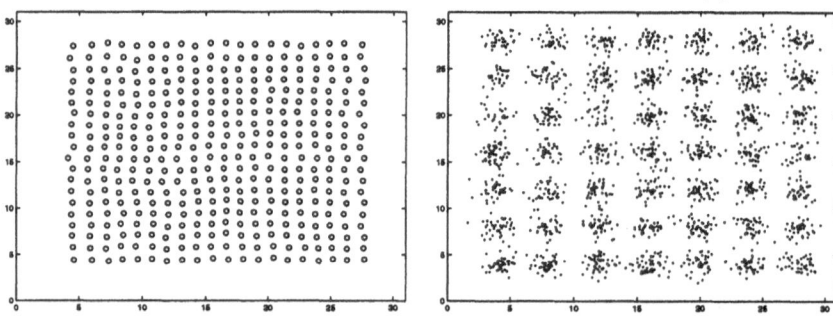

Figure 5: A 20 × 20 map trained with the RSOM approach in 2D input space. Weight vectors in the input space are drawn with small circles (o).

Figure 6: The training set of 49 different 2D inputs with additional Gaussian distributed ($N(0, 0.5)$) noise used to train the maps.

# References

[1] M. Varsta, J. Heikkonen, and J. del Ruiz Millán. A recurrent self-organizing map for temporal sequence processing. In *Proceedings of the ICANN'97*. Springer-Verlag Berlin Heidelberg New York, October 1997. ISBN 3-540-63631-5.

[2] M. Varsta, J. Heikkonen, and J. del R. Millán. Context learning with the self organizing map. In *Proceedings of WSOM'97, Workshop on Self-Organizing Maps, Espoo, Finland, June 4-6*, pages 197–202. Helsinki University of Technology, Neural Networks Research Centre, Espoo, Finland, 1997. ISBN 951-22-3589-7.

[3] G. J. Chappell and J. G. Taylor. The temporal Kohonen map. *Neural Networks*, 6:441–445, 1993.

[4] T. Kohonen. *Self-Organizing Maps*, volume 30 of *Lecture Notes in Inform. Sciences*. Springer, second edition, 1997.

[5] M. Cottrell. Theoretical aspects of the som algorithm. In *Proceedings of WSOM'97, Workshop on Self-Organizing Maps, Espoo, Finland, June 4-6*, pages 246—267. Helsinki University of Technology, Neural Networks Research Centre, Espoo, Finland, 1997.

# Multivariate Linear Regression on Classifier Outputs: a Capacity Study

Yann Guermeur and Hélène Paugam-Moisy

LIP, URA CNRS 1398, ENS Lyon

46 allée d'Italie, F-69364 Lyon Cedex 07, France

Helene.Paugam-Moisy,Yann.Guermeur@ens-lyon.fr

Patrick Gallinari

LIP6, UMR CNRS 7606, Université Paris 6

4, Place Jussieu, F-75252 Paris Cedex 05, France

Patrick.Gallinari@lip6.fr

### Abstract

We consider the problem of combining the outputs of several classifiers trained independently to perform a discrimination task, in order to improve the prediction accuracy of individual classifiers. We briefly describe the multivariate linear regression model which has already been implemented successfully for that purpose and we study its capacity, using generalizations of the notion of VC dimension.

## 1 Introduction

The idea of combining predictors instead of performing model selection has been intensively investigated in statistics during the last three decades [1]. These studies have given birth to various *ensemble methods*, characacterized either by the nature of the *combiner* [2] or the training strategy [3, 4, 5]. However theoretical evidence has been mainly developed for regression, whereas the specificities of discrimination have seldom been taken into account. In [6], we established how the Multivariate Linear Regression (MLR) model could be implemented in order to improve the class posterior probability estimates provided by a set of experts. It was observed that this combiner compared favourably with the methods currently used on a central open problem in predictive structural biology (see also [7]). This is because it represents a good compromise between capacity and robustness. Too simple combiners such as *linear opinion pools* [8] can fail to make the best of the data available whereas the complexity of neural network combiners may actually represent a drawback. In this article, we investigate further this problem by characterizing the capacity of our combiner using generalized definitions of the Vapnik-Chervonenkis (VC) dimension. Section 2 briefly outlines the implementation of the MLR model for class posterior probability estimates combination. Section 3 is devoted to the definition of the VC dimension and its extensions for multiclass discrimination. Section 4 deals with the influence of the nature of the predictors and the constraints on capacity. In Section 5, bounds on the MLR combiner capacity are derived for a standard extension of the notion of VC dimension. These bounds are tight enough to be compatible with the implementation of the Structural Risk Minimization (SRM) inductive principle [9].

## 2 Model implementation

Let us consider a $Q$-category discrimination problem and $P$ classifiers. Let $f_j$ denote the function computed by the $j^{th}$ classifier: $f_j(x) = [f_{jk}(x)] \in IR^Q$. The $k^{th}$ output $f_{jk}(x)$ approximates $p(C_k|x)$. Precisely, $f_j(x) \in U$ with

$$U = \left\{ u \in IR_+^Q / 1_Q^T u = 1 \right\} \tag{1}$$

$1_Q^T$ is the transpose of a column vector of $Q$ ones, and $1_Q^T u = 1$ thus defines a hyperplane of $IR^Q$. In other words, the outputs are non-negative and sum to 1. Under these hypotheses, the optimization problem corresponding to combining the experts by multivariate linear regression can be stated as follows:

**Problem 1** *Given a convex objective function $J$, find a function $g$ from $U^P$ to $U$, $g(x) = [g_k(x)]$ with $g_k(x) = F(x)^T v_k$, $(1 \leq k \leq Q)$ and $F(x) = [f_j(x)]$, which minimizes $J$.*

$v_{k,l,m}$, the general term of $v_k$, is the coefficient associated with the predictor $f_{lm}(x)$ in the regression computed to estimate $p(C_k|x)$. Let $v$ be the vector of all the parameters ($v = [v_k] \in IR^{PQ^2}$). We established in [6] that optimal solutions to Problem 1 were obtained by minimizing $J$ subject to $v \in V$ with

$$V = \left\{ v \in IR_+^{PQ^2} / \left\{ \begin{array}{l} \forall (l, m), \sum_{k=1}^{k=Q} (v_{k,l,m} - v_{k,l,Q}) = 0 \\ 1_{PQ^2}^T v = Q \end{array} \right. \right\} \tag{2}$$

This proposition holds irrespective of the nature of the convex objective function $J$. The training procedure thus consists in solving a simple convex programming problem. This can be done using standard algorithms such as the *gradient projection method* [10] (see [7, 11] for details).

## 3 VC dimension and extensions for multiclass

The VC dimension is a characteristic of the capacity of a set of functions.

**Definition 1** *The VC dimension of a set $H$ of indicator functions (values in $\{0,1\}$) is the maximum number $d$ of vectors that can be shattered, i.e. separated into two classes in all $2^d$ possible ways, using functions of $H$. If this maximum does not exist, the VC dimension is equal to infinity.*

This notion is of central importance in statistical learning theory. Vapnik has established [9] that, with high probability, the difference between the *risk* (i.e. the generalization performance) and the *empirical risk* (i.e. the training performance) is bounded above by an increasing function of the VC dimension. Loosely speaking, this implies that between two classifiers with the same training performance, it is safer to choose the one with the smallest VC dimension.

Definition 1 only applies to families of two-category discriminant functions. Several extensions have been proposed to deal with sets of real-valued or $\{0, \ldots, n\}$-valued functions (see [12] for a survey). Some of the latter can be used for families of $Q$-category discriminant functions. In the following sections, we consider two of them to assess the classifier obtained by applying Bayes' estimated decision rule on the outputs of the MLR combiner.

# 4 Incidence of the constraints on capacity

The first result holds for any extension of the notion of VC dimension.

**Theorem 1** *The VC dimension of the MLR combiner is the same whether the vector of parameters $v$ is constrained to belong to the convex set $V$ or not.*

**Lemma 1** *Each discriminant function which can be computed by the MLR combiner without restriction on the vector of parameters can still be computed by the MLR combiner under the constraint $v \in V$. This result remains true if the vectors of predictors $F(x)$ do not belong to $U^P$.*

The proof is straightforward. Let $w$ be any vector of $I\!R^{PQ^2}$. This vector characterizes a function $h$ corresponding to a particular discriminant function on the vectors of $I\!R^{PQ}$. Let $v$ denote the vector of $I\!R^{PQ^2}$ defined by:

$$\forall k \in \{1, \dots, Q\}, \; v_k = \epsilon \left( w_k - \frac{1}{Q} \sum_{n=1}^{n=Q} w_n \right) + (PQ)^{-1} 1_{PQ} \tag{3}$$

It is obvious that $v \in V$, provided $\epsilon > 0$ is small enough. Furthermore, we have $\forall (k, x)$, $g_k(x) = \epsilon h_k(x) + K(x)$, where function $K$ is independent of the index $k$ of category. Consequently, functions $g$ and $h$ correspond to the same discriminant function on $I\!R^{PQ}$ □.

Contrary to the constraints on $v$, the hypothesis on the inputs, namely that they live on $U^P$, has an incidence on the model capacity. To highlight this phenomenon, we need to choose a particular extension of the notion of shattered sample. For the sake of simplicity, we consider here the following naïve one:

**Definition 2** *A set $H$ of $Q$ − category discriminant functions shatters a set of $n$ vectors if and only if the vectors can be separated into $Q$ classes in all $Q^n$ possible ways using functions of $H$.*

With this definition, the following proposition holds, which highlights the influence of the hypothesis made on the inputs:

**Theorem 2** *The size of the largest sample of vectors of $I\!R^{PQ}$ that the MLR combiner can shatter according to Definition 2 is $PQ$. If the vectors are further required to belong to $U^P$, then this size decreases to $P(Q-1)+1$.*

Full proof of Theorem 2 is given in [11]. Here, we focus on the case where the vectors belong to $U^P$. Since the outputs of each classifier sum to unity, the dimension $d_E$ of the smallest subspace $E$ of $I\!R^{PQ}$ containing $U^P$ is $P(Q-1)+1$. Thus, any set of $P(Q-1)+2$ vectors in $U^P$ can be linearly combined with coefficients not all equal to 0 in order to generate the null vector. Furthermore, and this point is a specificity of our problem, the sum of the coefficients is necessarily 0. Assigning all the vectors with non-negative coefficients to class $C_1$ and the others to $C_2$, we exhibit a classification leading to a contradiction (non-empty intersection between the convex hulls of $C_1$ and $C_2$). This establishes that the size of a set shattered is at most $d_E$. Let $B_E$ be any basis of $E$. Its vectors are linearly independent. Consequently, for any separation of these vectors into the $Q$ categories, there exist $w = [w_k] \in I\!R^{PQ^2}$ and $\theta = [\theta_k] \in I\!R^Q$ such that: $\forall e_i \in B_E$, $e_i \in C_k \Longleftrightarrow e_i^T w_k + \theta_k > 0$. For each of these couples, let $w^* = [w_k^*] \in I\!R^{PQ^2}$ be given by: $\forall k$, $w_k^* = w_k + \theta_k \sum_{m=1}^{m=Q} b_{1,m}$, where $B = \{b_{l,m}\}$, $(1 \le l \le P)$, $(1 \le m \le Q)$, is the canonical basis of $I\!R^{PQ}$. The utility for using two indexes here will be clear later. It springs directly from the definition of $U^P$ that: $\forall e_i \in B_E$, $e_i \in C_k \Longleftrightarrow e_i^T w_k^* > 0$. $B_E$ is thus shattered by the hyperplanes of $I\!R^{PQ^2}$ and accordingly, from Lemma 1, by the constrained MLR combiner □.

# 5　The Natarajan dimension

The definition of shattering considered above has a drawback highlighted in [12]: it does not preserve the equivalence between finite VC dimension and *PAC learnability* [13]. This drawback falls with the definition due to Natarajan [14]:

**Definition 3** *A set $H$ of discrete-valued functions shatters a set $D$ of vectors $\Longleftrightarrow$ there exist two functions $h_1$ and $h_2$ belonging to $H$ such that:*
*(a) for any $z \in D$, $h_1(z) \ne h_2(z)$,*
*(b) for all $D_1 \subset D$, there exists $h_3 \in H$ such that $h_3$ agrees with $h_1$ on $D_1$ and with $h_2$ on $D \setminus D_1$, i.e. $\forall z \in D_1$, $h_3(z) = h_1(z)$, $\forall z \in D \setminus D_1$, $h_3(z) = h_2(z)$.*

This definition can be reformulated simply in the context of discrimination. Let us consider a set $D$ of cardinality $n$. Let $\Psi$ be the set of functions $\psi_{kl}$ $(1 \le k < l \le Q)$ from $\{C_1, \ldots, C_Q\}$ to $\{0, 1, *\}$, such that $\psi_{kl}(C_k) = 1$, $\psi_{kl}(C_l) = 0$, and $\psi_{kl}$ is undefined elsewhere (value $*$). The set $D$ is shattered by $H$ according to Definition 3 if and only if there exists a function $\psi^n$ in $\Psi^n$ such that $\psi^n \circ H$ shatters $D$ in the usual sense. Note that all the definitions considered up to now coincide in the degenerate case where $Q = 2$. For now on, we assume implicitly $Q \ge 3$. Using Natarajan's definition, we have established the following bounds:

**Theorem 3** *The Natarajan dimension $d_N$ of the MLR combiner satisfies:*
$\frac{1}{2}Q(Q-1)P + \lfloor \frac{1}{2}Q \rfloor \le d_N \le \frac{1}{2}Q(Q-1)^2 P$.

The line of argument used to establish the upper bound in Theorem 2 can be readily adapted to obtain the upper bound on $d_N$. This bound holds under the assumption $Q \ge 3$ (indeed, for $Q = 2$, $\frac{1}{2}Q(Q-1)^2 P < d_E$), and is tighter than the one resulting from theorem 7 in [12].

For lack of place, we only give the sketch of the proof of the lower bound. Let us consider the following matrix $A_1 \in M_{PQ,d_E}(I\!R)$ (see Figure 1):

$$A_1 = \frac{1}{2}\left( \left[ 1_{d_E}^T \otimes \sum_{l=1}^{l=P} b_{l,1} \right] + \left[ 1_{P+1}^T \otimes \sum_{l=1}^{l=P} b_{l,2}, 1_P^T \otimes \sum_{l=1}^{l=P} b_{l,3}, \ldots, 1_P^T \otimes \sum_{l=1}^{l=P} b_{l,Q} \right] \right)$$

where $\otimes$ denotes the Kronecker product. All column vectors of $A_1$ belong to $U^P$. Let $B_1 = \theta A_1 + (1 - \theta)\Delta_1$, with $\theta \in ]0, 1[$ and the column vectors of

$\Delta_1 = [\delta_{pq}^{(1)}]$ belong to $U^P$. Coefficients $\delta_{pq}^{(1)}$ are chosen adequately to ensure that the column vectors $F(x_i)$, $(1 \le i \le d_E)$ of $B_1$ constitute a basis of $E$. It has been shown in Section 4 that these vectors can thus be shattered according to Vapnik's initial definition by the family of hyperplanes of $I\!\!R^{PQ}$. We associate this basis with class $C_1$ and choose its first $P + 1$ vectors to be those that the combiner must be able to classify either in $C_1$ or in $C_2$. Similarly, the $P$ following vectors are those *shared* by $C_1$ and $C_3$ etc. Another basis, corresponding to all the vectors that the combiner must be able to classify in $C_2$, can then be built according to the same principle. Let matrix $A_2$ be given by:

$$A_2 = \frac{1}{2}\left(\left[1_{d_E}^T \otimes \sum_{l=1}^{l=P} b_{l,2}\right] + \left[1_{P+1}^T \otimes \sum_{l=1}^{l=P} b_{l,1}, 1_P^T \otimes \sum_{l=1}^{l=P} b_{l,3}, \ldots, 1_P^T \otimes \sum_{l=1}^{l=P} b_{l,Q}\right]\right)$$

Figure 1:
Matrices $A_1$ and $A_2$ :
PQ raws and
P(Q-1)+1 columns
*here: P=3, Q=4*
shaded boxes: term = 1/2
white boxes: term = 0
The first P+1 columns
(in black) are common
to $A_1$ and $A_2$

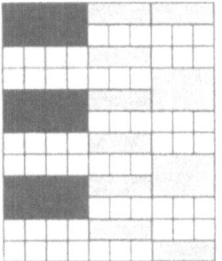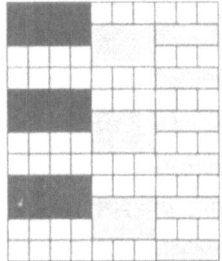

Let $B_2 = \theta A_2 + (1 - \theta)\Delta_2$, with $\Delta_2 = [\delta_{pq}^{(2)}]$. Coefficients $\delta_{pq}^{(2)}$ are chosen adequately to ensure that the column vectors $F(x_i)$ of $B_2$, with $i \in \{1, \ldots, P + 1\} \bigcup \{d_E + 1, \ldots, 2d_E - P - 1\}$, constitute a basis of $E$ sharing exactly its $P + 1$ first vectors with the former one. We have thus: $\forall (p, q)$, $(1 \le q \le P + 1)$, $\delta_{pq}^{(2)} = \delta_{pq}^{(1)}$. For any classification of these vectors into $C_2$ and $\neg C_2$, we can once more find a corresponding vector $w_2 \in I\!\!R^{PQ}$. Proceeding this way, we can generate overlapping sets of linearly independent vectors, each of which is shattered by the hyperplanes of $I\!\!R^{PQ}$. The total number of vectors finally generated depends on the parity of $Q$. If $Q$ is even, then each basis of odd index shares $P + 1$ vectors with the following one, and $P$ with each of the others. If $Q$ is odd, then $B_Q$ has only $P$ vectors in common with each of the bases. In any case, the number of vectors is thus equal to the lower bound announced.

To complete the proof, it remains to show that vector $w = [w_k]$ can always be chosen so as to satisfy the following property:

$$\forall (i, k, l), (F(x_i) \in B_k) \wedge (F(x_i) \notin B_l) \Longrightarrow F(x_i)^T (w_k - w_l) > 0$$

This is actually the case. A precise construction of matrices $\Delta_k$, $(1 \le k \le Q-1)$ even ensures a stronger result: any matrix $B_k$ can be chosen arbitrarily close to the corresponding matrix $A_k$ and, for any classification, vectors $w_k$ can be chosen arbitrarily close to $\frac{1}{P}\sum_{l=1}^{l=P} b_{l,k}$ (see [11] for details). Finally, using the transformation defined by equation (3), we can exhibit a vector $v$ for each desired classification of the $\frac{1}{2}Q(Q - 1)P + \lfloor\frac{1}{2}Q\rfloor$ vectors $F(x_i)$ generated, which completes the proof $\square$.

# 6 Discussion

The capacity of discrimination of the MLR combiner has been studied using extended definitions of the notion of VC dimension. Influence of both the specific nature of the predictors and the constraints has thus been highlighted. An attractive feature of our model is that many of the algorithms available to train it can be adapted easily to take into account complexity control. This is especially the case of the algorithms that use *active set methods*. An illustration of the possibilities offered can be found in [11], where we have modified with success the gradient projection method to incorporate a procedure of *early stopping*. The formula derived here can be used to go further by implementing the SRM principle. We are currently studying this subject, with the objective to increase the generalization ability of the model.

# References

[1] Bates, J.M. and Granger, C.W.J. The Combination of Forecasts, *Opl Res. Q.*, 1969, Vol. 20, 451-468.

[2] Xu, L., Krzyzak, A. and Suen, C.Y. Methods of Combining Multiple Classifiers and Their Applications to Handwriting Recognition, *IEEE Trans. on Systems, Man, and Cybernetics*, 1992, vol. 22, 418-435.

[3] Breiman, L. Stacked Regressions. *Machine Learning*, 1996, vol. 24, 49-64.

[4] Freund, Y. and Schapire, R.E. A decision-theoretic generalization of on-line learning and an application to boosting. *EuroCOLT'95*, 1995, 23-37.

[5] Sollich, P. and Krogh, A. Learning with ensembles: How over-fitting can be useful. *NIPS'8*, 1996, 190-196.

[6] Guermeur, Y., d'Alché-Buc, F. and Gallinari, P. Optimal Linear Regression on Classifier Outputs, *ICANN'97*, 1997, 481-486.

[7] Guermeur, Y. An Ensemble Method for Protein Secondary Structure Prediction. Submitted to the *Journal of Computational Biology*, 1998.

[8] Genest, C. and McConway, K.J. Allocating the Weights in the Linear Opinion Pool. *Journal of Forecasting*, vol.9, 53-73, 1990.

[9] Vapnik, V.N. *The Nature of Statistical Learning Theory*. Springer, N.Y., 1995.

[10] Rosen, J.B. The Gradient Projection Method for Nonlinear Programming. Part I. Linear Constraints. *J. SIAM*, 1960, vol. 8, $N^\circ$ 1, 181-217.

[11] Guermeur, Y. Combinaison de classifieurs statistiques, application à la prédiction de la structure secondaire des protéines. PhD thesis, Univ. Paris 6, 1997.

[12] Ben-David, S., Cesa-Bianchi, N., Haussler, D. and Long, P.M. Characterizations of Learnability for Classes of $\{0, \ldots, n\}$-Valued Functions. *Journal of Computer and System Sciences*, 1995, 50, 74-86.

[13] Valiant, L.G. A Theory of the Learnable. *Communications of the ACM*, 1984, vol. 27, 1100-1134.

[14] Natarajan, B.K. On learning Sets and Functions. *Machine Learning*, 1989, 4, 67-97.

# Poster Presentations:
# Applications

# Automated Statistical Recognition of Partial Discharges in Insulation Systems.

Massih-Reza AMINI, Patrick GALLINARI, Florence d'ALCHE-BUC

LIP6, *Université Paris 6*, 4 Place Jussieu, F-75252 Paris cedex 5.
{Massih-Reza.Amini, Patrick.Gallinari, Florence.dAlche}@lip6.fr

François BONNARD, Edouard FERNANDEZ

Schneider Electric, Research Center, A2 plant, 4 rue Volta, F-38050 Grenoble 09.
{francois_bonnard, edouard_fernandez}@mail.schneider.fr

### Abstract

We present the development of the successive stages of a statistical system for a classical diagnosis problem in the domain of power systems. It is shown how the different steps may be designed in a sound statistical way in order to develop a complete and efficient diagnosis system which provides the end user with a maximum of information about the system behaviour.

## 1  Introduction

We describe here the use of Neural Networks (NNs) and statistical techniques for a real life application in the domain of power systems. This application - the detection of electrical Partial Discharges (PD) in power apparatus - has been a challenging problem for more than 20 years. Efforts to automate PD detection began with expert systems [1]. Later on simple statistical models [2] and more recently Hidden Markov Models [3] and neural networks [4] have also been tested. However, most of this work has been exploratory and has addressed specific aspects of the problem. Some of the major statistical issues involved in the development of the successive processing steps were never considered. For example, in most approaches, feature selection relies on expertise, results validation is missing, data bases are too small to allow a sound comparison of different classifiers and feature sets. In the study presented here, a complete and valid approach ranging from feature extraction to statistical validation is developed. This allows us to compare both the performances and reliability of different systems and to show the interest of using flexible methods such as neural networks in real-world problems. In section 2, the PD phenomenon in insulating systems is described. Section 3 is devoted to feature extraction and selection and section 4 to the validation of the classifier. In section 5 we present experimental results.

## 2  Apparent charge detection

In high voltage power systems, small defects during manufacturing lead to the occurrence of Partial Discharges. Their accumulation is recognized as one of the major sources of deterioration for power apparatus and can trigger breakdown. Commercially available PD detectors allow to measure PD pulses, to process them and display the PD patterns on a time basis. Some of these detectors also offer basic diagnosis tools. The usual representation of these discharges is via 3-d *fingerprints* defined by the phase angle $\varphi_i$, apparent charge $q_j$ and pulse count $H_n(\varphi_i, q_j)$ coordinates (Figure. 1-left). Other useful statistics may be derived from this representation like (Figure 1-right): $H_{q\,max}(\varphi)$ the maximum pulse height distribution, $H_{qn}(\varphi)$ the mean pulse height distribution, $H_n(\varphi)$ the pulse count distribution, $H(q)$ the number of discharges vs. discharge magnitude and $H(p)$ the number of discharges vs. discharge energy.

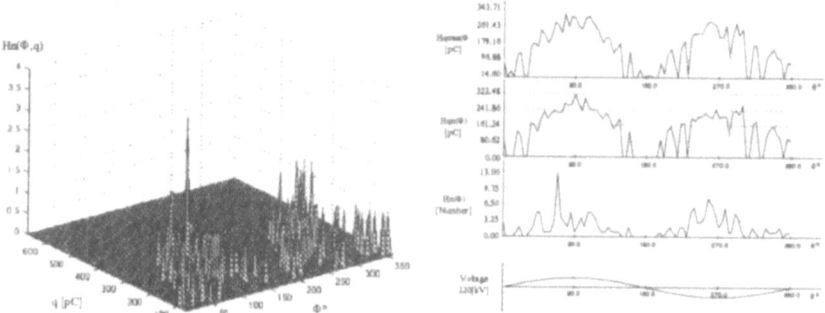

Figure 1 : left : 3-d representation of a PD, an entry $n_{ij}$ represents the number of PD pulses having magnitude $q_j$ and phase position $\varphi_i$. Right: some of the distributions which may be derived from the 3-d pattern, from top to bottom $H_{q\,max}(\varphi)$, $H_{qn}(\varphi)$ and $H_n(\varphi)$.

PDs characterize random phenomena and are influenced by several factors like aging, amplitude of voltage, frequency of voltage applications. PD measures are corrupted by interferences and intrinsic background noise. These 3-d patterns are thus complex and their characterization is a challenging task. The problem we tackle here is the detection and classification of PDs for apparatus which use SF6 gas for insulation. More precisely we are interested into classifying PD signals into 4 classes: corona discharges (sharp points in an electric field), surface discharges (surface irregularities), floating parts (small particles in an electric field) or background measurement noises.

# 3  Feature extraction and selection

In the dielectric literature, features derived from the 3-d representation are used to characterize PDs. We have considered here some of the most promising feature sets in order to evaluate them, we have also performed variable selection on each set and on a combination of all these feature sets. The features we have considered are: fractal characteristics of the 3-d patterns, simple statistics over $H_{q\,max}(\varphi)$, $H_{qn}(\varphi)$, $H_n(\varphi)$, $H(q)$ and $H(p)$ distributions, Fourier transform of $H_{qn}(\varphi)$. We first briefly describe these methods below and present after that variable selection.

## 3.1 Pre-processing

### *Fractal features*

Fractal measures allow to compute global characteristics of an image. We have used two fractal measures on the 3-d fingerprints, the fractal dimension and the lacunarity [5] which quantify respectively the irregularity and the denseness of an image surface.

### *Statistical operators*

Useful information from the distributions $H_{q\,max}(\varphi)$, $H_{qn}(\varphi)$, $H_n(\varphi)$, $H(q)$ and $H(p)$ may be inferred from their moments. The following features have been found useful for PD detection [2], $\mu_i$ denotes the $i^{th}$ order moment of variable $x$ for a given distribution: *Skewness*, $Sk = \mu_3/\mu_2^{3/2}$, which measures the degree of tilting of a distribution, *Kurtosis*, $Ku = \mu_4/\mu_2^2$-3, which measures the peakedness of a distribution, the number of modes of the distribution, and other features which characterize the asymmetry of the charges.

### *Fourier Analysis*

The normalized charge distribution can be evaluated by the average value of its spectral components in the frequency domain. The normalized average discharge is given by $\tilde{H}_{qn}(\varphi) = \dfrac{a_0}{2} + \sum_{k=1}^{\infty}\left[\overline{a}_k \cos(k\varphi) + \overline{b}_k \sin(k\varphi)\right]$ where $a$ and $b$ are the Fourier coefficients in the frequency domain, $\varphi$ is the phase angle. The first 8 spectral components $a$ and $b$ have been used here to approximate the charge distribution $H_{qn}(\varphi)$.

## 3.2 Variable selection

All measurements proposed by domain experts are not equally informative, it is thus useful to test automatic selection methods on these variable sets. Several methods have been proposed for variable selection with NNs. We have chosen here a method proposed in [6]. The relevance of a set of input variables is defined as the mutual information between these variables and the corresponding desired outputs. This dependence measure is well suited for measuring non linear dependencies as they are captured in NNs. For two variables $x$ and $d$, it is defined as: $MI(x,d) = \sum_{x,d} P(x,d)\log\dfrac{P(x,d)}{P(x)P(d)}$. Starting from an empty set, variables are added one at a time, the variable $x_i$ selected at step $i$ being the one which maximizes $MI(Sv_{i-1},x_i,d)$ where $Sv_{i-1}$ is the set of $i$-$1$ already selected variables and $d$ the desired output. Selection was stopped when MI increase falls below a fixed threshold. Densities were estimated by Epanechnikov kernels.

For the classification, we have used multilayer perceptrons trained according to a MSE criterion plus a regularization term.

# 4 Validation

The validation of classifier decisions is important for any real-world application. It includes both performance and confidence evaluation. We focus here on the latter. There are two aspects of interest regarding the confidence, the first, is the computation of global confidence intervals which allows to calibrate a classifier, the second is the computation of local confidence measures for each classifier decision.

For computing global confidence intervals, we have used both a classical parametric estimation which relies on the hypothesis of binomial classifier outputs and a bootstrap estimate [7]. For the latter, the quantity to be estimated is the percentage of correct classification for each class. We compute : for each bootstrap replicate $i$ and output $k$ the percentage of correct classification $\hat{\theta}_{ik}$, the mean of

these estimates $\hat{\theta}_k$ and their standard error $\hat{\sigma}_k$, $z_{ik} = \dfrac{\hat{\theta}_{ik} - \hat{\theta}_k}{\hat{\sigma}_k}$, $t^{(1-\alpha)}$ and $t^{(\alpha)}$ the

left and right $\alpha^{th}$ percentiles of the $\hat{\theta}_{ik}$ distribution. The *bootstrap-t confidence interval* is then $\left[\hat{\theta}_k - \hat{t}_k^{(1-\alpha)}\hat{\sigma}_k, \hat{\theta}_k + \hat{t}_k^{(\alpha)}\hat{\sigma}_k\right]$.

Note that the confidence could be computed in the same way for the global classification performances instead of class performances.

For the local confidence, several heuristic measures have been proposed in the literature. Most of them assume that the estimates of posterior class probabilities are accurate enough. We have used here the following two estimates :

- $\max_k y_k - \max 2_k y_k$, where $\max 2_k y_k$ is the second maximum output value.

- $1 - \dfrac{-\sum_{k=1}^{p} \hat{P}(k/x)\log \hat{P}(k/x)}{\log p}$ where $\hat{P}(k/x)$ is the $k^{th}$ output normalized so

that outputs sum to one.

# 5 Results

## 5.1 Experiments

Data were generated using an experiment design plan in order to insure a good representativity of the different classes and a minimum set of measures. The design factors for PD source discharges and background noise were : measure attenuation [0, 25] dB, gas pressure [1, 3] bar and test voltage [3.5 , 64.1] kV.

Simulations have been carried out to compare the different pre-processing methods and to show that the combination of these methods leads to a satisfactory discrimination. Variable selection allows to reduce the initial input dimension with a mean classification increase of about 1%. The number of variables is reduced from 28 to 12, 16 to 7 and 46 to 9 respectively for the statistical operators, Fourier components, and the combination of these two feature sets altogether with the 2 fractal measures. Performances are shown in table 1, the feature combination plus variable selection offers a significant performance increase compared to the other feature sets.

Table 1. Performances of different NN models retrained on the selected variables and for different space features.

| Space | NN Architecture | Performances (%) | |
|---|---|---|---|
| | | Training | Test |
| Fractal Dimension | 2-6-4 | 88.34 | 81.25 |
| Statistical operators | 12-16-4 | 97.25 | 93.12 |
| Fourier components | 7-16-4 | 98.75 | 94.38 |
| Combination | 9-16-4 | 100 | **96.25** |

## 5.2 Validation

Table 2, gives the confidence interval with 95% precision for the different classes using the 9 variables selected among the combination of the three feature sets. Intervals have been computed using the binomial assumption and the t-bootstrap interval method. With the t-bootstrap method, the confidence interval could be a singleton when a class has been recognized with a 0% or a 100% correct classification rate.

Table 2 : 95% confidence interval (on the test set) for the 9 input variables selected from the combination set. This interval has been computed with two methods (binomial assumption and t-bootstrap estimates).

| | binomial distribution | t-bootstrap interval |
|---|---|---|
| Corona discharge | [80.12%, 95.41%] | [84.31%, 97.18%] |
| Surface discharge | [96.62%, 100%] | {100%} |
| Floating parts | [98.17%, 100%] | {100%} |
| Noise | [96.62%, 100%] | {100%} |
| Global performance | [94.54%, 97.43%] | [94.16%, 97.18%] |

Additional information about the reliability of a classifier is provided by the confidence intervals computed directly on the outputs of the classifier instead of being computed on the classifier decision as in table 2. These intervals have been computed here via t-bootstrap and Figure 2 shows the corresponding box-plots for the combination of feature sets. In this case, the precision of the output values is high, it is significantly higher than for other feature sets.

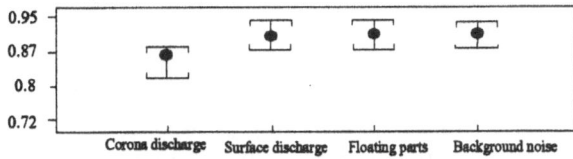

Figure 2 : Box-plots for the combination feature sets, The bold dot in each box indicates the median value and the lower and upper edges the 2.5 and 97.5% percentiles.

Local confidence intervals have been computed with the two methods described in section 4. Figure 3 shows the correct classification performances against the

706

percentage of reject. These experiments have been performed on the selected combination set. For both estimates of the local confidence, performances increase rapidly with the reject threshold to reach perfect recognition at 65% reject. These estimates allow the user to set up very easily the confidence level which fits the application.

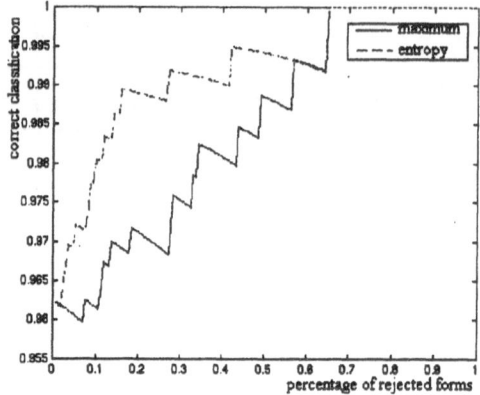

Figure 3: percentage of correct classification performances versus percentage of reject computed for two confidence measures.

# 6  Conclusion

We have described the development of the successive stages of a diagnosis system in the domain of power systems. Our goal was to illustrate how the different processing steps could be performed in order to develop, validate and calibrate as best as possible an operational system for a real world challenging problem. For this, we have used at the different steps (feature preprocessing and selection, classification and validation) methods which are well suited for the problem and complementary one to the other. Such software sensors could be easily implemented for several industrial diagnosis problems.

# References

[1] Wootton RE. Computer Assistance for the Performance and Interpretation of H.V. ac Discharge Tests. 5th Int. Conf. On H.V. Eng., Braunschweig, 1987 ; Journal 41.12

[2] Gulski E. Computer-Aided Recognition of Partial Discharges using Statistical Tools. PhD thesis, Delft University, 1991

[3] Satish L, Gururaj BI. Use of Hidden Markov Models for Partial Discharge Pattern Classification. IEEE trans. On Electrical Insulation, 1993 ; 28 :172-182

[4] Gulski E, Krivda A. Neural Networks as a Tool for Recognition of Partial Discharges. IEEE trans. On Electrical Insulation, 1993 ; 28 :984-1001

[5] Satish L, Zaengl WS. Can Fractal Features be Used for Recognizing 3-d Partial Discharge Pattern ? IEEE trans. On Electrical Insulation, 1995 ; 2 :352-359

[6] Bonnlander BV, Weigend AS. Selecting Input Variables Using Mutual Information and Nonparametric Density Evaluation. In Procedings of ISANN'94, Taiwan, 1994 ; 42-50

[7] Efron B, Tibshirani R. An Introduction to Bootstrap, Chapman Hall, London, 1993

# A Radar System with Phase-Sensitive Millimetric Wave Circuitry and Complex-Amplitude Neural Processing

Akira Hirose   and   Kazuyuki Sugiyama

Research Center for Advanced Science and Technology (RCAST)
The University of Tokyo
4-6-1 Komaba, Meguro-ku, Tokyo 153, Japan
e-mail: ahirose@ee.t.u-tokyo.ac.jp

## Abstract

A radar system based on complex-valued neural-network processing is proposed aiming at adaptive recognition and image reconstruction by utilizing complex information field. A prototype has been constructed as electronic circuitry and software. A recognition experiment has demonstrated that the radar system with the complex-valued neural network can learn different objects more effectively than those with the conventional (real-valued) networks. Radar images are detected by the system after coherent wave-propagation dynamics such as interference. It is observed that the wave physics is consistent with the neural dynamics that is constructed adaptively in the complex information field.

## 1   Introduction

Millimetric- / micro- wave image acquisition systems are widely used in various fields because of their wavelengths longer than visible lightwave, which lead to less absorption and less scattering, as well as a wavelength-dependent variation of detection objects. In such systems, detected raw image data are processed according to intended tasks to generate a desired output signal. Recently, the use of neural networks has been reported repeatedly for high performance recognition or image reconstruction by use of their adaptive and self-constructing behavior [1]-[4].

Generally speaking, the radar image acquisition systems employ a high-coherence electromagnetic wave. Therefore, they can intrinsically detect not only the amplitude but also the phase information [5]. In some aeroplane radar systems, for example, the amplitude contains the land-usage information whereas the phase the height of the landscape [6]. In these years, moreover, a use of phase-shifting filters has been proposed for an specific image processing [7].

In conventional neural network systems, the input data are processed in real-valued information space. Hence they do not have the processing ability for treating complex-amplitude (amplitude and phase) data in a consistent manner. If complex-amplitude processing neural networks are available, we can realize a radar image acquisition / recognition system in which both the amplitude and phase informations are processed profitably and adaptively by the learning and self-organizing neural-network features.

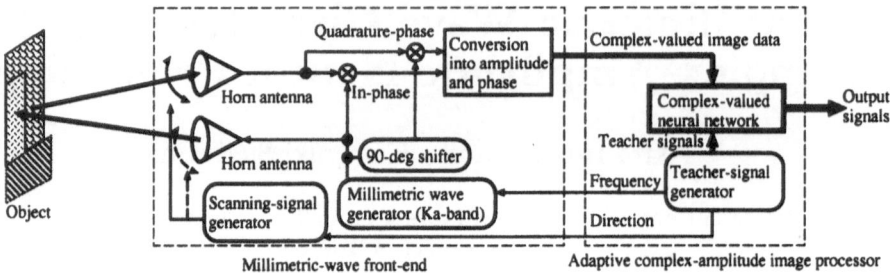

Figure 1: Construction of the complex-amplitude image-processing adaptive radar system using a complex-valued neural network.

In this paper, an adaptive radar system using a complex-valued neural network is proposed for processing the complex-amplitude millimetric-wave image data. The complex-valued networks deal with complex signals by constructing processing dynamics adaptively in complex-valued information space [8].

Because the raw radar images are constructed by the combination of coherent wave-propagation dynamics such as reflection, refraction, diffraction, and interference, the network can obtain consistent information-processing dynamics only in complex-valued (or phasor, in another word) information space. For instance, speckles are a typical interference noise observed often in the raw image data for land or sea observations from aeroplanes or satellites. It is expected that they are eliminated effectively by constructing an approximately reciprocal situation of the environmental world (propagation space) in the complex information field in the complex-valued neural networks.

A prototype of the radar system using complex-valued neural networks has been constructed at a Ka band (26.5~40[GHz]). Experimental results concerning image-dependent output-signal generation (recognition) are reported. It is demonstrated that the learning ability is higher than that of conventional systems using real-valued neural networks.

## 2  Construction

Figure 1 shows the construction of the adaptive image radar system with a complex-valued neural network for manipulating complex-amplitude raw data image. It consists of two parts; i.e., a millimetric-wave front-end and a complex-amplitude neural processor.

The millimetric-wave frequency range is Ka-band (26.5~40 [GHz]). In the present prototypal system, two horn antennas are mounted on direction-variable stages for the scanning of one of (or both of) the antennas to obtain the two-dimensional data, for simplicity, instead of a use of synthetic aperture antennas. The detected wave is mixed with two orthogonal reference waves to generate in-phase and quadrature-phase baseband signals These signals are finally converted into amplitude and phase outputs.

On the other hand, Fig.2 shows the construction of the neural network as

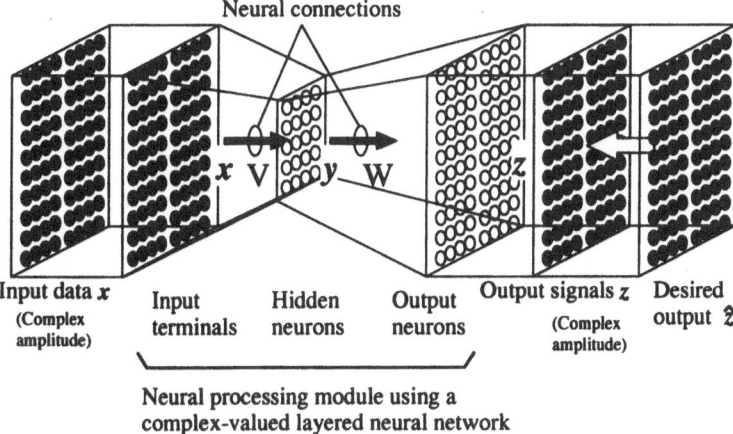

Neural connections

Input data $x$ (Complex amplitude)

Input terminals

Hidden neurons

Output neurons

Output signals $z$ (Complex amplitude)

Desired output $\hat{z}$

Neural processing module using a complex-valued layered neural network

Figure 2: Construction of the complex-valued neural network and the information flow.

well as a signal flow. Small open circles are neurons, whereas closed ones are input nodes or image pixels. The complex-valued neural network consists of two layers. Each layer deals with the complex-amplitude signals through complex-valued connection weights $\mathbf{W} = [w_{kj}]$ or $\mathbf{V} = [v_{ji}]$ and a nonlinear neuron activation function $f$. The forward process of the output layer is expressed as

$$z_k = f\left(\sum_j w_{kj}y_j\right) = A \, \tanh\left(g\left|\sum_j w_{kj}y_j\right|\right) \exp\left[i \arg\left(\sum_j w_{kj}y_j\right)\right] \quad (1)$$

where $\mathbf{y} = [y_j]$ and $\mathbf{z} = [z_k]$ denote input and output signals of the layer $\mathbf{W}$, respectively. The constants $A$ and $g$ are a saturation amplitude and a neuron gain, respectively. $i$ is the imaginary unit.

On the other hand, we define an error function $E$ as

$$E = \frac{1}{2A^2}|\mathbf{z} - \hat{\mathbf{z}}|^2 = \frac{1}{2A^2}\sum_k |z_k - \hat{z}_k|^2 \quad (2)$$

where $\hat{\mathbf{z}} = [\hat{z}_k]$ is teacher (desired) signals chosen for each input-data image. The steepest-descent learning rule to reduce the error function is derived by calculating partial derivatives $\partial E/\partial(|w_{kj}|)$ and $\partial E/\partial(\arg(w_{kj}))$ [8]:

$$\Delta(|w_{kj}|) = -K\left(1 - \frac{|z_k|^2}{A^2}\right)$$

$$\times \left\{\frac{|z_k|}{A} - \frac{|\hat{z}_k|}{A}\cos[\arg(z_k) - \arg(\hat{z}_k)]\right\} g \, |y_j| \quad (3)$$

$$\Delta(\arg(w_{kj})) = -K\frac{|z_k| \, |\hat{z}_k|}{A^2} \sin[\arg(z_k) - \arg(\hat{z}_k)]\frac{|y_j|}{|(1/N)\sum_j w_{kj}y_j|} \quad (4)$$

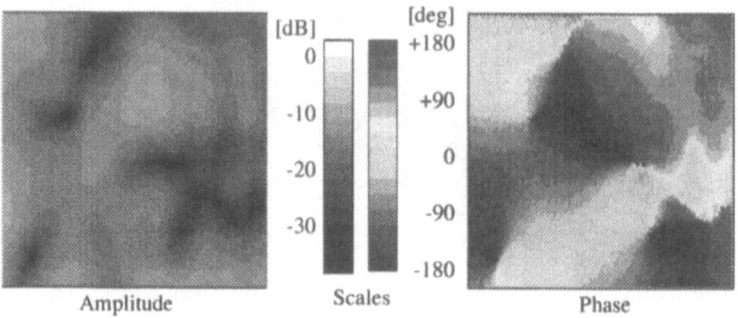

Figure 3: An example of obtained raw image data for an all-over foil reflection target (100×100 pixels).

where $K(> 0)$ is a learning gain constant. With these adjustive fractions, the connection gain (or loss) $|w_{kj}|$ and the connection phase shift $\arg(w_{kj})$ of a weight $w_{kj}$ are gradually changed in the learning process.

As to the backpropagation of the teacher signals for layer $\mathbf{V}$, the required teacher signal set $\hat{y} = [\hat{y}_j]$ is obtained as $\hat{y}^{\mathrm{T}} = f(\mathbf{W}^*\hat{z})$ where $\hat{y}^{\mathrm{T}}$ and $\mathbf{W}^*$ denote $\hat{y}$ transpose and the complex-conjugate transpose of $\mathbf{W}$, respectively.

The details of the steepest-descent method and the backpropagation learning in complex space is described in Ref.[8].

## 3  Experiment

In this experiment, five target patterns have been prepared by random-ragged aluminum foil placed in front of the antennas: The reflective areas are all-over, upper-half, lower-half, right-hand-half, and left-hand-half, respectively, which correspond to land or sea surfaces of various patterns in realistic observations. The foil is ragged at random with an average amplitude of around 20[mm], which is larger than the wavelength, so that the reflection becomes a random scattering. The situation is similar to a frosted glass illuminated by a laser beam.

Figure 3 shows an example of obtained raw image for the all-over reflective target. The reflection amplitude (Fig.3 left) is found inhomogeneous although a high reflectance is expected uniformly all over the observation area. Also we can find that the amplitude pattern has a correlation to the spatial variation of the phase data (Fig.3 right). The result suggests that the random-ragged reflecting surface generates the interference noise, which has to be distinguished from the pattern of the foil itself.

We take up a problem of image-dependent output-signal generation (recognition). In this experiment, the image size is 16×16. The output signals are chosen the same as the foil patterns of the target with the same number of output signals as that of input pixels. Such situation will directly lead to a higher-level tasks, i.e., the image reconstruction. The numbers of input ter-

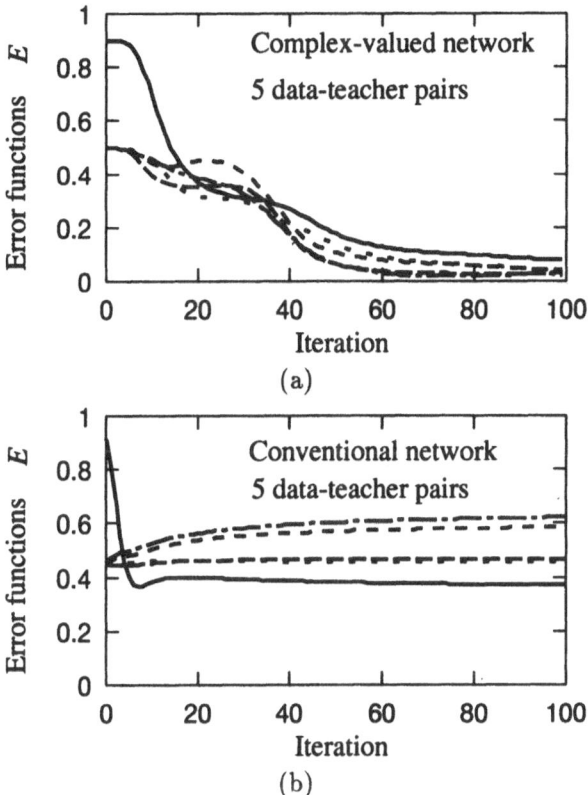

Figure 4: Error functions for (a)complex-amplitude and (b)real-valued-amplitude (conventional) processing systems, respectively, as functions of learning iterations.

minals, hidden neurons, and output neurons are 256 ( = 16×16 ), 64, 256, respectively.

Figure 4 shows typical learning curves observed in the backpropagation learning process for the five sets of target patterns and teacher signals. In Fig.4(a), five curves show the evolutions of the error functions for the five patterns when the complex-valued neural network deals with the complex-amplitude in complex-valued information space. It is found that the error values decrease almost monotonically.

Contrarily, in Fig.4(b), the curves show the error functions when the network is restricted to working as a real-valued neural network without phase terms of the weights and the activation functions. The phase data of the input images are also suppressed. Other situations are the same as in the above experiment. In this case, however, the error values have high floors, resulting in a failure of learning. This situation corresponds to a conventional neural radar system dealing only with the intensity of the reflected wave. Such difference of learning performance is observed in most cases using other obtained data.

It is found experimentally that the system using the complex-valued neural networks has a higher learning ability. The complex-valued networks are more suitable for processing the radar image data generated by wave phenomena. Performances for more complicated problems such as the image reconstruction will be reported elsewhere.

# 4 Conclusion

Adaptive radar systems using complex-valued neural networks for processing complex-amplitude millimetric-wave data have been proposed. A prototype system has been constructed as electronic circuitry and software. The experiment has demonstrated experimentally that the system has a higher learning ability than conventional systems using real-valued neural networks.

**Acknowledgment:** This work was partly supported by the Murata Science Foundation.

# References

[1] K.Watanabe, K.Shimizu, M.Yoneyama, K.Mizuno, "Post processing for millimeter wave radar images using feedforward neural network," IEICE Trans. on Electronics, J80-C-I (1997) 343-353

[2] K.S.Chen, W.P.Huang, D.H.Tsay, F.Amar, "Classification of multifrequency polarimetric SAR imagery using a dynamic learning neural network," IEEE Trans. on Geoscience and Remote Sensing, 34 (1996) 814-820

[3] H.Szu, B.Telfer, J.Garcia, "Wavelet transforms and neural networks for compression and recognition," Neural Networks, 9 (1996) 695-708

[4] S.Watanabe, M.Yoneyama, "An Ultrasonic Visual Sensor for Three-Dimensional Object Recognition Using Neural Networks," IEEE Trans. on Robotics and Automation, 8 (1992) 40-49

[5] L.C.Graham, "Synthetic interferometer radar for topographic mapping," Proc. IEEE, 62 (1974) 763-768

[6] S.N.Madsen, H.A.Zebker, J.Martin, "Topographic mapping using radar interferometry: Processing techniques," IEEE Trans. on Geoscience and Remote Sensing, 31 (1993) 246-256

[7] Q.Lin, J.F.Vesecky, H.A.Zebker, "New approaches in interferometric SAR data processing," IEEE Trans. on Geoscience and Remote Sensing, 30 (1992) 560-567

[8] A.Hirose, R.Eckmiller, "Coherent optical neural networks that have optical-frequency-controlled behavior and generalization ability in the frequency domain," Applied Optics, 35 (1996) 836-843

# Chances and Risks of Sensor Fusion with Neural Networks: An Application Example

Bernhard Sick

Chair of Computer Architectures (Prof. Dr. W. Grass), Univ. of Passau
Innstr. 33, 94032 Passau, Germany (email: sick@fmi.uni-passau.de)

## Abstract

Many real-world problems require multisensor solutions which generate a large amount of data. The transformation from complex and noisy inputs to a desired output is obviously a key element in the overall system performance. Very often, these transformations are nonlinear and hence neural networks are used to fuse the multisensor information. This paper demonstrates by means of a real-world example (online tool wear estimation in turning) that in practice, the fusion of sensor data with neural networks may actually produce worse results. Only if the data are pre-processed and their temporal development is considered in an appropriate way, the use of a multisensor system makes sense in this case. Additionally, the paper presents a generic sensor fusion architecture based on neural networks which may be used to describe sensor fusion systems.

## 1 Introduction

Multisensor fusion is a multilevel, multifaceted process dealing with the automatic detection, association, correlation, estimation, and combination of data from multiple sensors and other information sources (e.g. database systems or user inputs) [1]. Examples for sensor fusion applications are robotics, monitoring of machine tools, control of industrial production processes, weather forecasting, medical applications etc. The main advantages are an increased confidence (due to statistical advantages), reduced ambiguity and, therefore, a robust operational performance and an improved precision. In some applications an extended temporal and spatial coverage and an increased dimensionality (e.g. measuring different spectral bands) are also important [2].

Data from multiple sensors are fused using well-known techniques from different mathematical disciplines, e.g. signal processing, statistics, pattern recognition, soft computing (e.g. neural networks, fuzzy systems) etc. In this paper the focus will be on neural networks. This technique has been used in many applications in the last years [3]. However, results obtained with neural networks are often not validated in an appropriate way. It's a matter of common knowledge that neural networks are able to ignore disturbed or noisy information, detect fundamental interdependencies and approximate a sought nonlinear function. In practice, multisensor fusion with neural networks (as well as with other techniques) may produce worse results that could be obtained with a single sensor. One possible reason may be, that the network adapts learning patterns very well ("overfitting") and looses its generalization capability. This paper shows by means of an industrial application example (online tool wear estimation in turning), that it is important to pre-process data and to consider the temporal development of data in order to make multisensor fusion useful.

## 2  A Generic Sensor Fusion Architecture

Fig. 1 shows a generic sensor fusion architecture based on neural networks. Sensor fusion techniques may be applied in any level of the sensor fusion system. Additionally, information provided by a database system or user inputs may be processed. The proposed architecture is not a new fusion method, but may be used to describe hierarchical sensor fusion systems based on neural networks.

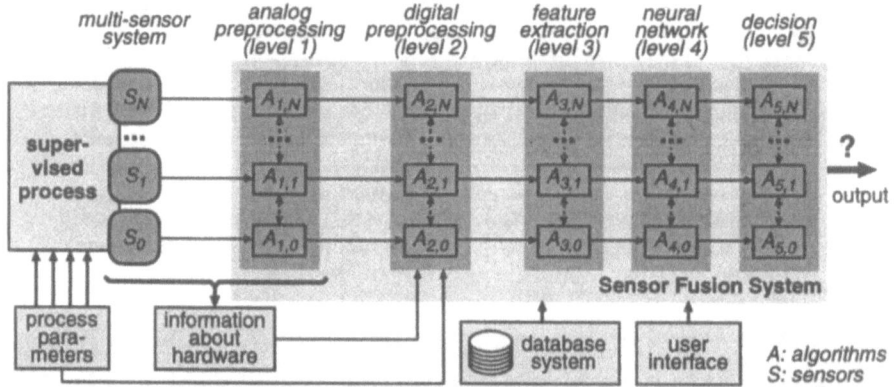

Figure 1: A Generic Sensor Fusion Architecture with Neural Networks

- *Analog preprocessing* deals with the preparation of raw data before digitization. For example, signals of one sensor are used to align the signals of another sensor (e.g. to eliminate undesired temperature influences).
- In the *digital preprocessing* level the information contents of the data is increased using available secondary information about the behaviour of analog or digital hardware (including the sensors) and about influences of process parameters on the sensor signals. An example is the linearization of a characteristic curve of a sensor or the alignment of the sensor signals by means of a physical model of the process which describes the influences of process parameters (which may be provided by additional sensors or by a database system) on the measured signals.
- The *feature extraction* condenses the remaining information in a few values which will be used as inputs for neural networks in the next level. Sensor fusion at this level deals for example with specific features like the principal components extracted from a set of signals (computed e.g. by means of a Karhunen-Loève-transformation).
- Different network paradigms may be used in the *neural network* level for data fusion (e.g. time-delay neural networks which consider the position of a single pattern in a pattern sequence).

- In the *decision* level methods like "mixture of experts" or "competing experts" are used to combine the results of several neural models.

The proposed generic sensor fusion architecture may be specialized to describe a specific sensor fusion system.

# 3 An Application Example

The real-world problem which will be investigated is online tool wear estimation in turning (some definitions are given in fig. 2). A lot of research has been done in the last years to solve this problem (see e.g. [4, 5] for lists of references), but due to insufficient generalization capabilities or simply a lack of precision even promising methods are not marketable up to now.

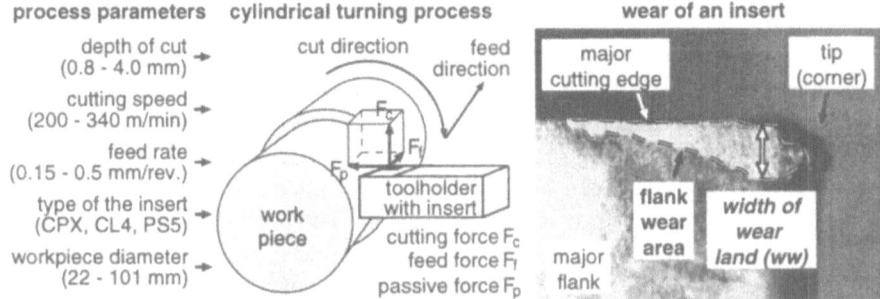

Figure 2: A Cylindrical Turning Process with Process Parameters and Forces

A turning process is influenced by many process parameters (five of them are varied here). Three sensors measure the forces in cutting, feed and passive direction (signals are extremely noisy in this and similar applications). There exist physical process models which describe the influences of some of the process parameters on the three forces [6]. A detailed description of metal cutting processes can be found in [7]. The wear of a tool (tools with throw-away inserts are used here) can be described for example by measuring the geometry of the wear areas on different faces of the insert. The width of wear land ($ww$) on the major flank (mainly caused by abrasion) will be estimated here (see fig. 2).

A solution using signals from all three sensors has been presented in [4, 5]. Two important characteristics distinguish this solution from others: The use of the mentioned process models to align the sensor signals (digital preprocessing level) and the consideration of the temporal development of the forces by means of specific neural network paradigms (time-delay neural networks, TDNN). TDNNs are feedforward networks which use more than one connection between two neurons in successive layers. Each connection is able to delay the propagation of values and possesses its own weight [3].

The question "Under what conditions does multisensor fusion improve the system performance in this application example?" will be answered here. Networks with one hidden layer (12 neurons, fully connected) have been trained 50000 epochs with the resilient backpropagation algorithm [3]. Inputs are the average values of the forces in a short time window. The activation function has been the non-linear sigmoid activation function. Multilayer perceptrons (MLPs) will be compared with TDNNs using delays of zero and one (in each layer of weights) corresponding to the current and one preceding cut with the tool. Additionally, experiments using aligned force signals will be compared with experiments using non-aligned force signals with or without process parameters as inputs. Variations of time delays and the number of neurons in the hidden layer are not investigated here. It could be shown, that the neural model is quite stable against smaller modifications of these parameters [5].

In order to show the (dis-)advantages of sensor fusion in detail, $5 \times 7$ sets of experiments have been carried out. To assess the generalization capability of a trained network, results for learning (training) patterns will be compared with results for unknown test patterns (extrapolation). Each experiment has been repeated 25 times (the overall number of experiments is 875) to show the repeatability of a result obtained with a learning algorithm starting with randomly initialized weights. With 25 repetitions, results of different experiments can be compared on a suitable level of significance. Therefore, the mean average error $\mu_\emptyset$ ("mean" refers to the 25 repetitions and "average" to the set of test patterns in one experiment) and the standard deviation of the average errors divided by the mean average error $\sigma_\emptyset / \mu_\emptyset$ have been determined. It must be stated, that the test patterns belong to three chipping experiments with combinations of process parameters which lie in an area of the parameter space covered quite well with chipping experiments. For the available data this measure turned out to be necessary as a result of the uneven and sparse distribution of parameter combinations in the parameter space. That's why results for test patterns are better than results for training patterns. However, relative improvements with different network paradigms and input parameters are obvious and the conclusions are valid in any way.

Tab. 1 gives the results of the following sets of experiments (**L**: learning, about 760 patterns; **T**: testing, about 220 "unknown" patterns):

- The first set uses non-aligned force values as inputs of a MLP.
- In the second set, five process parameters (see above) are used as additional inputs (the type of the insert has been coded in a binary notation at three input neurons).
- In the third set aligned force values are processed by means of a MLP.
- The temporal development of non-aligned force values is considered by a TDNN in set four.
- Set five finally uses aligned force values as inputs of a TDNN.

First, the single sensor experiments will be evaluated (column one, two and three). In almost every set the best results for learn **and** test patterns (i.e. the smallest mean average error) can be obtained using $F_f$. Only in the second set, where process parameters are used as additional inputs of the network, the best test results are achieved using $F_c$. In set two, the mean average error for learning patterns is reduced significantly compared with set one (but test results are only better for $F_c$). However, comparing the single sensor experiments of set one and two, the best test result is obtained using $F_f$ in set one. It might be concluded that the variation of process parameters doesn't have any influence on the precision of the wear estimation. Additionally, the standard deviations increase dramatically in set two. Due to the high degree of freedom of the neural model (large number of weights) the learning algorithm adapts the training patterns very well ("overfitting").

| inputs of the neural network | | | | | | | |
|---|---|---|---|---|---|---|---|
| cutting force $F_c$ | √ | | | √ | √ | | √ |
| feed force $F_f$ | | √ | | √ | | √ | √ |
| passive force $F_p$ | | | √ | | √ | √ | √ |
| **set 1: non-aligned signals, MLP** | | | | | | | |
| L: $\mu_\emptyset$ in $\mu m$ | 140.3 | 104.4 | 112.0 | 101.0 | 114.5 | 101.9 | 99.4 |
| $\sigma_\emptyset / \mu_\emptyset$ in % | 0.0 | 0.0 | 0.0 | 2.7 | 3.0 | 0.0 | 2.8 |
| T: $\mu_\emptyset$ in $\mu m$ | 97.3 | 53.1 | 66.6 | 62.2 | 86.6 | 55.5 | 70.0 |
| $\sigma_\emptyset / \mu_\emptyset$ in % | 0.5 | 0.0 | 0.2 | 2.7 | 2.6 | 0.1 | 4.3 |
| **set 2: non-aligned signals, with process parameters, MLP** | | | | | | | |
| L: $\mu_\emptyset$ in $\mu m$ | 60.1 | 46.1 | 57.1 | 42.4 | 51.3 | 40.8 | 33.9 |
| $\sigma_\emptyset / \mu_\emptyset$ in % | 5.4 | 5.6 | 4.3 | 6.3 | 5.5 | 8.7 | 6.9 |
| T: $\mu_\emptyset$ in $\mu m$ | 59.8 | 71.6 | 86.1 | 69.8 | 92.5 | 63.9 | 71.7 |
| $\sigma_\emptyset / \mu_\emptyset$ in % | 13.0 | 9.3 | 9.7 | 11.1 | 10.1 | 11.5 | 8.5 |
| **set 3: aligned signals, MLP** | | | | | | | |
| L: $\mu_\emptyset$ in $\mu m$ | 131.3 | 96.4 | 116.2 | 95.4 | 111.7 | 98.1 | 95.3 |
| $\sigma_\emptyset / \mu_\emptyset$ in % | 0.0 | 0.0 | 0.1 | 1.6 | 3.7 | 1.8 | 1.7 |
| T: $\mu_\emptyset$ in $\mu m$ | 80.3 | 52.7 | 66.9 | 52.9 | 63.4 | 53.2 | 54.1 |
| $\sigma_\emptyset / \mu_\emptyset$ in % | 0.1 | 0.0 | 0.2 | 1.6 | 3.9 | 0.8 | 2.3 |
| **set 4: non-aligned signals, TDNN** | | | | | | | |
| L: $\mu_\emptyset$ in $\mu m$ | 124.4 | 94.3 | 103.2 | 85.2 | 100.6 | 89.1 | 82.0 |
| $\sigma_\emptyset / \mu_\emptyset$ in % | 2.0 | 1.7 | 2.4 | 3.4 | 4.0 | 2.8 | 5.9 |
| T: $\mu_\emptyset$ in $\mu m$ | 97.4 | 57.2 | 66.2 | 58.3 | 80.1 | 62.9 | 64.3 |
| $\sigma_\emptyset / \mu_\emptyset$ in % | 1.2 | 2.0 | 4.6 | 5.1 | 7.5 | 5.3 | 6.8 |
| **set 5: aligned signals, TDNN** | | | | | | | |
| L: $\mu_\emptyset$ in $\mu m$ | 107.1 | 84.3 | 104.1 | 80.1 | 96.9 | 84.3 | 79.5 |
| $\sigma_\emptyset / \mu_\emptyset$ in % | 6.8 | 1.1 | 2.9 | 1.6 | 3.0 | 5.2 | 2.2 |
| T: $\mu_\emptyset$ in $\mu m$ | 66.4 | 47.9 | 63.7 | 45.1 | 55.5 | 47.1 | 43.7 |
| $\sigma_\emptyset / \mu_\emptyset$ in % | 11.7 | 1.0 | 4.8 | 1.0 | 4.3 | 4.1 | 2.2 |

Table 1: Results of the Sensor Fusion Experiments (Wear Estimation)

With aligned force signals (set three), noticeable improvements of the results for unknown test patterns can be obtained using $F_c$ (cf. set one). This is not surprising: It is more difficult to describe the influences of process parameters on $F_f$ or $F_p$ than on $F_c$ [6]. Using non-aligned signals and temporal information (set four), results for test patterns are comparable or worse. Finally, set five (combining process- **and** time-related information by means of aligned input values and TDNNs) shows the best test results in each column (apart from the experiment using $F_c$ in set 2).

If information from several sensors is fused (columns four up to seven), the learning task is more difficult and the learning algorithm may converge in a local minimum (causing an increase of the standard deviation). In any column of set five the mean average error for **learning** patterns in an experiment with two or three sensors is lower than in the corresponding experiments with single sensor data. This holds for almost all experiments in the other sets, too (apart from two exceptions which are not significant in a statistical sense). Fortunately, the statement is also true for results for **test** patterns in set five! That means, if a certain sensor combination is selected based on the results for learning patterns, this sensor combination shows the best results for test patterns, too. In any other set, the best learning results are achieved with three sensors and the best test results can be obtained using only one sensor! Obviously, $F_f$ contains the most important information for the wear estimation (even if process- and time-related information can be used better in experiments with $F_c$). Again, this is not surprising: The wear parameter $ww$ is measured on the major flank of an insert. The vector $F_f$ and the normal vector of the major flank distinguish only by a small angle. Finally, it must be stated that only the alignment of the force signals by means of a physical model of the forces (digital preprocessing level) **and** the use of temporal information by means of TDNNs makes the application of three sensors useful.

## 4   Conclusion

As a substantial result it can be claimed, that available secondary information about the observed or controlled process and about the measuring hardware should be used to improve the information content of the signals. Otherwise, the neural network treats some systematic disturbances of the signals like noise (even doing so, the signals are noisy enough). It is not sufficient to use this information as inputs of the neural models, because a neural network should be as small as possible in order to avoid overfitting. Therefore, secondary information should be used in a very early stage of the sensor fusion process. The proposed generic architecture may be used to describe specific fusion systems.

## References

[1] Klein LA: Sensor and Data Fusion Concepts and Applications. SPIE Optical Engineering Press, Bellingham (WA), 1993
[2] Hall DL: Mathematical Techniques in Multi-sensor Data Fusion. Artech House, Norwood (MA), 1992
[3] Zell A: Simulation Neuronaler Netze. Addison-Wesley, Bonn, Paris, Reading (MA), 1994
[4] Sick B: Wear Estimation for Turning Tools with Process-Specific Pre-Processing for Time-Delay Neural Networks. in: Dagli CH et al. (ed) Intelligent Engineering Systems Through Artificial Neural Networks - Vol. 7. ASME Press New York, 1997
[5] Sick B: Online Tool Wear Monitoring in Turning Using Time-Delay Neural Networks. In: Proc. of the 1998 International Conference on Acoustics, Speech, and Signal Processing (ICASSP '98), Seattle, 1998
[6] Paucksch E: Zerspantechnik. Verlag Friedr. Vieweg & Sohn, Braunschweig, Wiesbaden (9. ed.), 1990
[7] Shaw MC: Metal Cutting Principles. Oxford University Press, Oxford, 1989

# Support objects for domain approximation

Alexander Ypma and Robert P. W. Duin
Pattern Recognition Group
Faculty of Applied Sciences, Delft University of Technology
Lorentzweg 1, 2628 CJ, Delft The Netherlands
email: *ypma@ph.tn.tudelft.nl*

### Abstract

We propose a novel algorithm for extracting samples from a data set supporting the extremal points in the set. Since the density of the data set is not taken into account, the method could enable adaptation to novel (e.g. machine wear) data. Knowledge about the clustering structure of the data can aid in determination of the complexity of the solution. The algorithm is evaluated on its computational feasibility and performance with progressively more dissimilar data.

## 1 Introduction

Automatic recognition of machine wear and failure calls for methods that can deal with small sample sizes in high-dimensional spaces, undersampled fault classes and dynamically changing environments. Since normal machine behaviour is typically determined in a few calibration measurements of extremal operating conditions (e.g. when putting the machine into practice), an accurate but parsimonious description of the borders of the domain in the feature space indicating normal behaviour is expected to emerge. Failure detection can be performed by approximating the normal domain and rejecting samples not matching with this description. Ultimately, machine wear will be visible as a gradual shift of the borders of the admissible domain up to a point where a fault can be diagnosed (and this domain will consequently be labelled as faulty).

Conventional classification methods assume well-defined, static classes, while the problem of tailoring the complexity of the solution to a certain problem (e.g. choosing the number of hidden units in a neural network) is error-prone in cases with small sample sizes in high-dimensional spaces (*curse of dimensionality*). Recently proposed *support vector* methods circumvent these problems by editing the data set in a clever way, and expressing the solution in terms of the dot-product (i.e. *similarity*) of a new sample with the remaining (supporting) samples [5]. In addition one could use an arbitrary (non-analytic) similarity measure by looking upon the $m \times m$ distance matrix $D = d(\mathbf{x}_i, \mathbf{x}_j), i, j = 1, \ldots, m$ of a data set of size $m$ as a set of $m$ samples in an $m$-dimensional space [1]. When the rows in the distance matrix are considered as samples, and the columns as features, reducing the number of samples (data editing) results in a matrix with for each remaining sample (the *support objects*) the distances to each of the original objects. Inspired by these ideas, we propose a novel method for the extraction of objects in a data set that describe the domain of the set in a parsimonious manner, without taking the density of the set into account.

## 2  Domain approximation with support objects

In Vapnik's support vector classifier, the *optimal separating hyperplane* in the feature space specified by the nonlinear kernel $k(\mathbf{x}, \mathbf{y})$ for two separable classes of data $\{\mathbf{z}_i, i = 1, \ldots, l\}$ is given by [4, 5]

$$f(\mathbf{z}) = sgn(\sum_{i=1}^{l} \alpha_i y_i k(\mathbf{z}, \mathbf{z}_i) + b) \tag{1}$$

The coefficients $\alpha_i$ can be computed by a quadratic minimization procedure, and turn out to be nonzero only for the samples near the classification border (i.e. on the *margin*), the *support vectors* (dark objects in figure 1a). Hence, the solution can be found by taking only a limited subset of the available samples into account, whereas it is independent of the dimensionality of the data.

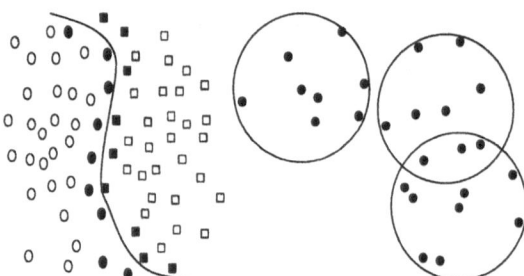

Figure 1: Support vectors in classification (l) and domain approximation (r)

The domain approximation problem can be stated in terms of equation (1) if we set the labels $y_i$ to 1 for samples in the domain and 0 otherwise, set the threshold $b$ to zero and choose as kernel function the rectangular kernel

$$k_r(\mathbf{x}, \mathbf{y}) = \begin{cases} 1, & \|\mathbf{x} - \mathbf{y}\| \leq r \\ 0, & otherwise \end{cases} \tag{2}$$

Since here there is no notion of a margin that is to be maximized and the labeling is different, the support vector learning algorithm determining the supporting samples in a dataset cannot be used any more. Using the fact that we approach a one-class problem, we develope an algorithm to minimize the radius $r$ of the kernel $k_r(\cdot)$ under the constraint that each training sample is decided by $f(\mathbf{z})$ as being *in the approximation domain*.

# 3 Domain approximation algorithm

Consider a data set $X = \{\mathbf{x}_i\}, i = 1, \ldots, m$. For *domain approximation*, every object in the support set $J = \{\mathbf{y}_j \in X\}, j = 1, \ldots, k, k \leq m$ is now given a *receptive field* in $R^m$ of radius $r$. A sample $\mathbf{x}_i$ in $X$ lies in the receptive field $R_p$ of a support object $\mathbf{y}_p$ when $p = arg\ min_j\ d(\mathbf{x}_i, \mathbf{y}_j)$ and $d(\mathbf{x}_i, \mathbf{y}_j) \leq r$. We took the Euclidean distance as distance measure $d(\mathbf{x}, \mathbf{y})$.

One could choose the set of $k$ support objects $J$ such that corresponding radius $r(J)$ is minimized, while all original objects are captured by some support object's receptive field (fig. 1b, dark objects in circle center), i.e.

$$J = arg\ min_{S_k} r(S_k) \qquad (3)$$

where $|S_k| = k, S_k \in 2^X$ i.e. $S_k$ a subset of $X$ of length $k$ and the corresponding receptive field radius $r(S_k)$ is given by

$$r(S_k) = max_i d(\mathbf{x}_i, \mathbf{y}_j), \mathbf{x}_i \in R_j, \mathbf{y}_j \in S_k \qquad (4)$$

We propose a variant of the *k-means clustering* algorithm adapted for the domain approximation problem (referred to as the *kcenters* algorithm) to cope with the subset selection problem that is still present. In the basic approach, different trials are repeated with different random choices for the initial subset. During the algorithm, each support object represents the center of the samples in its receptive field. If a better center can be found within its receptive field (i.e. such that the radius can be decreased, or *relaxed*), swap the former and the latter objects, until the subset can't be improved any more. Ultimately, the best subset over all trials is retained. Moreover, the support set size $k$ is also a parameter to be optimized, for which we use a *successive approximation* scheme: the support set size is increased from one to at most the number of samples, and the optimal subset at size $k$ is used as initial subset for size $k + 1$. The additional support object is initially taken to be the sample most remote to its receptive field center. As convergence criterion we choose the relative improvement in radius when increasing the support set from size $k$ to $k + 1$.

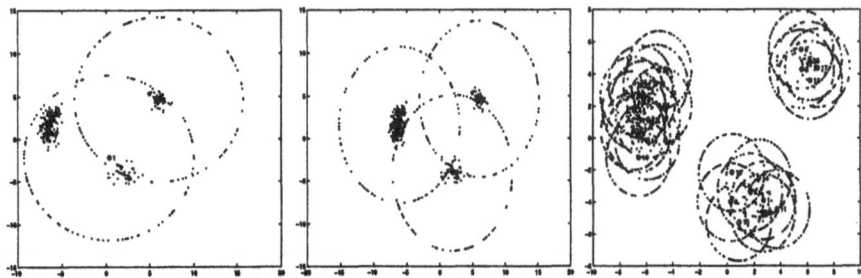

Figure 2: Trade-off between tolerance (left) and complexity (right)

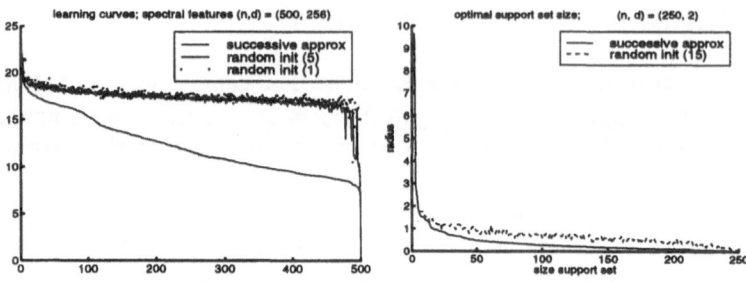

Figure 3: Successive approximation vs. random initialization

Note that there is a trade-off between complexity of the solution (support set size) and generalization capability (figure 2). Estimation of an appropriate support set size can be aided by using the data clustering structure. Moreover, the method assumes there are no serious outliers present in the data. For the application at hand (small number of well-defined calibration measurements) this assumption seems to be acceptable. Otherwise, an elementary outlier detection algorithm could be used prior to domain approximation.

# 4  Evaluation of the algorithm

The condition of mechanical machinery can be monitored by measuring the vibration behaviour of its rotating parts [2]. Measurements were taken from a submersible pump that was operating in three states: normal, presence of imbalance and presence of bearing failure (in the last case also at three different operating speeds). Nonlinear mapping of the data with multidimensional scaling revealed that the data consist of three clusters, where the cluster denoting bearing failure is composed of three subclusters. Bearing failure and imbalance may be diagnosed with dedicated vibration analysis techniques like envelope detection [2]. However, in [6] we showed that taking a few principal components of the acceleration spectrum already suffices for accurate recognition. This can be understood from the fact that a vibration signal $\mathbf{x}$ can be modeled as a harmonic series $\mathbf{s}$ plus additive white noise $\epsilon$. Now the principal eigenvectors of the correlation matrix $R_{xx} = E[\mathbf{x}\mathbf{x}^T]$ span the signal subspace [3], which is probably low-dimensional. Hence, a 256-points acceleration spectrum was obtained and normalized w.r.t. mean and standard-deviation.

## 4.1  Successive approximation algorithm

We compared the successive approximation algorithm to the *kcenters* algorithm with random initialization (1 and 5 trials) and arbitrary choice of support set size. Using the spectral features data set, we monitored the final radius of the surrounding spheres as a function of the number of spheres. In fig. 3a the successive approximation algorithm (the bottom line) can be seen to yield smaller final radius (vertical axis), while the radius is monotonically decreasing with

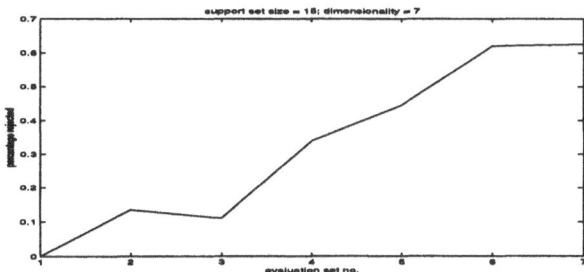

Figure 4: Rejection of novel data with successive approximation

support set size (horizontal axis). The radius obtained with 1 trial random initialization fluctuates randomly around the 5 trial random initialization results. It is clear that the stabilizing effect of more trials is at the expense of increased computing time, whereas successive approximation needs for each support set size only one run of the algorithm. Moreover, using more spheres leads to significant improvement with successive approximation, whereas the random case reaches a plateau. This illustrates that there is a relatively homogeneous distribution of distances in high-dimensional spaces, since the sphere radius can constantly be relaxed. With increasing support set size, the probability that a random guess for the initial subset is adequate decreases, hence the radius will be less improved using random initialization. When using data consisting of the first two principal components of the previous data set, again (fig. 3b) the successive approximation algorithm proved superior, but the difference is much smaller. Due to the three clusters in the data, combined with small dimensionality of the space, a sudden decrease in the radius occurs with three spheres (to a very small value). Moreover, the clusters are already represented adequately by the spheres, resulting in only marginal improvement in radius using more spheres.

## 4.2 Generalization to novel data

We analyzed the capability of the algorithm for incorporating gradually more dissimilar data by generating new samples out of the existing data. The offset of a new sample from its corresponding original sample is Gaussian distributed with zero mean and standard deviation $s$ times the mean signed difference between the point under consideration and its nearest neighbours in the original set. For $s$ running from 0 (original set) to 5, different validation sets were contructed. To track generalization, an independent test set of the same origin as the original set was used. Note that the use of multidimensional spheres can give rise to excessive tolerance in singular directions when the data lies in some (possibly nonlinear) subspace, hence we used the principal components of the spectral features data set in this experiment. It was observed that increasing feature size and support set size decreased generalization drastically (i.e. test samples and validation data for moderate values of $s$ were frequently rejected).

Results for reasonable values for dimensionality (7) and number of spheres (15) are shown in fig. 4. The first set is the train set, the second set the test set, and sets 3 to 7 are validation sets with $s = 1, \ldots, 5$. Since the domain is fitted as tight as possible, a test set already shows some rejections. Then the rejection rate increases with the amount of novelty, up to a point where always a certain fraction of the new samples lies somewhere on the original domain (the distribution of the offset has zero mean).

## 5  Discussion

We proposed a novel algorithm for the extraction of objects describing the domain of a data set using multidimensional spheres with minimal radius. The algorithm was demonstrated to be an adequate and efficient way for approximating the domain of a dataset. For more reliable detection of novelty the radii should not be uniform over the set, but vary with local resolution. Moreover, a predefined tolerance specifying the admissible amount of novelty should be incorporated in the method, since the proper rejection threshold will typically vary with each practical application. Since the density of points is never used in the algorithm, but the description is made on the basis of border points, the algorithm bears potential for on-line learning in problems with small sample sizes.

## 6  Acknowledgment

This work was partially supported by the Dutch Foundation for Applied Sciences (STW), project no. DTN-44.3584.

## References

[1] R. P. W. Duin. Relational discriminant analysis and its large sample size problem. *submitted to ICPR'98*, 1998.

[2] J. S. Mitchell. *An introduction to machinery analysis and monitoring - 2nd ed.* PennWell Publ. Comp., 1993.

[3] P. Pajunen, J. Joutsensalo, J. Karhunen, and K. Saarinen. Maximum likelihood estimation of equispaced sinusoids in rotating machine fault detection. In *Proc. of the ICSPAT'95, Boston, USA*, pages 1164–1168, 1995.

[4] B. Schölkopf. *Support vector learning.* PhD thesis, TU Berlin, 1997.

[5] V. Vapnik. *The nature of statistical learning theory.* Springer-Verlag, New York, 1995.

[6] A. Ypma, R. Ligteringen, E. E. E. Frietman, and R. P. W. Duin. Recognition of bearing failures using wavelets and neural networks. In *Proc. of TFTS'97*, pages 69–72. University of Warwick, Coventry (UK), 1997.

# A Comparison of Traditional and Soft-Computing Methods in a Real-Time Control Application

B. Sick, M. Keidl, M. Ramsauer, S. Seltzsam

University of Passau (Prof. Dr. K. Donner, Prof. Dr.-Ing. W. Grass)

Innstr. 33, 94032 Passau, Germany (email: sick@fmi.uni-passau.de)

## Abstract

The paper presents a performance comparison of traditional (controller based on a physical model) and soft-computing methods (neural and fuzzy controller) applied to a real-time control problem. The pros and cons of the controller paradigms will be investigated in this paper by means of an application example. In this example a ball has to be navigated through a maze mounted on a board. The board can be tipped in any direction using two step motors. An image of the board is supplied by a CCD-camera and two signal processors are used to process the image sequences and to generate motor step sequences, respectively. The results show that the choice of a suitable controller for a specific task mainly is a trade-off between the accuracy of the motion, the ball's speed and the execution time of the particular control algorithm.

## 1 Introduction

In the past years more and more soft-computing techniques (based on e.g. neural networks or fuzzy systems) have been developed for the solution of control problems. The reason is that many real-world control problems can hardly be solved by means of conventional techniques because needed information is not available or the systems under consideration are not well defined. The principle of soft-computing is: exploit the tolerance for imprecision, uncertainty, partial truth, and approximation to achieve tractability, robustness, low solution cost and even better rapport with reality [1]. In this sense, soft computing imitates certain human abilities. However, soft-computing techniques are not a panacea. Even if they are often very easy to handle, results have to be validated carefully. This paper presents a comparative evaluation of three different traditional and soft-computing techniques applied to a real-world control problem.

In this application example, a ball has to be moved through an arbitrary maze mounted on a board. The motion of the ball is achieved by tipping the board using two step motors. A video camera is placed above the board to observe the effect of the board's motion. At first sight the problem might appear playful, but similar or related problems (e.g. motion or balance problems) arise in many industrial applications. Fig. 1 shows the hardware components (left) and a photography of the board (right). The overall problem is solved in two subsequent steps. In the first step (pre-processing) it is necessary to detect the position of the ball and to find a passable way through the maze. The solution for this problem has been described in [2]. The result is a path from the starting to the target point of the maze in form of a polygonal. The second step is the real-time phase which is responsible to navigate the ball along the path by approaching the corner points of the polygonal one after the other. Initially, the new coordinates of the ball's center are estimated by considering some past positions of the ball. Then, the exact center is localized in an area ("sliding window") around the estimated center using e.g. a prototype fitting method (see [2]). The current position of the ball and the position of the next target point (corner of the polygonal) are used to calculate the motor step sequences. There are different possibilities to handle this task.

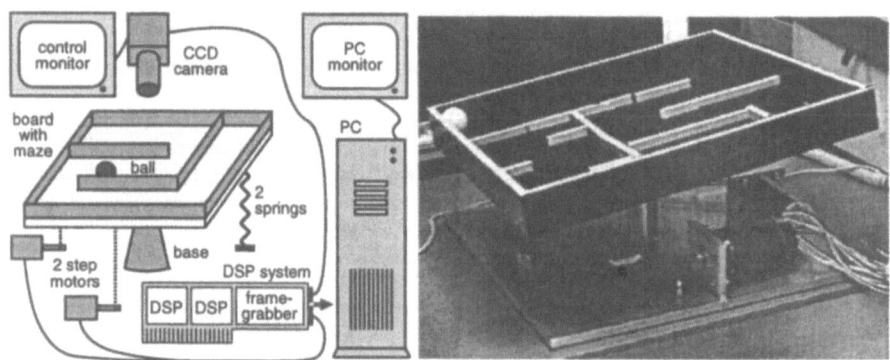

Figure 1: The hardware equipment of the application example

# 2 Controller Paradigms

Three types of controllers have been investigated: a fuzzy controller, a neural network and a conventional controller which is based on a physical model of the motion of the ball. These approaches are discussed in detail below.

## 2.1 Fuzzy Controller

The fuzzy controller doesn't need a precise physical model of the ball's behaviour. An intuitive description of the desired behaviour of the controller by means of linguistic variables, linguistic terms and fuzzy rules is sufficient [3]. The distance between the current position of the ball's center and the next corner of the polygonal and the current speed of the ball have been considered to be appropriate input variables for the controller. It is easy to see that the controller's behaviour in $x$- and $y$-direction has to be the same. Therefore, it is possible to use two identical controllers, one for each direction. In the same sense, the motion in positive direction is symmetrical to the motion in negative direction. For that reason, controllers with only one output variable (the signed number of steps in one of the two directions) have to be designed.

After defining the input and output variables of the control algorithm, it is possible to describe the transfer function of the controller by means of linguistic terms for the input and output variables and fuzzy rules. As described above, the linguistic variables are distance (in "pixels"), speed (in "pixels per image", i.e. per 1/50 *sec*) and number of motor steps. The corresponding linguistic terms (or fuzzy sets) are simple triangular functions. To enable the fuzzy controller to calculate the required number of motor steps, a set of fuzzy rules like "if low negative speed and high negative distance then many negative steps" has to be defined. The set of fuzzy rules describes the transfer function of the fuzzy controller. The controller can be considered to be a static, non-linear, and non-adaptive system [4]. We started with a more or less canonical definition of linguistic terms and rules and optimized the controller manually.

Further improvements of the controller's performance have been achieved by changing the way of tracking the path. With the original version of the fuzzy controller problems arise when the ball drifts away from the path, because the ball moves in the direction of the next target corner and not back to the path. To avoid this undesired behaviour the ball is led back to the path by calculating a virtual corner. If the center of the ball is far away from the path this virtual corner is the nearest point of the current path segment. The nearer the ball is to the path the nearer the virtual corner gets to the target corner. To prevent the ball from "lurching" from one side of the path to the other instead of moving in a straight line an additional PID controller has been connected with the output of the fuzzy controller. The PID variables have been optimized manually. With this PID, the overall control algorithm shows a dynamic behaviour.

## 2.2 Neural Controller

The main reason to use a neural controller is the possibility for an automatic adaption of the controller to other environmental conditions (e.g. another ball with different weight or other board surfaces; not evaluated here) and an improved performance due to a differentiable transfer function.

An offline pre-trained 3-layer feedforward neural network (multilayer perceptron, fully connected, four input neurons, 10 hidden neurons and 2 output neurons) is used as a starting point for a continuous, supervised online training [5]. Again, the inputs are the speed and the distance in $x$- and $y$-direction. The two output neurons give the number of motor steps. By using a pre-trained network, the duration of the training is shortened and the hardware is prevented from getting damaged in an early stage of the training (this would be the result of using randomly initialized weights). The learning patterns for the pre-training have been obtained by evaluating the transfer function of the fuzzy controller. Therefore, the neural controller and the fuzzy controller show quite the same behaviour after the offline training.

In order to improve the controller's performance two new input variables have been introduced: the inclination of the table and the deviation of the ball from the given path. Therefore four input neurons had to be added. The corresponding *additional* weights have been initialized with randomly selected values (unlike the original weights). In order to obtain "good" target output values for a supervised online training it is necessary to assess the obtained output values. By modifying the weights of the network in an appropriate way using the backpropagation learning algorithm, the behaviour of the neural controller can be improved continuously. The network should perform better in two ways: first, the speed of the ball should be increased and second, deviations from the given path should be minimized. For that reason, the target output values are calculated as a weighted sum of three values: the real output, a speed adjustment factor and a "lead the ball back to the path" adjustment factor. The speed adjustment is the difference between the desired speed and the real speed, and the lead-back values penalize major deviations of the ball from the given path. It is possible to determine whether the controller should be more accurate or faster by setting the weights in this sum in an appropriate way.

With the described online adaption of the synaptical weights, this controller type can be considered to be a static, non-linear, and adaptive system [4].

## 2.3 Conventional Controller

The conventional controller is based on a extremely simplified physical model of the ball/board system. Because an appropriate measuring hardware has not been available, only a few values have been determined experimentally: $\gamma_n$ (the inclination increment of the board when one impulse is sent to a step motor), $a_{max}$ (the maximum acceleration which can be achieved due to the limited inclination of the board), $v_{min}$ and $v_{max}$ (the minimum and maximum speed of the ball), and $\delta_{max}$ (a certain short distance to the target point where the ball should begin to slow down). The acceleration of gravity is $g_t$.

The control algorithm consists of two parts: First the desired speed $v_f$ is calculated and then the number of impulses which have to be sent to the board to achieve this speed is computed. The formulas described below are identical for each axis. The input values of the controller are the distance to the target point $\delta$ ($d = \text{sgn}(\delta)$ defines the direction), the current speed of the ball $v_c$, and the number of steps the table diverges from the starting position $t_c$. The desired speed of the ball $v_f$ can now be computed by

$0 \leq |\delta| \leq \delta_{max} : v_f := \delta \cdot \frac{1}{sec}$   (Slow down the ball.)

$\quad |\delta| > \delta_{max} :$ Adjust the ball's speed so that $v_f \in [v_{min}, v_{max}]$ :

$\qquad (0 \leq |v_c| < v_{min}) \vee (|v_c| > v_{max}) : v_f := d \cdot v_{min}$

$\qquad (v_{min} \leq |v_c| \leq v_{max}) :$ Don't move the board.

Finally, the number of motor steps $s$ is given by

$s := \frac{\gamma_f}{\gamma_n} - t_c,$

where

$a_c := \sin(\frac{t_c \cdot \gamma_n \cdot \pi}{180°}) \cdot g_t$   (current acceleration),

$a_f := \text{sgn}(a_f') \cdot \min\{|a_f'|, a_{max}\}$ with $a_f' := \frac{v_f - v_c}{1/50sec} - a_c$

$\qquad$ (required acceleration to speed up the ball from $v_c$ to $v_f$),

$\gamma_f := \arcsin(\frac{a_f}{g_t}) \cdot \frac{180°}{\pi}.$

The time to execute a certain number of steps is limited to $\frac{1}{50}$ sec, so only a limited number of steps may be executed. Therefore $s$ is scaled to $s \in [-s_{max}, s_{max}]$ where $s_{max}$ is the maximum number of motor steps.

As in the case of the fuzzy controller (and for the same reason), a PID controller is placed behind the conventional controller. With this PID the overall controller's behaviour changes slightly. The overall conventional controller is a dynamic, non-linear and non-adaptive system [4].

# 3 Results

To assess the different controller types, each controller has been tested using the same maze and the same path. To compare the controllers in a fair way, each test has been repeated twenty times and **mean** $\mu$ and **standard deviation** $\sigma$ of the following rating criteria have been determined (see tab. 1):

- the covered *distance* of the ball (measured in pixels),
- the *run time* between beginning (start point) and end (last target point) of the motion (measured in seconds),
- the *average deviation* (measured in pixels),
- the *max. deviation* of the ball (measured in pixels),
- the average number of processor *cycles* per calculation,
- the number of *motor steps*.

The path through the maze had been hard-coded in order to guarantee exactly the same path for each test and, therefore, comparable results. This "reference path" had a length of 1139 pixels. Results for the online trained network are shown **after** a certain time of online training (with online training the number of processor cycles is about 67000).

| paradigm | fuzzy controller | | pre-trained NN | | online trained NN | | conv. controller | |
|---|---|---|---|---|---|---|---|---|
| | $\mu$ | $\sigma$ | $\mu$ | $\sigma$ | $\mu$ | $\sigma$ | $\mu$ | $\sigma$ |
| distance | 1478 | 26.3 | 1459 | 27.3 | **1410** | 31.7 | 1560 | 24.0 |
| run time | 70.2 | 1.21 | 66.7 | 1.24 | **52.9** | 0.88 | 59.4 | 1.02 |
| avg. dev. | 2.50 | 0.21 | **1.88** | 0.16 | 3.47 | 0.19 | 3.24 | 0.22 |
| max. dev. | 12.1 | 1.40 | **10.4** | 1.89 | 13.9 | 1.36 | 16.6 | 4.02 |
| cycles | 3669 | 0.73 | 10992 | 0.07 | 10911 | 0.06 | **465** | 5.31 |
| mot. steps | 12316 | 177 | 11244 | 134 | **9048** | 193 | 9520 | 284 |

Table 1: Results of the controller experiments

It can be noticed, that the pre-trained neural network (which requires the transfer function of the fuzzy controller for the training) is the best to navigate the ball close to the given path (mean average dev. 1.88 pixels, mean maximum dev. 10.4 pixels). The online trained network guides the ball very fast through the maze (mean run time 52.9 seconds) and, therefore, the mean covered distance is the shortest (1410 pixels). Additionally, this controller needs the smallest number of motor steps (9048). Finally, the conventional controller has a very fast control algorithm (mean number of processor cycles 465).

# 4 Conclusion

It can be stated, that all controller types are able to control the motion of the ball in an excellent manner. The controllers (and the image processing algorithms, too) have to cope with a lot of disturbances in the control loop (see fig. 2). It is very difficult or even impossible to integrate these influences in a precise model of the system in order to consider them in the control algorithms.

The main advantage of the fuzzy controller is its robustness, e.g. dropped motor steps have only very little influence on good controlling results. On the

730

other side this controller is very hard to adapt to different environments (weight of the ball, friction of the board's surface, etc.). The linguistic terms and other parameters have to be optimized manually. The advantage of the neural network is its ability to learn an input/output relationship and, therefore, to adapt to different environments (not evaluated here). In this report the learning didn't start with randomly initialized weights. The network has been pre-trained offline by means of the transfer function of the fuzzy controller. After an offline training the user may specify whether the controller should train online on accuracy or speed or a combination of both. The main disadvantage of the online training is the execution time. Another problem is to control the learning parameters like momentum and learning rate carefully to guarantee convergence in learning. The conventional controller based on a extremely simplified physical model of the ball/board system may be used as well. The motor steps are calculated very fast and the average speed is excellent. On the other hand, adapting this controller to other environmental conditions is hard due to the necessity of adjusting the parameters of the underlying model manually. However, the period of development for this controller has been surprisingly short compared with the other controllers. The results show, that the overall system behaviour is a trade-off between accuracy of the motion, the ball's speed and the execution time for the control algorithm.

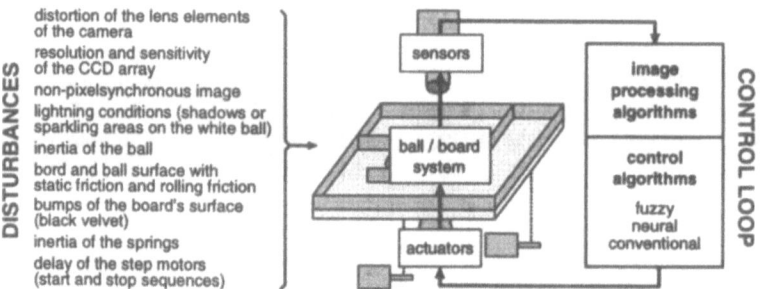

Figure 2: Disturbances in the control loop

**Acknowledgements:** The authors are grateful to the other members of the project team for the contribution of the interface hardware (R. Kickingereder, W. Maydl) and the image processing algorithms (G. Schichl, V. Vogelhuber).

# References

[1] Zadeh LA. What is soft computing?. in: Soft Computing, vol. 1 no. 1, 1997

[2] Keidl M, Kickingereder R, Maydl M, Ramsauer M, Schichl G, Seltzsam S, Vogelhuber V. Praktikum im Vertiefungsgebiet Technische Anwendungen der Informatik – Dokumentation. technical report, University of Passau, 1997

[3] Kahlert J. Fuzzy Control für Ingenieure. Verlag Friedr. Vieweg & Sohn, Braunschweig, Wiesbaden, 1995

[4] Oppenheim AV, Willsky AS. Signale und Systeme. VCH, Weinheim, 1992

[5] Rojas R. Neural Networks – A Systematic Introduction. Springer-Verlag, Berlin, Heidelberg, New York, 1996

# Efficient Local Subspace Construction for Neural Data Modeling

J. Bruske, J. Tröster, G. Sommer

Computer Science Institute, Christian-Albrechts-Universität Kiel

Kiel, Germany

### Abstract

In this article we show how optimally topology preserving maps (OTPMs) can be used for efficient local subspace construction and how existing local approximation networks can benefit from the local subspaces such constructed. The network models include radial basis function (RBF) networks, local linear maps (LLM) and local linear data modeling.

The local subspaces are constructed by local principal component analysis (PCA). Exploiting the OTPM, the local PCA can be shown to have only *linear time complexity* w.r.t. the dimensionality of the input space (in contrast to the prohibitive *cubic complexity of the conventional approach*), and hence the method becomes applicable even for very high dimensional input spaces as frequently encountered in computer vision, cf. [2]. Interesting on its own, we demonstrate the workability of the subspace approach by means of an appearance based robot grasping system.

## 1  Introduction

It has long been noticed that if input data stemming from some low $d$-dimensional manifold are embedded in a high $n$-dimensional input space, a local approximation scheme should work by projecting input data to the manifold and restrict approximation to within the manifold (hence eliminating noise orthogonal to the manifold). In a straightforward approach (e.g. [3]) a number of centers is distributed within the input manifold and an approximation to the manifold by local subspaces is obtained by local PCA of data in the Voronoi cells of the centers. Yet PCA has serial time complexity $O(n^3)$ independent of the intrinsic dimensionality $d$ of manifold, and hence the approach becomes prohibitive for high dimensional input spaces.

On ICANN'97, [2], we have shown how local subspace analysis by PCA becomes possible in time $O(n + m(d)^3)$ using optimally topology preserving maps (OTPMs), where $m(d)$ is a function of the intrinsic dimensionality of the manifold only. Hence by using OTPMs, local subspace analysis scales only linear with the input dimensionality and suddenly a large range of local subspace approximation schemes becomes applicable. For example, in the local linear data modeling approach [3], data are modeled within the local linear subspace. In RBF networks, stimuli can be projected to the local subspace prior to calculation of the activation function, resulting in Hyper Basis Function (HBF) networks, [5]. Finally, in Ritter's Local Linear Map (LLM), [7], the subspaces can be used for both projection and reduction of storage requirements for the Jacobian matrices.

## 2  Efficient local subspace construction

For the convenience of the reader, we will now briefly review the basic procedure for efficient local subspace construction with optimally topology preserving maps as presentend in [2]. Given a training set $T \subset \mathbb{R}^n$, it proceeds in four stages (batch-variant).

1. Generate a set of $N$ centers $S = \{c_1, \ldots, c_N\} \subset \mathbb{R}^n$ as the output of a vector quantization algorithm working on the training set $T$.

2. Calculate the graph $G = (V, E)$ by

   (a) associating each center in $S$ with a node in $V$, i.e.
   $$|V| = |S| \text{ and } c_i \in S \Leftrightarrow i \in V$$

   (b) for each $x \in T$, connecting the nodes associated with the best and second best matching centers, i.e.
   $$E = \{(i,j) \mid \exists x \in T \, \forall k \in V \backslash \{i,j\} : \max\{\| c_i - x \|, \| c_j - x \|\} \leq \| c_k - x \|\}$$

   $G$ is called the optimally topology preserving map, $OTPM_T(S)$, of $S$ w.r.t. $T$, cf. [2].

3. For each node $i \in V$ perform a principal component analysis of the set of $m_i$ difference vectors $\{c_{1_i} - c_i, \ldots, c_{m_i} - c_i\}$, with $(c_{j_i} - c_i)$ the difference vectors between $c_i$ and $c_{j_i}$, the center of its $j$-th direct topological neighbor in $G$.

4. Exclude eigenvectors corresponding to very small eigenvalues.

As a result of the vector quantization stage (step 1) the centers are placed within the data manifold $M \subseteq \mathbb{R}^n$ and noise orthogonal to $M$ is filtered out. In step 2, $OTPM_T(S)$ is constructed by simply connecting nodes corresponding to best and second best matching centers on presentation of $T$.

The central "trick" in step 3 is to use the difference vectors $(c_{j_i} - c_i)$ for PCA of each local subspace and not the data in a local region itself. First, the difference vectors have very low noise component orthogonal to $M$ (due to the noise reduction property of the vector quantizing stage), and second, the number of neighbors $m$ of a node in an OTPM does only depend on the intrinsic dimensionality $d$ and is small for small $d$. The latter property can be exploited to perform local PCA in time $O(m(d)^2 n + m(d)^3)$, [2], hence scaling only linearly (optimally) with the input dimensionality $n$.

Deciding in step 4 what size an eigenvalue $\mu_i$ as obtained by each local PCA must have to indicate an associated intra-manifold eigenvector (and not a noise component), amounts to determining a threshold $\alpha$ (significance level). In this work, we will regard an eigenvalue $\mu_i$ as significant if $\frac{\mu_i}{\max_j \mu_j} > \alpha$. If no prior knowledge is available, different values of $\alpha$ have to be tested.

# 3 Local subspaces for neural data modeling

Local subspace construction as described in section 2 supplies us with a set of (orthonormal) eigenvectors $e_1^i, \ldots, e_{l_i}^i$, $l_i \leq m_i$, spanning a local subspace for each center $c_i \in S$. In this section we show how knowledge of these subspaces can be used to improve existing local approximation schemes.

## 3.1 Local linear modeling

In [3], Kambhatla and Leen proposed an algorithm for local linear modeling of data based on clustering and straightforward PCA. They first generate a set of centers $S$ by clustering the data with a vector quantizer, then they use local PCA in each Voronoi cell to obtain the local eigenvectors. New data is coded as the index of the best matching unit, $bmu$, together with the projection coefficients to the local eigenvectors $(x - c_{bmu})^T e_1^{bmu}, \ldots, (x - c_{bmu})^T e_{l_{bmu}}^{bmu}$. The reconstruction (decoding) is given by

$$\hat{x} = c_{bmu} + \sum_{i=1}^{l_{bmu}} ((x - c_{bmu})^T e_i^{bmu}) e_i^{bmu}. \tag{1}$$

Utilizing the efficient local subspace construction procedure (section 2) we were able to apply local linear modeling to a sequence of $(64 \times 64)$-dimensional images in [2].

## 3.2 Subspace Local Linear Maps

The Local Linear Map (LLM) as introduced by Ritter et al., [7], has found widespread application for learning input - output mappings. The LLM rests on a locally linear (first order) approximation of the unknown function $f : R^n \to R^k$ and computes its output as (winner-take-all variant)

$$y(x) = A_{bmu}(x - c_{bmu}) + o_{bmu}. \tag{2}$$

Here $o_{bmu} \in R^k$ is an output vector attached to the best matching unit (zero order approximation) and $A_{bmu} \in R^{k \times n}$ is a local estimate of the Jacobian matrix (first order term). Centers are distributed by a clustering algorithm.

Due to the first order term, the method is very sensitive to noise in the input. With a noised version $x' = x + \eta$ the output differs by $A_{bmu}\eta$, and thus the LLM largely benefits from projecting to the local subspace, cancelling the orthogonal (w.r.t $M$) noise component of $\eta$. Equally important, instead of adapting and storing $k \times n$ parameters with each matrix $A_i$, by first projecting to the local $l_i$-dimensional subspaces only matrices $A_i' \in R^{k \times l_i}$ need to be stored. This results in much better scaling with the input dimension $n$ and, because of the reduced number of free parameters, better learning and generalization properties. The Subspace LLM (SLLM) proposed in this article hence takes the form

$$y(x) = A_{bmu}' E_{bmu}(x - c_{bmu}) + o_{bmu}, \tag{3}$$

where $E_i = [e_1^i, \ldots, e_{l_i}^i]$ denotes the local projection matrix as calculated by the efficient subspace construction procedure.

## 3.3 Subspace RBF networks

RBF networks for learning input - output mappings approximate the unknown function $f : R^n \to R^k$ by

$$y(x) = \sum_{c_i \in S} o_i h_i(\| x - c_i \|), \tag{4}$$

where $h_i : R^+ \to [0,1]$ denote strictly monotonically decreasing activation functions, e.g. $h_i(z) = \exp(-z^2/\sigma_i^2)$. Their netto input is usually calculated as the squared Euclidean distance, i.e. $\| x - c_i \|^2 = (x - c_i)^2$. With the squared Euclidean distance, some noise component $\eta$ orthogonal to the input manifold will result in a change of $\eta^2$ of all netto inputs and hence seriously affect approximation. Again, this noise component can be filtered out by projecting the input to the local subspaces. This amounts to setting $\| x - c_i \|^2$ to $(x-c_i)^T E_i E_i^T (x-c_i)$ where $E_i = [e_1^i, \ldots, e_{l_i}^i]$ is the local projection matrix. The basis functions now become Hyper Basis Functions (HBFs), cf. [5].

In general, the manifold can be curved and wrapped and hence we can not use all HBFs for approximation for a given stimulus (because they have lost any sensitivity orthogonal to the local eigenvectors). A solution to this problem is to use only the HBFs of the best matching unit and its direct topological neighbors w.r.t. the OTPM, $Nh^+(bmu)$, for output calculation:

$$y(x) = \sum_{i \in Nh^+(bmu)} o_i h_i((x - c_i)^T E_i E_i^T (x - c_i)), \tag{5}$$

Note that eq. (5) is just the Dynamic Cell Structure approximation scheme [1], except the additional projection to the local subspaces.

## 3.4 Normalized RBF networks and the effective projection property

In normalized RBF networks, the output of the RBF network is normalized with the total activation, i.e.

$$\bar{y}(x) = \sum o_i h_i / \sum h_i. \tag{6}$$

Interestingly, in case of Gaussian activation functions with equal widths $\sigma_i$ and all centers $c_i$ lying in a linear subspace $H \subset \mathbb{R}^n$, these nework possess an *effective projection property* [1]: They are insensitive to noise orthogonal to $H$ without explicitly projecting to $H$, i.e.

$$\bar{y}(x + \eta) = \bar{y}(x) \quad \text{for } x \in H \text{ and } \eta \perp H. \tag{7}$$

Hence in this particular case, explicit projection to $H$ has little effect.

# 4   The robot grasping application

Our demonstration concerns appearance based pose estimation of objects, cf. [4]. Fig. 1 (left) shows the experimental setup. Given an image of the object, the robot has to estimate its pose and to grasp it. The flat object has only one degree of freedom, its rotation around the z-axis, and hence images under different rotations lie on a 1-dimensional trajectory in image space.

Preprocessing of the image involves segmentation, scale normalization, taking the logarithm of grey values and convolving the image with 75 DC-corrected Gabor filters. The latter are distributed on a 5 × 5 grid, with 3 orientations on each position. As bandpass filters the Gabor filters serve two purposes: In conjunction with the logarithmically graded grey values they first achieve brightness invariance (filter out the DC component) and, second, they filter out high frequencies which would lead to discontinuous trajectories in image space. The 75 filter responses serve as input to the networks. We use 180 images and their corresponding grasping angles as a training set, 180 as a test set (differing by 1° rotation from those in the test set).

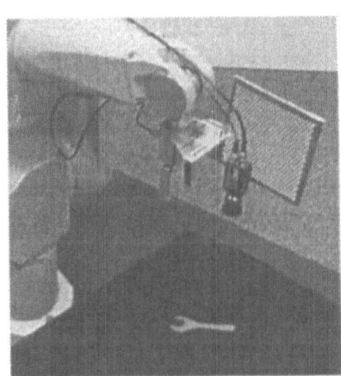 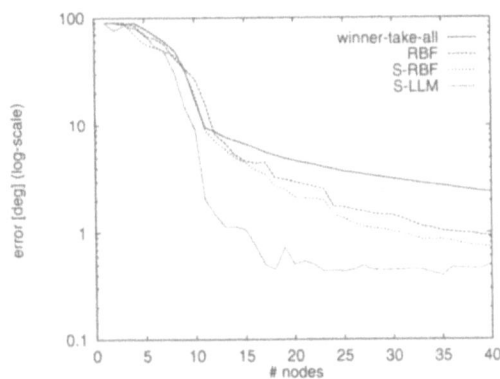

Figure 1: Left: Robot arm with attached camera for appearance based grasping of a spanner. Right: Pose error [°] for Subspace LLM on $\alpha = 0.2$ level, normalized RBF and Subspace RBF network and winner-take-all network.

Figure 1 (right) shows the grasping accuracy for a Subspace-LLM on the $\alpha = 0.2$ significance level, for a normalized Subspace-RBF network on the $\alpha = 0.2$ level, for a normalized RBF network and a winner take all scheme (averaged over 180 test images). All approximation schemes used the same center distribution (generated by an incremental LBG vector quantization algorithm). For the RBF networks we used the activation functions

$$h_i(z) = \exp(-z^2/\sigma_i^2) \text{ with } \sigma_i^2 \sim \tfrac{1}{|Nh(i)|} \sum_{j \in Nh(i)} (c_j - c_i)^2.$$

For each number of centers the output vectors $o_i$ (and Jacobian matrices $A_i$ in case of the Subspace-LLM) were optimized by singular value decomposition (SVD) [6]. Not surprisingly, the Subspace-LLM shows the best performance and achieves a pose estimation error of 1° with as few as 15 nodes in the network. The normalized Subspace-RBF network, however, performs only slightly better than the simple normalized RBF network. This can be attributed to the effective projection property of normalized RBF networks.

# 5 Conclusion

We have shown how to efficiently construct local subspaces utilizing OTPMs and how to use these subspaces for improved local approximation schemes. By projecting data to the relevant subspaces, othogonal noise is filtered out and, as in the case of the Subspace-LLM, the number of free parameters can be dramatically decreased.

Concerning our grasping application, we have succeeded in extending the demonstration presented in this article to an appearance based robot grasping system for multiple objects. In another project, we have equally successfully applied the local subspace approximation schemes for estimating the head posture in facial images. Based on these experiences we are confident that the presented subspace approach may become a valuable tool in pattern recognition applications characterized by high dimensional input spaces but low intrinsic dimensionality, in particular visual learning.

# References

[1] J. Bruske. *Dynamische Zellstrukturen - Theorie und Anwendung eines KNN-Modells*. PhD thesis, Inst. f. Inf. u. Prakt. Math., Christian-Albrechts-Universitaet zu Kiel, 1998. (submitted).

[2] J. Bruske and G. Sommer. Topology representing networks for intrinsic dimensionality estimation. In *Proc. ICANN'97*, Springer LNCS, Nr. 1327, pages 595–600, 1997.

[3] N. Kambhatla and T.K. Leen. Fast non-linear dimension reduction. In *Advances in Neural Information Processing Systems, NIPS 6*, pages 152–159, 1994.

[4] H. Murase and S. Nayar. Visual learning and recognition of 3-d objects from appearance. *International Journal of Computer Vision*, 14:5–24, 1995.

[5] T. Poggio and F. Girosi. Networks for approximation and learning. *Proc. of the IEEE*, 78(9):1481–1497, 1990.

[6] W.H. Press, S.A. Teukolsky, W.T. Vetterling, and B.P. Flannery. *Numerical Recipes in C - The Art of Scientific Computing*. Cambridge University Press, 1988.

[7] H. Ritter, T. Martinetz, and K. Schulten. *Neuronale Netze*. Addison-Wesley, 1991.

# Edge Detection of Multispectral Images Using the 1-D Self-Organizing Map

Pekka J. Toivanen, J. Ansamäki, S. Leppäjärvi, J. Parkkinen

Dept. of Information Technology, Lappeenranta Univ. of Technology

P.O. Box 20, FIN-53851, Finland

E-mail: Pekka.Toivanen@lut.fi

June 4, 1998

## Abstract

In this paper, a new method for edge detection in multispectral images is presented. It is based on the use of the Self-Organizing Map (SOM) and a conventional edge detector. The method presented in this paper orders the vectors of the original image in such a way that vectors that are near each other according to some similarity criterium should have scalar ordering values near each other. This is achieved using the 1-dimensional Self-Organizing Map. After ordering, the original vector image reduces to a gray-value image, and conventional edge detectors can be applied. In this paper, the Laplace and Canny edge detectors are used. It is shown, that using the Self-Organizing Map (SOM) in ordering the vectors of the original spectral image it is possible to find also those edges that the R-ordering based methods miss.

## 1 Introduction

It is not possible to define uniquely the ordering of multivariate data. A number of ways have been proposed to perform multivariate data ordering. They are usually classified into the following categories: marginal ordering (M-ordering), reduced or aggregate ordering (R-ordering), partial ordering (P-ordering), and conditional ordering (C-ordering). [3]

Conventionally, edge detection methods of multispectral images are based on gradient methods [5] or ordering the spectral vectors first using some ordering method, e.g. R-ordering [9]. Both of these methods have some drawbacks which can be overcome using the method presented in this paper. The gradient approach is unsatisfactory in cases where the image gradients show the same strength but in opposite directions. Then, the vector sum of the gradients would provide a null gradient [10]. On the other hand, if the vectors are first ordered using for instance R-ordering, we get a scalar for every vector and hence the vectors can be ordered. Unfortunately the R-ordering can give the same scalar value for spectra which are not similar and may either be the same color in RGB-space or may not. It will be shown in this paper, that using the 1-dimensional Self-Organizing Map (SOM) in ordering the vectors it is possible to find also those edges that the R-ordering based methods do not find.

# 2 Background

## 2.1 R-Ordering

In this paper, a color image is viewed as a vector field, represented by a discrete vector valued function $\mathbf{g} : Z^2 \to Z^p$, where $Z$ represents the set of integers and $p$ is an integer.

Let $\mathbf{x}$ represent a $p$-dimensional vector $\mathbf{x} = [x_1, x_2, ..., x_p]^T$, where $x_l$, $l = 1, 2, ..., p$ are the spectral components of a pixel and let $\mathbf{x}^j$, $j = 1, 2, ..., n$ be the pixel $j$ in the image $\mathbf{g}$. $n$ is the number of pixels in the image $\mathbf{g}$. Each $\mathbf{x}^j$ is a $p$-dimensional vector $\mathbf{x}^j = [x_1^j, x_2^j, ..., x_p^j]^T$. In R-ordering, each vector $\mathbf{x}^j$ is reduced to a scalar value $d_j$ in the following way:

$$d_j = \sum_{k=1}^{n} \|\mathbf{x}^j - \mathbf{x}^k\|, \tag{1}$$

where $\|.\|$ represents an appropriate vector norm. An arrangement of the $d_j$'s in ascending order $d_1 \leq d_2 \leq ... \leq d_n$, associates the same ordering to the multivariate $\mathbf{x}^j$'s: $\mathbf{x}^1 \leq \mathbf{x}^2 \leq ... \leq \mathbf{x}^n$. $\mathbf{x}^1$ is the vector median of the data samples [1].

## 2.2 The Self-Organizing Map

The basic idea of the Self-Organizing Map (SOM) assumes a sequence of input vectors $\{\mathbf{x}^j, j = 1, 2, ..., n\}$, where $n$ is the number of the vectors. $\{\mathbf{m}_i^j, i = 1, 2, ..., k\}$ denotes the set of representative neuron vectors in the SOM at the iteration phase $j$ which form the SOM. $k$ is the number vectors in the SOM. Every $\mathbf{m}_i^j$ is a $p$-dimensional vector.

It is assumed that the $\mathbf{m}_i^0$ have been initialized in some proper way; random selection will often do. Every input $\mathbf{x}^j$ is compared to all the $\mathbf{m}_i^j$. The input signal vector $\mathbf{x}^j$, the representative neuron vectors in the SOM $\mathbf{m}_i^j$, and best mathing unit $c$ are related by Equation (2),

$$\|\mathbf{x}^j - \mathbf{m}_c^j\| = min_i\{\|\mathbf{x}^j - \mathbf{m}_i^j\|\}, \tag{2}$$

where $\|.\|$ represents an appropriate vector norm. In this paper, the Euclidean norm is used [6].

Updating the Self-Organizing Map in the learning phase is done according to Equations (3) and (4),

$$\mathbf{m}_i^{j+1} = \mathbf{m}_i^j + \alpha^j[\mathbf{x}^j - \mathbf{m}_i^j] \quad \forall i \in N_c^j, \tag{3}$$

$$\mathbf{m}_i^{j+1} = \mathbf{m}_i^j \quad \forall i \notin N_c^j. \tag{4}$$

$N_c^j$ is a topological neighborhood which is centered around that representative neuron vector for which the best match with input $\mathbf{x}^j$ is found. The radius of $N_c^j$ is shrinking monotonically with time. $\alpha^j$ is a scalar parameter that decreases monotonically during the course of the process, $0 < \alpha < 1$ [6].

## 2.3   The Laplace and Canny Operator

The Laplacian edge detection operator is an approximation to the mathematical Laplacian $\frac{\partial^2 f}{\partial x^2} + \frac{\partial^2 f}{\partial y^2}$. In digital images it can be stated as follows: [2]

$$L(x,y) = f(x,y) - \frac{1}{4}[f(x,y+1) + f(x,y-1) + f(x+1,y) + f(x-1,y)] \quad (5)$$

The Laplace edge detector tends to doubly enhance any noise in the image. In this paper, though, this is not a problem because we assume that the original images are noiseless.

The Canny operator is defined as follows. Let us denote $G_n$ the first derivative of the 2-dimensional Gaussian function $G$ in the direction $\mathbf{n}$. $g = g(x,y)$ is the original image, as mentioned earlier.

$$G = e^{\frac{(x^2+y^2)}{2s^2}} \quad (6)$$

$$G_{\mathbf{n}} = \frac{\partial G}{\partial \mathbf{n}} \quad (7)$$

where $\mathbf{n}$ is the direction of the gradient and $s$ is the deviation. An estimate for the direction $\mathbf{n}$ is formed from the smoothed gradient direction:

$$\mathbf{n} = \frac{\nabla(G * g)}{|\nabla(G * g)|} \quad (8)$$

where $*$ denotes convolution and $\nabla$ the nabla operator. An edge point is defined as a local maximum in the direction $\mathbf{n}$ of the operator $G_{\mathbf{n}}$ applied to the image $g$,

$$\frac{\partial}{\partial \mathbf{n}} G_{\mathbf{n}} * g = 0. \quad (9)$$

Substituting for $G_{\mathbf{n}}$ from Equation (7) and associating Gaussian convolution, Equation (9) becomes

$$\frac{\partial^2}{\partial \mathbf{n}^2} G * g = 0 \quad (10)$$

# 3   Edge Detection with 1-Dimensional SOM

The basic idea of the new edge detection method proposed in this paper is to order the image spectra in such a way that vectors which are near each other according to some similarity measure will get scalar values which are near each other. This is achieved by first teaching a $32 \times 32$ Self-Organizing Map using the spectra of the whole Munsell book [7]. The spectra range from 400 nm to 700 nm with a step of 5 nm containing 61 components.

Let $\mathbf{M} = [\mathbf{m}_1, \mathbf{m}_2, ..., \mathbf{m}_k]$ be the 1-Dimensional Self-Organizing Map with $k$ vectors. In the teaching phase, every pixel vector of the original image $g(x, y)$ is taught to the 1-Dimensional SOM with the neighborhood $N_c$ defined as follows:

$$N_c = \{max(1, c - l), c, min(k, c + l)\} \tag{11}$$

where $l$ is a suitable positive integer. During the learning phase, $l$ decreases. After the teaching is completed, an input vector $\mathbf{x}$ of $\mathbf{g}$ is inserted into the SOM to find out the best matching unit $c$ as follows:

$$||\mathbf{x} - \mathbf{m}^c|| = min_i\{||\mathbf{x} - \mathbf{m}^i||\} \tag{12}$$

The scalar value of this best matching unit $c$ is inserted into a new image $f$ to replace the vector of the same location in $g$:

$$f(x, y) \leftarrow c \tag{13}$$

## 4    Results

The use of the 1-Dimensional Self-Organizing Map means that no Peano scan is needed. This is due to the fact that the 1-Dimensional Self-Organizing Map already orders the image vectors topologically. Figure 1 shows the original multispectral image in which every pixel is a 61-dimensional vector. Regions 3,4,5,7, and 8 hold different vectors. Regions 3 and 4 are green, regions 5 and 7 blue, and region 8 is red in RGB-space. Their scalar values given by the R-ordering method, Equation (1), are the same, which means, that they cannot be ordered using the R-ordering method:

$$d_3 = d_4 = d_5 = d_7 = d_8 \tag{14}$$

This means that is not possible to find edges between the regions 3,4,5,7, and 8 using an edge detection method in which the ordering of vectors is based on the R-ordering method.

Figure 2 shows the 1-dimensional Self-Organizing Map after teaching the whole Munsell book to it. The SOM consists of 20 neuron vectors. Figure 3 shows the corresponding gray-level image which is formed by replacong every pixel vector in the original image by its ordering scalar, which is given by the 1-D SOM. The edges found by first ordering the pixels by R-ordering and then using the Canny edge detector are depicted in Figure 4. Results of the new method are shown in Figures 5 and 6. Figure 5 shows the edges found after the 1-D SOM and the Laplace edge detector. Figure 6 shows an edge image given by the same SOM and the Canny edge operator.

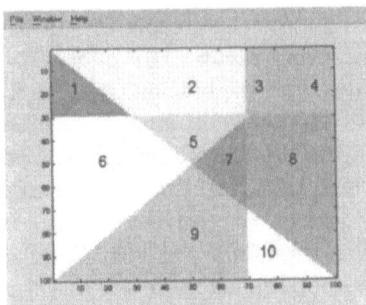

**Figure 1.** The original spectral image.

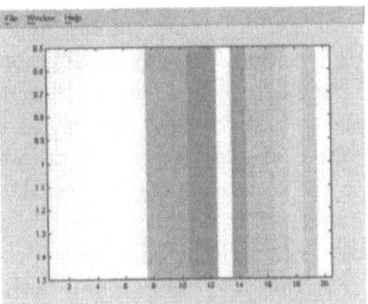

**Figure 2.** The 1-Dimensional SOM map after teaching.

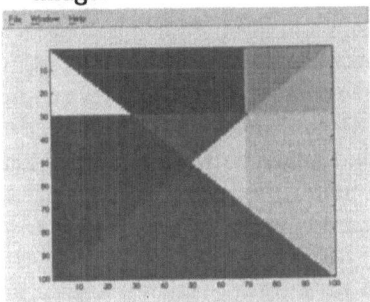

**Figure 3.** The gray-level image formed by the SOM. Every gray value is mapped to an RGB value.

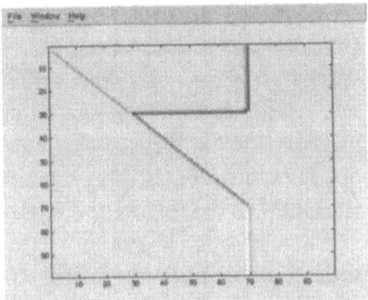

**Figure 4.** The edge image with R-ordering and Canny edge detector.

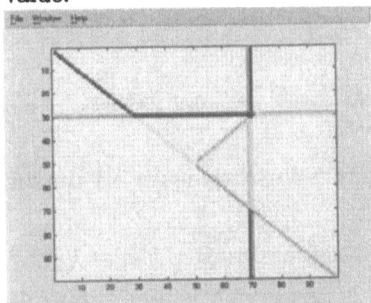

**Figure 5.** The edge image after Laplacian.

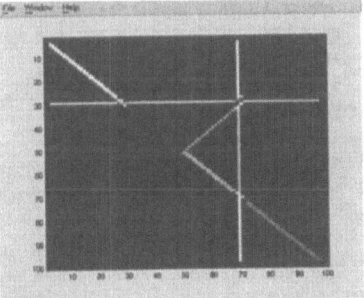

**Figure 6.** The edge image after Canny.

## 5 Discussion

It was shown in this paper, that using the 1-dimensional Self-Organizing Map (SOM) in ordering the vectors of the original multispectral image it is possible to find also those edges that the R-ordering based methods miss. The ordering is achieved by letting the Self-Organizing Map learn the whole vector space under consideration. In this paper, the whole Munsell book was taught to the 1-dimensional Self-Organizing Map. Those vectors which are topogically near each other in the original vector space are also near each other in the 1-dimensional Self-Organizing Map. After this, every vector of the original vector image is matched to the 1-dimensional SOM to find the best matching unit, i.e. the vector which is nearest to the vector of the original image. When this is found, the corresponding ordering scalar of the best matching unit is inserted to the original image to replace the vector and thus finally we get a gray-level image. The edge detection is easy to perform on the gray-level image using any conventional edge detector. In this paper, the Laplace and the Canny edge detectors are used. They both find all the possible edges in the test image. They are capable of finding edges between regions whose R-ordering values are different, and edges between regions whose R-ordering values are the same. In the latter case the vectors of the regions are not the same and also their RGB-values are not the same. Furthermore, with the method presented in this paper, it is also possible to find edges between regions whose R-ordering values are equal and whose RGB-values are the same although the vectors are not equal. By increasing the size of the Self-Organizing Map it is possible to improve the ability of the new method to find edges between vectors which are very near each other according to some similarity measure. If the size of the SOM is decreased the edge detector will become more robust by clustering more vectors to the node vector in the SOM, which would diminish its ability to find edges between vectors with only small variations.

# References

[1] J. Astola, P. Haavisto, and Y. Neuvo:
"Vector Median Filters", Proceedings of the IEEE, Vol. 78, pp. 678-689, 1990.

[2] D. H. Ballard and C. M. Brown:
Computer Vision, Prentice-Hall, 1982.

[3] V. Barnett:
"The Ordering of Multivariate Data", *Journal of the Royal Statistical Society*, A, 139, Part 3, pp. 318-355, 1976.

[4] J. Canny:
"A Computational Approach to Edge Detection", *IEEE Transactions on Pattern Analysis and Machine Intelligence*, Vol. PAMI-8, No. 6, 1986.

[5] A. Cumani:
"Edge Detection in Multispectral Images", *CVGIP: Graphical Models and Image Processing*, Vol. 53, No. 1, January, pp. 40-51, 1991.

[6] T. Kohonen:
Self-Organization and Associative Memory, Springer-Verlag, 3rd Edition, 1989.

[7] Munsell Book of Color-Matte Finish Collection. Munsell Color, Baltimore, Md., USA, 1976.

[8] E. A. Patrick, D. R. Anderson, and F. Bechtel:
"Mapping multidimensional space to one dimension for computer output display", *IEEE Transactions on Computers*, pp. 949-953, 1968.

[9] P.E. Trahanias and A.N. Venetsanopoulos:
"Color Edge Detection Using Vector Order Statistics", *IEEE Transactions on Image Processing*, Vol. 2, No. 2, pp. 259-264, 1993.

[10] S. Zenzo:
"A note on the gradient of a multi-image", *Computer Vision, Graphics and Image Processing*, Vol. 33, pp. 116-125, 1986.

# Design of Cellular Neural Networks for Binary and Gray Level Image Processing

David Monnin,* Lionel Merlat, Axel Köneke

French-German Research Institute of Saint-Louis

PO Box 34 - 68301 Saint Louis, France

monnin@lis.inpg.fr, monnin@ece.fr

Jeanny Herault

LIS Laboratory, INPG

Grenoble, France

**Abstract**

An original unified approach to the design of Cellular Neural Networks for many image processing tasks is proposed. It is shown that numerous traditional processing operators, among which are all convolution mask operators as well as morphological operators, can be successfully performed by a simple Cellular Neural Network with no feedback interconnection. The design of both gray and binary stable output operators is addressed, and the ability of a Cellular Neural Network to threshold or rescale image gray levels and to take advantage of the network initial state as a second input is highlighted.

## 1 Introduction

The real-time requirements involved in many artificial vision schemes necessitate the ability to deal, in a swift way, with a large amount of data. Deeply implicated in the early vision process, which aims to keep only the most significant components of the visual information perceived, the efficient implementation of low-level image processing algorithms is of great importance.

Cellular Neural Networks (CNNs) [1] are arrays of analog locally connected cells conceived for an implementation in VLSI technology. With cell-coupled light sensors, a CNN can be turned into a powerful artificial retina able of high-speed parallel image processing. As all analog cells are governed by the same set of parameters called cloning template, programming a CNN for a specific image processing task consists in finding those parameters.

A large number of common image processing algorithms are based on the principle of convolution masks. The present design method takes advantage of well-known results of digital image processing for providing a CNN equivalent to those algorithms. The method will be introduced through a cell dynamics based analysis, the case of gray scale and binary output image processing operators will then be discussed and sample applications will be given.

---

*David Monnin is also a member of the LIS Laboratory

## 2    Cell Dynamics Based Analysis

The operation of a cell (i, j) is described by the following equations [1]:

$$C\frac{dx_{i,j}}{dt} = -\frac{1}{R}x_{i,j} + A \otimes y_{i,j} + B \otimes u_{i,j} + I \tag{1}$$

$$y_{i,j}(x) = \frac{1}{2}(|x_{i,j} + 1| - |x_{i,j} - 1|), \tag{2}$$

where $\otimes$ denotes a two-dimensional discrete spatial convolution such that $A \otimes y_{i,j} = \sum_{k,l\in N(i,j)} A_{k,l}\cdot y_{i-k,j-l}$ for $k$ and $l$ in the neighborhood $N(i,j)$ of cell $(i,j)$. For convenience and without loss of generality positive real constants $R$ and $C$ will be set to unity in the rest of the paper.

Setting the feedback template $A$ to null values, except for its central element, inhibits the propagation effects due to the feedback interconnection between cells and reduces the complexity of the network. Benefits of the non-feedback interaction in terms of VLSI implementation were tackled in [2]. Further, as will be shown in the next sections, single feedback is sufficient for performing many common image processing tasks because no propagation effects are ordinarily needed for them. Equation (1) can thus be simplified and expressed as:

$$\frac{dx_{i,j}}{dt} = -x_{i,j} + a.y_{i,j} + B \otimes u_{i,j} + I \tag{3}$$

As such a cell does not receive any input from other cells and the external inputs and current constant are kept unchanged during the network transient phase, the cell convergence is guaranteed, which induces the stability of the whole CNN. The stability of a CNN was discussed in [1, 3, 4]. It must be noticed that the value of the feedback parameter $a$ is of great importance, because it leads to two distinct cell behaviors leading to two different families of operators. In order to establish this statement, (2) is replaced by its three linear constituents which are then introduced in (3). Thich yields:

$$\frac{dx_{i,j}}{dt} = \quad -x_{i,j} - a + B \otimes u_{i,j} + I \quad , \text{ for } x_{i,j} \in ]-\infty, -1] \tag{4a}$$

$$\frac{dx_{i,j}}{dt} = \quad (a-1).x_{i,j} + B \otimes u_{i,j} + I \quad , \text{ for } x_{i,j} \in [-1, 1] \tag{4b}$$

$$\frac{dx_{i,j}}{dt} = \quad -x_{i,j} + a + B \otimes u_{i,j} + I \quad , \text{ for } x_{i,j} \in [1, +\infty[ \tag{4c}$$

In the light of the characteristic representations of cell trajectories given in figure 1, it clearly appears that a drastically different trajectory is observed, depending on whether its slope is negative or positive in $[-1, 1]$.

In the case of a negative slope, i.e. when $a < 1$, there is only one possible equilibrium point located in $]-\infty, +\infty[$ and leading to an output value comprised in $[-1, 1]$. In the case of a positive slope, i.e. when $a > 1$, there are up to two possible equilibrium points located in $]-\infty, -1]$ and/or $[1, +\infty[$ and thus leading to the output value -1 or 1. Two primary categories of CNN image processing operators can hence be defined according to the value of parameter $a$, those providing a gray level output[1] and those providing a binary output. These two categories of operators are studied in sections 3 and 4.

---

[1] As Cellular Neural Networks are not digital but analog processors, "gray level output" actually means "continuously graded output" in this paper.

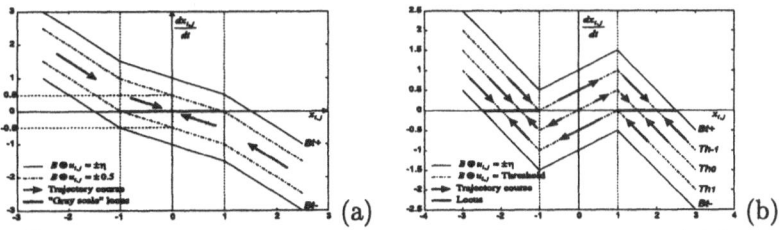

Figure 1: All possible cell trajectories, for a given set of parameters $a$, $B$ and $I$, are represented by the gray area between boundary trajectories $Bt_-$ and $Bt_+$, respectively obtained for $B \otimes u_{i,j} = -\eta$ and $B \otimes u_{i,j} = \eta$, where $\eta$ is the maximum convolution value defined in (7); $a < 1$ for (a) and $a > 1$ for (b)

# 3   Gray Level Output Processing Operators

Since the avowed objective of the design method discussed here is to devise linear filters based on convolution operations, it is convenient to restrict the cells locus to the $[-1, 1]$ domain where the output is a linear function of the internal state $x_{i,j}$, covering the whole range of output values. Hence, as $y_{i,j} = x_{i,j}$ for $x_{i,j} \in [-1, 1]$ and the steady state is reached for $\frac{dx_{i,j}}{dt} = 0$, the cell $(i, j)$ steady output value can be derived from (4b) and expressed as:

$$y_{i,j}^\infty = \frac{1}{1-a} B \otimes u_{i,j} + \frac{I}{1-a} \text{ or more generally as } y_{i,j}^\infty = \alpha . B \otimes u_{i,j} + \beta \quad (5-6)$$

According to assumptions made until now, it is obvious from (6) that a CNN can implement linear convolution filters using feedforward matrix $B$ as a convolution mask. Indeed, when CNN parameters are set in such a way that $\alpha = 1$ and $\beta = 0$, the CNN operator designed is no more than a linear convolution filter. But it is also possible to choose $\alpha$ and $\beta$ in a different way in order to rescale the result of the convolution operation. Hence, the original range $[m, n]$ of an image can be rescaled to a desired range $[M, N]$. Therefore, it is feasible to expand, reduce or translate the histogram function of a processed image. The present design method thus consists in finding parameters $a$ and $I$, given a convolution mask $B$, an original range $[m, n]$ and a desired range $[M, N]$. For convenience, let us define a normalization factor $\eta$ such that:

$$\eta = \sum_{k,l} |B_{k,l}| \quad (7)$$

Normalization factor $\eta$ is useful when a normalized convolution mask is wanted and feedforward matrix $B$ is not yet normalized. If $B$ is already normalized or should not be normalized, $\eta$ has to be set to unity. The expression of $\alpha$ and $\beta$ is then given by:

$$\alpha = \frac{\eta}{1-a} \quad , \text{ and } \quad \beta = \frac{I}{1-a} \quad (8a\text{-}b)$$

In order to find the expression of $a$ and $I$, let us express the range conversion as a set of two equations involving range boundaries:

$$\begin{cases} M = \alpha.m + \beta \\ N = \alpha.n + \beta \end{cases} \Leftrightarrow \begin{cases} M = \frac{\eta}{1-a}.m + \frac{I}{1-a} \\ N = \frac{\eta}{1-a}.n + \frac{I}{1-a} \end{cases} \quad (9)$$

Solving (9) for $a$ and $I$ finally gives:

$$a = 1 - \eta \frac{m-n}{M-N} \quad , \text{ and } \quad I = N.\eta \frac{m-n}{M-N} - n \quad (10a\text{-}b)$$

Furthermore, a "reverse video" effect can be obtained by simply reversing the sign of $B$ and using a new original range which is symmetrical to the old one with respect to the origin and yields a new current constant $I' = I + n + m$.

Figure 1a is given as an example in which the range $[-0.5, 0.5]$ of an image resulting from a normalized convolution operation is rescaled to $[-1, 1]$. If the original image has values in $[-1, -0.5]$ and/or $[0.5, 1]$, they will be respectively trimmed to $-1$ and $1$. As stated in section 2, when $a < 1$, each cell has only one equilibrium point which implies that, if the initial state of a cell is able to modify its transient trajectory, it has no influence on its final state. Consequently, when $a < 1$, the initial state is of no importance for CNNs designed to operate until steady state is reached.

## 4  Binary Output Processing Operators

Even if it is not prominent when $a > 1$, because of the binary nature of the outputs, the CNN intrinsic convolution mechanism highlighted in section 3 naturally remains here. So, it is still possible to devise linear filters based on convolution operations, but their outputs are now necessarily thresholded.

It is now helpful to observe figure 1b which exhibits interesting properties of the CNNs. Unlike the previous case, when $a > 1$, the initial state of a cell acts on its final state. This is perfectly illustrated by trajectory $Th_0$ which leads to $-1$ or $1$, according to whether the initial state is negative or positive. Moreover, $Th_0$ can be regarded as a threshold trajectory set to a threshold value of $0$ for a cell with an initial state equal to $0$. If the result of the convolution is greater than threshold $0$, the cell output is turned to $1$ or to $-1$ if it is less.

In a more general case, the threshold trajectory set to a threshold value $Th_x$ for a cell with an initial state $x_{i,j}(0) \in [-1, 1]$ can be defined as the trajectory for which $\frac{dx_{i,j}}{dt} = 0$ at point $x_{i,j} = x_{i,j}(0)$ and the result of the convolution $B \otimes u_{i,j}$ is equal to $Th_x$. According to (4b) this yields :

$$(a - 1).x_{i,j}(0) + Th_x + I = 0 \quad, \text{for} \quad x_{i,j}(0) \in [-1, 1] \tag{11a}$$

$Th_{-1}$ and $Th_1$ are also threshold trajectories of cells with initial states being respectively $-1$ and $1$ which both have a different threshold value. It is hence possible for the same convolution operation to set different threshold values to different cells according to their initial states. The design method now consists in finding parameters $a$ and $I$, given a convolution mask $B$ and one or more thresholds $Th_x$, depending on the initial state of cells they are applied to. The same principle is developed in several variants in the following subsections.

### 4.1  Single Threshold Processing

The purpose of the simplest variant of the method is to threshold the result of a linear convolution filter at a desired threshold value. As shown previously, if a unique threshold value is desired, it implies that the cells initial states are the same over the whole network. The value of the feedback parameter $a$ is here arbitrary, which yields the expression of current constant $I$:

$$I = (1 - a).x_{i,j}(0) - Th_x \quad, \text{for} \quad x_{i,j}(0) \in [-1, 1] \tag{11b}$$

For example, if the network initial state is chosen to be 0, $I$ is expressed as:

$$I = -Th_0 \tag{12}$$

An inversion effect can be obtained by simply reversing the sign of $B$ and $Th_x$.

## 4.2   Two Thresholds Processing

The aim of this second variant of the method is to threshold the result of a linear convolution filter with two different thresholds assigned to particular cells according to their input state. The CNN initial state is then used as a kind of second input which controls the threshold applied to each cell. For this method to be useful in the design of delay-type cloning templates [5], where a first cloning template would produce the initial state for a second one using two thresholds, it is important to adjust the thresholds for cell initial states being either $-1$ or $1$. This leads to the system of equations involving thresholds $Th_{-1}$ and $Th_1$ such that $Th_{-1} > Th_1$:

$$\left\{ \begin{array}{l} -1.(a-1) + Th_{-1} + I = 0 \\ \phantom{-}1.(a-1) + Th_1 \phantom{+} + I = 0 \end{array} \right. \tag{13}$$

Solving (13) for $a$ and $I$ gives:

$$a = 1 + \frac{Th_{-1} - Th_1}{2} \quad , \text{and} \quad I = -\frac{Th_{-1} + Th_1}{2} \tag{14a-b}$$

## 4.3   Single Threshold Processing and Boolean Operators

This is an adaptation of the previous method which permits to combine a binary initial state with the result of a thresholded convolution filter. "OR" and "AND" boolean operators are respectively obtained when:

$$Th_{-1} = \text{"threshold value"}, \text{ and } Th_1 \leq -\eta \tag{15}$$

$$Th_{-1} \geq \eta, \text{ and } Th_1 = \text{"threshold value"} \tag{16}$$

An inversion effect is obtained by reversing the sign of $B$ and "threshold value".

# 5   Applications

A few sample applications are presented here with their main characteristics and parameters. For each pair of images, the first image is the input and the second the output.

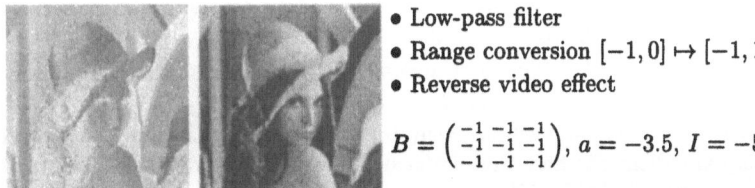

- Low-pass filter
- Range conversion $[-1, 0] \mapsto [-1, 1]$
- Reverse video effect

$$B = \begin{pmatrix} -1 & -1 & -1 \\ -1 & -1 & -1 \\ -1 & -1 & -1 \end{pmatrix}, a = -3.5, I = -5.5$$

Figure 2: Despeckling the image, reversing and enlarging the image histogram [§3]

- $Th_0 = -0.56$ and reverse video effect

$$B = \begin{pmatrix} 0 & 0 & 0 \\ 0 & -1 & 0 \\ 0 & 0 & 0 \end{pmatrix}, a = 2, I = 0.56$$

- $Th_{-1} = 0.40$ and $Th_1 = -0.81$

$$B = \begin{pmatrix} 0 & 0 & 0 \\ 0 & 1 & 0 \\ 0 & 0 & 0 \end{pmatrix}, a = 1.605, I = 0.205$$

Figure 3: Two thresholds segmentation operator, pass 1 : $X(0) = 0$, pass 2 : $X(0) = previous X(\infty)$ [§4.1, §4.2]

- $Th_{-1} = 10$ and $Th_1 = -8$
- Reverse video effect

$$B = \begin{pmatrix} -1 & -1 & -1 \\ -1 & -1 & -1 \\ -1 & -1 & -1 \end{pmatrix}, a = 10, I = -1$$

Figure 4: Morphological operator $U \cap (U \ominus B)^c$ for boundary extraction, $X(0) = U$ [§4.3]

- $Th_{-1} = 1$ and $Th_1 = -9$
- Four 90° rotations of $B$ are applied.

$$B = \begin{pmatrix} -1 & 0 & 1 \\ -2 & 0 & 2 \\ -1 & 0 & 1 \end{pmatrix}, a = 6, I = 4$$

Figure 5: Sobel operator for edge extraction, pass 1: $X(0) = -1$, passes 2,3,4 : $X(0) = previous X(\infty)$ [§4.3]

# 6   Conclusion

An original unified approach to the design of CNNs has been presented. The method permits to build complete image processing schemes including gray scale output operators to enhance and restore images and binary output operators for final image processing.

# References

[1] L. O. Chua and Yang, CNN Theory, IEEE T-CAS, vol. 35, pp. 1257-1272, 1988.

[2] L. Merlat, N. Sylvestre, and J. Mercklé, A Subclass of Cellular Neural Networks Cloning Templates, NEURAP'97, Marseille, France, 1997.

[3] L. O. Chua and T. Roska, Stability of a Class of Nonreciprocal Cellular Neural Networks, IEEE. T-CAS, vol. 37, pp. 1520-1527, 1990.

[4] L. O. Chua and C. W. Wu, On the Universe of Stable Cellular Neural Networks, in Cellular Neural Networks, J. Wiley & Sons, pp. 59-79, 1993.

[5] T. Roska and L. O. Chua, Cellular Neural Networks with Non-linear and Delay-type Template Elements and Non-uniform Grids, in Cellular Neural Networks, J. Wiley & Sons, pp. 31-43, 1993.

# Adaptive Illuminant Estimation
# Using Neural Networks

Vlad C. Cardei, Brian Funt and Kobus Barnard

School of Computing Science, Simon Fraser University
Burnaby, B.C., Canada
vcardei@cs.sfu.ca, funt@cs.sfu.ca and kobus@cs.sfu.ca

### Abstract

In this article we present results that show that a multi-layer neural network can be trained to estimate the chromaticity of the scene illumination. The network is trained with a set of artificially generated scenes and the corresponding illuminants that were used for generating the scenes. In the test phase, the network estimates the illuminant under which a given scene was taken. Tests with artificially generated as well as natural scenes show that the neural network outperforms many current color-constancy algorithms and that it is more stable and reliable for scenes that contain a small number of colors.

## 1  Color Constancy

### 1.1  Why Color Constancy?

Color constancy emerged as a very important and still open problem in many areas ranging from color indexing [1] to digital photography. In cognitive psychology, color constancy is the perceptual ability of the human visual system to discount variations in the color of the incident illumination and preserve the colors of the objects in a scene. Our goal is to design a neural network based system for machine vision, that can provide color-constant descriptors of objects in a scene. This will allow us to correct images taken with the wrong film balance, perform color based object recognition and perhaps contribute to a better understanding of human color constancy.

### 1.2  Previous work

The neural network presented in this paper will be compared to the algorithms mentioned below. All these algorithms estimate the illuminant under which a scene was taken.

Land's theory of Retinex [2] finds the maximum lightness and uses it to determine an illuminant invariant description of the scene. The version used for testing assumes that the highest sensor responses on each color channel correspond to that of a white patch. The chromaticity of the illuminant is assumed to be equal to the chromaticity of the white patch. Other algorithms are based on *a priori* assumptions about the composition of the scene. For example, Buchsbaum's *gray world* algorithm [3] assumes that the spectral average of the scene is gray and that any departure from that average is due to changes in the color of the illuminant.

One of the best performing color constancy algorithms was designed by Forsyth [4] and was extended by Finlayson [5]. Forsyth's algorithm estimates color constant descriptors for the objects in a scene, viewed under a standard "canonical" illuminant. Forsyth defines the "canonical gamut" as being the set of RGB camera responses obtained by perceiving a very large set of surfaces under a canonical illuminant.

## 2   A Neural Network Approach

### 2.1   Why a Neural Network?

Using a neural network instead of a well-defined mathematical model provides an alternative way for solving the color constancy problem and it also allows for a dynamic adaptation to a changing environment, since it has no built-in constraints, whereas classical algorithms would have to reconsider their very basic assumptions.

We thought of training a neural network [6] to learn the relationship between a scene and the chromaticity of its illumination. This would allow for the color correction of the scene [5], to make it look like it was taken under a canonical illuminant, thus providing color constancy.

### 2.2   The Neural Network

Each RGB pixel from a scene is projected into the rg-chromaticity space: $r=R/(R+G+B)$ and $g=G/(R+G+B)$. This space, which is bounded between 0 and 1, is uniformly sampled with a step S. All chromaticities that fall within the same sampling square of size S are considered equivalent and map to a distinct neuron in the input layer of the neural network. The input is set either to 0 or to 1, indicating whether there are any RGBs of chromaticity $(r,g)$ present in the scene. The size $N_I$ of the input layer is related to S: $N_I=(1/S)^2$. However, it is independent of the image size. We have made experiments with different sizes of the input layer (512, 1250, 2500 and 3600), with comparable color constancy results in all cases.

The neural network we used is a Perceptron [7] with two hidden layers. The input layer consists of a large number of binary inputs representing the presence or absence of a chromaticity in the scene. The first hidden layer contains around 400 neurons and the second layer, around 40 neurons. The output layer consists of only two neurons, corresponding to the chromaticity values of the illuminant. Our experiments show that the size of the hidden layers can vary in a wide range without affecting the performance of the network. All neurons have a sigmoid activation function.

The neural network was trained using the backpropagation algorithm [7]. The error function used for training and testing the network is the Euclidean distance in the *rg*-chromaticity space between the target and the estimated illuminant.

## 2.3 Optimizing the Neural Network

Initial tests performed with the 'standard' neural network architecture described above showed that it took a large number of epochs to train the neural network. To overcome this problem, a series of improvements have been developed and implemented [8].

The gamut of the chromaticities encountered during training and testing is much smaller than the whole (theoretical) chromaticity space. Thus, we modified the neural network's architecture, such that it will receive input only from the *active* nodes (the input nodes that were activated at least once). The *inactive* nodes (those nodes that were not activated at any time) are pruned from the neural network, together with their links to the first hidden layer. The network's architecture changes only during the first training epoch. The links from the first hidden layer are redirected only towards the neurons in the input layer that are active, i.e. those that correspond to existing chromaticities while links to inactive nodes are eliminated.

Due to the fact that the sizes of the layers are so different, we used different learning rates for each layer, proportional to the fan-in of the neurons in that layer [9]. Typical values for the learning rates are 0.1 for the output layer, 2 for the second hidden layer and 20 for the first hidden layer. This shortened the training time by a factor of more than 10, to about 5-6 epochs.

## 2.4 Training and Testing the Network

The results that we report in this paper were obtained with neural networks that have 3600 nodes in the input layer, 400 nodes in the first hidden layer, 40-50 nodes in the second hidden layer and 2 nodes in the output layer. Each node in the first hidden layer has 400 links to the input layer. The other layers are fully connected to the preceding ones. The neural networks were trained on 9800 artificially generated scenes. Each scene is composed of a variable number of patches (5 to 50) seen under one illuminant, randomly chosen from a database of 98 illuminants.

The patches correspond to matte reflectances, selected at random from a database of 260 surface reflectances. Therefore each patch has only one rg-chromaticity, derived from its RGB, which is computed by multiplying a randomly selected surface reflectance $S^j$ with the spectral distribution of an illuminant $E^k$ (selected at random, but the same for all patches in a scene) and with the spectral sensitivities of camera sensors $\rho$, over all visible frequencies $i$:

$$R = \sum_i E_i^k \cdot S_i^j \cdot \rho_i^R \ , \ G = \sum_i E_i^k \cdot S_i^j \cdot \rho_i^G \ , \ B = \sum_i E_i^k \cdot S_i^j \cdot \rho_i^B$$

# 3 Results

Tests were performed on synthesized scenes as well as on images taken with a CCD camera. The synthesized scenes were generated from the same databases used for generating the training sets. The results obtained using a neural network are compared to other color constancy algorithms.

## 3.1 Results for Synthetic Data

After training, the average error in estimating the illumination chromaticity for the training set data ranged from 0.0083 to 0.011. When tested on scenes that were not part of the training set, the average error then ranged from 0.012 to 0.022. These average errors are also a function of the distribution of the number of 'patches' in each scene, since scenes containing a smaller number of patches yield larger errors.

In the example given below, the testing set contained 100 scenes for each of the 98 illuminants. The performance of the neural network (NN) algorithm is compared with the 'white patch' (WP) version of the Retinex algorithm [2] which uses the maximum R, G and B values in the scene to estimate the illuminant and an implementation of the 'gray world' (GW) algorithm, which estimates the illuminant based on the average of all RGB values in the scene and on an *a priori* known average, computed from the whole surface reflectances database. The estimation errors of both the white patch and gray world algorithms converges to zero as the number of surfaces in the test scene approaches the size of the database.

| Method | 3 | 5 | 10 | 15 | 20 | 25 | 30 |
|--------|------|------|------|------|------|------|------|
| WP | .106 | .068 | .046 | .026 | .022 | .018 | .013 |
| GW | .082 | .063 | .041 | .033 | .027 | .025 | .023 |
| NN | .044 | .034 | .024 | .018 | 0.16 | .014 | .013 |

Table 1. Average error as a function of the number of patches in the scene.

## 3.2 Testing on Real Images

The network was tested on 48 images taken with a CCD camera under a wide range of light sources, from fluorescent tubes with blue filters, which emulate blue sky, to tungsten bulbs. The training set was the same one that was used for the network trained for tests on synthesized data. The neural network was compared with the white patch (WP) and gray (GW) world algorithms –described above– and with two implementations of Finlayson's version [5] of the gamut algorithm [4].

*Table 2* shows the results of the test performed on real data. The mean error represents the average distance between the estimated illuminant and the correct one and $\sigma$ represents the standard deviation. The errors are reported in the rg-chromaticity space as well as in CIE Lab space ($\Delta$Lab).

The mean variation of the illuminants was 0.09 in the *rg*-chromaticity space. This illumination chromaticity variation represents the average distance between a pre-determined canonical illuminant and the correct illuminants. This can be considered as a 'worst case' estimation, where we pick an *a priori* illuminant and consider that it represents our estimation of the illuminant, for all images.

The algorithms performed worse on real images than on synthesized data. The average errors were almost five times higher than the ones obtained for synthesized scenes. Noise, specularities, errors in camera calibration, etc. are some of the factors that might have affected the performance of the algorithms.

## 3.3 Modeling Specular Reflections

To improve the accuracy of the neural network illumination chromaticity estimate, we modeled the specular reflections in the training set [10], based on the dichromatic model of reflection [11]. The dichromatic reflection model states that the reflected light is an additive mixture of a specular and a body component. Therefore specularities were modeled in the training set simply by adding random amounts of the scene illumination's RGB to the matte component of the synthesized surface RGBs.

Each scene was generated by selecting $n$ surfaces at random and computing their RGB values, to which we added a random amount $r$ of the scene illumination. The value of $r$ for a scene $i$ was computed as the product between the maximum value of the specular component $S$ and a random, sub-unitary coefficient $p$: $r_i = S*p$.

Since surface specularities are not uniformly distributed in real images, we also created a non-uniform distribution of $p$ by squaring a uniformly distributed random function: $p=rnd()^2$. This model has an expected value for the specular coefficient $p$ of 33.3% and a standard deviation of 29.81%. This assures that generally only a few surfaces in the scene will be highly specular while still retaining a large variance. A random amount of white noise was also added to the data.

We generated training sets with different amounts of maximum specularity ($S$ ranging from 0% to 100%) and trained different networks for 10 epochs. The neural networks were tested using the same set of 48 real test images. All networks performed better than the ones trained without specular reflections.

Table 2 presents results obtained with a specular model containing at most 25% specularities ($S=25\%$). The results (in row #7) show that modeling specular reflections improved the performance of the neural network by around 25%. Statistical significance tests comparing the two neural network models yielded a confidence level of 94.1%.

| # | Method | Mean | $\sigma$ | $\Delta Lab$ |
|---|--------|------|----------|--------------|
| 1 | Illumination chromaticity variation | .090 | .062 | 22.38 |
| 2 | Gray World using average R, G, and B | .071 | .051 | 15.27 |
| 3 | Retinex (WP) using maximum R, G, and B | .075 | .049 | 16.36 |
| 4 | **Neural Network** | **.059** | **.043** | **15.03** |
| 5 | 2D gamut mapping using surface constraints only | .054 | .047 | 12.90 |
| 6 | 2D mapping using surface and illumination constraints | .047 | .039 | 12.67 |
| 7 | **Neural Network with 25% specularity model** | **.044** | **.032** | **12.13** |

Table 2. Results of colour constancy algorithms tested on real images.

The neural network also obtained more accurate estimates of the illumination chromaticity than any of the existing methods tested. However, much of the difference in error obtained for tests on real versus synthesized scenes, from around 0.044 to 0.010, remains unaccounted for, even when modeling the specular model.

# 4 Conclusions

This article presents a novel approach for achieving color constancy by estimating the chromaticity of the scene illuminant with the help of a neural network.

The neural network that we developed and tested is able to learn color constancy from synthesized data. On this type of scene, it outperformed the other color constancy algorithms. When tested on real images, it still performed well, although the errors were larger than for tests performed on synthesized scenes. Including specular reflections into the training set reduced the errors on real images.

Future work involves training the neural network on real images. The main impediment in training on real data is the very large number of images that are required to produce a training set.

# 5 References

[1] Swain M.J., Ballard D.H. Color Indexing. Intl. Journal of Computer Vision, 7:1, pp11-32, 1991

[2] Land E.H. The Retinex Theory of Color Vision. Scientific American, pp 108-129, 1977

[3] Buchsbaum G. A Spatial Processor Model for Object Colour Perception. J. Franklin Institute, 310 (1), pp 1-26, 1980

[4] Forsyth D.A. A Novel Algorithm for Color Constancy. Intl. Journal of Computer Vision, 5:1, pp 5-36, 1990

[5] Finlayson G. Color Constancy in Diagonal Chromaticity Space. IEEE Proc Fifth Intl Conf on Comp Vision, June 20-23, 1995

[6] Funt B., Cardei V. and Barnard K. Learning Color Constancy. Proc IS&T/SID Fourth Color Imaging Conference: Color Science, Systems and Applications, pp 58-60, Scottsdale, Arizona, November 1996

[7] Rumelhart D.E., Hinton G.E. and Williams R.J.: Learning Internal Representations by Error Propagation. In: Rumelhart D.E., McClelland J.L. and the PDP Research Group (ed). Parallel Distributed Processing: Explorations in the Microstructure of Cognition. Volume I: Foundations, pp 318-362, MIT Press, Cambridge, MA, 1986

[8] Cardei V., Funt B. and Barnard K. Modeling Color Constancy with Neural Networks. Proc Intl Conf. on Vision, Recognition, and Action: Neural Models of Mind and Machine, Boston, May 29-31, 1997

[9] Plaut D., Nowlan S., and Hinton G. Experiments on Learning by Back Propagation. Technical report, CMU-CS-86-126, Carnegie-Mellon University, Pittsburgh, USA, 1986

[10] Funt B., Cardei V. and Barnard K. Neural Network Color Constancy and Specularly Reflecting Surfaces. Proc AIC Color 97, Vol.II, pp 523-526, Kyoto, Japan, May 1997

[11] Shafer S.A. Using color to separate reflection components. Color Res Appl, vol 10, pp 210-218, 1985

# Automatic neural generalized font identification

A.M. González, J. Dorronsoro, C. Santa Cruz *
Department of Computer Engineering and
Instituto de Ingeniería del Conocimiento
Universidad Autónoma de Madrid, 28049 Madrid, Spain

## 1   Introduction

Neural methods are gaining a steady acceptance as powerful tools in a variety of pattern detection problems, OCR certainly being one of them. The concrete implementation of these neural OCR systems is of course a well guarded corporate secret, but in broad terms it can be said that in most of the cases, multilayer perceptrons (MLPs) are used. There are several reasons for the MLPs' success. To begin with, they are based in well understood mathematical and statistical principles and there are efficient tools and methodologies for their training and evaluation. Furthermore they have good generalization properties.

Nevertheless, MLPs have also some drawbacks. For instance, their correctness rates over individual characters, while very good from a broad point of view, are not usually good enough for what may be termed massive OCR tasks, involving processing jobs of hundreds of thousands or even millions of documents. Notice that a simple combinatorial argument shows that a fairly small error rate of 0.5 % per character translates into an unacceptable error rate of about 45 % in a ten field document with about 10 characters per field. Another drawback of MLPs is their relatively long training times, and more so in OCR, where a training set for recognition of large alphabets involving capital and lowercase letters, digits and some punctuation marks may well run into one million examples. Moreover, all this training effort can be partially undone if new samples are to be introduced for a better recognition rate.

These considerations would tend to suggest that MLP recognizers have to complemented with other tools for an effective use in massive OCR. A simple way to try to improve individual character recognition rates can be derived from the fact that very often massive OCR deals with printed data. Thus, the characters to be recognized can be assumed to concentrate in a relatively small number of fonts. Of course, to exploit this, a rather precise knowledge of the concrete font set involved is required. However such an a priori knowledge does not usually exist, and the sheer sample sizes in massive OCR make nearly

---
*With partial support of grant TIC 95–965 of Spain's CICyT.

impossible any manual font labeling of individual characters. In the following section we will briefly describe a general automatic approach based on radial basis function networks to what we may term "generalized font" detection. A strategy for the selection of the correct number of basis functions is discussed next, together with an illustration over a specific example.

## 2  RBF networks and generalized fonts

We will describe here an unsupervised approach for the identification of the fonts present on a sample of printed versions of a certain character, which is based on the estimation of its probability density. This task falls within the scope of both neural network methods and also classical statistical theory. A common ground between both approaches can be found if gaussian RBF networks are applied. Their very well known transfer function has the form $\sum_1^N w_i g(X, C_i, \Sigma_i)$, where $g(X, C, \Sigma)$ is a general multidimensional gaussian with a certain mean $C$ and covariance matrix $\Sigma$. If such a function is to represent a density $p(X, W)$, the normalizing conditions $w_i \geq 0$, $\sum_1^N w_i = 1$, $\Sigma_i$ positive definite have to be imposed. When doing so, $p(X, W)$ becomes what is called a *finite mixture distribution*. In our case we will use simpler, homogeneous gaussians, assuming that the covariance matrix is of the form $\Sigma = \sigma I$, with $I$ the identity matrix.

These networks have been extensively studied [1]. We will train them using the well known "Expectation–Maximization" (EM) algorithm, which seeks to maximize the log likelihood per single data of a $M$ character sample $\mathcal{L}(W) = \left(\log \Pi_{m=1}^M p(X_m, W)\right) / M = \left(\sum_{m=1}^M \log p(X_m, W)\right) / M$ with respect to the current weight set $W$ (see [2] for more concrete details of EM implementation and [4] for a thorough up to date analysis of EM convergence properties). Let us now discuss how to use this set of ideas to automatic font detection.

In our illustration, images are scanned at a 200 dpi black and white resolution, and once segmented, characters are normalized to a 32 by 32 bit matrix. Data space consists then of the first 40 Discrete Cosine Transform (DCT) coefficients derived from that matrix. Once a particular sample density function has been approximated by a gaussian mixture, each one of its components defines in a natural way an hyper-spherical influence region. We thus have an automatically constructed clustering of data space, which in our case can be naturally identified as *generalized fonts*.

Let's briefly explain what we mean. Ordinary typing fonts come in well defined families (courier, helvetica, times roman and so on), characterized by the precise shape and size of each character. However several factors (print quality, scanning effects, noise of various kinds) produce random variations of the originally defined font. In any case, if a gaussian RBF network approximates the sample density, each gaussian "attracts" a certain subset of the sample. It can be thus seen as capturing a general "font" around which randomly varying samples cluster. This approach has several clear advantages in OCR problems. First, classification is straightforward: individual samples are assigned to a con-

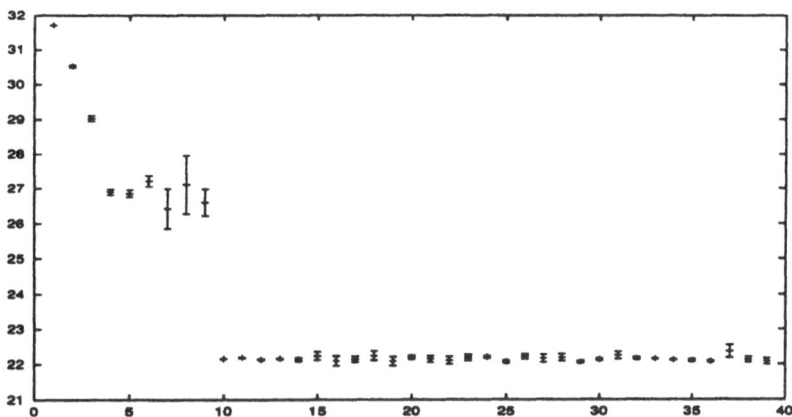

Figure 1: Approximate evolution of $N^2\Phi'(N)$ sample values derived from a 10 center gaussian mixture (the means and error bars given correspond to 10 experimental values).

crete character applying for instance likelihood ratio rules. Second, concrete a priori shape and size knowledge of font parameters is irrelevant: fonts are instead *defined* after training. Third, training times are much shorter since we deal with samples of individual characters and not with whole alphabets. Fourth, the addition of new samples does not imply a complete retraining: their own densities can be computed separately and merged with the previous ones after appropriate normalizations are performed. Fifth, "font removal" can be done in just the same way. In any case, a big question remains open: how to decide the number of such "generalized fonts" to be used or, equivalently, what is the appropriate number of gaussians in the RBF network. We will deal with this issue next, while considering a concrete example of such a generalized font identification.

## 3 Generalized font detection

Suppose we have a certain number $M$ of samples of a single character $C$. If all of them come from a well defined number $N_0$ of fonts, that sample would consist in gaussian noise perturbations of the corresponding "ideal characters" $C_1, \ldots, C_{N_0}$. If a RBF network with more than $N_0$ gaussians is used to cluster the sample, the network training procedure would ideally concentrate the sample characters around the "real" $N_0$ fonts, making negligible the contribution of the other $N - N_0$ gaussians. In other words, the sample likelihoods would be constant when $N \geq N_0$ gaussians are used. On the other hand, that likelihood would decrease rather sharply when the number $N$ of gaussians is well below $N_0$. This is precisely the situation depicted in figure 1. It shows the evolution of a numerical approximation to $N^2\Phi'(N)$ on a sample made up of gaussian noise

758

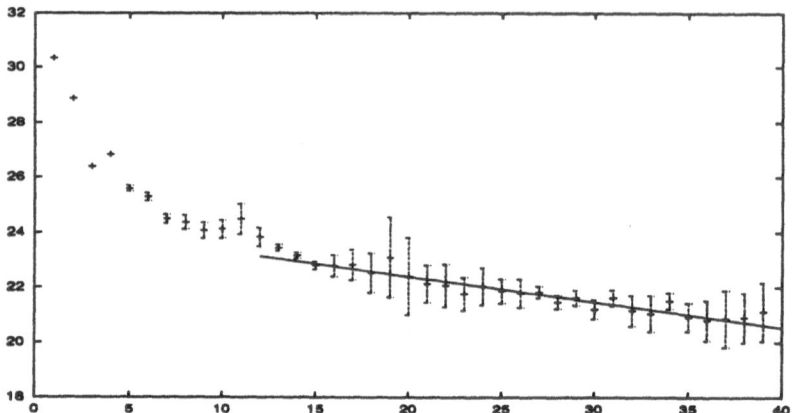

Figure 2: Approximate evolution of $N^2\Phi'(N)$ values obtained from a 6.000 character 3 sample (means and error bars given correspond to 10 experimental values). The straight line gives the best MSE fit in the interval [14,40].

added to 10 "ideal" characters 3, where $\Phi(N) = \mathcal{L}(N)/N$ is what we may call the "normalized" log–likelihood.

$\Phi(N)$ tries to capture the average "contribution" of the gaussians being used to the overall value of $\mathcal{L}(N)$. It also makes the dependence on $N$ of the values of $\mathcal{L}(N)$ more explicit. Here $\mathcal{L}(N)$ denotes the log–likelihood per single data after training of a $N$ gaussian RBF network. $\mathcal{L}$ is computed by the EM algorithm, starting with the centers obtained after the $K$–means clustering method is applied to randomly chosen initial centers. This seeks faster convergence of the subsequent EM iterations. As it can be seen in the figure, $N^2\Phi'(N)$ remains constantly equal to a value $\lambda$ while $N \geq N_0 = 10$, but this is no longer true once $N$ is below 10. This is what can be expected in the ideal situation. In fact, if $\mathcal{L}(N)$ is to be constantly equal to a given value $-\lambda = \mathcal{L}(N_0)$ for $N \geq N_0$, $\Phi(N)$ should then be essentially equal to $-\lambda/N$. We would have then $N^2\Phi'(N) = \lambda$.

However, when an actual character sample is used, things are different. Using a sample of 6.000 printed characters 3 obtained in a large scale OCR project [3], the corresponding log likelihoods $\mathcal{L}(N)$ show a slow decrease for large values of $N$ that accelerates for $N$ small. This is not surprising at all since a large number of gaussians may cause models to overfit. This may not happen when all the characters come from random sampling a certain gaussian mixture, as it was the case before, but overfitting is almost certain with the sample considered now. Figure 2 shows the corresponding evolution of $N^2\Phi'(N)$. Instead of constant values up to a given number of gaussians, $N^2\Phi'(N)$ increases first at a relatively constant rate, but this pattern breaks down for small values of $N$. A natural idea is to try to use these facts to fit a model to the values of $\mathcal{L}(N)$ above $N_0$. More precisely, we can consider then that $\mathcal{L}(N)$ has the form $\mathcal{L}(N) = -\lambda + \ell(N)$. $\ell$ is a positive, slowly increasing function with $\ell(N_0) = 0$ that captures the small "improvement" on $\mathcal{L}$ caused by model overfitting for $N \geq N_0$. For $N < N_0$ this

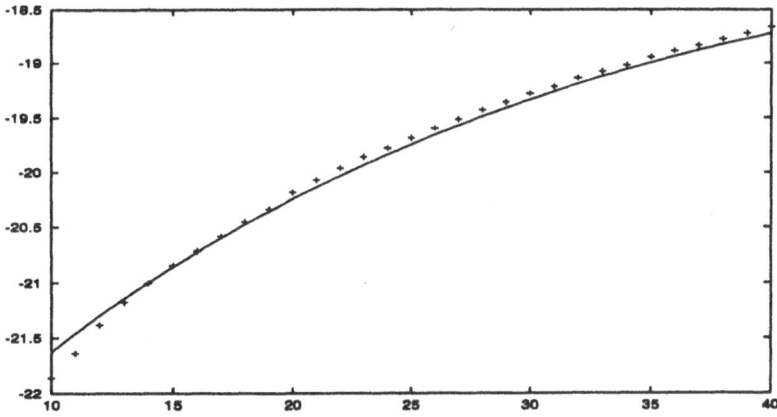

Figure 3: Actual values of $\mathcal{L}(N)$ for a 6.000 character 3 sample and values obtained from a model based on a 14 center gaussian mixture (continuous line).

model is not longer valid and $\mathcal{L}(N)$ decreases markedly faster. Thus, $N_0$ could be chosen then as the point where the models for $\ell(N)$ fail to apply.

Figure 2 gives support to this approach. It shows that the values $N^2\Phi'(N)$ have a linear structure for $N$ inside intervals starting around 10 and ending at 40, structure that is rapidly being lost for $N < 10$. This suggests to estimate the right number of gaussians (or the number of "generalized fonts") by choosing the best linear fit for $N^2\Phi'(N)$ and deriving a model of $\ell(N)$ and hence of $\mathcal{L}(N)$. The mathematical derivation is quite simple. We assume for the time being that $N \geq N_0$. The model $\mathcal{L}(N) = -\lambda + \ell(N)$ implies then that $\Phi'(N) = [N\mathcal{L}'(N) - \mathcal{L}(N)]/N^2$. Hence, $N^2\Phi'(N) = -\mathcal{L}(N) + N\mathcal{L}'(N) = \lambda - \ell(N) + N\ell'(N)$, which we approximate by a straight line with negative slope $\kappa - \gamma(N - N_0)$. It follows that

$$\Phi(N) = -\frac{\kappa + \gamma N_0}{N} - \gamma \log N + C,$$

where the value of $C$ is obtained making $N = N_0$ in this expression and recalling that $\Phi(N_0) = -\lambda/N_0$. Therefore, the final expression for $\Phi$ is

$$\Phi(N) = (\kappa + \gamma N_0)\left(\frac{1}{N_0} - \frac{1}{N}\right) - \gamma \log\frac{N}{N_0} - \frac{\lambda}{N_0},$$

and it follows that the fit for the sample values of $\mathcal{L}(N)$ when $N \geq N_0$ is

$$\hat{\mathcal{L}}(N) = -\lambda + \frac{\kappa + \gamma N_0 - \lambda}{N_0}(N - N_0) - \gamma N \log\frac{N}{N_0}.$$

For our character 3 sample, the best MSE linear model corresponds to $N_0 = 14$, giving a model for $\mathcal{L}$ in the range $[14, 40]$. The corresponding values of $\lambda$, $\kappa$ and $\gamma$ are then $\lambda = 21$, $\kappa = 23$ and $\gamma = 0.09$. Figure 3 shows a very close

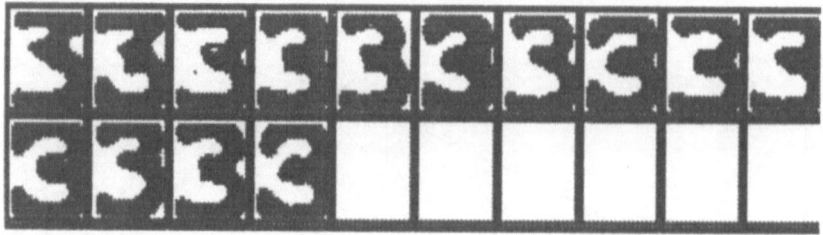

Figure 4: Generalized character 3 fonts deduced from a 6.000 character 3 from a 14 center RBF model.

fit between the model associated to these values and actual sample values of $\mathcal{L}(N)$. Finally, figure 4 shows the characters given by the inverse DCT of the 14 centers of the resulting gaussians. These are the "generalized fonts" associated to the final gaussian mixture.

## 4   Conclusion

Character recognition by RBF networks is an attractive and natural approach to be used in printed character OCR problems, provided that the right number $N_0$ of gaussian units (that is, of "generalized fonts") is chosen. We have briefly discussed its advantages and have proposed a simple method to estimate $N_0$ based on fitting the sample log–likelihoods $\mathcal{L}(N)$ for $N \geq N_0$. Of course, some open questions remain, such as the true nature of these generalized fonts and, more importantly, their recognition advantages against other RBF classifiers. These are currently under study. If successful, they could offer a complementary and competitive alternative to pure MLP recognizers.

## References

[1] F. Girosi, M. Jones and T. Poggio, "Regularization Theory and Neural Networks Architectures", Neural Computation 7 (1995), 219–269.

[2] R.A. Redner, H.F. Walker, "Mixture densities, maximum likelihood and the EM algorithm", SIAM Review 26 (1984), 195–239.

[3] A. Sierra, C. Santa Cruz, V. López, G. Fractman, J. Dorronsoro, C. Aguirre, J.M. Soto, A. Medina, R. López, "Neural networks in large scale bank effect recognition", Procs. SNN 1995 Symposium on Neural Networks, B. Kappen, S. Gielen (eds), 393–396, Springer Verlag, 1995.

[4] L. Xu, M.I. Jordan, "On convergence properties of the EM algorithm for Gaussian mixtures", Neural Computation 7 (1995), 129–151.

# An Approach to Blind Source Separation of Speech Signals

Shiro Ikeda, Noboru Murata

RIKEN Brain Science Institute

Wako, Japan

### Abstract

In this paper we introduce a new technique for blind source separation of speech signals. We focused on the temporal structure of signals which is not always the case in other major approaches. The idea is to apply the decorrelation method proposed by Molgedey and Schuster in time-frequency domain. We show some results of experiments with artificial data and speech data recorded in the real environment. Our algorithm needs considerably straightforward calculation and includes only a few parameters to be tuned.

## 1 Introduction

In this paper, we propose a blind source separation (BSS) method for speech signals recorded in a real environment. Speech signals have a temporal structure that it is stationary for a period shorter than 50~60msec but not stationary anymore if it is longer than 50~60msec and around 100msec. We use this time structure to build an algorithm.

The problem of BSS is defined as follows. Source signals are denoted by a vector $s(t) = (s_1(t), \cdots, s_n(t))^T$ and it is assumed that each component of $s(t)$ is independent to each other and mean 0. Recorded signals are $x(t) = (x_1(t), \cdots, x_n(t))^T$. We usually simulate a real-room recording with FIR filters, s.t. the observations are convolutive mixtures of source signals,

$$x(t) = A * s(t) = \left( \sum_k a_{ik} * s_k(t) \right), \quad a_{ik} * s_k(t) = \sum_{\tau=0}^{T_{max}} a_{ik}(\tau) s_k(t - \tau),$$

where $A(t)$ is a function of time, $a_{ik} * s_k(t)$ is the convolution of $a_{ik}(t)$ and $s_k(t)$, where $a_{ik}(t)$ is the impulse response from source signal $k$ to sensor $i$. The goal of BSS is to separate signals into the components which are mutually independent without knowing operator $A$ and source signals $s(t)$.

Basic BSS approaches have been developed for instantaneous mixtures. For convolutive mixtures, there are some trials[2]. We use the windowed-Fourier transform as in [3][5] to transform mixed source signals into the time-frequency domain. After that we apply Molgedey and Schuster's decorrelation algorithm[4] to the signals of each frequency independently. Most of the BSS approaches usually ignore the ambiguities of the amplitude and the permutation, but we have to remove these ambiguities to reconstruct the separated signals. Our idea is to use the inverse of the decorrelating matrices and the envelope of the speech signal.

## 2 Decorrelation Algorithm for Instantaneous Mixture

First we explain the decorrelation algorithm by Molgedey and Schuster [4] which was proposed for instantaneous mixtures, i.e. $x(t) = As(t)$ where $A$ is an $n \times n$ matrix.

The correlation matrix of observations is written as

$$\langle x(t)x(t+\tau)^T \rangle = R_{xx}(\tau) = A \langle s(t)s(t+\tau)^T \rangle A^T = AR_{ss}(\tau)A^T, \quad (1)$$

where $R_{xx}(\tau)$ and $R_{ss}(\tau)$ are correlation matrices. Since each component of $s(t)$ is independent, $R_{ss}(\tau)$ is diagonal for any $\tau$. Molgeday and Schuster showed that the BSS problem of finding $B$ is reduced to solve the eigenvalue problem

$$\left( R_{xx}(\tau_1)R_{xx}(\tau_2)^{-1} \right) B = B \left( \Lambda_1 \Lambda_2^{-1} \right). \quad (2)$$

This problem can also be solved by simultaneous diagonalization of matrices, where the number of the matrices doesn't have to be 2 but any number,

$$BR_{xx}(\tau_i)B^T = \Lambda_i, \quad i = 1, \ldots, r. \quad (3)$$

Although from the effect of the noise and small correlations among the source signals, (3) does not hold in practice. We implemented in the way to minimize the off-diagnal components of the matirices $BR_{xx}(\tau_i)B^T$. In order to obtain $B$, we use the algorithm which only needs straightforward calculations[6]. It consists of two procedures, sphering and rotation. (Fig.1)

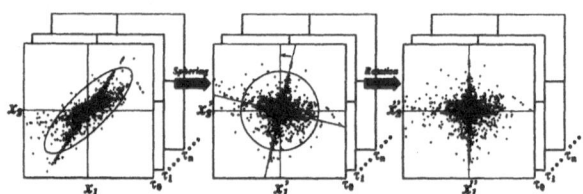

Figure 1: Sphering and Rotation

Sphering is a procedure to obtain a matrix $V$ which satisfies,

$$VR_{xx}(0)V^T = I \quad (4)$$

and rotation is a procedure to remove off-diagonal elements of correlation matrices with an orthogonal transformation. The implementation is to find an orthogonal matrix $C$ which minimizes

$$\sum_{l=1}^{r} \sum_{i \neq k} \left| (CVR_{xx}(\tau_l)V^TC^T)_{ik} \right|^2, \quad (5)$$

where $(*)_{ik}$ is the $ik$-element of a matrix. Cardoso and Souloumiac gave an implementation [1] with Jacobi-like algorithm to obtain $C$. Finally, matrix $B$ is given by $B = CV$. An advantage of this method is that it uses only the second order statistics and fixed amount of computation.

# 3 Proposed Method

In this section, the detail of the algorithm is shown along with the flow of it. First, the windowed-Fourier transform is applied to convolutive mixed signals,

$$\hat{x}(\omega, t_s) = \sum_t e^{-j\omega t} x(t) w(t - t_s),$$

$$\omega = 0, \tfrac{1}{N} 2\pi, \ldots, \tfrac{N-1}{N} 2\pi, \quad t_s = 0, \Delta T, 2\Delta T, \ldots \tag{6}$$

where $\omega$ denotes the frequency and $N$ denotes the number of points of the discrete Fourier transform, $t_s$ denotes the window position, $w$ is a window function (we used Hamming window) and $\Delta T$ is the shifting interval of moving windows. Let us redefine a $\hat{x}(\omega, t_s)$ for a fixed frequency $\omega$ as $\hat{x}_\omega(t_s) = \hat{x}(\omega, t_s)$. If the window length is long enough compared to the impulse response of $A(t)$, the relationship between observations and sources can be approximated as,

$$\hat{x}_\omega(t_s) = \hat{A}(\omega)\hat{s}(\omega, t_s), \tag{7}$$

where $\hat{A}(\omega)$ is the Fourier transform of operator $A(t)$, and $\hat{s}(\omega, t_s)$ is the windowed-Fourier transform of $s(t)$. This shows that for fixed $\omega$, a convolutive mixture is simply an instantaneous mixture. We extend the algorithm in §2 to complex values by substituting a Hermite matrix for a symmetric matrix, and apply it for each frequency. As the result, we have a separated time sequence,

$$\hat{u}_\omega(t_s) = B(\omega)\hat{x}_\omega(t_s). \tag{8}$$

Since BSS algorithms cannot solve the ambiguity of amplitude and permutation, even if we put each component of $\hat{u}_\omega(t_s)$ along with $\omega$, amplitudes are irregular and different independent sources will be mixed up. The problem of irregular amplitude can be solved by putting back the separated independent components to the sensor input with the inverse matrices $B(\omega)^{-1}$. Let us define $\hat{v}_\omega(t_s; i)$ as,

$$\hat{v}_\omega(t_s; i) = B(\omega)^{-1}(0 \ldots 0, \hat{u}_{i,\omega}(t_s), 0 \ldots 0)^T, \quad i = 1, \ldots, n \tag{9}$$

where $\hat{v}_{k,\omega}(t_s; i)$ represents the input of $i$-th independent component of $\hat{u}_\omega(t_s)$ into the $k$-th $(k = 1, \ldots, n)$ sensor. We applied $B(\omega)$ and $B(\omega)^{-1}$ to obtain $\hat{v}_\omega(t_s; i)$, therefore $\hat{v}_\omega(t_s; i)$ has no ambiguity of amplitude.

Remaining problem is permutation. We made an assumption that even for different frequencies, if the original souce is the same, the envelopes are similar. We utilize this idea for solving the permutation. Let us define an operator $\mathcal{E}$ to take the envelope as,

$$\mathcal{E}\hat{v}_\omega(t_s; i) = \frac{1}{2M} \sum_{t'_s = t_s - M}^{t_s + M} \sum_{k=1}^{n} |\hat{v}_{k,\omega}(t'_s; i)|, \tag{10}$$

where $M$ is a positive constant and $\hat{v}_{k,\omega}(t_s; i)$ denotes the $k$-th element of $\hat{v}_\omega(t_s; i)$. Inner product and norm are defined as

$$\mathcal{E}\hat{v}_\omega(i) \cdot \mathcal{E}\hat{v}_{\omega'}(k) = \sum_{t_s} \mathcal{E}\hat{v}_\omega(t_s; i)\mathcal{E}\hat{v}_{\omega'}(t_s; k), \tag{11}$$

$$\|\mathcal{E}v_\omega(i)\| = \sqrt{\mathcal{E}\hat{v}_\omega(i) \cdot \mathcal{E}\hat{v}_\omega(i)}, \tag{12}$$

and we define the similarity between two envelopes as the following,

$$\text{sim}(\omega) = \sum_{i \neq k} \frac{\mathcal{E}\hat{v}_\omega(i) \cdot \mathcal{E}\hat{v}_\omega(k)}{\|\mathcal{E}\hat{v}_\omega(i)\|\|\mathcal{E}\hat{v}_\omega(k)\|}. \tag{13}$$

Using these operations, we solve the permutation as follows:

**Solve the Permutation**

> $\omega = \text{sort}(\omega, \text{sim})$          *sorting $\omega$ to be $\text{sim}(\omega_1) < \cdots < \text{sim}(\omega_N)$*
>
> **for** $i = 1$ **to** $n$ **do**
>
>      $\hat{y}_{\omega_1}(t_s; i) := \hat{v}_{\omega_1}(t_s; i)$
>
> **done**
>
> **for** $l = 2$ **to** $N$ **do**
>
>      **for** $i = 1$ **to** $n$ **do**
>
>          $\sigma(i) := \text{argmax}_{i'} \left\{ \mathcal{E}\hat{v}_{\omega_k}(i') \cdot \left( \sum_{k=1}^{l-1} \mathcal{E}\hat{y}_{\omega_k}(i) \right) \right\}$
>
>          $\hat{y}_{\omega_l}(t_s; i) := \hat{v}_{\omega_l}(t_s; \sigma(i))$
>
>      **done**
>
> **done**

As a result, we obtain separated spectrograms as $\hat{y}_\omega(t_s; i)$. Applying inverse Fourier transform, finally we get a set of separated sources

$$y(t; i) = \frac{1}{2\pi} \cdot \frac{1}{W(t)} \sum_{t_s} \sum_\omega e^{j\omega(t-t_s)} \hat{y}_\omega(t_s; i), \quad i = 1, \ldots, n \tag{14}$$

where $W(t) = \sum_{t_s} w(t - t_s)$. Note that each $y_k(t; i)$ represents a separated independent component $i$ on sensor $k$, and $\sum_i y(t; i) = x(t)$ holds.

# 4 Experimental Results

## 4.1 Artificial Data

We show a result of an experiment on a set of data which were recorded separately and mixed on a computer. Since we wanted to simulate the general problem of recording sounds in a real environment, we built a virtual room as Fig.2 and calculated reflections and delays. We supposed that each wall, floor and ceil reflect the sound once and the strength of sounds varies in proportion to the inverse square of the distance. The strength of the reflection is 0.1 in power for any frequency.

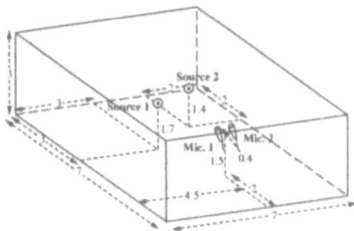

Figure 2: Virtual Room

Since we know the true sources and the mixing rates, we evaluate the performance with SNR (Signal to Noise Ratio) which is defined as,

$$\text{signal}_i(t;k) = a_{ik}s_k(t), \quad \text{error}_i(t;k) = y_i(t;k) - \text{signal}_i(t;k)$$

$$\text{SNR}_{ik} = 10\log_{10}\frac{\sum_t \text{signal}_i(t;k)^2}{\sum_t \text{error}_i(t;k)^2}.$$

We applied our algorithm, changing the window length from 8msec to 32msec.

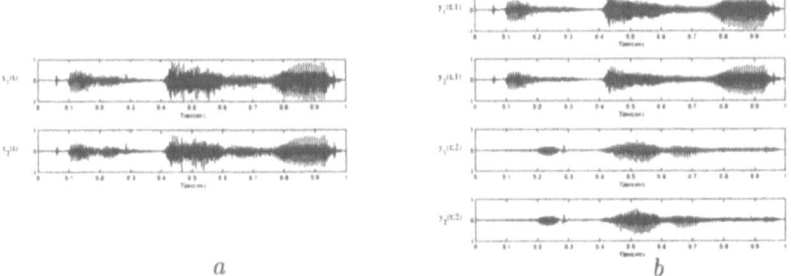

a                                                                    b

Figure 3: a: Mixed Signals in a Virtual Room b: Separated Signals

The SNRs of these results are shown in Tab. 1. Our approach with the window length of 32msec gave the best SNRs. Mixed inputs (Fig.3.a) and separated signals (Fig.3.b) are shown. Signals were clearly separated.

Table 1: SNRs (dB) for Separated Signals ($\Delta T$ is 1.25msec and $r$ in (3)is 40)

|  |  | SNR$_{11}$ | SNR$_{12}$ | SNR$_{21}$ | SNR$_{22}$ |
|---|---|---|---|---|---|
|  | 8msec | 4.36 | 6.32 | 11.94 | 11.66 |
| Window length | 16msec | 4.72 | 6.65 | 12.66 | 12.52 |
|  | 32msec | 6.47 | 7.30 | 14.40 | 13.19 |

## 4.2 Real-room Recorded Data

Finally, we applied the algorithm to the data recorded in a real environment. The data was provided by Prof. Kota Takahashi in the University of Electro-Communications. Two males were repeating different phrases simultaneously in a room and their voices were recorded with two microphones with 44.1kHz for 5sec then down-sampled to 16kHz. Inputs are shown in Fig.4.a.

We applied our algorithm to this data. Window length was 32msec (512 points), $\Delta T$ was 1.25msec and $r$ was 40. The result is shown in Fig.4.$b$. We show the separated signals in the graphs. We heard them and they were separated clearly.

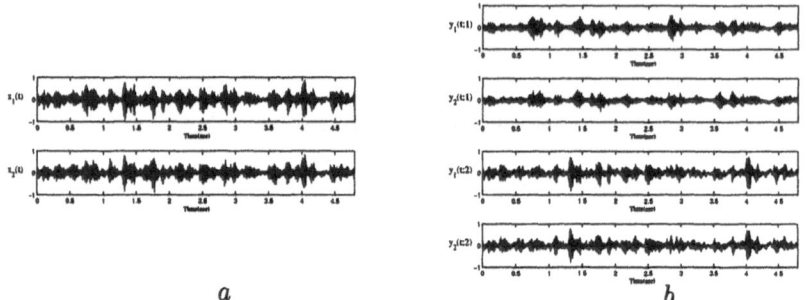

$a$ $b$

Figure 4: $a$: Recorded Signals in a Real Room $b$: Separated Signals

# 5 Conclusion

We proposed a BSS algorithm for speech signals. Our algorithm only uses straightforward calculation, and it includes only a few parameters to be tuned. Through the experiments, the algorithm worked very well for the data mixed on the computer and also for the real-room-recorded data. We haven't shown other results but they are available at

http://www.islab.brain.riken.go.jp/~shiro/blindsep.html

This algorithm is easy for hardware implementation, and we are now working for it. We are also working for realizing its on-line version. An on-line algorithm will make it possible to track walking speakers. In this article, we used the envelope of the signals to construct independent spectrograms form spectrograms form separated frequency components. It may be possible to use the continuity of de-mixing matrices between close frequency channels.

# References

[1] J.-F. Cardoso and A. Souloumiac. Jacobi angles for simultaneous diagonalization. *SIAM J. Mat. Anal. Appl.*, 17(1):161–164, 1996.

[2] S. C. Douglas and A. Cichocki. Neural networks for blind decorrelation of signals. *IEEE Trans. Signal Processing*, 45(11):2829–2842, 1997.

[3] T.-W. Lee, A. Ziehe, R. Orglmeister, and T. Sejnowski. Combining time-delayed decorrelation and ICA: towards solving the cocktail party problem. In *Proceedings of ICASSP'98*, 1998.

[4] L. Molgedey and H. G. Schuster. Separation of a mixture of independent signals using time delayed correlations. *Phys. Rev. Lett.*, 72(23):3634–3637, 1994.

[5] P. Smaragdis. Blind separation of convolved mixtures in the frequency domain. In *International Workshop on Independence & Artificial Neural Networks*, University of La Laguna, Tenerife, Spain, 1998.

[6] A. Ziehe. Statistische verfahren zur signalquellentrennung. Master's thesis, Humboldt Universität, Berlin, 1998. (in German).

# Data Fusion for Diagnosis
# in a Telecommunication Network

Philippe Leray and Patrick Gallinari

LIP6 - Pôle IA - Université Paris 6 - boîte 169

4, Place Jussieu - 75252 Paris cedex 05 - France

{Philippe.Leray, Patrick.Gallinari}@lip6.fr

### Abstract

We present a diagnosis system which combines the outputs of several classifiers performing local generation on the French Telephone Network. This data fusion process takes into account alarms occurring at different times and space locations. This allows to considerably improve upon the performances of a previous diagnosis system which relies only on local decisions.

## 1. Introduction

Industrial systems need to be monitored in order to prevent incidents, detect anomalies and maintain a high quality of service. The increasing complexity of these systems has motivated important efforts aimed at developping automatic monitoring and diagnosis methods.

A general diagnosis system relies on three tasks: alarm *generation* (i.e. detection of perturbed states in the system), alarm *filtering* and *understanding* (i.e. alarm processing for a better understanding of the phenomenon) and *command* (i.e. how to put the system back in its nominal state).

We have presented in [1,2,3] the first stages of a modular system for diagnosis in a Telephone Network where Neural Networks are used to implement each module. This diagnosis task is complex since the amount of data to process is tremendous and a large part of the information provided by the network is redundant or useless. In this system, alarm generation (i.e. the detection of overloads in the telephone network and the indication of the corresponding overload percentage) is performed locally in each transit centre. This diagnosis did not take into account spatial or temporal correlations existing in the network. We propose here to combine alarms, temporally - i.e. by considering previous states of a given centre - and spatially - i.e. by considering the states of neighbouring centres - in order to perform alarm filtering.

In section 2 we briefly describe the modular architecture we use for diagnosis in the telephone network and give some previously obtained results for the local diagnosis stage. We then present in section 3 and 4 new developments where neural networks are used to combine locally generated alarms so as to take into account temporal and spatial correlations. Finally we give results obtained with this global diagnosis architecture.

# 2. Local Diagnosis in the French Telephone Network

We consider a telephone network based on the French long-distance network. It consists of 5 Main Transit Centres (MTCs) and 68 Secondary Transit Centres (STCs). Nowadays, diagnosis is performed at the Network Management Centres and relies heavily on human operators. Measurements obtained from MTCs and STCs are aggregated into more explicit indicators in order to analyse the centre status, to detect abnormal conditions and to activate traffic controls so as to minimize the effects of disruptions.

The diagnosis problem we will study for each centre amounts to classifying 5 different situations denoted in the following $O_1$ to $O_5$ (they correspond respectively to a nominal situation, an outgoing overload, an incoming overload, an overall overload and a regional overload). Each of these situations can also be labelled with an overload level, expressed as a percentage of the nominal traffic conditions. In addition to the classification, our system will compute this overload level.

The modular system we proposed in [1,2,3] is built of two levels: a local level performs alarm generation for each transit centre, and a global level combines all these local alarms in order to perform the global diagnosis. At the local level, the situation of each transit centre is analyzed with different specialized modules:

- a first module (CLASSIF) determines the centre status (classes are denoted $O_1$ to $O_5$ for STCs and $O_1$, $O_4$ and $O_5$ for MTCs).
- a set of modules dedicated to each overload situation gives the corresponding overload percentage.

We briefly present in this section the performances of our system for local classification (CLASSIF modules) for one STC and one MTC. Our different modules are implemented with multilayer perceptrons with one hidden layer. The learning algorithm is a batch conjugate gradient. Learning is stopped either when a plateau is reached for the error on test data, or when overtraining is detected. Tables 1 and 2 give below CLASSIF confusion matrices for the two centres (the corresponding performances are 76.7 % and 73.3 % on test data). As can be noticed, $O_4$ and $O_5$ classes are ambiguous at STC or MTC level (see e.g. the confusion rates of 27 % and the 36 % in table 2). This confusion can be explained since $O_4$ and $O_5$ correspond respectively to an overload on the whole network and an overload on a whole region (also named main transit zone, ZTM). It is thus difficult to distinguish between these two events using the local information available at the transit centre.

|       | $O_1$ | $O_2$ | $O_3$ | $O_4$ | $O_5$ |
|-------|-------|-------|-------|-------|-------|
| $O_1$ | **88.3** | 0.0 | 0.0 | 11.5 | 0.2 |
| $O_2$ | 0.5 | **99.3** | 0.0 | 0.0 | 0.2 |
| $O_3$ | 1.2 | 0.0 | **98.8** | 0.0 | 0.0 |
| $O_4$ | 10.0 | 0.0 | 0.0 | **63.0** | 27.0 |
| $O_5$ | 3.2 | 0.0 | 0.0 | 36.0 | **60.8** |

Table 1: Confusion matrix (in %) on test data for one STC.

|       | $O_1$ | $O_4$ | $O_5$ |
|-------|-------|-------|-------|
| $O_1$ | **84.0** | 10.8 | 5.2 |
| $O_4$ | 20.8 | **65.0** | 14.2 |
| $O_5$ | 16.0 | 9.8 | **74.2** |

Table 2: Confusion matrix (in %) on test data for one MTC.

In order to improve the performances of systems which exploit local alarm generation, we propose in this paper to combine in our modular system, local alarms so as to take into account spatial and temporal correlations.

# 3. Using Temporal Correlations for Local Diagnosis

The telephone network is a dynamic system for which the state of a centre at a given time is highly dependent on its state at previous times. The classification modules described in section 2 are static. In order to improve upon that, we have developed a new module (COMBI) which makes use of the alarm temporal sequence at a given site. Several methods exist for that in the literature, for instance sequential hypothesis testing in statistic [4] or Markov models (used by [5] in a diagnosis system). We consider this task as a classifier combination problem. Let us denote $c(t)$ the output of a CLASSIF module at time t, several successive outputs are considered in order to compute $C(t)$ the new estimation of the centre status. We will denote $C(t) = F(c(t),...,c(t-n),W)$ where n is the number of past measures in the sequence, $F$ is the combination operator and $W$ its parameters.

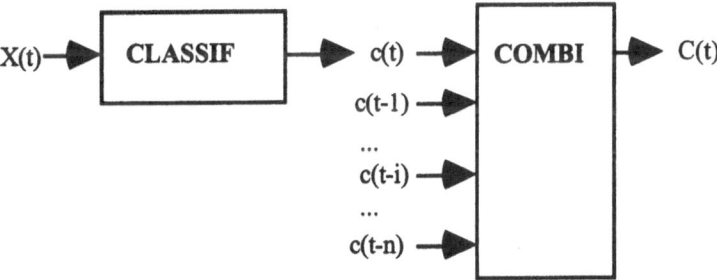

Figure 1: Use of temporal correlation between the outputs of CLASSIF module at different times to improve classification at time t. $c(t-i)$ is the CLASSIF module output at time t-i and $C(t)$ is the combination result.

Several approaches have been proposed for the combination of classifiers, a review may be found in [6]. We concentrate here on algebraic combinations of the classifier outputs, where coefficients **W** of the combination are learnt from data[1] and are thus specific for each centre.

Three different combination operators **F** have been trained: (LIN) is a simple linear combination, (PER) is a perceptron without hidden layer (i.e. a logistic regression), (MLP) is a multilayer perceptron with 5 hidden neurons. By varying n the length of the sequence considered in the combination we obtain the results presented in table 3. By comparing these results with the CLASSIF module performance (76.7 %), it is clear that the combination improves performances as soon as we take into account c(t) and c(t-1). In our database, only short perturbations were recorded, this is why considering longer sequences is not useful.

| | Number of delays in the combination | | | | |
|---|---|---|---|---|---|
| Architecture | 1 | 2 | 3 | 4 | 5 |
| LIN | 80.8 % | 80.2 % | 79.7 % | 79.6 % | 79.7 % |
| PER | 80.5 % | 80.2 % | 80.2 % | 80.3 % | 80.6 % |
| MLP | 82.3 % | 81.7 % | 82.1 % | 81.5 % | 82.0 % |

Table 3: Correct Classification Rate on test data for different combination architectures: (LIN) is a simple linear combination, (PER) is a perceptron without hidden layer, (MLP) is a multilayer perceptron with 5 hidden neurons. Numbers on the second line correspond to the value of n (number of previous valued added to c(t)).

# 4. Using Spatial Correlations for Local Diagnosis

By implementing separately the CLASSIF, EXPERT and COMBI modules for each centre, we still consider that the events at different locations are independent. This is not a realistic assumption for the global ($O_4$) and regional ($O_5$) overloads. At a given time, $O_4$ may be detected by any centre on the network and $O_5$ may be detected by all centres of the same region (a region will be denoted below MTZ).

In order to demonstrate how to take into account this spatial correlation and inspired by a technique used in [7], we have chosen different transit centres in two different regions. As shown in figure 2, $O_4$ and $O_5$ outputs for all the centres are combined in order to improve the detection of $O_4$ in the whole network (denoted Glob(network)) and $O_5$ in each region (denoted Reg(MTZi) for region i).

This new combination via the SYNTHESIS module (figure 2) is performed with another multilayer perceptron. Performances at the local levels of a centre and confusion matrices given in tables 4 and 5 were then re-computed using the following algorithm:

For each example on test set

- compute local alarms which take into account temporal correlations for each centre (CLASSIF and COMBI modules)

---

1 To limit the bias cause by this new training, we exchange our learning and validation databases for training the different combinations.

- compute global alarms which take into account spatial correlations for global and regional overload (SYNTHESIS module)
- use the global outputs of the SYNTHESIS module to refine the diagnosis at the local level of each centre by applying the following rule:
  - if Glob(network) is true then Glob(centre) is set true for each centre
  - if Reg(MTZi) is true then Reg(centre) is set true for each centre which belongs to this MTZi (region i).

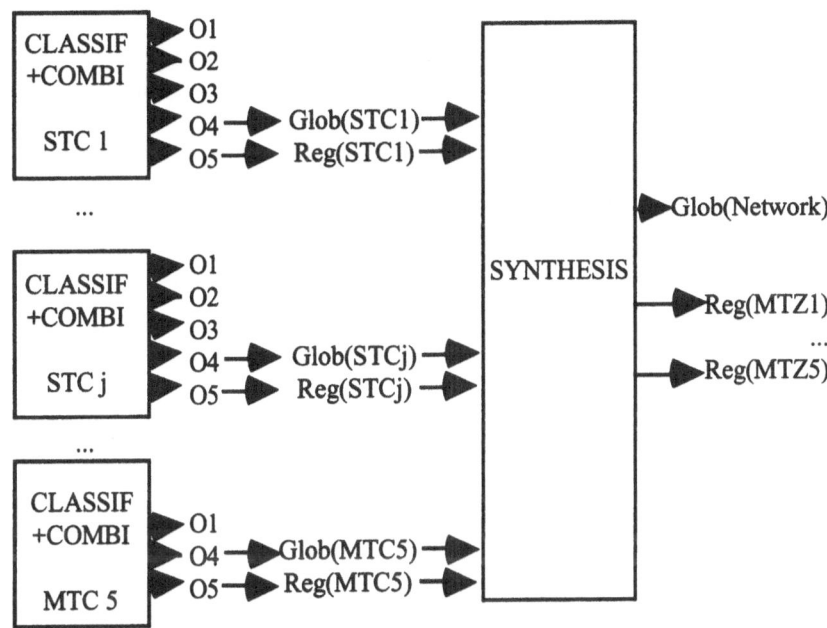

Figure 2: Detection of global and regional overloads through the fusion of local alarms: aggregation of the results obtained locally into one output "global overload" for the whole network and one output "regional overload" for each region.

|       | $O_1$ | $O_2$ | $O_3$ | $O_4$ | $O_5$ |
|-------|-------|-------|-------|-------|-------|
| O1    | **97.3** | 0.0 | 0.0 | 2.7 | 0.0 |
| O2    | 0.7 | **99.3** | 0.0 | 0.0 | 0.0 |
| O3    | 1.7 | 0.0 | **98.2** | 0.0 | 0.0 |
| O4    | 18.5 | 0.0 | 0.0 | **81.5** | 0.0 |
| O5    | 10.5 | 0.0 | 0.0 | 3.2 | **86.3** |

Table 4: Confusion matrix (in %) on test data for a STC using temporal and spatial correlations.

|       | $O_1$ | $O_4$ | $O_5$ |
|-------|-------|-------|-------|
| O1    | **97.3** | 2.7 | 0.0 |
| O4    | 17.2 | **82.8** | 0.0 |
| O5    | 10.3 | 3.2 | **86.5** |

Table 5: Confusion matrix (in %) on test data for a MTC using temporal and spatial correlations.

The use of temporal and spatial correlations for our modules yield excellent results. Performances are significantly increased: from 76.7 % to 92.6 % for the STC and from 73.3 % to 88.8 % for the MTC, which corresponds respectively to an error decrease of 68 % and 58 %. The confusion between global ($O_4$) and regional ($O_5$) overloads has practically disappeared (3.2 % in tables 4 and 5).

# 5. Conclusion

We have presented data fusion experiments for the last stages of a diagnosis system for the French telephone network. We have shown that, for this complex system, the fusion of local classifier outputs at different points in time can improve the performances. The fusion of classifier outputs at different space locations leads to a significant further increase in performances.

We are currently developing an improved diagnosis architecture using our neural modules for alarm generation and temporal filtering. This system makes use of a reject option with respect to a confidence measure on classifier outputs. Because of this reject, spatial filtering will have to be performed with an incomplete knowledge of local status and is implemented with a bayesian network.

**Acknowledgement:** This work has been performed in collaboration with Elisabeth Didelet, Luc De Bois and Daniel Stern from France Télécom - CNET under Grant 94 1B 003.

# References

[1]     Leray P, Gallinari P, Didelet E. Diagnosis Tool for Telecommunication Traffic Management, Proceedings of ICANN'96, 1996

[2]     Leray P, Gallinari P, Didelet E. Local Diagnosis for Real Time Network Traffic Management, Proceedings of the International Workshop on Applications of Neural Networks to Telecommunications 3, Alspector/Goodman/Brown ed., 1997

[3]     Leray P, Gallinari P, Didelet E. Neural Networks for Alarm Generation in the Telephone Network. Proceedings of DX'97, 1997

[4]     Fukunaga K. Introduction to Statistical Pattern Recognition, 2nd edition, Academic Press, inc, 1990.

[5]     Smyth P. Hidden Markov Models for Fault Detection in Dynamic Systems. Pattern Recognition 27(1):149-164

[6]     Tumer K and Gosh J. Theoretical Foundations of Linear and Order Statistics Combiners for Neural Pattern Classifiers, Technical Report TR-95-02-98, Computer and Vision Research Center, University of Texas, Austin, 1995

[7]     Wietgrefe H, Tuchs K, Jobmann K, Carls G, Fröhlich P, Nedil W and Steinfeld S. Using Neural Networks for Alarm Correlation in Cellular Phone Networks, Proceedings of the International Workshop on Applications of Neural Networks to Telecommunications 3, Alspector/Goodman/Brown ed., 1997

# Creating an Order in Distributed Digital Libraries by Integrating Independent Self-Organizing Maps

Andreas Rauber, Dieter Merkl

Institut für Softwaretechnik, Technische Universität Wien

Resselgasse 3/188, A–1040 Wien, Austria

{andi, dieter}@ifs.tuwien.ac.at

## Abstract

Digital document libraries are an almost perfect application arena for unsupervised neural networks. This because many of the operations computers have to perform on text documents are classification tasks based on "noisy" input patterns. The "noise" arises because of the known inaccuracy of mapping natural language to an indexing vocabulary representing the contents of the documents. A growing number of papers is dedicated to the usage of self-organizing maps to organize the contents of such digital libraries. These papers assume the central availability of the data; an assumption that is questionable given the massive amount of available information. In this paper we describe an approach for organizing distributed digital libraries based on a system of independent self-organizing maps each of which representing just a portion of the complete digital library. Furthermore, we argue in favor of integrating these independent maps in a hierarchical fashion, again by means of self-organizing maps. The integration is based on the trained low level maps.

## 1  Introduction

During recent years we have witnessed the appearance of an ever increasing flood of miscellaneous written information available in computer accessible form and originating from very different sources, culminating in the advent of massive digital libraries. As an example of such a digital library consider *The Internet Public Library* (http://www.ipl.org). Powerful methods for organizing, exploring, and searching collections of text documents are thus needed.

The map metaphor for displaying the contents of a document library in a two-dimensional display has gained considerable interest [1, 4, 5, 6, 7]. Maps are used to visualize the similarity between documents in terms of distance within the two-dimensional map display. Hence, similar documents may be found in neighboring parts of the map display.

The underlying assumption in all of the above mentioned papers is that the data items, i.e. the documents, are available locally for self-organizing map training. With ever growing information repositories, however, this assumption is questionable. One should rather expect that several repositories exist at different sites and a way for unified access should be provided for the users of such distributed digital libraries. Moreover, most of the reports on the application of self-organizing maps for the organization of document libraries propose the usage of a single map for document space visualization. It is only natural that with increasing size of the document library the self-organizing maps for representing this library grow larger, see e.g. [4]. This, however, does not necessarily improve the organization of the library. In particular, it is rather difficult for the user to find her orientation in large maps. To use an example from geography, it is much easier to find the way, say, from *Stockholm* to *Skövde* by using the map of *Sweden* than by using the map of *Europe* or, even worse, that of the *World*. In much the same way it is easier to locate a particular document using a map of a smaller portion of the library while relying on higher aggregated maps for overall orientation.

In this paper we suggest the utilization of individual self-organizing maps for organizing the different parts of a distributed digital library. Multiple maps may then be integrated to produce an aggregated visualization of the complete document space. We demonstrate the effects of such an approach based on a sample library of text documents.

The remainder of this paper is organized as follows. In Section 2 we give a brief review of self-organizing maps and describe their integration. Section 3 is dedicated to a description of our experimental document library. Experimental results of our approach to digital library organization are given in Section 4. Finally, our conclusions are contained in Section 5.

## 2 Hierarchies of self-organizing maps

The self-organizing map [2, 3], SOM, is one of the most prominent neural network models adhering to the unsupervised learning paradigm. The model consists of $p$ units, each of which is assigned an $n$-dimensional weight vector $m_i$, $1 \leq i \leq p$. Training is performed as a repetition of (i) input pattern presentation, (ii) selection of the unit with the closest weight vector, i.e. the *winner*, and (iii) adaptation of the weight vectors of the *winner* and of a number of units in the neighborhood of the *winner*. It is important to guarantee that both the strength of adaptation as well as the number of units that are adapted apart from the *winner* are decreasing in time. The distinctive feature of the self-organizing map is that similar input patterns will be arranged in neighboring regions of the resulting map. Hence, the similarity between input data items, i.e. the documents in our application, is mirrored in terms of the distance of the respective *winners* within the map.

In case of a digital library that exists only distributed over several sites, it might be more efficient to have independent self-organizing maps that represent the various parts of the digital library than transfering the whole information to one site for training. However, when some form of uniform access to the data is requested, the contents of the various sites has to be integrated. With our approach to digital library organization we suggest to utilize self-organizing maps to perform such an integration. In particular, the map that shall integrate different portions of the digital library may be trained by using the weight

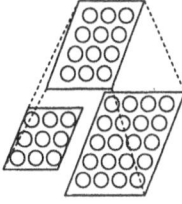

Figure 1: Integration of two self-organizing maps

vectors of the maps to be integrated. Such a strategy may be applied recursively in order to build hierarchies of arbitrary depth as shown in Figure 1. In this figure a 3 × 3 and a 4 × 5 map are integrated in a 3 × 4 map. Note that also selected parts of self-organizing maps may be integrated by using essentially the same architecture. The user simply selects areas of interest scattered across different maps for which an integration shall be performed. By this, the user may tie together pieces of information to build her own library fine-tuned to her particular interests.

The effect of such an integration, obviously, is that input data items that are separated in different low level maps are grouped together in the high level map. Input data that are mapped onto the same low level unit are represented together in the high level map.

# 3 An experimental document archive

For the experiments presented thereafter we used the 1990 edition of the *CIA Worldfactbook* (http://www.odci.gov/cia/publications/factbook/index.html) as a sample document archive. The *CIA Worldfactbook* represents a document collection containing information on countries and world regions. The information is split into categories such as *Geography, People, Economy, Defense Forces*, etc. In total, the 1990 edition of the *CIA Worldfactbook* consists of 245 documents.

The various documents are represented by means of simple histograms of word occurrences. We used all words occuring in more than 15 and less than 220 documents. Thus, 1056 distinct words, i.e. index terms, remained that were weighted according to a $tf \times idf$ weighting scheme [8], i.e. term frequency times inverse document frequency. Such a weighting scheme assigns high weights to index terms that occur frequently within a document but rarely within the whole document collection. Finally, the documents are represented by feature vectors where each feature corresponds to an index term and the specific value of the feature is derived by means of the $tf \times idf$ weighting scheme. These feature vectors are used as the input data during network training.

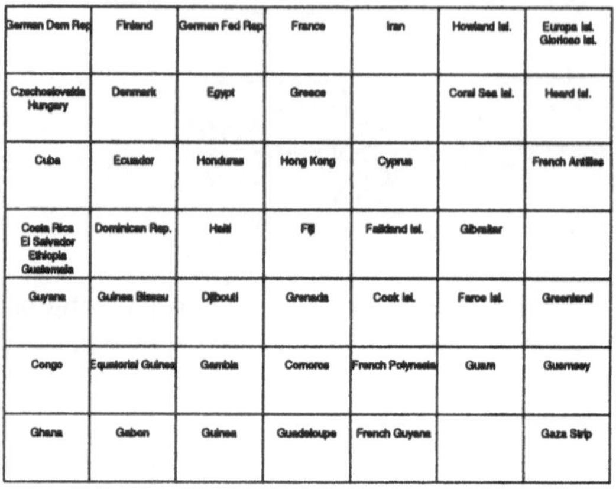

| German Dem Rep | Finland | German Fed Rep | France | Iran | Howland Isl. | Europa Isl. Glorioso Isl. |
|---|---|---|---|---|---|---|
| Czechoslovakia Hungary | Denmark | Egypt | Greece | | Coral Sea Isl. | Heard Isl. |
| Cuba | Ecuador | Honduras | Hong Kong | Cyprus | | French Antilles |
| Costa Rica El Salvador Ethiopia Guatemala | Dominican Rep. | Haiti | Fiji | Falkland Isl. | Gibraltar | |
| Guyana | Guinea Bissau | Djibouti | Grenada | Cook Isl. | Faroe Isl. | Greenland |
| Congo | Equatorial Guinea | Gambia | Comoros | French Polynesia | Guam | Guernsey |
| Ghana | Gabon | Guinea | Guadeloupe | French Guyana | | Gaza Strip |

Figure 2: Low level self-organizing map

# 4 Experimental results

In order to simulate a distributed digital library, the documents of the *CIA Worldfactbook* have been split randomly into five parts consisting of 50 documents each. Note that five documents are thus assigned to two parts at a time. Each part of the library was then used for training of a separate $7 \times 7$ self-organizing map. Each of these maps represents a topologically ordered portion of its subset of the document library. Due to space restrictions we can only show one of these maps in this paper, see Figure 2. The other four maps, however, are rather similar in spirit.

In a second step, we integrated the five independent maps into one single $7 \times 7$ self-organizing map which now represents the complete document library. The map that results from this integration is shown in Figure 3. Note that the main clusters are clearly visible from the map representation, with the information from the low level maps being arranged according to the organizational principles of the high level map. As examples consider the area containing African countries in the left upper part of the map, the area of oil producing countries, the area of countries from Latin America, or the area of countries usually referred to as the *first world* in the right upper part. With the latter area, please note the explicit distinction into countries belonging to the Western hemisphere and those belonging to the Eastern hemisphere[1]. Finally, it is worth noting that the countries that are contained twice in the library are pairwise assigned to the same unit in the integrated map. These countries are *Austria, Comoros, Iceland, Japan,* and *Mozambique.*

In a nutshell, the higher level map forms an orderly mapping based on the

---

[1] At this point it is important to recall that our document archive is the 1990 edition of the *CIA Worldfactbook*. Thus, the descriptions refer to a time before the "fall" of the Communist hemisphere.

| | | | | | | | |
|---|---|---|---|---|---|---|---|
| Kenya | Angola, Benin, Burkina Faso, Cameroon, Cape Verde, Centr Afric Rep, Chad, Congo, Equat Guinea, Gabon, Gambia, Guinea, Guinea Bissau, Ivory Coast, Madagascar, Malawi, Mali, Mauritania, Namibia, Niger, Sao Tome, Tanzania, Togo, Uganda, Zaire, Zambia, Zimbabwe | Haiti | Argentina, Bolivia, Brazil, Chile, Colombia, Costa Rica, Cuba, Dominican Rep, Ecuador, El Salvador, Ethiopia, Guatemala, Honduras, Mexico, Nicaragua, Panama, Paraguay, Peru, Uruguay, Venezuela | Fiji | United States | | Australia, Austria (2x), Belgium, Canada, Denmark, Finland, France, German Fed Rep, Greece, Italy, Japan (2x), Luxembourg, New Zealand, Norway, Portugal, South Korea, Spain, Sweden, Switzerland, United Kingdom |
| Ghana | Burundi, Rwanda | Mozambique (2x) | Guyana, Suriname | Iceland (2x), Ireland | Bhutan, Nepal | | Bulgaria, Czechoslovakia, German Dem Rep, Hungary, Poland, Romania, Soviet Union, Yugoslavia |
| Nigeria | Botswana, Lesotho, Swaziland | Seychelles | Mongolia, North Korea, Taiwan, Vietnam | Jordan | Algeria, Bahrain, Djibouti, Egypt, Iraq, Israel, Kuwait, Lebanon, Libya, Morocco, North Yemen, Oman, Qatar, Saudia Arabia, Somalia, South Yemen, Sudan, Syria, Tunisia, UAE | | Albania, Cambodia, Laos |
| Comoros (2x) | Antigua, Bahamas, Barbados, Belize, Grenada, Jamaica, Mauritius, St Kitts, St Lucia, St Vincent, Trinidad | Papua New Guinea, Solomon Isl, Tonga, Tuvalu, Western Samoa | Andorra | Afghanistan, Bangladesh, Brunei, Burma, China, Cyprus, India, Indonesia, Iran, Liberia, Malaysia, Maldives, Malta, Pakistan, Philippines, Singapore, Sri Lanka, Thailand, Turkey | South Africa | | South Georgia |
| Macau | Vanuatu | Americ Samoas, Guam, Marshall Isl, Micronesia, N Mariana Isl, Palau, Puerto Rico, Virgin Isl | | Hong Kong, Netherlands | Faroe Isl, Gibraltar, Greenland | Arctic Ocean, Atlantic Ocean, Indian Ocean, Pacific Ocean | Spratly Isl, Wake Isl |
| Kiribati, Nauru | | Liechtenstein, Monaco | Antarctica | Gaza Strip, West Bank | Ashmore Isl, Coral Sea Isl, French Antarctic Lands | Iraq-SA-Neutr Zone, Paracel Isl | Johnston Atoll, Midway Isl |
| Anguilla, Aruba, Bermuda, British Indian Ocean Terr, Cayman Isl, Christmas Isl, Cocos Isl, Cook Isl, Falkland Isl, Guernsey, Isle of Man, Jersey, Montserrat, Niue, Norfolk Isl, Pitcairn Isl, St Helena, Tokelau, Turks Isl | San Marino, Vatican | French Guiana, French Polynesia, Guadeloupe, Martinique, Mayotte, New Caledonia, Reunion, St Pierre, Wallis & Futuna | Svalbard, World | Jan Mayen | Basses da India, Bouvet Isl, Clipperton Isl, Europa Isl, Glorioso Isl, Heard Isl, Juan de Nova Isl, Kingman Reef, Navassa Isl, Palmyra Atoll, Tromelin Isl | Baker Isl, Howland Isl, Jarvis Isl |

Figure 3: Integration of the low level self-organizing maps

information contained in several low level maps, each of which representing a portion of the library. The higher level map thus is a convenient starting point for a user trying to find her orientation in a distributed digital library.

# 5    Conclusions

In this paper we presented a novel approach for organizing distributed digital libraries by using a system of self-organizing maps. In our approach, different portions of a digital library that may reside on different sites are independently organized by self-organizing maps. These individual maps may be integrated in order to provide an overview picture of the information contained in the digital library. We suggested that the integration of low level maps should be based on the various weight vectors representing groups of documents. Such an integration, however, is not restricted to complete maps. The user may equally well define areas of interest that cover just portions of low level maps. These areas can then be integrated by using the very same integration process. The benefit for the user of such a distributed digital library is that she may define her own personal library that reflects her particular interests.

# References

[1] S. Kaski, T. Honkela, K. Lagus, and T. Kohonen, Creating an order in digital libraries with self-organizing maps, In: Proceedings of the World Congress on Neural Networks, San Diego, CA, 1996.

[2] T. Kohonen, Self-organized formation of topologically correct feature maps, Biological Cybernetics 43, 1982.

[3] T. Kohonen, Self-Organizing Maps, Springer-Verlag, Berlin, 1995.

[4] T. Kohonen, S. Kaski, K. Lagus, and T. Honkela, Very large two-level SOM for the browsing of newsgroups, In: Proceedings of the Int'l Conference on Artificial Neural Networks, Bochum, Germany, 1996.

[5] X. Lin, D. Soergel, and G. Marchionini, A self-organizing semantic map for information retrieval, In: Proceedings of the Int'l ACM SIGIR Conference on Research and Development in Information Retrieval, Chicago, IL, 1991.

[6] D. Merkl, A connectionist view of document classification, In: Proceedings of the Australasian Database Conference, Adelaide, SA, 1995.

[7] D. Merkl, Exploration of text collections with hierarchical feature maps, In: Proceedings of the Int'l ACM SIGIR Conference on Research and Development in Information Retrieval, Philadelphia, PA, 1997.

[8] G. Salton, Automatic Text Processing: The transformation, analysis, and retrieval of information by computer, Addison-Wesley, Reading, MA, 1989.

# Competitive Learning
# for Binary Valued Data

Friedrich Leisch,* Andreas Weingessel* & Evgenia Dimitriadou

Institut für Statistik und Wahrscheinlichkeitstheorie, TU Wien

Vienna, Austria

Email: firstname.lastname@ci.tuwien.ac.at

## Abstract

We propose a new approach for using online competitive learning on binary data. The usual Euclidean distance is replaced by binary distance measures, which take possible asymmetries of binary data into account and therefore provide a "different point of view" for looking at the data. The method is demonstrated on two artificial examples and applied on tourist marketing research data.

# 1 Introduction

Most common clustering methods such as $k$-means, (hard and soft) competitive learning or neural gas minimize the usual Euclidean distance, i.e., perform least squares estimation [1]. Euclidean distance has a natural connection with normally distributed data, for normal distributions least squares and maximum likelihood estimation coincide. However, for non-normal data other distance measures may have advantageous properties, especially when the data are asymmetric and/or discrete.

In this paper we deal with data from tourist questionnaires. Vienna is one of the worlds largest destinations for city tourism, and marketing is of strategic importance. Homogeneous target groups are very important for advertising, it is desirable to segment tourists into groups which can be addressed separately.

People visiting Vienna are asked to fill out a form about their vacation preferences and general hobbies. These data are subsequently used to compute profiles of "prototypical" tourists. One of our datasets is about typical vacation activities such as tennis, golf, relaxing, shopping, concerts, theater or sightseeing. All answers are boolean, where a "yes" (encoded as 1) in tennis means that the corresponding person likes to play tennis when on vacation.

Clearly this data are not normally distributed and Euclidean distance need not be a good distance measure for this kind of data. Two persons both playing tennis have the same distance as two persons not playing tennis (in both cases the distance is 0), yet, two persons both playing tennis have more in common than two persons who both do not play tennis.

*This piece of research was supported by the Austrian Science Foundation (FWF) under grant SFB#010 ('Adaptive Information Systems and Modeling in Economics and Management Science').

## 2 Clustering binary data

Special distance measures for binary data have been used in statistics and cluster analysis for a long time [2], but mostly in combination with hierarchical cluster methods [3]. Hierarchical clustering is only feasible for small data sets, not for data mining in huge data sets containing several thousand cases.

Classic non-hierarchical methods such as $k$-means are hard to combine with binary distance measures because they need the explicit computation of cluster centers. Cluster centers are easy to compute for Euclidean or absolute distance, where they correspond to the mean or median of the cluster, respectively.

Adaptive methods such as online competitive learning have the advantage that they do not need the explicit computation of cluster centers, only the gradient of the distance measure is needed. Hence, neural network clustering methods can be used in combination with binary distance measures.

### 2.1 Distance measures for binary data

Numerous distance measures for binary data have been proposed in the statistical literature [2]. None of this can be considered to be "the right" distance for a given real world data set or application. It is up to the user to decide which features in the data set are more important or which differences he wants to find; and then to use an appropriate distance measure to extract these features.

Consider two $n$-dimensional binary vectors $x = (x_1, \ldots, x_n)'$ and $y = (y_1, \ldots, y_n)'$. We define the $2 \times 2$ contingency table $[\alpha, \beta; \gamma, \delta]$ where $\alpha = \#\{i : x_i = y_i = 1\}$ denotes the number of components where both $x$ and $y$ are one, $\beta = \#\{i : x_i = 0, y_i = 1\}$, $\gamma = \#\{i : x_i = 1, y_i = 0\}$, and $\delta = \#\{i : x_i = 0, y_i = 0\}$. We restrict ourselves to distance measures of type $D(x, y) = D(\alpha, \beta, \gamma, \delta)$. E.g., the well known Hamming distance (number of different bits of $x$ and $y$) can be written as $D(x, y) = \beta + \gamma$.

As mentioned in the introduction, we prefer asymmetric distance measures giving more weight to common ones than common zeros, because two common ones represent a common preference of two persons, whereas two zeros simply state that both persons do not like the respective activity. In the following we will concentrate on the following two (closely related) distances:

$$D_1(x, y) = \frac{\beta + \gamma}{\alpha + \beta + \gamma}, \qquad D_2(x, y) = \frac{\beta + \gamma}{2\alpha + \beta + \gamma}$$

$D_1$ is the famous Jaccard coefficient (see, e.g., [3]). Since both do not depend on $\delta$, questions where both subjects answered "no" are ignored. Distance $D_1$ is the percentage of disagreements in all answers where at least one subject answered "yes". Distance $D_2$ is similar, but puts more weight on answers where both subjects answered "yes".

### 2.2 Binary competitive learning

Hard competitive learning is a well-known online stochastic gradient descent algorithm for minimization of the average distance of a given set of data to its closest center. See, e.g., [4] for a survey on competitive learning methods. Let $X_N = \{x^1, \ldots, x^N\}$ denote the data set available for training and let $C_K = \{c^1, \ldots, c^K\}$ be a set of $K$ centers. Further let $c(x) \in C_K$ denote the center closest to $x$ with respect to some distance measure $D$. Then competitive learning tries to minimize $\sum_{n=1}^{N} D(x^n, c(x^n))$.

The simplest online hard competitive learning algorithm works as follows:

1. *Initialize $C_K$ at random (either by picking $K$ points from $X_N$ or from a random number generator). Set $t = 0$.*
2. *Pick a random point $x^i$ from $X_N$.*
3. *Let $c_t^j = c(x^i)$ be the center closest to $x^i$ at step $t$. Update the centers according to $c_{t+1}^j = c_t^j - \eta_t \nabla D(x^i, c_t^j)$ and $c_{t+1}^l = c_t^l$ for $l \neq j$ where $\nabla D$ is the gradient of $D$ and $\eta_t$ is a decreasing learning rate.*
4. *Stop if some convergence criterion (number of iterations, error) is fulfilled, else $t = t + 1$ and goto step 2.*

Two tasks have to be solved for using a binary distance $D$ with this algorithm: First, the gradient of $D$ with respect to the second argument $c^j$ has to be computed, which is straightforward. Second, a real-valued decreasing learning rate will result in non-binary (i.e., real valued) centers; hence, $D(x, c)$ must also be defined for non-binary centers $c$.

We overcome the second problem by interpreting a real-valued center $c = (c_i, \ldots, c_n)'$ with elements $0 \leq c_i \leq 1$ as a vector of probabilities, where $c_i$ is the probability, that the corresponding component is one. Note, that this approach is closely related to conventional Euclidean clustering, where the cluster centers are equal to the mean of the clusters and therefore equal to the probabilities of having a 1 in the corresponding component, if a variable is binary.

If a component $c_i^j$ of center $c^j$ after the update step 3 is larger than 1, we replace it by 1. Similarly, we replace it by 0 if $c_i^j$ is negative. In our experiments this lead to almost binary centers upon convergence, i.e., the $c_i^j$ of the final centers (almost) equal 0 or 1. Further investigation of this strategy is necessary.

Let $x$ be a (binary) data vector, and let $c$ be a real-valued center with components in $[0, 1]$, then the *expected values* of the contingency table entries are given by $\alpha = x'c$, $\beta = (1 - x)'c$, $\gamma = x'(1 - c)$, $\delta = (1 - x)'(1 - c)$. After some simple algebra we get the gradients of $D_1$ and $D_2$

$$\frac{\partial D_1}{\partial c_j} = \begin{cases} \frac{\alpha}{(\alpha+\beta+\gamma)^2}, & x_j = 0 \\ \frac{-1}{(\alpha+\beta+\gamma)^2}, & x_j = 1 \end{cases}, \quad \frac{\partial D_2}{\partial c_j} = \begin{cases} \frac{\alpha}{(2\alpha+\beta+\gamma)^2}, & x_j = 0 \\ \frac{-(\alpha+\beta+\gamma)}{(2\alpha+\beta+\gamma)^2}, & x_j = 1 \end{cases}$$

# 3 Experiments

## 3.1 Artificial Data

The questionnaires used by our research partners in tourism marketing use groups of questions concerning related questions. E.g., there are questions

| | $z_1 : x_1, x_2, x_3$ | $z_2 : x_4, x_5, x_6$ | $z_3 : x_7, x_8, x_9$ | $z_4 : x_{10}, x_{11}, x_{12}$ | m |
|---|---|---|---|---|---|
| Type 1 | high | high | low | low | 1000 |
| Type 2 | low | low | high | high | 1000 |
| Type 3 | low | high | high | low | 1000 |
| Type 4 | high | low | low | high | 1000 |
| Type 5 | low | high | low | high | 1000 |
| Type 6 | high | low | high | low | 1000 |

Table 1: Scenario 1: Symmetric distribution of 0s and 1s.

| | | $z_1 : x_1, x_2$ | $z_2 : x_3, x_4, x_5$ | $z_3 : x_6, x_7$ | $z_4 : x_8, x_9, x_{10}$ | m |
|---|---|---|---|---|---|---|
| I | Type 1 | high | high | low | low | 200 |
| | Type 2 | high | low | low | low | 800 |
| II | Type 3 | low | low | high | high | 200 |
| | Type 4 | low | low | high | low | 800 |
| III | Type 5 | low | low | low | low | 2000 |

Table 2: Scenario 2: Asymmetric distribution of 0s and 1s.

whether a person likes sports such as tennis, cycling, swimming, riding or water sports. Another group of questions is about cultural activities such as concerts, theater or museums. Obviously, answers inside such groups are correlated, i.e., a person generally interested in culture is more likely to visit both the theater and concerts than a person not so interested in culture.

This leads to the concept of latent variables, which cannot be measured directly, but through several observable variables. In the example given above, the latent variable would correspond to the feature "generally interested in culture"; the observable variables are concerts, theater and museum.

We compared clustering with Euclidean distance and the binary distances $D_1$ and $D_2$ on two artificial examples resembling this structure: We have 4 latent variables $z_1, \ldots, z_4$, which are measured through a varying number of observable variables $x_j$.

Table 1 shows a scenario where each latent variable $z_i$ is represented by three observable variables $x_j$. Six types of data are generated, each consisting of 1000 data points. Type 1 has a high probability (80%) that the variables corresponding to $z_1$ and $z_2$ are 1, and a low probability (20%) that the remaining variables are 1. Type 2 is exactly reverse, etc. [5]. Both competitive learning with Euclidean distance and with the two binary distances $D_1$ and $D_2$ found the six clusters without problems. We also tried $k$-means clustering, giving similar results.

Differences between binary and Euclidean distances should emerge if we break the symmetry in the distribution of 0s and 1s. Table 2 shows a asymmetric scenario with 5 clusters in 3 groups (I, II, III); the clusters have also different sizes.

In this examples, the Euclidean-based algorithms did not give stable results,

| | | | | | | | | | | | |
|---|---|---|---|---|---|---|---|---|---|---|---|
| $D_1$ | $c^1$ | 1 | 1 | 0 | 0 | 0 | 0 | 0 | 0 | 0 | 0 |
| | $c^2$ | 0 | 0 | 0 | 0 | 0 | 1 | 1 | 0 | 0 | 0 |
| E | $c^1$ | 0.09 | 0.09 | 0.12 | 0.10 | 0.11 | 0.92 | 0.91 | 0.38 | 0.36 | 0.34 |
| | $c^2$ | 0.37 | 0.36 | 0.18 | 0.18 | 0.19 | 0.10 | 0.08 | 0.09 | 0.09 | 0.10 |
| $D_1$ | $c^1$ | 1 | 1 | 0 | 0 | 0 | 0 | 0 | 0 | 0 | 0 |
| | $c^2$ | 1 | 1 | 1 | 1 | 1 | 0 | 0 | 0 | 0 | 0 |
| | $c^3$ | 0 | 0 | 0 | 0 | 0 | 1 | 1 | 1 | 1 | 1 |
| | $c^4$ | 0 | 0 | 0 | 0 | 0 | 1 | 1 | 0 | 0 | 0 |
| E | $c^1$ | 0.08 | 0.08 | 0.08 | 0.08 | 0.09 | 0.00 | 0.11 | 0.11 | 0.08 | 0.11 |
| | $c^2$ | 0.09 | 0.09 | 0.13 | 0.10 | 0.10 | 1.00 | 0.69 | 0.17 | 0.00 | 0.13 |
| | $c^3$ | 0.91 | 0.89 | 0.35 | 0.36 | 0.38 | 0.10 | 0.09 | 0.11 | 0.11 | 0.12 |
| | $c^4$ | 0.10 | 0.10 | 0.10 | 0.10 | 0.10 | 0.86 | 0.86 | 0.58 | 0.93 | 0.58 |

Table 3: Results for Scenario 2. $D_1$ denotes binary distance, E Euclidean distance.

i.e., different restarts of the algorithm typically gave different cluster centers. Using binary distances gave stable results. Table 3 shows typical results with two and four cluster centers: Euclidean distance always found one or more large clusters corresponding to type 5 (many 0s), and could not recover types 1–4 clearly. Using 5 cluster centers did not improve the situation. Binary distance $D_1$ always recovers types 1–4, but ignores type 5 due to its definition. In case of two cluster centers, groups I and II are recovered. Distance $D_2$ gave similar results, these are omitted due to space limitations.

## 3.2 Tourism Data

As described above we also clustered binary data from tourist questionnaires with 12 variables and 15066 cases. Results can be seen in Table 4. The binary clustering shows a clear structure. There is one big cluster (# 3) featuring "classical" tourist characteristics such as swimming, relax, shopping or sightseeing. Additionally there are 3 smaller clusters of almost the same size. # 4 can be seen as "typical tourist" similar to # 3, but with additional Viennese specialities like "Heurigen". # 2 is a more sportive kind of tourist (additionally cycling and water sports). Finally # 1 is a type of tourists "doing nothing but relax".

The results of Euclidean clustering are much more fuzzy. All clusters have almost the same size (3200-4800) and their profiles are not as distinctive as the results from binary clustering. Some clusters are similar to the results from binary clustering, but one needs additional postprocessing like thresholding to be able to read the results. However, by thresholding of the centers one alters the partition of the data set, hence the partition must not correspond to the used distance anymore.

| | tennis | cycling | riding | golf | swim | w.sport | relax | shop | sightsng | museum | theater | heuriger | size |
|---|---|---|---|---|---|---|---|---|---|---|---|---|---|
| $D_1$ | 0 | 0 | 0 | 0 | 0 | 0 | 1 | 0 | 0 | 0 | 0 | 0 | 2224 |
| | 0 | 1 | 0 | 0 | 1 | 1 | 1 | 1 | 1 | 0 | 0 | 0 | 2804 |
| | 0 | 0 | 0 | 0 | 1 | 0 | 1 | 1 | 1 | 1 | 0 | 0 | 7547 |
| | 0 | 0 | 0 | 0 | 0 | 0 | 1 | 1 | 1 | 1 | 1 | 1 | 2491 |
| E | .07 | .10 | .02 | .03 | .32 | .08 | .00 | .20 | .55 | .39 | .10 | .17 | 3268 |
| | .06 | .07 | .02 | .03 | .40 | .14 | .77 | .78 | .95 | 1.0 | .23 | .28 | 4803 |
| | .06 | .03 | .02 | .02 | .48 | .11 | 1.0 | .45 | .46 | .04 | .04 | .09 | 4090 |
| | .23 | .98 | .09 | .07 | .95 | .40 | .81 | .62 | .72 | .48 | .13 | .30 | 2905 |

Table 4: Clustering results for tourism data.

# 4  Summary

A—due to our knowledge—new approach for competitive learning with asymmetric binary distance measures has been proposed. This way, we provide a different "point of view" for looking at the data. For real world data, there is no way of determining which clustering algorithm is "best", because the data generating process is unknown. Typical clustering algorithms try to minimize a loss function depending on differences between cluster centers and data points. Using several different distance measures (and knowing about their special characteristics) can give valuable further insight into a data set.

Of course, one is not limited to the two binary distance measures proposed in this paper. A lot of different distance measures can be found in the literature, most of which can easily be adopted to our framework. Also, there are more popular competitive learning algorithms (e.g., soft competitive learning, neural gas or Kohonen maps), which could be generalized for binary distance measures in the same way. All these questions are currently under investigation.

# References

[1] Brian D. Ripley. *Pattern recognition and Neural networks*. Cambridge University Press, Cambridge, UK, 1996.

[2] Michael R. Anderberg. *Cluster analysis for applications*. Academic Press Inc., New York, USA, 1973.

[3] Leonard Kaufman and Peter J. Rousseeuw. *Finding Groups in Data*. John Wiley & Sons, Inc., New York, USA, 1990.

[4] Bernd Fritzke. Some competitive learning methods, April 5 1997. http://www.neuroinformatik.ruhr-uni-bochum.de/ini/VDM/research/gsn/.

[5] Sara Dolnicar, Friedrich Leisch, Andreas Weingessel, Christian Buchta, and Evgenia Dimitriadou. A comparison of several cluster algorithms on artificial binary data scenarios from tourism marketing. Working Paper Series 7, SFB "Adaptive Information Systems and Modeling in Economics and Management Science", http://www.wu-wien.ac.at/am/workpap.html, 1998.

# Neural Network-Based Inferential Sensing

Ressom Habtom and Lothar Litz

Institute of Process Automation, University of Kaiserslautern
PO Box 3049, 67653 Kaiserslautern, Germany
[habtom,litz]@e-technik.uni-kl.de

## Abstract

This paper addresses the problem of generating a neural-network based inferential sensor for difficult to measure process variables. An approach is proposed which is effective particularly in circumstances where it is not easy to model the relationship between the measurable process variables and the variables to be estimated. The method involves the modeling of an alternative relationship and the use of an a priori knowledge from which the unmeasured variables may be inferred.

To demonstrate the approach, a simulation model of an industrial drying drum which is commonly used in sugar industry is considered. The purpose of the drum is to decrease the water content of pressed pulp. We assume that the dry substance content of the pressed pulp at the inlet of the drum can not be measured on-line. A recurrent neural network is trained to predict a measurable variable, which is the dry substance content of the dried pulp at the outlet of the drum. The estimation of the unmeasured variable is carried out based on the prediction of the neural network and a knowledge from the physical insight about the effect of the unmeasured variable on the measured variable.

# 1  Introduction

The need for inferential sensing [1] becomes significant in circumstances where key variables cannot be measured on-line in real time but which, if included in the control strategy, would make a significant difference to process performance and safety. In some industries the laboratory analysis of samples is still widely used for control of process, but it is recognized that this is not genuine process control since the time constant of the measurement process are usually orders of magnitude greater than those of the process. The reason for the incapability of measuring the variables at the desired rate can be due to limited analyzer cycle times or a reliance upon off-line laboratory assays. In some cases the on-line measurement can be either costly or impossible. Hence, if the relationship between the difficult to measure variables (primary variables) and the on-line measurable process variables (secondary variables) can be captured, then the resulting model can be utilized to act as a virtual sensor during the on-line operation. In this regard, the use of neural networks as a model is appealing. This is due to their capability to learn relationships that can hardly be modeled efficiently by more usual statistical methods or from first principles.

However, depending on the nature of the relationship between the process variables, one may still find it difficult to train a neural network that can predict the desired process variables. In particular, if the primary variables are input variables to a process and all the secondary variables are inputs and outputs of the process, then the problem will be the training of an inverse model which is not usually as easy as the direct modeling. In [2], an approach is proposed that alleviates this problem by using the direct model and the extended Kalman filter (EKF) algorithm to infer unmeasured input variables. The present contribution addresses a more difficult problem by assuming that the primary variables can be measured neither on-line nor off-line except that a calibration procedure could be made. The calibration procedure is assumed to be carried out for constant values of the primary variables. An approach is proposed which employs an a priori knowledge to infer the unmeasured variable on-line.

## 2 On-line estimation of unmeasured input variables

A neural inferential sensor may be created, if there are sufficient accumulated process data representing the primary and secondary variables. However, if these variables (in particular the primary variables) can not deliver sufficient data for training, then obtaining a reasonable inferential sensor will be difficult. In such circumstances, we need to resort to other indirect methods. In this section, a method which involves the use of an a priori knowledge is proposed.

For the seek of illustration, we assume that the primary variables are input variables to a process and that these variables can not be measured except for the possibility of a calibration. Since the calibration procedure is assumed to allow only constant values of the secondary variables, the corresponding data can not be used to train a neural network with the primary variables as its output. As a result, we select from the secondary variables other variables which can be used as output variables. Let $y_u$ be a vector denoting the selected secondary variables and let $u$ be a vector containing the remaining secondary variables excluding those in $y_u$. Now, a neural network, which predicts $y_u$ from $u$, will be trained. Assuming that the variables have a dynamic relationship, a recurrent network will be used as a model. We consider a recurrent neural network which is described as a recurrent multilayer perceptron (RMLP) in [3]. In addition to the feedforward connections, the output of each unit is fed back to its input and is also connected to other units within the layer. The feedback connections are made after a unit delay. An RMLP network that has a single hidden layer and an external recurrence can be described mathematically as:

$$x(k+1) = f\left(W^y y(k) + W^u u(k) + W^h x(k)\right) \tag{1}$$

$$y(k+1) = g\left(W^x x(k+1) + W^o y(k)\right) \tag{2}$$

where $x(k)$ and $y(k)$ denote the output vector of the hidden units and output vector of the network at time k, respectively; $W^y$ and $W^u$ represent the weight matrices connecting the hidden units to the previous output of the network and to the input vector $u$, respectively; $W^x$ is the weight matrix between the hidden units and the output units; $\mathbf{W^h}$ and $\mathbf{W^o}$ are weight matrices representing the local and lateral connections for the hidden units and the output units, respectively. $\mathbf{f}(\cdot)$ and $\mathbf{g}(\cdot)$ are vector-valued activation functions for the hidden and the output units, respectively.

The prediction of the off-line trained RMLP network will be exact only if the primary variables are kept constant at the values used during calibration. The prediction error $(\mathbf{e})$ between the output vector of the neural network $(\mathbf{y})$ and that of the selected variables $(\mathbf{y_u})$ is attributed to the deviation of the primary variables from the constant values. Let $\mathbf{z}$ and $\mathbf{u_c}$ be vectors representing the estimate of the primary variables and the corresponding constant values used during calibration, respectively. As already pointed out, the objective is to infer the primary variables from the prediction error. This is possible if there is an a priori knowledge that provides us an information how we could incorporate the variable $\mathbf{z}$ in (1) and (or) (2). Considering that we know the relationship between the primary variables and some secondary variables, (1) and (2) can be written as:

$$x(k+1) = \mathbf{f}\left(\mathbf{W^y}y(k) + \mathbf{W^z}\mathbf{h}(\mathbf{u_z}(k), \mathbf{u_c}, \mathbf{z}(k)) + \mathbf{W^m}\mathbf{u_m} + \mathbf{W^h}x(k)\right) \tag{3}$$

$$y(k+1) = \mathbf{g}\left(\mathbf{W^x}x(k+1) + \mathbf{W^o}y(k)\right) \tag{4}$$

where $\mathbf{u_z}$ denotes a vector of the secondary variables whose relationship with the primary variables is known; $\mathbf{u_m}$ represents the remaining secondary variables in $\mathbf{u}$ excluding those in $\mathbf{u_z}$ ; $\mathbf{W^z}$ and $\mathbf{W^m}$ are the weight matrices which are extracted from $\mathbf{W^u}$ according to their connection to $\mathbf{u_z}$ and $\mathbf{u_m}$, respectively; $\mathbf{h}(.)$ is a vector valued function representing the a priori knowledge.

To use the EKF algorithm for the estimation of $\mathbf{z}$, we put (3) and (4) in a state space form and augment the resulting state space model by $\mathbf{z}$ as a constant plus a noise term, [4]. The resulting state model will be:

$$x(k+1) = \mathbf{f}\left(\mathbf{W^y}\mathbf{x_y}(k) + \mathbf{W^z}\mathbf{h}(\mathbf{u_z}(k), \mathbf{u_c}, \mathbf{z}(k)) + \mathbf{W^m}\mathbf{u_m}(k) + \mathbf{W^h}x(k)\right) + \mathbf{n_1}(k) \tag{5}$$

$$\mathbf{x_y}(k+1) = \mathbf{g}\left(\mathbf{W^x}\mathbf{f}(.) + \mathbf{W^o}\mathbf{x_y}(k)\right) + \mathbf{n_2}(k) \tag{6}$$

$$z(k+1) = z(k) + \mathbf{n_3}(k) \tag{7}$$

$$y(k) = \mathbf{x_y}(k) + \mathbf{v}(k) \tag{8}$$

where $\mathbf{n_1}$, $\mathbf{n_2}$, $\mathbf{n_3}$ are vectors of plant noise terms, while $\mathbf{v}$ is that of measurement noise terms. $\mathbf{x_y}$ represents the previous output vector of the network, which is fed back to the network's input. The state model can be represented in a matrix form:

$$\begin{bmatrix} x(k+1) \\ \mathbf{x_y}(k+1) \\ z(k+1) \end{bmatrix} = \begin{bmatrix} \mathbf{f}(\cdot) \\ \mathbf{g}(\cdot) \\ z(k) \end{bmatrix} + \begin{bmatrix} \mathbf{I} & \mathbf{0} & \mathbf{0} \\ \mathbf{0} & \mathbf{I} & \mathbf{0} \\ \mathbf{0} & \mathbf{0} & \mathbf{I} \end{bmatrix} \begin{bmatrix} \mathbf{n_1}(k) \\ \mathbf{n_2}(k) \\ \mathbf{n_3}(k) \end{bmatrix}$$

$$y(k) = \begin{bmatrix} \mathbf{0} & | & \mathbf{I} & | & \mathbf{0} \end{bmatrix} \begin{bmatrix} x(k) \\ \mathbf{x_y}(k) \\ z(k) \end{bmatrix} + \mathbf{v}(k) \tag{9}$$

where $\mathbf{f}(\cdot)$ and $\mathbf{g}(\cdot)$ represent the functions in (5) and (6), respectively; $\mathbf{I}$ and $\mathbf{0}$ are identity and null matrices, respectively. Now, the discrete EKF algorithm can be readily used to infer the unmeasured input vector $\mathbf{z}$ based on the prediction error, $\mathbf{e}$.

# 3 Simulation test and results

This example is based on a simulation model of an industrial drying process. In the course of extracting sugar from roots of sugar beet, there will be byproducts like scraps of beet and molasses. The purpose of the drying process is to decrease the water content of the byproducts. After a mechanical press the pressed pulp will be driven to a drying drum. A typical drying drum, which involves a high temperature drying process, is depicted in Fig. 1. The heat generated by the burner and flue gas coming from a boiler house pass through the drum creating high temperature in the drum. Upon rotation of the drum, the pressed pulp flow through the drum and come out at the outlet of the drum as dried pulp.

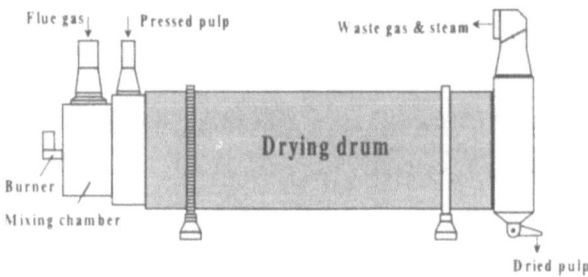

Figure 1: Schematic diagram of an industrial drying drum

The process variables, which are of interest to us are: the input feed flow rate of the pressed pulp ($u_1$), the dry substance content of the pressed pulp ($u_2$), the input oil quantity fed to the burner ($u_3$), and the dry substance content of the dried pulp ($y_o$). A block diagram representation of a drying drum's mathematical model, which is developed in [5] and adapted to a particular industrial drum in [6], is depicted in Fig. 2. A distinct feature of the model is that it contains high order linear dynamic transfer functions, transport delays, and that it is non minimum phase and nonlinear.

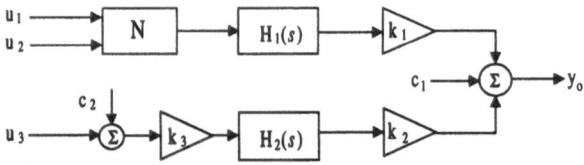

Figure 2: Block diagram representation of the mathematical model of the drying drum

$H_1(s)$ and $H_2(s)$ in Fig. 2 are linear transfer functions described as follows:

$$H_1(s) = \frac{-0.1143s^4 - 0.1504s^3 - 0.3519s^2 - 0.2834s^1 - 0.0075}{s^5 + 2.1543s^4 + 2.8923s^3 + 0.5515s^2 + 0.0593s^1 + 0.0013} e^{-9s} \tag{10}$$

$$H_2(s) = \frac{-0.1711s^4 - 2.441s^3 - 0.7681s^2 - 1.6166s^1 + 0.1721}{s^5 + 1.7922s^4 + 2.9957s^3 + 1.1311s^2 + 0.2902s^1 + 0.0139} e^{-3s} \tag{11}$$

while the constants and the nonlinear function N in the figure are given by:

$$k_1 = 0.84 \quad k_2 = 1.19 \quad k_3 = 0.00095 \quad c_1 = 91 \quad c_2 = -2200$$
$$N = 0.31\big(u_1(1 - 0.01\,u_2) - 36.6\big) . \tag{12}$$

In a closed loop operation, the dry substance content of the dried pulp is controlled through the oil quantity to achieve a value of around 90%, which is only 36 to 46% at the inlet of the drum. The input feed flow rate ($u_1$) and the dry substance content of the pressed pulp ($u_2$) act as disturbances. Since the latter is not measurable, the controller cannot get direct information about this disturbance.

Hence, the primary variable in this example is $u_2$. To demonstrate the approach proposed in the previous section, we consider that we know how $u_1$ and $u_2$ are related. In other words, we use the a priori knowledge that the two input variables appear in the form $u_1(1 - 0.01u_2)$, see (12). Among the secondary variables, $y_0$ is used as an output variable, while $u_1$ and $u_3$ are used as input variables to an RMLP network. Thus, the vectors $\mathbf{y_u}$, $\mathbf{u_z}$, $\mathbf{u_m}$, and $\mathbf{z}$ in the previous section correspond to this example's variables $y_0$, $u_1$, $u_3$ and the estimate of $u_2$, respectively. Note that, in the sequel, $y_0$, $u_1$, $u_3$, and the estimate of $u_2$ are denoted as $y_u$, $u_z$, $u_m$, and $z$, respectively.

5000 samples are generated by simulating Fig. 2 for $u_z$ ($u_1$ in the figure) over the range 58 to 72 t/h; for a constant value of $u_2$ ($u_c = 41\%$); and $u_m$ ($u_3$ in the figure) over the range 2000 to 3000 l/h with a sampling rate of 1 min. Such data may also be gained from the real process by using standard pulp whose dry substance content is known. 80% of the total samples are used for training the RMLP network while the remaining samples are used for validation. The RMLP network has 8 hidden units and one output unit. It is trained using the real time recurrent learning algorithm, which involves the Levenberg-Marquardt approach. Fig. 3 shows the output of the simulation model compared to the prediction of the off-line trained RMLP network for the test samples. Note that this result does not show a one-step ahead prediction, but the response of the network over the whole range of time where the output of the network is fed back to its input.

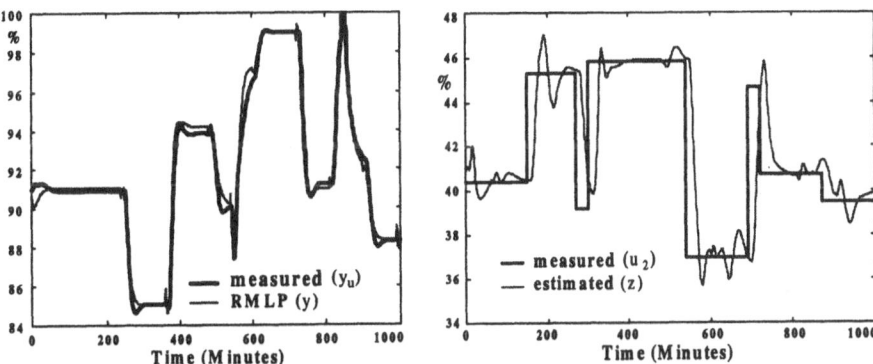

Figure 3: Measured and predicted values of $y_u$    Figure 4: Measured and estimated values of $u_2$

After having obtained the neural model, the knowledge about the relationship between $u_1$ and $u_2$ is used to put the required equations in the form of (5)-(8). Using the mentioned a priori knowledge, the function h in (5) for this problem is given by:

$$h\big(u_z(k), u_c, z(k)\big) = \frac{u_z(k)\big(1 - 0.01\,z(k)\big)}{1 - 0.01\,u_c} \tag{13}$$

where the denominator $(1-0.01u_c)$ is included so as to get the correct prediction when the unmeasured variable is equal to the constant value $u_c$. Finally, a state space model of the form (9) is constructed and the estimation of the unmeasured variable together with the state variables $\mathbf{x}$ and $\mathbf{x}_y$ is carried out using the EKF algorithm. Fig. 4 shows the measured and estimated dry substance contents of the pressed pulp.

# 4 Conclusions

This paper presents that the use of a neural network and the EKF algorithm is an attractive alternative to other more traditional approaches. It is shown that the approach can be employed to generate an inferential sensor not only when the available data can be readily used to train a neural network that predicts the unmeasured data from other measurements but also in circumstances where it becomes impossible to come up with a reasonable network due to scarcity of data or difficulty to train the relationship. Of course the need for a calibration procedure and an a priori knowledge are a requirement to establish the required relationships and eventually to carry out the estimation. The simulation result has demonstrated the effectiveness of the proposed method. The estimation approach can play a significant role in achieving a better performance of a closed loop operation by providing the estimated disturbance information to the controller, [7].

# References

[1] Willis M. J, Montague G. A, Di Massimo C, Tham M. T, Morris A. J. Artificial neural networks in process estimation and control. Automatica 1992; 28,1181-1187.

[2] Habtom R, Litz L. Estimation of unmeasured inputs using recurrent neural networks and the extended Kalman filter. In Proceedings of IEEE International Conference on Neural Networks, Houston 1997, pp. 2067-2071.

[3] Fernandez B, Parlos A. G, Tsai W. K. Nonlinear dynamic system identification using artificial neural networks (ANNs). In Proceedings of International Joint Conference on Neural Networks, San Diego 1990, vol. II. pp. 131-141.

[4] Williams R. J. Training recurrent networks using the extended Kalman filter, in Proc. of the International Joint Conference on Neural Networks, Baltimore 1992, vol. IV, pp. 241-246.

[5] Mann W. Identifikation und digitale Regelung eines Trommeltrockners. In Kernforschungszentrum Karlsruhe: Projekt Prozeßlenkung mit Datenverarbeitungsanlagen (PDV), Forschungsbericht KfK-PDV 189, Karlsruhe 1980.

[6] Leibrock G. Ein Konzept zur Automatisierung einer industriellen Trocknungsanlage für Zuckerrübenschnitzel mit Fuzzy Control. Diplomarbeit D49, Institute of Process Automation, University of Kaiserslautern, Kaiserslautern 1995.

[7] Habtom R, Litz L: Neurocontrol of nonlinear dynamic systems subject to unmeasured disturbance inputs. In: Gerstner W (ed) Artificial Neural Networks-ICANN '97. Springer-Verlag, Heidelberg, 1997, pp 855-860 (Lecture notes in computer science no. 1327)

# Retina Encoder Inversion for Retina Implant Simulation

M. Becker, M. Braun, R. Eckmiller

Informatik VI (Neuroinformatik), Universität Bonn

D-53117 Bonn, F. R. Germany

email: mb@nero.uni-bonn.de, URL: http://www.nero.uni-bonn.de

## Abstract

To test tuning methods for the Retina Encoder (RE) of a Retina Implant (RI) as a visual prosthesis for blind subjects with retinal degenerations a suitable simulation of the patient's evaluative response to RE state alterations must be provided. RE simulates real time retinal information processing and consists of several hundreds of spatio-temporal receptive field (RF) filters to generate electrical signals for ganglion cell (GC) stimulation. We propose a neural network to reconstruct the RE input from a number of consecutive RE output frames. The network can be interpreted as a simulation of a part of the central visual system with GC signals as input and perception visualization as output. We present first results using Evolution Strategies for neural network weight optimization.

## 1  Problem Formulation

Retina Implants [1] for blind subjects with retinal degenerations are currently under development [3]. In the future the blind subject wears a Retina Encoder (RE) in a frame of glasses to observe the visual environment and to process image signals in real time according to a suitable model of retinal information processing. Intact retinal ganglion cells (GC) are stimulated electrically with the simulation output signals to elicit visual perceptions to regain a modest amount of vision. RE consists of receptive field filters (RF), one for each GC to be stimulated. Each RF has several spatial and temporal parameters and can be tuned continuously within a wide range of GC-types observed in retina physiology. The set of RF parameters forms the RE state space. During stimulator implantation the actually contacted GC-types remain unknown and must be determined in a dialog with the implant carrying subject. Learning algorithms have been proposed to adjust the RE from the patient's evaluative feedback to consecutive RE state alterations [2] to optimize the patient's visual perception. For the purpose of testing these algorithms patient feedback must be provided. At present it has to be simulated since real feedback from implant carrying patients is not yet available. Investigations so far have shown that suitable feedback is crucial for successful learning [2]. Therefore we are developing *man in loop* techniques that apply human evaluative feedback for

---

[1]Supported by Federal Ministry for Education, Science, Research, and Technology (BMBF).

patient simulation. Here we present an implementation based on the idea of inverting the RE output signals to reproduce the according RE input image.

## 2 Retina Model

The implemented retina model is an intermediate version towards the DSP based RE *Mark II* [4], which is currently being developed in our group. Figure 1 explains the implemented RF input-output model which was adopted from physiological investigations of the primate retina [5]. Each RF of RE is ap-

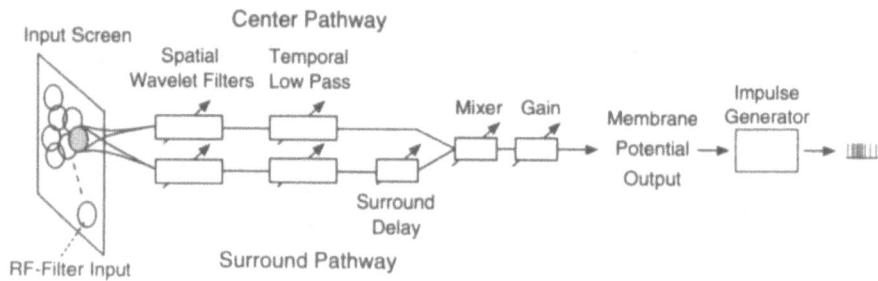

Figure 1: *Receptive field filter (RF) of RE. RF receives input data from the photosensor array. Data is processed via two distinct spatio-temporal pathways for center and surround computation. Pathways converge at the mixer-stage. The resulting signal is finally multiplied with a gain factor. Each component is individually tunable to achieve a wide range of GC-types observed in retina physiology.*

portioned a spatial receptive field as two-dimensional image input. Inseparable spatio-temporal processing is achieved by two distinct filter pathways, one for center computation and the other for the surround. Each pathway performs a spatial scalar product of the image pixel data and a rotationally symmetric two dimensional Gaussian (G) with the corresponding widths $\sigma_c$, $\sigma_s$. The sigmas determine the spatial extent of its RF. The resulting scalar signals then each pass a temporal first order low pass ($LP$), with (inverse) time constants $\lambda_c$, $\lambda_s$. Additionally, the surround pathway signal can be delayed ($D$) by a number of RE-calculation cycles $n\tau$. The signals from both pathways converge at the *mixer* component which determines center vs. surround contribution. Finally, a gain factor ($g$) enables range adaptation and switching between on-off and off-on GC properties. The resulting signal simulates the GC membrane potential ($MP$) and is used to generate stimulation signals. Equations 1 - 4 summarize the model operators.

$$Path_c = LP(\lambda_c)\,G(\sigma_c) \tag{1}$$
$$Path_s = D(n\tau)\,LP(\lambda_s)\,G(\sigma_s) \tag{2}$$
$$Mixer = mPath_c - (1-m)Path_s \quad m \in [0,1] \tag{3}$$

$$MP \;\; = \;\; g \cdot Mixer \qquad\qquad (4)$$

The model is able to simulate a wide range of primate ganglion cell types observed in retina physiology. There are 2 basic types of GC: sustained P-cells and transient M-cells. Cells in each of these classes show antagonistic center-surround weighting with either the center positively and the surround negatively weighted (on-off), or vice versa (off-on). Thus, RE states that actually match physiological findings will be located around 4 focal points in the RE state space: P-on, P-off, M-on, M-off. Note that states of cells in each class can deviate significantly from the corresponding focal point state.

Since the first stage of RE learning procedures will concentrate on optimizing the four base classes the goal of the present investigation is to develop RE inversion approximators for the so called *four-class* RE. It consists of RFs attached to one of the four base classes. For this study, RF centers are located at hexagonal grid points on the RE input screen, but the implementation has no general restrictions for RF center positioning.

## 3 Patient simulation with RE inversion and subjects with normal vision

RE learning procedures [2] employ patient feedback to generate RE state alterations in order to optimize the patient's visual perception. One approach to simulate the patient is to define a plausibly chosen RE state to correspond to a fictitious patient's optimal perception (Fig. 2). Then create an inversion

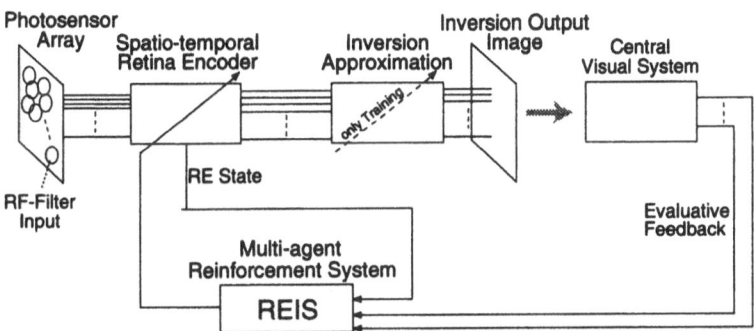

Figure 2: *Scheme of RE adjustment with Reinforcement System (REIS) processing feedback from a normal sighted humans to test REIS Learning. IM approximates the inversion of RE.*

module (IM) which approximately generates RE input images from RE output for the defined optimal RE state. The reconstructed images are presented to a normal sighted test person. When RE state is then tuned apart from the optimal state the IM output image will no longer correspond to the RE input. The test person's perceived IM output image is taken as a substitute for the future implant carrying subject's perception. The test person evaluates consecutive RE state alterations. This feedback is processed by the learning system [2]. It changes RE state which yields alterations at the displayed IM output. Iteratively, the feedback loop of IM, IM output image, normal sighted test person, and learning system adjusts RE towards its (pre-defined) optimal state. Note that IM output images of non-optimal RE states are not necessarily correlated with the implant carrying subject's perception in the future. Only the optimal state RE produces IM images comparable to the patient's perception with her personal optimal RE state.

# 4   RE Inversion Approximation

The inversion module (IM) consists of a tunable neural network and operators to change image resolution. We used a RE with 256 RFs for this study. In this case IM input resolution is $16 \times 16$ elements. Since RF function is non-invertible an inversion approximation has to take into account several RFs to reconstruct RE input pixels. Besides IM topology the degree of overlapping of RFs is crucial to approximation quality. An input screen pixel typically spreads to 2 to 4 adjacent RFs of RE. IM stores a certain number of previous RE

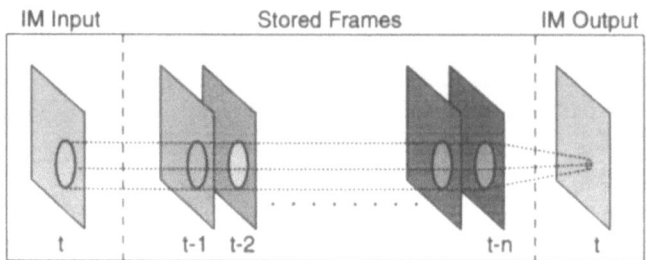

Figure 3: *Scheme of inverter module (IM). At each time step* t *it receives the current RE output frame as input and calculates an approximation of RE input images. A number of past RE output frames is stored to generate IM output with a neural network of asymmetric pair interactions of spatially adjacent elements which is motivated by internal feedback connections along the visual pathway.*

output frames to utilize RFs temporal filtering properties (Fig. 3). An element of IM output is calculated from spatially neighboring elements of temporally consecutive RE output frames. Within a generalized view this corresponds to the observed feedback-loops along the visual pathway [6]. IM state space consists of asymmetric pair interaction neural network weights. The weight between two elements depends on the affiliation of each element to one of the four GC base classes and its location along the visual pathway feedback loop. The weights are optimized by an Evolution Strategy (ES) [1]. In contrast to local neural networks used in previous studies [7] a global network concept is applied here.

# 5 Experiments and Results

IM is trained with a spatio-temporal stimulus sequence. A moving ring with horizontal sinusoidal speed profile turned out to be well suited. The stimulus is projected onto RE input screen and the corresponding RE output is recorded. In the training phase IM is fed with the RE output sequence while measuring IM output which depends on the current IM state. A scalar error value (*fitness* in terms of Evolution Strategy) measures the difference between the IM output images and RE input images as desired IM output.

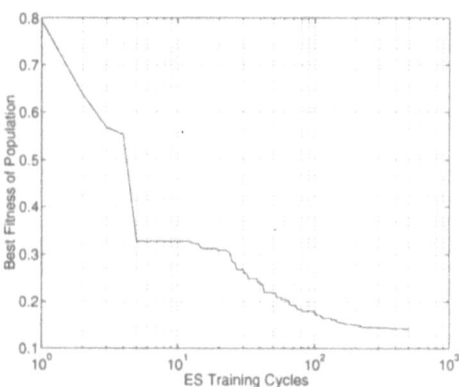

Figure 4: *Typical fitness curve for IM training.* $(\mu + \lambda)$ *ES was applied with* $\mu = 150$, $\lambda = 121$.

The training phase is repeated while optimizing IM state with ES. Figure 4 shows a typical fitness curve for IM training with the fitness value of the population's best individual at each training cycle. $(\mu + \lambda)$ strategy with $\mu = 150$, $\lambda = 121$ was applied here with IM state space dimension of 261. Figure 5 shows first results of a recall situation with the same stimulus that was used for training. The ring on the left is the RE input image (transformed to IM output resolution). The ring in the right image was reconstructed from RE output displayed in the middle figure. Obviously, RE output signals bear low resemblance with the corresponding RE input image. Generalization was tested with a bar stimulus which is moving horizontally (Fig. 6). The right image is typical for a reconstructed image with the input image not included with the training image sequence.

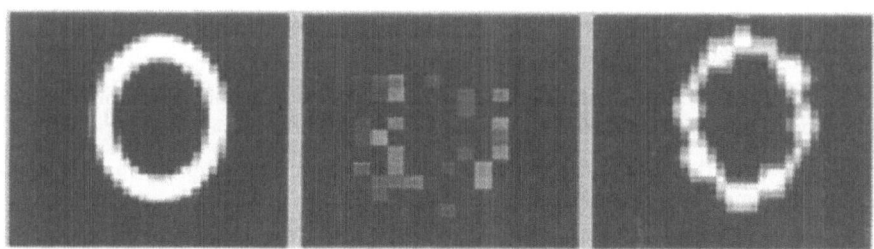

Figure 5: *Screen shots of corresponding (resized) RE input image, RE output, and IM ouptut image. The input stimulus moves horizontally with sinusoidal speed profile.*

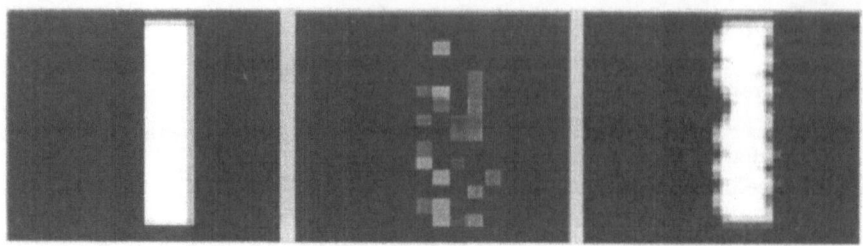

Figure 6: *Generalization to stimuli not trained. RE input image on the left, RE output in the middle, and reconstructed image on the right.*

# 6 Conclusions

The functional structure of the presented IM with GC signals as input and a simulation of visual perception as output is motivated by observed feedback loops along the visual pathway. With IM we are now able to investigate RE learning systems with evalutions of humans with normal vision. Eventually, the usefulness of IM on learning tests will show with normal sighted test persons. From current estimates the quality of IM images will suffice for test persons to perceive image changes on RE state alteration. The study shows that a specialized network topology with an ES learning algorithm is likely to perform superior to general approaches (e. g. MLP with Backprop [7]) when special knowledge is available.

# References

[1] Bäck T, Schwefel HP. Evolution Strategies I: variants and their computational implementation. In: Winter G et. al. (eds) Genetic Algorithms in Engineering and Computer Science. Wiley, 1995, pp 111-125

[2] Becker M, Eckmiller R. Spatio-temporal filter adjustment from evaluative feedback for a retina implant. In: Gerstner W et. al. (eds) Artificial Neural Networks - ICANN'97. Springer, 1997, pp 1181-1186 (Lecture notes in computer science no. 1327)

[3] Eckmiller R. Learning retina implants with epiretinal contacts. Ophthal. Res. 1997; 29:281-289

[4] Eckmiller R, Hünermann R, Becker M. Dialog-based tuning of a Retina Encoder for Retina Implants. Invest Ophthal & Vis Sci 1998; (Suppl.)39 (in press)

[5] Fleet DJ, Hallett PE, Jepson AD. Spatiotemporal inseparability in early visual processing. Biol Cybern 1985; 52:153-164

[6] Ghose GM, Freeman RD. Intracortical connections are not required for oscillatory activity in the visual cortex. Vis. Neurosci. 1997; 14:1192-1208

[7] Walther R. Simulation system for perception based, neuronal optimization of a learning retina encoder for retina implants. Diploma thesis, University of Bonn, August 1997 (in German)

# The Automated Identification of Tubercle Bacilli using Image Processing and Neural Computing Techniques

K. Veropoulos, C. Campbell

Department of Engineering Mathematics, University of Bristol
Bristol, United Kingdom
E-mail: veropoul@cs.bris.ac.uk

G. Learmonth, B. Knight

Department of Anatomical Pathology, University of Cape Town
Cape Town, South Africa
E-mail: learmonth@compuserve.com

J. Simpson

Provincial TB Laboratory, South African Institute for Medical Research
Western Cape Province, South Africa
E-mail: joannj@mail.saimr.wits.ac.za

### Abstract

Tuberculosis is currently the world's leading cause of adult death from a single infectious disease. Sputum examination remains the cornerstone of diagnosis in epidemic situations. To improve the diagnostic process we are developing an automated method for the detection of tubercle bacilli in clinical specimens, principally sputum smears. A preliminary investigation is presented here, which makes use of image processing techniques and neural network classifiers for the automatic identification of TB bacilli on Auramine stained sputum specimens. Currently, the developed system shows a sensitivity of 93.5% for the identification of individual bacilli. As there are usually fairly numerous TB bacilli in the sputum of patients with active pulmonary TB, the overall diagnostic accuracy for sputum smear positive patients is expected to be very high. Potential benefits of automated screening for TB are rapid and accurate, diagnosis, increased screening of the population, and reduced health risk to staff processing slides.

## 1   Introduction

Tuberculosis (TB) is currently the world's leading cause of adult death from a single infectious disease and in developing countries it is responsible for more adult deaths than AIDS and malaria combined. The disease is now on the increase in both developing and developed countries with a projected increase in annual mortality of about one million by 2004. This increase has stemmed from a number of factors such as the increased multiple drug resistance of the disease, inadequate control programmes, co-infection with HIV, increased migration, and the increased number of young adults in the world - the age group with the highest mortality from the disease. Globally, more than 50 million individuals may now be infected with multi-drug resistant TB. In 1993, due to the increasing incidence of multi-drug resistance and the highly infectious nature of TB, WHO declared tuberculosis to be a global emergency, the only declaration of this kind in the history of WHO.

Since TB remains largely treatable the key to controlling the disease is through correct initial diagnosis and subsequent monitoring. WHO guidelines suggest diagnosis of TB by viewing and screening Ziehl-Neelsen (ZN) stained specimens under a light microscope or Rhodamine/Auramine stained specimens under a fluorescence microscope [5]. Fluorescence microscopy is more sensitive and thus considered to be superior to light microscopy. However, manual screening for TB is labour intensive and has a high false negative rate.

This investigation is the first attempt to automatically identify TB bacilli in sputum smears and its aim is the detection of tubercle bacilli using image processing and recognition techniques.

## 2 Method

The method is mainly divided into three steps: image capture, image processing and analysis, and image recognition using neural network methods.

### 2.1 Data Preparation & Image Capture

The first stage involves visual identification of the TB bacilli on Auramine stained sputum specimens, examined under a fluorescence microscope. The bacilli are then captured using a digital camera attached to the microscope. The data used for this investigation was drawn out of a set of 1000 randomly chosen Auramine stained slides, prepared at the TB Control Laboratory at the South African Institute for Medical Research (SAIMR) in Cape Town, South Africa. The bacilli were identified firstly at low power magnification (x100) and high power magnification (x400). For image capture x630 magnification was used. The digital camera used for capturing the images contained a CCD chip with a resolution of 720x512 pixels (72 dpi).

The image in Figure 1 shows numerous fluorescent TB bacilli in a background that is free of debris and contaminants. However, in the majority of smears there are only a few bacilli scattered over the entire slide (Figure 3 - left). Moreover, the bacilli may be faint, occluded, obscured by cells or remnants, or inside macrophages – this imparts a hazy outline to the bacilli, which may cause oversights in recognition. Also, the background can be complex due to debris and other features in the sputum, making recognition even more difficult. During the course of infection the number of bacilli in the sputum may fluctuate close to the estimate of $10^4$/ml needed for a smear-positive result.

A few of the images used for this study contained a very high density of bacilli (Figure 3 - right). These images were discarded from the training and test sets in the investigation, because the bacilli are so numerous that they are frequently superimposed on each other or indistinguishable from each other. However, this was not regarded as a drawback since these images generally contain a sufficient number of bacilli for reliable identification of tuberculous infection.

## 2.2 Image Processing & Analysis

Image processing and analysis is the second stage of the method. Recognition of TB bacilli is solely based on the shape of bacilli. The image processing techniques used in this investigation help in enhancing the images and highlighting features needed for the shape description of each bacillus in the image.

Separate entities in the images, such as TB bacilli, will be referred to as objects (each object being considered as a separate region).

*1. Edge Detection.* A number of edge detection algorithms have been investigated [8,10] and it was found that the most appropriate edge detector for the captured images is the Canny operator [2,10]. The sensitivity of the Canny filter must be adjusted according to the amplification gain and resolution used during the original image capture.

*2. Region Labelling & Region Removal.* The next step is to label regions and remove some of them based on their size. The region removal process enables the system to examine only regions that belong to a certain size-range excluding regions that could not possibly be bacilli due to their size – this eliminates some unnecessary processing. In this step and the boundary tracing below, 8-connectivity is used [10].

*3. Edge Pixel Linking.* After region labelling, edge pixel linking [4] needs to be applied on objects with possible open edges. All points (pixels) in the image - that are similar in terms of magnitude and angle - are linked, forming a closed boundary of pixels. This step is necessary to improve segmentation of the image, since none of the existing edge detection operators yields optimal segmentation.

*4. Boundary Tracing.* When segmentation is complete and nearly optimal, boundary tracing techniques are used to find the boundaries of each object in the image (Figure 2). In this investigation the inner boundary tracing algorithm is used which results in a shape closer in size to the size of the object being examined [10]. This step is necessary for the derivation of shape descriptors needed for the identification of each object.

*5. Shape Description.* Having found the regions and their boundaries, shape descriptors can then be derived from the boundary curve of each region. After experimentation, a total of 15 Fourier descriptors was found to be sufficient to represent each region [4,7,9,10].

## 2.3 Image Recognition

The shape descriptors were fed into a classifier to identify any relevant regions as bacilli. A number of classification techniques were used at this stage, e.g.

discriminant methods, neural networks, etc [6]. The approach that has performed optimally in terms of overall accuracy is a multi-layered neural network with one hidden layer, trained by using standard Back-Propagation.

## 3 Results

A training set consisting of 900 image objects, a validation set of 100 objects and a test set of 147 objects (bacilli and non-bacilli) were used to train and test a neural network. There were 15 input values corresponding to the 15 Fourier descriptors, and one output value (positive/negative) for labelling each image object. 10-fold cross validation was used in our experiments and the average overall accuracy on the test set is shown in Table I.

|  | KNN(5) | RBF | BP | KR |
|---|---|---|---|---|
| Accuracy | 91.80% | 88.06% | 97.57% | 95.24% |

Table I: The expected class is the category assigned to regions by manual identification. The detected class is the category found by the computer.

The best performing algorithm was back-propagation (BP) which showed little variability for 0-5 hidden nodes (the figure quoted is for no hidden nodes). K-nearest neighbours (KNN) and RBF networks were poorer overall. The Kernel Regression (KR) is a new algorithm, which emulates Support Vector Machines [1].

For back-propagation the sensitivity was 93.53% and the specificity 98.79% where sensitivity is the ratio of true positive decisions against the number of positive cases and specificity is the ratio of true negative decisions against the total number of negative cases [11].

Table I shows the detection rate for individual bacilli present. However, on most positive smears, there will be more than one bacillus present. Positive diagnosis of TB is not usually made unless three or more bacilli are detected [3]. Thus the diagnostic accuracy for these smears will be very high.

## 4 Discussion

In this paper it is shown that automated detection of TB bacilli is certainly feasible and it could be a cost effective approach in areas with high incidence rates such as the Western Cape region of South Africa [12]. It has been claimed [13] that current screening methods may miss 33-50% of active cases and automation may reduce this error rate since a greater number of fields will be investigated by the machine. Automation would also enable the screening of a larger proportion of the population and the increased monitoring of patients currently receiving treatment.

Recognition rates using ZN (light microscopy) and Papanicolaou (fluorescence microscopy) stained specimens are currently being investigated. An extension of the work presented here is to introduce a confidence measure for the classification

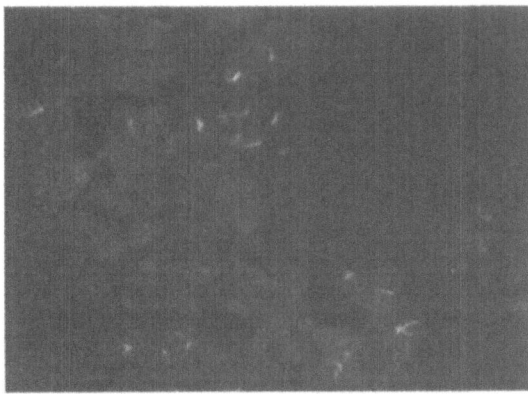

Figure 1: An image with a number of distinct bacilli present.

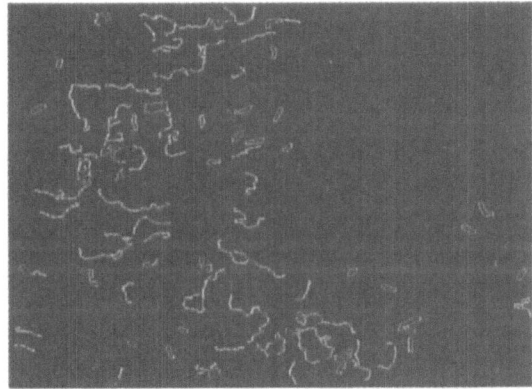

Figure 2: The image of Figure 1after edge detection, region labelling/removal, edge pixel linking, and boundary tracing have been applied.

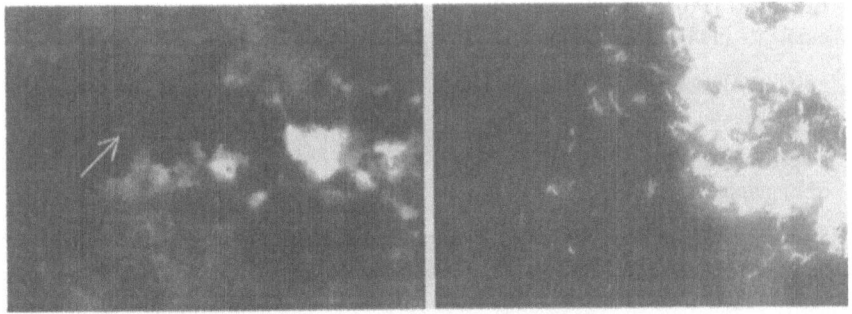

Figure 3: A few bacilli may be scattered among a large number of fields (left); some fields have a very high densities of bacilli (right).

of image objects. The use of bacilli counts and time elapsed since onset of treatment is also being investigated as a means of distinguishing multi-drug resistant and non-resistant infection. These and other results will be presented elsewhere.

# 5 Acknowledgements

The authors wish to thank Dr. Mark Jepson and the staff of the Cell Imaging Facility at the Biomedical Sciences School (Bristol) and Ms. Marlein Bosman and the staff at SAIMR (Cape Town) for their technical assistance.

# References

[1] Campbell C, Cristianini Nello, Veropoulos K. (preprint in preparation)

[2] Canny JF. A computational approach to edge detection. IEEE Trans Patt Anal & Mach Intell 1986; 8:679-698

[3] Collins CH, Grange JM, Yates MD. Organisation and practice in tuberculosis bacteriology. Butterworths, London, 1985

[4] Gonzalez RC, Woods RE. Digital image processing. Addison-Wesley, USA, 1992

[5] Kupper TH, Steffen K, Wekle G, Richartz G, Pfitzer P. Morphological study of bacteria of the respiratory system using fluorescence microscopy of Papanicolaou stained smears with special regard to identification of Mycobacteria in sputum. Cytopath 1995; 6:388-402

[6] Mitchie D, Spiegelhalter DJ, Taylor CC. Machine learning, neural and statistical classification. Ellis Horwood, New York, 1994

[7] Persoon E, Fu K. Shape discrimination using Fourier descriptors. IEEE Trans Sys, Man, and Cyber 1977; 7(3):170-178

[8] Phillips D. Image processing in C: analyzing & enhancing digital images. R&D Publications, Kansas, 1994

[9] Reeves AP, Prokop RJ, Andrews SE, Kuhl FP. Three-dimensional shape analysis using moments and Fourier descriptors. Proc 7$^{th}$ Int Conf Patt Rec 1984; 447-450

[10] Sonka M, Hlavac V, Boyle R. Image processing, analysis and machine vision. Chapman and Hall, London, 1993

[11] Veropoulos K. The Probabilistic Neural Network. MSc thesis, University of the West of England, Bristol, 1995

[12] Wilkinson D, de Cock KM. Opinion: tuberculosis control in South Africa – time for a new paradigm? South African Med J 1996; 86:33-35

[13] WHO Tuberculosis Diagnostics Workshop (ed. Foulds J). Product development guidelines. WHO, Cleveland, Jul 1997

# Black-Box Software Sensor Design for Environmental Monitoring[*]

S. Canu[1], Y. Grandvalet[2], M.H. Masson[2]

[1] PSI, INSA de Rouen, France

[2] HEUDIASYC, UMR CRNS 6599, Université de Technologie de Compiègne

### Abstract

Software sensor design consists in building a model to estimate an unknown quantity, with error bars, using other available measurements. In the environmental domain, due to a lack of physical model, non-linearities, and unknown time dependencies, black-box modelling is required. An application in river water quality monitoring illustrates a neural network based methodology. All stages of the method are described from data cleaning, and model selection, predictor estimation and prediction validity assessment. The originality of the approach is that it provides automatically an estimation of inputs relevance in merging the input selection and prediction estimation steps.

## 1  Introduction

Environmental monitoring requires valid measurements but sensors are often either expensive or unreliable. The development of software sensors is a major issue for the next generation of monitoring devices. The problem is to build a model capable of giving an estimate of the quantity of interest, with a confidence interval.

Environmental data are characterized by the lack of physical models. Moreover the relations are often non linear, with unspecified temporal dependencies. This requires the use of black-box models such as kernel regressors, regression splines, SVM, or neural networks. This paper illustrates a general methodology of software sensor design with an application developed within the $EM^2S$ Esprit project (Environmental Monitoring and Managing Systems). The data was provided by the French water supplier Lyonnaise des Eaux. To monitor the water quality and assess the performance of a waste water treatment plant, observation stations provide the following on-line measurements: temperature, pH, conductivity, oxygen, and ammonium ($NH_4$). The ammonium is one of the key parameters used to assess the overall quality of the water. Because of the cost of the physical $NH_4$ sensor, the feasibility of a virtual sensor has been studied. The main steps in the design, sample selection, data splitting, black-box modelling, and confidence interval determination are described in the following sections.

---

[*]This work has been done within the $EM^2S$ Esprit Project P-22442. The $EM^2S$ consortium is: Suez - Lyonnaise des Eaux, F, VKI Water Quality Institute, DK, Danfoss System Control, DK, Hitec, N, Computas, N and Heudiasyc CNRS, F.

# 2 Sample selection and data splitting

Before building a predictor, a preliminary data selection step has to be carried out, since sensor failures must not be modelled. The first part of the $EM^2S$ project involved the development of on-line sensor data validation procedures [1]. The software was used to validate the available raw database. The latter is composed of 25000 measurements taken over a period of five months (April-August) standardized to a sampling period of 6mn. The result of the validation procedure is a set of time contiguous data blocks made up of 3200 valid data. To take into account short and long-term dependencies, it was chosen to retain 44 explicative variables gathering past values of each possible measurement (pH, temperature, oxygen, and conductivity) recorded at time $t - \alpha \Delta t$ with $\alpha = 0, 1, ..., 10$ and $\Delta t = 10$. This choice resulted from a compromise between a good description of the process dynamics and a moderate number of covariates.

Data splitting refers to the problem of dividing the data set into learning and test sets. For dependent data it is not possible to split the sample randomly because of the correlations between sequential data. Since the beginning and the end of the data set correspond to different situations (season, state of the river), it is not appropriate to split the sample by taking the first half for training and the last part for testing. Each block of contiguous data provided by the validation procedure is assumed to be stationary. Therefore, each block was split into two blocks one for training the other one for test purposes.

# 3 Black-box modelling

In black-box modelling, prediction requires four steps [2]: 1) definition of a structure of (possibly) nested models of increasing complexity; 2) estimation of a predictor for each model; 3) estimation of the generalization error for each predictor; 4) selection of the predictor minimizing the generalization error.

## 3.1 Structure definition

The software sensor predicts a variable from a set of covariates. The first step in its construction is thus the determination of this set. It should include all relevant features, and rule out the other ones. A relevant feature may be characterized as being likely to improve the prediction. Thus, it depends on 1) the relationship between the covariates and the response variable; 2) the predictor used to capture this relationship.

To take into account these two dependencies, features should not be selected independently of the predictor in a pre-processing step. However, irrelevant features perturb prediction. Thus, feature selection should be made possible in the nested models, as a means to tune complexity.

We use here a version of adaptative ridge regression [3] penalizing/pruning the covariates, according to their relevance, while controlling the smoothness of the input-output mapping. The chosen structure of nested models is thus defined by a fixed MLP architecture and a hyper-parameter $\lambda$, tuning the overall penalization applied to the network. The value of $\lambda$ indexes the models in the structure, hence the model selection step is equivalent to the estimation of $\lambda^*$ minimizing the generalization error.

## 3.2 Predictor estimation

The chosen architecture is a one-hidden-layer perceptron with 20 hidden units. As there are 44 input variables, the number of free parameters is about 1000, for about 1600 points in the training sample. The version of adaptive ridge regression used here penalizes differently 45 groups of variables, as illustrated in fig. 1: 44 groups gather the

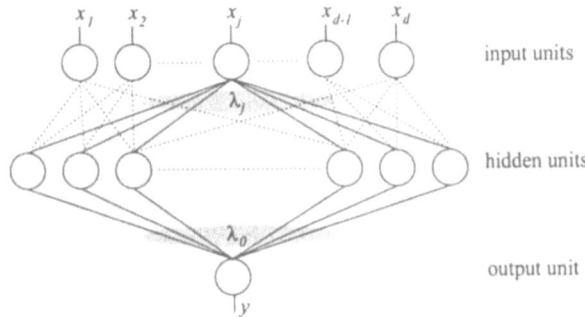

Figure 1: weight groups used by adaptive ridge penalisation.

outcoming weights $w_j$ of the input units, $j = 1, \ldots, 44$, and one group the incoming weights $w_0$ of the output unit, excluding all bias terms which are not penalized. The first 44 groups are used to penalize irrelevant features, and the last one applies smoothness constraints on the mapping. The estimate $\hat{f}_\lambda$ is defined by

$$
\begin{cases}
\hat{f}_\lambda = \underset{f \in \mathcal{F}}{\text{Argmin}} \frac{1}{\ell} \sum_{i=1}^{\ell} (f(x_i) - y_i)^2 + \sum_{j=0}^{d} \lambda_j \|w_j\|^2 \\
\text{subject to} \quad \frac{1}{d+1} \sum_{j=0}^{d} \frac{1}{\lambda_j} = \frac{1}{\lambda} \quad , \lambda_j > 0 \ ,
\end{cases}
\tag{1}
$$

where $\mathcal{F}$ is the set of MLPs with 20 hidden units, and where the value of $\lambda$ is determined by estimation of the generalization error.

## 3.3 Generalization error estimation/predictor selection

Tibshirani [4] compared various analytic estimates of generalization error to the bootstrap for MLP. He concludes that the resampling scheme performs better. We do not use bootstrap here, since the $(x_i, y_i)$ pairs are recorded at close time intervals. They are thus dependently drawn from some distribution, and we have no model of this dependency. Block-bootstrap [5] could have been used, but it was simpler to use K-fold cross-validation [5], using blocks of contiguous data. Breiman [6] recommends the number of blocks to be between 5 and 10. We chose to use 5 blocks because of the data correlation. The resulting estimate is variable, but reliable enough to select a correct model.

## 3.4   Results

Of the 44 covariates, 22 are estimated to be irrelevant during the cross-validation procedure. They are thus ruled out of the training sample when estimating $\widehat{f}_\lambda$ on the whole training set. The final prediction on the test set is given in fig. 2. The predictor performance is compared for reference to three other predictors in table 1. The results for MLPs are significantly improved over the ones of linear prediction, supporting the existence of non-linearities in the dependency. The benefits of adaptive ridge in terms of prediction accuracy are also significant.

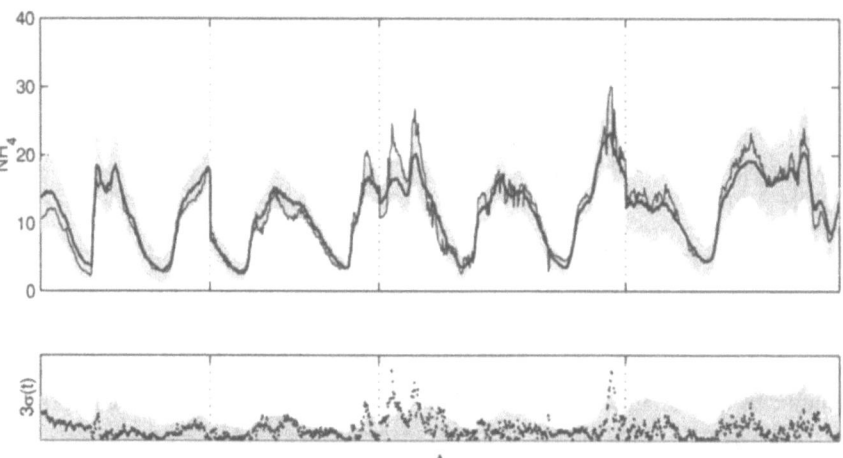

Figure 2: top: $NH_4$ measurement (thin line) and prediction (thick line) on test data with 90% confidence interval (shaded region); bottom: absolute errors (points) and 90% confidence interval (shaded region).

Another benefit of adaptive regularization is that the interpretation of results is eased by the computation of the relevance index (proportionally to $1/\lambda_j$ (1)). This index is given in table 2 for the most significant covariates. Oxygen and conductivity are by far the main explicative variables. This result is surprising for the chemist, who would expect pH to be more important, but pH is highly correlated with oxygen, and the pH measurement is less accurate.

# 4   Prediction accuracy and validity domain

For practical use, the prediction provided by the software sensor is not sufficient. The end user needs a measurement of the prediction accuracy such as a confidence interval. Furthermore, since black-box modelling is based on empirical data, predictions are valid only within a certain neighborhood of the sample. Prediction cannot be assumed to be valid for new inputs far from previously seen data. Actually these two quantities (accuracy and validity) are linked since invalid data can be associated with a very large confidence interval (if correctly defined).

Table 1: Prediction error for ridge regression (RR) and adaptive ridge regression (ARR), for a linear model and multilayered perceptron. The given intervals are estimated from the dispersion of the validation set errors.

| Method | Prediction error |
|--------|------------------|
| linear + RR | 4.0±0.2 |
| linear + ARR | 3.8±0.2 |
| MLP + RR | 3.7±0.2 |
| MLP + ARR | 3.1±0.2 |

Table 2: Relevance (computed from $\lambda_j$) for the top 5 selected covariates.

| Variable | Relevance index |
|----------|-----------------|
| oxygen(t) | 33% |
| conductivity(t) | 21% |
| temperature(t-100) | 8% |
| pH(t-40) | 5% |
| conductivity(t-10) | 5% |
| pH(t) | 5% |

Following [7, 8] a neural network was designed with the absolute residuals as targets. This method suits our problem well since it does not require any strong assumption about the nature of the uncertainty. The algorithm used for prediction was re-used with the absolute residuals on the cross validation sets (jackknife samples) as output targets. Following stacking regression ideas [9], new potentially explicative variables were integrated in the the the set of learning descriptors. These new input variables are the predicted value $\hat{f}_\lambda(x)$ and distances between the new input and the sample. Several distances have been used (to the nearest neighbor or the gravity center) and included as potential explicative variables.

In our approach, confidence interval estimation is used to define the prediction validity domain. Above a certain level of confidence, the prediction is presumed invalid, even if this uncertainty may be caused by the variability of the $NH_4$ measurements.

The results of the algorithm on the test set are displayed in figure 2. The estimation error is not constant. It varies in a ratio of one to four. Points with a large error estimation can be rejected as invalid. The relevance index shows that the main explicative variable is the nearest neighbor distance. These results support the approach of determining the validity domain from prediction accuracy.

Even if the results are relevant for the end user some questions remain unanswered. For instance it is not clear yet how to asses the quality of the confidence interval. Furthermore, stacking could also have been used to improve prediction. The examination of the residuals shows that they are asymmetrical. A method to deal with this asymmetry could be useful.

# 5 Conclusion

This application shows how black-box modelling can be successfully used when little is known about the dependency between variables. In such applications, due to temporal dependencies, a major issue is to retrieve the relevant explicative variables. A feature is relevant according to the dependence and to the predictor used to model this dependence. Thus feature selection should be integrated into the modelling process. The adaptive ridge algorithm is well-suited since it performs input selection while tuning the predictor complexity. Thanks to this selection mechanism, non-linearities are exhibited in our application although the ratio of sample size to input dimension is low.

A usual criticism of black-box models is their failure to provide a confidence interval. The proposed approach allows a data-based estimation of such intervals to be built. The same supervised algorithm is applied using both the cross-validation residuals as targets and additional explicative variables. Among the latter, the distance between the current input and its nearest neighbor in the training set appears to be the most relevant, justifying the use of training set characteristics as additional variables. More than solving a particular problem, this paper provides general guidelines for the design of software sensors capable of providing a prediction together with a confidence interval.

# References

[1] Heudiasyc diagnosis group. Sensor data validation. Technical Report CNRS/EM$^2$S/310/12-96, U.T.C., 1996.

[2] V.N. Vapnik. *The Nature of Statistical Learning Theory*. Springer Series in Statistics. Springer-Verlag, New York, 1995.

[3] Y. Grandvalet. Least absolute shrinkage is equivalent to quadratic penalization. In *ICANN'98*. Springer-Verlag, 1998.

[4] R.J. Tibshirani. A comparison of some error estimates for neural networks models. *Neural Computation*, 8(1):152–163, 1996.

[5] B. Efron and R.J. Tibshirani. *An Introduction to the Bootstrap*, volume 57 of *Monographs on Statistics and Applied Probability*. Chapman & Hall, New York, 1993.

[6] L. Breiman. Bagging predictors. *Machine learning*, (26):123–140, 1996.

[7] T. Heskes. Practical confidence and prediction intervals. In *Advances in Neural Information Processing Systems 9*, pages 176–182. MIT press, 1997.

[8] J. N. Fidalgo, M. A. Matos, and M. T. Ponce de Leão. Assessing error bars in distribution load curve estimation. In *Artificial Neural Networks - ICANN'97*, pages 1017–1022. Springer-Verlag, 1997.

[9] D. Wolpert. Stacked generalization. *Neural Networks*, 5:241–259, 1992.

# Classifying Regional Seismic Signals Using TDDN-alike Neural Networks

Arno J. Klaassen & Xavier Driancourt

Neuristique S.A., 28 Rue des petites Ecuries, F-75010 Paris, France

[Arno.Klaassen|Xavier.Driancourt]@neuristique.fr

Stéphanie Muller & Jean-Denis Muller

CEA-DAM, B.P. 12, F-91680 Bruyères-le-Châtel, France

[mullerst|muller]@bruyeres.cea.fr

### Abstract

We present a shared-weights neural architecture for classifying spectrograms from regional seismic events. This approach does not use any high-level information available (like epicenter, magnitude, time origin etc.) on the event other than the estimated arrival times for the P-wave and eventually the S-wave. This severe restriction is to make sure that the system continues to function in the absence of high-level information. In spite of the very noisy character of the spectrograms, the classification rates obtained are comparable to those obtained by methods which do depend on the availability of high-level information ($\approx 90 - 98\%$), though the rejection rate on non-trained events is rather high ($\approx 20\%$).

## 1 Introduction

At the French territory, a seismometers network continuously records seismic events. There are multiple possible reasons for these events: natural earthquakes, quarry blast, rock bursts, etc.. Most events are recorded at multiple stations. This study concerns the classification of low-power (i.e. small magnitude) events. This power, together with the sensibility of the seismic sensors, makes that the events are detected by all stations within a radius of some hundreds of kilometers from the event's epicenter.

Seismic signals can roughly be cut in two parts: first comes the so-called P-wave which is primarily caused by longitudinal propagation along the earth's crust. Transversal propagation is less quickly and thus arrives later at a recording station. When it arrives, we say the S-wave starts. Since the P-phase essentially is characterized by a sudden increase in spectral energy, its automatic detection is rather straightforward and reliable. Reliable automatic labeling of the S-phase is more difficult.

The CEA uses inverse physical models (combining multiple records of the same seismic event) to determine a multitude of characteristics like its epicenter, origin time, magnitude, etc.. These high-level characteristics contain valuable information about the class of the seismic event. Together with other characteristics obtained by simple signal processing like the spectral amplitude or slope ratio of the S- and P-waves, they provide enough information for classifying systems to obtain very good results [1, 2, 3].

The present study is meant as being complementary to this approach: the inverse model might fail and the raw seismic signal might contain more valuable information than used up to now. So we decided to stay as low-level as possible, and not to blindly rush for maximum performance (the system resulting from this study will be part of a much greater system in which data fusion techniques finally determine the class). We decided to apply a proven neural network approach from speech processing [4, 5] to seismic signal classification.

## 2 Preprocessing

Typical examples of the raw recordings for each kind of seismic event are shown in Figure 1. The sampling frequency is $50Hz$. Apart from the raw signal we also use the estimated start of the P-wave as computed by an inverse physical model. Our data base contained 53 earthquakes making up for 352 recordings, 411 explosions resulting in 1659 recordings and 40 rock bursts yielding 139 recordings. We randomly choose a test-set consisting of 20 earthquakes, 20 explosions and 10 rock bursts (typically making up for some 300 recordings). Obviously, these recordings are not part of the training set.

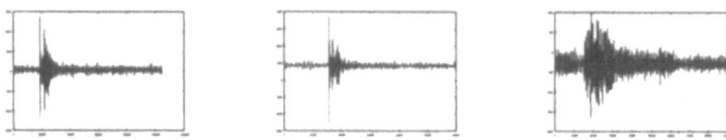

Figure 1: *Typical examples of the raw recordings for each kind of seismic event; from left to right: earthquake, explosion, rock burst*

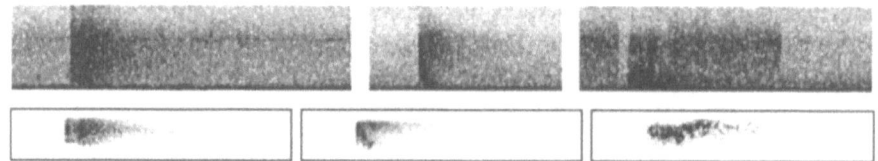

Figure 2: *Raw and denoised spectrograms of the signals from Figure 1.*

The approach now is to cut the raw signal in multiple overlapping frames, to perform a spectral decomposition of each frame and to design a shared weight neural architecture which classifies a series of frames. Given the sampling frequency of $50hz$ we decided to use a window of a 100 samples with 50 samples overlap. This is in the same range as values found in literature. To each window we applied the following Hamming function $H(x) = 0.54 - (0.46 * \cos(\frac{2\pi}{100}x))$, $H(0) = H(99) = 0$. We then applied a Fourier transform of which we ignore the DC-component. We calculated the energy in all other frequency bands, added an offset of 1 and took the logarithm of it. In Figure 2 we show the so obtained spectrograms for the three signals of Figure 1. Clearly, there is a lot of noise in each recording and at first sight the spectrograms seem even more uniform than the raw signals. To increase the contrast in the spectrograms, we applied a method which has already successfully been used in classification of German earthquake data by Joswig [6]: for each frequency band we determined the mean value and the standard deviation of the values of the frames preceding the P-wave. Then we subtracted this mean value and one time the standard deviation from all frames. Thus obtained negative values were set to zero.

Further, in accordance with observations found in literature, there is a lot of energy in the low frequency bands before, after and during the event, and at higher frequencies there is less and less variation from one frequency-band to another. So first we squeezed the 50 frequency bands into 15, by starting from the lowest frequency band and then summing the energy in the $n$ following bands, where $n$ is given by [1]: $\inf((\frac{3}{2}\sqrt{3})^i)$, $0 < i < 16$. We then aligned the spectrograms so that the supposed P-onset is at the same frame for all spectrograms (see Figure 2). Finally, we assumed the relative time-frequency pattern being independent of the power of the seismic event [7]

---

[1] This equation yields (1 1 1 1 2 2 2 3 3 4 4 5 6 7 8).

and normalized all spectrograms individually to have a mean value of 0 and a standard deviation of $\frac{2}{3}$. As we will see later on, our results suggest that this assumption might not be true.

# 3 Network architecture

We inspired our neural network architectures on TDNN-like ones which successfully have been applied to speech recognition [4, 5]. Essentially a connection-mask of $n$ frames repeatedly is shifted $m$ frames until the end of the spectrogram has been reached. Between the input layer and the first hidden layer we took a mask of $n = 4, m = 2$ reducing the number of frequency bands from 15 to 10, between the first and second hidden layer we took $n = 9, m = 5$ further reducing the number of frequency bands from 10 to 5. Due to the shared weights, this kind of architecture is robust against slight pattern shifts and/or partially correct or missing frames. For the first two layers we used a column-shared bias (i.e. par frequency band). We used a stochastic gradient descent learning algorithm of which the learning rate was lowered manually from time to time. We experimented with conjugate gradient learning algorithms as well, but this lead to overtraining.

However, due to the slower propagation speed of the S-wave, the number of frames in between the arrival of the P-wave and the arrival of the S-wave, increases with the distance from the epicenter to the recording station. This yields different activation patterns in the last hidden layer for the very same event which has been recorded by different stations with different distances to the epicenter. As a first remedy against this, we designed a second network with basically the same architecture but this time using the pre-determined arrival of the S-wave as well: the P-arrival is aligned on the 10th frame, the S-arrival on the 60th. The 4 frames before the S-arrival are used with decreasing amplitude as the end of the P-wave and with increasing amplitude as the start of the S-wave.

# 4 Training algorithm

Our data base contained significantly more examples of explosions than of earthquakes and rock bursts. Further, individual earthquakes are recorded at more stations than individual explosions or rock bursts. Furthermore, of each event it is not clear *a priori* which recordings best characterize the event (remember that we did not want to use high-level information like the distance to the epicenter). All this taken into account, we came up with the following learning strategy: 1) define some kind of signal-level estimator for each recording; 2) during one learning epoch only present recordings with signal-level estimated above a certain threshold and randomly choose them in equal proportions; 3) from epoch to epoch gradually lower the threshold.

The signal-level estimator has been defined as follows: we took the mean value of all frames of twice the P to S interval before and after denoising the raw spectrogram (see Section 2). The signal-level is defined as the post-denoise mean divided by the pre-denoise mean. The clearest recordings had signal levels of about 50%; recordings below 1.2% have been discarded. Since it turned out that even with this method the explosions were over-represented, we set the threshold for earthquakes and rock bursts at 0.6 times the threshold for explosions. Finally, to artificially increase the number of examples, at each presentation of a spectrogram to the network we applied a shift randomly chosen from $[-s, s]$. Given our network architecture the theoretical maximum value of $s = 5$, in practice we most often used $s = 1$ or $s = 2$.

# 5 Results: unsegmented architecture

For the results presented here we used an initial signal-level threshold of 27%. When each pattern of the epoch had been presented twice to the network, we multiplied this threshold by 0.85 and filled in the next epoch. In Figure 3 we show the performance and mean square error for individual recordings. The network quickly learns to discriminate perfectly the recordings with a high signal-level from its training set. However, on a individual recording base, generalization is rather poor (using a max classifier, typically some 85% of all signals in the training set are correctly classified and some 75% of the ones in the test set).

On the other hand, the curves of Figure 3 suggest that generalization for strong signals only, might be a lot better, but our test set was much too small to make any sensible conclusion using *exclusively* strong signals. So we decided to combine the network outputs for all recordings of an event taking into account the signal-level estimation.

Suppose $m$ is a matrix of which the rows contain the outputs of the network for each recording, and $sl$ is a vector containing the signal-level estimation of each recording. Then we multiply all values in row $i$ of $m$ by $f(sl_i)$, where $f$ either is the identity function, or $\log(x + 1.59)$. Then a vector $res$ is obtained by summing up over all columns; $res$ is normalized by dividing by $|res|$; we then apply the operator $classify$ to $res$. In any case $res$ is rejected if its max value is smaller than $thr+$. The following combinations of $f$, $classify$ and $thr+$ were successful most often:

| Merge Method | $f$ | $classify$ | $thr+$ |
|---|---|---|---|
| 4 | ∞ | max | 0.01 |
| 5 | ∞ | max | 0.1 |
| 8 | $\log(x + 1.59)$ | max | 0.01 |
| 9 | $\log(x + 1.59)$ | max | 0.1 |
| 18 | ∞ | quad | 0.01 |
| 19 | ∞ | quad | 0.1 |
| 22 | $\log(x + 1.59)$ | quad | 0.01 |
| 23 | $\log(x + 1.59)$ | quad | 0.1 |
| 30 | $\log(x + 1.59)$ | max | -1.0 |
| 31 | $\log(x + 1.59)$ | quad | -1.0 |

The results are shown in Table 1. The upper left shows the confusion matrix for individual signals (q=quake, e=quarry blast, b=burst, r=rejected); the upper right shows the overall classification and rejection rates for the different merge methods M; the lower part show confusion matrices for a selection of four merge methods.

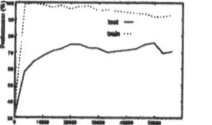 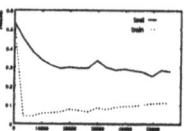

Figure 3: *Performance and normalised mean square error (NMSE) on individual signals*

# 6 Results: aligned on P-onset and S-onset

The results for the architecture having the P-arrival aligned on the 10th frame and the S-arrival on the 60th are shown in Table 2. A striking difference with the results for the unsegmented architecture is that the merge methods on the current training epoch longtime yield classification rates near to a 100% when gradually lowering the signal-level threshold. The generalization performance on individual signals is slightly better, and merging the individual network outputs in one event classification clearly yields better results than for the unsegmented architecture. The rejection rates are somewhat lower as well, but still stay rather high.

|        |   | quad |      |     |    |     | max |      |    |
|--------|---|------|------|-----|----|-----|-----|------|----|
|        |   | q    | e    | r   | b  | r   | q   | e    | b  |
|        | q | 101  | 20   | 18  | 7  | 8   | 112 | 29   | 13 |
| test   | e | 17   | 61   | 9   | 4  | 11  | 27  | 69   | 6  |
|        | b | 2    | 3    | 4   | 24 | 3   | 3   | 8    | 25 |
|        | q | 143  | 9    | 12  | 0  | 7   | 154 | 15   | 2  |
| train  | e | 63   | 1113 | 100 | 17 | 89  | 124 | 1221 | 37 |
|        | b | 0    | 2    | 2   | 95 | 0   | 0   | 4    | 96 |

| M  | Epoch                    | Training set             | Test Set              |
|----|--------------------------|--------------------------|-----------------------|
| 4  | 93.4% (296/317 [5.1%])   | 98.3% (416/423 [3.6%])   | 95.0% (38/40 [20%])   |
| 5  | 94.2% (291/309 [7.5%])   | 98.3% (413/420 [4.3%])   | 94.7% (36/38 [24%])   |
| 8  | 93.7% (296/316 [5.4%])   | 97.9% (415/424 [3.4%])   | 94.9% (37/39 [22%])   |
| 9  | 94.8% (291/307 [8.1%])   | 98.3% (412/419 [4.6%])   | 97.2% (35/36 [28%])   |
| 18 | 94.5% (291/308 [7.8%])   | 98.3% (414/421 [4.1%])   | 95.0% (38/40 [20%])   |
| 19 | 95.0% (286/301 [9.9%])   | 98.3% (411/418 [4.8%])   | 94.7% (36/38 [24%])   |
| 22 | 94.5% (290/307 [8.1%])   | 97.9% (413/422 [3.9%])   | 94.6% (35/37 [26%])   |
| 23 | 95.3% (285/299 [10.5%])  | 98.3% (410/417 [5.0%])   | 97.1% (33/34 [32%])   |
| 30 | 92.8% (310/334 [0.0%])   | 96.8% (425/439 [0.0%])   | 86.0% (43/50 [0%])    |
| 31 | 94.5% (290/307 [8.1%])   | 97.9% (414/423 [3.6%])   | 94.6% (35/37 [26%])   |

|    |   | epoch |     |    |    |    | train |     |    |    |    | test |    |   |   |   |
|----|---|-------|-----|----|----|----|-------|-----|----|----|----|------|----|---|---|---|
|    |   | q     | e   | r  | b  | r  | q     | e   | r  | b  | r  | q    | e  | r | b | r |
| 18 | q | 28    | 1   | 2  | 0  | 2  | 29    | 1   | 3  | 0  | 0  | 14   | 0  | 3 | 0 | 1 |
|    | e | 9     | 234 | 14 | 6  | 8  | 0     | 356 | 13 | 0  | 2  | 1    | 16 | 2 | 1 | 0 |
|    | b | 0     | 1   | 0  | 29 | 0  | 0     | 1   | 0  | 29 | 0  | 0    | 0  | 2 | 8 | 0 |
| 19 | q | 26    | 1   | 2  | 0  | 4  | 29    | 1   | 3  | 0  | 0  | 13   | 0  | 3 | 0 | 2 |
|    | e | 9     | 231 | 14 | 4  | 13 | 1     | 353 | 13 | 0  | 5  | 1    | 16 | 2 | 1 | 0 |
|    | b | 0     | 1   | 0  | 29 | 0  | 0     | 1   | 0  | 29 | 0  | 0    | 0  | 2 | 7 | 1 |
| 30 | q | 31    | 2   | 0  | 0  | 0  | 31    | 2   | 0  | 0  | 0  | 17   | 2  | 0 | 1 | 0 |
|    | e | 16    | 250 | 0  | 5  | 0  | 9     | 365 | 0  | 2  | 0  | 1    | 18 | 0 | 1 | 0 |
|    | b | 0     | 1   | 0  | 29 | 0  | 0     | 1   | 0  | 29 | 0  | 0    | 2  | 0 | 8 | 0 |
| 31 | q | 28    | 1   | 3  | 0  | 1  | 28    | 1   | 4  | 0  | 0  | 14   | 0  | 6 | 0 | 0 |
|    | e | 10    | 233 | 15 | 5  | 8  | 6     | 357 | 10 | 1  | 2  | 0    | 14 | 3 | 1 | 2 |
|    | b | 0     | 1   | 0  | 29 | 0  | 0     | 1   | 0  | 29 | 0  | 0    | 1  | 2 | 7 | 0 |

Table 1: *Results for unsegmented spectrograms; upper left: confusion matrix for individual signals (q=quake, e=quarry blast, b=burst, r=rejected); upper right: overall classification and rejection rate for different merge methods M; lower: confusion matrices for four differents merge methods.*

Figure 4: *Performance and normalised mean square error (NMSE) on individual signals*

# 7 Conclusions

We described a methodology to classify raw regional seismic signals. The only high-level information needed is a rough guess about the arrival of the P-phase which most often is trivial to obtain. We used TDDN-alike neural networks. Combining the classification of individual signals into a single event classification, yields classification rates of about 95% with $0 - 5\%$ rejection on the training set and $0 - 30\%$ on the test set. Aligning on the S-wave as well improves these results to about 97% with $0 - 5\%$ rejection on the training set and $0 - 20\%$ on the test set. These results did not vary much for different training and test sets from our data base. Application to a set of recordings from a different data base (with much more earthquakes and few blasts), show almost similar results for the segmented architecture (slightly higher confusion between earthquakes and explosions, see Table 3). Performance of the unsegmented architecture drops to about 80%.

One of the assumptions was that the relative time-frequency spectrum is constant per event class. The fact that favorizing signals with a higher signal-level yields significantly better results, indicates that this assumption might not be true because probably high signal-level means nearby and/or high-magnitude events. However, the signal/noise ratio of all recordings is very low, which might explain this as well. On the other hand, it is known that the signal is colored by the earth it propagates through from the epicenter to the recording site. Future work has to show whether the use of an estimation for the epicenter (for the normalization/correction of spectrograms) significantly improves our results.

| | | quad | | | | | max | | |
|---|---|---|---|---|---|---|---|---|---|
| | | q | e | r | b | f | q | e | b |
| test | q | 97 | 14 | 20 | 9 | 10 | 110 | 25 | 15 |
| | e | 7 | 73 | 14 | 1 | 7 | 12 | 86 | 4 |
| | b | 1 | 6 | 3 | 22 | 4 | 3 | 8 | 25 |
| train | q | 141 | 12 | 10 | 2 | 6 | 150 | 18 | 3 |
| | e | 98 | 1072 | 97 | 31 | 84 | 150 | 1183 | 49 |
| | b | 3 | 2 | 5 | 90 | 0 | 4 | 3 | 93 |

| M | Epoch | Training set | Test Set |
|---|---|---|---|
| 4 | 96.5% (274/284 [4.4%]) | 96.7% (410/424 [3.4%]) | 100% (42/42 [16%]) |
| 5 | 96.8% (271/280 [5.7%]) | 97.3% (400/411 [6.4%]) | 100% (39/39 [22%]) |
| 8 | 96.8% (275/284 [4.4%]) | 97.2% (415/427 [2.7%]) | 100% (40/40 [20%]) |
| 9 | 96.8% (270/278 [6.4%]) | 97.8% (405/414 [5.7%]) | 100% (36/36 [28%]) |
| 18 | 96.8% (270/279 [6.1%]) | 97.6% (405/415 [5.5%]) | 100% (42/42 [16%]) |
| 19 | 97.1% (267/275 [7.4%]) | 98.3% (395/402 [8.4%]) | 100% (39/39 [22%]) |
| 22 | 97.1% (271/279 [6.1%]) | 97.1% (407/419 [4.6%]) | 100% (40/40 [20%]) |
| 23 | 97.4% (266/273 [8.1%]) | 97.8% (397/406 [7.5%]) | 100% (36/36 [28%]) |
| 30 | 94.6% (281/297 [0.0%]) | 96.8% (426/439 [0.0%]) | 90% (45/50 [0 %]) |
| 31 | 97.1% (271/279 [6.1%]) | 97.1% (409/421 [4.1%]) | 100% (41/41 [18%]) |

| | | epoch | | | | | train | | | | | test | | | |
|---|---|---|---|---|---|---|---|---|---|---|---|---|---|---|---|---|
| | | q | e | r | b | f | q | e | r | b | f | q | e | r | b | f |
| 18 | q | 30 | 2 | 1 | 0 | 0 | 28 | 2 | 2 | 0 | 0 | 16 | 0 | 4 | 0 | 0 |
| | e | 5 | 211 | 12 | 2 | 5 | 6 | 348 | 11 | 1 | 10 | 0 | 18 | 1 | 0 | 1 |
| | b | 0 | 0 | 0 | 29 | 0 | 1 | 0 | 0 | 29 | 0 | 0 | 0 | 2 | 8 | 0 |
| 19 | q | 29 | 1 | 1 | 0 | 2 | 27 | 1 | 2 | 0 | 3 | 14 | 0 | 4 | 0 | 2 |
| | e | 5 | 209 | 12 | 2 | 7 | 4 | 339 | 11 | 1 | 21 | 0 | 18 | 1 | 0 | 1 |
| | b | 0 | 0 | 0 | 29 | 0 | 1 | 0 | 0 | 29 | 0 | 0 | 0 | 2 | 7 | 1 |
| 30 | q | 31 | 2 | 0 | 0 | 0 | 28 | 5 | 0 | 0 | 0 | 17 | 1 | 0 | 0 | 0 |
| | e | 9 | 221 | 0 | 5 | 0 | 7 | 368 | 0 | 1 | 0 | 0 | 20 | 0 | 0 | 0 |
| | b | 0 | 0 | 0 | 29 | 0 | 1 | 0 | 0 | 29 | 0 | 0 | 0 | 8 | 0 | 0 |
| 31 | q | 30 | 2 | 1 | 0 | 0 | 27 | 4 | 2 | 0 | 0 | 14 | 0 | 6 | 0 | 0 |
| | e | 4 | 212 | 11 | 2 | 6 | 6 | 353 | 8 | 1 | 8 | 0 | 19 | 1 | 0 | 0 |
| | b | 0 | 0 | 0 | 29 | 0 | 1 | 0 | 0 | 29 | 0 | 0 | 0 | 2 | 8 | 0 |

Table 2: *As table 1 using segmented spectrograms.*

| | | quad | | | | | max | | |
|---|---|---|---|---|---|---|---|---|---|
| | | q | e | r | b | f | q | e | b |
| | q | 828 | 259 | 163 | 116 | 92 | 947 | 333 | 178 |
| | e | 77 | 131 | 34 | 12 | 17 | 96 | 155 | 20 |
| | b | 18 | 8 | 20 | 203 | 16 | 26 | 13 | 226 |

| | | quad | | | | | max | | |
|---|---|---|---|---|---|---|---|---|---|
| | | q | e | r | b | f | q | e | b |
| | q | 795 | 184 | 143 | 100 | 87 | 913 | 257 | 139 |
| | e | 43 | 139 | 34 | 6 | 18 | 65 | 165 | 10 |
| | b | 5 | 3 | 16 | 207 | 17 | 13 | 5 | 230 |

| M | signal-level > 27 | All |
|---|---|---|
| 4 | 78.4% (149/190 [13.6%]) | 77.8% (214/275 [16.7%]) |
| 5 | 79.3% (142/179 [18.6%]) | 80.2% (202/252 [23.6%]) |
| 8 | 78.6% (151/192 [12.7%]) | 79.4% (212/267 [19.1%]) |
| 9 | 79.6% (144/181 [17.7%]) | 81.8% (202/247 [25.2%]) |
| 18 | 77.9% (141/181 [17.7%]) | 78.7% (210/267 [19.1%]) |
| 19 | 79.1% (136/172 [21.8%]) | 80.8% (198/245 [25.8%]) |
| 22 | 78.3% (144/184 [16.4%]) | 80.0% (208/260 [21.2%]) |
| 23 | 79.3% (138/174 [20.9%]) | 82.6% (199/241 [27.0%]) |
| 30 | 73.2% (161/220 [0.0%]) | 73.0% (241/330 [0.0%]) |
| 31 | 78.0% (145/186 [15.5%]) | 80.2% (210/262 [20.6%]) |

| M | signal-level > 27 | All |
|---|---|---|
| 4 | 89.6% (173/193 [10.2%]) | 88.1% (238/270 [17.7%]) |
| 5 | 92.7% (166/179 [16.7%]) | 90.7% (234/258 [21.3%]) |
| 8 | 89.6% (173/193 [10.2%]) | 89.1% (236/265 [19.2%]) |
| 9 | 92.8% (168/181 [15.8%]) | 89.5% (231/258 [21.3%]) |
| 18 | 90.2% (166/184 [14.4%]) | 88.0% (235/267 [18.6%]) |
| 19 | 93.5% (159/170 [20.9%]) | 90.6% (231/255 [22.3%]) |
| 22 | 90.2% (166/184 [14.4%]) | 88.9% (232/261 [20.4%]) |
| 23 | 93.6% (161/172 [20.0%]) | 89.4% (227/254 [22.6%]) |
| 30 | 87.9% (189/215 [0.0%]) | 82.3% (270/328 [0.0%]) |
| 31 | 89.7% (166/185 [14.0%]) | 88.6% (233/263 [19.8%]) |

| | | signal-level > 27 | | | | | All | | | | |
|---|---|---|---|---|---|---|---|---|---|---|---|
| | | q | e | r | b | f | q | e | r | b | f |
| 18 | q | 86 | 23 | 21 | 14 | 7 | 134 | 38 | 41 | 11 | 7 |
| | e | 3 | 6 | 3 | 0 | 1 | 8 | 17 | 7 | 0 | 0 |
| | b | 0 | 0 | 4 | 49 | 0 | 0 | 0 | 6 | 59 | 2 |
| 19 | q | 83 | 20 | 21 | 13 | 14 | 122 | 32 | 41 | 9 | 27 |
| | e | 3 | 5 | 3 | 0 | 2 | 6 | 17 | 7 | 0 | 2 |
| | b | 0 | 0 | 4 | 48 | 0 | 0 | 0 | 6 | 59 | 2 |
| 30 | q | 101 | 31 | 0 | 19 | 0 | 158 | 54 | 0 | 19 | 0 |
| | e | 6 | 7 | 0 | 0 | 0 | 10 | 21 | 1 | 0 | 0 |
| | b | 2 | 1 | 0 | 53 | 0 | 5 | 0 | 0 | 62 | 0 |
| 31 | q | 90 | 23 | 19 | 14 | 5 | 133 | 37 | 47 | 9 | 5 |
| | e | 4 | 6 | 3 | 0 | 0 | 6 | 18 | 8 | 0 | 0 |
| | b | 0 | 0 | 4 | 49 | 0 | 0 | 0 | 6 | 59 | 2 |

| | | signal-level > 27 | | | | | All | | | | |
|---|---|---|---|---|---|---|---|---|---|---|---|
| | | q | e | r | b | f | q | e | r | b | f |
| 18 | q | 102 | 14 | 19 | 3 | 8 | 147 | 25 | 50 | 3 | 4 |
| | e | 1 | 10 | 2 | 0 | 0 | 4 | 23 | 4 | 0 | 1 |
| | b | 0 | 0 | 0 | 54 | 2 | 0 | 0 | 2 | 65 | 0 |
| 19 | q | 98 | 9 | 17 | 3 | 8 | 143 | 19 | 50 | 1 | 16 |
| | e | 0 | 8 | 2 | 0 | 3 | 4 | 23 | 4 | 0 | 1 |
| | b | 0 | 0 | 0 | 53 | 3 | 02 | 0 | 2 | 65 | 0 |
| 30 | q | 121 | 20 | 0 | 5 | 0 | 178 | 36 | 0 | 15 | 0 |
| | e | 1 | 12 | 0 | 0 | 0 | 6 | 26 | 0 | 0 | 0 |
| | b | 0 | 0 | 0 | 56 | 0 | 1 | 0 | 0 | 66 | 0 |
| 31 | q | 102 | 14 | 18 | 5 | 7 | 146 | 23 | 53 | 3 | 4 |
| | e | 0 | 10 | 3 | 0 | 0 | 4 | 23 | 5 | 0 | 0 |
| | c | 0 | 0 | 0 | 54 | 0 | 0 | 0 | 3 | 64 | 2 |

Table 3: *Test results on another data set; left: unsegmented; right segmented*

# References

[1] S. Muller, P. Garda, J.-D. Muller, R. Crusem, and Y. Cansi, "A neuro-fuzzy coding for processing incomplete data: Application to the classification of seismic events," *Neural Processing Letters*, vol. 8, August 1998.

[2] S. Muller, J.-F. Legrand, J.-D. Muller, Y. Cansi, R. Crusem, and P. Garda, "Seismic events discrimination by neuro-fuzzy data merging." Submitted to Geophysical Research Letters.

[3] S. Muller, J.-F. Legrand, P. Garda, J.-D. Muller, R. Crusem, and Y. Cansi, "Seismic events discrimination by a neuro-fuzzy merging of incomplete data," in *Annales Geophysicae, 23rd General Assembly of the European Geophysical Society*, vol. 16, (Nice, France), April 1998.

[4] C. Dugast and L. Devillers, "Incorporating acoustic-phonetic knowledge in hybrid tdnn/hmm frameworks," in *ICASSP92*, vol. I, (San Francisco, USA), p. 421, 1992.

[5] X. Driancourt and L. Bottou, "Tdnn-extracted features," in *Neuro-Nimes 90*, (Nimes, France), EC2, 1990.

[6] M. Joswig, "Automated classification of local earthquake data in the bug small array," *Geophys. J. Int.*, 1995.

[7] G. Romeo, F. Mele, and A. Morelli, "Neural networks and discrimination of seismic signals," *Computers & Geosciences*, vol. 21, pp. 279–288, March 1995.

# A hierarchical self-organizing map model for sequence recognition

Otávio Augusto S. Carpinteiro

IEE, Escola Federal de Engenharia de Itajubá

Itajubá, MG, 37500–000, Brazil

otavio@iee.efei.br

## Abstract

The paper proposes a novel neural model made up of two self-organizing map nets — one on top of the other. The model makes an effective use of context information, and that enables it to perform sequence classification and discrimination efficiently. It was trained and assessed on a four-part fugue of J. S. Bach. The model has application in domains which require pattern recognition, or in particular, which demand recognizing either a set of sequences of vectors in time or sub-sequences into a unique and large sequence of vectors in time.

## 1   Introduction

Many researchers have developed artificial neural models to classify sequences in time. Several models based on Kohonen's self-organizing feature map [1], nonetheless, have faced some well-known flaws, such as high computational cost, loss of context, and inability to recognize sub-sequences inside a large and unique input sequence [2, 3, 4].

The model introduced here is also based on Kohonen's map. However, thanks to its hierarchical topology, it makes an efficient use of context information which, in its turn, prevents it from suffering from those mentioned flaws.

The model is trained and evaluated on a real sequence in time — a four-part fugue of J. S. Bach. In the following sections, we shall describe the representation, model, and the experiment on musical sequences.

## 2   Representation for the sequences

The concept behind the representation is the division of a musical piece into equal size time intervals. The small figure in a musical sequence is set to be the *time interval (TI)*. Thus, all other figures in the sequence are multiples of TI. For example, if TI is an eighth note, a quarter note lasts two TIs, a half note lasts four TIs, and so on.

A *time interval counter (TIC)* may also be defined. One TIC lasts one TI. TIC is the unit in which the musical sequence is measured. Therefore, at each TIC, either there is a rest, or a note onset, or a note sustained.

The input data in the experiment consists of a sequence of musical intervals, which corresponds to a Bach's fugue. Data is input one TIC at a time.

Fifteen neural units are used in the representation. Each unit represents one musical interval ranging from an octave down to an octave up. We assume here, therefore, that listeners are able to separate out voices if intervals between notes in the voices are greater than a fifteenth. When there is a rest, none of the input units receives activation. Otherwise, when a note is onset or sustained, the unit corresponding to the interval receives activation.

The representation for musical sequences takes into consideration all musical voices, and therefore, is complex because the voices interact. The representation assumes three facts. First, any note onset occurring in a TIC makes up an interval with all notes onset or sustained which occurred in the TIC immediately before. Second, the representation does represent the intervals which occur in a TIC, but does not represent multiple instances occurring through the voices in the TIC. Third, at any given TIC, an interval corresponding to a note onset masks any occurrence of the same interval corresponding to a note sustained.

## 3 The model

The model is made up of two self-organizing maps (SOMs), as shown in figure 1. Its features, performance, and potential are better evaluated in [5, 6].

The input to the model is a sequence in time of $m$-dimensional vectors, $\mathbf{S_1}$ = $\mathbf{V}(1)$, $\mathbf{V}(2)$, ..., $\mathbf{V}(t)$, ..., $\mathbf{V}(z)$, where the components of each vector are non-negative real values. The sequence is presented to the input layer of the bottom SOM, one vector at a time. The input layer has $m$ units, one for each component of the input vector $\mathbf{V}(t)$, and a time integrator. The activation $\mathbf{X}(t)$ of the units in the input layer is given by

$$\mathbf{X}(t) = \mathbf{V}(t) + \delta_1 \mathbf{X}(t-1) \tag{1}$$

where $\delta_1 \in (0,1)$ is the decay rate. For each input vector $\mathbf{X}(t)$, the winning unit $i^*(t)$ in the map[1] is the unit which has the smallest distance $\Psi(i, t)$. For each output unit $i$, $\Psi(i, t)$ is given by the Euclidean distance between the input vector $\mathbf{X}(t)$ and the unit's weight vector $\mathbf{W}_i$.

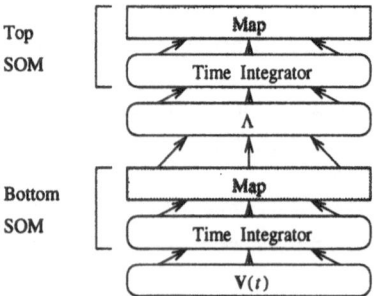

Figure 1: The model

[1] Also known as array, grid, or output layer.

Each output unit $i$ in the neighbourhood $N^*(t)$ of the winning unit $i^*(t)$ has its weight $\mathbf{W}_i$ updated by

$$\mathbf{W}_i(t+1) = \mathbf{W}_i(t) + \alpha \Upsilon(i)[\mathbf{X}(t) - \mathbf{W}_i(t)] \tag{2}$$

where $\alpha \in (0,1)$ is the learning rate. $\Upsilon(i)$ is the *neighbourhood interaction function* [7], a Gaussian type function, and is given by

$$\Upsilon(i) = \kappa_1 + \kappa_2 e^{-\frac{\kappa_3[\Phi(i,i^*(t))]^2}{2\sigma^2}} \tag{3}$$

where $\kappa_1$, $\kappa_2$, and $\kappa_3$ are constants, $\sigma$ is the radius of the neighbourhood $N^*(t)$, and $\Phi(i, i^*(t))$ is the distance in the map between the unit $i$ and the winning unit $i^*(t)$. The distance $\Phi(i', i'')$ between any two units $i'$ and $i''$ in the map is calculated according to the maximum norm,

$$\Phi(i', i'') = max\{|l' - l''|, |c' - c''|\} \tag{4}$$

where $(l', c')$ and $(l'', c'')$ are the coordinates of the units $i'$ and $i''$ respectively in the map.

The input to the top SOM is determined by the distances $\Phi(i, i^*(t))$ of the $n$ units in the map of the bottom SOM. The input is thus a sequence in time of $n$-dimensional vectors, $\mathbf{S_2} = \Lambda(\Phi(i, i^*(1))), \Lambda(\Phi(i, i^*(2))), \ldots, \Lambda(\Phi(i, i^*(t))), \ldots, \Lambda(\Phi(i, i^*(z)))$, where $\Lambda$ is a $n$-dimensional *transfer function* on a $n$-dimensional space domain. $\Lambda$ is defined as

$$\Lambda(\Phi(i, i^*(t))) = \begin{cases} 1 - \kappa\Phi(i, i^*(t)) & \text{if } i \in N^*(t) \\ 0 & \text{otherwise} \end{cases} \tag{5}$$

where $\kappa$ is a constant, and $N^*(t)$ is a neighbourhood of the winning unit.

The sequence $\mathbf{S_2}$ is then presented to the input layer of the top SOM, one vector at a time. The input layer has $n$ units, one for each component of the input vector $\Lambda(\Phi(i, i^*(t)))$, and a time integrator. The activation $\mathbf{X}(t)$ of the units in the input layer is thus given by

$$\mathbf{X}(t) = \Lambda(\Phi(i, i^*(t))) + \delta_2\mathbf{X}(t-1) \tag{6}$$

where $\delta_2 \in (0,1)$ is the decay rate.

The dynamics of the top SOM is identical to that of the bottom SOM.

## 4  The experiment

The experiment was on recognition of the instances of a theme occurring in a musical sequence — the sixteenth four-part fugue in G minor of the first volume of The Well-Tempered Clavier of Bach. The fugue had 544 TICs, and TI was a sixteenth note.

The input data consisted of two sets, hereafter referred to as *input set I* and *input set II*. Input set I consisted of a large and unique sequence of musical intervals, which corresponded to the fugue. Input set II contained many sequences, which were formed by segmenting the fugue whenever there were rests.

We wanted the model to be tested on musical sequences, because the musical domain set three strong conditions on the model. First, that the model were able to recognize both a set of input sequences and a set of sub-sequences within a large and unique input sequence. The model was required to recognize a set of input sequences when the whole sequence was segmented. Otherwise, the entire piece consisted of a unique input sequence, and the model was thus required to recognize sub-sequences of that sequence.

Second, that the model classified sequences (or sub-sequences) properly in the presence of noise. The reason followed from the fact that any two sequences which differed slightly had to achieve similar classifications.

Third, that the model recognized sequences (or sub-sequences) in a very precise form. The reason for the latter is that any two sequences which shared either some intervals, or even all intervals, but in an alternative order or rhythm, were musically different, and as a consequence, had to be recognized as distinct.

The experiment pursued two aims. Firstly, to determine whether the model recognizes all instances of the theme in the fugue. Secondly, to determine whether any other sequence (or sub-sequence), which was not an instance, was not misclassified as theme.

The training of the two SOMs of the model took place in two phases — coarse-mapping and fine-tuning. In the coarse-mapping phase, the learning rate and the radius of the neighbourhood were reduced linearly whereas in the fine-tuning phase, they were kept constant. The bottom and top SOMs were tested and trained respectively with map sizes of $15 \times 15$ in 700 epochs, and $18 \times 18$ in 850 epochs. The initial weights were given randomly to both SOMs.

The experiment comprised five studies. In the last four, in order to study the effect of noise on the classifications, reinforcement in activation was given to input units when representing instances of theme. Table 1 shows the activation values of notes onset and sustained, whether reinforced or not, as well as the input set employed in each study[2].

Table 1: Parameter values of the studies

| Study | Input Set | Reinforcement Value | Note Onset | Note Sustained | N. Onset (Reinforced) | N. Sustained (Reinforced) |
|---|---|---|---|---|---|---|
| I | I | 1 | 0.1 | 0.07 | 0.1 | 0.07 |
| II | I | 5 | 0.1 | 0.07 | 0.5 | 0.35 |
| III | I | 10 | 0.1 | 0.07 | 1.0 | 0.7 |
| IV | I | 100 | 0.1 | 0.07 | 10.0 | 7.0 |
| V | II | 100 | 0.1 | 0.07 | 10.0 | 7.0 |

A sequence (or sub-sequence) $S_a$ is said to have the same classification as that of the theme $S_t$ if the distance $\Phi(i_a^*(z), i_t^*(z)) < 2$, where $i_a^*(z)$ and $i_t^*(z)$ are the last winning units of $S_a$ and $S_t$. In case of $S_a$ be also an instance of the theme, the error of the instance $S_a$ is then given by calculating the sum

---

[2]Reinforcement was provided from the seventh common TIC between the theme and any of its instances.

of the distances between each winning unit of $S_a$ and its corresponding in $S_t$[3]. The mean error is given by the sum of the errors of each instance divided by the number of instances.

Table 2 displays the classifications and misclassifications of the studies. We divided the cases of misclassifications into two groups — minor and major misclassifications. We considered a case of minor misclassification when the model kept on classifying as theme the next few TICs which followed the theme. All other cases were otherwise referred to as major.

Table 2: Classifications and misclassifications of the model

| Study | Classifications | | | Misclassifications | |
|-------|-----------------|--|--|--------------------|--|
|       | No. Hits | No. Failures | Mean Error | No. Minor Miscl. | No. Major Miscl. |
| I     | 0  | 16 | 162.44 | 2  | 5  |
| II    | 3  | 13 | 85.51  | 0  | 6  |
| III   | 4  | 12 | 72.19  | 0  | 14 |
| IV    | 11 | 5  | 43.66  | 11 | 0  |
| V     | 13 | 3  | 52.92  | 6  | 0  |

Some conclusions may be drawn from the results. First, as displayed in table 2, the model held a high number of misclassifications in the third study. Such a high number was due to the fact that the model classified an intermediate part of theme as its final part, and consequently, kept on misclassifying intermediate parts of instances of theme as their final parts as well.

Second, by analysing the results displayed in the table 2, one may observe that the model was fault tolerant to errors. It classified properly several instances which differed slightly from the theme, whether in the pitch or in the duration of one single note. The model performed classification efficiently in the presence of noise as well. When instances of theme occurred concurrently with other polyphonic voices, the degree of noise was so high that it caused the model not to classify instances correctly. However, when reinforcement was given to instances, the remaining polyphonic voices started playing roles of noisy backgrounds, and then, the model started classifying rightly instances of theme.

Third, the model failed, in all studies, in recognizing three instances of theme. It succeeded, however, in recognizing two other instances of theme in the last four studies. These related instances, which occur between TICs 265 and 282, TICs 273 and 290, TICs 441 and 456, TICs 449 and 466, and TICs 457 and 460, overlapped through the voices, making up the two cases of *stretto*[4] present in the fugue. One may conclude, therefore, that the recognition of *strettos* was not performed reasonably by the model.

Finally, by comparing studies IV and V in the table 2, one may verify that there is not a significant difference between their results. The straightforward conclusion which may be drawn is thus, that the segmentation of the fugue on rests, i.e., the resetting of the model on inputs which corresponds to rests, does not yield any improvement in terms of classification.

---

[3]The error and the distances of each instance are computed from the seventh common TIC between the instance and the theme.

[4]*Stretto* is a musical passage where two or more instances of theme overlap.

# 5 Conclusion

An original representation for musical sequences, and an original artificial neural model for sequence classification is presented. The model has a topology made up of two self-organizing map networks, one on top of the other. It encodes and manipulates context information effectively, and that enables it to perform sequence classification and discrimination efficiently. The model has application in domains which demand classifying either a set of sequences of vectors or sub-sequences into a unique and large sequence of vectors in time.

The results obtained have shown that the artificial neural model was able to perform efficiently, even in the presence of noise, sequence classification and discrimination. The model was able to classify properly most of the instances of theme occurring in the musical piece. It is worth noticing that the model performed classification even in the presence of noise, i.e., even when instances occurred modified, in different past contexts, and amidst different polyphonic voices in the musical piece.

The model could also discriminate instances of theme from sequences that shared some similarity with the theme. Very many of these pseudo-instance sequences occurred in the fugue. Rightly yet, the model did not classify them as instances of theme.

We intend to do further research on the model, applying it to other domains to better assess its potentiality for performing classification and discrimination of sequences in time.

# References

[1] T. Kohonen. *Self-Organization and Associative Memory*. Springer-Verlag, Berlin, third edition, 1989.

[2] J. Kangas. *On the Analysis of Pattern Sequences by Self-Organizing Maps*. PhD thesis, Helsinki University of Technology, Finland, 1994.

[3] G. J. Chappell and J. G. Taylor. The temporal Kohonen map. *Neural Networks*, 6:441–445, 1993.

[4] D. James and R. Miikkulainen. SARDNET: a self-organizing feature map for sequences. In *Proceedings of NIPS*, volume 7. Morgan Kaufmann, 1995.

[5] O. A. S. Carpinteiro. A hierarchical self-organizing map model for pattern recognition. In L. Caloba and J. Barreto, editors, *Proceedings of the Third Brazilian Congress on Artificial Neural Networks*, pages 484–488. UFSC, Florianópolis, SC, Brazil, 1997.

[6] O. A. S. Carpinteiro. A hierarchical self-organizing map model for sequence recognition. To appear in *Neural Processing Letters*, 1998.

[7] Z. Lo and B. Bavarian. Improved rate of convergence in Kohonen neural network. In *Proceedings of the IJCNN*, volume 2, pages 201–206, 1991.

# Discrete Time Backpropagation and Synaptic Delay Based Artificial Neural Networks in Chaotic Time Series Prediction[*]

## R. J. Duro[1] and J. Santos[2]

[1]Dpto. Ingeniería Industrial, [2]Dpto. Computación
Universidade da Coruña, Spain
e-mail: richard@udc.es, santos@dc.fi.udc.es

## Abstract

This paper is concerned with the application of a new training algorithm for delay based neural networks to the prediction of future values in chaotic time series. In the networks we employ, the transmission of information through synapses is delayed by a trainable amount. The main application of these structures is in training to perform operations that require reasoning with events occurring in different instants of time without any time windowing process. We test the validity of the approach to the prediction of future values in chaotic time series using iterative multistep prediction.

## 1. Introduction

It is very important in signal processing and prediction to consider the temporal aspects involved, as most of this type of problems are intrinsically dynamic in nature. Traditionally, this has been achieved through the presentation of appropriately selected windows of the signal to static networks as parallel inputs. This procedure is obviously not very well suited for the description of time dependent phenomena, unless a perfectly good window may be found. The windowing approach presents many well known drawbacks, and to avoid them we require networks that can learn the appropriate state space representation of the mechanisms involved straight from the signal, without any previous processing or windowing.

There are basically two approaches to the problem of adding to neural networks the capability of handling data produced in different instants of time. On one hand we find recurrent neural networks, which take into account previous information through the use of feedback loops [3][4]. The problem with recurrent networks is that they are very hard to train and their scalability is very limited. On the other hand, we may consider networks based on delayed links, that is, that contain an explicit representation of time. An example of these are TDNNs, proposed by Waibel [5]. But these networks become huge when anything but a few instants of

[*] Supported by the Xunta de Galicia (project XUGA-16602A96) and the Universidade da Coruña.

time are considered.

Another example is continuous time temporal backpropagation with adaptable time delays considered by S.P. Day and M. R. Davenport [1]. They propose an algorithm that, using a network with delays in the different synapses, trains the values of these delays considering their future effect on the outputs and obtaining a backpropagated error term from this so as to respect the principle of causality. The problem with this strategy is that it requires future information in order to be trained.

Other approaches involve FIR [9] or IIR [8] filter type synapses, based on tapped

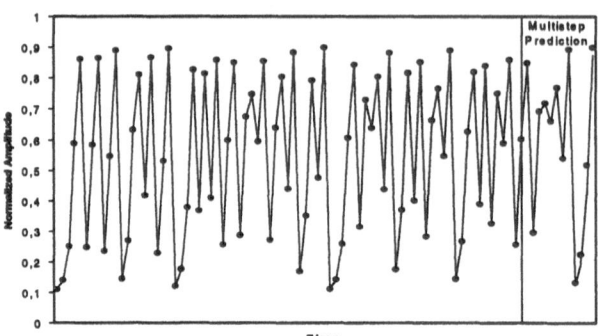

Figure 1: Logistic time series data (circles) and prediction by a 1-10-10-1 synaptic delay based network (solid line). After the dividing line, the predictions are obtained using the network's previous predictions as inputs for the next predictions.

delay lines. The structures of the networks, and thus their training, in these cases become quite complicated. In fact, in order to run a FIR filter synapses type neural network it is necessary to store a weight vector for each synapse that is as long as the number of steps in the tapped delay line corresponding to the synapsis.

As a first step in directly training the delays found in a generalized synaptic delay network, using a gradient descent method, we have developed an extension of backpropagation that provides really adequate results [1]. In previous work by our group [2], this methodology has been successfully applied to the design and training of a neural network for the detection and classification of QRS complexes in ECG signals, and in this work we are going to present another possible application in the realm of the prediction of chaotic time series, as they are in general a paradigmatic problem when demonstrating the ability of a structure to model the behavior of a time series.

## 2. The Artificial Neural Network Structure and Training

The artificial neural network we consider for training consists of several layers of neurons connected as a Multiple Layer Perceptron (MLP). The only difference with traditional MLPs is that the synapses are represented by a delay term in addition to the classical weight term. That is, now the synaptic connections between neurons are described by a pair of values, $(W, \tau)$, where W is the weight describing the ability of the synapse to transmit information and $\tau$ is a delay, which in a certain sense provides an indication of the length of the synapse.

The justification of the training procedure employed, that is Discrete Time Backpropagation, may be found in [2]. This algorithm permits training the network through variations of synaptic delays and weights, in effect changing the length of the synapses and their transmission capacity in order to adapt to the problem in hand.

## 3. Chaotic Time Series: Multistep Prediction

In order to test the algorithm proposed in [2] in the field of time series prediction, we have chosen two chaotic time series often found in the literature as benchmarks.

A) Logistic series: $$x(t) = 4x(t-1)[1 - x(t-1)]$$

B) Mackey-Glass [10]: $$\frac{\partial x(t)}{\partial t} = a\frac{x(t-\tau)}{1 + x(t-\tau)^{10}} - bx(t).$$

We are going to choose values for $\tau$ for which the series becomes chaotic. In order to be as standard as possible in the series we employ, we have chosen values for $a$ and $b$ of 0.2 and 0.1 respectively, which are the values we have found to be most often used in the literature [6][9]. As in the case of [1] we have taken as boundary conditions $x(t)=0.8$ for $t<0$. In every case we scaled the signal in order to fit it into the [0,1] interval but did not remove the offset, as some other authors do.

## 4. Results

In the first test, the network was a two hidden layer perceptron type synaptic delay based network. There was a single input neuron through which the signal was input one sample per instant of time and one output neuron that predicted the value of the time series in $t+n$, being $t$ the current instant of time. The hidden layers consisted of 10 sigmoid type neurons in the case of the logistic function and 20 for the case of the Mackey Glass time series. The initial weights and delays were chosen randomly, from the intervals [-10,10] and [0,10] respectively.

Figure 1 displays the predictions for t+1 in a logistic time series. The prediction values obtained for the right hand side of the dividing line correspond to iterative multistep prediction, that is, to using the output generated by the predicting network as input for the prediction of the next value. It is clear that the network performs a very good prediction for the next ten values, without any appreciable error. The result is a first indication that the network learns the underlying mechanism that generates

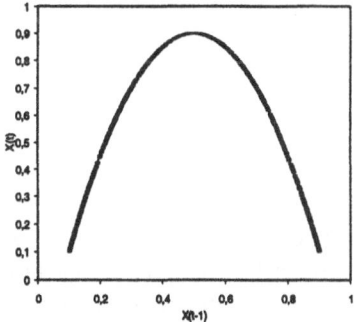

Figure 2: State Space representation of the logistic times series (solid line) and the data obtained by the same network as figure one (dashes).

the series. This can be further seen in figure 2, where we display the state space plot for the logistic time series (line) and the values obtained by the network straight from the signal (small horizontal lines).

In figure 3 we display the results of the application of this network to a Mackey-Glass type series with $\tau=17$ (circles) and its prediction of the value of the series in time $t+100$ in an iterative multistep fashion (solid line). In the bottom of the figure we include the fractional prediction error.

The data corresponding the normalized mean squared error (NMSE) for iterated multistep predictions of one of our networks are displayed in figure 4 as a function of the depth into the future considered. The 1-20-20-1 network was trained to predict one instant of time ahead and used to predict different temporal distances into the future by means of using as input for the prediction of the value in the next instant the predictions made by the network for previous instants. This will provide a good measure of how well the network is able to learn the underlying mechanism that generates the time series.

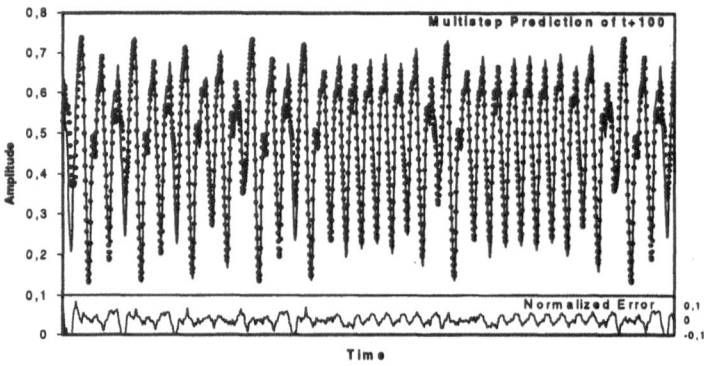

Figure 3: Iterative multistep prediction (solid line) of the Mackey Glass Series data (circles) for 100 instants into the future. In the bottom part of the figure we display the normalized error for this prediction.

The error values obtained for one step ahead prediction using a 2000 point test set after training the network on-line on the signal, (MSE=$10^{-5}$, NMSE=$8*10^{-4}$) compare quite favorably with those of other authors such as McDonnell [8] whose predictions are obtained using an evolved recurrent non linear IIR filter network (MSE =$5*10^{-5}$, about four times ours), or Chng and Chen [7], who predict the same Mackey Glass series using a GRBF network and obtain, in their best effort, a NMSE of around $10^{-3}$, an order of magnitude larger than the one obtained by our network. Other authors that provide error results for the same time series include Chow and Leung [6] who obtain results after a non-linear preprocessing stage of around MSE=$10^{-5}$ in the case of one step prediction, and different values provided by authors utilizing recurrent neural networks with different learning algorithms, such as Williams and Zipser [11] (NMSE=$5.4*10^{-3}$), Peralmutter and Barak [12] (NMSE=$2.2*10^{-3}$) or Sun, Chen and Lee (NMSE=$1.6*10^{-2}$) [13].

In the above we have considered, as most authors do, a one step ahead prediction, which is a fairly easy problem as in many areas of the signal, the Mackey Glass series varies "predictably" that is, there are not that many transitions or brusque changes in the values, which are the problems that would require more knowledge on the part of the predictor. The problem becomes noticeably more complicated when predictions must be made of values for the signal several instants ahead and, even harder when these predictions must be iteratively obtained using as inputs previous predictions made by the network. This leads to an accumulation of. the error which usually increases very rapidly. Also, it must be taken into account that when training for one step ahead prediction, there is certainly a risk that in the quest for the minimum possible error in a training or test set, the network will become incapable of performing multistep prediction through an overfitting problem, thus reducing the generalization capabilities and henceforth the possibility of recuperating from small errors which can always arise in multistep prediction. For this reason, we have only compared our results to authors that included multistep prediction in their work. Some other authors, such as Yu and Bang [9] with their FIR synapse OLL trained network, do obtain better NMSE error results than our networks for one step prediction, but do not include multistep prediction results, and therefore are not comparable to ours.

It can be clearly appreciated in figure 4, that the NMSE certainly increases as errors are compounded when looking deeper into the future, but this increase is fairly controlled allowing for predictions to be made quite deeply into the future and still preserve a satisfactory degree of accuracy. This can be seen in figure 3, where the predictions obtained for t+100 by a 1-20-20-1 network operating in an iterative multistep fashion are shown. Actually most other authors tend to obtain unbearably large error values in iterative multistep prediction, some of them after just a few steps, as is the case of Mcdonnell, [8] with a MSE of 5.11 after just three steps. Many authors do not even report error values for more than three or four steps and almost none report more than 10 steps into the future. In the case of recursive networks, these values become large for 8 steps: NMSE=0.22 for Williams and Zipser [11], 0.29 for Pearlmutter [12] and 1.37 in the case of Sun, Chen and Lee [13]. One of the best values we have found is a NMSE of around $6*10^{-2}$ obtained by Chng

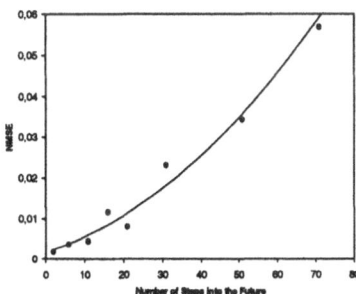

Figure 4: NMSE as a function of number of steps in iterative multistep prediction for a 1-20-20-1 network predicting the Mackey Glass chaotic time series.

and Chen [7] using their second order GRBF model in a 10 step prediction.

In this regard, the network presented here performs really well. Compared to the GRBF case, it can iteratively predict more than 70 steps into the future before the performance of the network degrades down to Chng and Chen´s value for 10 time steps, and, compared to the best recursive network for 8 time steps, our network can predict with a lower value for the error for more than 160 steps into the future.

# 5. Conclusions

In this paper we have presented an application of the Discrete Time Backpropagation training algorithm for neural networks that include trainable synaptic delays. The application of these techniques to the prediction of future values of chaotic time signals indicates that the networks easily learn the models underlying the time series. This has been shown in the case of the logistic time series and the Mackey Glass series. It has also been shown that the networks perform very well in iterative multistep prediction, which requires a temporal generalization capability that is not always obtained when trying to simply minimize the one step ahead prediction error.

# References

[1] Day, S. P. and. Davenport, M. R.. Continuous Time Temporal Backpropagation with Adaptable Time Delays. *IEEE Trans. on Neural Networks*. Vol 4, N°2. pp 348-354. 1993.

[2] Duro, R.J. and Santos, J.. Fast Discrete Time Backpropagation for Adaptive Synaptic Delay Based Neural Networks. *Submitted to IEEE Trans. ANN*.

[3] Elman J. L.. Finding Structures in Time. *CRL Technical Report N°8801*, La Jolla University of California, San Diego. 1988.

[4] Jordan, M.I.. Attractor Dynamics and Parallelism in a Connectionist Sequential Machine. *In Proceedings of the 1986 Cognitive Science Conference*. L. Erlbaum, Hillside N. J. pp 531-546, 1986.

[5] Waibel A. Hanazawa., T, Hinton, G.. Lang J, and Shikano K.. Phoneme Recognition Using Time-Delay Neural Networks. *IEEE Trans. Acoust. Speech Signal Processing* 37, pp. 328-339, 1989.

[6] Chow, T., Leung, C.T. Performance Enhancement Using Nonlinear Preprocessing. *IEEE Trans ANN*, V7 N4, pp 1039-1042, July 1996.

[7] Chng E.S., Chen S., Mulgrew, B. Gradient Radial Basis Function Networks for Nonlinear and Nonstationary Time Series Prediction . *IEEE Trans ANN*, V7 N1, pp 190-194, Jan 1996.

[8] McDonell, J. R. Evolving Recurrent Perceptrons for Time Series Modeling. *IEEE Trans ANN*, V5 N1, pp24-38, Jan 1994.

[9] Yu, H.Y. and Bang, S. Y. An Improved Time Series Prediction by Applying the Layer-by-Layer Learning Method to FIR Neural Networks. *Neural Networks*, Vol 10, N. 9. Pp 1717-1729, 1997.

[10] Mackey, M. C. and Glass, L., "Oscillation and Chaos in Physiological Control Systems", *Science*, vol 197, pp. 287-289, July 1977.

[11] Williams, R. J. and D. J. Zipser. Experimental Analysis of Real Time Recurrent Learning Algorithm. *Connection Science*, Vol. 1, No. 1, 1989.

[12] Peralmutter and Barak. Dynamic Recurrent Neural Networks, *Technical Report* CMU-CS-90-196, 1990.

[13] Sun, Guo-Zehng, Chen and Lee, Green's Function Method for Fast On Line Learning Algorithm of Recurrent Neural Network, *Advances in Information Processing Systems 4*, 1992.

# A Learning Method of Fuzzy Inference Rules Using Vector Quantization

Kazuya KISHIDA

Control Engineering, Kagoshima National College of Technology

Kagoshima, Japan

kishida@kctmgw.kagoshima-ct.ac.jp

Hiromi MIYAJIMA

Faculty of Engineering, Kagoshima University

Kagoshima, Japan

miya@eee.kagoshima-u.ac.jp

## Abstract

Some models using self-organization systems of neural networks are proposed in recent studies. These models show good results in point of the number of fuzzy rules in high dimensional problems. However, most of these models determine a distribution of initial fuzzy rules by considering only input data. In this paper, we propose a method considering not only input data but also output data. In order to demonstrate the validity of the proposed method, some numerical examples are performed.

## 1 Introduction

In recent studies, some fuzzy models using self-organization or vector quantization are proposed[2, 3, 4, 5, 6, 7]. It is effective to use self-organization or vector quantization for constructing fuzzy rules in high dimensional problems. Because self-organization and vector quantization approximate input data without influence by the increase of input dimensions. That is, by using self-organization or vector quantization, fuzzy rules are created directly in input space and distributed in input domain where input data exist. Most of above models[2, 3, 4, 5] are applied for pattern recognition or pattern classification problems. Umano[2] and Horikawa[5] give good results by using self-organization. Further we have proposed a self-creating method of fuzzy rules using vector quantization[7], that is applied for function approximation problems. Our method is effective to construct fuzzy rules in high dimensional problems. However, in the conventional methods like these, only input data is used to determine distribution of initial fuzzy rules. When self-organization and vector quantization are used to construct fuzzy rules, we have to consider not only input data but also output data. By considering output data, fuzzy rules are created more properly. That is, reference vectors(fuzzy rules) are distributed over input domain where input data exist and which influences change of inclination of output strongly.

In this paper, we propose a learning method of fuzzy rules using one of methods of vector quantization, neural gas network. First, we introduce a method that determines the appearance frequency of input data like probability density using output data, then distribution of fuzzy rules is made properly by vector quantization. Next, the number of fuzzy rules to an objective value, inference error, is determined by a constructive method. In order to demonstrate the validity of the proposed method, some numerical examples are performed.

## 2 Self-tuning Method of Fuzzy Rules by Descent Method

When the input data are expressed by $x_1, \cdots, x_m$ and the output is expressed by $y^*$, the rules of simplified fuzzy inference can be expressed as the following:

$$R^i) \; if \; x_1 \; is \; M_{i1} \; and \; x_2 \; is \; M_{i2} \cdots x_m \; is \; M_{im} \; then \; y^* \; is \; \omega_i \tag{1}$$

where $i$ $(i = 1, \cdots, n)$ is a rule number, $j$ $(j = 1, \cdots, m)$ is a variable number, $M_{i1}, \cdots, M_{im}$ are membership functions of the antecedent part, and $\omega_i$ is a real number of the consequent part.

A membership value of the antecedent part $\mu_i$ is expressed as the following:

$$\mu_i = M_{i1}(x_1) \cdot M_{i2}(x_2) \cdots \cdot M_{im}(x_m) \tag{2}$$

$$M_{ij}(x_j) = exp(\frac{-(x_i - c_{ij})^2}{b_{ij}}) \tag{3}$$

where $M_{ij}$ is membership function of antecedent part, and the center value $c_{ij}$ and the width $b_{ij}$ are parameters.

The output of fuzzy inference $y^*$ can be derived from the Eq.(4).

$$y^* = \frac{\sum_{i=1}^{n} \mu_i \cdot \omega_i}{\sum_{i=1}^{n} \mu_i} \tag{4}$$

A learning method based on the descent method is described as follow.

The objective function $E$ is defined to evaluate the inference error between the desirable output $y^r$ and the output of fuzzy inference $y^*$:

$$E = \frac{1}{2} (y^* - y^r)^2 \tag{5}$$

In order to minimize the objective function $E$, the parameters $c_{ij}$, $b_{ij}$, and $\omega_i$ are updated according to following equations.

$$\Delta c_{ij}(t) = K_c \cdot \frac{\mu_i}{\sum_{g=1}^{n} \mu_g} \cdot (y^* - y^r) \cdot (\omega_i - y^*) \cdot \frac{2(x_j - c_{ij})}{b_{ij}} \tag{6}$$

$$\Delta b_{ij}(t) = K_b \cdot \frac{\mu_i}{\sum_{g=1}^{n} \mu_g} \cdot (y^* - y^r) \cdot (\omega_i - y^*) \cdot \frac{(x_j - c_{ij})^2}{b_{ij}^2} \tag{7}$$

$$\Delta \omega_i(t) = K_\omega \cdot \frac{\mu_i}{\sum_{g=1}^{n} \mu_g} \cdot (y^* - y^r) \tag{8}$$

where $K_c$, $K_b$ and $K_\omega$ are constants.

A general flow of learning algorithm for fuzzy rules performing self-tuning method is omitted. The reference [1] describes the detailed explantion.

# 3  A Proposed Learning Method Using Vector Quantization

Some models using self-organization or vector quantization are proposed. These models determine an initial assignment of fuzzy rules as the result of vector quantization or self-organization. Then fuzzy rules are created over input domain where input data exist. These methods are effective in high dimensional problems. However, when determining an initial assignment of fuzzy rules by the conventional method using vector quantization, only input data is considered.

In this paper, so as to determine more proper initial distribution of fuzzy rules, we propose a method determining an initial assignment of fuzzy rules, by considering not only input data but output data. We use neural gas network to determine the positions of initial fuzzy rules. Neural gas network is one of methods used for vector quantization and shows outstanding approximation accuracy. An explanation of neural gas network is omitted. The reference [8] describes the detailed one.

## 3.1  Introduction of the Appearance Frequency of Input Data

In Fig.1, we assume that input data exist in whole input space. Then initial fuzzy rules are distributed homogeneously in input space by the conventional method as shown in Fig.1(a). We determine the width of membership function of antecedent part of each fuzzy rule so that the summation of membership values of fuzzy rules is 1. The circle in Fig.1 shows a real value of consequent part of each fuzzy rule. Then the output of fuzzy inference is shown as a broken line. A solid line shows the output of system. In Fig.1(a), a broken line does not approximate a solid line not much. In Fig.1(b), a broken line approximates a solid line more exactly compared with Fig.1(a). From Fig.1(b), it is needed that fuzzy rules are created over input domain which influences change of inclination of the output. So as to make a distribution of initial fuzzy rules as shown in Fig.1(b), it turns out that what is necessary is to give input data in black parts repeatedly when we use vector quantization algorithm. So we introduce an appearance frequency to input data. We determine an appearance frequency of input data by considering both input and output data in the following:

1. For some input data $x_i$, other input data $x_j$ are ranked by the distance between $x_i$ and $x_j$ as follows:

$$x_{il} = \min_l \{\|x_i - x_j\|\} \tag{9}$$

, where $l$ varies from 1 to $M$ and $\min_l$ means to calculate the $l$-th smallest value. $x_{iM}$ shows the $M$-th nearest input data to $x_i$ in input space.

2. Determine $H(x_i)$ which shows the degree of change of inclination of the output around output data to input data $x_i$, by the following equation:

$$H(x_i) = |\sum_{l=1}^{M} \frac{y_i - y_{il}}{D_{il}}| \tag{10}$$

, where $D_{il}$ shows the distance between $x_{il}$ and $x_i$.

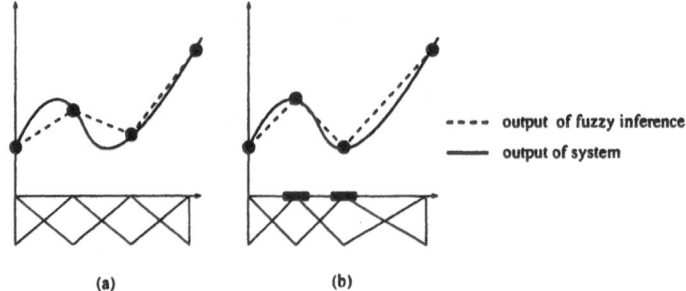

(a)              (b)

---- output of fuzzy inference
—— output of system

Figure 1: Relation between the positions of initial fuzzy rules and output of fuzzy inference

Determine the appearance frequency $p(x_i)$ for $x_i$ by normalizing $H(x_i)$.

$$p(x_i) = \frac{H(x_i)}{\sum_{j=1}^{P} H(x_j)} \tag{11}$$

## 3.2 A Proposed Algorithm

We describe the proposed algorithm using vector quantization as follows:

[STEP 1] **Initialization :**
Determine the number of initial reference vectors $L$, the threshold $T_1$ of inference error. The appearance frequency for input data is determined according to 3.1.

[STEP 2] **Approximation of input space :**
Input data with apperance frequency are approximated by neural gas network.

[STEP 3] **Creation of fuzzy rules :**
Reference vectors in STEP 2 are used as initial center values of fuzzy rules. The number of created fuzzy rules is as same as the number of reference vectors. The width of each fuzzy rule is determined by the distance between the center value of inference rule and the center values of fuzzy rule which is the nearest to each fuzzy rule in input space. That is, the width of fuzzy rules increases when the distance between fuzzy rules increases.

[STEP 4] **Learning :**
Fuzzy rules are tuned based on self-tuning method. The inference error $E(t)$ is calculated from the Eq.(12).

$$E(t) = \frac{1}{P} \sum_{p=1}^{P} |y_r^p - y^p| \tag{12}$$

[STEP 5] **Termination :**
The $E(t)$ and the threshold $T_1$ are compared. Then
if $E(t) < T_1$ then the algorithm terminates, else go to STEP 2 with $L \leftarrow L+1$.
□

# 4  Numerical Simulations

The systems are identified by fuzzy rules. We use two method, self-creating method[7] and the proposed method. In order to show the advantage of the proposed algorithm, two methods are made comparison. Here are three systems

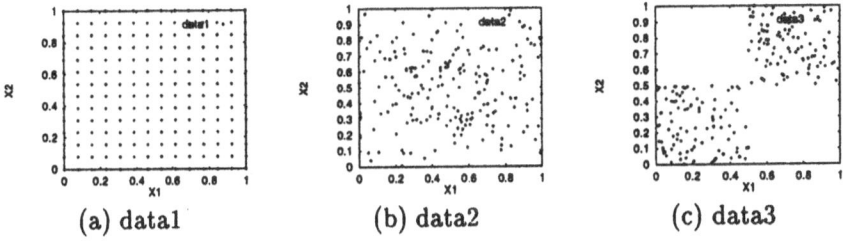

(a) data1  (b) data2  (c) data3

Figure 2: Distribution of three different type of input data.

each of which is specified by each function. Every system has two inputs $x_1, x_2$ and one output $y$. We use three different type of

1) $y = \sin(\pi x_1^3) \times x_2$

2) $y = \dfrac{1.9(1.35 + e^{x_1}\sin(13(x_1 - 0.6)^2)e^{-x_2}\sin(7x_2))}{7.0}$

3) $y = \dfrac{\sin(10(x_1 - 0.5)^2 + 10(x_2 - 0.5)^2 + 1.0)}{2}$

input data. Data 1 is one hundred ninety six data taken regularly in input space on $[0, 1]^2$. Data 2, 3 consist of two hundred random data. Fig.2 shows the distribution of three type of input data. The threshold $T_1$ is $3 \times 10^{-3}$. We use the same set of test data for performance assessment, namely sample of size 2500 on a regularly spaced grid. Inference error to test data is derived from Eq.(12).

As the result, Table 1 shows the number of fuzzy rules, the inference error to test data. Fuzzy system based on data 3 are not evaluated using inference error. Because input data does not exist in whole input space. From Table 1, the proposed method has constructed fuzzy systems which have the small number of fuzzy rules compared with that constructed by self-creating method.

By cosidering both input data and output data, fuzzy rules are distributed in input space more properly. Fuzzy rules are tuned easily by descent method. Then fuzzy systems which has the small number of fuzzy rules are constructed.

# 5  Conclusion

In this paper, we proposed the learning method of fuzzy rules using vector quantization and introduce a method determining the appearance frequency of input data. The appearance frequency is given by considering output data. So the proposed method makes more effective distribution of reference vectors(fuzzy rules) in input space than the conventional one. And fuzzy systems which

### Table 1: The result of numerical simulations

#### (a) The result of self-creating method

| systems | evaluations | data1 | data2 | data3 |
|---------|-------------|-------|-------|-------|
| (1) | the number of rules | 12 | 6 | 9 |
| | inference error | 0.0028 | 0.0056 | |
| (2) | the number of rules | 29 | 24 | 19 |
| | inference error | 0.0040 | 0.0100 | |
| (3) | the number of rules | 55 | 21 | 10 |
| | inference error | 0.0134 | 0.0095 | |

#### (b) The result of a proposed method

| systems | evaluations | data1 | data2 | data3 |
|---------|-------------|-------|-------|-------|
| (1) | the number of rules | 11 | 6 | 8 |
| | inference error | 0.0031 | 0.0052 | |
| (2) | the number of rules | 23 | 22 | 18 |
| | inference error | 0.0050 | 0.0099 | |
| (3) | the number of rules | 44 | 20 | 9 |
| | inference error | 0.0068 | 0.0123 | |

have the suitable number of fuzzy rules are constructed. We demonstrated the validity of the proposed method by performing some numerical examples.

# References

[1] H.Nomura, I.Hayashi and N.Wakami, "A Learning Method of Fuzzy Inference Rules by Descent Method", FUZZ-IEEE/IFES'92, pp.203-210.

[2] M.Umano, T.Fukunaka, I.Hatono and H.Tamura, "Extraction of Fuzzy Rules Using Fuzzy Neural Network with Learning Vector Quantization and Backpropagation with Forgetting", 11th Fuzzy System Symposium, pp.815-818, 1995.

[3] S.Horikawa and A.Tsukamoto, "Composition Method of Fuzzy Classification Systems Using Self-Organizing Feature Map", 12th Fuzzy System Symposium, pp.639-642, 1996.

[4] T.Nomura and T.Miyoshi, " An Adaptive Rule Extraction with the Fuzzy Self-Organizing Map and a Comparison with Other Methods", Journal of Japan Society for Fuzzy Theory and Systems, Vol.8, No.2, pp.347-357, 1996.

[5] S.Horikawa, "Composition Method of Fuzzy Classification Systems Using Self-Organizing Feature Map (II)", 13th Fuzzy System Symposium, pp.659-662, 1997.

[6] A.Ishihara, T.Takagi and S.Nakanishi, "Constitution of Fuzzy Inference by a Kohonen Network", 9th Fuzzy System Symposium, pp.425-428, 1993.

[7] K.Kishida, H.Miyajima, M.Maeda and S.Murashima, A Self-tuning Method of Fuzzy Modeling using Vector Quantization, FUZZ-IEEE'97, pp397-402.

[8] T.Martinetz, S.Berkvich and K.Schulten, "Neural-Gas Network for Vector Quantization and its Application to Time-Series Prediction", IEEE Transactions on Neural Networks, Vol.4, No.4, pp558-569, 1993.

# Mixed Fuzzy-system and Artificial Neural Network Approach to the Automated Recognition of Mouth Expressions.

J.C. Wojdeł        L.J.M. Rothkrantz

Knowledge Based Systems,
Faculty of Information Technology and Systems,
Delft University of Technology
Delft, The Netherlands
(J.C.Wojdel, L.J.M.Rothkrantz)@cs.tudelft.nl

### Abstract

One of the most important parts of the automatic recognition of facial expression is recognition of the mouth expression. This paper describes a new approach to this task. Contrary to the works already done in this fields, it is knowledge- rather than graphically- based. We use here a fuzzy-system in combination with an artificial neural network to obtain the recognition of the mouth shape.

## 1   Introduction

At this moment there is an ongoing project in the Knowledge Based Systems group at the Delft University of Technology, which is aimed at the development of the Integrated System for Facial Expression Recognition (ISFER). The multimedia workbench [1] is currently the basis for the development of the system. Some modules of the ISFER system are already developed. For example the interpretation part of the system was presented in [2]. Such systems for facial expression recognition are being developed not only at TU Delft, but also for example at MIT Media Laboratory [3].

One of the parts of facial expression recognition is the detection of the exact expression of the mouth region in the face. In fact, this region is the most changing part of the face and so should be considered as a very important one. Apart from the applications of the whole facial expression recognition, the recognition of the mouth expression alone may be used to develop an automated lip-reading system. Such a system can then be used in a bimodal speech recognition application to improve the effectiveness of the speech recognition [4] or to support just a segmentation of the speech signal in a noisy environment.

## 2   Problems with finding the mouth contour

Apart from the problems common to any of the image recognition tasks, such as e.g. the presence of noise, the recognition of the mouth shape puts the researcher to some additional difficulties. Firstly, the mouth has to be located in the image. Each of the regions matching some properties be taken as *mouth*

834

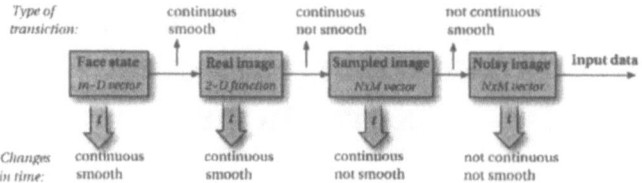

Figure 1: Data flow from the state of the face to the recognition system.

*location hypothesis.* All the hypothesis must then be evaluated, and the proper one must be chosen. The evaluation of hypothesis is based on such properties of found regions as its size, its relative position to other regions (as e.g. eyes) and face symmetry axis. Description of a system performing such a task can be found in [5].

After locating the mouth region one has to recognize correctly its shape features. So far most of the works in recognizing mouth shape concerns finding the mouth contour [6, 5]. Unfortunately, there is no significant change in the image intensity at the contour of the mouth (except for women wearing lipstick). In fact, it is also hard to define correctness of the contour which was found. It seems that the only criterion that can be used is whether effect corresponds to a human perception of the mouth image.

## 3   A new representation of the mouth shape

The first thing that has to be understood is the nature of the data that is processed in mouth feature recognition. The process that generates mouth image-sequence is continuous and smooth in time. Yet the system never gets directly a state of the face as an input (Figure 1). As it can be seen the system should be able to construct a smooth and continuous response on the basis of data that is neither smooth nor continuous.

The information from a mouth expression can be shown using only a single drawing line and it still remains perfectly well recognizable for all the people. How does it come ? After a closer look we can observe that the edge information (gradient) in a simple drawing is just an exaggerated approximation of that derived from real-life photo. The information about whether a mouth is "smiling" or "sad" is passed through the average edge direction in the corners of the mouth. If the edge is on average "going up" mouth is interpreted as "smiling" if it "goes down" mouth is interpreted as "sad". So, it seems that, after all information about average edge direction is more important than information about an edge in some specific point. The above considerations led us to the conclusion that the appropriate representation of the mouth shape may be the information about the average edge intensity and direction in an image.

## 4   Fuzzy-system for edge information acquisition

We present now a new tool for obtaining information about the average edge direction in a mouth image. The basic idea of this tool comes from [7]. In this article a system based on fuzzy-reasoning for edge detection was proposed. The proposed system performs reasoning based on two linguistic labels:

- local level of symmetry in the image in a specific point
- local image steepness in the direction of the symmetry axis

which correspond to two features of the gradient:

- the function is locally symmetrical along the gradient direction
- the gradient value corresponds to local steepness of the function

Based on those two features we can perform reasoning about the edges in the image (e.g. if the symmetry level and steepness are high then the gradient is high and can be accounted for an edge). A difference in our approach in comparison to the referred article [7] is that our main information is rather the direction of the gradient than its value (contrary to normal edge detection problem).

The numerical values representing symmetry and steepness level are converted by a *fuzzifier* part of the fuzzy-system into the labels. Those labels describe whether each of the values is *low, medium* or *high*. Such statements are then passed to the *reasoning engine*. This part of the fuzzy-system is based on nine rules. Those rules correspond to our assumptions about the edges (e.g. if the steepness is *high* and the symmetry level is *high* then the edge intensity in this point is *high*). The reasoning part produces then also three statements (*low, medium* and *high*) about the edge intensity in a given point .

The greatest advantage of using the fuzzy system is that we can freely use common sense to design the reasoning rules and then use them in a fuzzy system to obtain the tool which deals with the numerical values.

To obtain the information about both: edge intensity and direction , we use the information about the direction of the symmetry axis which was computed in an intermediate step of the processing. Combining the information about the direction and the intensity of the edge in a given point we obtain a vector. This vector-field is then averaged and only 50 average vectors are the output of this stage of image processing.

## 5 Classification of mouth expression by ANN

This part of the system is just an implementation of a simple feed-forward, back-propagation trained Artificial Neural Network (ANN). Only the network architecture is a little bit changed in order to reflect the specific features of the task.

The thing which is reflected in the network layout is that the mouth has the vertical symmetry axis. Of course it may happen that the left-hand side of the mouth is different than the right-hand one, as both sides may act almost independently. The symmetry means in this case that both sides can perform the adequate actions. There is no possible action of the one side of the mouth which couldn't be performed by the other one. That means that there are some shape features of the mouth which are the same for both of the sides. The network which was implemented in this work is therefore divided into two

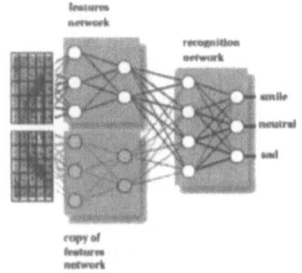

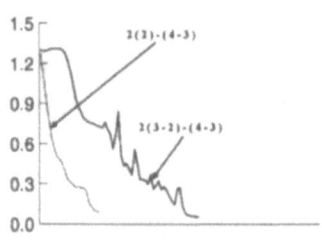

Figure 2: The architecture of the implemented ANN.

Figure 3: Average error in epoch during training process for two network architectures.

parts:

- shape features extraction part, which is built of two symmetrically responding networks (features networks),
- reasoning part which on the basis of outputs from both of the features networks produces some network output.

This architecture is shown in Figure 2. One of the important things is that both features networks should perform the same tasks and so they can be implemented as two copies of the same network. Of course then the error must be propagated not only within the single network but also from the recognition network to both of the features networks. It can be seen that in this case, for each training sample the features network is updated twice – once for the left side and once for the right side. Such a construction of the network should reflect in both: the speed-up of the training process, and better generalization properties

# 6  Experimental results on single pictures

To evaluate the method we chose a set of about 100 pictures of the mouth area which were manually classified to one of the three categories: *smiling*, *neutral* and *sad*. In each experiment, the 10 examples were randomly chosen as a test set and the network was trained on the remaining 90 samples. We show here the results obtained with two different network architectures. Both architectures are shown in the Table 1.

In the first case, each copy of the features network had two layers, and respectively three and two neurons in each layer. The recognition network had also two layers with four neurons in the first layer, and three neurons in the second one.

This network achieved full 100% recognition level for both training and test sets. The training took about 60 epochs. Changes of the average error of the

|  | Features network | | Recognition network | |
|---|---|---|---|---|
|  | 1st layer | 2nd layer | 1st layer | 2nd layer |
| Architecture 1 | 3 | 2 | 4 | 3 |
| Architecture 2 | 2 | – | 4 | 3 |

Table 1: Number of neurons in different network architectures

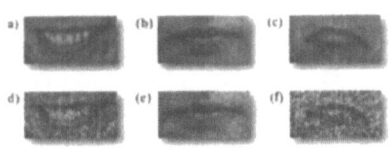

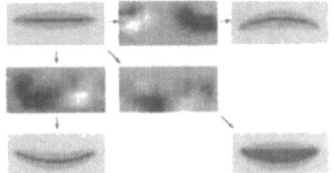

Figure 4: Noiseless images (a), (b), (c), and noisy images which were correctly recognized (d), (e), (f).

Figure 5: The image sequences used and the visualized output of the fuzzy-system.

network response during the training process can be seen in the Figure 3. An average response error of this network on the testing set was 0.08.

The next network which was tested had only one layer of neurons in the features network. The results obtained with this one were similar (the average response error 0.11), but the training took only about 20 epochs. The progress of the training is also displayed in Figure 3.

Another thing which was tested, was the sensitivity of the system to noise. On some representative images Gaussian noise was added with ascending noise level. The achieved results are at least satisfactory (see Figure 4); the presented method is not sensitive for the noise in a sufficiently wide range.

# 7 Experimental results on artificially made image sequences

The method described in the previous sections is aimed at the processing of a sequence of images. Therefore we decided to visualize the changes in the output of the fuzzy-system in some example sequences. For the beginning we chose the artificially made sequences of the simplified mouth actions. Three sequences of the mouth changing from *neutral* to *smile*, *open smile* and *sad* (Figure 5) were investigated.

The areas in which the edges are rotating clockwise are depicted in dark grey contrary to the areas with the edges rotating counter-clockwise which are bright. The areas with no significant change of the edge direction are in neutral grey color. For this visualization we took the average change of the fuzzy-system output in three following frames of the sequence. The patterns obtained from such a visualization can be seen in Figure 5. In Figure 5 we can see also the *neutral mouth* image which was the beginning of the sequence and three images to which this first one was morphed.

It can be well seen that the obtained patterns are very easily recognizable. The sequences with the same "emotional contents" (*smile* and *open smile*) give also similar patterns with bright areas in the right-hand, bottom part of the image and dark areas in the left-hand bottom part of it. This may help the neural network which will classify those patterns to generalize the output.

# 8 Conclusions

The proposed method proved to be valid at least in some basic cases. As it was derived on the basis of the nature of a video signal rather than special properties of the mouth image, it can be easily applied to other face regions. In such a case only some special considerations about the nature of the gradient field in a specified region must be made in order to construct an appropriate tool for the shape classification.

Both steps (fuzzy-system and ANN) of the presented method can be well parallelized. The first one because it repeatedly and independently uses the same fuzzy-system for all the pixels of the image. The second because the ANN consists of independently working neurons and even in this case full networks. This property of the proposed method may be crucial in a real-time applications such as e.g. lip-reading. Also the small size of the data passed from the fuzzy-system to the neural network results in robustness of this approach.

It was not yet proved whether the proposed method is sufficiently sensitive. Some broader research should be done to check whether the method is sensitive enough to allow for example the recognition of the Action Units from the FACS set [8]. The method uses only some average properties of the image and therefore it may destroy subtle differences between mouth expressions which are crucial for proper recognition of all the AUs.

# References

[1] L. J. M. Rothkrantz, M. R. van Schouwen, F. Ververs, and J. C. M. Vollering. A multimedia workbench for facial expression analysis. In *Proceedings of Euromedia-'98*.

[2] M. Pantic, L. J. M. Rothkrantz, and H. Koppelaar. Automation of non-verbal communication of facial expressions. In *Proceedings of Euromedia-'98*.

[3] I. A. Essa and A. P. Pentland. Facial expression recognition using a dynamic model and motion energy. In *Proceedings of the Fifth International Conference on Computer Vision*, Cambridge, Massachusetts, 1995. IEEE Computer Society Press.

[4] S. Nakamura, R. Nagai, and K. Shikano. Improved bimodal speech recognition using tied-mixture HMMs and 5000 word audio-visual synchronous database. In *Proceedings of ECSA, Eurospeech97*, Rhodes, Greece, 1997.

[5] M. J. T. Reinders. *Model Adaptation for Image Coding*. PhD thesis, Delft University of Technology, Delft, The Netherlands, Dec. 1995.

[6] P. Juran and P. Krsek. Mouth segmentation using active contours. Isfer report, Delft University of Technology, Delft, The Netherlands, Oct. 1996.

[7] T. Law, H. Itoh, and H. Seki. Image filtering, edge detection and edge tracing using fuzzy reasoning. *IEEE Transactions on Pattern Analysis and Machine Intelligence*, 18(5), 1994.

[8] P. Ekman and W. V. Friesen. *Facial Action Coding System (FACS): A Technique for the Measurement of Facial Action*. Consulting Psychologists Press, Palo Alto, CA, USA, 1978.

# COMVIS: A communication framework for computer vision

Alex Cozzi, Florentin Wörgötter

Institute of Physiology, Department of Neurophysiology, Ruhr-University
Bochum, Germany

### Abstract

We present an advanced computer vision system based on the integration
of different visual modules. The data fusion procedure is based on a
representation of the scene in terms of planar patches. The applications
of state-of-the-art algorithms allows to fuse correlated information while
guaranteeing conservative error estimates.

Extensive test on synthetic and real images show that our system
produces a consistent and marked improvement over the results of single
visual modules, with error reduction up to a factor of ten and with typical
reduction of a factor 2–4.

## 1 Computer Vision

A visual module is an algorithm to analyze a particular kind of information from
one or more images, specialized to detect particular features in the images and,
on that basis, to infer the presence of significant structure in the scene (e.g.,
edge detectors, stereo modules, etc.).

The specialization of visual modules is necessary to keep visual processing at
a manageable level of complexity. On the other hand, isolated visual modules
have problems in dealing with the huge variability of real scenes, producing
erroneous or incomplete results.

The solution that we propose consists of the development of more advanced
computer vision systems based on the integration of heterogeneous visual modules. The aim of this work is the design of a communication and data fusion
process able to merge visual information generated by independent visual modules with the goal to produce an improved, consistent reconstruction of the
viewed scene.

## 2 The Biological Background

In order to arrive at a consistent interpretation of a scene, different features that
apparently do not belong together have to be linked and interpreted as parts
of the same object. It has been suggested that synchronization processes can
underly such "feature binding" in the nervous system [3]. Nerve cells respond
with temporal activity patterns to a stimulus and it is known that certain
oscillatory patterns prevail during optimal stimulation. Much in the line of
the example above it has been shown that groups of nerve cells synchronize

their oscillations during stimulation with a common object, even if this object is partly obscured.

Such a synchronization mechanism, therefore, provides a common communication scheme, which in principle is understood throughout the whole nervous system. In this way modules that deal with quite different aspects of the visual scene can be bound together as soon as their activity is synchronous.

The application of such synchronization process in a computer vision system is possible and has been done, still it is so inefficient and uncontrollable to make it inconvenient for such applications.

# 3 The Communication Mechanism

The representation chosen as basis of the communication mechanism has to fulfill several requirements: (1) It has to be global in the sense that all modules must be able to use it. (2) It has to be optimal in the sense that it should encode the information with little redundancy. (3) It has to be unambiguous and (4) it has to be efficient, not requiring complicated pre-processing to be attained or complicated post-processing to be decoded.

A structure that fulfills all these requirement is the *planar patch*: a delimited planar surface. A planar patch consists of the parameters of its 3-D plane equation ($aX + bY + cZ = 1$) and the region that defines the points of the plane that belong to the planar patch. The planar patch is completed by a covariance matrix that describes the imprecision of the plane parameters, $\Sigma$. The availability of an estimate of how good the data are is critical to devise a rigorous data fusion mechanism.

Most of the visual modules produce information that is easily expressed in terms of 3-D plane equations. The covariance matrix is more difficult to derive. Nevertheless, it is possible to derive the covariance matrix even when the mathematical model of each algorithmic step performed by the visual module is not available [4].

The goal of the communication scheme is to merge the lists of planar patches generated by the different visual modules and produce an unified, minimal, consistent, complete and correct representation of the scene from the planar patches produced by different visual modules or by the same visual module on different inputs.

The central component of the COMVIS system is a communication mechanism between visual modules. In this perspective, two kind of information are to be communicated: (1) the parameters of the plane equations and (2) the image regions.

The communication process is divided into four main steps:

1. Collect all the planar patches generated by the different visual modules and transform the planes to the same coordinate system.

2. Identify the planar patches to be merged together.

3. Merge the plane equations.

4. Merge the image regions.

The first step is performed applying the 3-D geometry of coordinates transforms. To perform the second step—identify the planar patches to be fused—we need a way to measure how "near" are two planar patches. This is done defining a distance measure between planar patches. The distance measures that are used in COMVIS take in account only the parameters of the plane equation and the covariance matrix of a planar patch, ignoring the form or the size of its region. A distance measure that takes in account the covariances of the planes is the Mahalanobis distance:

$$(\mathbf{a} - \mathbf{b})^{\mathrm{T}} (\Sigma_{\mathbf{a}} + \Sigma_{\mathbf{b}})^{-1} (\mathbf{a} - \mathbf{b})$$

where $\mathbf{a}$ and $\mathbf{b}$ are the vectors of the plane parameters and $\Sigma_{\mathbf{a}}$ and $\Sigma_{\mathbf{b}}$ the respective covariance matrices.

The third step is the fusion the information about $\mathbf{a}$ and $\mathbf{b}$ together to yield a new estimate $\{\mathbf{c}, \Sigma_{\mathbf{cc}}\}$.

The most common data fusion algorithms compute a new estimate by a linear combination of the means and then analytically determine its covariance. The Kalman filter, for example, uses the linear update rule in the form

$$\mathbf{c} = \mathbf{W_a a} + \mathbf{W_b b} \tag{1}$$

Where $\mathbf{W_a}$ and $\mathbf{W_b}$ are the two matrices that weights the importance given to $\mathbf{a}$ and $\mathbf{b}$, respectively. Assuming $\mathbf{a}$ and $\mathbf{b}$ to be uncorrelated and choosing the weight matrices to minimize the trace of $\Sigma_{\mathbf{c}}$, the usual Kalman filter equations [6] are obtained.

The problem is that in real situations the measures $\mathbf{a}$ and $\mathbf{b}$ are usually correlated and their correlation grade can not be measured. A general solution to this problem has been recently proposed as extension of the Kalman filter paradigm: the *covariance intersection* algorithm [7].

Covariance intersection performs a convex combination of the means and covariances in inverse covariance space. The covariance intersection estimate is computed as

$$\begin{aligned} \Sigma_{\mathbf{c}} &= (\omega \Sigma_{\mathbf{a}}^{-1} + (1 - \omega) \Sigma_{\mathbf{b}}^{-1})^{-1} \\ \mathbf{c} &= \Sigma_{\mathbf{c}} (\omega \Sigma_{\mathbf{a}}^{-1} \mathbf{a} + (1 - \omega) \Sigma_{\mathbf{b}}^{-1} \mathbf{b}) \end{aligned} \tag{2}$$

where $\omega \in [0, 1]$. The estimate described by Eq. 2 can be proved to be consistent, for all possible cross-correlation grades and choices of $\omega$ [7].

This work represent to our knowledge the first application of covariance intersection in computer vision.

## 3.1 The Segmentation of the Regions

Markov random field (MRF) theory [5] is a powerful tool to model complex joint multivariate distributions that have some locality properties.

The goal of the segmentation step is to assign each pixel of the reference image to one and only one planar patch. The basis of the mechanism is the modeling of the problem in terms of the MRF framework by defining an appropriate potential functions that define a "good" segmentation.

A potential function limited to only to the first order potential can already produce good results if the luminance in the images of the scene changes enough to distinguish between true correspondences and casual ones. In most of the real scenes this is not the case, which leads to two kind of problems: either different assignments have the same potential, or their differences are under the noise threshold, and thus should not be taken as significant.

The solution is to exploit the correlation between neighboring pixels. Adjacent pixels belong most of the time to the same planar patch. This can be modeled by the the second order potential. It expresses the correlation of neighboring pixels by adding a cost when neighboring pixels are assigned to different patches.

An example of segmentation of a real scene is reported in Fig. 1.

An extensive testing of the techniques described in this paper has been performed with synthetic and real data [1, 2]. The results of these tests show that the COMVIS system is able to deliver precise and reliable measurements, producing a consistent and marked improvement over the results of single visual modules, with error reduction up to a factor of ten and with typical reduction of a factor 2–4 (Fig. 2).

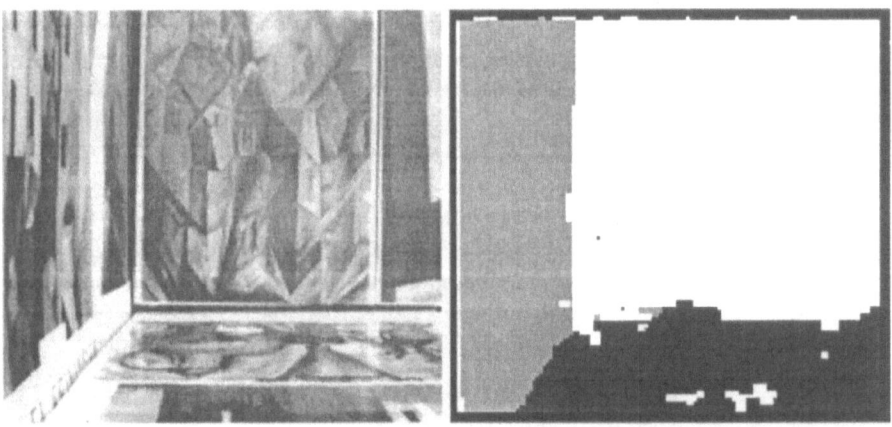

**Figure 1:** A real scene and the planar patches extracted by COMVIS.

# 4   Conclusions

This study extends in an evolutionary manner the standard approaches used in computer vision, building on the classical works of the field and proposing new solutions that still well fit into the current context of computer vision research. The approach here advocated integrates many recent research developments

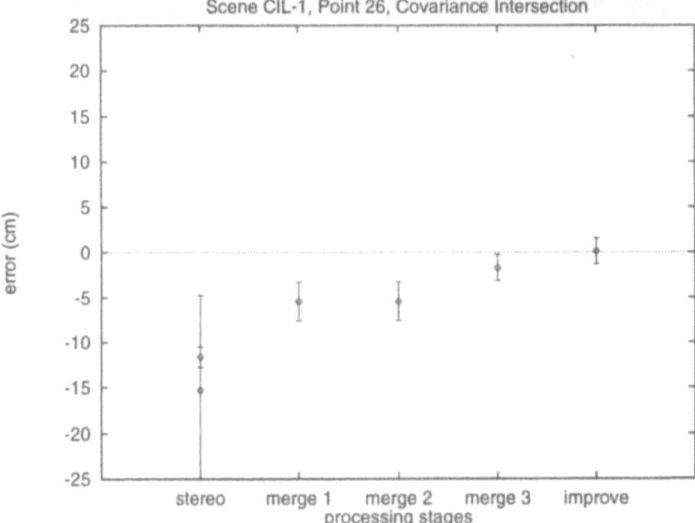

**Figure 2:** One image of the CIL-1 scene with the points whose real distance is known and the graph of the errors of the point 26 in the various processing phases. In phase "stereo" are reported the data points produced by three stereo modules, the "merge" phases are the results after merging the results of one, two and three modules, the "improve" phase minimize the mismatch error of a planar patch by changing the plane parameters.

844

(e.g. covariance intersection, error propagation, Markov random field theory) in a powerful and coherent unity. A goal of this work is to show the potential and the flexibility of the communication approach, providing a starting base for the development of more complex communication schemes in the form of a class library implemented with the Java programming language (it is freely available on the Internet at http://www.neurop2.ruhr-uni-bochum.de/java).

The planar patch has been previously used as a primitive to represent 3-D scene models. New in this work is its application to the problem of communication between visual modules and the pursuit of the possibilities offered by its simplicity to develop rigorous solutions to the problems of data fusion and error estimation in a computer vision system. In consequence it was possible not only to propose rigorous solutions to the problems, but also to prove their optimality—i.e., that no better solutions are available.

A characteristic of the communication schema proposed in this work is its generality: only very general assumptions are made about the nature of the visual modules and the scene being analyzed: any visual module that encodes its results in terms of planar patches can be easily integrated into the system. Most computer vision modules fit this description.

# References

[1] A. Cozzi, B. Crespi, F. Valentinotti, and F. Wörgötter. Performance of phase-based algorithms for disparity estimation. *Machine Vision and Applications*, 9(5-6):334–340, 1997.

[2] A. Cozzi and F. Wörgötter. Reclustering techniques improve early vision feature maps. *Pattern Analysis and Application*, 1998. In press.

[3] R. Eckhorn, R. Bauer, W. Jordan, B. M., W. Kruse, M. Munk, and H. Reitböck. Coherent oscillations: a mechanism of feature linking in the visual cortex? *Biological Cybernetics*, 60:121–130, 1988.

[4] R. M. Haralick. Propagating covariance in computer vision. In H. I. Christensen, W. Förstner, and C. B. Madsen, editors, *Workshop on Performance Characteristics of Vision Algorithms*, pages 1–12, Cambridge, U.K., April 1996. ECVNet.

[5] S. Li. *Markov Random Field Modeling in Computer Vision*. Springer-Verlag, Berlin, 1995.

[6] P. S. Maybeck. *Stochastic Models, Estimation, and Control*, volume 1. Academic Press, New York, 1979.

[7] J. K. Uhlmann. Covariance intersection home page, 1997. http://www.ait.nrl.navy.mil/people/uhlmann/CovInt.html.

# Implementing a Hybrid Architecture for Artificial Neural Network Applications

Sean Hill and Maia Wentland[1]
INFORGE - HEC, University of Lausanne
Lausanne, Switzerland
shill@iphysiol.unil.ch, mwentland@hec.unil.ch

## Abstract

We have constructed a prototype hybrid architecture consisting of a core neural network simulator tightly coupled with knowledge-based systems. An engine for performing numerical analysis is integrated with the knowledge-based systems, allowing a dynamic, computer-guided analysis. The user interactively defines the domain, problem and goal, while the software progressively refines the task using feedback, questions and data analysis. This process, while not requiring knowledge of neural networks, can help the user build a successful neural network application.

## 1  Introduction

With an increase in computerization and automation of both the corporate and research industries, ever-increasing quantities of complex data are generated. It may be helpful to use this data to identify patterns, categories or make predictions of a system. For example, the financial markets are complex dynamic systems whose behavior is not well understood. Making a prediction about the behavior of these markets could guide intelligent investment decisions [2]. As another example, categorizing electrophoresis gels to distinguish a sick patient from one who is healthy may rely on an analysis of complex patterns of proteins [1,6]. Many experts in a particular field, such as finance or medicine, do not have an expertise in data modelling with artificial neural networks (ANN). Our goal is to provide a program which offers a non-specialist the possibility of designing an ANN application without having ANN expertise.

This can be done by providing a tool, in the form of a hybrid symbolic-connectionist architecture, which contains domain knowledge for an application as well as neural network modelling. The system then guides the user through the whole design process, from data pre-processing to ANN outputs interpretation, as well as the ANN configuration and learning validation. It queries the user and automatically tunes itself according to the answers obtained.

For example, the user has a small numerical data set which he wants to use to build a neural network model. The tool identifies the number of dimensions and the statis-

---

[1] The authors gratefully acknowledge the support of the Swiss National Science Foundation (FNRS) Grant # 2129-47.026.96/1.

tical characteristics of the data automatically. Subsequently by using rules and heuristics gathered from the neural network literature and community, the appropriate neural network structure and parameters are estimated by an expert system. Following training, the performance of the model is tested and the user is queried to determine if the performance is satisfactory. If necessary the process is repeated with the user interacting with the system to answer basic questions about the data (such as: is this truly relevant data?).

We have constructed this hybrid architecture (HANNA) using a core ANN simulator tightly coupled with expert or knowledge-based systems (KBSs) [4]. Each KBS has a specific role in helping the user: choose input datasets, tune ANN parameters, interpretation of the ANN outputs, etc. The current program implements the architecture described in this paper, however the rules which populate the knowledge-bases remain to be developed. The knowledge to be formalized in the KBSs will stem from three principal sources: results and experience accumulated during the HAREM project [9]; existing literature; and new applications with which we will experiment.

# 2   Implementation

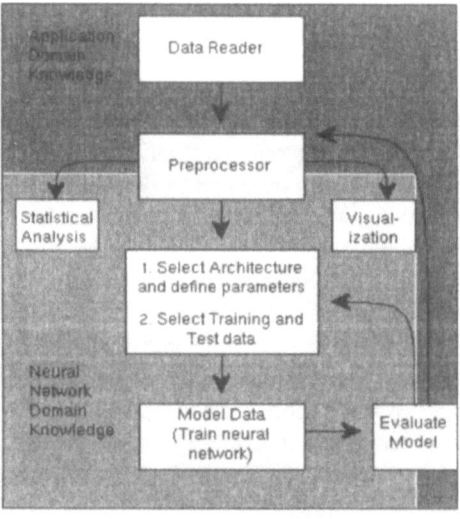

Figure 1: The general structure of the HANNA program. Each of the five main stages use knowledge-based systems to guide the user and the data processing. The gray regions show the influence of two separate knowledge domains on a neural network application.

The application (figure 1) consists of integrated KBSs, data analysis tools, and neural network engine (SNNS) [11]. As necessary, the KBS presents the user with a question to determine the next course of action. The responses to these questions as well as the analyses performed on the data build a global "context" dictionary on which further decisions (by the KBSs) are based.

The initial reading and cleaning of the data is guided by an application domain KBS (Reader) along with user input. After the data is cleaned and converted into numeric data, a series of analyses are completed which build a statistical description of the data for the next expert system (Preprocessor) to determine what preprocessing is necessary and which features of the data will be useful for discriminating classes or making predictions. This stage can profit from user input in identifying if the data is appropriate for the chosen learning task. The next stage is to design the actual neural network architecture (Modeler) to model the data using the SNNS neural network engine. After a training with a given set of parameters and neural network architecture, the Evaluator must choose an appropriate validation technique (at this time cross-validation is used) and measure the network performance.

An important part of this process is user-interaction, feedback and decision-making. This program alternately guides and is guided by the user during the preprocessing, training and evaluation phases. The expert systems can augment the domain knowledge of the user. The knowledge of the user and the ability to assist the computer in identifying good and bad design decisions is critical to the application of the program.

## 2.1 Automatic data-parsing and identification

The DataReader reads data from an ASCII plain text file and converts it into numerical vectors for further processing. It maintains information about the type of data and characteristics which are useful for the post-processing stage. For example, it is important to convert responses such as Yes, No, Maybe into numbers for processing. In this case the domain expert system "knows" about this set and converts the values to 1, -1 and 0 respectively. Similarly, the expert system recognizes and converts known date formats into numerical representations. If the system does not have a rule for converting a particular type of data, it prompts the user to guide it in assigning numerical values to the data.

An internal representation of the data is constructed at this stage which will be used to determine the preprocessing techniques to be applied. This representation records the history of transformations and user analysis of the data. For example when a piece of data is marked as a temporal value it can then be used for specialized preprocessing for detecting trends and cycles. Values such as dates are converted into numerical values by converting them to the number of seconds since a reference date. Other values can be explicitly identified as temporal dimensions as needed. Symbolic data is converted into user-defined values as well. Customized readers for other data types (such as images or databases) can be added.

## 2.2 An analytical expert-system guides preprocessing

An expert system guides the user and Preprocessor while cleaning the data and filling in missing values. The expert system can take advantage of the computational abilities of this program to perform preprocessing, analysis, and visualization of the data to aid the user in determining the preprocessing strategy. A new attribute may be specified (by a rule) in terms of a mathematical manipulation of one or more other attributes.

These results are added to the internal representation for each data attribute.

A high-performance versatile numerical mathematics engine, MATLAB from Math-works [5], has been integrated into this program. An object-oriented interface has been wrapped around Matlab allowing this functionality to be supplemented or replaced by a different computational engine.

Domain level rules, within the expert system, can implement the pre-processing strategy directly by calling Matlab. In the example that follows, the expert system has determined that creating a new dimension for the data, indicating whether a trade would be profitable or not, could be useful. This additional information may help train the neural network to predict when a stock should be bought or sold.

The two rules that were applied here exist in the application domain expert system in the following form:

```
BuySellRule =
{
    Condition = (diff(data) > 15 | diff(data) < -15);
    Action = (
        newdata(diff(data) > 15) =
            BuyValue.*ones(size(find(diff(data) > 15))),
        newdata(diff(data) < -15)) =
            SellValue.*ones(size(find(diff(data) < -15))),
    addDimension(newdata));
}
```

This rule determines if the numerical derivative (or difference) of data exceeds a certain magnitude. If so, then it recommends a buy or sell. It then adds a new dimension to the dataset consisting of this derived data. The parameter such as the threshold magnitude (here it is equal to 15) could be user specified to agree with the trading strategy of the user, or it could be provided by the output of another analysis rule. This rule has been applied to stock data from the Swiss Market Index and a graph is displayed to the user (figure 2). This oversimplified example illustrates the integration of preprocessing techniques directly into an expert-system. This same type of methodology can be applied to identify clusters of data, estimate the complexity of samples, and use heuristics to define a neural network architecture and parameters. Visualization can be a very useful part of preprocessing and analysis.

The expert system is able to perform the preprocessing itself as described by rules written to include data processing directives. In the above example, the Matlab language is used to describe the conditions and data manipulations. Additional methods may be easily added by writing routines in Matlab or adding a pre-written toolbox of code. In order to aid in visualization and the user decision-making process, statistical routines provide feature detection and visualization of the data. These methods consist of several class separation algorithms discussed by Ripley [7], and a nonlinear dimension-reducing Sammon mapping [8]. These will be extended to include projection pursuit and principal component methods [7].

## 2.3 Building a neural network model

The KBS in Modeller uses the global "context" to make decisions about the type of neural network architecture and how many inputs, hidden units, and outputs are to be

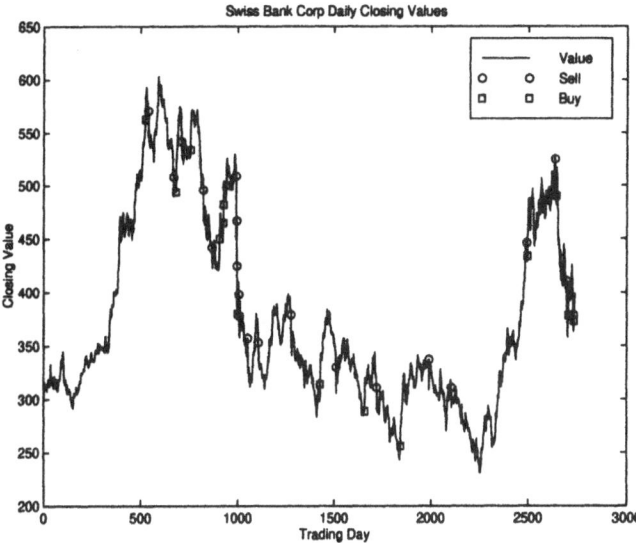

Figure 2: Stock market values superimposed with signals generated by the rules in the expert system. *Buy* indicates that the system will increase by at least 15 points in the next step. *Sell* shows that the price will fall by at least that much.

used. It can suggest initial parameters or parameter ranges for training. A sample of the data is selected and set aside for testing purposes. The SNNS neural network engine gives the flexibility to use the most common neural network architectures by defining a code module which will create the appropriate architecture via calls to the SNNS kernel [11]. The knowledge-base contains information concerning the particular neural network architecture, the types of categorization and modelling it can perform and the required inputs and format of the outputs. In our case we use two different types of networks. The first is a backpropagation network with which one can define the number of inputs, hidden units and outputs. The other is a Kohonen network where the dimension of the input and the number of units used to form a representation map is required [3]. These different types of network require different heuristics which are provided by knowledge-bases associated with each code module.

## 2.4 Evaluating performance

The evaluation stage uses a combination of questions to the user, a knowledge-base and cross-validation [7] to determine if the performance level of the model is adequate for the current application. Depending on the range of errors produced, the network may need to be adjusted (e.g. add or remove hidden units), parameters changed (e.g. alter the learning rate) or if the model is not working at all, the program should return the user to the original preprocessing stage where the preprocessing strategy and selected features and available data can be re-evaluated by the KBS.

This module is a key part of a revision loop where both the program and the user provide feedback about the performance of the model and make revisions to the architecture, data and learning process as needed. The interface presents the questions graphically showing the performance of the network and posing questions to the user which help in determining how to revise the network or training procedure.

# 3 Future Work and Conclusion

The goal now is to develop knowledge bases for two specific applications. The first application is in financial market prediction [9,10]. We will use five years of stock market data for several companies listed on the Swiss exchange. With this data, the knowledge bases will be tuned for predicting profitable trades. The second application is to recognize the presence or absence of a particular protein or complex of proteins in electrophoresis gels [1,6]. The development of these knowledge bases are critical to the success of each application. However, this tool provides a means of supplementing the experience of the user with knowledge accumulated in an expert system.

# References

[1] Appel R, Hochstrasser D, Funk M, Vargas R, Pellegrini C, Muller, AF, Scherrer J-R. The MELANIE project - From Biopsy to Automatic Protein Map Interpretation by Computer. Electrophoresis 1991;12:722-735

[2] Doboeck GJ, editor. Trading on the edge: Neural, Genetic, and Fuzzy Systems for Chaotic Financial Markets. John Wiley & Sons, Inc, Finance Editions, USA, 1994

[3] Kohonen T. Self-Organization and Associative Memory, 3rd edition. Springer-Verlag, Paris, 1989

[4] Michie D, Spiegelhalter DJ, Taylor, CC, editors. Machine Learning, Neural and Statistical Classification. Ellis Horwood, Series in Artificial Intelligence, UK, 1994

[5] Matlab, see http://www.mathworks.com

[6] Pun T, Hochstrasser D, Appel R, Funk M, Villars V, Pellegrini C. Computerized classification of two-dimensional gel electrophoretograms by correspondence analysis and ascendant hierarchical clustering. Applied and Theoretical Electrophoresis 1988;1:2-9

[7] Ripley BD. Pattern Recognition and Neural Networks. Cambridge University Press: Cambridge, 1996

[8] Sammon, JW. A nonlinear mapping for data structure analysis. IEEE Transactions on Computers 1969; C-18(5):401-409

[9] Simillion F, Hennebert J, Wentland M. From Prediction to Classification: The Application of Pattern Recognition Theory to Stock Price Movements Analysis. In: Proceedings of the 2nd SIGEF Congress, Spain, 1995

[10]Simillion F, Wentland M. Towards a Hybrid Architecture for Neural Network based Applications in Finance. University of Lausanne, submitted for publication in the Polish Journal of Operations Research (special issue on Financial Applications of Neural Networks), 1996

[11]Stuttgart Neural Network Simulator, User Manual, Version 4.1 Report No. 6/95, Institute for Parallel and Distributed High-Performance Systems (IPVR), University of Stuttgart, Breitwiesenstrasse 20-22, 70565 Stuttgart, Fed. Rep. of Germany

# Application of ANN to the Selection of a Valve from the Catalogue

Szymon Grymek

Department of Machine Design, Technical University of Gdansk
Gdansk, Poland
sgrymek@sunrise.pg.gda.pl

Tomasz Kiczkowiak

Department of Control Systems, Technical University of Koszalin
Koszalin, Poland
kiczko@lew.tu.koszalin.pl

**Abstract**

For many years we have worked on CAD software in our departments. In 1988 there was completed a software package for pneumatic systems [1], and in 1995 we started to novel it. It has been financially supported by the Polish State Committee for Scientific Research.

Within the system, there is a module for aiding calculations and supporting selection of pneumatic elements from catalogues. Analysis that has been done at a start on traditional computing algorithms has proven, that they are not adequate and not useful, especially those concerned control valves. To the contrary, it seems possible to use ANN to augment the selection of control valves from catalogue [2].

Below, an assessment of the traditional algorithms is presented, on requirements imposed by CAD systems. Problems derived by computer implementation of traditional methods are shown and it is discussed how they were coped. What more, first results of ANN applications to the selection of control valves are presented. Further goals are discussed as well.

## 1 Methods of calculation and selection of control valves

Table 1 Range of validity and parameters calculated for considered methods

| Algorithm | NM | F | P. diameter [m] | | Pressure [MPa] | | Velocity [mm/s] | | |
|---|---|---|---|---|---|---|---|---|---|
| | | | min | max | min | max | min | max | |
| Festo | Y | N | 0.012 | 0.25 | for 0.6 | | 6.2 | 1e3 | dk |
| CPOAC[1] | Y | N | | | | | | | Kv, dk |
| Combined | N | Y | | | | | | | Qn |
| Lucifer | Y | N | 0.025 | 0.7 | 0.1 | 2.0 | 0.2 | 1e5 | Kv |
| Martonair | N | Y | | | | | | | Kv |

---

[1] In the CPOAC algorithm, the range of input data is not directly given, but the stroke time estimated on the average velocity and the stroke is not to exceed 20 s.

For various methods, different parameters are used for selection, and this may be [3]: manifold orifice diameter *dk*, volume flow rate of an air at the standard conditions *Qn*, valve flow coefficient *Kv* (VID/VDE 2173) and coefficient of flow *C* (dcm³/s/bar).

Under consideration there have been algorithm of MARTONAIR, FESTO, CPOAC, LUCIFER, as well as the COMBINED algorithm [3]. In the present paper, details of methods are not included, as they may be found in the handbooks or other publications, for example [3]. The ranges of validity, as well as parameters used, are given in Table 1.

## 1.1   How the methods meet requirements of CAD systems

Traditional methods of calculations of pneumatic elements are of various modes, as nomograms (NM), tables, formulas (F) or combinations (see Table 1). They were devised for simple and easy computations of necessary parameters. However, such approach may be contradicted to their usefulness for CAD programmes.

Computer needs mathematical formulas, and it can calculate an algebraic expression, with a great but limited accuracy. In the same time, nomograms and tables may be adopted in CAD systems if data code can be transformed into analytical formulas.

Draw–backs of existing methods are:

1. Variety of types of resulted parameters, what makes that results obtained of various methods are not directly comparable.

2. In general, the coefficient *Kv* is not given in catalogues, which one is mostly calculated in algorithms (Table 1). The only one common variable is the manifold orifice diameter *dk*. On catalogues data, the flow rate *Qn* may be calculated (sometimes it is given).

3. Different types of input variables for calculations (Table 1) and of ranges, what makes that the effective control of a given data is difficult in CAD systems.

Only two algorithms: COMBINED and that of the MARTINAIR Co. are based on mathematical formulas, so they are capable to be directly used in CAD systems. Remaining algorithms use nomograms and tables of data, thus they need some approximations, for example:

1. In the FESTO algorithm to determine economy of the selection there are tables linked by nomograms. In the module that we propose the method point–to–point calculations is applied, where the nearest greater values of the input parameters are searched. Additionally, the average velocity of the piston is computed in regard to the supplying pressure.

2. An experiment of approximation of nomograms of the LUCIFER Co. with application of the fluid flow theory was not successful, thus in a computer algorithm a geometrical approximation of diagrams was used.

3. An experiment of approximation of nomograms of the CPOAC Co. with application of the fluid flow theory was not successful. A numerous set of complicated curves makes that the accuracy of an approximation was not satisfied.

## 1.2 Assessment of the calculation methods

Starting to the computer implementation of the calculation methods and the selection of control valves, the question has arisen: are these methods capable for the calculation and the selection of control valves from catalogues? To answer, next uncertain points have arisen:

1. The diameter of the manifold orifice $dk$ does not describe the flow characteristic of a valve; hence it seems, that this variable is not to be searched for.

2. Nomograms in vendor's catalogues, which take into account the flow characteristics, are not capable to be approximated by the flow theory. A judgement may be made, that the nomograms code experimental data, but there are no information about it.

3. Majority of nomograms has been devised long time ago, so a question of adequacy for new types of valves is valid, especially with different diameter $dk$.

4. The nomograms were prepared for specific data, for example for a concrete value of the supplying pressure. Usually there is no information how data may be extended for different values.

5. It is observed that for common conditions the orifice diameter of a control valve is similar to that of a cylinder.

Of another nature is a problem of methods accuracy, which may be only by real experiments verified, as well as the convergence of results obtained by various methods. There are reports [3] about substantial divergence, for some situations. Perhaps the reason is that the methods were used out of their ranges of validity.

## 1.3 Conclusions

A conclusion may be derived, that a new method for calculations and selection of control valves is needed, which is easy to be applied in CAD systems. The method should take under consideration the gas flow characteristic of a valve and as many parameters concerned as possible. It should be valid in a broad scope of parameters values.

## 2 Conception of ANN application

Having analysed the real calculations and selection methods, as well as the vendor's catalogues, the variables are proposed, as follows:

1. **Independent variables** (input signals to the ANN): cylinder diameter, piston diameter, force load on the piston, mass of moving elements (of a piston, of a rod and an additional one), stroke of a piston, supplying pressure and time of a stroke (as a result of simulation).

2. **Output variables** (output signals from the ANN): coefficient of the gas mass flow rate [4], supplying orifice diameter, time delay of the valve.

## 2.1 Training data set and selection of ANN structure

It has been decided that learning data set is to be created by the computer simulation, based on the E. W. Gerc model. The model is described in details in [4]. In the current phase of research, a heat flow between the gas in chambers and in ambient atmosphere is omitted, as well as the influence of the temperature on the displacement of the piston. From the extended model, three differential equations were adapted: for gas pressure in the first chamber, in the second one and for the displacement of the piston and additional elements. The mathematical model was implemented in the MatLab/Simulink package environment. Input data do not cover all the scope, but only possible and existing links of ranges.

The basic objective of ANN is an approximation of relations between the control valve parameters and the cylinder parameters, and of specifications for the elementary system (the cylinder with the valve). Thus the feed–forward ANN structure (the multi––layer perceptron) has been applied [5].

From the analysis of theoretical relations, it was derived that the function, which is to be approximated, is a continuous one. So the ANN has one hidden layer with non––linear neurons of hyperbolic tangent function, and the output layer with linear neurons.

As a result of simulation experimentation a 4571 vectors set of data was obtained. 173 vectors were removed because of errors. A test subset was created of 48 randomly chosen vectors. Finally, subset of 4350 vectors was a real training data. The set was normalised and scaled into the scope (0—1).

## 2.2 Training process

A goal of the preliminary training was to assess the hidden layer neurons number.

1. The experiment started with the classical error back propagation procedure, but no reasonable result was obtained: the learning process did not even start; modifications of the learning rate coefficient did not result.

2. Learning procedure was extended by the momentum term technique, but it did not resulted, either.

3. For the Quick Propagation algorithm, it was successful to push the net out the starting point, but the convergence of the learning process was poor, anyway.

4. Satisfying learning process was obtained for the error back propagation algorithm, based on the Levenberg–Marquardt method.

5. Watching the process for 30 epochs, it was stated that in the hidden layer 15 to 18 neurons is necessary.

6. The learning process was extended to 100 epochs; after analysis 16 neurons in the hidden layer were taken.

7. Training was completed of 700 epochs, then the testing was done.

Table 2 illustrates the progress of learning process. For the starting 100 epochs SSE (Sum Square Error) and MSE (Mean Square Error) decrease rapidly.

Table 2 Changes of error during learning of chosen ANN

| Epoch | 0 | 100 | 200 | 300 | 400 | 700 |
|---|---|---|---|---|---|---|
| SSE | 4082.90 | 3.70 | 2.66 | 2.23 | 2.22 | 2.06 |
| MSE | 0.93860 | 0.00085 | 0.00061 | 0.00051 | 0.00051 | 0.00047 |

## 2.3 Results

Some tests were completed to determine the quality of the ANN, especially its capability for generalisation. The total Square Error and the Mean Square Error for the test data set is given in the Table 3. It is shown that the error decreases for the testing set, also, what suggests a possibility of an effective generalisation capacity of the net.

Table 3 Errors for test set

| Epoch | 100 | 200 | 300 | 400 | 700 |
|---|---|---|---|---|---|
| SSE | 0.0504 | 0.0402 | 0.0343 | 0.0336 | 0.0294 |
| MSE | 0.00105 | 0.00083 | 0.00071 | 0.00070 | 0.00061 |

Percentage vectors fractions in the learning and in the testing sets were computed, where the absolute value of the error exceeds 3, 5 and 10%, for sequential stages of the training process. It may be concluded, that a small number of elements in the testing set and a greater number of errors (above 5%) suggest an inferior generalisation capacity of the net. However, there are not errors exceeding 10%. In the same time it may be stressed, that:

- In the testing set there are vectors very much different that those in the training set, so while they have been correctly interpreted it is a proof of a good generalisation capacity of the net.

- The 6.2% of errors (for the trained net) are only 3 cases within 48.

While assessing results, a 5% error was accepted as a satisfying, because in the training set there was a considerable noise of data, which is a result of a rounding up the values and of the numerical integration procedure at the end of the piston stroke. Due to that the piston stops a bit further then the proper declared stroke. The difference was related to values of cylinder parameters.

Table 4 Percent participation of vectors with different absolute error

| Error limit | Relative (%) vector number of epochs after number of epochs | | | | | | | | | |
| | 100 | | 200 | | 300 | | 400 | | 700 | |
|---|---|---|---|---|---|---|---|---|---|---|
| 3% | 25.60 | 29.1 | 18.80 | 18.7 | 14.70 | 16.6 | 14.40 | 16.6 | 13.40 | 12.5 |
| 5% | 8.70 | 12.5 | 5.20 | 10.4 | 4.20 | 8.3 | 4.10 | 8.3 | 3.60 | 6.2 |
| 10% | 0.55 | 0.0 | 0.25 | 0.0 | 0.22 | 0.0 | 0.25 | 0.0 | 0.160 | 0.0 |

# 3 Conclusions

From an analysis of the traditional algorithms of calculations and selection of the control valves it may be concluded, that a new method is necessary, which meets requirements listed in the paragraph 1.3.

The authors believe, that the ANN (Artificial Neural Network) may be an adequate tool, what was proved by preliminary results reported in this paper. Assuming the mathematical model error does not exceed 3%, and the computing error of the ANN does not exceed 5%, it may be stated that even now at the beginning development phase, the proposed technique is better than any commercial method.

As opposed to the classic methods, this technique:

- Considers substantial parameters and requirements of the basic subsystem.

- Takes into consideration the distribution of the velocity, and not the average value of the velocity.

- Does not require any additional calculations to determine the real time of displacement.

Development efforts are continued. What is planned, is:

1. To increase the training set of data gained from computer simulation.

2. To device our own implementation of numeric procedures for integration to reduce the error of a rounding up the values.

3. To apply RBF net (Radial Basic Function) and ALN net (Adaptive Logic Network); it may be noted, that for the selection problem a net is taught not on the data derived from simulation but the reverse relation, and an application of the ALN type net should make the training easier.

4. To check the possibility of a classical approach.

5. An experimental verification on real basic subsystems (cylinders with control valves).

# References

[1] Tarnowski W. Foundations of engineering design, WNT, Warszawa, 1997 (in Polish)

[2] Grymek Sz, Kiczkowiak T. Artificial neural networks as a tool in design of pneumatic driving systems, ZN Faculty of Mechanical Engineering Technical University of Koszalin No: 22, Koszalin, 1998

[3] Milanowski J, Kiczkowiak T. Pneumatic driving and control systems, Technical University of Koszalin, Koszalin, 1991 (in Polish)

[4] Gerc E.W. Dinamika pnevmaticzeskich sistiem maszin, Maszinostroienie, Moskwa, 1985

[5] Kosko B. Neural network and fuzzy systems, Prentice–Hall International Editions, Englewood Cliffs, USA, 1992

# A Neural Network Approach to Functional MRI Pattern Analysis — Clustering of Time-Series by Hierarchical Vector Quantization

Axel Wismüller[1], Dominik R. Dersch[2],
Bernadette Lipinski[3], Klaus Hahn[1], and Dorothee Auer[3]

[1]Institut für Radiologische Diagnostik,
Ludwig-Maximilians-Universität München,
Klinikum Innenstadt, Ziemssenstr. 1, D-80336 München, Germany
email: Axel.Wismueller@physik.uni-muenchen.de

[2]Integral Energy Corp., Sydney, Australia

[3]Max Planck Institute of Psychiatry, Munich, Germany

## Abstract

In this paper, we present a neural network approach to hierarchical unsupervised clustering of functional magnetic resonance imaging (fMRI) time-sequences of the human brain by self-organized fuzzy minimal free energy vector quantization (VQ). In contrast to conventional model-based fMRI data analysis techniques, this deterministic annealing procedure does not imply presumptive knowledge of expected stimulus-response patterns, and, thus, may be applied to fMRI experiments in which the time course of the stimulus is unknown like in spontaneously occurring events, e.g. hallucinations, epileptic fits, or sleep. Moreover, as minimal free energy VQ represents a hierarchical data analysis strategy implying repetitive cluster splitting, it can provide a natural approach to the subclassification task of activated brain regions on different scales of resolution with respect to fine-grained differences in pixel dynamics.

## 1  Introduction

fMRI experiments induce spatio-temporal patterns of changing imaging properties in the human brain. Interpretation of these patterns as a response to a given experimental stimulus is the key problem of fMRI data analysis. Model-based approaches like cross-correlation techniques are commonly used to perform this task. However, as they imply presumptive knowledge of expected stimulus-response patterns, they may sometimes fail in unveiling complex signal changes, thus discarding valuable information about the fMRI signal. Moreover, in fMRI studies of spontaneously occurring events like hallucinations, epileptic fits, or sleep, even the exact time course of the stimulus is unknown.

Unsupervised clustering techniques offer a powerful strategy to overcome these problems. In this context, different vector quantization (VQ) algorithms have been proposed for a wide scope of biomedical signal processing problems including fMRI data analysis [11]. Here, the time-sequences of pixel grey values obtained from fMRI experiments can be interpreted as feature vectors representing a multidimensional probability distribution. VQ procedures map a data space onto a finite set of prototypical feature vectors, a so-called codebook. Examples of this class of algorithms are Kohonen's self-organizing maps (SOMs) [8], minimal free energy VQ [10], [5], [4], [2], and the 'neural gas' algorithm [9].

The mathematical properties of these algorithms, their motivation from statistical mechanics, as well as their strengthes and weaknesses in the field of biosignal analysis have been thoroughly investigated in the literature (see e.g. [2], [11]). SOMs have already been applied to fMRI data analysis [7]. Minimal free energy VQ is a deterministic annealing procedure minimizing the free energy of a multiparticle system in analogy to a canonical ensemble tending towards thermal equilibrium. In contrast to SOMs, this algorithm offers a specific advantage for practical data analysis problems: it provides a *hierarchical* clustering scheme *on different scales of resolution* [2], [3].

In the field of fMRI time-sequence analysis, this offers a convenient method for subclassification of activated areas according to similarities in signal time-sequences. At the same time, heuristic manual merging of pixel clusters belonging to different codebook vectors like in the SOM approach can be avoided. Merging into larger meta-clusters can be replaced by back-tracking the codebook hierarchy tree to an earlier stage of the annealing procedure. Thus, the coarse-grained structure of the data set can be explored in a natural manner.

## 2  Theory

Let $n$ denote the number of subsequent scans in a fMRI experiment. The dynamics of each voxel $i$, i.e. the sequence of grey values $x_i(t)$ over all scan acquisition time spots $t$ can be interpreted as a vector $\mathbf{x}_i \in \mathbb{R}^n$ in the $n$-dimensional feature space of possible fMRI signal time-sequences. Clustering identifies groups $k$ of pixels with similar dynamics. These groups are represented by prototypical time-sequences called codebook vectors $\mathbf{w}_k$. Soft-competing VQ procedures determine these cluster centers by an iterative adaptive update according to

$$\mathbf{w}_k(t+1) = \mathbf{w}_k(t) + \epsilon a_k(\mathbf{x}_i(t); W(t), \kappa)(\mathbf{x}_i(t) - \mathbf{w}_k(t)), \tag{1}$$

where $\epsilon$ denotes a learning parameter, $a_k$ a so-called cooperativity function which, in general, depends on the codebook $W(t)$, a cooperativity parameter $\kappa$, and the presented feature vector $\mathbf{x}_i$ itself. In the fuzzy clustering scheme proposed by Rose, Gurewitz, and Fox [10], the cooperativity function $a_k$ reads

$$a_k(\mathbf{x}_i; W, \kappa \equiv \rho) = \frac{\exp(-E_k(\mathbf{x}_i)/2\rho^2)}{\mathcal{Z}} \tag{2}$$

Here, the 'energy' $E_k(\mathbf{x}_i) = \|\mathbf{w}_k - \mathbf{x}_i\|^2$ measures the distance between the codebook vector $\mathbf{w}_k$ and the data vector $\mathbf{x}_i$. $\mathcal{Z}$ denotes a partition function given by $\mathcal{Z} = \sum_k \exp(-E_k(\mathbf{x}_i)/2\rho^2)$ and $\rho$ is the cooperativity parameter of this model. That so-called 'fuzzy range' $\rho$ defines a length scale in data space and is annealed to repeatedly smaller values in the VQ procedure. The learning rule (1) with $a_k$ given by (2) describes a stochastic gradient descent on the error function

$$F_\rho(W) = -\frac{1}{2\rho^2} \int P(\mathbf{x}) \ln \mathcal{Z} d^n x, \tag{3}$$

which ia a free energy in a mean-field approximation [3]. Here, $P(\mathbf{x})$ denotes the probability density of feature vectors $\mathbf{x}_i$. For the minimal free energy VQ procedure, the codebook vectors mark local centers of this multidimensional probability distribution. Thus, for the application to fMRI signal analysis, the codebook vector $\mathbf{w}_k$ is the weighted average fMRI signal of all the time-sequences $\mathbf{x}_i$ belonging to group $k$ with respect to a fuzzy tesselation of the feature space.

In contrast to SOMs, minimal free energy VQ

(i) can be described as a stochastic gradient descent on an explicitly given energy function (see (3)), [10],

(ii) preserves the probability density without distortion [5], [6], and, most important for practical applications,

(iii) allows hierarchical data analysis on different scales of resolution [4].

In the beginning of the VQ process, there is only one cluster representing the center of the whole data set. As the deterministic annealing procedure continues, phase transitions occur and large clusters split up into smaller ones marking increasingly smaller regions of the feature space. Tracing this repetitive cluster splitting through the whole VQ procedure leads to a 'genealogy' of cluster centers, i.e. a resemblance tree of codebook vectors. Thus, the manual merging of cluster centers into larger meta-clusters like in the SOM approach to fMRI analysis can be avoided. At the same time, the scope of resolution can be adapted according to the observer's needs. The similarity of different codebook vectors can easily be derived by back-tracking the clustering tree. The procedure can be monitored by various control parameters like the free energy, entropy, reconstruction error etc. which allow an easy detection of cluster splitting [5].

# 3 Methods

Functional imaging was performed on a 1.5 T system (Signa, General Electrics, Milwaukee) using a GI-EPI sequence (TR/TE = 4,000/66 msec) with 8 slices and 64 images per experiment. Resolution was 3x3x4 mm, and three periods of photic stimulation (8 Hz alternating checkerboard, central fixation point) were interleaved by four control periods (dark background, central fixation point). The first scan was discarded from analysis for remaining saturation effects. Movement artifacts were compensated by automatic image alignment (AIR software, [12]).

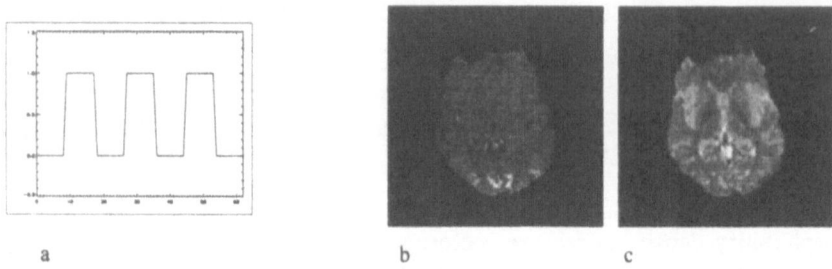

Figure 1: (a) Stimulus. (b) Cross-correlation image. (c) Anatomical image.

Average-corrected time-sequences of each pixel were clustered by minimal free energy VQ employing 30 codebook vectors. The results were compared with classical cross-correlation images (e.g. [1]).

# 4 Results

Fig.2 shows a part of the hierarchical clustering tree covering two subsequent VQ steps of the deterministic annealing procedure. Fig.2a presents one of 17 cluster centers present at the observed stage of VQ. Note the apparent similarity compared with the stimulus (fig.1a). Fig.2d shows all the pixels belonging to this cluster center according to a minimal distance criterion in the metric of the time-sequence feature space. The highlighted regions can be attributed to the visual cortex. They clearly correspond to the activated regions in the cross-correlation image (fig.1b).

Now a phase transition occurs in the subsequent VQ step, and the cluster of fig.2a splits up into two descendant clusters representing smaller regions of the visual cortex with different pixel dynamics. They are presented in the lower part of fig.2. Note that the sum of the activated areas in fig.2e and fig.2f is greater than the area of the cluster in fig.2b. This is based on a reduction in local reconstruction error due to the fact that the new codebook structure better fits the underlying local probability density. Thus, the descendant clusters can take over pixels which formerly were attributed to adjacent codebook vectors.

# 5 Discussion and Conclusion

The study shows that deterministic annealing by the minimal free energy VQ is a useful strategy for the analysis of fMRI data sets without presumptive knowledge of stimulus-response models or the stimulus function itself. In contrast to Kohonen's SOM algorithm, it realizes a hierarchical clustering procedure unveiling the structure of the data set with gradually increasing resolution. Therefore, heuristic manual merging of pixel clusters belonging to different codebook vectors like in the SOM approach can be avoided. Merging into larger meta-clusters can be replaced by back-tracking the codebook hierarchy

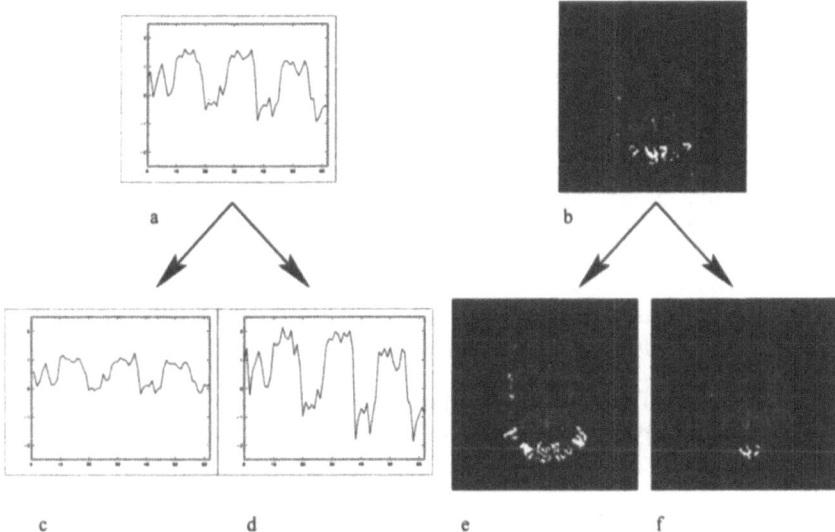

Figure 2: Part of the hierarchical clustering tree demonstrating cluster separation during deterministic annealing by minimal free energy VQ. (a) Cluster center before phase transition, i.e. cluster separation. (b) Corresponding pixel cluster before phase transition according to a minimal distance criterion. (c,d) Cluster centers after phase transition. (e,f) Corresponding pixel clusters after phase transition.

tree to an earlier stage of the annealing procedure.

Thus, the structure of the data set can be explored in a natural manner. Therefore, we recommend minimal free energy VQ as an alternative to SOMs for unsupervised fMRI data analysis. Especially, it may be helpful in situations where subclassification of activated brain regions on different scales of resolution is focused with respect to fine-grained differences in pixel dynamics.

## Acknowledgements

This work has been funded by grants from the Hanns-Seidel-Foundation and the German Federal Ministry of Science and Technology (BMBF).

# References

[1] P.A. Bandettini, A. Jesmanowicz, E.C. Wong, and J.S. Hyde. Processing strategies for time-course data sets in functional MRI of the human brain. *Magn. Reson. Med.*, 30:161–173, 1993.

[2] D.R. Dersch. *Eigenschaften neuronaler Vektorquantisierer und ihre Anwendung in der Sprachverarbeitung.* Verlag Harri Deutsch, Reihe Physik, Bd. 54, Thun, Frankfurt am Main, 1996. ISBN 3-8171-1492-3.

[3] D.R. Dersch, S. Albrecht, and P. Tavan. Hierarchical fuzzy clustering. In A. Wismüller and D.R. Dersch, editors, *Symposion über biologische Informationsverarbeitung und Neuronale Netze - SINN '95, Konferenzband.* Hanns-Seidel-Stiftung, München, 1996.

[4] D.R. Dersch and P. Tavan. Control of annealing in minimal free energy vector quantization. In *Proceedings of the IEEE International Conference on Neural Networks ICNN'94*, pages 698–703, Orlando, Florida, 1994.

[5] D.R. Dersch and P. Tavan. Load balanced vector quantization. In *Proceedings of the International Conference on Artificial Neural Networks ICANN*, pages 1067–1070. Springer, 1994.

[6] D.R. Dersch and P. Tavan. Asymptotic level density in topological feature maps. *IEEE Transactions on Neural Networks*, 6(1):230–236, 1995.

[7] H. Fischer, M. Buechert, and J. Hennig. Assessing the dynamics of fMRI data using self-organizing map clustering. In *Proceedings of the 5th SMR meeting*, 1997.

[8] T. Kohonen. The self-organizing map. *Proceedings of the IEEE*, 78(9):1464–1480, 1990.

[9] T.M. Martinetz and K. Schulten. A 'neural gas' network learns topologies. In *Proceedings of the International Conference on Artificial Neural Networks ICANN*, pages 397–402, Amsterdam, 1991. Elsevier Science Publishers.

[10] K. Rose, E. Gurewitz, and G.C. Fox. Vector quantization by deterministic annealing. *IEEE Transactions on Information Theory*, 38(4):1249–1257, 1992.

[11] A. Wismüller and D.R. Dersch. Neural network computation in biomedical research: chances for conceptual cross-fertilization. *Theory in Biosciences*, 116(3), 1997.

[12] R.P. Woods, S.R. Cherry, and J.C. Mazziotta. Rapid automated algorithm for aligning and reslicing PET images. *Journal of Computer Assisted Tomography*, 16:620–633, 1992.

# Penalized Training for Serially Correlated Data

R. A. Dunne

Victoria University of Technology

Melbourne, Australia

N. A. Campbell

CSIRO Mathematical and Information Sciences

Perth, Australia

### Abstract

This paper treats the problem of selecting appropriate penalization training methods in the case where the data consists of curves sampled at many grid points. Several methods are presented and discussed for treating such data and an illustrative example is presented.

## 1  The Multi-layer Perceptron (MLP) Model

The standard MLP model with two layers of adjustable parameters, $p$ inputs, $h$ hidden–layer units and $q$ output units, can be described by

$$\mathrm{mlp}(x_i, \Upsilon, \Omega) = f_q(\Upsilon[1, \{f_h(\Omega x_i)\}^t]^t), \qquad (1)$$

where $\Upsilon$ (of size $q \times h + 1$) and $\Omega$ (of size $h \times p + 1$) are the two matrices of adjustable weights or parameters, $f_q : R^q \to (0,1)^q$ and $f_h : R^h \to (0,1)^h$ apply the same one–variable "squashing" function to each of their coordinates.

A set of training data, $D = \{x_n, t_n\}_{n=1}^N$, is assumed to be available, where each $x_n$ is a feature vector of length $p$, augmented by the addition of a 1 in the first coordinate position and $t_n$ is an encoding of the class label as a target vector of length $q$.

For convenience, we write $y = \Omega x$ and $y* = [1, \{f_h(y_1, \ldots, y_h)\}^t]^t$, the $y$ vector being an argument to the function $f_h$, augmented by a 1 in the first coordinate position. This now forms the data input to the next, and in this case final, layer of the network. We then write $z = \Upsilon y*$ and $\mathrm{mlp}(x, \Upsilon, \Omega) = z* = f_q(z)$. Fitting the MLP model involves minimizing a penalty function $\rho\{z^*(x_n), t_n\}$, for which the derivatives with respect to the weights $\Upsilon$ and $\Omega$ are generally required.

## 2  Penalization

Due to the flexibility of MLP models and the ill-posed nature of many problems, additional constraints, in the form of an appropriate prior on the class of approximation functions, may be used to control the fitting process.

An additional assumption that is often made is one of smoothness, that is, it is assumed that $z^*$ is a smooth function and a term is imposed on $\rho$ to penalize roughness. In the context of MLP models, the term is often one that penalizes the magnitudes of the weights.

We explore particular constraints applicable in the case where the individual observations are spectra sampled at many points. Instead of regarding a datum as a point, $x_i$, in $R^p$, we model it as a smooth function, $X(t)$, plus additive noise. More complex models involving sums of orthogonal curves are considered in [6]. [4] discusses a discriminant function classifier based on similar arguments. Generally, when the grid is equally spaced, we map the range of $X$ to the integers $1, \ldots, p$ so that $X : N \to R$ and $x_i = X(i)$.

Consider the limiting case where we have two smooth functions of one variable, $s_1(x)$ and $s_2(x)$ – spectral curve 1 and spectral curve 2. We are looking for a function $\psi$ that discriminates between $s_1(x)$ and $s_2(x)$ in the sense that

$$\| \int \psi(x) s_1(x) dx - \int \psi(x) s_2(x) dx \|,$$

the magnitude of the difference of the inner products $\int \psi(x) s(x) dx$, is a maximum. A reasonable assumption about $\psi$ is that it will be a piecewise smooth function.

Now we would expect, for a sample from $X$, that the weights $\omega_i = \psi(i)$ should reflect this smoothness. While it is possible to draw a $C^\infty$ function through a given set of weights, $\omega_i$, $i = 1, \ldots, p$, we expect the weights to be visually smooth and not have the white noise appearance that may be associated with fitting $p$ independent parameters.

However, for some of the commonly used penalization terms, the rationale may be quite different to this. Weight decay [5] is the MLP equivalent of ridge regression; its effect is to smooth the boundaries of the decision region rather than to smooth $\psi(x)$. Weight elimination [8] is a variable selection technique, and assumes an (improper) prior of a Gaussian contaminated with a uniform distribution. It essentially divides the weights into two groups, those from the uniform distribution that are not equal to 0, and those from the zero-centered Gaussian that can be considered to be equal to zero and removed. This is likely to result in $\psi$ being a spikey function.

A better penalization procedure when the data are curves would be to ensure that the weight vector is smooth. In order to incorporate the smoothness constraint into the MLP, the unknown function $\psi$ can be modeled as

$$\psi \approx \sum_{k=1}^{K} \alpha_k B_k,$$

where $\{B_k\}_{k=1}^{K}$ is some set of basis functions. Basis functions examined to date include splines and truncated Gaussians, but clearly any set of convenient basis functions could be used. It appears preferable to use basis functions that have compact support so that the fitting procedure (iterative function minimization) only has to make local adjustments.

In this way, the number of estimated parameters will be reduced from $p+1$ to $K+1$ per hidden layer unit, where $K$ is the number of basis functions used to model $\psi$. We fit the parameters $\alpha_k$ by a function minimization scheme so that the model is still described by (1) but now

$$w_{ji} = \sum_{k=1}^{K} \alpha_{jk} B_{jk} \Bigg|_{\text{evaluated at } i.}$$

The $\alpha$'s are now the parameters optimized in the fitting procedure and a further penalty term, such as a weight decay term, can be imposed on them if necessary.

Another way to incorporate a serial smoothness constraint on the weights is to impose a penalty term incorporating the second forward difference operator. Let $D_1$ be a matrix that applies the first difference operator to a vector. Then

$$D_{1ij} = \begin{cases} 1 & i = j \\ -1 & i+1 = j \\ 0 & \text{otherwise,} \end{cases}$$

and $D_2$, the second difference matrix, is formed by

$$\underset{(n-2)\times n}{D_2} = \underset{(n-2)\times(n-1)}{D_1} \underset{(n-1)\times n}{D_1}$$

and

$$\underset{n\times n}{\Omega} = D_2^T D_2.$$

The appropriate penalty term is $\rho + \lambda w^T \Omega w$. Note that this generalizes ridge regression to give a finite approximation to

$$\rho + \lambda \sum_n \int \left( \frac{\partial^2 w(x)}{\partial x} \right)^2 dx. \tag{2}$$

[1] uses

$$\sum_n \| z_n^* - t_n \|^2 + \lambda \sum_n \int \left( \frac{\partial^2 z_n^*}{\partial x} \right)^2 dx \tag{3}$$

for a regression network with linear output units. [7] makes the comment that (3) might be used in classification problems by applying it to the total input to the function $f_q$.

Note that different and unrelated quantities are being smoothed in (2) and (3). In (2) the differential operator is applied to the weight vector and in (3) it is applied to the outputs of the MLP. It is possible for the output of the MLP to be smooth without the weights, considered as a sequence, being in any way smooth.

Based on this discussion, four suggested procedures are: fit a standard MLP with no weight decay penalty term (as the initial weights are selected close to zero, this produces a set of weights that are serially correlated, much like the data); smooth the weights after fitting the model; use a basis function expansion; and use the second forward difference penalty term. We investigate these strategies in the following example.

# 3    Example – AVIRIS data

AVIRIS is an acronym for the Airborne Visible Infrared Imaging Spectrometer. The spectrometer delivers calibrated images of the spectral radiance from 0.4 to $2.5 \times 10^{-6}$ meters in 224 contiguous spectral bands. Being contiguous, these bands define a continuous spectrum of sufficient detail to allow differentiation of the spectral signatures of many earth surface materials.

Given the number of bands and the fact that the training sets should be of unambiguous class ascription, training sets are likely to contain fewer pixels than the number of spectral bands. In this case, the classification task presents certain difficulties since MLPs and other flexible discriminant techniques may achieve a spurious 100% correct classification on the training data.

The data set is a $512 \times 614$ image of an irrigated agricultural region of the San Luis Valley, Colorado. A visual inspection reveals that potato and alfalfa have a significant overlap throughout their spectral range and so these two classes were chosen for a small-scale study. Because of the correlation between the spectral values of neighboring pixels [2], the correct units of data to consider when doing cross–validation are the sites rather than the observations. Each pair of sites (potato versus alfalfa) was used as training data in turn and tested on the remaining sites. The results, with a number of classifiers, are summarized in table 1.

Many of the comparator techniques require a parameter to be set. This is usually done by cross-validation or by integrating it out as a hyper-parameter. In order to remove the effect of estimating this parameter from the comparison, we have selected the $\lambda$ value that performed best on the test data. Clearly no parameter estimation procedure on the training data can do better than this.

As expected, linear discriminant analysis performs quite poorly and quadratic discriminant analysis fails as some of the training sites lead to singular covariance matrices. Nearest neighbor does somewhat better and the performance improves with the number of neighbors until $k = 55$. (However, the computational burden of checking that many neighbors makes this an unattractive technique.)

The standard MLP gives an error of 13% and the use of a weight decay or weight elimination penalty causes an increase in the error rate. However, use of any of the suggested smoothing techniques causes a drop in error rates. The post-training smoothing was done with "super-smoother" [3]. The basis function MLP, which was fitted using a set of truncated Gaussians with $\sigma = 2$ (ie two spectral bands) and 25 basis functions, gives an error rate of 4%.

# 4    Discussion and Conclusion

We note that all of the techniques achieved a zero error rate on the training data. In addition, we note that there is no obvious similarity between the sets of optimized weights shown in figure 1.

It is easy to find sets of weights that separate the training data, weight elimination appears to be using only two weights. However, due to the ill-posed nature of the problem, these sets of weights may not generalize well. This sort of problem would usually be approached by imposing a penalty term on the fitting process; however, as can be see from table 1, two of the commonly used penalty terms lead to a worse error rate than the standard MLP.

This illustrates the general point that penalization methods must be chosen to reflect the characteristics of the data set. In the case of serially correlated data it is argued that imposing a smoothness constraint, rather than a magnitude constraint, on the weights will lead to improved generalization.

The presented example demonstrates the utility of approaches based on smoothing the weight vector.

# References

[1] C. M. Bishop. Curvature-driven smoothing: a learning algorithm for feed-forward networks. *IEEE Transactions on Neural Networks*, 4:882–884, 1993.

[2] N. A. Campbell and H. T. Kiiveri. Neighbour relations and remotely sensed data. Internal report, Division of Mathematics and Statistics, CSIRO, Western Australia, 1988.

[3] J. H. Friedman. A variable span smoother. Tech. Rep. No. 5, Laboratory for Computational Statistics, Dept. of Statistics, Stanford Univ., California, 1984.

[4] H. T. Kiiveri. Canonical variate analysis of high–dimensional spectral data. *Technometrics*, 34(3):321–331, 1992.

[5] D. C. Plaut, S. J. Nowlan, and Hinton G. E. Experiments on learning by backpropagation. Tech. Rep. CMU-CS-86-126, Carnegie Mellon University, Pittsburg, 1986.

[6] John A. Rice and B. W. Silverman. Estimating the mean and covariance structure nonparametrically when the data are curves. *Proceedings of the Royal Society of London B*, 53(1):233–243, 1991.

[7] B. D. Ripley. *Pattern Recognition and Neural Networks*. Cambridge University Press, Cambridge, 1996.

[8] A. S. Weigend, D. E. Rumelhart, and B. A. Huberman. Generalization by weight-elimination with application to forecasting. In R. P. Lippmann, J. E. Moody, and D. S. Touretzky, editors, *Advances in Neural Information*

*Processing Systems 3. Proceedings of the 1990 Conference*, pages 875–882, San Mateo, 1991. Morgan Kaufmann.

| classifier | Training | Test |
|---|---|---|
| Linear discriminant functions | 0 | 0.4503704 |
| Nearest neighbor (k=1) | 0 | 0.2809877 |
| Nearest neighbor (k=52) | 0 | 0.2088889 |
| standard MLP | 0 | 0.1274074 |
| standard MLP (weight elimination, ($\lambda = 0.00004$) | 0 | 0.1308642 |
| standard MLP (weight decay, $\lambda = 1.2$) | 0 | 0.1471605 |
| smoothed MLP | 0 | 0.1182478 |
| second difference ($\lambda = 0.01$) | 0 | 0.1120988 |
| basis MLP ($K = 25$) | 0 | 0.0454321 |

Table 1: A comparison of several classification techniques on potato versus alfalfa classes. The error rates on the test and training sets are given.

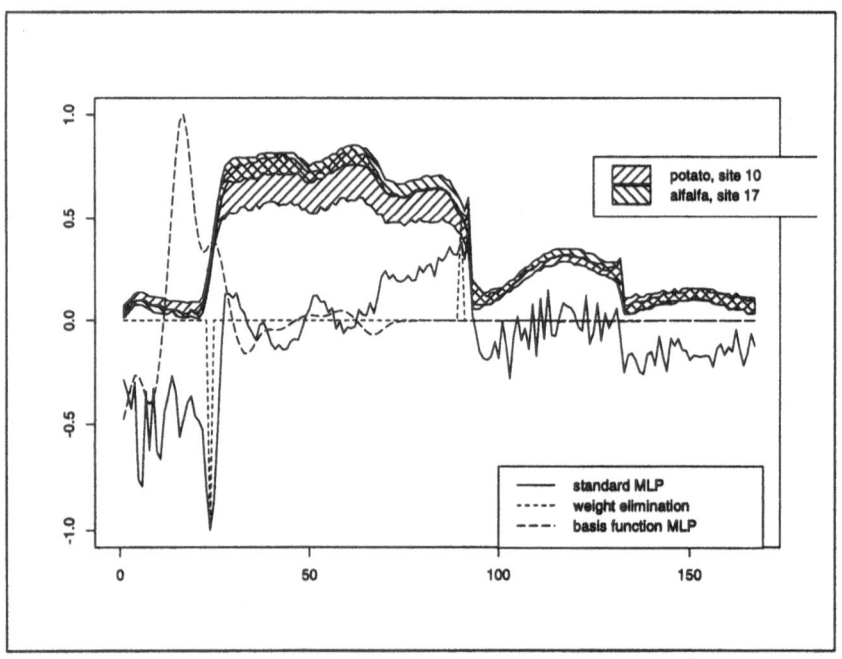

Figure 1: *This plot combines an overlay of the spectral envelopes for site 10 (potato) and site 17 (alfalfa) with the weights from: a standard MLP; an MLP fitted with weight elimination ($\lambda = 0.00004$); and a basis function MLP (K=25). The weights are not shown at their true scale. As can be seen, weight elimination pushes all but two significant weights to 0.*

# Gaussian Mixture and Kernel Based Approaches to Blind Separation of Sources Using Neural Nets

Mats G. Gustafsson

Signal and Systems, Uppsala University

Uppsala, Sweden

### Abstract

This work presents revised results from the work reported in [1] where a rigorously derived algorithm for blind separation of sources (BSS) based on Gaussian mixtures was presented and evaluated. Experimental results using the two first rigorously derived BSS algorithms are included for comparison and to illustrate the possibility to develop algorithms which allow more general source distributions.

## 1 Introduction

The standard blind separation of sources (BSS) problem is to recover (except for scaling and permutation) the original $N$ signals $s_n(t), n = 1, 2, \ldots, N$ when observing the $N$ signals $x_n(t) = \sum_j a_{nj} s_j(t), n = 1, 2, \ldots, N$. $A$ $(a_{ij})$ is the so called $N \times N$ dimensional mixing matrix. The first rigorously derived neural network based algorithms for BSS presented recently [2, 3] do not allow arbitrary source distributions. This work presents revised results from the work reported in [1] where a rigorously derived BSS algorithm based on Gaussian mixtures was presented and evaluated. During the anonymous review of the present paper, it has come to our knowledge that very closely related but independent work recently has been presented [4].

Using a single layer neural net with weight matrix $W$, and output $\mathbf{y}$ the new algorithm is derived by minimizing the mutual information $I(\mathbf{y})$ as in [3]. However, instead of approximating the marginal probability density functions (pdfs) using the so called Gram-Charlier expansion, they are approximated using Gaussian mixtures. Using the mixture model $f_i(y_i) = \sum_{k=1}^K a_{ik} \phi_k(y_i)$ where $\phi(y_i)$ is a Gaussian pdf the new algorithm can be summarized as

$$W(t+1) = W(t) + \eta_w (I - \mathbf{g}(\mathbf{y})\mathbf{y}^{\mathbf{T}})(W^{\mathbf{T}})^{-1} \tag{1}$$

where the components of $\mathbf{g}(\mathbf{y})$ can be written

$$g_i(y_i) = \frac{\sum_k a_{ik} \phi_k'(y_i)}{f_i(y_i)} \tag{2}$$

The mixture coefficients $a_{ik}$ are updated using a simplified maximum likelihood procedure as

$$a_{ik}(t+1) = a_{ik}(t) + \eta_a \phi_k(t) \tag{3}$$

A Parzen window approach has also been considered to avoid the need for adjustments of the mixture coefficients $a_{ik}$. Here the marginal pdfs are approximated directly using $P$ consecutive samples as $f_i(y_i) = (\sum_{p=1}^{P} \frac{1}{V_p} \psi((y_i - m_{iy})/v_p))/P$ where $V_p$ and $v_p$ are parameters to be set and $\psi(x)$ is typically a bell-shaped function. In compariosn with the Gaussian mixture approach, the main advantage is that the density estimates are obtained directly without adjustments of mixture coeffcients. The approach has been successfully tested on real data (music and speech) using a window with 200 consecutive samples (not reported here).

Here follows the results from two simple but illustrative experiments where the new method was compared with the algorithms in [1,2].

## 2   Experiment1

In experiment 1, we considered the problem of separating the sources $s_1(t) = n(t)$, $s_2(t) = 0.6\sin(600t)\cos(70t)$, and $s_3(t) = \mathrm{sign}[\sin(500t + 9\cos(50t))]$ where $n(t)$ is white noise (independent samples) with the amplitude uniformly distributed on the interval $[-1, +1]$. The signals are presented in Figure 1.

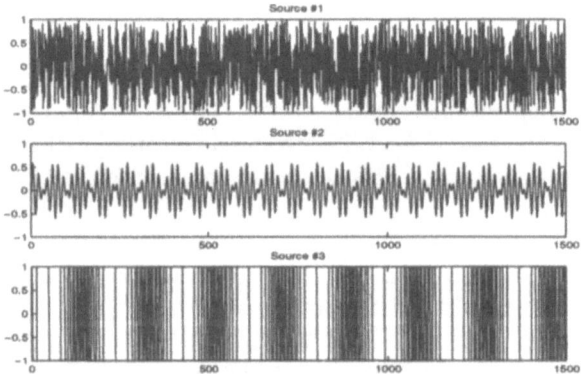

Figure 1: The source signals used in experiment 1.

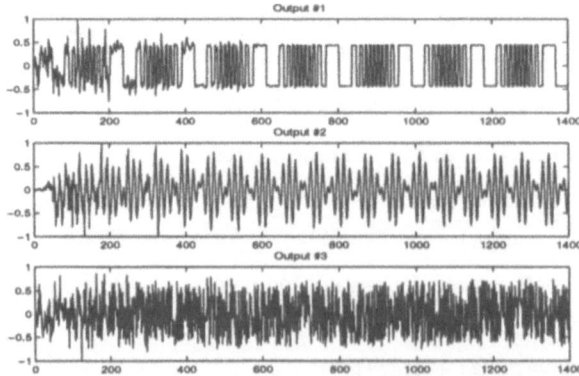

Figure 2: Results from experiment 1 using the algorithm presented in [3].

The sources were mixed using a matrix $A$ with elements drawn from the uniform distribution on $[-1, 1]$. These test signals were also used in [3]. One should note that $s_3(t)$ corresponds to a degenerate pdf consisting of a sum of two Dirac impulses centered at $-1$ and $+1$, $f_3(y_3) = 0.5\delta(y_3 + 1) + 0.5\delta(y_3 - 1)$. This makes the differential entropy $h_3(y)$ undefined (the integral diverges).

Moreover, such pdfs are difficult to model by means of a Gaussian mixture with few components. One should also note that the samples are not statistically independent since they have a common reference, the sampling time instant $t$. In particular, knowledge about the sample $s_3(t)$ from the third source reduces the possible values for the corresponding sample $s_2(t)$. Although not satisfying the assumptions used in the derivations, this is an interesting case which resembles a practical application for example in speech processing or in ultrasonics.

In Figure 2, results from experiment 1 using the algorithm suggested in [3] are presented. Since the sources have unusual pdfs which do not fit the assumptions behind the algorithm in [2], that algorithm was not successful. In contrast, the algorithm presented in [3] shows impressive results in Figure 2.

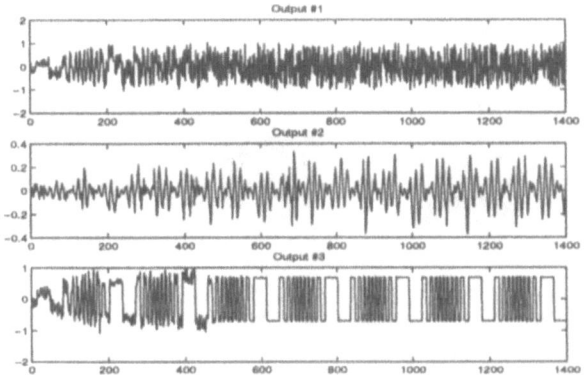

Figure 3. Results from experiment 1 using the new Gaussian mixture based algorithm. The sources are separated after about 500 time steps.

In Figure 3, results using the new algorithm are presented. The number of terms in the Gaussian mixtures were $K = 81$ with standard deviation $\sigma_k = 0.02$ and mean $m_k = -1.00 + 0.025k$, $k = 1, 2, ..., K$. The learning rate $\eta$ was 0.0005 for the updating of the mixture coefficients $a_{ij}$ and 0.001 for the weights. For this particular realization the performance is comparable with the those in Figure 2. Unfortunately, the new algorithm is not robust, the convergence is very much dependent on the initial weight matrix $W$ and the randomly selected mixing matrix $A$.

# 3 Experiment 2

In the second experiment, the two sources $s_1(t)$ and $s_2(t)$ were white noise processes (independent samples) with uniform amplitude distributions on the interval $[-1, 1]$. The mixing matrix rotated every sample 45° ($\pi/4$ radians); the elements were $A(1, 1) = A(2, 2) = A(2, 1) = 2^{-1/2}, A(2, 1) = -(2)^{-1/2}$.

In Figure 4, results from the second experiment using the algorithm in [3] are presented. The Figure consists of 6 subpictures where the fist one (upper left) is the samples observed by the network. In the following 5 sub-pictures the network output after every 200 pattern presentations are presented which shows how the network successively removes the dependencies between the two outputs. The results are again very good, the network converges within about 500 iterations.

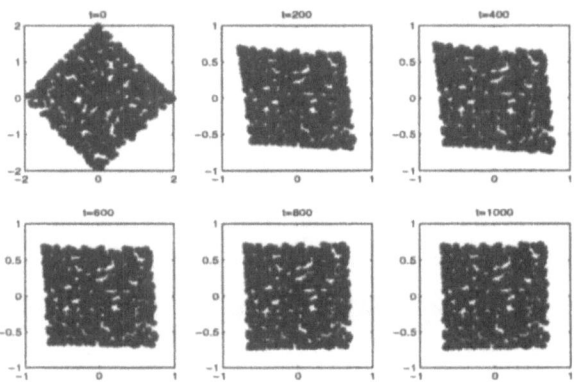

Figure 4. Results from experiment 2 using the algorithm in [3]. In the first subpicture (top left), samples observed by the network are presented. In the following subpictures, the outputs after every 200 samples are presented. Clearly, the net converges after a few hundred iterations.

As in experiment 1, there was no success with the algorithm in [2]. However, by modifying the distributions of the sources to have more kurtosis, the separation succeeded as shown in Figure 5. Instead of uniform pdfs for the noise (source) generator n(t), the results in Figure 5 were obtained using the source generator $[n(t)]^6 \text{sign}(n(t))$ which yields a pdf with most value near zero as can be seen from the sample distribution. In Figure 6, results form the new algorithm on the rotated uniform distribution are presented. Here the number of components in the Gaussian mixtures were only 20. The learning rates were 0.001 and 0.002 for the mixture coefficients and the weights respectively. The learning rate for the weights was successively decreased after 200 iterations to decrease the variances. The network converges after a few hundred iterations and shows a much more robust behaviour than in experiment 1.

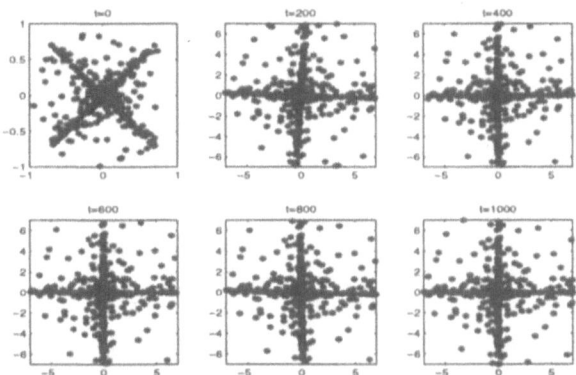

Figure 5. Results from experiment 2 using the algorithm in [2] but with the pdfs of the sources modified to have a high kurtosis. In the first subpicture (top left), samples observed by the network are presented. In the following subpictures, the results of processing the samples in the first subpicture by means of the converging network after every 200 iterations are presented. The net converges after a few hundred iterations.

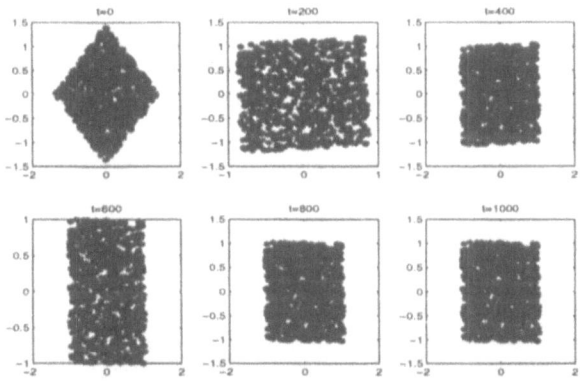

Figure 6. Results from experiment 2 using the new algorithm. In the first subpicture (top left), samples observed by the network are presented. In the following subpictures, the results of processing the samples in the first subpicture by means of the converging network after every 200 iterations are presented. The net converges after a few hundred iterations.

## 4   Summary and Conclusions

From the above experiments, we find that the algorithm in [2] certainly has problems with signals where the pdfs of the sources do not match the nonlinearities of the neurons. In contrast, the algorithm in [3] seems more robust as it can handle both examples presented here. In fact, this algorithm can handle much more demanding cases. The new algorithm presented here is not as robust as the one in [3]. One explanation for the limited performance of the new algorithm in the first experiment presented here might be that two of the source signals were deterministic and statistically dependent. Since the new method has difficulties to obtain reasonable estimates of the marginal pdfs for such sources and the independence assumption is violated it is not unexpected that the performance is limited. In order to handle badly scaled signals, one possibility is to introduce automatic tuning of the mean values and variances of the components in the Gaussian mixtures. Presently they are fixed which means that all outputs must have approximately the same dynamics, if not some of the marginal pdfs will be very badly modeled.

This work demonstrates the possibility to obtain more universal blind separation of sources. Whereas the earlier approaches are not easily generalized to more realistic conditions including for example nonlinearities or different number of dimensions, the new approach seems to be easily extended to handle such cases. Unfortunately, the current version of the new algorithm is not very robust and requires computationally demanding nonlocal matrix inversions. How to eliminate such limitations is of course an important issue for future work

Finally, it is important to note that the new approach presented in this work has several disadvantages and limitations. Besides the limited robustness observed, the new method is slower than the two presented in section 2, especially when the number of components in the Gaussian mixtures is large. Updating the estimates of the marginal pdfs is relatively time consuming in the present version but will be improved in future work. Another improvement would be to have adaptive means and variances in the Gaussian mixtures to make this approach more flexible and more sensitive to badly scaled problems.

# References

[1] M.G. Gustafsson. An exploratory step towards more universal blind separation of sources using neural nets. UPTEC 96042R, Uppsala University, 1996.

[2] A.J. Bell and T.J. Sejnowski. An information-maximization approach to blind separation and blind deconvolution. *Neural Computing*, 7:1129–1159, 1995.

[3] S. Amari, A. Cichocki, and H.H Yang. A new learning algorithm for blind signal separation. In *Advances in Neural Information Processing Systems, NIPS 7*. MIT Press, 1996.

[4] L. Xu, C. Cheung, H. yang, and S. Amari. Independent component analysis by the information-theoretic approach with mixture of denisty. In *Proc. 1997 IEEE Intl Conf on Neural Networks*, pages 1821–1826. 1997.

# The Adaptive Setback Thermostat

Olle Gällmo & Patrik Lögdahl

Department of Computer Systems, Uppsala University
Uppsala, Sweden
neuron@docs.uu.se

## Abstract

We present an adaptive setback thermostat (AST) which switches between two temperature setpoints – one optimized for user comfort and one for saving energy. The AST operates locally in an office room and makes its decisions based on how the room is (expected to be) used. Core issues, in decreasing order of importance, are user comfort, user friendliness (ease of installation and use) and to reduce energy costs. It is argued why a reinforcement learning approach may not be the best solution, and then shown how to reformulate the problem using a simple heuristic where reward maximization is replaced by explicit prediction of user arrivals, using temporal difference learning.

## 1 Introduction

Conventional thermostats are user-friendly in that they maintain a constant (comfortable) room temperature and are easy to use, but it is not (energy) cost efficient to keep the room temperature on a comfortable level in the user's absence. An alternative is to use *setback thermostats*, that can be set to different temperatures for different time intervals. This may save energy but is not as user-friendly. In a home environment, the time intervals must be set by the user. In an office, they are normally fixed and shared with other rooms (possibly the whole building), beyond control for the user.

Decisions made by the Adaptive Setback Thermostat (AST) proposed in this paper are based on local usage of the room, not only on time. The task is to maintain a comfortable temperature, $T_c$, in the room when the user is present (detected by a motion detector) and some energy saving temperature, $T_e$, when the user is absent. Heating/cooling of the room takes time, due to the room's thermal resistance and capacitance, so the AST must anticipate the user's arrivals and departures.

The goal is to save energy without being more difficult to handle than a conventional thermostat. User friendliness is of utmost importance, both in terms of user comfort and ease of operation.

A similar application can be found in [1], where a more general temperature control problem (controlling the furnace in a home environment) is described. Our approach is to generate temperature setpoints for a conventional thermostat (which may also be controlled by the user), rather than controlling the furnace directly. Thus, the control problem to maintain a constant temperature in a room is avoided, and the AST can focus on higher level decisions.

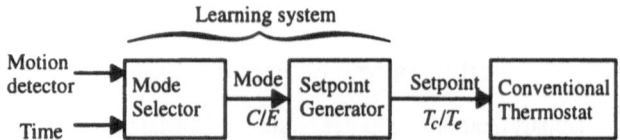

Figure 1: The Adaptive Setback Thermostat includes a conventional thermostat, controlled by a learning system which generates temperature setpoints.

Every $\Delta$ minutes (here, $\Delta$ = 15 minutes), a *mode selector* sets the *setpoint generator* in one of two possible modes, user comfort mode (*C*-mode) or energy saving mode (*E*-mode). This decision is based on data from a motion detector and the current time. The setpoint generator then selects a suitable temperature level to strive for (i.e. setpoints for the thermostat).

The desired temperature when in *C*-mode ($T_c$) is, of course, subjective to the user. Therefore, a reasonable (and simple) definition of $T_c$ is the temperature last requested by the user (using the thermostat).

The corresponding temperature when in *E*-mode ($T_e$) is a function of many variables[1]. For example, one should consider the temperature outdoors and in surrounding rooms and the expected length of the current *E*-mode period. If the room has air-conditioning and the surrounding temperatures are greater than $T_c$, it could be cost efficient to *increase* the temperature when the user is absent, i.e. set $T_e > T_c$. On the other hand, if the user is expected to be back soon, it may be better to keep the temperature constant at $T_c$ degrees (i.e. set $T_e = T_c$) than to change it.

This suggests that the task of the setpoint generator is a learning problem in itself, involving predictions of user arrivals. Though the focus of this paper is on the first step in the decision chain – the mode selector – the observation that the setpoint generator requires such predictions will be exploited.

## 2 An *ad hoc* reinforcement learning approach

At first, it seemed natural to define mode selection as a reinforcement learning problem [2] where, in each time step, $k$, an *agent* is given a state, $s_k$, selects an action $a_k$ and, in the next time step, receives a new state $s_{k+1}$ and a reward $r_{k+1}$. The agent's environment is a discrete time Markov decision process (MDP) and the objective is to find an optimal *policy*, i.e. a rule for action selection, which maximizes future rewards.

In this case, there are only two selectable actions in each state; "*C*-mode" or "*E*-mode". The state $s_k$ at time step $k$ is a pair ($P_k$, $k$), where $P_k$ is some estimate of the probability that the user is present in the room, e.g. a moving average of past motion detector values. Note that the time step, $k$, is included in the state. The mode selector needs this to (implicitly) anticipate future arrivals and departures.

A peculiarity of this MDP is that the sequence of states encountered by the agent is independent of its previous actions; The temperature in the room, decided by the agent's previous actions, is very unlikely to affect the user's future presence in the room and the time step $k$ is of course independent of previous decisions. The effect on the MDP formulation is that the state transition probabilities, $P^a_{ss'}$ (the probability that selecting action $a$ in state $s$ leads to state $s'$), normally defined as:

1. Since the controlled device is a thermostat, not the furnace, the AST can not simply decide to turn of heating (or cooling) to save energy.

$$P^a_{ss'} = P(s_{t+1} = s' \mid s_t = s, \ a_t = a) \tag{1}$$

can be rewritten:

$$P^a_{ss'} = P_{ss'} = P(s_{t+1} = s' \mid s_t = s) \tag{2}$$

Furthermore, many of these probabilities will be zero, since the state description includes the time step itself. If $s = (P_k, k)$ and $s' = (P_k', k')$, then:

$$P_{ss'} > 0 \text{ iff } k' = k + 1 \tag{3}$$

Depending on how the other half of the state description (the probability $P_k$) is computed, the number of states reachable from state $s$ in one time step can be restricted even further (many state transition probabilities are close to 0). For example, if $P_k$ is an average of past motion detector values, then $P_{k+1}$ is very likely to be close to $P_k$.

The immediate reward, $r_k$, given to the agent each time step $k$, must reflect the objective – to minimize energy costs to and keep the user comfortable. Energy costs are measurable, but user comfort not obviously so.

One way to measure user comfort is to treat manual adjustments to the thermostat as a sign of discomfort. For example, one could include a term in the immediate reward, which is –1 if the user has moved the thermostat sometime between time steps $k-1$ and $k$ and 0 otherwise.

However, this is a very crude measure of user comfort. If the thermostat has been set (by the AST) to $T_c$ and the user arrives before the room has had time to reach that temperature, it is not certain that she will move the thermostat (observing that it is already set to $T_c$). Furthermore, not all manual adjustments to the thermostat should be treated as complaints to the agent. If the room is $T_c$ degrees, but the user feels chilly that day and sets the thermostat to $T_c+1$ degrees, it is not a fault of the mode selector.

A better measure, then, would be a term $M_k$, which is 1 if the temperature is 'too far' (subjective) from $T_c$ when the user is present, and 0 otherwise. Thus, the immediate reward could be defined by:

$$r_t = - E_t - aM_t \tag{4}$$

where $E_k$ is the energy cost since the last time step and $a$ is a trade-off constant.

## 2.1 Discussion

There are two potential problems with the reinforcement learning approach. One is that reinforcement learning implies trial-and-error search which, in this case, can be very uncomfortable to the user if not restricted in some way.

The second problem is based on the observation that the two terms in the reward definition correspond to two tasks – to reduce energy costs and to keep the user comfortable. If the AST is to be accepted by the user the latter must have higher priority than the former. Minimizing the energy costs (i.e. selecting $E$-mode) should only be tried in the user's absence. Likewise, when the user is present, the system should have no choice but to select $C$-mode.

Neither of these observations disqualify the use of reinforcement learning methods for this problem, but the second suggests that there is another, simpler, approach based on the following deterministic 'policy' or heuristic:

> If the user is in the room, or is expected to arrive within $\delta$ time periods, select C-mode, else select E-mode.

The value of $\delta$ should be set to a value close to, but slightly greater than, the time it takes to change room temperature from the current temperature to $T_c$.

The learning problem is no longer a reinforcement learning problem (to maximize rewards) but a prediction problem, estimating the time until the user's next arrival[2]. If the user is uncomfortable it is not because of a bad policy, but because of bad predictions (or bad values of $\delta$). As illustrated in Section 1, the setpoint generator may have use of such a predictor as well (when deciding the value of $T_e$).

# 3 TD-AST – a temporal difference learning approach

In the mode selector, called TD-AST, shown in Figure 2 a *counter* simply counts the number of time steps, $\Delta_{prev}$, since the latest registered movement. A *movement predictor* takes $\Delta_{prev}$, the time of day and day of week, and estimates $\Delta_{next}$ – the time until the motion detector will react next. The selector, then, selects $C$-mode if motion has been detected recently or if the estimated time until next movement is less than $\delta$.

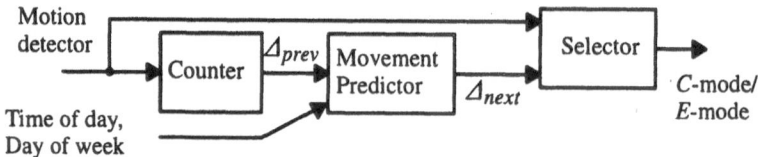

Figure 2: A mode selector, based on predictions of the time
until the next movement in the room.

The movement predictor is implemented as lookup table, trained on-line by temporal difference learning [3] to approach the ideal predictor, where $\Delta_{next}$ is 1 if motion is detected in the following time period and equal to 1 + the next estimate otherwise.

But, the absolute value of $\Delta_{next}$ is not important – only whether it is greater or less than $\delta$. All $\Delta_{next}$ values greater than $\delta+1$ could therefore be set to be equal to $\delta+1$ or $\Delta_{next}$ can be discounted by a factor $\gamma \in [0,1[$ to keep it finite:

$$\Delta_{next} = \begin{cases} 1, & \text{if } \Delta'_{prev} = 0 \\ 1 + \gamma\Delta'_{next}, & \text{otherwise} \end{cases}$$

(5)

where $\Delta'_{prev}$ and $\Delta'_{next}$ denote the counter and movement predictor values at the next time step. The predictor is trained to minimize the temporal difference error, i.e. the difference between the right hand and the left hand side of (5), by successive updates of the corresponding table values:

$$\Delta_{next} \leftarrow \Delta_{next} + \eta \begin{cases} 1 - \Delta_{next}, & \text{if } \Delta'_{prev} = 0 \\ 1 + \gamma\Delta'_{next} - \Delta_{next}, & \text{otherwise} \end{cases}$$

(6)

where $\eta \in [0,1]$ is the step-length (learning rate).

## 3.1 Discussion

TD-AST is an on-line learner, without the potentially bad (uncomfortable) effects of unrestrained trial-and-error search. A design based on off-line supervised learning is possible [4] but would typically have to be trained for an initial period of several weeks

---

2. The heuristic may be extended to include estimates of user departures as well. Doing so may save some energy, but may also have negative effects on user comfort.

and then retrained with regular intervals (which may require supervision from an operator to ensure convergence).

In contrast, once the value of $\delta$ has been set (measured) the TD-AST can be installed and will be operational the first day. If the table entries are initialized to something less than $\delta$, the TD-AST will initially behave as a conventional thermostat, making small adjustments to the table entries according to the observed behaviour of the user but without affecting her. After a while, when some of the table entries exceed the value of $\delta$, it will start changing the temperature in the user's absence, but still tend to underestimate the time until she returns (which is to the user's advantage).

# 4 Experiments

TD-AST has been tested and compared to a supervised approach (a committee of multi layer perceptrons) and a reinforcement learning approach in a simulated office environment. Space considerations make it impossible to include all the details of these simulations here, but see [4] for a complete description and also some results from experiments in a real office environment.

The thermal model of the simulated room is very simple. To heat the room 1°C takes 15 minutes, and the temperature change is linear. Cooling the room takes 30 minutes/°C, also linear. A more realistic thermal model, such as the resistance-capacitance model used in [1], would perhaps increase the credibility of the simulation results, but it should be noted that the proposed method does not depend on the model.

The setpoint generator is also simple; $T_c$ is initially 20°C and then equal to the latest setpoint chosen by the user on the thermostat. $T_e$ is a constant (15°C).

The AST makes a decision every 15 minutes and the TD-AST learning constants are $\gamma=0.98$ and $\eta=0.2$. $\delta=10$ is also a constant, corresponding to the worst case scenario of changing from $T_{min}=15$°C to $T_{max}=25$°C, which would take 2.5 hours = 10 time periods. The table used to represent the $\Delta_{next}$ values[3] is initialized by setting all elements equal to $\delta-1 = 9$ time periods. The effect is that the room will have constant temperature of $T_c$ degrees during the first $(1/\eta) - 1 = 4$ weeks of operation, i.e. that the system initially behaves as a conventional thermostat.

The simulated user has an ideal temperature of $20\pm0.5$°C. Outside this interval, she will 'complain' by setting the thermostat to 20°C, unless the AST has already done so. She arrives to work between 6 and 8 a.m. and leaves between 3 and 8 p.m. (drawn from truncated Pascal distributions). Some week days she does not arrive at all and never on Saturdays, nor Sundays. The user takes a lunch break (out of office) for an hour, at the same time every day.

The motion detector outputs a 1 if motion has been detected sometime during the latest 15 minutes. The real motion detector is very sensitive, but in simulation there is a probability $q=0.4$ that the motion detector fails to detect the user when she is at work. This is to simulate the user leaving the room now and then during the day.

Figure 3 shows how the weekly average of the temperature changes with time over a simulated 30 week period. As predicted, the AST maintains a constant temperature of 20°C during the first 4 weeks. From week 5, when the temperature starts to move, the average temperature is 16.8°C. The average temperature, only counting times

---

3.   The table size is $96 \times 7 \times 4$. There are 96 (15 minute) periods in a day, 7 days in a week and the $\Delta_{prev}$ values are represented by four levels; $\Delta_{prev} \leq 1$, $1 < \Delta_{prev} \leq 2$, $2 < \Delta_{prev} \leq 8$ and $\Delta_{prev} > 8$.

880

when the user is at work (i.e. when the temperature is supposed to be 20°C) is 20.0°C with a standard deviation of 0.2. The corresponding average (and std.dev.) in the user's absence is 16.1°C (1.9).

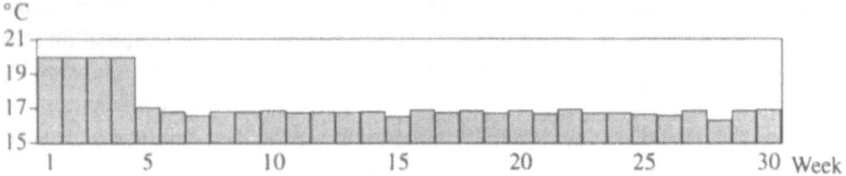

Figure 3: The weekly average temperature for all thirty weeks.

The user makes manual adjustments to the thermostat on 5 occasions over the full 30 week period. Though this is a crude measure of user comfort, it can be compared to the 307 such 'complaints' from the same user (same arrival/departure sequence) when testing a reinforcement learning approach in [4].

## 5 Conclusion

The Adaptive Setback Thermostat (AST), presented in this paper, is similar to a conventional (time based) setback thermostat in that it switches between two temperature setpoints, one comfortable, $T_c$, and one energy preserving, $T_e$. Normally a setback thermostat switches to/from $T_c$ and $T_e$ at fixed, predetermined, times that must be set by hand. The AST is more flexible in that the time intervals are not fixed, and autonomous in that the intervals are automatically adjusted based on how the room is used.

It seems natural to formulate the task of the AST as a reinforcement learning problem in a Markovian environment, but this may not be the best solution. If the AST is to be accepted by the user it must be user-friendly, which in this case seems easier to achieve by defining hard (but reasonable) constraints on the objective function.

The suggested design, TD-AST, is therefore not based on reinforcement learning, but on a simple heuristic combined with explicit prediction of user arrivals. A table of predictions is initialized to be on the 'safe' (user-friendly) side and then trained online using temporal difference learning. Experiments in a simulated environment (only summarized here) show that TD-AST allows for saving energy with very little negative effects on the user.

## Acknowledgements

This work was sponsored by Vattenfall Utveckling AB. Thanks also to Jakob Carlström, Lars Asplund and Ernst Nordström for their comments and suggestions.

## References

[1] Mozer MC, Vidmar L, Dodier RH. The Neurothermostat: Predictive optimal control of residential heating systems. In: Mozer MC, Jordan MI, Petsche T (Eds), Advances in Neural Information Processing Systems 9, MIT Press, 1997, pp 953–959

[2] Sutton RS, Barto AG. Reinforcement Learning: An Introduction, MIT Press, 1998.

[3] Sutton RS. Learning to Predict by the Methods of Temporal Difference, Machine Learning 1988; 3:9-44, Kluwer Academic Publishers

[4] Lögdahl P. The Adaptive Setback Thermostat: Experiments in simulated and real office environments. MSc Thesis, Dept. of Computer Systems, Uppsala University, Sweden, 1998

# Architecture Optimization in Feedforward Connectionist Models

## M. YACOUB and Y. BENNANI

LIPN-CNRS, Institut Galilée-Université Paris 13

Avenue J.B. Clément 93430 Villetaneuse France

{yacoub,younes}@lipn.univ-paris13.fr

### Abstract

Given a set of training examples, determining the number of free parameters is a fundamental problem in neural network modeling. The number of such parameters influence the quality of the solution obtained. This paper deals with the problem of adapting the effective network complexity to the information contained in the training data set, and the task's difficulty. The method we propose consists of choosing an oversized network architecture, training it until it is assumed to be close to a training error minimum then selecting the most important input variables and pruning irrelevant hidden neurones. This method is an extension of our previous one used for input variables selection, it is simple, cheap and effective. We show its effect experimentally through one classification and one regression problem.

## 1 Introduction

Choosing an appropriate architecture for a learning task is an important issue in training neural networks. Two different approaches exist for determining a suitable architecture: constructive and destructive methods. In the first approach, the learning algorithm gradually increases the number of free parameters by adding hidden units to the network. The algorithm stops adding hidden units when some validation criterion indicates that the network performance is good enough. Most methods based on this approach seem to be very "greedy" of neurones. They have tendency to favour rote learning and consequently generalise badly. In the second approach, based on pruning and/or regularization, we start rather with a relatively large network and gradually remove either connections or complete units. Regularisation involves adding an extra term to the cost function to penalize complex structures. Pruning methods are based on the following general procedure. First, a relatively large network is trained using one of the standard training algorithms. Then the network is examined to estimate the relative importance of the weights or neurons, and the least important are removed. Several methods have been proposed. For a survey of pruning methods for feedforward neural networks, see for example [1].

In this paper, we propose the use of HVS measure [2] for pruning excess hidden neurones. It is an extension of our previous work on variables selection [2, 3]. We present experimental results on two problems: a 3 classes discrimination problem and a univariate time series regression problem.

# 2    Pruning Procedure

The method we propose for architecture selection consists to eliminate redundant variables and those which do not contain enough relevant information, and prune the non-relevant hidden neurones. To do so, we will use all input variables that may have an effect on the output and train a relatively large three layered fully connected MLP. After that, we will first select the most important input variables using our heuristic measure HVS [2]. Then we considere the hidden layer as the input layer of a sub-network made of this hidden layer and the output layer. We will then use HVS measure for hidden units selection.

# 3    Experimental Validation

We present two validation problems: the Breimann waveforms discrimination problem and the well known Sunspots regression problem. For the two problems, we use a three layered fully connected MLP. The two networks are trained using the back-propagation algorithm. We will use the following notation to represent the neural networks' architecture: $< n_I|n_H|n_O >$, where $n_I$ represents the dimension of the input layer, $n_H$ the number of hidden neurons, and $n_O$ the number of neurons in the output layer.

## 3.1    Discrimination Task

In order to show the effect of our method on discrimination problem, we choose the "Waveform" Classification Problem. It was proposed by Breiman et al.[4] and a noisy version was used by De Bollivier et al.[5]. It is a three class problem. The examples are real vectors of dimension 40 including 19 noise components. The representation of the Waveform problem in the two first principal components space is shown in figure 1. Polygons show the hull for each class. We can notice the high overlapping between all classes.

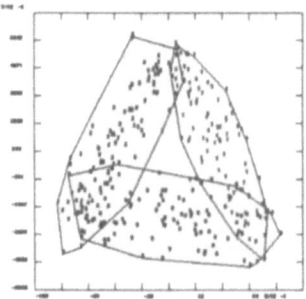 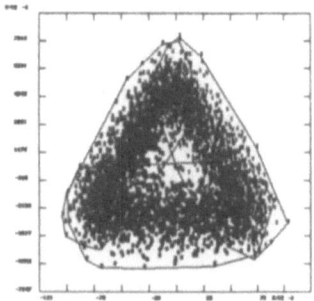

Figure 1: Representation of the Waveform problem in the two first principal components space for the learning (left) and test set (right). Polygons show the hull for each class.

For this problem, our experimental design is as follows: for learning we use a set $D^l$ of 300 instances and a set $D^v$ of 700 instances for validation. For the test, we use a set $D^t$ of 4300 instances. We have used the initial architecture $< 40|10|3 >$. HVS procedure has selected the set of variables $S^v$(HVS) = $\{4, 5, 6, 7, 8, 9, 10, 11, 12, 13, 14, 15, 16, 17, 18, 19\}$, and rejected variables 1 to 3 and 20 to 40. It may be noticed that the HVS procedure selected the most important variables in the first part of the signal, and rejected all variables in the "noise" part of the signal. As shown in Table 1, after pruning input variables, performances on $D^t$ are increased from 82.8372 to 85.3721. For comparison of our method to others methods concerning this problem see [2]. Notice that after each pruning step the network is retrained.

| | Architecture | performances on | | |
| --- | --- | --- | --- | --- |
| | | $D^l$ | $D^v$ | $D^t$ |
| Before pruning | $< 40|10|3 >$ | 92.00 | 83.8571 | 82.8372 |
| | (443 parameters) | [88.37 , 94.57] | [80.95 , 86.40] | [81.68 , 83.93] |
| After variables | $< 16|10|3 >$ | 88.66 | 85.5714 | 85.3721 |
| selection | (203 parameters) | [84.58 , 91.78] | [82.77 , 87.98] | [84.28 , 86.40] |
| After hidden | $< 16|6|3 >$ | 89 | 85.7143 | 85.4651 |
| neurons pruning | (123 parameters) | [84.95 , 92.06] | [82.93 , 88.11] | [84.38 , 86.49] |

Table 1: Number of parameters and performances before pruning, after variables selection, and after variables selection and hidden neurons pruning.

After variables selection, we have used HVS measure for hidden units selection. We present on figure 2 the results by drawing the performances on $D^v$ during hidden neurons pruning. The best performance on $D^v$ is obtained after pruning 4 hidden neurons. We can notice that the time needed for network's stabilization depends on the disturbance's importance due to the pruning process. As shown in Table 1, using HVS for input and hidden neurons pruning allows to cut 320 parameters, which corresponds to a percentage of deletion of 72.23%, and performances are increased from 83.8571 to 85.7143 on $D^v$ and from 82.8372 to 85.4651 on $D^t$. Notice that the result of our approach is 0.54 from the Bayesian theoritical limit which is approximately equal to 14% error's rate on $D^t$.

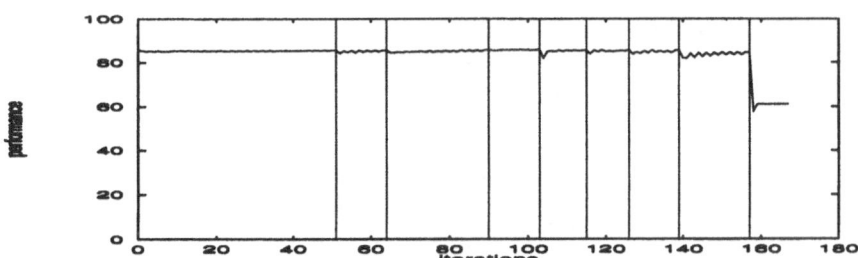

Figure 2: Performances evolution on $D^v$ during hidden neurons pruning. Notice that intervals delimited by vertical lines correspond to different pruning phases.

## 3.2 Regression Task

In this experiment, we use a real-world time series of limited record length, the sunspot data. It contains the annual averaged sunspot activities from 1700 to 1979 (280 points). The series is shown in figure 3.

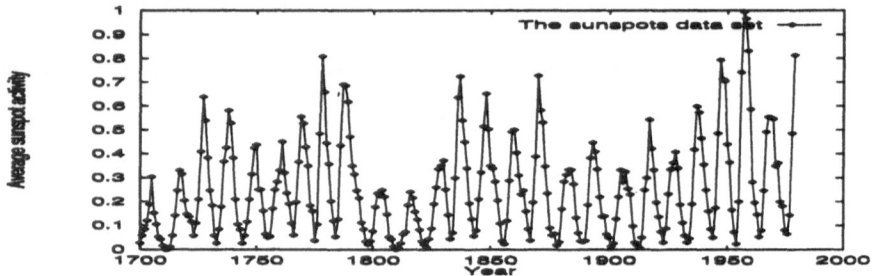

Figure 3: The sunspots data for years 1700 to 1979.

For this problem, we attempt to predict one value using the twelve previous ones, leading to a network with 12 input units and one output unit. We use the data from 1712 to 1920 for learning (a set $D^l$ of 209 examples) and the data from 1921 to 1955 for validation (a set $D^v$ of 35 examples). For the test, we use the data from 1956 to 1979 (a set $D^t$ of 24 examples). We have used the initial architecture $< 12|10|1 >$ with the hyperbolic tangent and identity as transfert functions of the hidden and output layer, respectively. We start with 12 inputs neurones, because Weigend et al [6] notice that increase from 12 to 25 input units did not lead to significant further improvement.

Performance is measured in terms of average relative variance (arv), which is the ratio between the average squared error of the model and the variance of the data. Notice that the definition of the arv quantity would relate the error of the model and the variance of the data calculated on the same set D (taken from the entire set S).

$$arv(D) = \frac{\sum_{i \in D} (y^{(i)} - f(x^{(i)}))^2}{\sum_{i \in D} (y^{(i)} - \mu_D)^2}$$

Where $y^{(i)}$ is the true value of the time series at time i, $f(x^{(i)})$ is the output of the network at time i, and $\mu_D$ the average of the target value in D.

However, the widely used definition of the arv scales by the overall variance of the data.

$$arv(D) = \frac{\frac{1}{|D|} \sum_{i \in D} (y^{(i)} - f(x^{(i)}))^2}{\frac{1}{|S|} \sum_{i \in S} (y^{(i)} - \mu_S)^2}$$

In order to compare our arv with established results, we will stick to the widely used definition.

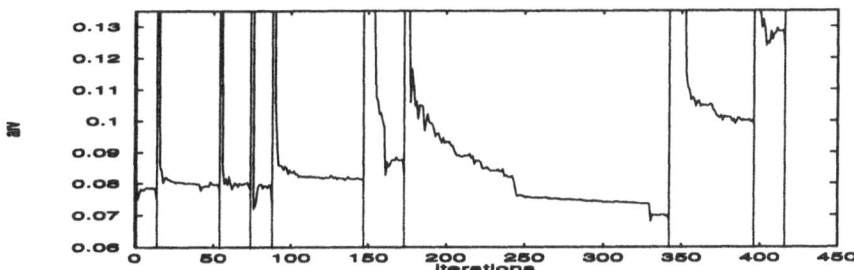

Figure 4: Arv evolution on $D^v$ during hidden units pruning for one experiment. The intervals delimited by vertical lines correspond to different pruning phases.

First, we have used HVS measure for input variables selection. We have done 30 experiments. For all the experiments, we have used the initial architecture $< 12|10|1 >$. The inputs : $x_{t-12}, x_{t-11}, x_{t-8}, x_{t-3}, x_{t-2}$, and $x_{t-1}$ have been selected in the majority of our experiments.

After input units pruning we proceed to pruning hidden units, this process can be seen as a hidden dimensions discovering process: the number of hidden units can be interpreted as an upper limit of the dimension of the dynamics of the time series. Figure 4 shows the evolution of the arv on $D^v$ during the hidden neurons pruning procedure for one experiment. As we can see, for this experiment, the best arv on $D^v$ is obtained after pruning 7 hidden neurons. Thus, for this experiment, using HVS measure to input variables and hidden neurons allows to cut 116 parameters, which corresponds to a perecentage of deletion of 82.27%.

Reported in Table 2 are our results, averaged over 30 experiments, after input variables selection and pruning hidden neurons, and some results obtained using different methods. Notice that our method allows not only to reduce the effective complexity of the model but also to obtain predictions with better precisions compared to the prediction qualities of the other approachs.

| Model/arv | $D^l$(1712-1920) | $D^v$(1921-1955) | $D^t$(1956-1979) |
|---|---|---|---|
| Yacoub & Bennani | 0.090 | 0.071 | 0.318 |
| Goutte[7] | 0.082 | 0.082 | 0.357 |
| Svar & al.[8] | 0.090 | 0.082 | 0.35 |
| Weigend & al.[6] | 0.082 | 0.086 | 0.35 |
| Linear | 0.131 | 0.128 | 0.36 |

Table 2: Comparison of the results obtained using different methods. Our results are averaged over 30 experiments.

# 4 Conclusion

We have presented a units pruning measure, and have tested it on one relatively hard discrimination problem and one real world time series regression problem of limited length. Results show that we can improve or preserve the performances on the validation and test set while decreasing the effective complexity of the model by selecting the most important input variables and pruning irrelevant hidden neurones using HVS measure which is simple and cheap to implement.

# References

[1] Reed R.: Pruning algorithms-A survey, IEEE Trans. Neural Networks, vol. 4, no. 5, pp. 740-747 (1993).

[2] Yacoub M. and Bennani Y.: HVS: A Heuristic for Variable Selection in Multilayer Artificial Neural Network Classifier. In Intelligent Engineering Systems Through Artificial Neural Networks, Vol 7: C. Dagli, M. Akay, O. Ersoy, B. Fernandez and A. Smith (Editors), pp. 527-532, (1997).

[3] Yacoub M. and Bennani Y.: A Neural Network Methodology for Machines' Class Identification. Proc. IEEE International Joint Conference on Neural Networks (IJCNN'98), Vol. 1, pp. 322-325 (1998).

[4] Breiman L., Freidman J., Olshen R., Stone C.: Classification and regression trees. Wadsworth Int. Group. (1984).

[5] De Bollivier M., Gallinari P., Thiria S.: Cooperation of neural nets and task decomposition. International Joint Conference on Neural Networks (IJCNN'91), Vol. 2, pp. 573-576 (1991).

[6] Weigend A.S., Huberman B.A. and Rumelhart D.E.: Predicting the future: a connectionist approach. Int. Journal of Neural Systems, Vol. 1, No 3, pp. 193-209 (1990).

[7] Goutte C.: On the use of pruning prior for neural networks. In Neural Network for Signal Processing VI, pp. 52-61 (1996).

[8] Svarer C., Hansen L.K., Larsen J.: On design and evaluation of tapped-delay neural networks architectures. In IEEE International Conference on Neural Networks, pp. 46-51 (1993).

# Neural Velocity Force Control for Industrial Manipulators Contacting Rigid Surfaces

M. Dapper, R.Maass, R. Eckmiller

University of Bonn, Dept. of Comp. Science VI
Römerstr. 164, D-53117 Bonn, F.R. Germany
Tel.: +49-228-73-4168, FAX: +49-228-73-4425
e-mail: dapper, maass, eckmiller@nero.uni-bonn.de

**Abstract.** We present a new neural velocity force control scheme for a 6 DOF industrial manipulator ensuring tracking of end effector positions along unconstrained directions and tracking of contact force along the constrained direction to significantly expand the range of manipulator applications. Neural velocity force control is actually feasible even in the case of an extreme stiff environment, which is a quite common situation in industrial applications. A cascaded velocity controller (CVC) ensures the precise approach with a tender impact to the unknown surface. The neural controller is of inverse dynamics type with a force feedforward action and performs an adaptive computation of the inverse manipulator model. Simulation results for a 6 DOF industrial manipulator are reported and show the convenience of the approach for demanding tasks like dismantling or surface tracking inclusive establishing contact to rigid objects with precisely bounded impact forces[1].

## 1 Introduction

In many industrial/manufacturing applications, a robot manipulator is required to establish contact with its environment (e.g. contour following, grinding, scribing, deburring and disassembly related tasks). In these applications, simultaneous control of the end effector position and the interaction force is required [2][3][4]. Many robotic applications involve intentional interaction between the manipulator and the environment where contact stability to moving objects and surface tracking with defined contact force typify demanding tasks [1][5]. Establishing stable contact to rigid objects as well as tracking surfaces with unknown shape and elasticity [8] are the prerequisites of a manipulator handling its environment naturally[6][9].

---

[1] This work was supported in part by Federal Ministry for Education, Science, Research and Technology (BMBF) under grant "DEMON"

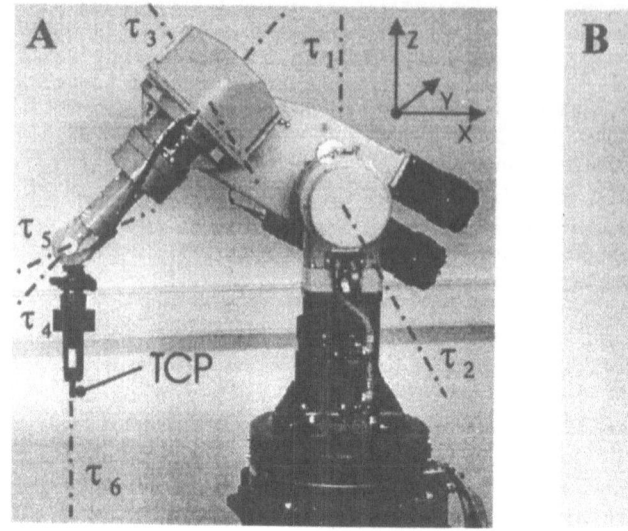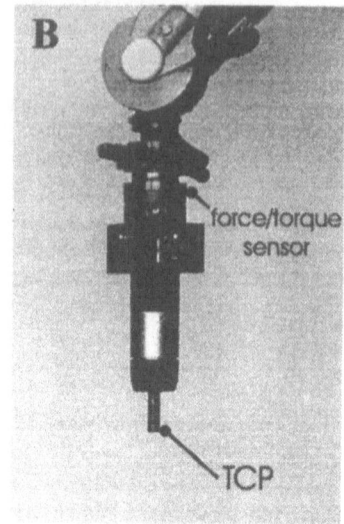

**Fig. 1.** A: Industrial manipulator (Siemens Manutec r2) equiped with a 6D force/torque sensor(ATI 100/5) which will be controlled with the simulated neural CVC in the near future. B: Zoom of manipulators endeffector with force/torque sensor

## 2 Results

### 2.1 Velocity Force Control System

The velocity force control system consists of several components to either guarantee a tender impact while establishing contact to the unknown constraining surface and to maintain a certain contact force [4][7].

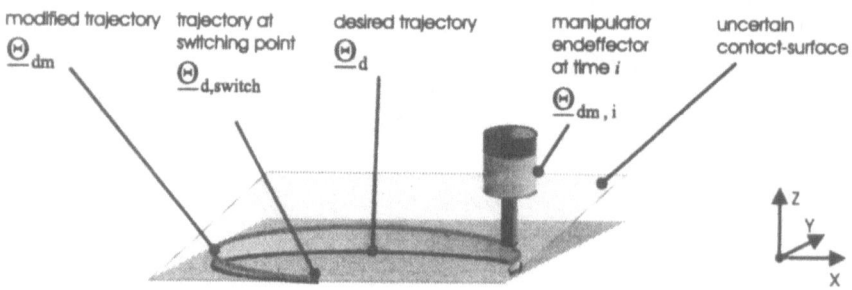

**Fig. 2.** Manipulators TCP contacting and tracking uncertain surface with precisely defined contact force.

*CVC:* An essential role is played by the 'Cascaded **Velocity Controller**' (**CVC**), which controlls the cartesian velocity $\mathbf{\dot{X}}$ of the endeffector (fig. 1) via an accelerating or decelerating force $\mathbf{F}_d$, fed into the neural force/position controller while having no contact to the uncertain surface. CVC automatically raises $\mathbf{F}_d$ to the

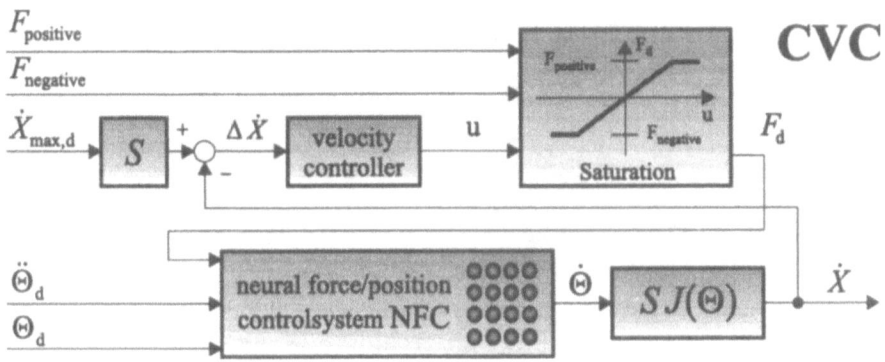

**Fig. 3.** Cascaded Velocity Control System (**CVC**) with subordinated neural force/position controller

specified force $\mathbf{F}_{negative}$, when contacting the static or moving surface. Thus CVC is automatically turned off and only the neural force/position controller remains active.

*TMO:* The neural force/position controller (fig. 4) has been extended with the essential module 'Trajectory Modification for Moving Objects' (**TMO**), which allows contacting **moving** objects. The orthogonal position separation law (eq. 1) naturally requires TMO, to keep the control error $\mathbf{\Delta\theta}$ sufficiently small.

$$e_p = \mathbf{J}^{-1}(\theta) \cdot (\mathbf{I} - \mathbf{S}) \cdot \mathbf{J}(\theta) \cdot \mathbf{\Delta\theta} \tag{1}$$

TMO modifies the trajectory $\theta_d$ in the force controlled direction to make position deviations in the force controlled direction invisible for the position controller. Equations 2, 3 and 4 show the control law used in TMO with the Jacobian $\mathbf{J}_i$ of the actual- and the Jacobian $\mathbf{J}_{d,i}$ of the desired trajectory at time step i. S is the separation matrix of the neural force/position control.

$$\theta_{T,i} = \theta_{T,i-1} + \mathbf{J}_i^{-1} \cdot [\mathbf{I} - \mathbf{S}] \cdot \mathbf{J}_{d,i} \cdot [\theta_{d,i} - \theta_{d,i-1}] \tag{2}$$

$$\theta_{F,i} = \theta_{F,i-1} + \mathbf{J}_i^{-1} \cdot [\mathbf{S}] \cdot \mathbf{J}_i \cdot [\theta_i - \theta_{i-1}] \tag{3}$$

$$\theta_{dm,i} = \theta_{d,switch} + \theta_{T,i} + \theta_{F,i} \tag{4}$$

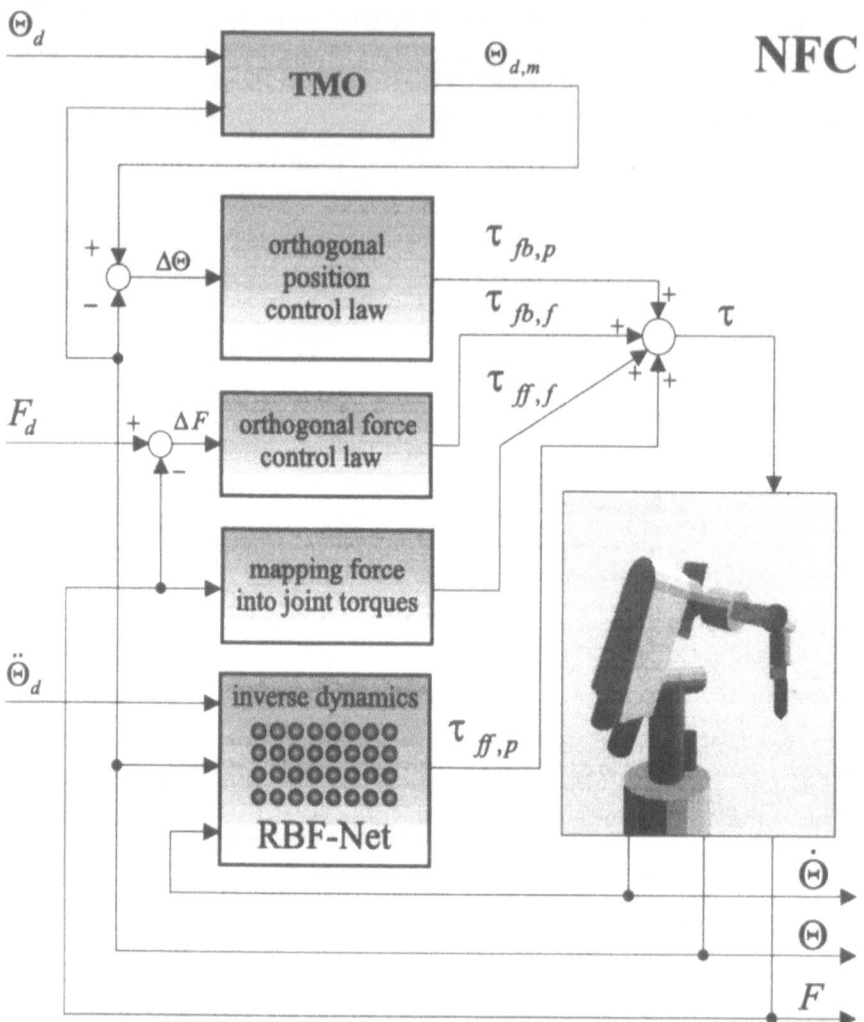

**Fig. 4.** Neural force/position control system with 'Trajectory Modification for Moving Objects' (**TMO**) and neural RBF-based inverse dynamics

At the switching time $T_{switch}$ to neural force/position control [3] the angle $\theta_{d,switch}$ is saved and used in equation 4. The angles $\theta_{T,i}$ and $\theta_{F,i}$ are initially set to zero. Figure 4 shows the augmented neural force/position control system with the TMO module and the neural inverse dynamics, which consist of a RBF type network with joint angles $\theta_2$ and $\theta_3$ as inputs. The neurons are placed in a distance of 10 degrees which leads to a net of 598 neurons, taking into account joint angle boundings. The training of the inverse dynamics net is performed during free movement of the manipulator executing a sufficient number of fast adaptation trajectories [3][6].

## 2.2 Simulations

Figure 5 shows the simulation results for the 6 DOF velocity force control system. The dynamic simulations of the industrial manipulator were performed on a pentium II dual 266 MHz system with a sampling rate of 2 milliseconds in the control loop corresponding to the conditions of the realtime control loop inclusive the real industrial manipulator. The figures 5 A,B and C exhibit the

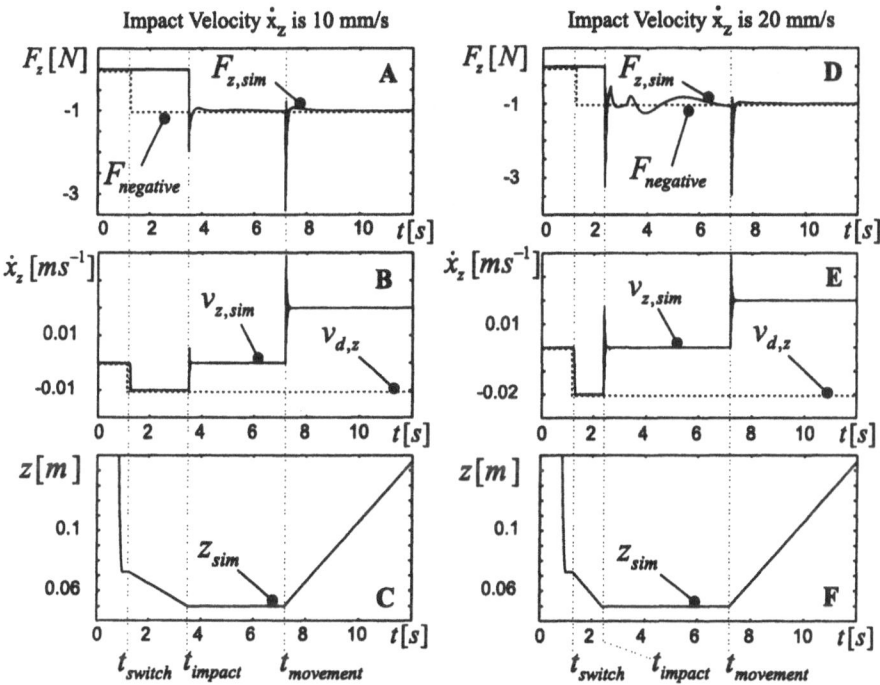

**Fig. 5.** Simulation results for contacting and tracking a moving rigid object. A and D: Cartesian TCP force in z-direction, B and E: Cartesian TCP velocity in z-direction, C and F: cartesian z component of the TCP trajectory

simulated contact force, velocity and position of the TCP in z-direction for a desired cartesian velocity of 10 mm/s. Figure 5 D,E and F indicate the same values for a desired velocity of 20 mm/s. Comparing figure 5A with figure 5D shows a lower impact force at time step $t_{impact}$ caused by the reduced cartesian impact velocity in figure 5B. Thus via a sufficient small impact velocity the maximum impact force can be bounded. At $t_{movement}$ the surface starts to move in positive z-direction with 20mm/s and generates the same force peak at figure 5A and figure 5D at time step $t_{movement}$.

# 3 Discussion

Simulations and first experiments have pointed out the performance of neural velocity force control CVC as a demanding extension to neural force/position control. CVC is capable of precise tracking and smooth contacting uncertain rigid surfaces inclusive maintaining a desired force trajectory. TMO as a fundamental enhancement of neural force/position control has been developed and proved to work excellent in simulations. It will be implemented into the real control loop in the near future.

# 4 Conclusions

Neural velocity force control for industrial manipulators significantly improves the phase before establishing contact and ensures a properly bounded impact force via an adequate approaching velocity. Simulations for a 6 DOF industrial manipulator have pointed out the feasibility of neural velocity force control. Especially demanding robot control problems like dismantling, deburring or contacting uncertainly defined rigid surfaces with precise contact forces are the superior features of our neural velocity force controller.

# References

1. B. Bona and M. Indri. Friction compensation and robustness issues in force/position controlled manipulators. *IEE Proc.-Control Theory Appl.*, 142(6):569–574, 1995.
2. S. Chiaverini, B. Siciliano, and L. Villani. Force and position tracking: Parallel control with stiffnes adaptation. *IEEE Control Systems Magazine*, 18(1):27–33, 1998.
3. M. Dapper, R. Maaß, V. Zahn, and R. Eckmiller. Neural force control (nfc) for complex manipulator tasks. In *Springer Berlin, Lecture Notes in Computer Science 1327, Artificial Neural Networks - ICANN'97, Lausanne, Switzerland, October*, pages 787–792, 1997.
4. M. Dapper, R. Maaß, V. Zahn, and R. Eckmiller. Neural force control (nfc) applied to industrial manipulators in interaction with moving rigid objects. In *IEEE Proc. Int. Conf. Robotics and Automation, Leuven, Belgium, May(in press)*, 1998.
5. D.M. Dowell, G. Irwin, G. Lightbody, and G. Connel. Hybrid neural adaptive control for bank to turn missiles. *IEEE Transactions on Control Systems Technology*, 5(3):297–308, 1997.
6. J. Kalkkuhl, K. Hunt, R. Zbikowski, and Dzielinski A. *Applications of Neural Adaptive Control Technology*. World Scientific, Series in Robotics and Intelligent Systems (17), Singapore, 1997.
7. R. Maaß, V. Zahn, and R. Eckmiller. Neural force/position control in cartesian space for a 6 dof industrial robot: Concept and first results. In *IEEE Proc. Int. Conf. Neural Networks, ICNN97, Houston, USA, June*, pages 1744–1748, 1997.
8. F. Pfeiffer and C. Glocker. *Multibody Dynamics With Unilateral Contacts*. John Wiley and Sons Inc., New York, 1996.
9. C.C. de Wit, B. Siciliano, and G. Bastin. *Theory of Robot Control*. Springer, Berlin, 1997.

# Neural Trajectory Optimization (NTO) for Manipulator Tracking of Unknown Surfaces

R. Maaß, M. Dapper, R. Eckmiller

Department of Computer Science, University of Bonn, F. R. Germany

e-mail: maass@nero.uni-bonn.de

## Abstract

We developed a novel approach for surface tracking and force/position control based on neural networks. A new concept of neural trajectory optimization NTO will be presented as a part of the neural force/position control NFC and as a very capable and versatile tool for the generation of natural manipulator movements (fast, flexible and smooth). The NTO concept is based on DRBF neural networks, an extension of the RBF type network, to be proposed in this paper. Experimental results of the realtime implementation of NTO and simulation results of its combination with the neural force/position control NFC will be presented. As a testbed we use a 6 DOF industrial manipulator executing demanding tasks such as surface tracking with defined normal force. [1]

## 1 Introduction

Most of current industrial manipulator tasks include grinding, deburring, polishing and assembly operations. Such surface tracking applications generally need to control dynamically the tangential velocity along an uncertain work surface as well as a normal distance or contact force to the workpiece (fig.1) [1]. In recent years some novel concepts were proposed to control the contact parameter while tracking and working along the workpieces surface [2][3][6]. A challenge concerning these approaches is to add a modular trajectory tool to the controller, that enables the manipulator to perform constraint motions with naturally high speed, while guaranting the stability of the contact parameter (distance or contact force).

## 2 Results

### 2.1 Concept of DRBF-Neurons

Optimization of trajectories concerning physical properties of the manipulator, and taking into account the manipulator's hardware constraints for its

---

[1] This work was supported in part by Federal Ministry for Education, Science, Research, and Technology (BMBF) under grant "DEMON"

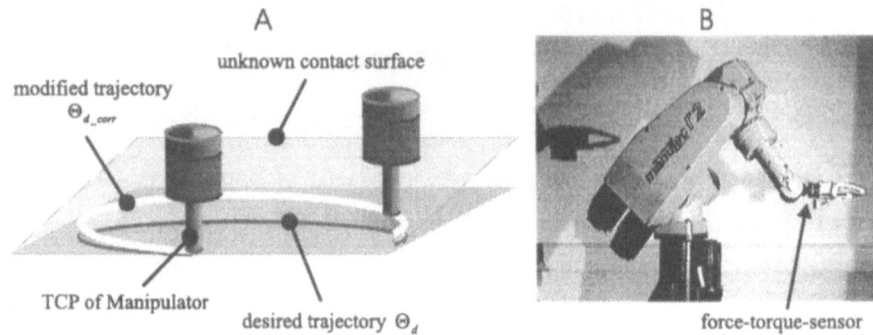

A

B

modified trajectory
$\Theta_{d\_corr}$

unknown contact surface

TCP of Manipulator

desired trajectory $\Theta_d$

force-torque-sensor

Figure 1: A: Surface tracking with defined contact to uncertain surfaces,
B: Realization for a 6DOF Industrial robot (Siemens Manutec) with 6D-wrist
force/torque sensor (ATI)

movement, require modification of joint angel velocity, -acceleration and -jerk
while maintaining the desired course. Many applications in computer science
and robotics deal not only with the approximation of a function itself, but also
with the optimization of higher derivatives of the desired function (e.g. generating of time optimal movements). In this paper, an extension to the RBF-type
net is proposed, the 'Derivable Radial Basis Function Net' (**DRBF**) which
delivers not only the common net outputs with the weighted activities

$$y_j = \sum_i w_{ij} a_i \quad , \text{ with } \quad a_i = exp(\frac{\sum_k (x_k - c_{ki})^2}{\sigma_i^2}) = exp(C), \quad (1)$$

but also the derivatives of the single, weighted neuron activities:

$$\frac{\partial y_j}{\partial x_k} = \sum_i w_{ij} \delta_i a_i \quad , \text{ with } \quad \delta_i = \frac{\partial C}{\partial x_k} \quad (2)$$

$$\frac{\partial^2 y_j}{\partial x_k^2} = \sum_i w_{ij} \delta\delta_i a_i \quad , \text{ with } \quad \delta\delta_i = \frac{\partial^2 C}{\partial x_k^2} + (\frac{\partial C}{\partial x_k})^2 \quad (3)$$

The 'exponential function' - character of the used RBF net functions allows
generating the derived net outputs via the factor $\delta_i$ and $\delta\delta_i$ (fig.2 A). The complete derivative of the common net outputs is build by the sum of the new
weighted activities.

Higher dimensional input vectors are derived with ease for any valid function. Thus DRBF nets are very convenient for realtime applications like surface
tracking and force/position control, where higher derivatives of the robots joint
angles are required. The modulation facilities of DRBF nets can be enhanced
by adding a second hidden layer. The structure, used in the NTO module consists of two one-dimensional neural DRBF nets $N_1$ and $N_2$ with time $t$ as input
of the first and $t^*$ as input of the second net. Thus the resulting joint angle

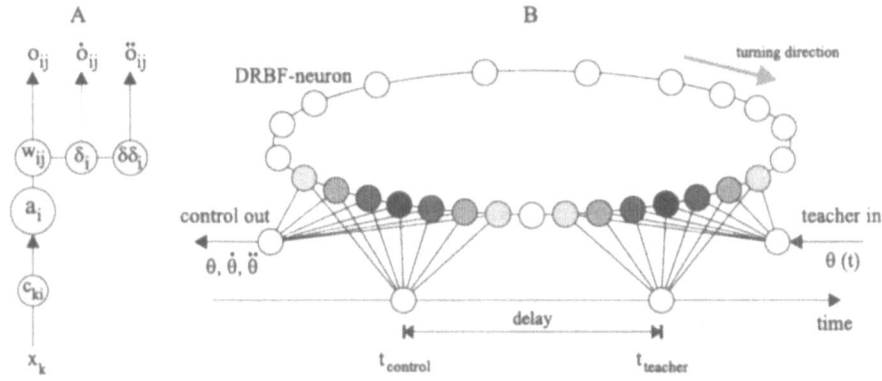

Figure 2: A: DRBF Neuron with output $o_{ij}$ and its derivatives $\dot{o}_{ij}$ and $\ddot{o}_{ij}$, B: Schema of Turning DRBF Trajectory Net for realtime processing of viapoints

positions for a 6DOF manipulator appear in the form $\theta(t^*(t)) = N_2(N_1(t))$. The weights of net $N_1$ are modified, taking temporary violations of velocity and acceleration constraints of the manipulator into account.

## 2.2 Online Training and Generation of Manipulator Trajectories NTO with Turning DRBF-Network

In many cases manipulator movements have to be generated or modified quickly, thus it is necessary to have a tool that generates smooth and kinematically valid trajectories from viapoints in realtime. With DRBF-Neurons a circular structure was designed, the Turning DRBF-Network (fig.2 B). During a realtime process viapoint sequences are used as teacher input and desired control values $(\theta, \dot{\theta}, \ddot{\theta})$ as output. The time difference $\Delta t_{delay}$ between teacher input $t_{teacher}$ and control output $t_{control}$ depends on the expected velocities and the desired smoothness of the trajectory. Typical values are: Input of new viapoint every $20msec$, sampling time of $2msec$ and a delay $\Delta t_{delay}$ of 0.1 to 0.4 seconds.

The functionality of the NTO module as a part of a robot control is described as follows:

The NTO Modul receives Cartesian viapoint sequences from a planning instance, describing a desired manipulator movement. Then the viapoints are trained into the DRBF net offline or online. Output values are handed down to a computed torque controller in the closed loop ($2msec$). Kinematical mappings, generating the transformation between the Cartesian and the joint angle space are delivered by an optional neural or analytical module [4][5]. Two feedbackloops allow adaptations and optimizations (offline or online) in joint space, Cartesian space or in the time function $t^*$. E.g. in surface tracking tasks a desired contact trajectory can be adapted to an uncertain surface during the first run and improved by iterated processing.

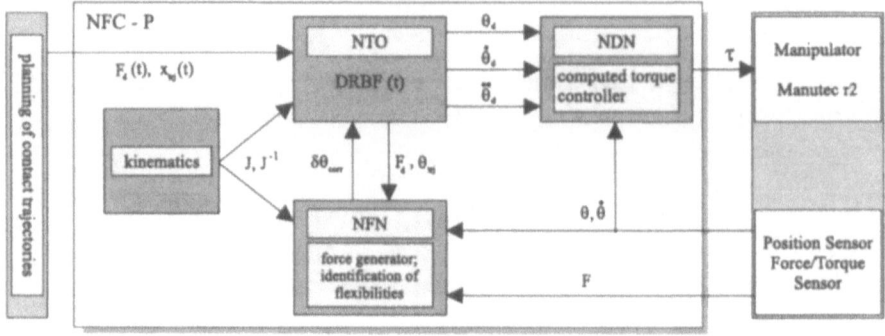

Figure 3: Schema of position based neural force control NFC-P including the NTO module, a kinematic and dynamic (NDN) module and the neural force net NFN module for processing the force data in the control loop.

## 2.3 Combination of NTO with Hybrid Force/Position Control NFC-P

The neural trajectory module NTO can be included into a force/position (f/p) controller to solve surface tracking tasks dynamically and adaptiv. Fig.3 shows the schema of position based neural f/p control NFC-P with the extension of the NFN (Neural Force Net) module in the closed loop [5]. This module has two functions: The influence of the tool (screw driver, pen) on the force sensor data, depending on the manipulator's position $\theta$, can be trained and deducted. Secondly a virtual force parameter can be adapted online:

$$c_{virtual}(t) = \frac{\Delta F_{act}}{SJ\Delta\theta_{des}(t - \delta t)}. \tag{4}$$

From this follows a new desired offset value to the trajectory joint value $\Delta\theta_{des}(t)$ to control the contact force $F$. The matrix S separates force and position controlled components of the trajectory. J is the Jacobian matrix.

## 2.4 Experimental Results

A realtime implementation of the NTO module was tested in several experiments with our 6DOF Manipulator (Siemens Manutec r2). Fig.4 (e.g.) displays the influence of the trajectory modifications of NTO to the joint acceleration: kinematically invalid (figures above) trajectories given by viapoints during realtime process are changed into kinematically valid movements (figures below). These modifications have no significant influence on the position errors of the trajectory.

Other experiments have shown, that NTO enables the manipulator to follow viapoints with a delay $t_{delay}$ of only $0.2sec$ to $0.4sec$ even if the desired joint velocities and accelerations are near or above the manipulator's limiting values.

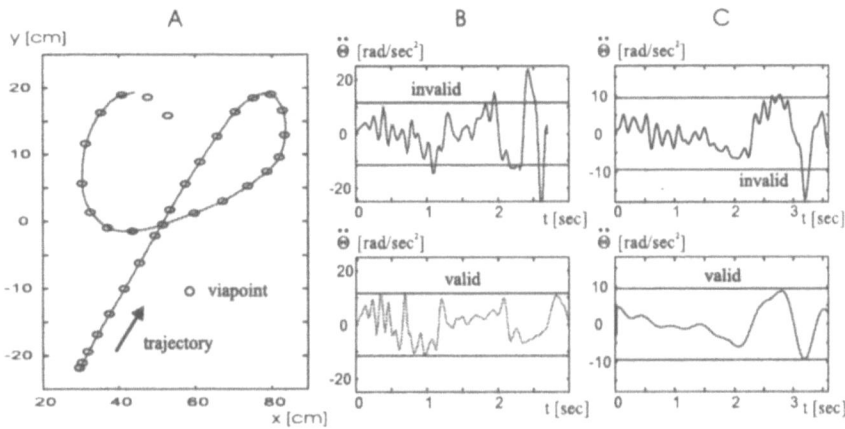

Figure 4: Online trajectory optimization NTO: A: Viapoints of desired trajectory and actual manipulator trajectory, B: Time-modulated trajectory, and C: Modulation of smoothness to kinematically valid joint accelerations $d^2\theta/dt^2$

## 2.5 Simulations

The simulation results for surface tracking (fig.5) with the neural f/p controller NFC-P suppose a quit stiff surface ($c = 50N/cm$) with uncertain shape in z-direction. The tangential velocity of the TCP is quit high ($\approx 30cm/sec$) even so the controller holds steady the desired contact force of $10N$ and follows the sudden change to $20N$ during process.

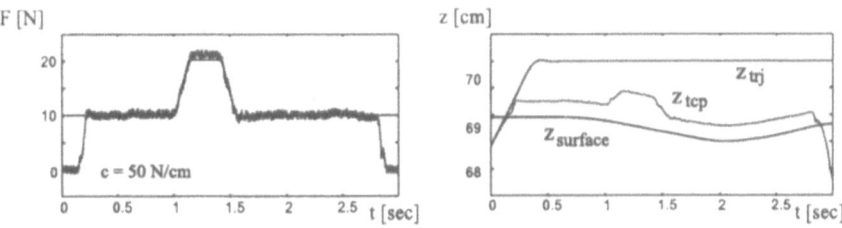

Figure 5: Simulations: neural force/position control NFC-P with NTO included: Tracking the trajectory $z_{tcp}$, given by the viapoints $z_{trj}$ of fig.4A, in contact with an uncertain surface $z_{surface}$. The force response $F$ holds the desired value of $10N$ or temporary $20N$.

## 3 Discussion

The advantages of trajectory generation based on DRBF neural networks compared to common methods like cubic splines are:

- The trained trajectories are infinitely often continuously derivable.
- Optimizations, regarding time and path behavior of trajectories, can be realized easy and fast.
- Simple changes of neuron distances or widths allow a continuous adjustment of trajectory smoothness versus exactness.
- The required time for the training and output calculation of the DRBF network is short enough to be processed in realtime ($2msec$) during fast manipulator movements or movements with defined contact to a surface (f/p control).

These specified features of NTO were successfully tested in several realtime experiments.

## 4 Conclusions

We presented a novel approach for manipulator control during surface tracking applications NTO. Based on Derivable Radial Basis Function (DRBF) networks, this flexible module calculates and optimizes trajectory data during real time processes and is therefore appropriate for several different control principles (e.g. NFC-P, NFC [4]). NFC-P with the proposed NTO expands the range of manipulator applications to various difficult tasks, such as grinding, deburring, polishing and assembly operations.

## References

[1] B.H. Park, J.S. Lee and S.Y. Park. An Adaptive Learning Control Method for Constrained Motion of Uncertain Robotic Systems. *IEEE Proc. Int. Conf. Robotics and Automation, Albuquerque, New Mexico, April*, pages 531–536, 1997.

[2] S. Chiaverini, B. Siciliano, and L. Villani. Force and position tracking: Parallel control with stiffnes adaptation. *IEEE Control Systems Magazine*, 18(1):27–33, 1998.

[3] M.S. Querioz, D.M. Dawson and H.Canbolat. Adaptive Position/Force Control of BDC-RLED robots without Velocity Measurements. *IEEE Proc. Int. Conf. Robotics and Automation, Albuquerque, New Mexico, April*, pages 525–530, 1997.

[4] M. Dapper, R. Maaß, V. Zahn, and R. Eckmiller. Neural force control (nfc) for complex manipulator tasks. In *Springer Berlin Lecture Notes in Computer Science 1327, Artificial Neural Networks - ICANN97, Lausanne, Switzerland, October*, pages 787–792, 1997.

[5] R. Maaß, V. Zahn, and R. Eckmiller. Neural force/position control in cartesian space for a 6 dof industrial robot: Concept and first results. In *IEEE Proc. Int. Conf. Neural Networks, ICNN97, Houston, USA, June*, pages 1744–1748, 1997.

[6] C.C. de Wit, B. Siciliano, and G. Bastin. *Theory of Robot Control.* Springer Berlin, 1997.

# Computer Network User Behaviour Visualisation Using Self Organising Maps

Albert J. Höglund

Nokia Research Center, Nokia Group

Helsinki, Finland

albert.hoglund@research.nokia.com

Kimmo Hätönen

Nokia Research Center, Nokia Group

Helsinki, Finland

kimmo.hatonen@research.nokia.com

## Abstract

Computer systems are vulnerable to abuse by insiders and to penetration by outsiders. The amount of monitoring data generated in computer networks is enormous. Tools are needed to ease the work of system operators. Anomaly detection attempts to recognise abnormal behaviour to detect intrusions.

A prototype Anomaly Detection System has been constructed. The system provides means for automatic anomaly detection and user behaviour visualisation. The system consists of a data gathering component, a user behaviour visualisation component, an automatic anomaly detection component and a user interface. This paper is focused on the user behaviour visualisation component. This component uses large Self Organising Maps as a basis. The construction and the usage of the component is presented. Some discussion on comments from the test usage of the Anomaly Detection System is also provided.

## 1 Introduction

Computers and computer networks are becoming more and more important. The computer networks are normally protected from unauthorised usage by security mechanisms, such as passwords and access controls. However, if an abuser or intruder manages to bypass these security mechanisms and gains access to vital information, the potential loss is enormous.

The loss can be decreased by detecting intruders or abusers at an early stage. The two main intrusion detection techniques are rule-based misuse detection and anomaly detection. Rule-based misuse detection attempts to recognise specific behaviours that are known to be improper. That is, if a user follows certain intrusive patterns, he is classified as an intruder. Anomaly detection, on the other hand, attempts to recognise anomalous or abnormal user behaviour to detect intrusions. Anomalous or abnormal behaviour is suspected if the current behaviour deviates sufficiently from the previous behaviour, which is assumed normal. Anomaly detection deals with behaviour that is not known in advance, while rule-based misuse detection deals with predefined violations. In this difficulty also lies the advantage of anomaly detection: it can namely be used to detect types of intrusions that have never occurred before. References on anomaly detection and misuse intrusion detection research can be found in [1, 2, 3, 4].

A prototype Anomaly Detection System for the UNIX environment has been constructed [5]. The system provides means for automatic anomaly detection and user behaviour visualisation. The system consists of a data gathering component, a user behaviour visualisation component, an automatic anomaly detection component and a user interface. This paper presents the user behaviour visualisation component.

The objective with the user behaviour visualisation was to visualise the user behaviour during a certain period in a simple way. Such a component is useful, since the amount of data generated by users in a computer network is enormous.

The amount of information is reduced by selecting a set of features that characterises the behaviour of the users in the network. This set of features should form a daily fingerprint of the network user, which means that it has to be selected carefully. Although the amount of data is reduced in the feature selection process, it is still difficult to compare and analyse the user behaviour. The visualisation problem is tackled using an approach, in which the Self Organising Map is used to visualise the user behaviour in two dimensions. The Self Organising Map is a neural network based on unsupervised learning and it is suitable for visualisation and interpretation of large high-dimensional sets of data.

## 2 Methods and Implementation

### 2.1 Software and Environment

The prototype Anomaly Detection System was built in the UNIX environment. The data gathering and the data processing are performed on separate servers. This enhances security and makes it more difficult to disturb the operation of the system.

The routines that build the prototype Anomaly Detection System have been coded using C and Perl and the Self Organising Maps are made using the procedures of SOM_PAK [6]. The user interface uses Netscape[1] to view the html pages generated by the system.

### 2.2 Data Gathering and Scaling

For a period of 400 days the user account logs of more than 600 users have been stored. The user account logs give information on the processes performed by the users. This information includes CPU-times, characters transmitted and blocks read. The selection of features describing the user behaviour is discussed in Section 2.3.

---

[1] The Netscape browser can be downloaded from http://www.netscape.com/

Since the magnitude of the features varies greatly, logarithmic or linear scaling was considered necessary. The features were scaled according to (1) and (2), where $f$ is the feature in question. The division by the maximum scales the parameters to the range [0, 1]. Histograms were studied in order to determine which one of the two scalings was more suitable for the feature in question.

$$f_{i\_Log-scaled} = \frac{\ln(f_i + 1)}{\max_i[\ln(f_i + 1)]}, \quad f_{i\_Lin-scaled} = \frac{f_i}{\max_i[f_i]} \qquad (1), (2)$$

## 2.3 Feature Selection

Feature selection provides a means of reducing the enormous amount of data generated by computer network users. The feature selection problem can be stated as follows: Objects are described with a large set of features. The objective is to find a subset of features that distinguishes the objects from each other as well as possible.

Features characterising the user behaviour during a period of 24 hours are used in the user behaviour visualisation component. Firstly an initial feature set with 34 features was derived. Features describing CPU-time and transmitted characters for different services were included, but also session, process and login information. The initial feature set was reduced to a set of 16 features. This was achieved by omitting features with strong linear dependency to other features and by omitting very noisy features. The linear correlation was checked using a correlation test and the noisy features were found by closely examining the variances. Careful consideration was used in the feature selection process.

## 2.4 User Behaviour Visualisation

The idea is to use Self Organising Maps (SOM) to visualise the user behaviour during a certain period in a simple way. The SOM is an effective tool for visualisation of high-dimensional data [7, 8]. The principal goal of the SOM is to transform an incoming signal pattern of arbitrary dimension into a one- or two-dimensional discrete map, and to perform this transformation in a topologically ordered fashion. The algorithm and detailed theory on the SOM can be found [7].

The user behaviour visualisation component uses two-dimensional SOM:s. The maps are constructed using the whole data set for the whole period, which means the data for all the users for all the days in the period. A large number of tests indicated that maps of size 18x14 give sufficient accuracy. The type of lattice used is hexagonal (six neighbours) and the neighbourhood function type used is bubble [6, 7]. In this paper the map has been labelled with the user number and the number of "hits" on the neuron. These labels are separated by a "_". The maps are visualised using the U-matrix method [9, 10, 11]. In this method the neurons are marked with dots and the distances between them are described with greyscales. The darker the cell between two neurons, the greater the distance between them. In addition, the user behaviour visualisation component provides real values for the neurons of the map, a connection to the real data and feature statistics.

# 3   Using the User Behaviour Visualisation Component

The user behaviour visualisation component of the Anomaly Detection System prototype has been in test usage during a period of several months. The following user behaviour clustering examples illustrate the use of the user behaviour visualisation component.

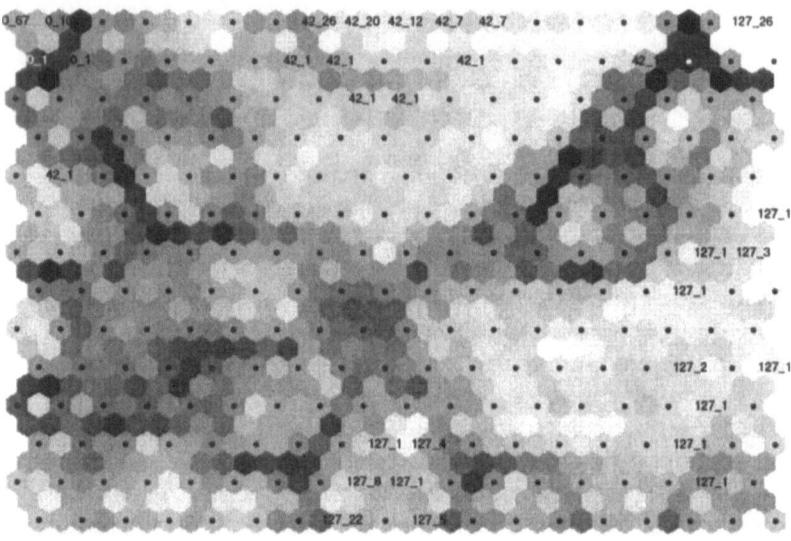

Figure 1  Map of size 18x14 trained with the whole set of data and labelled with the usage of user 0, 42 and 127

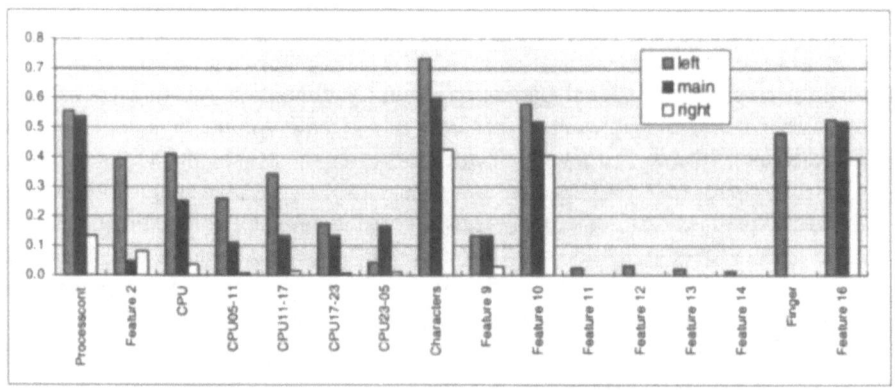

Figure 2  Feature distribution for user 42 main usage cluster compared with the deviation on the right and the deviation on the left.

The behaviour of three users during a period of 79 days is visualised in Figure 1. The behaviour of user 42 is quite nicely clustered. Figure 1 shows that user 42 has one main usage cluster in the upper middle of the map, but one notices that there are two deviations from this cluster, one on the right and one on the left. An explanation to these deviations is given in Figure 2. In this figure the feature distribution of the neuron with 20 hits from the main usage cluster is compared with the feature distribution of the deviations on the left and on the right. The network usage of the deviation on the right is lighter than normal with fewer processes, less CPU-time and fewer characters transmitted. This deviation can be explained with a breakdown in the network. The deviation on the left, on the other hand, is a bit anomalous. The usage is heavier than normal, especially the CPU-times in the morning and in the afternoon are bigger than normal. Another anomalous thing with the deviation on the left is the high usage of the finger service compared with no use at all for the normal behaviour.

The behaviour of hundreds of users have been analysed during the test usage. There is for example a well bounded no usage cluster in the upper right corner of the map in Figure 1. Users that rarely use the computer network are mostly mapped to this cluster. Figure 1 shows that user 127 has 26 days with no usage during the period of 79 days. Another well-bounded cluster is located in the upper left corner of the map. A system demon id with user number 0 is mapped to this cluster (see Figure 1). The behaviour of this id can be considered very regular since its behaviour is mapped to only four neurons on the map. A misuser or intruder using the id of the system demon would be very critical, since it has greater privileges in the network than the normal users. Deviations in the behaviour of the system demon id can easily be noted using the user behaviour visualisation component.

There are also users that are mapped to two clearly separated usage clusters, which means that the users have two working modes. An example of this is user 127. Figure 1 shows that the behaviour of user 127 is clustered to three clearly separated clusters. One is the no usage cluster and the two others are the normal working modes. These modes can be analysed further using the same procedures as with user 42 above. The phenomenon with two working modes may, for example, originate from the fact that the users work in more than one project. There are of course some users whose behaviour is very irregular and their behaviour is therefore not so nicely clustered on the map.

# 4   Conclusions and Discussion

The initial feedback from the test usage of the Anomaly Detection System has been quite positive. Comments like "The user behaviour visualisation component gives a quick overview of the user behaviour" were quite encouraging. The test usage feedback also indicated that the component is practical when analysing user behaviour that has been reported anomalous. Section 3 showed how the Self Organising Map in the user behaviour visualisation component can be used to analyse user behaviour. The user feedback also included comments on necessary improvements. Better connections to the real data for further analysis were suggested and have now been implemented. Improvements in the labelling of the map and map region classification have also been suggested and will be implemented in the future.

The general impression of the authors is that the Self Organising Map provides a good method for reducing the dimensions of the data and for comparing and visualising the behaviour of network users. Examples and test usage give just indications of the performance of the user behaviour visualisation component, though. Simulation experiments for further evaluation of both the user behaviour visualisation component and the automatic anomaly detection component will therefore be performed. A publication on the automatic anomaly detection component is also under preparation.

# References

[1] Javitz H S, Valdes A, Lunt T F, Tamaru A, Tyson M, Lowrance J. Next generation intrusion-detection expert system (NIDES): Statistical algorithms rationale and rationale for proposed resolver. Technical report, Computer Science Laboratory, SRI International, Menlo Park, California, The USA 1993.

[2] Kumar S, Spafford EH. A pattern matching model for misuse intrusion detection. In Proceedings of the 17th National Computer Security Conference, October 1994, pp 11-21.

[3] Lankewicz L B, A nonparametric pattern recognition approach to anomaly detection. Doctoral Thesis, Tulane University, 1992.

[4] Lunt T. F. A survey of intrusion detection techniques, Computers and Security 1993; 12(4):405-418

[5] Höglund A. An Anomaly Detection System for Computer Networks. Master's thesis, Helsinki University of Technology, Helsinki, 1997.

[6] Kohonen T, Hynninen J, Kangas J, Laaksonen J. Manual of SOM_PAK, The Self-Organising Map Program Package, Version 3.1, April 7, 1995, http://nucleus.hut.fi/nnrc.html.

[7] Kohonen T. Self Organising Maps. Second Edition. Springer-Verlag, Heidelberg 1997.

[8] Neural Networks Research Centre & Laboratory of Computer and Information Science. Triennal Report 1994-1996. Helsinki University of Technology 1997.

[9] Iivarinen J, Kohonen T, Kangas J, Kaski S. Visualising the clusters on the Self Organising Map. Multiple Paradigms for Artificial Intelligence (SteP94), 122-126. Finnish Artificial Intelligence Society, 1994.

[10] Kraaijveld M A, Mao J, Jain A K. A non-linear projection method based on Kohonen's topology preserving maps. Proceedings of the 11th International Conference on Pattern Recognition (11ICPR), 41-45, Los Alamitos, CA. IEEE Comput. Soc. Press, 1992.

[11] Ultsch A, Self organised feature maps for monitoring and knowledge acquisition of a chemical process. Gielen S, Kappen B, editors, Proceedings of the International Conference on Artificial Neural Networks (ICANN93), London. Springer-Verlag, 1993, pp 864-867.

# Neural Control of a Virtual Prosthesis

Lars Eriksson

Dept. of Electrical Measurement, Lund Institute of Technology
Lund, Sweden
cie91lme@student1.lu.se

Fredrik Sebelius

Dept. of Electrical Measurement, Lund Institute of Technology
Lund, Sweden
cie91fse@student1.lu.se

Christian Balkenius

Lund University Cognitive Science
Lund, Sweden
christian.balkenius@fil.lu.se

## 1 Introduction

The abilities of the currently existing hand prostheses are typically limited to opening or closing the hand. This limits the usefulness of the prosthesis considerably compared to the many degrees of freedom in an intact hand. In order to develop more advanced hand prostheses two main problems have to be solved. The first is to develop more advanced mechanical solutions that allows for more degrees of freedom. The second, that we address below, is to devise a way of controlling the additional dexterity of such a prosthesis. Before the second problem is solved, the development of more advanced prostheses will be severely hindered.

The most natural way to control a prosthesis would be in the same way as the hand is controlled: by neural control mediated by the nerves intended for the amputated hand or arm. The system we present below is an implementation of this type of control that exploits that the muscle activity of the remaining part of the arm is correlated with the desired action of the prosthesis. By recording from several surface mounted electrodes on the arm, we are able to predict the corresponding motion of the hand using an artificial neural network.

Figure 1 shows the principle of the system. Eight electrodes record EMG signals from the limb. These are subsequently fed into a preprocessing stage and then into a variant of the self-organizing feature map that learns to recognize categories of muscle patterns. These categories are then associated with the appropriate motor action recorded from a data glove. The signal processing and classification methods are based on our earlier work with signals recorded from rats [1, 2].

At the current state of the project, we are using surface mounted electrodes because they do not require surgery and are easy to manage. While the surface mounted electrodes seem to provide enough information, implanted electrodes have better signal conditions and would therefore enhance performance. The benefits of surface mounted electrodes would however come about to the expense of more complex filtering and pre-processing. Since no prosthesis with the desired mechanical properties exist, we use a simulated hand generated on a computer screen in an OpenGL environment. This hand shows the movement of the virtual prosthesis corresponding to the EMG patterns sent to the neural network.

The following sections describe the processing of the EMG signals and the architecture of the neural network used. In the last sections we describe preliminary results with a patient with a congenital amputation of the left hand. After training the neural network, the patient was able to control the virtual prosthesis by will.

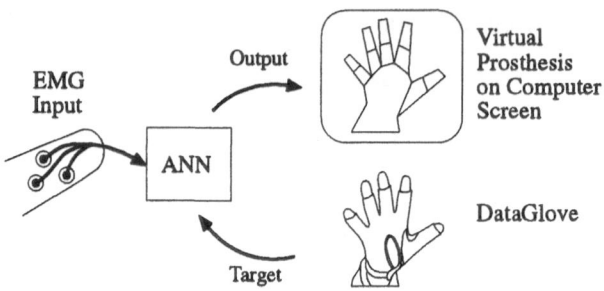

Figure 1: A virtual prosthesis can be controlled by EMG signals that are interpreted by a neural network.

## 2 Recording of EMG Signals

The EMG signals were recorded using sixteen electrodes and sent to a specially built eight channel bipolar amplifier. The amplifier uses a switched notch filter at 50Hz. The signals were then digitized at 8 kHz on a standard analogue to digital converter card on a PC.

Figure 2 (top) illustrates the signals picked up from electrodes mounted on a patients lower limb. The index finger was moved back and forth two times during the measurement.

We only want to retrieve the important information in the muscle signals. If this is not achieved then the network will, of course, give poor results. Thus, we reduce the data as much as possible to create a smooth function with a small time delay (Figure 2 bottom).

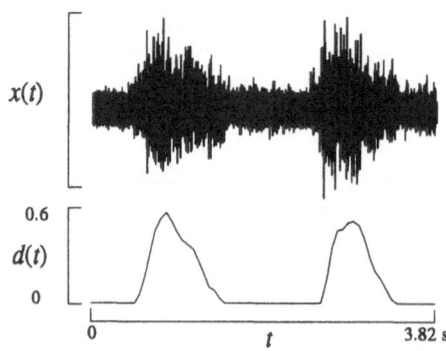

Figure 2: (top) One of the raw muscle signal as recorded from the patient's left limb. From left to right, the picture shows the measured signals derived when moving the finger twice. (bottom) The signal after pre-processing.

# 3 Preprocessing

In order to reveal the true signals, we have employed a custom made preprocessing methods. These are in order (figure 3): (1) rectification of the signal, (2) correlation between individual channels are used to remove inter-channel crosstalk, (3) a special band-pass filter that uses time versus amplitude differences in the signal, (4) downsampling to remove redundant information and reduce memory storage. These methods are applied to each of the eight channels. The following sections describe them in detail.

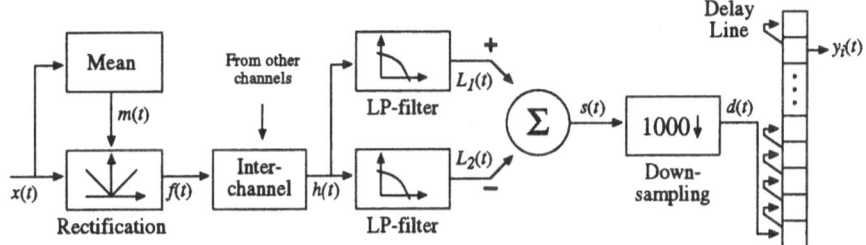

Figure 3: The preprocessing of the EMG signals

## 3.1 Rectification

To rectify the recorded signals, we let a floating value represent the mean of the signal, and then mirror the activity round it. This is necessary since the muscle signals contain both a positive and a negative peak [3]. With this method we achieved an adequate amplitude for the signal. The formula is:

$$f(t) = |x(t) - m(t)|$$ (3.1)

where x(t) is the sampled value and m(t) is the floating mean of the original signal.

## 3.2 Correlation Between Channels

In order to inhibit inter-channel cross talk we used the following algorithm:

$$g(t) = (1 - \alpha)g(t-1) + \alpha \sum_{k=0}^{K} f_k(t-1)$$ (3.2)

$$h_k(t) = f_k(t) \left( \sum_{k=0}^{K} f_k(t) \right)^{-1} g(t)$$ (3.3)

where K is the number of channels, $\alpha$ is a constant and g(t) represents the floating mean of the total activity of the signal.

## 3.3 Band-Pass Filtering

The next step is to band-pass filter the signals. This is done using two third order low-pass filters as follows:

$$L^0(t) = h(t)$$
$$L^{i+1}(t) = \gamma L^i(t) + (1 - \gamma)L^i(t-1) \tag{3.4}$$

The signal is divided into two parts with different parameters in the filters. The difference between the outputs from the two low-pass filters are then calculated:

$$s(t) = L^3_{\gamma_1}(t) - L^3_{\gamma_2}(t) \tag{3.5}$$

The two time constants $\gamma_1$ and $\gamma_2$ generate differences in the two low pass filters and we thus achieved a band pass filter. In order to decrease the storage space needed it is now possible to down-sample the data set without losing any important information.

$$d(t) = \frac{1}{N} \sum_{i=0}^{N-1} s(tN + i) \tag{3.6}$$

Figure 2 (bottom) shows the outcome of the preprocessing stage. Notice how the essentials of the signal are revealed. We can see that a great clarification has been achieved through the amplitude differentiation algorithm. Because of the dramatic decrease in density, the storage space required is also far less than that of the original signal, which in turn boosts network performance. Finally we normalize the signals before we send them to a delay line which constructs the input vectors for the categorization network.

## 4 Classification of EMG Patterns

For classification, we use the self-organizing feature map [4] modified with the conscience mechanism suggested by DeSieno [5]. Additionally, we bias each input signal to the network with the estimated variance of that signal in such a way that a signal with small variance is allowed to make a larger contribution to the category. The variance is calculated locally at each node for the signals that are categorized as belonging to its class.

The basis for the conscience mechanism is that all nodes in the feature maps should be used. To make this possible, the selection of the winning node is biased by how much it has been previously activated. Nodes that have not been activated very much gain an advantage in the competition. To accomplish this, we calculate an estimate, f, of how much a node has won the competition,

$$f_i(t+1) = f_i(t)(1 - \beta) + \beta z_i(t) \tag{4.1}$$

where $z_i(t)$ is 1 when node i wins at time t, and 0 otherwise. The parameter $\beta$ sets the time horizon for the conscience mechanism. This estimate is used to calculate a bias for node i as,

$$bias_i(t) = \lambda\left(\frac{1}{N} - f_i(t+1)\right)$$

(4.2)

$N$ is the number of nodes in the network and $\lambda$ is a constant. The calculation of the distance to the winning node is biased in the following way [5],

$$dist_i(t) = \sqrt{\sum_{i=0}^{n-1}(x_i(t) - w_i(t))^2(1 - v_i(t))} - bias_i(t)$$

(4.3)

where $v_i$ is a floating variance estimate,

$$v_i(t+1) = \gamma(x_i - w_i)^2 + (1 - \gamma)v_i(t)$$

(4.4)

In this way each node determine how substantial each input is. Changes to the weights are subsequently done in the standard way [4]. Finally, each node in the network is associated with the corresponding hand posture recorded by the data glove. On recall, these stored hand signals are used to predict the movement of the virtual prosthesis. The inclusion of the conscience and the variance mechanisms improves the learning speed and the correctness of the classification compared to the standard SOFM-network [1].

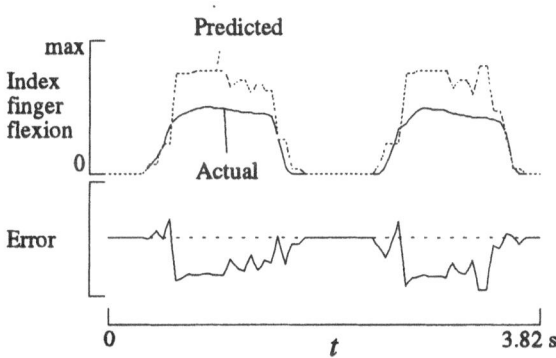

Figure 4: The actual finger position recorded from the data glove (black) compared to the predicted finger position (gray) generated by the neural network and the difference between actual and predicted position.

## 5 Experiment and Results

A 24 year old patient who was born without a left hand tried to imagine moving her left hand in the same way as the intact right hand. The position of the intact right hand was recorded by a data glove simultaneously with EMG signals measured at her left limb. Eight bipolar surface mounted electrodes were used.

The patient performed five different movement patterns with her hand. Each finger was flexed individually and each movement was performed twice. In all, signals were recorded in four sets at two different occasions. Data from the first session were used to train the network, and data from the second session were used to verify the performance of the network.

Figure 4 shows the behavior of the system for the index finger during two movements. It can be seen that the neural network closely follows the actual movement of the finger. Similar results were obtained for the other fingers and for other movements.

The difference between actual and predicted finger position is less than what is required to control a prosthesis. Also note that in this experiment, the patient is not allowed to learn the behavior of the virtual prostheses. If this were allowed, even better performance could be expected.

# 6 Conclusion

Our experiment shows that neural networks are feasible for categorizing patterns of EMG signals. The signals recorded by the surface electrodes are sufficient to control the movements of a virtual prosthesis. The presented method offers great potential for the development of future hand prostheses.

# Acknowledgements

This research was performed at and partly supported by the Dept. of Hand Surgery at Malmö University Hospital. We would like to thank Göran Lundborg for initiating and promoting the project, Styrbjörn Lindberg for the effort he put into constructing the measurement hardware, and Birgitta Rosén for help with the measurements. Earlier stages of this project were achieved with the kind help of Ingmar Rosén, Dept. of Clinical Neurophysiology, Lund University and Lars Montelius, Dept. of Solid State Physics, Lund University. We also thank Nissho Electronics, Japan, for lending us the SuperGlove used in the experiments.

# References

[1]   Eriksson L. Sebelius F. Pattern recognition of nerve signals using artificial neural networks. M.Sc. thesis, Department of Solid State Physics, Lund University, 1996

[2]   Montelius L, Sebelius F, Eriksson L, Holmberg H, Schouenbourg J, Danielsen N, Wallman L, Laurell T, Balkenius C. (1996). Pattern recognition of nerve signals using an artificial neural network. Proceedings of the 18th Annual International Conference of the IEEE Engineering in Medicine and Biology Society , 1996

[3]   Deutch, S. and Deutch, A. Understanding the Nervous System. IEEE Press, New York, 1992

[4]   Kohonen, T. Self-organization and associative memory. Springer-Verlag, Berlin, 1984

[5]   DeSieno, D. Adding a conscience to competitive learning. IEEE International Conference on Neural Networks, IEEE Press, New York, 1988, vol. 1, pp 117--124

# Adaptive Clustering and Multidimensional Scaling of Large and Highdimensional Data Sets

Friedhelm Schwenker, Hans Kestler and Günther Palm

Department of Neural Information Processing, University of Ulm

D-89069 Ulm, Germany

## Abstract

We describe a algorithm for exploratory data analysis which combines the adaptive c-means clustering and the multi-dimensional scaling procedure (ACMMDS). ACMMDS is an algorithm for the online visualization of clustering processes and may be considered as a alternative approach to Kohonen's self organizing feature (SOM). Whereas SOM is a heuristic neural network algorithm, ACMMDS is derived from multivariate statistic algorithms. The possible implications of ACMMDS are illustrated through two different data sets.

## 1   Introduction

In many practical applications one has to explore the underlying structure of a large set objects. Typically, each object is represented by a feature vector $x \in \mathcal{X}$, where $\mathcal{X}$ is the feature space endowed with a distance measure $d_{\mathcal{X}}$. This data analysis problem can be tackled utilizing *clustering methods* (see [2] for an overview). A widely used clustering algorithm is *c-means clustering* [6, 5, 4] where the aim is to reduce a set of $M$ data points $X = \{x_1, \ldots, x_M\} \subset \mathcal{X}$ into a few, but representative, cluster centers $\{c_1, \ldots, c_k\} \subset \mathcal{X}$.

A neural network algorithm for clustering is Kohonen's *selforganizing feature map (SOM)* [3]. SOM is similar to the classical sequential c-means algorithm (see section 2) with the difference that in SOM the cluster centers are mapped to a *display space* $\mathcal{Z}$ with a distance measure $d_{\mathcal{Z}}$. Each cluster center is mapped to a fixed location of the display space. The idea of this display space $\mathcal{Z}$ is that cluster centers corresponding to nearby points in $\mathcal{Z}$ have nearby locations in the feature space $\mathcal{X}$. Typically, $\mathcal{Z}$ is a 2-dimensional (or 3-dimensional) grid. Therefore it is often emphasized that Kohonen's SOM is able to combine clustering and visualization aspects. In this context it has to be mentioned that SOM is a heuristic algorithm — not derived from a objective function which incorporates both a clustering criterion and some kind of topological or neighborhood preserving measure [3, 8].

Another approach for getting an overview over a high-dimensional data set $X \subset \mathcal{X}$ are *visualization methods*. In multivariate statistics several linear and nonlinear techniques have been developed. A widely used nonlinear visualization method is *multidimensional scaling (MDS)* [10, 9]. MDS is a class of distance preserving mappings from the data set $X$ into a low-dimensional *projection space* $\mathcal{Y}$ which is endowed with some distance measure $d_{\mathcal{Y}}$. Each feature vector $x_\mu \in X$ is mapped to a point $y_\mu := p(x_\mu) \in \mathcal{Y}$ in such a way that the distance matrices $D_X := (d_X(x_i, x_j))_{1 \leq i,j \leq M}$ in feature space $X$ is approximated by the distance matrix $D_Y := (d_Y(y_i, y_j))_{1 \leq i,j \leq M}$ in projection space $\mathcal{Y}$.

In this paper an algorithm, which we call ACMMDS, is described. It combines c-means clustering and MDS. This procedure is able to be utilized for the online visualization of clustering processes and should be considered as a alternative approach to Kohonen's self organizing feature (SOM).

Throughout this paper we restrict our considerations to the feature space $X = \mathbb{R}^n$ and the projection space $\mathcal{Y} = \mathbb{R}^r$. The Euclidean distance $d$ is used as distance measure in both spaces.

The paper is organized as follows. The three basic algorithms *c-means*, *SOM*, and *MDS* are discussed in section 2. In section 3 we describe the *ACMMDS* algorithm; which is then applied to two different data sets an artifical data set and a real world data set. We discuss these numerical experiments in section 4.

# 2    Components of ACMMDS

The *c-means clustering* algorithm moves a fixed set of $k$ cluster centers into the centers of gravity of the accumulations of data points [6, 5]. For the interpretation of the clustering result it is important to choose the right number of cluster centers $k$. If the choosen number of clusters is different from the actual number of clusters hidden within the data, the result of the clustering process has to be reconsidered.

During the c-means clustering process the current data point $x \in X$ is classified to the closest center $c_{j^*}$, i.e. the data point $x$ belongs to cluster $C_{j^*}$ if $d(x, c_{j^*}) = \min_i d(x, c_i)$. The quantization error $H(c_1, \ldots, c_k)$ defined by

$$H(c_1, \ldots, c_k) = \sum_j \sum_{x \in C_j} d^2(x, c_j)$$

is minimial, if each cluster center $c_j$ is equal to the corresponding center of gravity of cluster $C_j$, e.g. if for all $j = 1, \ldots, k$ hold

$$c_j = \frac{1}{|C_j|} \sum_{x \in C_j} x$$

where $|C_j|$ denotes the size of cluster $C_j$. This type of algorithm is called *batch c-means* [6].

We concentrate on the *sequential c-means clustering* method realized by the updating rule

$$\Delta c_{j^*} = \frac{1}{|C_{j^*}| + 1}(x - c_{j^*}) \tag{1}$$

where $c_{j^*}$ is the closest cluster center to the data point $x \in X$ [5].

Sequential c-means clustering is closely related to learning in artificial neural networks in particular, to the paradigm of *competitive neural networks* [7]. *Kohonen's selforganizing feature map (SOM)* is a competitive learning scheme

[3]. During SOM learning cluster centers $c_j$ that are close in the display space $\mathcal{Z}$ will be adapted to the same input $x \in X$:

$$\Delta c_j = \eta_t h(r_j, r_{j\bullet})(x - c_j) \tag{2}$$

where $h : \mathcal{Z} \times \mathcal{Z} \to \mathbb{R}_+$ is a neighborhood function with $h(r_j, r_{j\bullet}) \to 0$ for increasing distance of $r_j$ and $r_{j\bullet}$. For the special case of $h(r_j, r_{j\bullet}) = 1$ for $j = j^*$ and $h(r_j, r_{j\bullet}) = 0$ otherwise, Kohonen's SOM updating rule is identical to sequential c-means clustering.

Given a set of data points $X \subset \mathcal{X}$ and a transformation $p : \mathcal{X} \to \mathcal{Y}$ a natural evaluation criterion for the distance preservation of $p$ is some kind of difference between the matrices $D_\mathcal{X} := (d_\mathcal{X}(x_i, x_j))_{1 \le i,j \le M}$ in $\mathcal{X}$ and $D_\mathcal{Y} := (d_\mathcal{Y}(y_i, y_j))_{1 \le i,j \le M}$ in $\mathcal{Y}$.

MDS is a multivariate statistics procedure that start with a distance matrix $D_{\mathbb{R}^n}$ of $M$ data points $X = \{x_1, \ldots, x_M\} \subset \mathbb{R}^n$ and generates a set of corresponding representation points $Y = \{p(x_1), \ldots, p(x_M)\} \subset \mathbb{R}^r$ [2, 9]. This projection $p : X \to \mathbb{R}^r$ is calculated in such a way that the distance matrices $D_{\mathbb{R}^n}$ and $D_{\mathbb{R}^r}$ are similar. The difference is measured through *stress functions* defined by

$$S(x_1, \ldots, x_M) = \alpha \sum_{j,i=1}^{M} \left( \Phi[d^2(x_i, x_j)] - \Phi[d^2(y_i, y_j)] \right)^2 \tag{3}$$

where $\alpha > 0$ and $\Phi : \mathbb{R} \to \mathbb{R}$ is a strictly increasing differentiable function, i.e. $\Phi(x) = x$, $\Phi(x) = \sqrt{x}$ or $\Phi(x) = \log(x + 1)$. The consecutive points $y_j$ may be calculated by a gradient descent algorithm

$$\Delta y^j = \eta_t \cdot \alpha \sum_{i \neq j}^{M} \delta_{ji}(y_i - y_j), \tag{4}$$

where $\delta_{ji}$ is the weighted difference between $d(x_i, x_j)$ and $d(y_i, y_j)$ given by

$$\delta_{ji} = \Phi'[d^2(y_i, y_j)] \left( \Phi[d^2(x_i, x_j)] - \Phi[d^2(y_i, y_j)] \right)$$

and $\eta_t > 0$ with $\eta_t \to 0$ as $t \to \infty$.

## 3 The ACMMDS–Algorithm

In this approch we combine the sequential c-means clustering procedure, to detect the cluster structure in feature space, with a MDS algorithm, to get a low-dimensional representation of the cluster centers. To achieve this, the cluster centers $c_j \in \mathbb{R}^n$ move according to the sequential c-means iteration rule (1) and simultanously a set of low-dimensional representation centers $p_j := p(c_j)$ move in $\mathbb{R}^r$. These representation centers move in such a way that the distances $d(p_i, p_j)$ are close to the distances $d(c_i, c_j)$ of the cluster centers. This is realized by a gradient descent algorithm,

$$\Delta p_j = \eta_t \alpha \sum_{i \neq j}^{k} \delta_{ji}(p_i - p_j) \tag{5}$$

where $\delta_{ji}$ is

$$\delta_{ji} = \Phi'[d^2(p_i, p_j)]\Big(\Phi[d^2(c_i, c_j)] - \Phi[d^2(p_i, p_j)]\Big)$$

which minimizes the stress function

$$S(p_1, \ldots, p_k) = \alpha \sum_{i,j=1}^{k} \Big(\Phi[d^2(c_i, c_j)] - \Phi[d^2(p_i, p_j)]\Big)^2 \qquad (6)$$

---

**ACMDS algorithm**

```
estimate thresholds θ_new and θ_merge
set k = 0 (no prototypes)
    choose a data point x ∈ X
    calculate d_j = d(x, c_j), j = 0, ..., k
    detect the winner j* = argmin_j d_j
    if (d_j* > θ_new) or k = 0
        c_k := x and p_k according to (5)
        k := k + 1
    else
        adapt c_j* by (1) and p_j* by (5)
        calculate D_l = d(c_l, c_j*), l = 0, ..., k
        detect l* := argmin_{l≠j*} D_l
        if (D_l* ≤ θ_merge)
            merge(c_l*, c_j*), k := k - 1
    goto:  choose data point
```

---

In addition we incorporate into the clustering algorithm a scheme for adjusting the number $k$ of cluster centers in order to address the problem of finding a good choice for number of cluster centers and initial locations of cluster centers. We use a variant of c-means clustering in the ACMMDS procedure, which allows the creation of new cluster centers and the merging of close clusters and call it *adaptive c-means clustering*. For this clustering algorithm parameters $\theta_{merge}$ and $\theta_{new}$ have to be derived from the data subset $X$ in advance.

When a new cluster center is inserted, the location of the corresponding representation center $p_j$ has to be determined, by setting the inital position to be a linear combination of its two nearest neighbors and then adapting $p_j$ by the iteration rule (5).

# 4   Results and Discussion

To illustrate the behavior of the ACMMDS procedure we present the result of a subset of the 3-dimensional helix $H := \{(sin(t), cos(t), t) \mid t \geq 0\}$. The first 8 loops of $H$ are shown in Fig. 1. A data set $X$ containing $M = 1001$ data points was equidistantly sampled from $H$. This data set has been analyzed by the

ACMMDS algorithm. The set of representation centers $p_j \in \mathbb{R}^2$ is depicted in Fig. 1, the 8 loop 3D-helix is projected into a sine-like curve with 8 periods.

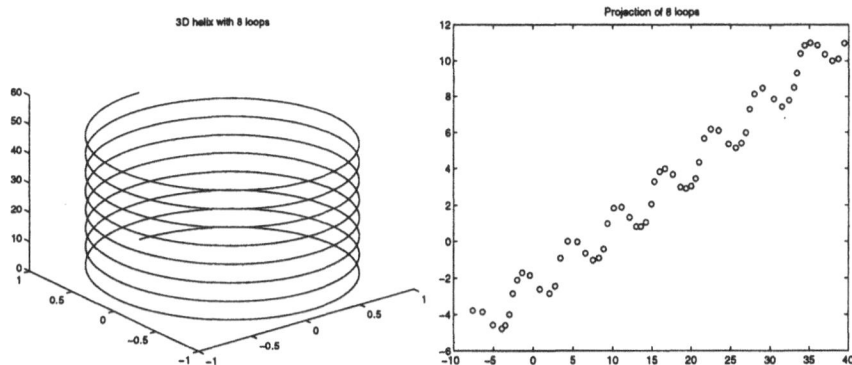

Fig.1: The 3D-helix $H := \{(sin(t), cos(t), t) \mid 0 \le t \le 16\pi\}$ with 8 loops and its 2D-projection.

The next data set stems from the problem domain of classifying ECG signals in order to predict sudden cardiac death. Three features are used [1]: Duration of the filtered QRS (one heart-beat), rootmean square of the last 40 ms and duration of the terminal part of the QRS below 40 $\mu$V, the $QRSD$.

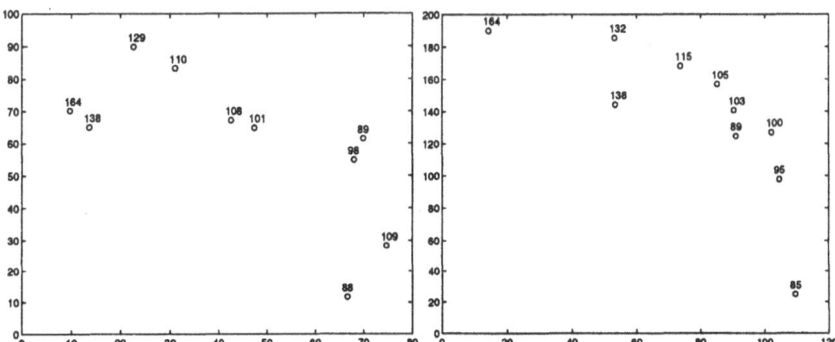

Fig.2: The 2D-projection of the cluster centers with the annotaion of the duration of the QRS-complex after the 1. (left) and 10. (right) learning epoche.

In Fig.2 the projection centers after the first and tenth learning epoche are shown. Each projection center is labeled with die mean of the $QRSD$. Whereas the representation centers in the beginning of the training procedure were only roughly ordered in the 2D-map, after the tenth epoche (final) these centers were grouped around a 1-dimensional curve, e.g. the chain of cluster centers ordered by the $QRSD$-labels 164-132-115-105-100-95-85. The distances between all cluster centers were preserved in the final 2D-map. We conclude that the 3-dimensional data set is redundant and can be embedded roughly into a 1-dimensional feature space, which can be spanned more or less by the duration of the QRS-complex.

The ACMDS algorithm is an adaptive data analysis procedure for clustering and visualization of large and high-dimensional data sets. The adaptivity of this clustering and representation procedure makes it useful for many applications, where the clustering itself is part of a larger program in a changing enviroment. In such cases, where one wants the system to be able to learn without throwing away the accumulated experience and under human supervision, the procedure described here may turn out to be very useful.

# References

[1] G. Breithardt, M. .E. Cain, N. El-Sherif, N. Flowers, V. Hombach, M. Janse, M. B. Simson, and G. Steinbeck. Standards for analysis of ventricular late potentials using high resolution or signal-averaged electro-cardiography. *European Heart Journal*, 12:473–480, 1991.

[2] A.K. Jain and R.C. Dubes. *Algorithms for Clustering Data*. Prentice Hall, Englewood Cliffs, New Jersey, 1988.

[3] T. Kohonen. *Self-Organizing Maps*. Springer, 1995.

[4] Y. Linde, A. Buzo, and R.M. Gray. An algorithm for vector quantizer design. *IEEE Transactions on Communications*, 28(1):84–95, 1980.

[5] S.P. Lloyd. Least squares quantization in PCM. *IEEE Transactions on Information Theory*, 28(2):129–137, 1982.

[6] J. MacQueen. Some methods for classification and analysis of multivariate observations. In L.M.LeCam and J.Neyman, editors, *Proceedings of the Fifth Berkeley Symposium on Mathematical Statistics and Probability*, volume I, pages 281–297. Berkeley University of California Press, 1967.

[7] J. Moody and C.J. Darken. Fast learning in networks of locally-tuned processing units. *Neural Computation*, 1(2):281–294, 1989.

[8] H. Ritter and K. Schulten. Convergence properties of Kohonen's topology converving maps:fluctuations,stability, and dimension selection. *Biological Cybernetics*, 60:59–71, 1988.

[9] J.W. Sammon. A nonlinear mapping for data structure analysis. *IEEE Transactions on Computers*, C-18:401–409, May 1969.

[10] D.W. Scott. *Multivariate Density Estimation*. John Wiley & Sons, New York, 1992.

# Behavior of the Weights of a Support Vector Machine as a function of the Regularization Parameter $C$

Rodrigo Fernández

LIPN, Institut Galilée-Université Paris 13

Avenue J.B. Clément 93430 Villetaneuse France

rf@lipn.univ-paris13.fr

### Abstract

It is known that the parameters defining a Support Vector Machine (SVM) are obtained by solving a constrained Quadratic Programming (QP) problem. When the restrictions of the original Support Vector problem cannot be satisfied, Vapnik introduces the use of a regularization parameter $C$ in order to bound the size of the solution. In this paper we derive two results on constrained quadratic optimization that allow us to understand the behavior of the solution as $C$ grows to infinite. Based on these results we propose a simple method to stop the cross-validation process when we are searching for the optimal $C$ and to learn whether a given SV problem is realizable or not. We illustrate our results by examples on regression and classification.

## 1 Introduction

We consider the problem of computing the parameters of a SVM, that is, support vectors and their associated weights. The central idea of Vapnik's SVM is to map the entries into a high dimension feature space implicitly defined by a Hilbert-Schmidt kernel function, in the feature space the algorithm computes a hyperplan, called *optimal hyperplan*, minimizing a risk functional following the *structural risk minimization* induction principle. A detailed description of SVM can be found in [1] (chapter 5) and [2]. Basic notions on continuous optimization can be found in [3].

From the implementation point of view, training a SVM is equivalent to solving a linearly constrained QP problem. The weights vector defining the optimal hyperplan corresponds to the QP problem solution. Every SV application (classifiers, regressors, operator inversors) leads to a slightly different QP problem, nevertheless, all these problems could be viewed as a particular case of a more general one. The aim of this paper is to characterize the solution of a SV problem in this general context.

Consider the following QP problem in $I\!R^n$:

$$(P_C) \quad \min_{\alpha \in D_C} q(\alpha)$$

$$D_C = \{\alpha \in I\!R^n \ s.t. \ 0 \le \alpha_i \le C, a^t \alpha = 0\}$$

918

where $q = \alpha^t A\alpha + \alpha^t b$ is a quadratic function with $A$ a positive semi-definite $n \times n$ matrix; $a, b \in \mathbb{R}^n$ and $C \in \mathbb{R}^+ \cup \{+\infty\}$. Remark that we explicit dependence between the problem and the upper bound for the admissible region $C$, when the domain is unbounded for above, the problem will be noted as $P_\infty$ and the admissible region as $D_\infty$.

Let us consider a situation where the learning set and all the other parameters defining a SVM are fixed. We are interested by the solution of $P_C$ only as a function of $C$. When $P_\infty$ has an optimum, geometrical constraints of the original SV problem are fully satisfied and we say the SV problem is *realizable*. In the opposite case $P_\infty$ does not have a solution and we have to comply with the solution of $P_C$ for some, usually sub-optimal, $C$. In the next section we present some results that help to understand the role of this parameter.

# 2 Behavior of the QP problem solution as a function of $C$

Two propositions characterizing the behavior of the QP problem solution when $C$ grows to infinite are presented. The first proposition applies in the realizable case and the second one applies in the non realizable case.

## 2.1 Realizable case

**Proposition 1** *The vector $\tilde{\alpha}$ is solution of $P_\infty \iff \exists C1 < C2 \in \mathbb{R}$ such that $\tilde{\alpha}$ is solution of $P_{C1}$ and $\tilde{\alpha}$ is solution of $P_{C2}$.*

**Proof:**

($\Rightarrow$) Trivial.

($\Leftarrow$) Take $C1 < C2 \in \mathbb{R}$ such that $\tilde{\alpha}$ solves $P_{C1}$ and $\tilde{\alpha}$ solves $P_{C2}$. We define the quantity $q_C = \min q(\alpha)$, $\alpha \in D_C$. Clearly $q_C$ is a decreasing function of $C$, since $q_{C1} = q_{C2} = q(\tilde{\alpha})$, we deduce that $q_C$ is constant over $[C1, C2]$.

Now consider an arbitrary $\gamma \in D_\infty$. There are two possibilities:

(i) If $\gamma \in D_{C2} \Longrightarrow q(\gamma) \geq q(\tilde{\alpha})$.

(ii) If $\gamma \notin D_{C2}$, since $\tilde{\alpha} \in D_{C1}$, there exists $\mu \in (0,1)$ such that $(\mu\gamma + (1-\mu)\tilde{\alpha}) \in D_{C2}$[1] and the following inequality is direct

$$q(\mu\gamma + (1-\mu)\tilde{\alpha}) \geq q_{C2} = q(\tilde{\alpha}) \tag{1}$$

Since $q$ is convex, we have

$$q(\mu\gamma + (1-\mu)\tilde{\alpha}) \leq \mu q(\gamma) + (1-\mu)q(\tilde{\alpha}). \tag{2}$$

---

[1]Notice that the restriction $a^t\alpha = 0$ is linear, then $[a^t\alpha = 0] \Longrightarrow [a^t(K\alpha) = 0]$ is always true. So, we don't have to care about it.

Combining (1) and (2) we get

$$\begin{aligned}
q(\tilde{\alpha}) \le\ & q(\mu\gamma) + (1-\mu)\tilde{\alpha}) \le & \mu q(\gamma) + (1-\mu)q(\tilde{\alpha}) & \Rightarrow \\
& q(\tilde{\alpha}) & \le\ \mu q(\gamma) + q(\tilde{\alpha}) - \mu q(\tilde{\alpha}) & \Rightarrow \\
& q(\tilde{\alpha}) & \le\ q(\gamma). &
\end{aligned}$$

In both cases $\tilde{\alpha}$ solves $P_\infty$ $\square$.

## 2.2 Non-realizable case

When $P_\infty$ does not have an optimum, the behavior of $P_C$ solution is analyzed when $C \rightsquigarrow \infty$.

We start with some remarks about $q$: The quadratic term $\alpha^t A\alpha$ is not negative, it will be noted as $\alpha^t A\alpha = q^+(\alpha)$. If $\tilde{\alpha}$ solves $P_C$ then $\tilde{\alpha}^t b \le 0$ since $\min_{\alpha \in D_C} q(\alpha) \le 0$. We note $\alpha^t b = q^-(\alpha)$.

**Proposition 2** *Let be $\tilde{\alpha}$ a solution of $P_C$, then*
$q^+(\tilde{\alpha}) = 0 \Longleftrightarrow K\tilde{\alpha}$ *solves* $P_{KC}$ $\forall K > 1$.

**Proof:** We will prove both implications.

($\Longrightarrow$) From the above remarks we have that $q(\tilde{\alpha}) = q^+(\tilde{\alpha}) + q^-(\tilde{\alpha}) = q^-(\tilde{\alpha})$ since $q^+(\tilde{\alpha}) = 0$. And also

$$\begin{aligned}
q(K\alpha) =\ & K^2 q^+(\alpha) + K q^-(\alpha) \\
=\ & K q^-(\alpha) \\
\le\ & q^-(\alpha) & \text{since } (K > 1) \text{ and } (q^-(\alpha) \le 0)
\end{aligned}$$

Let be $\alpha$ some element of $D_{KC}$. By definition of $D_{KC}$ there exists $\alpha_0 \in D_C$ such that $K\alpha_0 = \alpha$. We have then

$$\begin{aligned}
q(\alpha) =\ & q(K\alpha_0) \\
=\ & K^2 q^+(\alpha_0) + K q^-(\alpha_0) \\
\ge\ & K(q^+(\alpha_0) + q^-(\alpha_0)) & (K > 1) \\
\ge\ & K q(\tilde{\alpha}) & \alpha_0 \in D_C \\
=\ & q(K\tilde{\alpha})
\end{aligned}$$

that means that $\forall \alpha \in D_{KC}$ $q(\alpha) \ge q(K\tilde{\alpha}) \Rightarrow K\tilde{\alpha}$ solves $P_{KC}$.

($\Longleftarrow$) Let be $\tilde{\alpha}$ a solution of $P_C$ and $K > 1$ such that $K\tilde{\alpha}$ solves $P_{KC}$. Since $D_C \subset D_{KC}$ we have $q(K\tilde{\alpha}) \le q(\tilde{\alpha})$, therefore

$$\begin{aligned}
q(\tilde{\alpha}) \ge\ & q(K\tilde{\alpha}) & \Leftrightarrow \\
q^+(\tilde{\alpha}) + q^-(\tilde{\alpha}) \ge\ & K^2 q^+(\tilde{\alpha}) + K q^-(\tilde{\alpha}) & \Leftrightarrow \\
-q^-(\tilde{\alpha})\frac{K-1}{K^2-1} \ge\ & q^+(\tilde{\alpha}) & \text{since } K > 1
\end{aligned}$$

but $K$ is as large as we like and $\lim_{K \rightsquigarrow \infty} \frac{K-1}{K^2-1} = 0$. It's easy to conclude that $q^+(\tilde{\alpha}) = 0$. $\square$

## 2.3 Connection with SVMs

Using the above statements, we can deduce some things about the original SV problem solution. Suppose we are training a SVM and we find $C1 < C2$ such that the solutions of $P_{C1}$ and $P_{C2}$ are the same, from Proposition 1, it follows that $P_{\infty}$ has a solution. On the other hand, suppose we find out that, for large enough values of $C$, the solution $\tilde{\alpha}$ behaves like $C\alpha_0$. Using Proposition 2 we are able to infer that $q^+ \approx 0$ (easy to check) and that behavior of the solution will be similar to $C\alpha_0$ for larger values of $C$.

The following rule to stop the cross-validation process on $C$ summarizes the ideas exposed in this section: *If the solution is stable, stop, the SV problem is realizable and the optimum is already available. If the solution behaves as $C\alpha_0$ for large values of $C$, stop, the SV problem is non realizable and, in terms of the original problem, the solution will not be improved since $\forall k$ positive $\alpha_0$ and $k\alpha_0$ realize the same hyperplan.*

# 3 How does it work in practice?

We illustrate our results by two examples: a regression problem introduced by Smola[4] showing realizable cases and a non-realizable separation of 2 overlapped classes in $\mathbb{R}^2$.

## 3.1 Smola's regression example

In [4] Smola uses a simple real valued function as a toy problem for testing Support Vector Regression Machines. The example is $f(x) = \frac{sin(x)}{x} + \psi$ where $\psi \rightsquigarrow N(0, br)$.

The work interval [-1.5,1.5] has been sampled with a step of 0.02. In all experiences we use the $\epsilon - insensitive$ loose function

$$L(\eta, x_i) = \begin{cases} 0 & if \ |\eta| < \epsilon \\ |\eta| - \epsilon & otherwise \end{cases}$$

and a Radial Basis Function kernel $K(x, y) = e^{-\frac{\|x-y\|^2}{2\sigma^2}}$. A detailed description of the use of SVMs in regression tasks can be found in [4] [5] and [6], those who are interested by Hilbert-Schmidt kernels should consult [7]. Let us only say that the parameter $\epsilon$ is the desired precision of the SV regressor since the risk functional is minimized when the maximum gap between regressed data and input data is $\epsilon$.

We use a noise free regression problem to illustrate a realizable case (figure 1). Three SVM are presented. Different SVMs are determined by their kernel parameter $\sigma$, here we present SVMs with $\sigma = 0.03, 0.06$ and $0.1$. Notice how, for $C$ greater than 100, requested precision is reached by the three machines (right curves). On the other hand, we find that the solution is stable, that is, the number of support vectors (NBSV) becomes constant and the number of support vectors whose associated weight is $C$ (NBFIX) becomes zero (left curves). This behavior is predicted by Proposition 1.

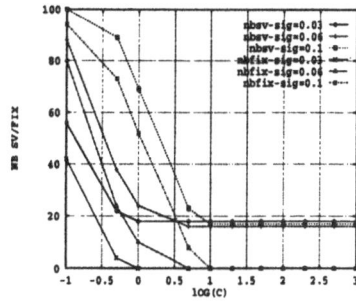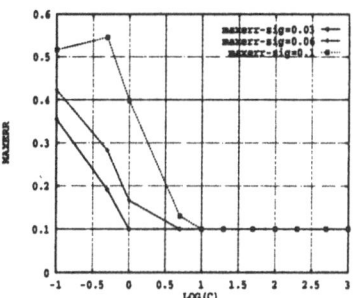

Figure 1: Noise free regression with $\epsilon = 0.1$ for 3 different SVMs ($\sigma =$ 0.03, 0.06 et 0.1). Left: NBSV is the number of support vectors, NBFIX is the number of support vectors whose associated weight is $C$. Right: MAXERR is the maximal gap between original and estimated curves. This quantities are plotted against $log_{10}(C)$. The regressions are realizable. Notice how for every $\sigma$ the requested precision is reached and how for $C$ bigger than 100 the solution does not change.

## 3.2 A non realizable classification example

In this section we present a non realizable classification problem on $\mathbb{R}^2$. Two classes are drawn from uniform distributions over 2 unitary circles centered at (-0.8,0) and (0.8,0) respectively. Since the distributions have overlapping supports, the classes cannot be separated. We trained SV classifiers with homogeneous polynomial kernels $K(u, v) = \left(\frac{u \cdot v}{2} + 1\right)^d$ where $d$ is the degree of the polynomial. The training set contains 100 examples, 50 for each class.

Results are shown in figure 2. Several quantities characterizing the quality of the solution of the original SV problem and the related QP problem are plotted against $C$. Polynomial SV classifiers with $d = 2$ and 3 are presented. Notice that NBERR does not decrease to 0, that is, training set cannot be separated without errors. The number of support vectors remains stable for $C > 250$ (figure 2, left), but the solution itself is not stable (figure 2, right). Notice how as $q^+$ decreases to 0, the ratio $\frac{\|\tilde{a}\|}{C}$ remains constant , this behavior is predicted by Proposition 2. Even when the norm of the weights grows in direct proportion to $C$ when $C \rightsquigarrow \infty$, in terms of the original SV problem the quality of the solution is not being improved.

# 4 Conclusion

We derived two results on linearly constrained quadratic optimization caracterizing the behavior of the solution of QP problems associated to the training of Support Vector Machines when the regularization parameter $C \rightsquigarrow \infty$. These results allow us to decide whether a proposed SV problem is realizable and provide a simple and reliable tool to stop the cross-validation process over $C$. Examples of both realizable and non-realizable cases confirm our assertions and show how they can be used in pactice.

922

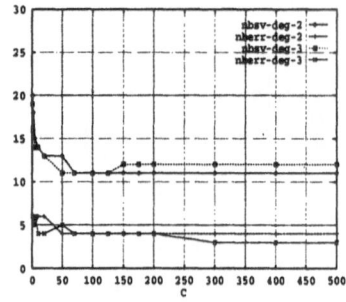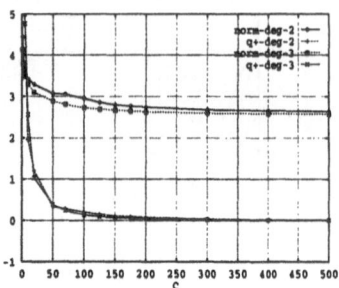

Figure 2: A non realizable classification problem, two overlapped circles in $R^2$ must be separated. Always against $C$, the following quantities caracterizing SV classifiers are plotted: number of support vectors (nbsv), number of training classification errors (nberr), ratio between the norm of the weights vector and C (norm), and the positive term of the quadratic objective function ($q^+$, see Proposition 2). SVMs with homogeneous polynomial kernel of degree 2 and 3 are shown.

# References

[1] V. Vapnik. The Nature of Statistical Learning Theory. Springer. 1995.

[2] C. Cortes and V. Vapnik. Support-Vector Networks. Machine Learning, vol 20, 273-297. 1995.

[3] H. Eiselt, G. Pederzoli, C-L. Sandblom. Continuous Optimisation Models. Walter de Gruyter. 1987.

[4] A.J. Smola. Regression estimation with support vector learning machines. Technishe Universität München, Fakultät für Physik. 1996.

[5] V. Vapnik, S. Golowich, A.J. Smola. Support Vector Method for Function Approximation, Regression Estimation and Operator Inversion. Advances in Neural Information Processing Systems, vol 9. San Mateo, CA. 1997.

[6] A.J. Smola, B. Schölkopf, K.-R. Müller. General Cost Functions for Support Vector Regression. Proceedings ACNN'98, Australian Congress on Neural Networks. 1998, forthcoming.

[7] A.J. Smola, B. Schölkopf. On a Kernel-based Method for Pattern Recognition, Regression, approximation and Operator Inversion. Technical Report 1064, GMD First. 1997.

# Convergence Properties of a Modified Temporal Anti-Hebbian Model

Robert Geary

Department of Computer and Information Systems, University of Paisley
Paisley, Scotland

Colin Fyfe

Department of Computer and Information Systems, University of Paisley
Paisley, Scotland

### Abstract

This paper proposes a modification to Girolami and Fyfe's temporal variant of Foldiak's first model. In the original temporal model asymmetric memory based lateral weights are adapted using a cross-correlation criterion and compared with information maximisation applied to the blind separation of sources which have experienced convolved mixing. In our new model the weights are adapted using a linearised discontinuous function which shares the same quadrature sign behaviour as the decorrelation function. The model provides separation and restoration of convolved mixes of signals where network weights converge very closely to ideal FIR deconvolving filter coefficients. We compare the convergence properties of a 'linearised correlation' model with that of the original model and examine potential for improvement of convergence behaviour.

## 1 Introduction

A two input model of anti-Hebbian learning as proposed by Foldiak [1] is show in Figure 1, The model behaviour is governed by the following equation of neuron dynamics as

$$\tau \, dy_i/dt = -y_i + x_i + \sum_{j=1}^{N} w_{ij} \, y_i \tag{1}$$

where $w_{ij}$ is the synaptic strength between pre- and post- synaptic synaptic activities and the law of repulsion is governed by the anti-Hebbian rule $\Delta w = -\eta \, y_i \, y_j \; : \forall \, i \neq j$

The adiabatic approximation from (1) gives

$$y_i = x_i + . \sum_{j=1}^{N} w_{ij} \, y_i \tag{2}$$

and in matrix form is

$$\mathbf{y} = (\mathbf{I} - \mathbf{W})^{-1} \mathbf{x} \qquad (3)$$

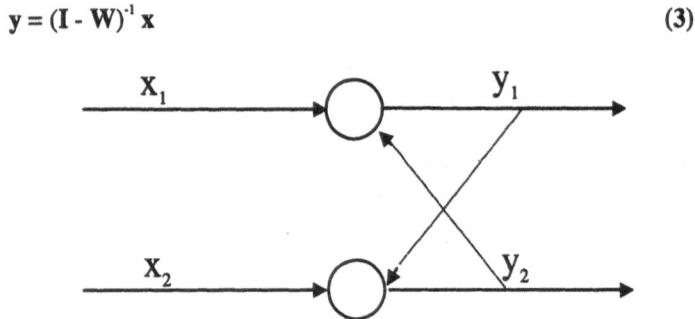

Figure 1: Two neuron lateral inhibition network

Girolami and Fyfe [2] employ memory based, synaptic lateral connections to the temporal model is depicted in Figure 2.

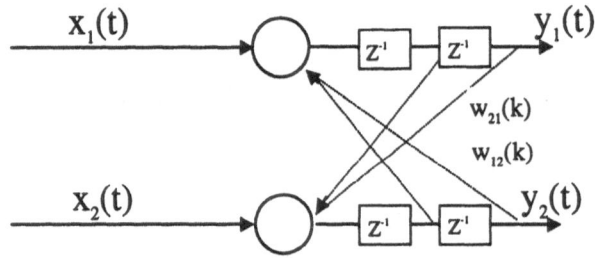

Figure 2: Two neuron, two tap temporal model

Girolami and Fyfe showed that for causal convolutive mixtures of sources, temporal anti-Hebbian learning yields comparable results to the entropy maximisation algorithm developed by Torkkola [3].

The network output is given as

$$y_i(t) = x_i(t) + \sum_{j=1}^{N} \sum_{k=0}^{M} w_{ij}(k) \, y_j(t-k) \qquad (4)$$

and the anti-Hebbian learning for each of the tapped delay weights is

$$\Delta w_{ij}(k) = -\eta \, y_i(t) y_j(t-k) \qquad \forall \; i{\neq}j \, \wedge \, k{\in}\{1..M\} \qquad (5)$$

The cross weights grow in an inhibitory fashion if there is correlation between output $y_i(t)$ and each of the tapped outputs of the adjacent neuron $y_j(t-k)$ until the following steady state condition is satisfied

$$E\{y_i(t)y_j(t-k)\}=0 \qquad \forall \ i{\neq}j \ \wedge \ k{\in}\{1..M\} \qquad (6)$$

In general the weight matrix will not be symmetrical i.e. $\Delta w_{ij}(k){\neq}\Delta w_{ji}(k)$ since in all but the most artificial situations the original mixing multipath convolution transfer-functions will have been different. This may have implications on the convergence behaviour of this network and will be discussed further in this paper.

## 2 Modification to the Temporal Model

It is perhaps useful to think of the cross-correlating adaptation rule as a detector which identifies when output $y_i(t)$ shares the same sign as the delayed output value $y_j(t-k)$. With this view in mind the cross-correlation function is a simple mechanism which measures in some sense the similarity between two functions or process realisations. In our network the deconvolving filter weights grow rapidly if $y_i(t)$ and $y_j(t-k)$ share the same sign often and are both large. In the early stages of adaptation only large sign coincident and non-coincident values have an effect on the growth dynamics of the FIR filter weights. We consider that, especially in the early stages of learning it is useful to use all the available sign information to improve the network adaptation behaviour.

We propose that the cross-correlation in the adaptation rule for temporal anti-Hebbian learning is replaced by a quasi-linear adaptation function which shows the same quadrature sign behaviour as cross-correlation. The new 'linearised' function is depicted in Figure 3

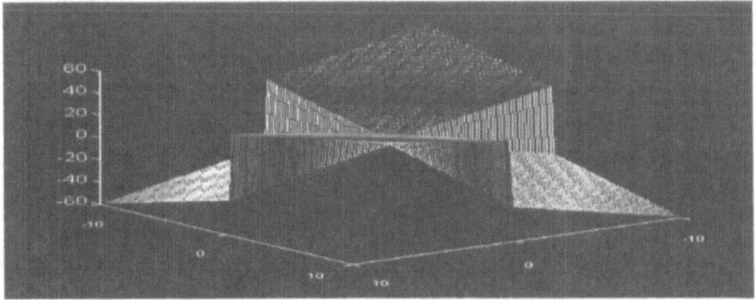

Figure 3: The new adaptation function

The calculation of $\Delta w_{ij}(k)$ is now given by

$$\Delta w_{ij}(k) = -\text{SGN}(y_i(t), y_j(t-k))\, \eta\, (|y_i(t)| + |y_j(t-k)|) \quad \forall\; i \neq j \;\wedge\; k \in \{1..M\} \quad (7)$$

where
$$\begin{aligned}
\text{SGN}(y_i(t), y_j(t-k)) &= +1 \;\mid\; y_i(t)y_j(t-k) > 0 \\
&= -1 \;\mid\; y_i(t)y_j(t-k) < 0 \\
&= \;\;0 \;\mid\; y_i(t)y_j(t-k) = 0
\end{aligned} \qquad (8)$$

This new rule allows sign information to drive learning since $\Delta w_{ij}(k)$ is not as dependent on larger valued terms. It was considered that the new rule might lead to some instability, but in practice this has been no evidence of this. When the network is near convergence the lower level weights do appear to carry small fluctuations. It was straightforward to reduce this effect by employing a search-then-converge strategy such as the approach suggested by Darken and Moody [4] where in the initial phase of learning the rate parameter $\eta$ is kept almost constant and in the second phase it is exponentially decreased to zero. There are many popular functions that can be used. A simple one, which was found to be effective, is given by

$$\eta(t) = \eta_0/(1+t/t_0) \qquad (9)$$

where $t_0$ is the break time and is chosen to make the decay of $\eta(t)$ suitably slow.

With the new rule, cross weights grow in an inhibitory fashion until the following steady state condition is satisfied

$$\mathbf{E}\{\; \text{SGN}(y_i(t),y_j(t-k))(|y_i(t)| + |y_j(t-k)|) \;\} = 0 \qquad (10)$$

Since in general the weight matrix will not be symmetrical, convergence will not occur simultaneously at all neurons. (This is found to be the case in our simulations.) The cross-coupled nature of the network means improvement in one neuron gives improvement in the other and so it may beneficial to choose asymmetric learning rates. Generalising further since particular tap values corresponding to stronger path mixing take longer to reach their ideal deconvolving filter values, a localised learning strategy could be employed which will allow the FIR filters to grow with more degrees of freedom. Recently Almeida et al. [5] developed a step size adaptation technique for accelerating stochastic gradient optimisation. Their simulations show a convincing gain in performance over fixed step size learning. The application of such approaches to this network is a subject for further study.

# 3 Simulation

Two unknown male speech input signals sampled at 12KHz were mixed using the following polynomial mixing matrix

$$\begin{bmatrix} x_1 \\ x_2 \end{bmatrix} = \begin{bmatrix} 1-0.45z^{-10}+0.3z^{-15}-0.2z^{-35}-0.1z^{-60} & 0.5z^{-30}-0.3z^{-53}-0.1z^{-63} \\ 0.5z^{-15}+0.25z^{-26}-0.15z^{-32}-0.1z^{-50} & 1-0.2z^{-25}-0.2z^{-34}+0.15z^{-58} \end{bmatrix} \begin{bmatrix} s_1 \\ s_2 \end{bmatrix}$$

Signal separation was performed using both algorithms. Neuron cross-weights were captured at 200 sample intervals recording the network growth history. The final converged state for one half of the network is depicted in figures 4 and 5. A set of time stacked histograms were also obtained which give detailed dynamic growth information.

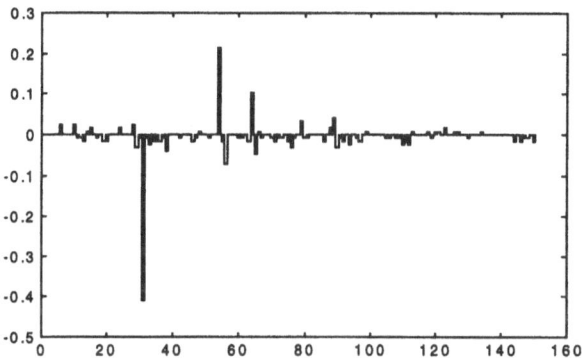

Figure 4 Converged state of $w_{12}(k)$ for new algorithm

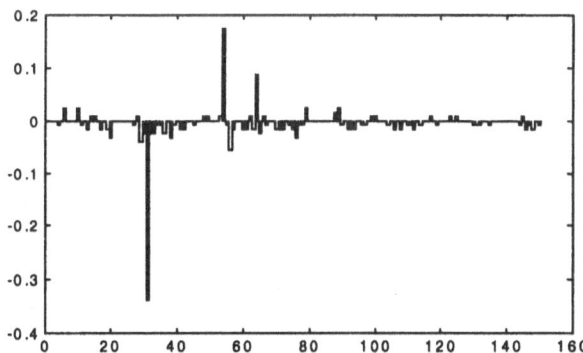

Figure 5 Converged state of $w_{12}(k)$ for original decorrelation

# 4 Conclusions

Both 150 tap networks were trained with a choice of learning parameters, which were the best compromise between fast growth and potential instability. The new network converged faster and as can be seen from the simulation results, converged to a causal stable solution. The steady state neuron weights for $w_{12}(k)$ give near perfect results when compared with computed idealised inverting filter weights [2]. Dynamic growth information not shown here indicates that individual filter weights can indeed grow at quite different rates throughout training and so a localised learning strategy would be likely to be beneficial.

# References

[1] Foldiak P. Adaptive network for optimal linear feature extraction. Artif Intell, IEEE/INNS International Joint Conference on Neural Networks, Vol 1, pp.401-405,Washington DC:Institute of Electrical and Electronic Engineers, SanDiego, 1989

[2] Girolami M, Fyfe C: A temporal model of linear anti-Hebbian learning, Neural Processing Letters, vol 4, No 3, pp.139 - 147, 1996

[3] Torkkola K. Blind separation of convolved sources based on information maximisation. IEEE Workshop on Neural Networks Ellis for signal processing, NNSP'96, Seika, Kyoto, Japan, September 4 - 6, 1996

[4] Darken C, Moodey J: Towards faster stochastic gradient search. In: Advances in Neural Information Processing Systems 4. Morgan Kaufman, San Mateo, 1991, pp 1009 - 1016

[5] Almeida L, Langlois T, Amaral J: On-line step size adaptation, Technical Report INESC RT07/97, INESC/IST, Lisbon, Portugal, 1997

# Volatility Prediction with Mixture Density Networks

Christian Schittenkopf, Georg Dorffner
Austrian Research Institute for Artificial Intelligence
Dept. of Medical Cybernetics and Artificial Intelligence,
University of Vienna, Austria
chris@ai.univie.ac.at, georg@ai.univie.ac.at

Engelbert J. Dockner
Dept. of Business Administration,
University of Vienna, Austria
dockner@finance2.bwl.univie.ac.at

### Abstract

Despite the lack of a precise definition of volatility in finance, the estimation of volatility and its prediction is an important problem. In this paper we compare the performance of standard volatility models and the performance of a class of neural models, i.e. mixture density networks (MDNs). First experimental results indicate the importance of long-term memory of the models as well as the benefit of using non-gaussian probability densities for practical applications.

## 1 Introduction

Stock market returns typically exhibit the following time series characteristics. While the returns are uncorrelated, the squared returns show a rich structure that can be approximated by linear and non-linear models. Especially the appearance of volatility clustering renders the assumption of a constant variance (homoscedasticity) doubtful. This assumption is usually made when feedforward networks are trained to fit a given time series by gradient descent on the standard error function (mean squared error).

In this paper we apply the concept of mixture density networks (MDNs) [1] to estimate and predict the volatility of the Austrian stock market index ATX. Consequently, our neural models are *heteroscedastic*. Furthermore, the multidimensional networks (several gaussian distributions in the output) are able to also approximate non-gaussian, typically leptocurtic (fat tailed) distributions. We measure the performance of the MDNs and of some standard models of volatility with respect to the likelihood function evaluated on test sets and with respect to a prediction error of change of volatility. We find that small errors on the test sets do not necessarily imply good predictions and a profitable application of volatility models in terms of trading strategies.

Standard models for volatility estimation are briefly described in Section 2. The architecture and training of MDNs is summarized in Section 3. Section 4 includes our preliminary results on the ATX. We discuss planned and partially implemented extensions of our MDN architecture in Section 5.

## 2 Classical Models

Basic to standard models and our models of volatility is the notion that the financial time series $\{x_t\}$ under study can be decomposed into a predictable component $\mu_t$ and an unpredictable component $e_t$, which is assumed to be zero mean gaussian noise of variance $\sigma_t^2$: $x_t = \mu_t + e_t$. The models are thus characterized by *time-varying* conditional variances $\sigma_t^2$ and are thus well suited to explain volatility clusters. The most widely used (standard) models of volatility are ARCH/GARCH models and the GJR approach [2, 3, 4]. Financial time series often exhibit means close to zero and negligibly small correlations. In these cases the corresponding models can be forced to predict a conditional mean $\mu_t = 0$. If there are reasons to believe that the conditional mean is significantly different from zero, an extra component for $\mu_t$ should be provided in the model. The classical ARCH($q$) model [2] is given by

$$x_t \sim N(\mu_t; \sigma_t^2), \sigma_t^2 = \alpha_0 + \sum_{i=1}^{q} \alpha_i e_{t-i}^2, \tag{1}$$

where $N(\mu_t; \sigma_t^2)$ denotes a Gaussian random variable of mean $\mu_t$ and of variance $\sigma_t^2$. To ensure that the variance $\sigma_t^2$ is positive for each $t$ the restrictions $\alpha_0 > 0, \alpha_i \geq 0, i = 1, \ldots, q$, are imposed on the parameters. A GARCH($p, q$) model [3] is an extension of an ARCH($q$) model because the variance is calculated recursively by

$$\sigma_t^2 = \alpha_0 + \sum_{i=1}^{q} \alpha_i e_{t-i}^2 + \sum_{i=1}^{p} \beta_i \sigma_{t-i}^2. \tag{2}$$

Additionally to the constraints of the ARCH model we require that $\beta_i \geq 0$, $i = 1, \ldots, p$. This specification implies that the conditional variance $\sigma_t^2$ follows an autoregressive process for which stationarity is guaranteed, if the sum of coefficients $\sum_{i=1}^{q} \alpha_i + \sum_{i=1}^{p} \beta_i < 1$. In many applications it is sufficient to choose $p = q = 1$. Finally, the GJR model [4], which is an extension of the GARCH(1,1) model, has been successfully applied to financial time series. It incorporates asymmetric effects, and it is defined by

$$\sigma_t^2 = \alpha_0 + \alpha_1 e_{t-1}^2 + \alpha_2 s_{t-1} e_{t-1}^2 + \beta_1 \sigma_{t-1}^2 \tag{3}$$

where $s_{t-1} = 1$ if $e_{t-1} < 0$ and $s_{t-1} = 0$ otherwise. The use of the GJR model has been proposed because stock returns are characterized by a leverage effect, i.e. volatility increases as returns for stocks decrease.

## 3 Network Architecture

Within the last years MDNs [1, 5] have turned out to be a very useful tool to model conditional probability density functions (pdfs) in different fields such as non-linear inverse problems [1] and time series analysis [6]. The main idea of MDNs is to use multi-layer perceptrons (MLPs) to predict the parameters of the pdf of the next observation $x_t$ in dependence of the past observations $x_{t-1}, \ldots, x_{t-m}$. A very natural way to approximate the conditional pdf of $x_t$ is to choose a weighted sum of $n$ gaussian densities, i.e.

$$x_t \sim \sum_{i=1}^{n} \alpha_{i,t} N(\mu_{i,t}; \sigma_{i,t}^2), \tag{4}$$

$$\alpha_{i,t} = s(\tilde{\alpha}_{i,t}), \tilde{\alpha}_{i,t} = \text{MLP}_j(x_{t-1}, \ldots, x_{t-m}), 1 \le j \le n, \tag{5}$$

$$\mu_{i,t} = \text{MLP}_j(x_{t-1}, \ldots, x_{t-m}), n+1 \le j \le 2n, \tag{6}$$

$$\sigma_{i,t}^2 = \exp(\text{MLP}_j(x_{t-1}, \ldots, x_{t-m})), 2n+1 \le j \le 3n. \tag{7}$$

The softmax function

$$s(\tilde{\alpha}_{i,t}) = \frac{\exp(\tilde{\alpha}_{i,t})}{\sum_{j=1}^{n} \exp(\tilde{\alpha}_{j,t})} \tag{8}$$

ensures that the priors $\alpha_{i,t}$ are positive and that they sum up to one, which makes the right hand side of Eq. (4) a pdf. The exponential function in Eq. (7) guarantees positive conditional variances. As a result the MDN receives the $m$-dimensional input $x_{t-1}, \ldots, x_{t-m}$ and produces a $3n$-dimensional output. The first $n$ components $\text{MLP}_j, 1 \le j \le n$, are used to calculate the priors, the outputs $\text{MLP}_j, n+1 \le j \le 2n$, are the conditional means, and the components $\text{MLP}_j, 2n+1 \le j \le 3n$, are squared to estimate the conditional variances. The parameters of the MDN (the MLP) and of the models of Section 2 are updated according to scaled gradient descent on the negative logarithm of the likelihood function [1]. To test the performance of the models on independent test sets, the same function applied to the test data can be used as a loss or generalized error function.

## 4 Experimental Results

The time series $\{x_t\}$ of the Austrian stock market index ATX from 7 January 1986 until 14 June 1996 (2575 measurements) was preprocessed using the transformation $r_t = 100(\log x_{t+1} - \log x_t)$. The resulting time series of returns $r_t$ and the autocorrelation functions of $r_t$ and $r_t^2$ are depicted in Fig. 1. There is an obvious change in structure in the time series at time $t \approx 950$ when the trading conditions at the stock exchange in Vienna were changed. Several volatility clusters (accumulations of large positive and negative returns) are clearly visible. The two horizontal lines on the right hand side of Fig. 1 indicate the 95% confidence interval for an identically and independently distributed (i.i.d.) process (white noise). Consequently, only the first autocorrelation of $r_t$ should be assumed to be statistically significant. The squared returns $r_t^2$, however, show a very regular structure which is significant for all lags $k$ ($1 \le k \le 25$). The

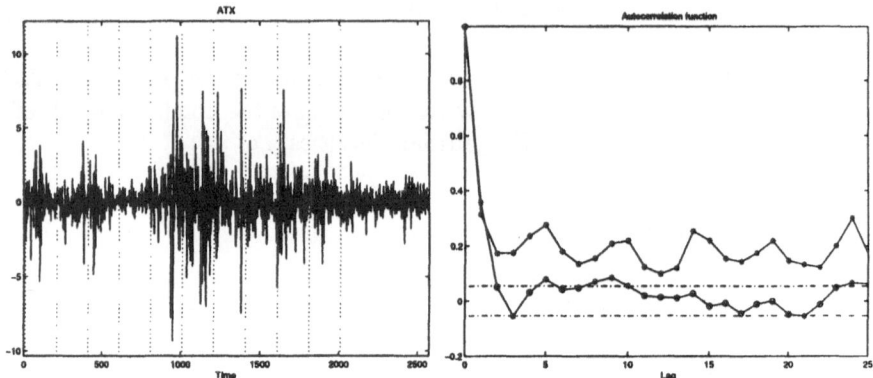

Figure 1: (Left) The returns $r_t$ of the ATX from 7 January 1986 until 14 June 1996 and the division into training and test sets (dotted lines). (Right) The autocorrelation function of $r_t$ (lower curve) and $r_t^2$ (upper curve) and the 95% confidence intervals for white noise.

quasiperiodicity of period five might indicate that the volatilities of identical days of the week are particularly correlated.

First, we fitted an ARCH(1), a GARCH(1,1) and a GJR model to the time series of returns $r_t$. Due to the correlation analysis the mean component $\mu_t$ was modelled by an autoregressive process of first order, i.e. $\mu_t = a x_{t-1}$. In order to evaluate the performance of the models we used the concept of cross validation. More precisely, the time series was divided into ten subsequent intervals of equal size: $I_1 = (r_{11}, \ldots, r_{210}), \ldots, I_{10} = (r_{1811}, \ldots, r_{2010})$. The rest of the data $T = (r_{2011}, \ldots, r_{2575})$ was used as an independent test set (see Fig. 1). Then each model was trained on nine of these ten intervals and the error $E_j$ on the missing interval $I_j$ was calculated ($1 \leq j \leq 10$). Additionally, each model was trained on the whole training data set $I = (r_{11}, \ldots, r_{2010})$ and evaluated on the test set $T$. The first set $I_1$ starts with $r_{11}$ since we wanted to present the same training sets to the models (of different order $m$).

The mean value and the standard deviation of the errors $E_j$ are summarized in Table 1. We see that the GARCH(1,1) model has the best performance of the standard models. Table 1 also gives the results for the trained MDNs. A network with two inputs $(x_{t-1}, x_{t-2})$, three hidden neurons and two gaussian distributions is denoted MDN(2-3-2), for instance. The best network is MDN(2-3-2) which is also better than the best standard model GARCH(1,1). The performance of the largest network MDN (5-4-3) is slightly worse which could be the result of insufficient training. We emphasize that the standard deviation is very large for all models (in comparison to the mean) because of the change in structure at $t \approx 950$. In fact, there are subintervals $I_j$ which can be easily modelled ($j = 4$, for instance), whereas some periods are characterized by large returns which are hard to predict ($j = 5$, for instance). If the models are trained on the whole training data set $I$ and tested on the independent set $T$, the GARCH(1,1) model performs best.

| Model | Parameters | mean | std. | $T$ | correct |
|---|---|---|---|---|---|
| ARCH(1) | 3 | 1.617 | 0.415 | 1.138 | 53.9% |
| GARCH(1,1) | 4 | 1.505 | 0.411 | 0.925 | 67.6% |
| GJR | 5 | 1.514 | 0.404 | 0.934 | 65.4% |
| MDN(1-3-1) | 18 | 1.601 | 0.375 | 1.157 | 51.6% |
| MDN(2-3-2) | 33 | 1.448 | 0.368 | 0.994 | 52.0% |
| MDN(5-4-3) | 69 | 1.458 | 0.414 | 1.002 | 56.6% |

Table 1: Overview of models fitted to the Austrian stock market index ATX.

Another test for the quality of volatility forecasts is the analysis of the profitability of trading strategies based on the predicted volatilities. More precisely, the volatility forecast based on historical returns gives us the information if the volatility is going to increase or decrease in the next period. This information can be interpreted as a buying or selling signal for a straddle [7]. If the predicted volatility is lower than the current one (volatility decreases) we go short, and if the volatility increases we take a long position. Therefore the quality of a volatility model can be measured by the percentage of correctly predicted directions of change of volatility from this period to the next (increase or decrease).

The first part of the ATX data set, i.e. $I$, was used to train the classical and neural models which were evaluated on the independent test set $T$ afterwards. The performance of the models concerning the correctly predicted directions of change of volatility is summarized in the last column of Table 1 (the concrete implementation of trading strategies is planned for the future). Strictly speaking, the squared returns $r_t^2$ are considered the "true" volatility and compared to the forecasted volatility $\sigma_t^2$. A prediction is thus classified as correct if and only if $(\sigma_t^2 - r_{t-1}^2)(r_t^2 - r_{t-1}^2) > 0$. For the MDNs with several gaussian distributions the "accumulated" variance of the distribution [1] is used. The best model is again the GARCH(1,1) model with impressive 67.6%. From Table 1 we also learn that the predictive quality of the MDNs increases with the number of inputs (past values).

## 5   Discussion and Conclusion

These results indicate the importance of long-term memory of the models if they are implemented in trading strategies. For the GARCH(1,1) model the conditional variance $\sigma_t^2$ is heavily influenced by the previous conditional variances $\sigma_{t-1}^2, \ldots$ owing to the parameter $\beta_1 \approx 0.918$ (for the training set $I$). A promising idea is thus to include *recurrent structures* into the MDNs. Our new architecture, which is currently investigated, consists of three MLPs which estimate the priors, the means and the variances separately. Following the GARCH(1,1) specification the MLP estimating the variance $\sigma_t^2$ receives a two-dimensional input: the squared error $e_{t-1}^2$ and the previous variance $\sigma_{t-1}^2$. We think that a comparison of the performance of standard and neural models is only fair if this extended MDN architecture is considered.

Furthermore, the implementation and evaluation of different trading strategies might provide further valuable insights into the behavior and predictive power of standard and neural volatility models.

# Acknowledgements

The MDNs were implemented using the NETLAB neural network software written by I. Nabney and C. Bishop (`http://neural-server.aston.ac.uk/`). This work was supported by the Austrian Science Fund (FWF) within the research project "Adaptive Information Systems and Modelling in Economics and Management Science" (SFB 010). The Austrian Research Institute for Artificial Intelligence is supported by the Austrian Federal Ministry of Science and Transport. The authors want thank A. Weingessel and F. Leisch for valuable discussions.

# References

[1] Bishop CM. Mixture density networks, Neural Computing Research Group Report: NCRG/94/004, Aston University, Birmingham, 1994

[2] Engle RF. Autoregressive conditional heteroscedasticity with estimates of the variance of U.K. inflation. Econometrica 1982; 50:987-1008

[3] Bollerslev T. A generalized autoregressive conditional heteroscedasticity. Journal of Econometrics 1986; 31:307-327

[4] Glosten LR, Jagannathan R., Runkle DE. On the relation between the expected value and the volatility of the nominal excess return on stocks. Journal of Finance 1993; 48:1779-1801

[5] Neuneier R, Finnoff W, Hergert F, Ormoneit D. Estimation of conditional densities: a comparison of neural network approaches. In: Marinaro M, Morasso PG (ed) ICANN 94 - Proceedings of the International Conference on Artificial Neural Networks. Springer-Verlag, Berlin, 1994, pp 689-692

[6] Schittenkopf C, Deco G. Testing nonlinear Markovian hypotheses in dynamical systems. Physica D 1997; 104:61-74

[7] Noh J, Engle RF, Kane A. Forecasting volatility and option prices of the S & P 500 index. Journal of Derivatives 1994; 17-30

# Introducing a Clustering Technique into Recurrent Neural Networks for Solving Large-Scale Traveling Salesman Problems

Kunikazu Kobayashi

Faculty of Engineering, Yamaguchi University

2557 Tokiwadai, Ube, Yamaguchi 755-8611, Japan

e-mail: k@csse.yamaguchi-u.ac.jp

### Abstract

It is difficult to solve large-scale traveling salesman problems (TSPs) using recurrent neural networks (RNNs). Because local solutions increase remarkably as the number of cities does and then computational cost increases dramatically. In this paper, introducing a clustering technique, it is shown that RNNs could apply to large-scale TSPs. At first, a large-scale TSP is divided into some small-scale TSPs, which can be easily solved by RNNs. If necessary this process is continued recursively. Then, such small-scale TSPs are solved by RNNs. Finally, a total tour of the large-scale TSP is derived from connecting the tours of small-scale TSPs. Through computer experiments, it is verified that the proposed algorithm is effective for solving large-scale TSPs.

## 1 Introduction

It is well known that traveling salesman problem (TSP) is one of the NP-hard problems. The problem is to find a closed tour which visits each city once, returns to the starting city and has a shortest total path length. A lot of efforts have been devoted to solve TSP because of its various applications in engineering such as a wiring of LSI substrates, scheduling and so on.

Hopfield and Tank proposed a recurrent neural network (RNN), so-called Hopfield model, for solving combinatorial optimization problems [1]. In their model, it is guaranteed that the network gradually changes its state so as to decrease a cost function defined for the problem. They have applied it to TSP and have shown that an approximate solution is obtained. As the number of cities increases, however, the network tends to be trapped into local minima because they are dramatically increased. In general, this problem has been escaped using annealing techniques like the Boltzmann machine [2]. On the other hand, regarding computational cost, it becomes much heavier.

Nozawa added a negative self-feedback to a discrete-type Hopfield model [3]. Since this allows each neuron to have a chaotic state, the network converges to a global minimum not trapping into local minima. Furthermore, Chen and Aihara improved convergence property in Nozawa model, where the value of the self-feedback is gradually decreased [4]. This realizes a chaotic simulated annealing. However, both models still have a problem on computational cost for large-scale TSPs. Then it is difficult to obtain a good approximate solution for the problems with more than 50 cities.

In the present paper, we focus on the reduction of computational cost for the large-scale TSP and propose an effective algorithm using a clustering technique. In our method, a large-scale TSP is divided into small-scale TSPs by clustering the cities according to their coordinates. Here, the small-scale TSP refers to one which can be easily solved by RNNs. Then, each small-scale TSP is solved by RNN. Finally, a total tour is given by connecting the tours of small TSPs.

# 2  Solving large-scale TSP using RNN

In this section, transiently chaotic neural network (TCNN) proposed by Chen and Aihara, which is one of the TSP solvers using RNNs, is described (2.1). Then, a new method for solving large-scale TSPs is proposed (2.2).

## 2.1  Transiently chaotic neural network

The dynamics of TCNN with $N$ neurons is characterized by the following equations [4].

$$x_i(t) = \frac{1}{1 + \exp(-y_i(t)/\epsilon)}, \tag{1}$$

$$y_i(t+1) = ky_i(t) + \alpha \left\{ \sum_{j=1, \, j \neq i}^{N} w_{ij} x_j(t) + I_i \right\} - z_i(t)\{x_i(t) - I_0\}, \tag{2}$$

$$z_i(t+1) = (1 - \beta)z_i(t), \tag{3}$$

where $x_i$ and $y_i$ are an output and internal state of neuron $i$, respectively, $\epsilon$ is a slope parameter of sigmoid function, $k$ is a decay parameter, $\alpha$ is a positive scaling parameter for inputs, $w_{ij}$ is a connection weight between neuron $i$ and $j$, $I_i$ is an external input to neuron $i$, $z_i$ is a self-feedback connection weight, $I_0$ is a positive constant, and $\beta$ is a decay parameter of $z_i$.

TCNN is characterized as follows. At the initial state, $z_i$ is so large that a network state is chaotic and the network searches a global solution. Then, $z_i$ is gradually decreased with time. The network is also gradually changed from chaotic to steady state. An optimization process of TCNN is regarded as a chaotic simulated annealing (CSA) [4].

In TCNN, an energy function and connection weights for solving TSP is same as Hopfield model [1]. Chen and Aihara have shown that global solutions for 10- and 48-city problems were obtained using TCNN [4]. However, they focus on the error rate, which is a measure of quality of the approximate solution, but not on computational cost. For example, when the number of cities becomes twice the number of neurons and computational cost per neuron also increases twice. As a result, total computational cost becomes four times. In fact, it is difficult to obtain a good approximate solution using TCNN for the problems with more than 50 cities.

## 2.2 Dividing large-scale TSP

To overcome the problems explained in the previous section, computational cost is reduced using a clustering technique. The basic idea is that a large-scale TSP is divided into some small-scale ones and then each small-scale TSP is solved using TCNN.

The proposed algorithm is described as follows.

1. Divide a large-scale TSP into some small-scale ones according to their coordinates. In the present paper, we used $K$-mean method as a clustering technique because it is simple and requires no heavy computation.

2. Determine a visiting order among the clusters. Finding the visiting order is regarded as a TSP, called reference TSP. The coordinates of cities in the reference TSP are assumed as mean vectors in each cluster. Then solve the reference TSP.

3. Select two cities in each cluster, which have a shortest distance between the adjacent clusters and correspond to the starting and last cities, respectively. In the later experiment, this is realized that we set the external input for such cities to a higher value than the other cities. As a result, the two neurons corresponding to such cities tend to fire.

4. Solve each small-scale TSP using TCNN.

5. Connect every two cities selected in step 3 to obtain a total tour of the large-scale TSP.

If the number of cities in a cluster is so large that TCNN cannot solve the TSP our algorithm is applied recursively.

Our algorithm will be illustrated using an example, which is 101-city problem, eil101.tsp, from TSPLIB [1]. At first, 101 cities are divided into 9 clusters, which have 20 to 30 cities as shown in Fig.1. In this figure, the same symbol denotes the cities in the same cluster. Next, a reference TSP composed of 9 mean vectors is solved using TCNN. The tour is illustrated by the dashed line in the figure. Then, 9 small-scale TSPs are solved using TCNN, respectively. Finally, by connecting the 9 tours we can obtain a total tour of 101-city problem, which is illustrated by the solid line in the figure.

# 3 Computer experiment

The first experiment (Experiment 1) confirms that the proposed method is effective for reduction of computational cost. The second experiment (Experiment 2) shows that it is possible to reduce computational cost and obtain a good approximate solution compared with greedy method and genetic algorithm (GA).

---

[1]http://www.iwr.uni-heidelberg.de/iwr/comopt/soft/TSPLIB95/TSPLIB.html

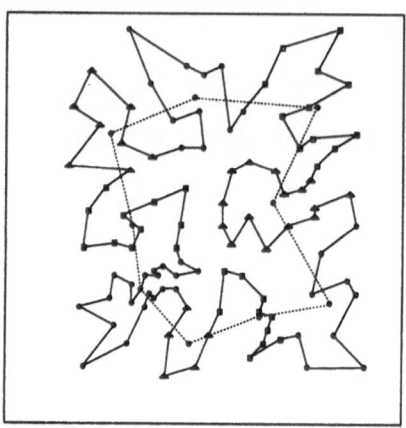

Figure 1: An example

## 3.1   Experiment 1

In this experiment, we compared our method with the original TCNN using the former 101-city problem, eil101.tsp. At first, we have to determine the number of clusters because of $K$-mean method. Table 1 shows the numerical result of both methods. In this table, CPU time of our method refers to the relative value when that of TCNN is assumed 100.

Table 1: Comparison of TCNN with and without clustering

| TCNN without clustering | | |
|---|---|---|
| No. of clusters | Error rate [%] | CPU time |
| - | 15.6 | 100.0 |
| TCNN with clustering | | |
| 6 | 10.0 | 4.31 |
| 7 | 7.8 | 4.00 |
| 8 | 7.2 | 2.62 |
| 9 | 5.7 | 1.78 |
| 10 | 8.4 | 1.49 |
| 11 | 8.4 | 1.49 |
| 12 | 7.5 | 1.49 |

As seen in this table, our method could dramatically reduce computational cost compared with TCNN without clustering. The error rate and computational cost depend on the number of clusters. The more the number of clusters is the lower computational cost is. Of course, the distribution of the cities and the number of cities in a cluster affect the quality of the solutions. Obviously, the optimal number of clusters for this problem is nine.

## 3.2 Experiment 2

In this experiment, our method was compared with two other TSP solvers, i.e. greeding method and genetic algorithm (GA) [5]. The three problems, pr76.tsp, eil101.tsp, and ch130.tsp in TSPLIB, were used for evaluation. The number of generations in GA was 1,000, 1,500, and 2,000 for pr76.tsp, eil101.tsp, and ch130.tsp, respectively.

Table 2 shows the numerical result of the experiments. In this table, GM, GA, and TC refer to greedy method, genetic algorithm, and our method, respectively. The result of GM represents the best solution in 10 trials and our method and GA are the average of 10 trials. Regarding CPU time, the time using GA is assumed 100.

Table 2: Comparison with other two methods, greedy method and GA

| 76-city problem (pr76.tsp) | | | |
|---|---|---|---|
| Method | Error rate [%] | CPU time | No. of clusters |
| GM | 21.0 | 0.83 | - |
| GA | 3.0 | 100 | - |
| TC | 7.9 | 9.2 | 7 |
| 101-city problem (eil101.tsp) | | | |
| GM | 14.6 | 1.72 | - |
| GA | 6.0 | 100 | - |
| TC | 5.7 | 5.17 | 9 |
| 130-city problem (ch130.tsp) | | | |
| GM | 17.8 | 3.29 | - |
| GA | 9.1 | 100 | - |
| TC | 9.1 | 5.32 | 11 |

As seen in this table, the error rate is changed according to the distribution of cities. The greedy method is very fast because it uses only local information. But it cannot escape the local minima. Therefore, the quality of the solutions is worst. And the solution using GA is better but computational cost is high. On the other hand, the error rate of our method is almost same with that of GA but CPU time is lower than GA. Of course, the error rate depends on the number of clusters.

# 4   Applying to larger-scale TSP

The merit of the proposed method is that it can apply to larger-scale TSPs using recursive clustering.

Figure 2 shows an approximate solution for 1000-city problem, dsj1000.tsp, in TSPLIB using our method. In this case, the initial number of clusters is 20. Then the recursive clustering is applied. As a result, the final number of clusters is 63.

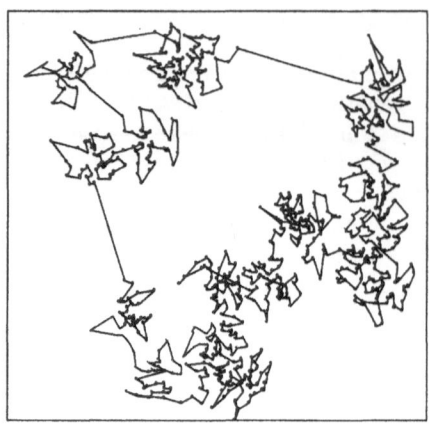

Figure 2: An approximate solution of 1000-city TSP (dsj1000.tsp)

## 5 Conclusions

In the present paper, we have proposed the TSP solver for large-scale problem using RNN. It is shown that our method could solve large-scale problems with more than 50 cities, which cannot be solved by other RNNs. Using the clustering technique, the computational cost is dramatically reduced.

The quality of solution and computational cost highly depends on the number of clusters and cities in a cluster. Moreover, the quality of the solution of reference TSP affects the final tour length.

The optimal clustering and clustering using neural networks are future work. The application to other combinatorial optimization problems should be investigated.

## References

[1] Hopfield J. J., Tank D. W. "neural" computation of decisions in optimization problems. Biological Cybernetics 1985; 52:141-152

[2] Ackley D. H., Hinton G. E., Sejnowski T. J. A learning algorithm for Boltzmann machines. Cognitive Science 1985; 9:147-169

[3] Nozawa H. A neural network model as a globally coupled and applications based on chaos. Chaos 1992; 2:377-386

[4] Chen L., Aihara K. Chaotic simulated annealing by a neural network model with transient chaos. Neural Networks 1995; 8:915-930

[5] Pai K. F. Genetic algorithm for the traveling salesman problem based on a heuristic crossover operation. Biological Cybernetics 1993; 69:539-546

# Poster Presentations:

# Computational Neuroscience and Brain Theory

# Comparing Different Measures of Spatio-Temporal Patterns in Neural Activity

Gustavo Deco

Siemens AG, Corporate Technology, 81730 Munich, Germany

Laura Martignon

Max Planck Institute for Human Development, 14195 Berlin, Germany

Kathryn Blackmond Laskey

Dept. of Systems Engineering, George Mason University, Fairfax, VA 22030, USA

### Abstract

Advances in the technology of multi-unit recordings have created the need for new statistical approaches to detect the presence of spatiotemporal patterns in neuron spike train data. We present statistical approaches to detect synchronisation in the activity of three or more neurons. These phenomena must be modelled as higher order correlations that cannot be reduced to a simple combination of second order correlations. We examine three measures for the presence of higher-order patterns of neural activation: coefficients of log-linear models, connected cumulants and redundancies. We present arguments in favour of the coefficients of log-linear models. We introduce the Constraint-Perturbation-Procedure (CPP) as an alternative to Iterative proportional Fitting to construct test statistics for detecting the presence of higher order interactions. The methods are applied to experimental data drawn from of multi-unit recordings from the frontal cortex of behaving monkeys.

## 1 Introduction

Hebb conjectured that information processing in the brain is achieved through the collective action of groups of neurons, which he called *cell assemblies* [1]. Evidence for collective phenomena confirming the "cell assembly" hypothesis has only recently begun to emerge as a result of the progress achieved by multi-unit recording technology. In pursuit of experimental evidence for cell assembly activity in the brain, physiologists thus seek to observe the activities of many separate neurons simultaneously, preferably in awake, behaving animals. These "multi-neuron activities" are then inspected for possible signs of interactions between neurons. Of interest are patterns involving more than three neurons, and which cannot be described in terms of pair-correlations. Such patterns are genuine *higher order phenomena*. The models reported in this paper were developed for the purpose of describing and detecting such higher order correlations. In the data we analyse, the spiking events (in the 1 msec range) are encoded as sequences of 0's and 1's, and the activity of the whole group is described as a sequence of binary configurations. This paper presents a family of statistical models for analysing such data. In our models, the parameters represent spatio-temporal firing patterns. We develop statistical tests for detecting the presence of a genuine order-$n$ correlation and distinguishing it from an artefact that can be explained by lower-order interactions. The tests compare observed firing frequencies on the involved neurons with frequencies predicted by a distribution that maximises entropy among all distributions consistent with observed information on synchrony of lower order. We introduce the Constraint-Perturbation-Procedure (CPP) to construct test statistics for detecting the presence of higher order interactions. We compare our approach based on log-linear models with two other candidates for detecting the presence of higher order correlations approaches drawn from statistical physics and information theory. Finally, results are presented to multi-unit recordings.

# 2 Detecting Higher Order Synchronisation

## 2.1 Coefficients of Log-Linear Models

Consider a set of $n$ neurons. Each neuron is modelled as a binary unit that can take on one of two states: 1 ("firing") or 0 ("silent"). The state of the $n$ units is represented by the vector $\underline{x} = (x_1,...,x_n)$, where each $x_i$ can take on the value zero or one. There are $2^n$ possible states for the $n$ neurons. If all neurons fire independently of each other, the probability of configuration $\underline{x}$ is given by: $p(x_1,...,x_n) = p(x_1) \cdots p(x_n)$. Methods for detecting correlations look for departures from this model of independence. Neurons are said to be correlated if they do not fire independently. A correlation between two neurons is expressed as: $p(x_1,x_2) \neq p(x_1)p(x_2)$. Extending this idea to larger sets of neurons introduces complications. It is not sufficient simply to compare the joint probability $p(x_1,x_2,x_3)$ with the product $p(x_1)p(x_2)p(x_3)$ and declare the existence of a triplet when the two are not equal. This would confuse authentic triplets with overlapping doublets. That is, neurons 1, 2, and 3 may fire together more often than the independence model would predict because neurons 1 and 2, and neurons 2 and 3, are each involved in a binary interaction. The probability distributions for the three pair configurations are $p(x_1,x_2)$ $p(x_2,x_3)$ and $p(x_1,x_3)$. A genuine third-order interaction among the three neurons would occur if it was impossible to determine the joint distribution for the three-neuron configuration using only information in these pair distributions. A natural way to construct a joint distribution on three neurons from the pair distributions would be to maximise entropy subject to the constraint that the two-neuron marginal distributions are given by $p(x_1,x_2)$ $p(x_2,x_3)$ and $p(x_1,x_3)$. This is the distribution that adds the least information beyond the information contained in the pair distributions and that can be written as:

$$p(x_1, x_2, x_3) = e^{\{\theta_0 + \theta_1 x_1 + \theta_2 x_2 + \theta_3 x_3 + \theta_{12} x_1 x_2 + \theta_{13} x_1 x_3 + \theta_{23} x_2 x_3\}}, \tag{1}$$

where the $\theta$'s are real-valued parameters, and where $\theta_0$ determined from the other $\theta$'s and the constraint that the probabilities of all configurations sum to 1. Not all joint probability distributions on three neurons can be written in the form (1). A general joint distribution for three neurons can be expressed by including one additional parameter:

$$p(x_1, x_2, x_3) = e^{\{\theta_0 + \theta_1 x_1 + \theta_2 x_2 + \theta_3 x_3 + \theta_{12} x_1 x_2 + \theta_{13} x_1 x_3 + \theta_{23} x_2 x_3 + \theta_{123} x_1 x_2 x_3\}} \tag{2}$$

Holding the other parameters fixed and increasing the value of $\theta_{123}$ increases the probability that all three neurons fire simultaneously. Parametrizations of the general form (1) and (2) are called *log-linear models* because the logarithm of the probabilities can be expressed as a linear sum of functions of the configuration values. It seems natural, then, to define a genuine order 3 correlation as a joint probability distribution that cannot be expressed in the form of Eq. (1) because $\theta_{123} \_ 0$. This idea can be extended naturally to larger sets of neurons and to temporal patterns. Consider a set $\Lambda$ of $n$ binary neurons and denote by $p$ the probability distribution on sequences of binary configurations of $\Lambda$. Assume that the sequence of configurations $\underline{x}_t = (x_{1t}, \cdots, x_{nt})$ neurons forms a Markov chain of order $r$. Let $\delta$ be the time step, and denote the conditional distribution for $\underline{x}_t$ given previous configurations by $p(\underline{x}_t | \underline{x}_{(t-\delta)}, \underline{x}_{(t-2\delta)}, \cdots, \underline{x}_{(t-r\delta)})$. We assume that all transition probabilities are strictly positive and expand the logarithm of the conditional distribution as:

$$p(\underline{x}_t | \underline{x}_{(t-\delta)}, \underline{x}_{(t-2\delta)}, \cdots, \underline{x}_{(t-r\delta)}) = e^{\{\theta_0 + \sum_{A \in \Xi} \theta_A \underline{x}_A\}} \tag{3}$$

In this expression, each A is a subset of pairs of subscripts of the form $(i, t - s\delta)$ for $r \geq s \geq 0$ that includes at least one pair of the form $(i, t)$. The variable $\underline{x}_A = \Pi_{1 \leq j \leq k} x_{(i_j, t - m_j \delta)}$ is equal to 1 in the event that all neurons in A are active and zero otherwise. The set $\Xi \subseteq 2^\Lambda$ for which $\theta_A \_ 0$ is called the *interaction structure* for the distribution $p$. The parameter $\theta_A$ is called the *interaction strength* for the interaction on subset A. Clearly, $\theta_A = 0$ means $A \notin \Xi$ and is taken to indicate absence of an order $|A|$ interaction among neurons in A. We denote the structure-specific vector of nonzero interaction strengths by $\theta_\Xi$.

**Definition 1:** We say that neurons $\{ i_1, i_2, \ldots, i_k \}$ exhibit a *spatio-temporal pattern* if there is a set of time intervals $m_1 \delta, m_2 \delta, \ldots, m_k \delta$ with $r \geq m_j \geq 0$ and at least one $m_i = 0$, such that $\theta_A \neq 0$, where $A = \{(i_1, t - m_1 \delta), \ldots, (i_k, t - m_k \delta)\}$.

**Definition 2:** A subset $\{ i_1, i_2, \ldots, i_k \}$ of neurons exhibits a *synchronisation* or *spatial correlation* if $\theta_A \neq 0$ for A$=\{ (i_1, 0), \ldots, (i_k, 0) \}$.

In the case of absence of any temporal dependencies the configurations are independent and we drop the time index: $p(\underline{x}) = \exp\{\theta_0 + \sum \theta_A \underline{x}_A\}$ where each A is a nonempty subset of $\Lambda$ and $\underline{x}_A = \prod_{i \in A} x_i$ .

## 2.2 Other Measures for Higher Order Correlation

We consider two alternate approaches to measuring the existence of higher order correlations in a set of binary neurons: Connected Cumulants and Redundancy. Connected cumulants, a measure of higher order correlation used in statistical mechanics, is a candidate parameter for spatio-temporal patterns of neural activation. Let us define the partition function $Z = \sum_{x_1,\ldots,x_N} \exp'(\,\theta_1 x_1 + \ldots + \theta_{1,2,\ldots,N} x_1 x_2 \ldots x_N + \sum_j J_j x_j)$. The n-point correlation functions are defined by the equation

$$\left\langle x_{i_1} \ldots x_{i_N} \right\rangle = \frac{1}{Z} \frac{\partial^N Z}{\partial J_{i_1} \ldots \partial J_{i_N}} \; . \tag{4}$$

This is also a correlation measure. However, this expression may be decomposable as a sum of lower order correlations. In order to measure the amount of correlation which is not decomposable as a sum of smaller groups of neurons, statistical physics introduces the concept of connected cumulant given by,

$$C_{i_1 \ldots i_N} = \frac{\partial^N \log Z}{\partial J_{i_1} \ldots \partial J_{i_N}} = \left\langle x_{i_1} \ldots x_{i_N} \right\rangle - \sum \left[ \begin{array}{cccc} products & of & connected & cumulants \\ & C_{i_1 \ldots i_M} & where & M < N \end{array} \right] \tag{5}$$

The connected cumulant is the contribution to the n-point correlation that remains when all other contributions of lower order have been substracted off. Thus it provides a candidate for measuring higher-order correlation. In the case of two neurons, the presence of correlation defined by means of cumulants is equivalent to the presence of correlation established by the effects of log-linear models. That is, the connected cumulant is equal to zero if and only if the neurons exhibit no correlation as defined above. For three neurons the concepts are not equivalent. It can be shown that the connected cumulant may be nonzero when there exist overlapping couples. Such overlapping couples, we argue, constitute a second-order phenomenon and not a true third-order interaction.

Another approach to measuring higher-order correlations is drawn from information theory by using the concept mutual information. For the case of 2 neurons $x_1$ and $x_2$ the redundancy, i.e. the mutual information $I(x_1;x_2)$, is an information-theoretical measure of statistical correlation that takes into account non-linear dependencies. For the case of 3 neurons, the redundancy can be decomposed as $I(x_1;x_2;x_3) = I(x_1;x_2) + I(x_2;x_3) + I(x_1;x_3|x_2)$. Let us define $I_3 = I(x_1;x_2;x_3) - I(x_1;x_2) - I(x_1;x_3) - I(x_2;x_3)$ which, as is clear, is a measure of higher order correlation. As in the case of connected cumulants the presence of second order correlations defined by redundancy is equivalent to the one established by effects.

## 2.3 The Constrained Perturbation Procedure

We adopt a frequentist approach. This section is devoted to constructing the adequate Null-Hypotheses for detecting higher order correlations suggested by Good's theorem. There exists a famous algorithm for this construction in the general case, called the Iterative Proportional Fitting Procedure, or I.P.F.P. We introduce a more general method for constructing the null-hypotheses corresponding, for each of the three measures defined in section 2, to the unique distribution consistent with the lower order marginals of the data and annulling the measure to be tested. Assume again that $P$ is a probability distribution defined on the configurations of a set $A$ of vertices. Consider the following problem: find a distribution $P^*$ such that the marginals of order less than $|A|$ coincide with those of $P$ and the corresponding measure $\theta_A^*, C_A^*, I_A^*$, is equal to 0. If $B$ is a nonempty subset of $A$, denote by $\chi_B$ the configuration that has a component $1$ for every index in $B$ and 0 elsewhere. Define

$P^*(\chi_B) = P(\chi_B) + (-1)^{|B|}\Delta$, where $\Delta$ is to be determined by solving for $\theta^*_A \equiv 0$,

where $\theta^*_A$ is the coefficient corresponding to $A$ in the log-expansion of $P^*$, i.e

$\theta^*_A = \sum_{B \subset A} (-1)^{|A-B|} \ln p^*(\chi_B)$. Furthermore, define $P^*(0) = 1 - \sum_{\phi \neq B \subset A} P^*(\chi_B)$.

This distribution $P^*$ maximises entropy among those with the same marginals of $P$ on proper subsets of $A$. This follows from Good's Theorem and the fact that, as remarked above, there is exactly one distribution with the same marginals on subsets of $A$ satisfying the additional condition that its A-coefficient in the log-expansion is 0. Observe that our construction contains only one step where a numerical approximation becomes necessary, namely solving for $\theta^*_A = 0$ in terms of $\Delta$. This is done by using Newton's approximation method.

*Example*: Suppose that $A = V = \{1,2,3\}$ and that $P$ is a distribution on $A$. We want to construct the distribution $P^*$ that maximises entropy among all those consistent with Marg($\{\{1,2\},\{1,3\},\{1,3\}\}$). Define: $P^*(1,1,1) = P(1,1,1) - \Delta$; $P^*(1,1,0) = P(1,1,0) + \Delta$; $P^*(1,0,1) = P(1,0,1) + \Delta$; $P^*(1,0,0) = P(1,0,0) - \Delta$; $P^*(0,1,1) = P(0,1,1) + \Delta$; $P^*(0,1,0) = P(0,1,0) - \Delta$; $P^*(0,0,1) = P(0,0,1) - \Delta$; $P^*(0,0,0) = 1 - \sum_{B \subset \{1,2,3\}} P(\chi_B)$. It is possible to see at a glance that all second order marginals of $P^*$ coincide with those of $P$. The only additional step we need to undertake, for the measures $\theta_A^*, C_A^*, I_A^*$, is to solve for $\Delta$ in the equations $\theta_A^*, C_A^*, I_A^*$, equal to zero, which is done by Newton's method in the case of effects and redundancy. In the case of the cumulant, there is no need for Newton's method, because choosing $\Delta$ as $C_A$ provides the solution.

## 2.4 Monte-Carlo Surrogate Method

The simultaneous firing of a group of neurons is by rejecting a null hypotheses produced by the surrogate method. The rejection of a null hypothesis is based on the computation of the discriminating statistics for the original data $D_o$ and the discriminating statistics $Ds_i$ for the $i$-th surrogate generated under the null hypothesis. The null hypothesis is rejected if the significance $S = |D_o - \mu_S|/\sigma_S$ is greater than $\hat{S}$ corresponding to a p-value $p = erfc(\hat{S} / \sqrt{2})$ ($p$ being usually 0.05 which corresponds to $\hat{S} = 1.645$), and being $\mu_s$ and $\sigma_s$ are the mean value and standard deviation of the discriminating statistics. The surrogates data used for testing synchronization of order k are generated by different realizations of the probability distribution $P*$ via the Monte Carlo method. The discriminating statistic D is therefore given by the higher-order measure $M_k$. The generated surrogates data have the same marginals than the original data and correspond to the null hypothesis that syn-firing of order k does not exists, i.e. $M_k = 0$. In the analysis we did of the neural data for this paper we used 1000 surrogates.

# 3    Application: Data from Multi-Unit Recordings

We applied our program to data from an experiment in which spiking events among groups of neurons were analysed through multi-unit recordings of 6-16 units in the frontal cortex of Rhesus monkeys (from E. Vaadia, Dept. of Higher Brain Function, Hadassah Medical Hospital, Hebrew University of Jerusalem, March 1995). The monkeys were trained to localise a source of light and, after a delay, to touch the target from which the light blink was presented. The spiking events (in the 1 msec. range) of each neuron were encoded as a sequence of zeros and ones, and the activity of the whole group was described as a sequence of configurations or vectors of these binary states. The experimenter provided a data-set of 188,000 msec of a subset of 6 neurons. Table 1 shows the obtained results.

| Neural Cluster | Effects (Surrogates) | Cumulants (Surrogates) | Redundancy (Surrogates) |
|---|---|---|---|
| 1,2 | 0.27 | 0 | 0 |
| 1,4 | 0.007 | 0.007 | 0.013 |
| 2,4 | 0.004 | 0.004 | 0.008 |
| 3,4 | 0.0013 | 0.029 | 0 |
| 2,4,6 | 0.015 | 0.004 | 0.005 |

Table 1: Synchronizations

By performing temporal shifts between neurons of up to 100 msec and applying the synchronization method at each instance of the resulting shifted data we checked for temporal correlations in sets of two and three neurons.

# References

[1] Hebb, D. (1949) *The Organization of Behavior.* New York: Wiley, 1949.

# Influence of Recurrent Excitation and Inhibition on Receptive Field Size and Contrast Sensitivity in Layer 4C of Macaque Striate Cortex

U. Bauer[1], M. Scholz[3], J. B. Levitt[2], J. S. Lund[2], and K. Obermayer[3]

[1] Technische Fakultät, Universität Bielefeld, 33501 Bielefeld, Germany

[2] Institute of Ophthalmology, UCL, London EC1V 9EL, U.K.

[3] FB Informatik, TU Berlin, 10587 Berlin, Germany

e-mail: ubauer@TechFak.Uni-Bielefeld.DE

## Abstract

Neurons in layer 4C in macaque striate cortex show an increase in receptive field size and achromatic contrast sensitivity from the bottom to the top of the layer. Using a computational model which is based on realistic anatomical and physiological data we demonstrate that part of the observed changes can arise from differences in the overall balance between recurrent excitation and lateral inhibition from two different neuron types. The model predicts that - given the above hypotheses - lateral recurrent excitation must come from an increasingly wider range with rise in depth of layer 4C, and lateral inhibition must have higher threshold and gain in upper 4Cα. The anatomical substrate of recurrent excitation are the stepped projections of spiny stellate cells. As the possible anatomical substrate of differential inhibition we suggest the "clutch" cell in lower and mid 4C and the α-6 [1] cell in upper 4Cα which replaces the clutch cell as a somatic inhibitor.

## 1 Introduction

Layer 4C in macaque striate cortex can be divided into two subdivisions 4Cα and 4Cβ which correspond to the fully segregated termination zones of the P- and the M-type thalamic afferents. In two studies [2, 3] it was reported that basic response properties, i.e. receptive field size and achromatic contrast sensitivity of cortical neurons, do not follow this step in the type of input, but rather show a gradual increase from P- to M-like properties from the bottom to the top of the layer.

Recently, we have performed a model study which showed that differential convergence of P- and M-inputs onto spiny stellate cells, whose dendritic arbors intrude into both termination zones, may form the substrate of the gradual change observed in lower and mid 4C. The almost exponential gradient of the basic response properties in upper 4Cα, however, led us to predict a sub-population of LGN-M cells with higher sensitivity and field size and which preferentially project to upper 4Cα [4, 5]. In the present contribution we investigate another potential source of increased sensitivity and field size in upper 4Cα: differences in the local circuitry of recurrent excitation of

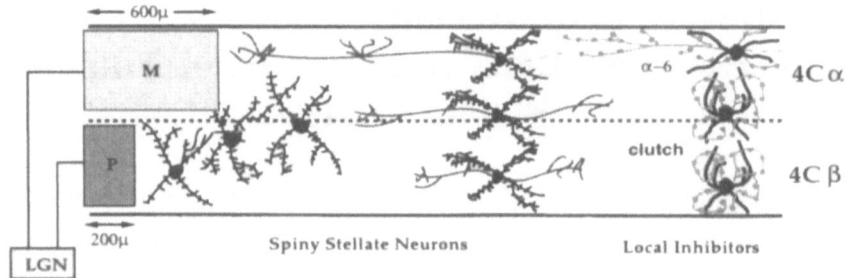

Figure 1: Anatomical organization of layer 4C. The left side shows the projection territory of a thalamic P- and M- axon; note that the dendrites of spiny stellate cells (black) in mid 4C overlap into both P- and M- termination zones. Each spiny stellate cell has a local axonal field (250 $\mu$m, not shown) which is slightly larger in lateral spread than its dendritic field (200 $\mu$m). The local axons of excitatory cells in mid-4C make step-like projections (grey) reaching 350-550$\mu$m from their somata. Axons of spiny stellate cells at the top of layer 4C$\alpha$ make a second lateral sidestep at 1000-1200$\mu$m. The right side shows the inhibitory cells. The "clutch" cell is prominent in the lower 2/3 of the layer and characterized by a dense dendritic and axonal field (both 175$\mu$m). The $\alpha$-6 cell is restricted to upper 4C$\alpha$ and has a dendritic field of 200 $\mu$m but shows a larger and more stratified axonal arborization (1000-1400$\mu$m) [1].

spiny stellate cells and a change of inhibitory strategy in upper layer 4C. For a detailed comparism of both models see the discussion.

The anatomically segregated termination zones of the physiologically distinct P- and M-pathways within depth of layer 4C are associated with different width of lateral projection patterns of the excitatory spiny stellate cells [6] and different types of local GABA-positive interneurons [1] (see Fig. 1). Only a small portion (10%) of the constant number of spines of a spiny stellate cell is occupied by thalamic synapses. The rest of the excitatory input comes from local collaterals of other spiny stellate cells within the layer and recurrent collaterals of layer 6 pyramids which we do not consider in this study. Unpublished observations from our own laboratory have revealed that the number of somatic inhibitory (type-2) synapses is approximately constant throughout the depth of the layer. Lower 4C$\beta$ and mid 4C contain a small basket neuron called a "clutch" cell [7] or $\beta$-1/$\alpha$-1 cell [1]. The clutch cell has been shown to contact the somata of spiny stellate cells within layer 4C and it might be activated by all three sources of excitatory input provided to layer 4C [7]. In upper 4C$\alpha$ the clutch cell is replaced by an $\alpha$-6 wide arbor basket cell [1]. The targets of the $\alpha$-6 cell are not known but its "basket" morphology suggests it also may contact the somata of the local spiny stellate cells. We, therefore explore the hypotheses that the difference in the anatomical organization in depth of layer 4C could be responsible for the almost exponential increase of basic response properties in upper 4C$\alpha$.

## 2 Methods

Our computational model consists of a visual field layer, two LGN layers for the P- and the M-populations, and eight "sublayers" representing neurons at eight different

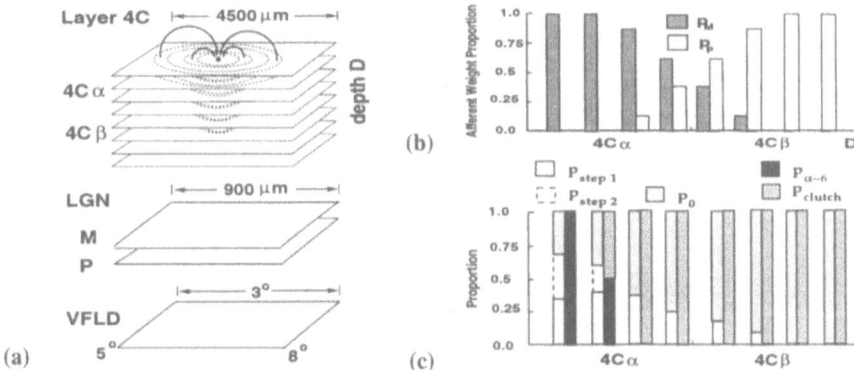

Figure 2: (a) Neural network architecture. The concentric annuli in layer 4C indicate the region from which a cortical cell gets lateral recurrent input via stepped projections. (b) Proportion $p_P(D)$ and $p_M(D)$ of afferent LGN-P and LGN-M input to cortical cells at different depth $D$ in layer 4C. (c) Proportion of lateral excitation and inhibition at different depth in layer 4C. Excitation comes from different spatial origin i.e. from local neighbours ($p_0$), from the first ($p_{step_1}$) and second ($p_{step_2}$) sidestep of lateral projections of spiny stellate cells. Inhibition comes from different cell types ($p_{clutch}$ and $p_{\alpha-6}$).

depths $D$ in 4C (Fig. 2a). Only monocular ON-cells are considered and model parameters correspond to $5° - 8°$ eccentricity (for references see [8]). Cells in different layers are connected in a topographic fashion and are modelled within a connectionist framework (for details see [4]).

The receptive field profiles $S$ of geniculate P- and M-cells are given by a Difference-of-Gaussians model. The geniculate output is calculated by convolving the activity $L(x)$ of the visual field layer with the DoG profile $S$ and applying a transfer function $f, O = f(S*L)$. The parameters of the transfer function $f(x) = a(x-s)/(b+x-s)$ are determined via a fit of $f$ to the experimental response vs. contrast function of P- and M-cells (for details [4]).

Each sublayer in 4C consists of 80% excitatory cells ($e$) which correspond to spiny stellate neurons and 20% inhibitory cells ($i$) which correspond to either the clutch (D=1,...,6) or the $\alpha$-6 cells (D=7,8) [1]. The afferent input to a cortical cell at location u and depth $D$ is given by

$$I_{aff}(\mathbf{u}, D) = \sum_{T\in\{M,P\}} p_T(D) \int_{\Re^2} d\mathbf{x}\ w_T(|\mathbf{u} - \mathbf{x}|)\ O_T(\mathbf{x}), \qquad (1)$$

where $p_T(D)$ is the proportion of the geniculate P and M input to a cortical cell at depth $D$ (Fig. 2b). The terms $w_T(\mathbf{x})$, $\int d\mathbf{x}\ w_T(\mathbf{x}) = W_{aff}$, are normalized circular symmetric weight kernels which are proportional to the areal overlap between the afferent axonal arbor of type $T$ at position x and the dendritic field of a cortical cell at position u. The cortical cells within each sublayer are connected via excitatory and inhibitory connections. Each neuron is characterized by a "membrane potential" m whose dynamics is given by

$$\frac{d}{dt}m^P(\mathbf{u}, t, D) = -m^P(\mathbf{u}, t, D) + I_{lat}^P(\mathbf{u}, t, D) + I_{aff}(\mathbf{u}, D), \qquad (2)$$

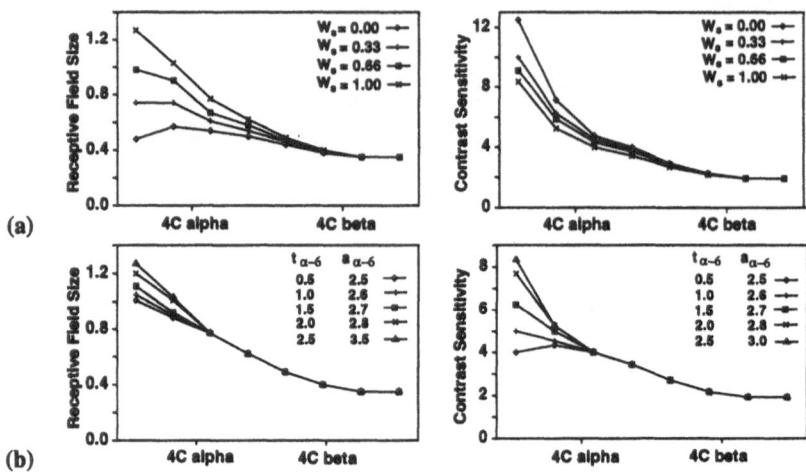

Figure 3: The curves connect receptive field size (left) and contrast sensitivity (right) measures of our model cortical cells at eight different depth of layer 4C for varying efficacy $W_s$ of the lateral stepped projections (a) and for varying properties of the $\alpha$-6 cell transfer function (b).

where $P \in \{e, i\}$ is the cell type. The lateral recurrent input is given by

$$I_{lat}^{P}(\mathbf{u}, t, D) = \sum_{P' \in \{e,i\}} \int_{\Re^2} d\mathbf{x} \, w_{P' \to P}(|\mathbf{u} - \mathbf{x}|, D) \, g_{P'}(m^{P'}(\mathbf{x}, t, D)) \qquad (3)$$

where $w_{P' \to P}(|\mathbf{u} - \mathbf{x}|, D)$, $\int d\mathbf{x} \, w_{P' \to P}(\mathbf{x}, D) = W_{P' \to P}$, are normalized symmetric weight kernels. The normalization constants $W_{aff}$, $W_{e \to P}$ and $W_{i \to P}$ which represent the excitatory and inhibitory synaptic load of a cortical cell are set to 0.08, 0.72 and 0.2, respectively. Because the geometry of axonal and dendritic fields as well as neural composition changes within depth of layer 4C, individual weight kernels $w_{P' \to P}$ are chosen for each depth D of layer 4C. Let $d =: |\mathbf{u} - \mathbf{x}|$ denote the distance between two cortical cells:

$$w_{e \to P}(d, D) = \sum_{j=1,2} W_s \, p_{step_j}(D) \, w_{e \to P}^{step_j}(d) + W_0 \, p_0(D) \, w_{e \to P}^{0}(d) \qquad (4)$$

$$w_{i \to P}(d, D) = p_{clutch}(D) \, w_{i \to P}^{clutch}(d) + p_{\alpha-6}(D) \, w_{i \to P}^{\alpha-6}(d). \qquad (5)$$

The circular symmetric weight kernels $w_{e \to P}^{0}, w_{i \to P}^{clutch}, w_{i \to P}^{\alpha-6}$ and $w_{e \to P}^{step_j}$ which give the weights of $j$th stepped axonal projection are calibrated by realistic anatomical data (for details see Introduction and Fig. 2a). All weight kernels in eqs. (4) and (5) are normalized to 1.0. Thus weight portions $p_{step_j}$, $p_0$, $p_{clutch}$ and $p_{\alpha-6}$ give the excitatory and inhibitory synaptic load of a cortical cell at depth $D$ arising from different spatial origin or different inhibitory cell types (for numerical values see Fig. 2c). The parameters $W_s$ and $W_0 = 3 - 2\,W_s$ control the efficacy of lateral excitation from the stepped projections versus excitation from the axonal projections of local neighbours.

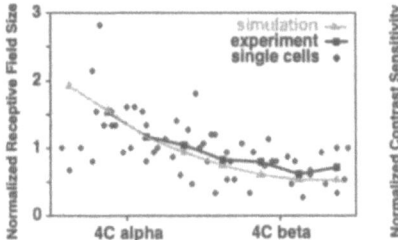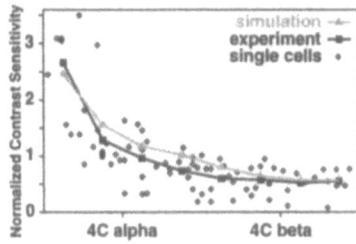

Figure 4: Normalized receptive field size and contrast sensitivity as a function of depth in layer 4C. Dots denote normalized experimental data ([2, 3]) from single units; the black line connects their mean values for eight equally spaced depth intervals. The grey line indicate best predictions of the recurrent model. Parameters were $W_s = 1.0$ and $t_{\alpha-6} = 2.5, a_{\alpha-6} = 3.0$.

The output of a cortical cell is calculated by applying a rectifying, piecewise linear transfer function $g_P(x) = \max(0, a_P(x - t_P))$ to the membrane potential. The threshold and gain for spiny stellate cells, inhibitory clutch cells and $\alpha$-6 cells are set to $t_e = 0.1$, $a_e = 1.0$, $t_{clutch} = 0.5$, $a_{clutch} = 2.5$ and $t_{\alpha-6} = 2.5$, $a_{\alpha-6} = 3.0$ (but see Results).

# 3 Results and Discussion

Figure 3a shows how the *efficacy of the lateral stepped projections* which is controlled by the model parameter $W_s$ affects the response properties of the model cortical cells. If there is no input from stepped projections ($W_s = 0.0$) the receptive field size curve shows only moderate increase with rise in depth of layer 4C and a plateau at the top of layer 4C$\alpha$, due to strong recurrent inhibition from the $\alpha$-6 cell. If $W_s$ is increased, receptive field size curves become steeper, however, contrast sensitivity decreases in upper 4C$\alpha$. The reason for reduced contrast sensitivity is a lower firing rate of the presynaptic LGN cells for large stimuli. In summary, recurrent local excitation may explain the increase of receptive field size in upper 4C$\alpha$ but it effectively decreases contrast sensitivity in the same layer. The decrease in contrast sensitivity however can be balanced by an increase in *threshold and gain of the somatic ($\alpha$-6) inhibitor* (Fig. 3b). Higher threshold and gain of the $\alpha$-6 cell is in particular consistent with the observation, that direction selectivity which is GABA mediated and emerging first in upper 4C$\alpha$ requires strong lateral inhibition [9]. Figure 4 shows the *best predictions* of the recurrent network model compared to the experimental data. Note, that both stepping connections and differential inhibition with depth are needed to explain the experimental data. We conclude that within a biological realistic parameter regime the intracortical recurrent model presented above can account for the exponential gradient of the basic response properties in depth of layer 4C.

This "intracortical" study explores an alternative to our recently published "afferent" hypotheses which predicts higher sensitivity and field size for an anatomically dentified special population of LGN-M cells [4, 5] – the so-called M1 cells – which project preferentially to upper layer 4C$\alpha$; we have suggested that the "afferent" hypothesis can be tested by re-examining the physiological properties of LGN-M cells together with intraaxonal filling to examine their axon morphology. The "intracortical" model presented above predicts substantial excitation from the lateral sidestepped projections of spiny stellate cells and considerable differences in the physiological properties or postsynaptic effects of two anatomically-identified basket cell populations. Both hypothesis are consistent with the current data, hence new experiments are needed to distinguish between them. The physiological properties of the clutch cell and the $\alpha$-6 cell might be testable experimentally by observing the IPSPs recorded intracellularly for cells in 4C, but it might be difficult to test the efficacy of the stepped connections. Thus both models make clear physiological predictions for anatomically identified cell classes and the key experiment which ultimately distinguishes between both hypothesis will be the re-examination of the physiological properties of LGN-M cells. Thus experimental tests must show if either of the hypotheses are likely and, if they are both confirmed, modeling may show how they combine to produce the actual biological condition.

**Acknowledgements** Supported by DFG (Ob 102/2-1), MRC G9409137, Wellcome Trust, and by a DFG scholarship (GK 231) to U.B.

# References

[1] J. S. Lund. Local circuit neurons of macaque monkey striate cortex: I. Neurons of laminae 4C and 5A. *J Comp Neurol* 1987; 147:455–496

[2] G. G. Blasdel and D. Fitzpatrick. Physiological organization of layer 4 in macaque striate cortex. *J Neurosci* 1984; 4:880–895

[3] M. J. Hawken and A. J. Parker. Contrast sensitivity and orientation selectivity in lamina IV of the striate cortex of Old World monkeys. *Exp Brain Res* 1984; 54:367–372

[4] U. Bauer, P. Adorján, M. Scholz, J. B. Levitt, J. Lund, and K. Obermayer. On the anatomical basis of receptive field size, contrast sensitivity, and orientation selectivity in macaque striate cortex: A model study. In W. Gerstner, A. Germond, M. Hasler, and J.-D. Nicoud, editors, *Artifical Neural Networks – ICANN'97*. Springer-Verlag, Heidelberg, 1997, pp. 213–218

[5] M. Scholz, U. Bauer, J. B. Levitt, K. Obermayer, and J. S. Lund. Merging of P and M properties: A model for geniculocortical information processing in macaque striate cortex. *Soc Neurosci Abstr* 1997; 23:2059

[6] T. Yoshioka, J. B. Levitt, and J. S. Lund. Independence and merger of thalamocortical channels within macaque monkey primary visual cortex: Anatomy of interlaminar projections. *Vis Neurosci* 1994; 11:467–489, 1994

[7] Z. F. Kisvarday, A. Cowey, and P. Somogyi. Synaptic relationships of a type of gaba-immunoreactive neuron (clutch cell), spiny stellate cells and lateral geniculate nucleus afferents in layer IVc of the monkey striate cortex. *Neuroscience* 1986; 19:741–61

[8] J.S. Lund, Q. Wu, P. T. Hadingham, and J. B. Levitt. Cells and circuits contributing to functional properties in area V1 of macaque monkey cerebral cortex: Bases for neuroanatomically realistic models. *J Anat* 1995; 187:563–581

[9] H. Sato, N. Katsyama, H. Tamura, and Y. Hata, and T. Tsumoto. Mechanisms underlying direction selectivity of neurons in the primary visual cortex of macaque. *J Neurophysiol* 1995; 74:1382–1394

# Self-Organization of Shift-Invariant Receptive Fields through Pre- and Post-Synaptic Competition

Kunihiko Fukushima,    Kazuya Yoshimoto

Graduate School of Engineering Science, Osaka University

Toyonaka, Osaka 560-8531, Japan

fukushima@bpe.es.osaka-u.ac.jp

## Abstract

This paper proposes a new learning rule by which cells with shift-invariant receptive fields are self-organized. During the learning, straight lines of various orientations sweep across the input layer. With this learning rule, cells similar to simple and complex cells in the primary visual cortex are generated in a network.

To demonstrate the new learning rule, we simulate a three-layered network, that consists of input layer (or the retina), a layer of S-cells (or simple cells) and a layer of C-cells (or complex cells). S-cells are created through competition among postsynaptic S-cells depending on their instantaneous outputs. For the self-organization of C-cells, however, not only postsynaptic C-cells but also presynaptic S-cells compete. Although presynaptic cells compete depending on their instantaneous outputs, postsynaptic C-cells compete depending, not on their instantaneous outputs, but on the traces (or sustained aftereffect) of their outputs.

## 1 Introduction

This paper proposes a new learning rule by which cells with shift-invariant receptive fields are self-organized. Namely, cells similar to simple and complex cells in the primary visual cortex are generated in a network trained by the new learning rule.

The author proposed previously a neural network model of the visual system, called a *neocognitron* [1][2]. The neocognitron has a hierarchical multi-layered architecture similar to the classical hypothesis by Hubel and Wielsel [3], and acquire an ability to robustly recognize visual patterns through unsupervised learning. It consists of layers of S-cells, that resemble simple cells in the visual cortex, and layers of C-cells, that resemble complex cells. S-cells are feature-extracting cells. C-cells, similarly to complex cells, exhibit approximate invariance to position of the stimuli presented within their receptive fields. Although input connections to S-cells are variable and are modified through unsupervised learning, input connections to C-cells are fixed and unmodifiable in the conventional neocognitron.

This paper proposes a new learning rule by which input connections to C-cells, as well as S-cells, can be created through learning. During the learning, straight lines of various orientations sweep across the input layer. With this learning rule, shift-invariant receptive fields can be automatically generated in a network.

The conventional neocognitron has architecture called cell-planes from the beginning before the learning starts. The cells in each cell-plane share the same set of input connections. In contrast to the conventional learning algorithm used in the neocognitron, the new learning algorithm proposed in this paper does neither premise the architecture of cell-planes nor shared connections to realize shift-invariant receptive fields of C-cells.

## 2  Network Architecture

We simulate a three-layered network, that has an input layer ($U_0$) consisting of photoreceptor cells, a layer of S-cells ($U_S$) and a layer of C-cells ($U_C$).

Each S-cell is accompanied by an inhibitory V-cell as in the case of the neocognitron [1][2][4]. The S-cell receive variable excitatory connections from photoreceptor cells of the input layer (or the retina), and a variable inhibitory connection from the V-cell. The location of the *connectable area* of each S-cell is determined so that the retinotopy can be maintained. (A cell receives variable connections only from the cells in the *connectable area*.) The accompanying V-cell has fixed excitatory connections from the same set of photoreceptor cells as the S-cell, and always responds with the average (root mean square) intensity of the outputs of the photoreceptor cells. Besides these connections, a mechanism of recurrent lateral inhibition is built in among S-cells.

Each C-cell receives variable excitatory connections from S-cells of the preceding layer. The connectable area of each C-cell, like that of an S-cell, is also predetermined so that the retinotopy can be maintained. The output of a C-cell is proportional to a weighted sum of the outputs of the presynaptic S-cells, but is depressed if the sum of excitatory input connections is larger than a certain value. In other words, C-cells with a larger number of excitatory input connections generally yields a smaller output for the same input signals. This depression is useful in competitive learning for hindering C-cells that happened to have undesired input connections from becoming winners. Layer of C-cells also has recurrent lateral inhibition.

The built-in architecture of cell-planes, which the conventional neocognitron has, does not exist in the network. The self-organization can start from a homogeneous network. Shift-invariant receptive fields are automatically generated without using an architecture like cell-planes.

Initial values of excitatory input connections to each S- or C-cell have a weak and almost flat spatial distribution in its connectable area.

## 3  Learning Rule

Training patterns that are presented during the learning phase are moving lines that sweep across the input layer. The lines are long enough to cover the input layer, and have various orientations.

S-cells are created through competition among postsynaptic S-cells depending on their instantaneous outputs. For the self-organization of C-cells, however, not only postsynaptic C-cells but also presynaptic S-cells compete. Although presynaptic cells compete depending on their instantaneous outputs, postsynaptic C-cells compete depending, not on their instantaneous outputs, but on the traces (or sustained aftereffect) of their outputs. These processes will be discussed below in more detail.

## 3.1 Learning Rule for S-cells

The learning rule of the S-cells is the same as that for the conventional neocognitron [4]. Each cell competes with other cells in its vicinity (called the *competition area* of the cell), and the competition depends on instantaneous activity of the post-synaptic cells (S-cells). (We will use the term *activity* almost in the same meaning as *output of a cell* in this paper.) Winners of the competition have their input connections increased. The increment of each connection is proportional to the presynaptic activity. Therefore, the input connections of a winner cell will come to form a template that exactly matches the stimulus presented to its receptive field. Since the variable inhibitory connection from the accompanying V-cell is also increased, the winner S-cell acquire the selective responsiveness to this line stimulus.

If a straight line is presented during the learning phase, winners in the post-synaptic competition are generally distributed along the line with approximately equal intervals, because of the limited size of the competition areas, and also because of the recurrent lateral inhibition among the S-cells. As the result of modification of the input connections, the winners come to have receptive fields selectively responsive to this line [4]. When the line is moved in parallel and comes to another location, another set of winners is chosen, and the new winners come to have receptive fields of the same preferred orientation but at different locations. Thus, after finishing a sweep of the line across the input layer, S-cells whose receptive fields have the same preferred orientation as the line are generated and become distributed over the layer $U_S$. S-cells of other preferred orientations are generated by sweeps of lines of other orientations.

## 3.2 Learning Rule for C-cells

For the self-organization of C-cells, not only postsynaptic C-cells but also presynaptic S-cells compete.

Postsynaptic C-cells compete with each other based, not on their instantaneous activities, but on the *traces* (or sustained aftereffects) of their activities. A trace is a kind of temporal average (or moving average). Incidentally, use of traces in self-organization was suggested previously by Földiák [5]. He proposed a modified Hebbian rule, in which the amount of modification is proportional to the presynaptic activity and the trace of the postsynaptic activity.

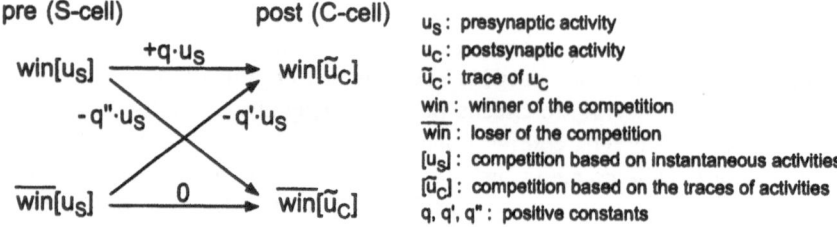

Figure 1: Learning rule for C-cells: the amount of increment or decrement of a synapse from S-cell to C-cell.

The competition among presynaptic S-cells, however, is based on their instantaneous activities.

Figure 1 illustrates the learning rule, and roughly shows how a synaptic connection from presynaptic S-cell to postsynaptic C-cell increases or decreases. The connections from presynaptic winners to postsynaptic winners increase, but the connections from presynaptic losers to postsynaptic winners decrease. The connections from presynaptic winners to postsynaptic losers decrease, but the connections from presynaptic losers to postsynaptic losers are unchanged. The increment or decrement of each connection is proportional to the instantaneous presynaptic activity, but the connections have a characteristic of saturation and do not exceed a certain boundary value.

We will now discuss how the self-organization of the C-cell layer progresses under this learning rule. To simplify the explanation, we assume here that the self-organization of the preceding S-cell layer has already been finished.

If a line stimulus sweeps across the input layer, S-cells whose preferred orientation matches the orientation of the line become active. The timings of becoming active, however, are different among S-cells. For the creation of shift-invariant receptive fields of C-cells, it is desired that a single C-cell comes to receive excitatory connections from all of these S-cells in its connectable area.

To prevent an occurrence of a situation in which many C-cells come to receive connections from the same set of S-cells, competition among postsynaptic C-cells, which have receptive fields nearly at the same location, is required. In this competition, the same C-cell has to continue to be a winner throughout the period when the line is sweeping across its receptive field. This condition can generally be satisfied by our learning rule, by which winners are determined based on traces of outputs of C-cells, and not on instantaneous outputs. If a C-cell once becomes a winner, the C-cell will be apt to keep winning for a while after that, because the trace of an output lasts for some time after the extinction of the output. Thus, the C-cell will finally come to receive connections from all relevant S-cells.

Now let us consider how the presynaptic competition is useful. A desired state after the learning is that a C-cell comes to receive excitatory connections from all S-cells of a particular preferred orientation, but not from S-cells of any other preferred orientations.

If a line sweeps across the input layer, S-cells whose preferred orientation matches the orientation of the line become active. At the same time, S-cells of slightly different preferred orientations will also become active, but weakly. Among these activated S-cells, only the ones whose preferred orientations exactly match the orientation of the line will become winners, and S-cells of other preferred orientations will become losers. Hence, by the proposed learning rule, postsynaptic winner C-cells increase connections only from the S-cells whose preferred orientations exactly match the orientation of the line, and decrease connections from S-cells of slightly different preferred orientations. A C-cell thus comes to receive excitatory connections from S-cells of a particular preferred orientation, but not from S-cells of any other preferred orientations. If presynaptic competition is not used and all active S-cells are allowed to make connections to a C-cell, however, the C-cell might come to have connections from S-cells of every orientation, when, for example, training lines of slightly different orientations are presented sequentially.

According to the proposed learning rule, postsynaptic loser C-cells decrease input connections from presynaptic winner S-cells, if the loser C-cells are not silent. This rule is useful for eliminating undesired connections that happen to have been generated during the learning. Suppose a C-cell happens to have had strong connections from S-cells of two different preferred orientations. Let these orientations be $\alpha$ and $\beta$. Also suppose there is another C-cell that has connections only from S-cells of preferred orientation $\alpha$. When a line of orientation $\alpha$ is presented to the input layer, the latter C-cell will usually yield a larger output than the former C-cell. This is because C-cells with a larger number of excitatory input connections generally yields a smaller output for the same input signals, as was discussed in Section 2. As a result, the former C-cell will become a loser, and reduce the connections from S-cells of preferred orientation $\alpha$, which are presynaptic winners. The former C-cell will thus lose selectivity to orientation $\alpha$ and come to have only one preferred orientation, $\beta$.

# 4 Computer Simulation

A three-layered network is simulated on a computer. It has input layer $U_0$, S-cell layer $U_S$, and C-cell layer $U_C$, connected in a cascade. Layer $U_0$ has $27 \times 27$ photoreceptors, layer $U_S$ has $51 \times 51$ S-cells, and layer $U_C$ has the same number of C-cells. The density of the cells is different in the three layers: it is highest in $U_C$ and lowest in $U_0$.

Training patterns are moving lines of various orientations. The width of the lines used for the simulation is 1. Each photoreceptor has a square face of $1 \times 1$ in size, and its output is proportional to the degree of overlap between the stimulus line and the photoreceptor face. Lines of eight different orientations chosen with an interval of 22.5° are used. The moving speed of a line is 0.3. In other words, the location of a line is shifted by 0.3 at each discrete time unit.

Figure 2 shows responses of the network that has finished self-organization. The responses of photoreceptors of $U_0$, S-cells of $U_S$ and C-cells of $U_C$ are represented by the size of the cells.

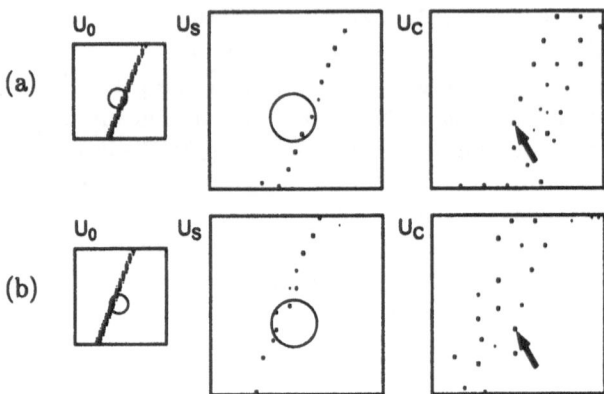

Figure 2: Responses of the network that has finished self-organization.

If a line is presented to a location on the input layer as shown in Fig. 2(a), several numbers of S-cells respond to the line, and consequently C-cells that receive excitatory connections from these S-cells become active. Let us watch, for example, an active C-cell marked with an arrow in layer $U_C$. The circles in $U_S$ and $U_0$ show the connectable area and the effective receptive field of the C-cell, respectively. (When comparing the sizes of these areas, you should note that layers $U_S$ and $U_C$, that have a larger number of cells, are displayed in this figure with a larger scale than layer $U_0$.)

If a line is shifted to a new location as shown in Fig. 2(b), other S-cells become active because S-cells are sensitive to the location of the stimulus. In layer $U_C$, however, the same C-cell, which is marked with an arrow again, is active. This shows that the C-cell exhibits a shift-invariance within the receptive field. When the line is rotated to another orientation, this C-cell, of course, is silent, even if the line is presented in its receptive field. We can thus observe that other C-cells also have shift-invariant, and orientation-sensitive receptive fields.

# References

[1] Fukushima K. Neocognitron: a hierarchical neural network capable of visual pattern recognition. *Neural Networks* 1988; 1:119-130

[2] Fukushima K, Miyake S. Neocognitron: a new algorithm for pattern recognition tolerant of deformations and shifts in position. *Pattern Recognition* 1982; 15:455-469

[3] Hubel DH, Wiesel TN. Receptive fields and functional architecture in non-striate areas (18 and 19) of the cat. *J. Neurophysiology* 1965; 28:229-289

[4] Fukushima K. Analysis of the process of visual pattern recognition by the neocognitron. *Neural Networks* 1989; 2:413-420

[5] Földiák P. Learning invariance from transformation sequences. *Neural Computation* 1991; 3:194-200

# A Cortical Interpretation of ASSOMs

N. Mayer, M. Herrmann, H.-U. Bauer, T. Geisel

Max-Planck-Institut für Strömungsforschung
Bunsenstraße 10, 37073 Göttingen, Germany

## Abstract

Self-organizing maps have been successfully used to model map formation in the visual cortex of mammals. When applying natural images as stimuli, properties of the maps obtained for low-dimensional input manifolds, such as retinotopy, are not equally well reproducible. The present study points to the virtues of the adaptive subspace self-organizing map (AS-SOM) in modeling neural maps. Since the representation of position and orientation and that of stimulus phase are automatically mapped to different hierarchical levels of the ASSOM, topography is established for orientation and position, but not for phases. This agrees to evidence for the absence of smooth phase maps. Further, we show that some biologically implausible conditions of the ASSOM rule can be relaxed.

## 1   Introduction

Feature maps in visual cortex are often considered to result from a self-organizing process, which has been modeled in various ways such as self-organizing maps (SOMs) [1], cf. e.g. [2, 3, 4]. In high-dimensional SOMs stimuli transferred from the retina via LGN are represented by vectors containing activities of individual neurons or small groups of LGN neurons as entries. If the stimulus set consists of elongated blobs with varying retinal position, orientation maps can be produced that in many aspects coincide with those obtained experimentally. Taking into account that cortical maps show some degree of adaptiveness also postnatally, the question arises how the statistical features of visual stimuli received in a natural environment affect the map layout.

Studies based on independent component analysis [5] emphasize the role of edges in natural images. Band-pass filtering of the images (describing the effect of the lowest levels of the visual system) transforms edge-like features into Gabor patches. On the other hand, Gabor-shaped receptive fields (RFs) in cells of the primary visual cortex of primates and cats have been revealed using reverse correlation techniques [6, 7]. Finally, in [8] the correspondence between stimuli taken from filtered natural images and the development of Gabor-like RFs has been investigated using a sparse-coding scheme.

In the SOM framework, where afferent synaptic weights from LGN ON-center cells are described by positive, those from OFF-center cells by negative numbers, the Gabor-type stimuli cause a formal problem. Namely, considering

Euclidean distances between RF vectors suggests that RFs with similar position and orientation but opposite phases are more different than those referring to distant retinal positions. This is because inhibitory and excitatory subfields are exchanged in the first case, leading to twice the numerical distance value compared to the case of nonoverlapping RFs, cf. Fig. 1. Since SOMs are topographical, i.e. tend to map nearby positions to close cortical locations, and are based on distances (or scalar products) of the RF vectors, the retinotopic structure may deteriorate when using complex stimuli.

The problem may be solved by using different algorithms, such as the approach in [9], where a correlation-based model relying essentially on binocularity was applied. We propose here to retain the conceptual simplicity of the SOM approach, but to use a hierarchical version similar to the adaptive subspace SOM (ASSOM) [1]. The algorithm is reviewed in the following section, where also its formal conditions are relaxed for achieving a more plausible neural implementation. The third section presents results from simulations of both ASSOM and the modified version of this algorithm. The final section discusses the implications of the ASSOM-like model for the representation of phases in the visual cortex.

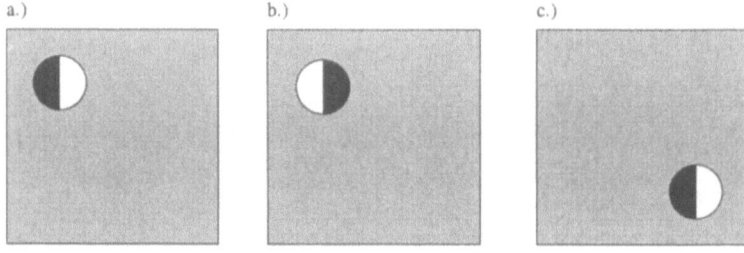

Figure 1: *Schematic receptive fields represented in retinal coordinates. Black and white regions correspond to inhibitory or excitatory subfields. RFs a) and c) have different positions, but equal phase. For a) and b) positions coincide, but the phase is shifted. The pixelwise distance of representative input vectors a) and b) is twice as large as that for the distant RFs a) and c).*

## 2 The ASSOM

The adaptive subspace SOM [1] is a variant of SOM that matches projections of data vectors instead of the data themselves. It has thus capabilities of retrieving invariant data features. It can be considered as a two-level hierarchy, where the first level resembles the original SOM and a second level extracts local principal components within regions defined by the first stage. The SOM level contains an $M$-dimensional array of units i, each of which contains a $K$-dimensional local subspace, i.e. a set of basis vectors $\mathbf{w}_{ik}$, $k = 1, \ldots, K$. These vectors are adapted in order to locally match the data manifold $\mathcal{V}$.

The response of a unit when presenting a data vector $\mathbf{v} \in \mathcal{V} \subseteq \mathcal{R}^d$ is defined as a $K$-dimensional vector containing the scalar products

$$r_{ik}(\mathbf{v}) = \langle \mathbf{v}, \mathbf{w}_{ik} \rangle \tag{1}$$

where both $\mathbf{v}$ and $\mathbf{w}_{ik}$ are Euclidean normalized vectors. The best-matching unit is determined by the squared length of the projection of $\mathbf{v}$ onto the subspace spanned by the $\mathbf{w}_{ik}$.

$$\mathbf{i}^* = \arg\max_{\mathbf{i}} \sum_k r_{ik}^2 \tag{2}$$

The vectors $\mathbf{w}_{ik}$ are updated analogous to the SOM rule.

$$\Delta\mathbf{w}_{ik} = \epsilon h_{\mathbf{i}\mathbf{i}^*} (r_{ik}(\mathbf{v})\, \mathbf{v} - \mathbf{w}_{ik}), \tag{3}$$

where $\epsilon$ is the learning rate and $h_{\mathbf{i}\mathbf{i}^*} = \exp(-\|\mathbf{i} - \mathbf{i}^*\|^2 / 2\sigma_k^2)$. After the presentation of a stimulus (or a stimulus epoch consisting of a sequence of inputs) the $\mathbf{w}_{ik}$ are orthogonalized separately for each unit $\mathbf{i}$.

SOMs have proven to serve as relevant, though abstract models of cortical map formation. Determination of the winning unit and interaction of units in the learning dynamics can be understood as being produced by collective dynamics of groups of neurons. Similar arguments may be applicable, but shall not be of interest in the present paper, for the neural implementation of Eq. (2). The orthonormalization step of the ASSOM is, however, more suspicious from a neurobiological perspective. In order to achieve local low-dimensional representations also mechanisms different from the theoretically more stringent orthogonalization are possible. For this purpose we considered the vectors in a unit as individual neurons, whose RFs are now supposed to fill the local subspace rather than to parametrize it. For the two-dimensional subspace in the simulations below, instead of two basis vectors, $K \gg 2$ neurons per unit are used, cf. Fig. 2. All neurons of a unit contribute to the determination of the winning unit according to Eq. (2), which now does not refer to a projection of the stimulus onto the local subspace, but to a combined effect of the stimulus on the unit. An inhibitory interaction among the neurons of a unit serves to distribute their RFs across the local region, which not necessarily will form a subspace. In this way, the RF vectors evolve according to

$$\Delta\mathbf{w}_{ik} = \epsilon h_{\mathbf{i}\mathbf{i}^*} (\eta_{kk^*} \mathbf{v} - \mathbf{w}_{ik}), \tag{4}$$

where $k^* = \arg\max_k \langle \mathbf{v}, w_{ik} \rangle$ is the best matching neuron of a unit and $\eta_{kk^*}$ governs the intra-unit adaptation. We tested two choices for $\eta_{kk^*}$ and $K = 5$. In order to achieve results in tractable times maps of $32 \times 32$ units were used with a set of parameters that matched the size of the map. For an essentially inhibitory interaction we set $\eta_{kk^*} = 1$ for $k = k^*$ and $\eta_{kk^*} = -1/K$ otherwise. The results given below are, however, based on $\eta_{kk^*} = \cos(2\pi(k - k^*)/K)$.

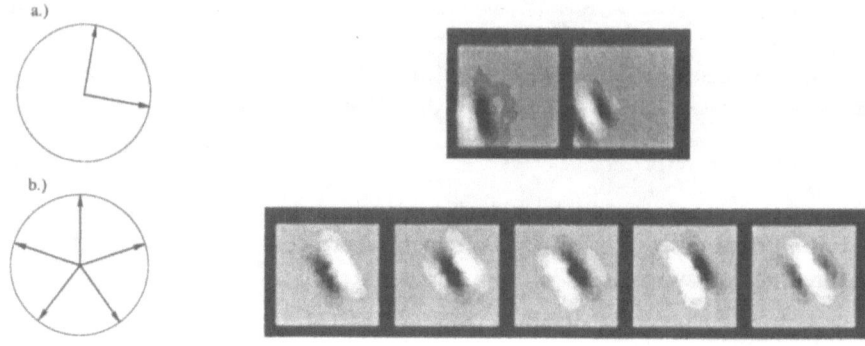

Figure 2: **a**) *Left: Orthogonal neural pointers as obtained by the ASSOM are displayed schematically in the local subspace. Right: example of corresponding receptive fields exhibiting a phase shift of 90° due to orthonormalization.* **b**) *Modified ASSOM with $K = 5$ neural pointers per unit (left). While position and orientation are similar, spatial phase differences of multiples of $360°/K$ are observed among the neurons within this unit.*

## 3    Results

We have performed simulations of the original ASSOM consisting of $128 \times 128$ units each consisting of $K = 2$. Inputs are patches of natural images taken from forests, lake sceneries etc. [8]. Ten images of $512 \times 512$ pixels are used. The images are low-pass filtered by the filter function $f(k) = k \exp(-(k/k_0)^4)$ with $k_0 = 200$ cycles (cf. [8]). Input vectors are image patches of $d = 16 \times 16$ pixels , which are sampled at random positions and are randomly rotated. The patches are multiplied with a Gaussian mask of size of $\sigma_g = 4.0$ centered at a random position inside the margins of the patch of four pixels width. During an initial period of $5 \times 10^4$ of a total of $6 \times 10^5$ steps $\sigma_k$ is decreased linearly from 6.0 to 2.5 and kept constant thereafter. The learning rate $\epsilon$ decayed exponentially from 0.2 to 0.01 throughout the learning process. The maps are initialized by retinotopically ordered Gaussian distributions, multiplied with white noise. At the end of each iteration the $\mathbf{w}_{ik}$ are kept multiplicatively normalized to $\langle \mathbf{w}_{ik}, \mathbf{w}_{ik} \rangle = 1$ .

The RFs in the developed maps are shaped like Garbor filters. The AS-SOM produces two orthogonal components at each unit that correspond to RFs with virtually identical positions and orientation, but with a 90° phase shift, cf. Fig. 2 a). Stimuli with similar position and orientation and an arbitrary phase are matched by linear superposition of the basis vectors. The modified algorithm leads to units whose neurons have RFs of similar retinal position and close orientation preferences. A unit generally contains different spatial phases distributed across the full 360°-range of phases, cf. Fig. 2 b).

Fig. 3 displays a general view of the representation of the positions of the

Figure 3: *Receptive fields obtained by the ASSOM algorithm. Displayed is one of the RFs of every eighth unit from a cortical patch of* $128 \times 128$ *units.*

RFs in the input space obtained by the ASSOM. Each unit of the ASSOM develops a position- and orientation-sensitive, but phase-independent representation of a patch of the input space. Further, like high-dimensional SOMs other features as position and orientation are arranged in a topology preserving manner, i.e. the map is retinotopic (RFs centered bottom left are displayed in the bottom left box of the map, etc.), smooth orientation variation across the map except at a few pinwheels.

# 4   Conclusion

The results reported here demonstrate that adaptive subspace SOMs retain similar modeling properties as high-dimensional SOMs even in the case of realisitc stimuli. The fact that the algorithm encodes orientation and position by unit location and represents the phase within units is due to the linearity of phases, which is lacking for orientation and position for the high-dimensional stimuli. This agrees with the suggestion in [10] that in contrast to orientation, ocular dominance or position no smooth phase map exists. High-dimensional SOMs, however, tend to produce smooth phase maps since similarities of the individual stimuli are to be represented topographically rather than similarities of stimuli classes as in the ASSOM. Hence we are suggesting that ASSOMs or more generally hierarchical SOMs could provide a framework for modeling neural maps in the cortex. These can be extended to include ocular dominance maps, direction maps, and spatial frequency maps.

# References

[1] T. Kohonen, *Self-organizing maps*. Springer, Berlin (1995). (For ASSOM cf. pp. 161-173).

[2] K. Obermayer, G. G. Blasdel, K. Schulten, A principle for the formation and for the spatial structure of cortical feature maps. *Proc. Natl. Acad. Sci. USA* **87** (1990) 8345-8349.

[3] M. Riesenhuber, H.-U. Bauer, and T. Geisel, Analyzing phase transitions in high-dimensional self-organizing maps. *Biol. Cyb.* **75** (1996) 397-407.

[4] M. Riesenhuber, H.-U. Bauer, D. Brockmann, T. Geisel, Breaking rotational symmetry in a self-organizing map model for orientation map development. *Neur. Comp.* **10**:3 (1998) 717-730.

[5] A. J. Bell, T. J. Sejnowsky, The 'Independent components' of natural scenes are edge filters. *Vis. Res.* **37**:23 (1997) 3327-3338.

[6] J. Daugman, Two-dimensional spectral analysis of cortical receptive field profiles. *Vis. res.* **24**:9 (1984) 847-856.

[7] J. Jones, L. Palmer, An evaluation of the two-dimensional Garbor-filter model of simple receptive fields in cat striate cortex. *J. Neurophys.* **58**:6 (1987) 1233-1258.

[8] B. Olshausen, D.J. Field, Emergence of simple-cell receptive field properties by learning a sparse code for natural images. *Nature* **78** (1996) 315-333.

[9] T. Burger, E. Lang, A CBL network model with intracortical plasticity and natural image stimuli. In: W. Gerstner, A. Germond, M. Hasler, J.-D, Nicoud (eds.), *Artificial Neural Networks – ICANN'97*, Springer, Berlin (1997) 225-230.

[10] R. D. Freeman, G. C. Angelis, G.M. Ghose, I. Ohzawa, Clustering or response properties of neurons in the visual cortex. *Soc. Neurosci. Abstr.* **23**:1 (1997) 567.

# Decoding Population Responses in Short Epochs

S. Panzeri[1], A. Treves[2], S. Schultz[1], E.T. Rolls[1]

[1]Department of Experimental Psychology, University of Oxford

Oxford OX1 3UD, UK

Email: stefano.panzeri@psy.ox.ac.uk

[2]SISSA, 34013 Trieste, Italy

### Abstract

We study, in the limit of short time epochs, the effectiveness of stimulus reconstruction procedures in decoding the information contained in the responses of a neuronal population. We show, both analytically and with computer simulations, that, in the limit of short times, if the information extracted in the stimulus reconstruction is calculated by taking into account only the most likely stimulus in each trial, then almost all the information in the neuronal responses can be decoded. If instead the information extracted with decoding is computed using all the probabilities that each of the possible stimuli in the set was the actual one, then in this limit almost all the information in the responses is lost.

## 1   Introduction

Understanding the way in which environmental stimuli are represented in the responses of a population of neurons requires the ability to reconstruct (or decode) the stimuli from the action potentials. Decoding is very useful first in providing insight into how the brain might use the information encoded in the neuronal responses in solving computational problems, and second, in providing a tool to quantify the accuracy with which the stimuli can be estimated from the observation of the activity of groups of neurons. There is substantial evidence that the time scale with which information is transmitted by the neuronal activity could be quite short, suggesting that it may also be decoded in short times [1, 2]. In this paper we study the accuracy of the decoding procedures in time periods shorter than the mean interspike interval. The novelty of our study is that, instead of considering simple quantification of the decoding performance, like the mean squared error [3], we address the decoding accuracy directly in terms of information quantities. We compare, for short time intervals, the information about the stimuli that can be gained by any decoding of the neuronal responses to its upper bound, i.e. the information which is actually transmitted (encoded) by the responses themselves. This is particularly of interest because calculating the information carried by neuronal populations is made difficult by the large amount of data needed to sample adequately the response space of a set of cells. Decoding can compress the dimensionality of the response space efficiently, and lead to precise estimates of the information carried by populations of cells. The information estimation can be done by considering the mutual information between the stimuli and the most likely stimulus in each trial (we call this information $I^{ml}$). A slightly more complex variant often used is a step that extracts from the responses in each trial not only the single most likely stimulus, but all the probabilities that each of the possible stimuli in the set was the actual one. (we call this information $I^p$). Of course the information quantities $I^{ml}$ and $I^p$, calculated using the decoding step, cannot be higher than the information contained in the neuronal responses, because the decoding cannot add new information. But, if we want to use $I^{ml}$ or $I^p$ as a reliable approximation to the full information contained in the neuronal responses, we require the difference between true information in the responses and decoded information to be very small. It also very interesting to investigate how the information loss due to decoding depends on the number of neurons in the ensemble. These points are addressed in the rest of the paper, where we develop an analytical formalism to study in detail the decoding procedure and the related information quantities in short time windows. We use computer simulations to illustrate the results.

## 2 The information carried by population responses in short times

The neuronal responses are quantified by the number of spikes simultaneously emitted by each of $C$ neurons in the time window $[t_0, t_0 + t]$. This gives us a vector $\mathbf{r}$, each component of $\mathbf{r}$ being the number of spikes emitted by the respective neuron. Stimuli are taken from a discrete set $S$ of $S$ elements, each occurring with probability $P(s)$. The probability of events with response $\mathbf{r}$ is denoted as $P(\mathbf{r})$, and the joint probability as $P(s, \mathbf{r})$. The information that the neuronal responses convey about the set of stimuli is a function of response probabilities and of the epoch length $t$:

$$I(t) = \sum_{s \in S} \sum_{\mathbf{r}} P(s, \mathbf{r}) \log_2 \frac{P(s, \mathbf{r})}{P(s)P(\mathbf{r})} \tag{1}$$

Since one has to compute $I(t)$ from (1) using the experimental frequency tables instead of the true probabilities $P(s, \mathbf{r})$, the information can be overestimated due to limited sampling. With the amount of data typically obtained from a cortical recording session the use of finite sampling corrections [4] allows a direct calculation of the information for ensembles comprised of no more than a few cells.

An alternative approach to the study of the instantaneous rate at which information accumulates from time $t_0$ is to consider the time-derivatives of information at $t_0$. $I(t)$ is approximated by its first order Taylor expansion, $I(t) = t\, I_t + O(t^2)$, where $I_t$ is the first time-derivatives of $I(t)$ calculated at $t_0$. Here we study the short epoch expansion for the true information contained in the neuronal responses, and in the next section we apply the same formalism to the study of the informations extracted by decoding procedures, comparing the results. We assume that the firing rate distribution reflects a stationary random process, so that the mean firing rate $\bar{r}_{s;c}$ of cell $c$ to stimulus $s$ is a well defined quantity. We also assume that the probability of observing one spike emitted by a cell $c$ in the time window $[t_0, t_0 + t]$ conditional upon the emission of a different spike by any neuron(s) in the population, when a stimulus $s$ is presented, is proportional to $t$. The $t$ expansion of response probabilities becomes an expansion in the total number of spikes emitted by the population. The only events with non-zero probability are to first order in $t$ those with no more than one spike emitted in total :

$$p(0|s) = 1 - t \sum_{c=1}^{C} \bar{r}_{s;c} + O(t^2); \quad p(\mathbf{e}_c|s) = t\, \bar{r}_{s;c} + O(t^2) \tag{2}$$

where $\mathbf{0}$ is the response vector with zero spikes emitted by each cell; $\mathbf{e}_c$ is the response vector with one spike in the $c$-th cell component and zero in the other ones. Substituting (2) into (1), and defining $\bar{r}_c \equiv \sum_s P(s)\bar{r}_{s;c}$, we obtain:

$$I_t = \sum_{c=1}^{C} \sum_{s \in S} P(s)\bar{r}_{s;c} \log_2 \frac{\bar{r}_{s;c}}{\bar{r}_c} \tag{3}$$

## 3  Decoding Responses from Short Epochs

The optimal way to transform any actual response into a prediction of the stimulus that elicited the response is to make use of Bayes' rule[1]:

$$P(s'|\mathbf{r}) = \frac{P(\mathbf{r}|s')P(s')}{P(\mathbf{r})} . \tag{4}$$

This of course requires knowledge of the response probabilities $P(\mathbf{r}|s)$. In practice, this means fitting $P(\mathbf{r}|s)$ to a model function. However, it will be seen that in the short epoch limit the choice of response probability model is not important, as the response probabilities in this limit depend only upon the mean firing rates. To avoid biasing the estimation of conditional probabilities, the training responses (used in estimating $P(\mathbf{r}|s)$) should not include the particular test trial for which $P(s'|\mathbf{r})$ is going to be derived. Summing over different test trial responses to the same stimulus $s$, we can extract the probability that by presenting stimulus $s$ the neuronal response is interpreted as having been elicited by stimulus $s'$,

$$P(s'|s) = \sum_{\mathbf{r} \in test} P(s'|\mathbf{r})P(\mathbf{r}|s) , \tag{5}$$

Having estimated the relative probabilities that the test trial response have been elicited by any one stimulus, the stimulus $s' = s^p$ for which this likelihood is maximal can be said to be the stimulus *predicted* on the basis of the response. In general $s^p$ will not coincide with the true $s$, and the accuracy in the decoding

---

[1]For a discussion of decoding strategies not based on Bayes rule see e.g. [3]

can be quantified by the fraction of correct decodings, or alternatively by the mutual information extracted from the probability table $Q(s^p|s)$,

$$I^{ml} = \sum_{s,s_p \in \mathcal{S}} P(s)Q(s^p|s)\log_2 \frac{Q(s^p|s)}{Q(s^p)} , \tag{6}$$

where $Q(s^p|s)$ is the fraction of times an actual stimulus $s$ elicited a (test) response that led to a predicted (most likely) stimulus $s^p$. Thus $I^{ml}$ measures the information in the predictions based on *maximum likelihood*, and as such it does not only reflect, the number of times the decoding is exact, but also the distribution of wrong decodings. To first order in $t$, all we need to know for decoding are the conditional probabilities of posited stimuli $P(s|\mathbf{r})$ for the $C+1$ possible first order responses $\mathbf{0}, \mathbf{e}_1, \cdots, \mathbf{e}_C$. $P(s|\mathbf{r})$ depend only on the mean firing rates of the cells in response to the different stimuli, and on the probability of occurrence of the stimuli[2]. The most likely stimulus $s^p$ for response $\mathbf{0}$ is the stimulus which elicits the smallest population response, i.e. the stimulus $s$ that minimizes $\sum_c \bar{r}_{s;c}$. We call this stimulus the "worst stimulus". The most likely stimulus $s^p$ for the response $\mathbf{e}_c$ is instead the stimulus that maximizes the mean response $\bar{r}_{s;c}$ of the cell $c$ that fired. We call this stimulus the "preferred" stimulus for cell $c$. It is important to note that the stimuli decoded by the $C+1$ events $\mathbf{0}, \mathbf{e}_1, \cdots, \mathbf{e}_C$ may not all be different. So the number of the stimuli that have a non-zero probability to be decoded is a number that we call $K+1$, where $K$ is the number of stimuli which are predicted by any of the $\mathbf{e}_c$ responses, and which are distinct from one another and from the worst stimulus. Since the ordering of the stimuli is arbitrary, we assign to the worst stimulus the index $s = 0$. Similarly, we call $s = 1, \cdots, K$ the $K$ distinct stimuli predicted by an $\mathbf{e}_c$ response. The set of cells which have $s = k(k = 0, \cdots, K)$ as a preferred stimulus is denoted $\mathcal{C}(k)$. Expanding the maximum likelihood information as a power series in $t$, $I^{ml} = t\, I_t^{ml} + O(t^2)$, we can compare the information rate $I_t^{ml}$ estimated through maximum likelihood information with the full information rate $I_t$ contained in the neuronal responses (3). $I_t^{ml}$ can be computed by first calculating, by means of (2) and (4), the fractions of decodings $Q(s^p|s)$, and then expanding (6) in powers of $t$:

$$I_t^{ml} = \sum_s P(s) \sum_{k=1}^D \left( \sum_{c \in \mathcal{C}(k)} \bar{r}_{s;c} \right) \log_2 \left[ \frac{\left(\sum_{c \in \mathcal{C}(k)} \bar{r}_{s;c}\right)}{\left(\sum_{c \in \mathcal{C}(k)} \bar{r}_c\right)} \right] \tag{7}$$

One can show that, due to the log-sum inequality, $I_t^{ml} \leq I_t$. The difference between $t\, I_t$ and $t\, I_t^{ml}$ quantifies the information loss due to the decoding procedure to first order in $t$. One can show from (7) that all the information in the neuronal responses is preserved, to first order in $t$, by the decoding procedure when each of the events $\mathbf{0}, \mathbf{e}_1, \cdots, \mathbf{e}_C$ predicts a different stimulus. When there is overlap between stimuli predicted by the events $\mathbf{0}, \mathbf{e}_1, \cdots, \mathbf{e}_C$, then all the information is fully decodable *if and only if* there is no overlap

---

[2]We suppose in the following that the stimuli are equiprobable, and that the predicted stimuli are non-degenerate (the general case is fully considered in a work in preparation).

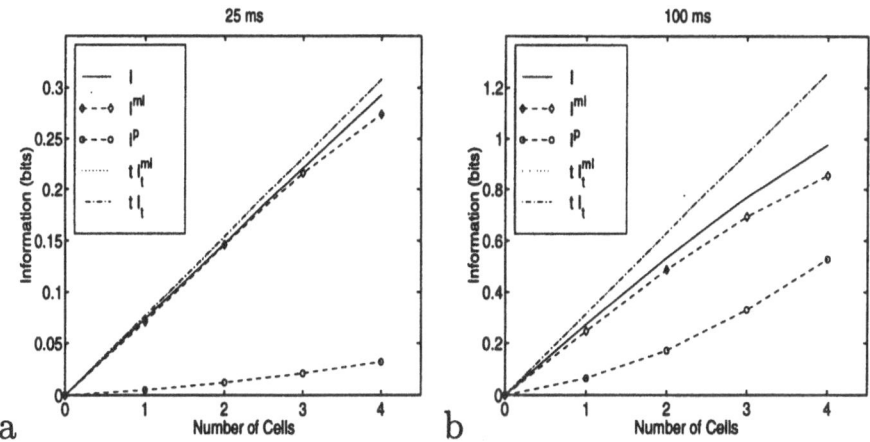

Figure 1: Comparison of estimates of information for Poisson data from four cells. The epoch length is 25 ms (a) and 100 ms (b).

between the preferred stimuli of some of the cells and the worst population responses *and*, if two or more cells share the same preferred stimulus, they have the same response profile (up to a proportionality constant) to all the different stimuli in the set. Therefore the difference between the true and the maximum likelihood information $I^{ml}$ is in general expected to be very small for fewer cells, and to progressively increase as the number of cells $C$ increases: with many cells, overlapping between predicted stimuli by different cells becomes more likely.

A further quantity is the mutual information

$$I^p = \sum_{s,s' \in \mathcal{S}} P(s, s') \log_2 \frac{P(s, s')}{P(s)P(s')} \tag{8}$$

obtained from the *probability* $P(s'|s)$ of confusing $s$ with $s'$, which is given by averaging $P(s'|\mathbf{r})$ over the responses to $s$, eq (5). This second information measure reflects, unlike the first, also the degree of certainty with which each single trial has been decoded. Both information measures have been used in the literature [5, 6, 7]. It can be shown, by using eqs. (5) and (2), that the first derivative of $I^p$ is *always* zero: $I^p(t) \approx O(t^2)$. This means that $I^p$ cannot estimate information transmission rates, and it gives poor estimates of information for relatively short epochs. We note that, by expanding $I^p$ in powers of the number of cells $C$ instead in powers of $t$ one would obtain $I^p \propto C^2$. Therefore a use of $I^p$ for short times may lead to a substantial imprecision also in measures of the redundancy of the messages carried by different cells.

To confirm these results, numerical simulations were performed by generating Poisson responses for each of four cells independently with a stimulus-dependent mean firing rate. Firing rates in response to stimuli ranged in the same regime as real hippocampal spatial view cells [7]. 100 trials were generated for each of the five stimuli in the set. The stimuli were then decoded from the responses with a Bayesian decoding based on a Poisson model of responses. Then $I^{ml}$ and $I^p$ were calculated, as were their first order approximations $I_t^{ml}$ and $I_t$. The true information $I$, Eq. (1), was also computed from the true probabilities. Finite sampling corrections were applied. Each cell had a different preferred stimulus, and in addition there was another stimulus which elicited the worst population response. In this case, $I_t$ and $I_t^{ml}$ coincide, and to first order in $t$ no information should be lost due to decoding. It can be seen from Fig. 1 that $I^{ml}$ is an almost perfect reconstruction of the true information up to the point where the first order approximation fails; small losses in $I^{ml}$ are due to second order effects. In contrast, $I^p$ is a poorer estimator of the true information, and grows super-linearly with the number of cells, as expected by the analysis.

In conclusion our study clearly shows that, for relatively short time windows, $I^{ml}$ is a very good estimate of the true information; $I^p$, however, cannot be used to estimate information transmission rates, and is thus an extremely poor estimate of the true information.

S.P. is supported by an EC Research Training Grant ERBFMBICT972749.

# References

[1] Tovée M.J., Rolls E.T., Treves A., Bellis R.J. Information encoding and the responses of single neurons in the primate temporal visual cortex. J. Neurophysiol. 1993; 70: 640-654

[2] Rieke F., Warland D., de Ruyter van Steveninck R.R., Bialek W. Spikes: exploring the neural code. MIT Press, Cambridge MA,1996

[3] Salinas E., Abbott L. Vector reconstruction from firing rates. J. Comput. Neurosci. 1994; 1: 89-107

[4] Panzeri S., Treves A. Analytical estimates of limited sampling biases in different information measures. Network 1996; 7: 87-107

[5] Gochin P.M., Colombo M.,Dorfman G.A., Gerstein G.L., Gross C. G. Neural ensemble encoding in inferior temporal cortex. J. Neurophysiol. 1994; 71: 2325-2337

[6] Gawne T.J., Kjaer T.W., Hertz J.A., Richmond B.J. Adjacent visual cortical complex cells share about 20% of their stimulus-related information. Cerebral Cortex 1996; 6: 482–489

[7] Rolls E.T., Treves A., Robertson R.G., Georges-Francois P., Panzeri S. Information about spatial views in an ensemble of primate hippocampal cells. J. Neurophysiol. 1998; 79: 1797-1813

# Modeling Reward Dependent Activity Pattern of Caudate Neurons

Thomas Trappenberg and Hiroyuki Nakahara

RIKEN Brain Science Institute, Lab. for Information Synthesis

2-1 Hirosawa, Wako-shi, Saitama 351-0198, Japan

{thomas,hiro}@brain.riken.go.jp

Okihide Hikosaka

Department of Physiology, School of Medicine, Juntendo University

2-1-1 Hongo, Bunkyo, Tokyo 113-0033, Japan

hikosaka@med.juntendo.ac.jp

### Abstract

A hypothesis on the function of the basal ganglia was recently proposed based on reinforcement learning, however, only at conceptual level [1]. Our ongoing project is to quantify this hypothesis in cooperation with the experiment of reward-modulated activities of caudate neurons by Kawagoe et al. [2]. This paper, as our initial effort, aims to summarize the followings: (1) predictions of experimental results drawn from a minimal model, (2) comparison between these predictions and currently-obtained experimental results, (3) some extensions of a minimal model, (4) requirement for further experimental and computational studies.

## 1  Introduction

Inspired by the experiment on dopaminergic (DA) neurons in the substantia nigra pars compacta (SNc) [3], a recently-proposed hypothesis on the basal ganglia[1, 4, 5] is that the spiny neurons (Sps) in the striatum perform reinforcement learning based on the actor-critic scheme (see [5]) by use of a prediction error carried by DA neurons in the SNc. This hypothesis, however, is not yet quantitatively examined, particularly with respect to neural activities in the striatum. Our ongoing project is, based on this hypothesis, to have a quantitative analysis of neural activities in the striatum in cooperation with the ongoing experiment of reward-modulated neural activities in caudate (CD) nuclei (a part of the striatum) of monkeys by Kawagoe et al. [2]. In the following we first briefly introduce the experiment in [2] and outline some of their currently obtained results. Second, to start a quantitative analysis of those we introduce a minimal model based on the above hypothesis and state predictions given by this model. Third, we show that several modifications of the minimal model can explain some additional details of the experimental results. Finally, we discuss requirements for experimental and computational studies for further quantitative investigations.

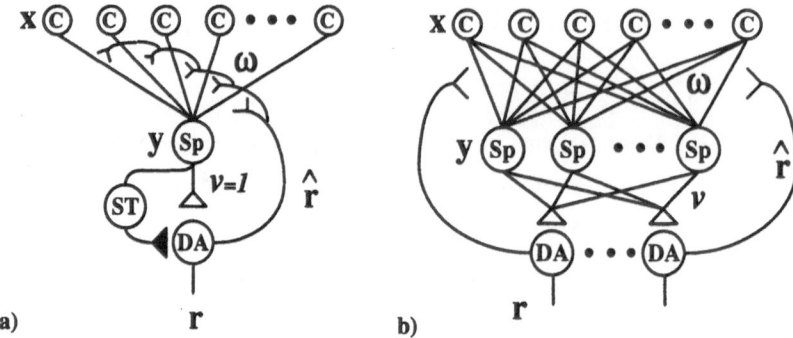

Figure 1: Model architecture of the caudate-SNc loop. Cortical neurons are denoted by c, spiny neurons by Sp, dopaminergic neurons by DA, and the subthalamus by ST.

## 2  Reward dependent caudate activities

Kawagoe et al. [2] have studied the activity pattern of Sps in CD in a memory-guided saccade task with several reward conditions. In each trial, one of four possible target locations was randomly illuminated for a short duration, and a monkey was required to saccade to the memorized target when the fixation point was removed. In each block of trials, a successful saccade to all of four possible targets was rewarded with liquid in ADR condition. In exclusive-1DR condition, only one of four targets comes with a reward, and in relative-1DR condition, one of four comes with a bigger reward than the other three targets. Kawagoe et al. [2] found a variety of neural activity patterns in these different reward conditions. Most salient results are (1) that many Sps in CD showed the biggest response for rewarded directions in each block of all four possible targets in exclusive-1DR and (2) that the reversed behavior of (1) was also observed, that is, some Sps responded increasingly only for the non-rewarded direction. Let us call such Sps of (1) and (2) reward-dependent (R-dependent) Sps and, particularly, such Sps of (2) reversed Sps for convenience. The activities of R-dependent Sps were also consistent in ADR in that the enhancement or suppression of responses occurred with all the rewarded directions. We will mention further details of results in [2] during the course of this paper.

## 3  Towards a quantitative description

In the actor-critic scheme (see [5]) the 'critic' works to evaluate a current state while the 'actor' works to choose an optimal action given the state. Figure 1a shows a possible neural implementation of one critic module based on some known anatomy of the basal ganglia [1]. A simplification we made in our minimal model is to neglect the subthalamic (ST) loop, which is suggested to convey information of the predicted value of the previous states to the DA as in the temporal difference (TD) learning framework [4]. Even though that is very helpful for delayed reward in general, we neglect it in our minimal model because there is not much delayed reward aspect nor varying timing of reward in the experiments of Kawagoe et al [2]. Also, our minimal model is just a linear perceptron for sake of simplicity. Then, provided a cortical input vector $\mathbf{x}$, the output, $\mathbf{y}$, of the Sp in CD can be given by

$$y = \sum_j \omega_j x_j, \tag{1}$$

where $\omega$ denotes the cortico-striatal synaptic efficacy (weights).

The Sp project in an inhibitory way to a DA in the SNc, which will also receive an excitatory input $r$ related to the physical liquid reward. These two inputs, $y$ and $r$, determine the output $\hat{r}$ of the DA at the dopamine receptors of the Sp,

$$\hat{r} = (\hat{r}_0 + r - \sum_i v_i\, y_i)\Theta(\hat{r}_0 + r - \sum_i v_i\, y_i). \tag{2}$$

This rule differs from the one in [1] only slightly in that we restricted $\hat{r}$ to positive values with the step function $\Theta$ and included $\hat{r}_0$ to represent a constant background activity of DA. The cortico-striatal weights are only changed if the dopamine level is different from this constant background amount according to

$$\omega_{ij}^{new} = \omega_{ij}^{old} + \alpha(\hat{r} - \hat{r}_0)x_i, \tag{3}$$

where $\alpha$ is a learning rate. Note that we do assume here for simplicity that the timing of the reward is synchronized with the arrival of cortical activity.

## 3.1 The basic predictions of the minimal model

A major aim of this paper is to outline the basic predictions of this minimal model. We start here with the additional simplifying choice of using orthogonal input vectors representing the targets. The simulation of two blocks in exclusive-1DR is shown in Figure 2 (a & b), where we used input vectors which have an entry 1 for the illuminated location and 0 otherwise (e.g. $\mathbf{x} = (0, 1, 0, 0)$ for the target at location $j = 2$) and used the reward value $r = 1$ for a rewarded direction. The curves for the Sp activity of the model (Figure 2a) are similar, as a first approximation, to typical R-dependent Sp activities (not reversed ones) in [2]. The DA activity (Figure 2b) for an unexpected reward is large and converges to the background activity once the conditioning is established, which is in general agreement with cell recording data from Schultz et al. [3].

When the amount of reward is varied in exclusive-1DR, the saturated Sp response of the minimal model is linearly proportional to that amount, simply because the increase of Sp response stops when it correctly predicts this amount, i.e. $y = r$ (Figure 2c). We can not quantify this relation directly with the data in [2] since we do not have a direct measure for the reward input to DAs, $r$, but only indirectly as the amount of liquid in the experiment. However, the magnitude of $r$ is expected to monotonically increase as the amount

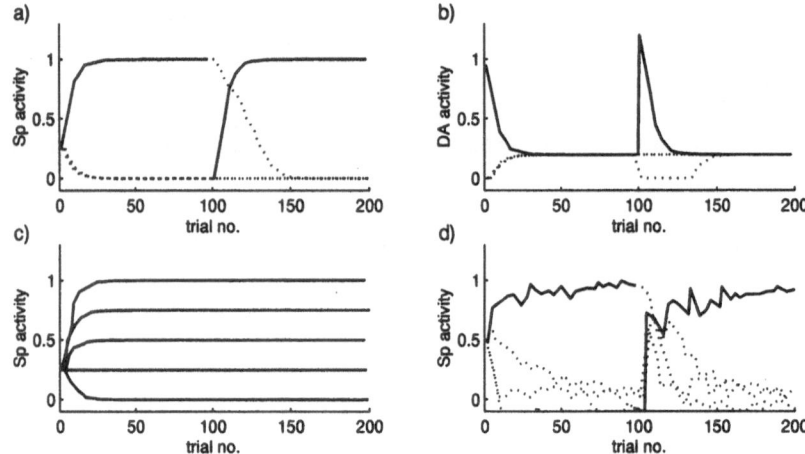

Figure 2: Simulated Sp and DA responses (rewarded as solid and non rewarded as dotted lines) in exclusive-1DR by the minimal model. (a,b) 2 blocks with 100 trials with orthogonal input vectors. (c) 1 block of 200 trials only for rewarded direction, superimposed for different reward values $r$ between 0 and 1 with $\Delta r = 0.25$. (d) Same as (a) with partial overlapping input vectors.

of liquid increases in a certain range, which was also generally consistent in exclusive-1DR [2]. Then, the prediction of the minimal model should be a monotone relation between R-dependent Sp activity and the amount of liquid for exclusive-1DR and also for ADR and relative-1DR. However, this does not seem to hold for some data of the relative-1DR experiments and should be investigated further.

## 3.2 Partial overlapping input vectors

The choice of orthogonal input vectors seems rather special. Hinted by the topographical organization of the cortical areas that also have a topographical projection to CD, one way to relax orthogonal condition is to introduce partially overlapping input vectors, while the input vectors are still separable by our perceptron model (eq. 1). The example of such input vectors used in this study is

$$x_i^j = \begin{cases} 1 & \text{if } i = j \\ 0.5 & \text{if } i = j \pm 1 \quad \text{with} \quad i = 0, ..., 5 \\ 0 & elsewhere \end{cases} \tag{4}$$

where $j = 1, ..., 4$ denotes four possible targets.

Figure 2d Shows the Sp responses with these input vectors in the same setting as of Figure 2a. Note that fluctuation of Sp responses in Figure 2d is induced by the overlapping components of input vectors for different directions. The data in [2] do show fluctuations, and it should be explored if some parts of those fluctuations can be related to overlapping input vectors.

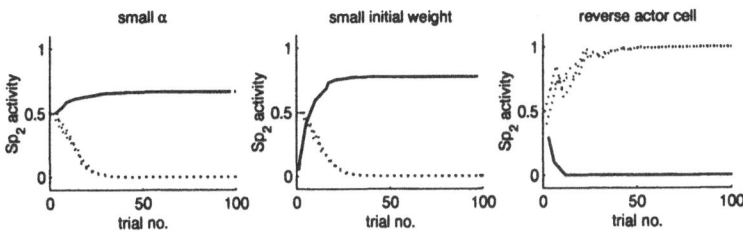

Figure 3: Some varieties of Sp activities in response to the rewarded target (solid line) and non-rewarded targets (dotted lines) in the model simulations. (a) Critic cell with small learning rate for rewarded target; (b) Critic cell with small initial weight to rewarded target; (c) Actor cell with reverse input.

## 3.3 Including multiple Sps

The minimal model with a single Sp and DA can be extended to include several Sps and DAs as shown in Figure 1b. Here, we only discuss the case of several Sps and one DA. Note that we have included synaptic weights $v$ for the Sp-DA projections in Figure 1b. This architecture does include critics and actors in the sense that actors do not contribute to the reward prediction and can be simulated by zero weights to the DAs. There are different ways to combine the predictions (outputs) of different critic Sps. In the following we used simply a geometrical average of critic Sp predictions by setting $v_i = 1$.

We showed a typical response of R-dependent (not reversed) critic Sps by the minimal model in Figure 2a. A variation, however, exists in the degree of such enhancement for rewarded targets in Kawagoe et al. [2], though the response for rewarded targets was still the biggest in each block. There are several ways to achieve such responses in our model including (1) varying learning rates for some synapses, (2) varying the initial synaptic strength, (3) varying the magnitude of input, and (4) limiting the range of synaptic strength. We demonstrated the first two scenarios in Figure 3 (a&b) with a setting similar to the first block in Figure 1a except that we let $r = 2$ and included a second Sp (critic node with a unit projection to DA). The learning rate of the synapse for the target in $Sp_2$ was smaller than other synapses of $Sp_1$ and $Sp_2$ in Figure 1a, whereas the initial value of the same synapse in $Sp_2$ was smaller than others in Figure 3b. Both modifications result in a weakened activity ($y < 1$).

As mentioned in Section 2, Kawagoe et al. [2] found the reversed Sp behavior which enhanced their activity only for non-rewarded targets. This behavior can be simulated with a minimal model by treating such a Sp as an actor. We demonstrate this in Figure 3c with an actor cell (no projection to DA) which receives reverse input $1 - x$. This reversed response is very interesting to consider as actor in relation to the direct and indirect pathways scheme of the basal ganglia. In this scheme, the inhibition of Sps in CD results in facilitating movement in the direct pathway, whereas the excitation of Sps results in suppressing movement in the indirect pathway. Then, it is intriguing to ask whether the reverse Sps are related to the direct pathway.

# 4 Discussion

Based on the actor-critic hypothesis of the basal ganglia, this paper has outlined a minimal model which can be compared to the experiments on spiny (Sp) neurons in the caudate (CD) of monkeys under various reward conditions by Kawagoe et al. [2]. We first demonstrated that our minimal model exhibits a typical reward-dependent behavior of Sps found in [2] with qualitative feature resembling dopamine (DA) neuron responses in SNc similar to that found in [3].

We also outlined some extensions to capture more details of their experimental data, necessary for a more quantitative analysis. It was shown that some overlaps in the cortical input representation lead to fluctuations in Sp responses and we also noted that various patterns of reward-dependent responses found in [2] could be realized by considering several modifications such as different learning rates and initial preferences in synaptic projections. It was also possible to show reversed Sp response by simple modifications of the minimal model.

Many important questions remain to be investigated experimentally and computationally. The relation between DA responses and physical rewards should be quantitively investigated in the on-going experiments. It should be noted that a monotone relation between Sp responses in CD and the physical rewards seems not to hold in relative-1DR in the preliminary experiments. We need more experiment data, which are under way. Also, we should explore different models of several Sps, e.g., competitive scheme between Sps. We pointed out that reverse Sp responses may work as actor particularly in the direct pathway. A further experimental and computational analysis is required to relate their responses with behavior, or saccades. This paper only treated one DA but the effects of multiple DAs (see Figure 1b) on reinforcement learning in CD should be investigated with different models of Sp-DA projections.

# References

[1] J.C. Houk, J.L. Adams and A.G. Barto, A Model of How the Basal Ganglia Generate and Use Neural Signals That Predict Reinforcement, in *Models of Information Processing in the Basal Ganglia*, (eds. Houk, Davis and Beiser), MIT Press, Cambridge 1995

[2] R. Kawagoe, Y. Takikawa and O. Hikosaka., Basal ganglia translate motivation into oculomotor action., to be published.

[3] W. Schultz, P. Apicella and T. Ljungberg, Response of Monkey Dopamine Neurons to Reward and Conditioned Stimuli during Successive Steps of Learning a Delayed Response Task, J. of Neuroscience, 1993, 13(3):900-913

[4] R.S. Sutton, Learning to predict by the method of temporal differences, Machine Learning, 1988, 3:9-44

[5] A.G. Barto, Adaptive Critics and the Basal Ganglia, in *Models of Information Processing in the Basal Ganglia*, (eds. Houk, Davis and Beiser), MIT Press, Cambridge 1995

# Neural Signalling: It's a Gas!

Andrew Philippides, Phil Husbands and Michael O'Shea

Centre for Computational Neuroscience and Robotics, University of Sussex

Brighton, U.K., BN1 9SB

andrewop@cogs.susx.ac.uk, philh@cogs.susx.ac.uk, M.O-Shea.susx.ac.uk

### Abstract

This paper presents a computational model of Nitric Oxide (NO) diffusion in nervous systems. The discovery that this freely diffusing molecule acts as a neurotransmitter has recently had a major impact on our thinking about fundamental neural mechanisms. For the first time, we have modelled diffusion from biologically realistic structures and results from our investigation into spatial and temporal aspects of NO spread are described. It is explained how this study links to more abstract modelling of the modulatory affects of NO in neuronal networks that may have important consequences for both neuroscience and engineering.

## 1 Introduction

The discovery that the gas Nitric Oxide (NO) is a neuronal signalling molecule has radically altered our thinking about how information is transmitted in the brain [1]. Traditionally neurotransmission is thought to be spatially and temporally restricted and from the pre-synaptic to the post-synaptic neuron. However the release of NO does not require specialized point-to-point synaptic contacts and unlike traditional neurotransmitters, NO can diffuse through cell membranes. NO may therefore act without the need for conventional synaptic connectivity and its action is not necessarily locally confined to the immediate post-synaptic neuron [2].

In the brain NO is generated by the calcium-activated enzyme NO synthase, or NOS [3]. Once synthesized, NO diffuses in three dimensions away from the site of synthesis regardless of intervening cellular or membrane structures. Synthesis and diffusion are not the only parameters which define the temporal and spatial dynamics of the spread of NO in the nervous system. NO is a highly reactive molecule which oxidizes readily to produce biologically inactive derivatives. The half-life and therefore the spread of NO in the brain is influenced by NO decay which depends importantly on the oxidizing environment in which it is diffusing.

Crucial to an understanding of how NO functions as a neuronal signalling molecule are the temporal and spatial dynamics of its spread. A model of NO spread therefore could potentially provide a theoretical framework for evaluating the signalling capacity of NO in the brain. Although models of NO diffusion have been published ([4], [5]) almost all of this work has concentrated on modelling the instantaneous activation of NO synthesis and the spread of NO from point sources which have no physical dimension. Although this may give some insight, serious mathematical anomalies and biologically implausible results are introduced by this approach. For example, at the source upon instantaneous activation of synthesis, the concentration of NO is infinite. In this paper we have dealt with this and other serious problems associated with point source models by modelling diffusion from continuous structures with actual and realistic dimensions. In this way and for the first time, we are able to model the spread of NO away from biologically interesting morphologies providing a new insight into the properties of the NO signalling system in the brain. Moreover, the use of different conditions and different structures give rise to qualitative changes in the kinetics of the diffusional process. However, space does not allow full presentation of these results and so only a subset have been presented here, with a complete account to follow.

## 2 Method

The dynamics of diffusion are governed by the modified diffusion equation:

$$\frac{\partial C}{\partial t} - D\nabla^2 C = -\lambda C \tag{1}$$

where $C$ is concentration and $D$ is the diffusion coefficient [6]. The term on the right hand side is an inactivation function, used to model the loss of NO through various chemical reactions. This has been taken to be exponential decay since there is no real data to suggest any other function. Thus: $1/2life = ln\,(2/\lambda)$.

### 2.1 Instantaneous symmetrical sources

Some solutions to (1) taken from the field of thermodynamics [7] are readily applicable to modelling diffusion. The main technique used in this field is to build up solutions for complicated structures from summation of contributions from point source solutions. The solution for a point-source in 3 dimensions is:

$$C_{inst}(r,\theta,t) = \frac{S_0}{8(\pi Dt)^{3/2}} e^{\left(\frac{-r^2}{4Dt}-\lambda t\right)} \tag{2}$$

where $C_{inst}(r,\theta,t)$ is the concentration of NO at time $t$ at a point $(r,\theta)$ due to a quantity $S_0$ of NO appearing instantaneously at the origin[1]. Hence for another source morphology, a ring of radius $r'$ in 2D for instance, the method is to sum the contributions from point sources uniformly distributed at a distance $r'$ from the origin, i.e. around the ring. In this case, the concentration at $(r,\theta)$ and time $t$ due to the ring-source is:

$$C_{ring}(r,r',t) = \int_0^{2\pi} C_{pt}(r,r',\theta,t)d\theta \tag{3}$$

where $C_{pt}(r,r',\theta,t)$ is the concentration at position $(r,0)$ and time $t$, due to a point source in 2D at position $(r',\theta)$. Similarly, to obtain the concentration at time $t$ and distance $r$ from the origin, $C_{ann}(r,t)$, due to an annulus in 2D of

---

[1]Notice that here, as in 3 and 4, there is radial symmetry and thus, no dependence on $\theta$

inner radius $R_1$ and outer radius $R_2$, centred on the origin, the contributions from rings of radii: $r \in [R_1, R_2]$ are summed as follows:

$$C_{ann}(r,t) = \int_{R_1}^{R_2} C_{ring}(r,r',t)dr' \tag{4}$$

This approach is valid since the diffusion equation is linear and thus appeal to the principle of superposition of linear solutions can be made. However, it relies on symmetry of the structure and generally, radial symmetry is needed for tractable solutions. If this is not the case, other techniques (e.g. difference equations) must be used.

## 2.2 Solutions for continuous sources

The solution for a source which emits NO continuously is derived from the solution for an instantaneous source in the natural way, again via the principle of superposition [6]. Thus if the concentration at time $t'$ and distance $r$ from the origin, due to an instantaneous source is: $f(r,t')$, if the source emits continuously, we have:

$$C_{cont}(r,t) = \int_0^t f(r,t')\,dt' \tag{5}$$

This can be understood by seeing that the contribution at time $\tilde{t} < t$ is due to an instantaneous pulse of NO $\tilde{t}$ seconds previously. Thus, in (5), the most recent pulses of NO are responsible for the lower limit of the integration, whilst the oldest pulses account for the upper limit. Similarly, we can derive the solution $t_1$ seconds after a source with instantaneous solution $f(r,t')$, which emitted continuously for $T$ seconds, has stopped synthesising, with:

$$C_{burst}(r,t_1+T) = \int_{t_1}^{t_1+T} f(r,t')\,dt' \tag{6}$$

## 2.3 The models used

In order to investigate the differences in solutions which one obtains when using a structure rather than a point source, a hollow sphere, inner radius 50 microns, outer radius 100 microns, has been used. This is intended to approximate the proportions of a large spherical cell, whose cytoplasm synthesises NO and whose nucleus does not. The instantaneous solution is [7]:

$$C_{sphere}(r,t') = \frac{\rho S(t)}{2r\sqrt{\pi Dt}} \int_{50}^{100} r'e^{-\lambda t}(e^{\frac{(r'-r)^2}{4Dt}} - e^{\frac{(r'+r)^2}{4Dt}})dr' \tag{7}$$

where $\rho$ is the density of the sources and $S(t)$ the source strength. $S(t)$ has been modelled as a square wave with sigmoidal rise and fall, with maximum source strength of $2.1 \times 10^{-17} mol/sec$, i.e the concentration due to one NO producing molecule [8].

For the point source we have:

$$C_{inst}(r, \theta, t) = \frac{S'(t)}{8(\pi D t)^{3/2}} e^{\left(\frac{-r^2}{4Dt} - \lambda t\right)} \tag{8}$$

which is the same as given in (2), with $S_0$ replaced with $S'(t)$, a square wave with sigmoidal rise and fall, which has a maximum value of $21 M/sec$. It should be noticed that the units of $S'(t)$ are different to those of $S(t)$. This change has occurred since we are interested in the concentration of NO, but all we know is the amount of NO produced i.e. the strength. To get the concentration we must multiply the strength by the density of the sources, as in (7). However, this is extremely problematic for a point source as, by definition, it occupies zero volume. Therefore, when we come to calculate the density, i.e. the number of sources per volume, though we may decide that a point source is a single source, what volume should we say it occupies. If we give it zero volume, $\rho$ is infinite, which is clearly impossible as we get an infinite concentration everywhere! Thus, paradoxically, we have to assign dimensions (volume) to an implicitly dimensionless entity - an inherent problem of the point source model. In this case, it was decide to assume a density of 1 source per $\mu m^3$.

These solutions can then be integrated over the appropriate time interval to obtain the solution either during, or after, a burst of synthesis of a particular length, as in (5) and (6). In all cases, the solutions required numerical integration. This was performed by the 'quad8' function in Matlab, which is based on the mathematical concept of quadrature (e.g. [9]), to a relative accuracy of 0.1%.

The values of the other constants that were used are as follows: $D = 3300 \mu m^2/sec$ [8] and $\rho = 2.38 \times 10^{-21} \mu m^{-3}$ [4].

# 3 Results and Discussion

The evolution of the concentration solutions for point-source and sphere was examined for a burst of synthesis of duration 0.5 seconds, with results shown in figure 1. As can be seen, they are qualitatively different in both the synthesis and post-synthesis stages. Very little attention has been paid to the effects of having a continuous source structure. Most research has assumed that solutions for such morphologies can be extrapolated from the point-source solution. However, this approach is severely limited: firstly, the point source solution is handicapped by its singular nature; secondly, as demonstrated below, a much richer range of dynamics is obtainable using a continuous structure.

For the point-source, the concentration rises rapidly but soon seems to approach an equilibrium - the steady-state solution (fig.1a). As a consequence of this, much of the work in modelling of NO has focussed on the steady-state solution, [4], [5]. However, such an approximation would not be valid for the sphere, as from fig.1c, we see that steady-state is not approached.

Once the period of synthesis is over, the point-source peak concentration dies away very rapidly, dropping to 20% of the maximum attained in 0.25 secs (fig.1b). This fall is accompanied by a swift spreading of NO to the surrounding area. In contrast, for the sphere, we see that during the synthesis period the

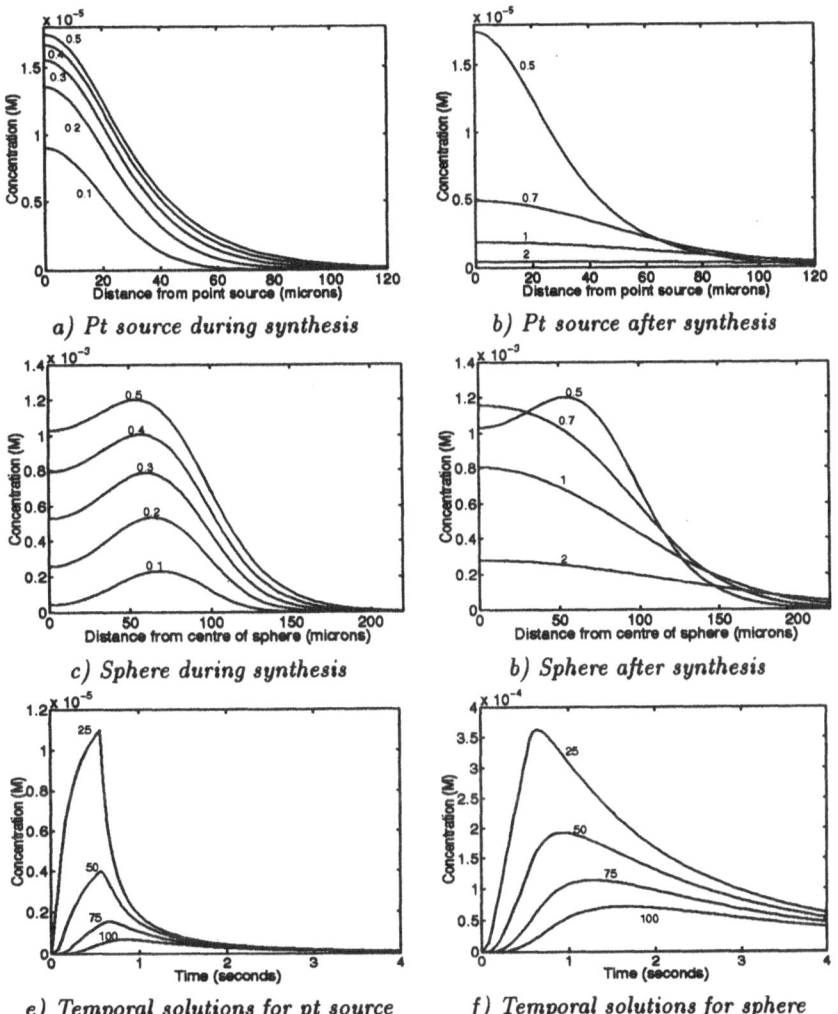

a) Pt source during synthesis

b) Pt source after synthesis

c) Sphere during synthesis

b) Sphere after synthesis

e) Temporal solutions for pt source

f) Temporal solutions for sphere

Figure 1: *Solutions for the sources for a burst of synthesis of length 0.5 secs. a)- d) are profiles of the concentration of NO at various times (given in secs) after synthesis is initiated. e) and f) are temporal concentration profiles at various distances (given in μms) measured from the outer edge of the source. N.B. Here, it is the relative dynamics of the solutions which are important and not the values of the concentrations*

concentration in the area surrounding the source rises relatively slowly, whilst in the centre of the sphere (equivalent to the nucleus of the cell), a 'reservoir' of NO builds up (fig.1d). This is due to the fact that the NO which has diffused into the centre, from sources in the cytoplasm, cannot be dissipated due to the higher concentration present at these sources. Thus, when synthesis ceases and the cytoplasmic concentration falls, we see this reservoir spreading away from the cell, in a wave of high concentration. The different dynamics are caused by shallower concentration gradient in the centre of the sphere, which is itself a consequence of the larger volume over which the concentration is high.

Examination of the concentration at 100 microns from the source (fig.1f), shows that it remains low until about 0.5 seconds after synthesis has stopped. It peaks shortly afterwards and stays relatively high for a relatively long period. This has implications for the temporal dynamics of NO-signalling in a neurobiological context. Thus if an NO-responsive neuron with an NO threshold of 0.05 mM was 100 microns from the source neuron, it would be affected after a delay of 0.5 secs, for a period of about 3 secs. Such a process could be used to introduce a time delay in NO-mediated neural signalling. The dynamics of this delay process can be seen to vary when other source morphologies are analysed.

We hope that computational modelling will prove an important tool in better understanding the spatial and temporal characteristics of NO diffusion in nervous systems. Current invasive neuroscience does not allow measurement of NO diffusion in the brain. However, a combination of the kind of detailed modelling presented here with more abstract models of the modulatory affects of NO in neuronal networks, may give very useful insights. We have abstracted general principles about the way the diffusion of NO triggers long-term and short-term changes in network properties and incorporated these principles in artificial nervous systems used to control autonomous robots [10]. We believe that this enterprise, at the interface of neuroscience and engineering, will not only give us a better understanding of how brains work, but will also spawn a new generation of more powerful adaptive machines.

**Acknowledgement** Funded by BBSRC and BT.

# References

[1] Hölscher C. Nitric oxide, the enigmatic neuronal messenger: its role in synaptic plasticity. TINS 1997; 20:298-303

[2] Hartell NA. Strong activation of parallel fibers produces localized calcium transients and a form of LTD that spreads to distant synapses. Neuron 1996; 16:601-610

[3] Bredt DS, Snyder SH. Isolation of nitric oxide synthetase, a calmodulin-requiring enzyme. Proc Natl Acad Sci USA 1990; 87:682-685

[4] Wood J, Garthwaite J. Models of the diffusional spread of nitric oxide: Implications for neural nitric oxide signalling and its pharmacological properties. Neuropharmacol 1994; 33:1235-1244

[5] Lancaster J. Simulation of the diffusion and reaction of endogenously produced nitric oxide. Proc Natl Acad Sci USA 1994; 91:8137-8141

[6] Crank J. The mathematics of diffusion. Oxford University Press, 1980

[7] Carslaw HS, Jaeger JC. Conduction of heat in solids. Oxford University Press, 1959

[8] Malinski et al. Biochem. Biophys. Res. Commun. 1993;193:1076-1082

[9] Press WH, Teukolsky SA, Vetterling WT, Flannery BP. Numerical recipes in C: the art of scientifc computing. Cambridge University Press 1971; 4:129-164

[10] Husbands P, Smith T, Jakobi N, Philippides AO, O'Shea M.Brains, Gases and Robots. In: Proc. ICANN '98. Springer Verlag 1998; (invited paper)

# Fast Learning of Dynamic Compensation in Motor Control

Pietro G. Morasso, Francesco Frisone and Sergio Bruni

DIST (Dept. of Informatics, Systems, Telecommunication), University of Genova
Via Opera Pia 13, I-16145 Genova, Italy
V: +39-10-3532749; F: +39-10-3532154; E: morasso@dist.unige.it

### Abstract

In the framework of the theory equilibrium-point control, a model for learning the compensation of dynamic loads is presented. It is self-supervised but non Hebbian and can compensate unexpected load variations in 1-2 repeated trials. A preliminary study is presented as regards the generalisation across tasks and the role of the cerebellar circuitry is discussed as a dynamic co-processor capable to implement part of the required computations.

## 1 Introduction

Human motor control is a complex system which has to solve a number of non-linear problems, namely direct and inverse kinematics, direct and inverse dynamics. Different learning models have been studied for tackling such problems which are inspired, in most cases, by the Piagetian concept of circular reaction which follows a strategy of self-supervised learning. The problem with this paradigm, as it is the case in general with most processes of self-organisation, is that it is intrinsically *slow*, in the sense that it requires a statistically dense sampling of the parametric space before any structure can emerge. Empirical knowledge about motor learning, particularly during development, is not in contrast with this kind of learning paradigm as regards the different aspects of the sensorimotor transformation which underlay motor planning and control [1,2]. An exception is the unique human capacity to quickly adapt to unexpected load changes (consider the full-bottle/empty-bottle experiment) not in the course of the current action, which typically results in a strong overshoot/undershoot, but in the subsequent few repetitions. What is remarkable of the human performance is that it can achieve a good compensation while keeping a rather high level of muscle compliance, which is an essential component of the smoothness and dexterity of human movements. A standard feedback controlled robot would indeed achieve a similar dynamic compensation, even on-line, but at the expense of an extremely high stiffness which is incompatible with dexterity.

Two weapons are available to the biological machinery for solving the load compensation problem: one is related to the intrinsic viscous-elastic properties of the muscles, augmented by the segmental stretch reflex, and the other takes advantage of the unique features of the cerebellar circuitry to deal with dynamics and, in particular, store and recall sequences [3]. The load-compensation strategy investigated in this paper is based on the repeated-trial paradigm and requires the storage and computation of whole sequences. A realistic muscle model is included, which takes into account the essential non-linearities of muscle contraction (recruitment, tetanic fusion, Hill's law, feedback delay), as well as a model of the Lagrangian dynamics of a realistic arm, in order to evaluate the interaction between the learning method and the complexity of the musculo-skeletal apparatus.

## 2   The model

The simulated musculo-skeletal system is shown in figure 1a: it is 2-dimensional and gravity is ignored for simplicity. The geometric and physical parameters agree with those used by [4][1]. The model has 4 single-joint muscles and 2 double-joint muscles. The parameters of the muscle models were tuned in such a way to agree with the experimental data about the stiffness ellipses of the arm [5].

### 2.1   Muscle model.

The following muscle model has been used, which is a particular form of the family of $\lambda$-models:

$$\lambda(t) \rightarrow \alpha(t) = l(t-\delta) - \lambda(t) + \mu \, dl(t-\delta)/dt \rightarrow f_r(t) = \rho \, (e^{c\,\alpha(t)}-1) \rightarrow f_s + \tau \, df_s/dt = f_r(t) \rightarrow$$

$$f_m(t) = f_s(t) \, h(dl/dt)$$

where $\lambda(t)$ is the centrally generated motor command (the threshold of the segmental reflex); this command is combined with the segmental (tonic and phasic) feedback (with a delay of $\delta=15$ ms and a phasic gain of $\mu=0.06$ s) thus yielding the $\alpha$ muscle-control signal; the recruited force $f_r$ has an exponential growth according to the size principle ($\rho$ is proportional to the cross-sectional-area of the muscle[2] and $c=0.112$ $m^{-1}$ is a constant characteristic of the contractile tissue); the transformation from the recruited force $f_r$ to the tetanically fused force $f_s$ is approximated with a low-pass filter (time constant $\tau = 50$ ms) and finally the output muscular force $f_m$ is modulated according to the Hill's law (approximated with a sigmoid). Figure 1b shows the block diagram of the Simulink© implementation of the model.

### 2.2   Control.

The motor commands $\lambda(t)$ sent to all the muscles are composed from the combination of three parts:

$$\lambda(t) = \lambda_R(t) + \lambda_C(t) + \lambda_D(t)$$

---

[1] Link mass (1.59 Kg, 1.44 Kg); link length (0.3 m, 0.35 m); moment arms: shoulder flexor and extensor (4 cm, 0 cm), elbow flexor and extensor (0 cm, 2.5 cm); double-joint flexor and extensor (2.8 cm, 3.5 cm).

[2] $\rho$ is computed as 4 Kg times the cross-sectional area (in $cm^2$) of each muscle: shoulder flexor and extensor (14.9); elbow flexor (11); elbow extensor (12.1); double joint flexor (2.1); double joint extensor (6.7).

respectively a *reciprocal* command (which reflects the desired/planned trajectory), a *co-activation* command (to allow the modulation of the muscle stiffness), and a *load-compensation* command (which implements a computation of inverse dynamics). Piagetian learning is only applicable to the first two components of the command vector, which are load-independent and only depend on the non-linear geometry of the musculo-skeletal system [1,2]. The last component is the load-sensitive one, including the *internal loads* which depend on the Lagrangian dynamics of the system:

$$\text{Internal-load} + \text{External-load}^3 = J_m^T \, f_m(\lambda \ )$$

where $J_m$ is the moment-arm matrix (constant in our geometric model) and $f_m$ is the output of the muscle model. In the simulations the reciprocal component of the motor command was computed by means of an explicit inverse kinematic transformation and the coactivation component was kept constant, at a 10% level.

## 2.3   Learning.

In the framework of the classic equilibrium-point models (see [6] and [7] for a review) only the first two components of the control vector are deemed necessary, under the assumption that muscle stiffness is sufficient to keep the discrepancy between the real and desired trajectory within physiological limits. However this is not true in general (the issue has been the topic of a lively debate) and we must think of an additional brain mechanism for generating the load compensation signal. In particular, we are concerned with plausible models of learning such signal (or its modifications with respect to an initial bias) as a response to unexpected load changes. In the simulations we assume that there was no initial bias in the command and so what is learnt is the inverse dynamic model of the arm. The method is iterative, each iteration requiring a new trial movement. The first goal is the approximate estimate of the total load and this can be carried out locally (i.e. independently for each actuator subsystem) by comparing the desired shortening/lengthening curve - coded by $\lambda_R(t)$ - with the actual curve $l_{ms}(t)$ measured by the muscle spindles[4]. In fact, by definition the difference of the two is the load-compensation command $\lambda_D(t)$ that would have been necessary for achieving perfect compensation of a desired trajectory equal to the real one:

Total-load $\leftrightarrow \lambda_D(t) = \lambda_R(t) - l_{ms}(t)$

The problem is that the load is dependent upon the trajectory and thus the load-compensation command computed directly as above is certainly inaccurate. A better

---

[3] In the experiment there was no external load. The internal load was given by the Lagrange equations of the arm:

$$Q(t) = I(q, dq/dt) \, d^2q/dt^2 + C(q, dq/dt) \, dq/dt + G(q)$$

where Q is the vector of joint torques, q is the vector of joint rotations, I is the inertial matrix, C is the matrix of Coriolis, and G is the gravity component.

[4] In the simulation model, for simplicity, the kinematic variables are mapped to the x-y plane and the learning equations are expressed in the Cartesian coordinates.

approximation can be obtained by parametrising this command as a function of the temporal structure of the desired trajectory. In particular, we used a linear approximation in terms of the first two derivatives of the desired curve, which yields a 3-parameter compensation vector $p_D=[\alpha,\beta,\gamma]$:

$$\lambda_D(t) = \alpha\, l(t) + \beta\, dl/dt + \gamma\, d^2l/dt^2$$

where $l(t)$ is determined by the desired law of motion. The parameters are estimated by means of a least squares approximation of the whole simulated sequence. The linear model is then applied to the originally planned curve in order to compensate it. The procedure can be iterated, thus producing a sequence of repeated trials, which eventually converge to a perfect compensation of the load. In practice 1 or 2 trials were sufficient to achieve the goal, particularly as regards the elimination of the internal delays due to inertia and transmission. The procedure is robust and mimics the ability of humans to compensate unexpected loads in 1-2 repeated trials. Consider for example a desired straight movement with duration of 0.5s and a bell-shaped velocity profile. The panel T1 of figure 2 shows the variation over time of the travelled distance: the dashed line is the desired curve, the continuous line is the actual curve, and the star-dotted line is the curve after the first trial. The first panel shows the planned and compensated trajectories in the Cartesian plane. Significant load variations or even parametric variations of the muscle model can be compensated effectively in the same way.

## 2.4 Generalisation across tasks.

The proposed model of load compensation is fast and local. The basic computational units or neural assemblies (one for each muscle or group of motor units) operate independently on the basis of individual efferent and afferent information and the coupling among the units of such neural network is carried out by the physical interactions in the real world. The learning approach is self-supervised and requires an explicit behavioural strategy (trial repetition) to be contrasted with the random exploration of the Piagetian strategy. The problem is that what is learnt is strictly task-specific, i.e. the parameter vectors $p_D$ are functions of the various characteristic components of the task: (i) the position in the configuration space, (ii) the timing of the movement, (iii) the shape, size, orientation of the trajectory, (iv) the nature and structure of the load or load field. So the question can be asked if the repeated-trials learning paradigm is capable to achieve some degree of generalisation across tasks without having to re-learn each task separately. A pilot exploration of this topic was performed by applying the same parameter vectors $p_d$ learnt for the dynamic compensation of a specific task, say the trajectory T1 of the figure, also to other tasks which differ for various task components. For example, the different panels of figure 2 show the effect of varying the direction ($10^{\circ}$ or $20^{\circ}$ on both sides: T2 to T6). Although the performance degrades, it does so in a graceful way, particularly as regards timing. Similar experiments of task manipulation yielded similar results which motivate the assumption that the repeated trial paradigm might not be incompatible with generalisation across tasks as regards load compensation. Much of the merit can be probably attributed to the mechanical properties of the muscles: they are insufficient to carry out the full compensation of the loads but are sufficient to attenuate the residual error due to the approximated compensation model, thus widening the region of applicability of any given compensation template. In the future we shall investigate a more general learning model in which the fast learning feature of the repeated trial paradigm described in this paper is integrated with a task-oriented computational framework, which builds some kind of map of such compensation templates.

# 3   Discussion

The proposed model of load compensation is in agreement with the opinion expressed by [3] that "the cerebellum takes care of the physics that is implied but not explicitly contained in the simple motor commands emanating from the cortex", thus exploiting the unique characteristics of the micro-zones of the cerebellar circuitry for the processing of whole sequences of efferent/afferent sensorimotor signals. The $\lambda_D(t)$ profiles of our model are the "sculpting signals", in the terminology of J. Eccles, which adapt the cortically generated commands to the dynamics of the world. We wish also to emphasise the importance of the mechanical properties of the peripheral apparatus for allowing the functional convergence of the learning procedure, thus pointing out the *ecological* (non-hierarchical) integration, from the computational point of view, among the processes occurring in the cortex, the cerebellum, the muscles, and the environment.

# References

[1] Morasso, P., and Sanguineti, V. (1995)   Self-organizing body-schema for motor planning. Journal of Motor Behavior, 26 : 131-148.

[2] Sanguineti, V., Morasso, P. and F. Frisone (1997)   Cortical maps of sensorimotor spaces,   in "Self-organization, Computational Maps, and Motor Control" (P. Morasso and V. Sanguineti (eds), , Elsevier Science Publishers, Amsterdam, 1-36.

[3] Braitenberg, V., Heck, D. and F. Sultan (1997) The detection and generation of sequences as a key to cerebellar function: experiments and theory. Behavioral and Brain Sciences, 20.

[4] Kawato end Gomi, H. and M. Kawato (1992) Adaptive feedback control models of the vestibulocerebellum and spinocerebellum, Biological Cybernetics, 68, 105-114.

[5] Tsuji, T., Morasso, P., Goto, K. and K. Ito (1995) Human hand impedance characteristics during maintained posture. Biological Cybernetics, 72, 475-485.

[6] Bizzi, E., Hogan, N., Muss Ivaldi, F.A. and S. Giszter (1992) Does the nervous system use equilibrium-point control to guide single and multiple joint movements?, Behavioral and Brain Sciences, 15, 603-613.

[7] Feldman, A.G. and M.F. Levin (1995) The origin and use of positional frames of references in motor control. Behavioral and Brain Sciences 18, 723-745.

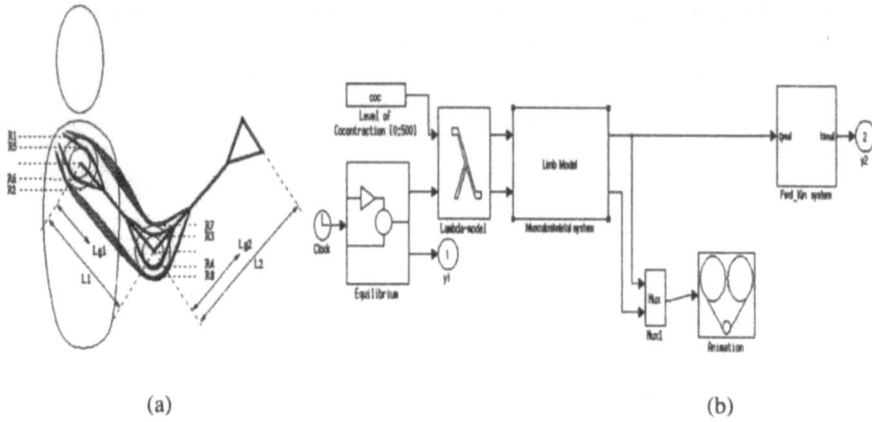

(a)  (b)

Figure 1

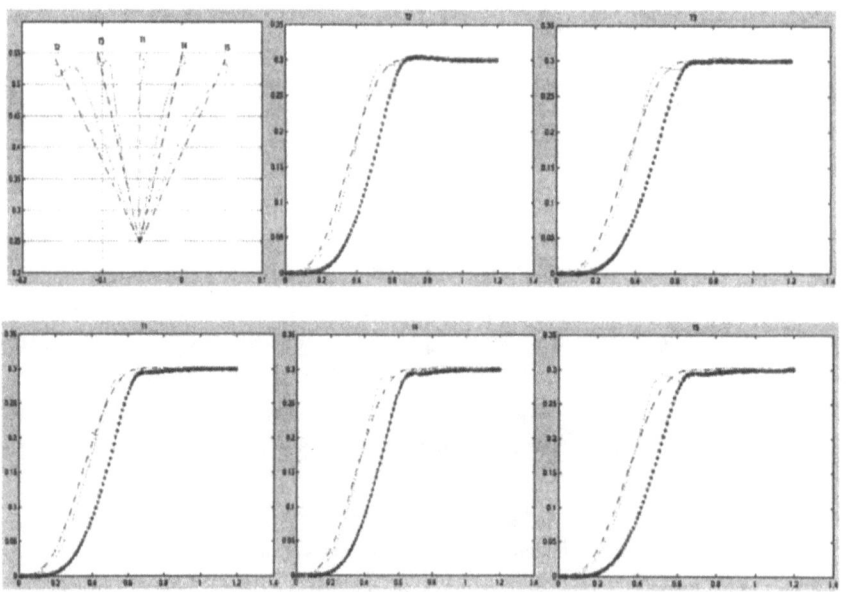

Figure 2

# A One-dimensional Frequency Map Implemented using a Network of Integrate-and-fire Neurons

Leslie S. Smith

Department of Computing Science and Mathematics, University of Stirling
Stirling FK9 4LA, Scotland. email: lss@cs.stir.ac.uk

### Abstract

A network of integrate-and-fire units (consisting of five excitatory units and one inhibitory unit) is shown to implement a one dimensional frequency map over one octave (80 - 160Hz). The network has a biologically plausible structure, conforming to Dale's law, and using plausible synaptic and axonic timings.

## 1 Aims

The aim of this work is to demonstrate a one dimensional frequency map (i.e. a network in which an ordered sequence of neurons respond to different frequencies of input) using a biologically plausible network of integrate-and-fire neurons. Biological plausibility here means obeying Dale's law, using EPSP and IPSP timings which are plausible, and using inter-neuron delays of the correct order of magnitude.

## 2 Background

This work was motivated by a problem in audition. For an auditory stimulus which consists of many harmonics of a (low) fundamental frequency, the resultant auditory nerve signal from those parts of the cochlea which cannot resolve each harmonic is amplitude modulated at the fundamental frequency [6]. Amplitude modulation is amplified by stellate cells in the cochlear nucleus (choppers) [9], and there are cells in the inferior colliculus which form a map ordered by amplitude modulation frequency [11]. We sought a neural network solution to the problem of forming a map ordered by frequency. There are a wide variety of different neurons in the early auditory system [10], and one option would have been to model a system of these, based directly on the neurophysiology [3, 13]. We decided instead to attempt to produce a frequency-sensitive system using the simplest form of neuron which is sensitive to time-varying signals at all, namely the integrate-and-fire neuron [7], exploiting their parameters and interconnection to produce differential frequency selectivity.

Integrate-and-fire units have been used for grouping signals [2, 8], and for clustering [12]. Using them in a frequency mapping system emphasises their importance as a useful abstraction of real neurons. The work reported here builds on earlier work by the author's student [1] (in which pulsatile input was used, and the excitatory units directly inhibited other excitatory units).

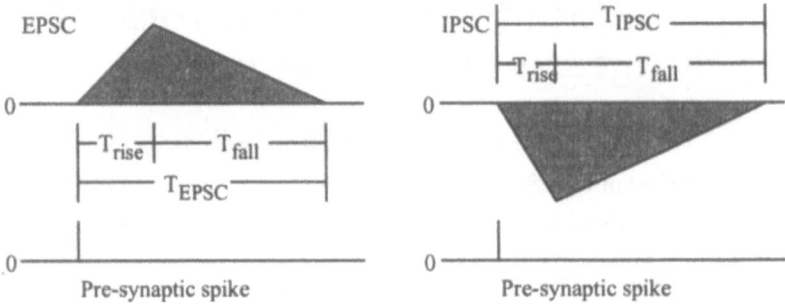

Figure 1: Excitatory and inhibitory post-synaptic currents for a single pre-synaptic spike. For a weight of 1 (-1 in the inhibitory case) the total charge transferred (over $T_{EPSC}$ ($T_{IPSC}$)) is 1 (-1). $T_{EPSC} = 3.7$ms, and $T_{IPSC} = 3.3$ms, with $T_{rise} < T_{fall}$ in both cases.

# 3 Methods

The experiments were carried out on a simulator for networks of integrate-and-fire neurons developed by the author and Chumbo [1] which runs under the NeXTStep operating system.

## 3.1 The neurons

All the neurons are integrate-and-fire units, characterised by their voltage-like state variable, $V$, between spikes:

$$\frac{dV}{dt} = -\alpha V + I(t)$$

where $\alpha$ is the dissipation. If $\alpha = 0$, this is simple integration, and, for $\alpha > 0$ leaky integration. When $V$ hits some threshold ($\theta$, initially 1 in this case), the unit fires: that is, it emits a spike. There then follows a refractory period, during which $\theta = \infty$, at the end of which $V$ is reset to 0. This is followed by a relative refractory period, which starts with $\theta = 10$, followed by a quadratic (parabolic) decay back to the original threshold.

Input to the neurons takes two forms. External input to the network is considered to be like injected current. From an auditory perspective (i.e. where the "real" input was an amplitude modulated signal), this input would be made up from many auditory nerve fibers or outputs from primary cells in the cochlear nucleus. These will have arrived nearly simultaneously at a number of synaptic sites, resulting in (i) half-wave rectification of the amplitude modulated signal and (ii) smoothing of high frequency content of the signal [4].

Input from other neurons is modelled using simple triangular EPSPs and IPSPs as shown in figure 1. This is a simplification of real EPSP and IPSP shapes, similar to that discussed in [5].

## 3.2 The network

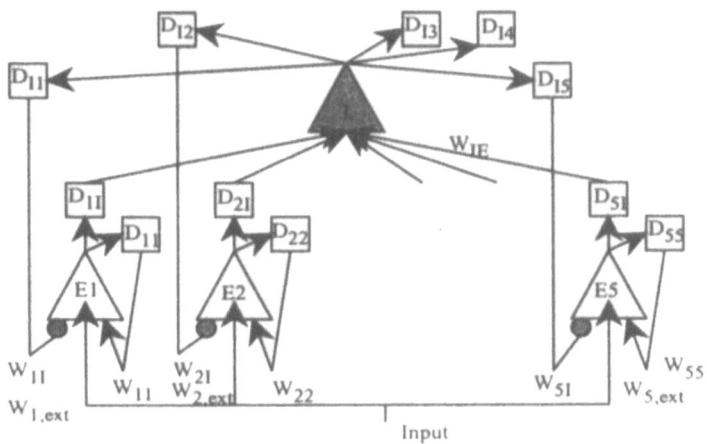

Figure 2: The network used. Only 3 of the 5 excitatory neurons are shown. Neurons $Ei$ are excitatory, and neuron $I$ is inhibitory. The weights $(W_{ij})$ and delays $(D_{ij})$ are discussed in the text.

The network consists of five excitatory neurons each with a recurrent excitatory connection, all exciting an inhibitory interneuron, which in turn inhibits all the excitatory neurons (see figure 2). Each has either all excitatory or all inhibitory outputs, so that the network satisfies Dale's law. There is a delay associated with each connection between neurons: this delay is crucial to network design.

## 3.3   Design of the frequency mapping network

We set the dissipation of each excitatory neuron so that it would be maximally sensitive to half-wave rectified input at its "best" frequency. For this to occur, the activation should peak near the end of each positive half-cycle. We chose the dissipation to be $\frac{1}{T_{\text{half-cycle}}} = 2f_{\text{best}}$, where $f_{\text{best}}$ is the desired "best" frequency, then adjusted the strength of the input to each neuron $(W_{i,\text{ext}})$ so that it just fired in response to input of strength 1 at its best frequency.

The basic ideas for the weights and delays in the filter were taken from digital filtering. There, bandpass filters are produced using multiple weighted delayed signals which are fed back to the input of an amplifier. However, using a neuron whose output was fed back through many synapses, each with different delays, some with positive and some with negative weights is not biologically plausible. Instead, we used one excitatory recurrent connection, with a total delay (including the EPSC delay) of one period of the "best" frequency. We set the refractory period (RP) on each of the excitatory neurons to 2ms, and the relative refractory period (RRP) to a value such that $RP + RRP = \frac{2}{f_{\text{best}}}$. The idea was that the excitatory feedback should be needed in order to make the neuron fire for a second time. The strength of the recurrent excitatory feedback $(W_{ii})$ was then set so that input at $f_{\text{best}}$ resulted is a stream of spikes. This results in each excitatory neuron being a low-pass filter (LPF), in the sense that it fires once per cycle for $f \leq f_{\text{best}}$, and less often for $f > f_{\text{best}}$.

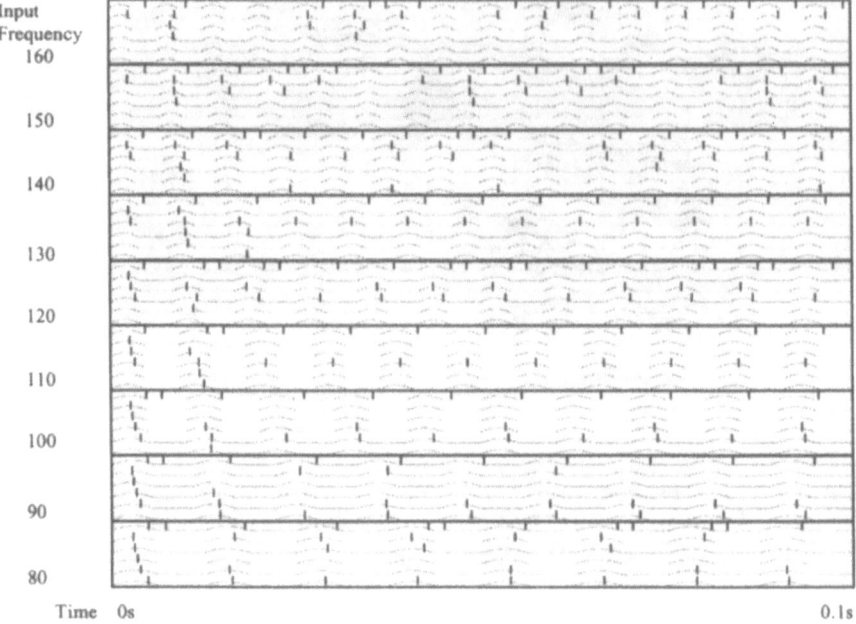

Figure 3: Network response to half-wave rectified signals from 80 to 160Hz. Grey lines are input, black vertical lines are spike outputs. In each small rectangle, the bottom 5 lines are the input and output from each of the excitatory neurons (lowest frequency at bottom, highest at top), and the top line is the inhibitory interneuron.

Turning the LPF characteristic of each neuron into a frequency map requires inhibition. To provide this without sacrificing biological plausibility, we used an inhibitory interneuron. This integrate-and-fire neuron is excited by each excitatory neuron through a fast synapse with short delay. All the weights to this neuron ($W_{IE}$) and all the delays ($D_{iI}$) are identical. The weights were set so that any single spike from an excitatory neuron will cause the inhibitory interneuron to fire. The RP and RRP were set to small values (1ms each), and the dissipation to 200 (altering this seems to have little effect). The delays from the inhibitory neuron back to the excitatory neurons ($D_{Ii}$) were set so that the round-trip delay from output of the excitatory neuron back to inhibition of the excitatory neuron was $1.5 * \frac{1}{f_{best}}$. Inhibition occurs when the excitatory neuron should not be firing if it is responding to its $f_{best}$. The inhibitory weight to the excitatory neuron, ($W_{iI}$) was set to the same value for all the neurons.

# 4  Results

A network with 5 excitatory neurons, and one inhibitory neuron was designed, and set up for $f_{best}$ of 80Hz, 100Hz, 120Hz, 140Hz, and 160Hz. It was tested with input (of strength 1) at frequencies between 80 and 160 Hz, in intervals of

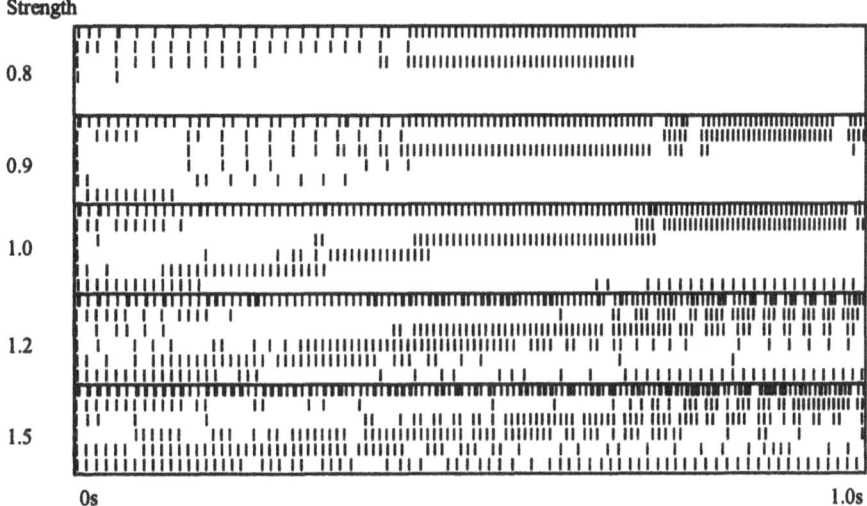

Figure 4: Network response to chirp input (see text) at different strengths. Top line is inhibitory interneuron.

10Hz, The results are shown in figure 3. Initially, each neuron works as a LPF, but then the inhibitory interneuron takes effect, and the network settles down with the appropriate neurons responding most strongly. The exception to this occurs at 80Hz, where the highest frequency neuron also responds.

The network was tested with a "chirp" input (a half-wave rectified sine wave, whose frequency varied linearly from 80Hz to 160Hz over 1 second). This was applied at varying strengths, and the results are shown in figure 4. When the signal has strength 0.8, it is not strong enough to make the network behave correctly. At higher strengths, the appropriate neurons fire. As the input strength is increased, more excitatory neurons fire, although the inhibitory neuron also fires more often. The response is centred on the correct neuron, although there is additional noise as the strength increases. The exception is near 160Hz, where the 80Hz unit also fires, though only on every second pulse.

# 5 Discussion

We have provided constructive proof that one can build a frequency mapping network using only integrate-and-fire neurons. Real neurons are more complex, and it may be that the neurons of the early auditory system have structures which make this task easier: however, we have demonstrated a biologically plausible neural circuit for producing a frequency map similar to that in the inferior colliculus. The network is designed, not adapted: it is not clear how one would produce an adaptive network to perform this task. We have not experimented with scaling this network up: we suspect that larger clusters of excitatory neurons controlled by one inhibitory neuron would be less stable.

# References

[1] P.S.C. Chumbo. Integrate and fire neural network simulator and bandpass filter. Technical report, Centre for Cognitive and Computational Neuroscience, University of Stirling, Stirling UK, 1997.

[2] U. Ernst, K. Pawelzik, and T. Geisel. Multiple phase clustering of globally pulse coupled neurons with delay. In M. Marinaro and P.G. Morasso, editors, *ICANN94: Proceedings of the international conference on artificial neural networks, Sorrento, Italy, 26-29 May 1994*, pages 1063–1066. Springer-verlag, 1994.

[3] M.J. Hewitt and R. Meddis. A computer-model of amplitude-modulation sensitivity of single units in the inferior colliculus. *Journal of the Acoustical Society of America*, 95(4):2145–2159, 1994.

[4] M.J. Hewitt, R. Meddis, and T.M. Shackleton. A computer model of a cochlear nucleus stellate cell: responses to amplitude-modulated and pure-tone stimuli. *Journal of the Acoustical Society of America*, 91(4):2096–2109, 1992.

[5] W. Maass. Networks of spiking neurons: The third generation of neural network models. *Neural Networks*, 10(9):1659–1671, 1997.

[6] M.I. Miller and M.B. Sachs. Representation of voice pitch in discharge patterns of auditory nerve fibers. *Hearing Research*, 14:257–279, 1984.

[7] R.E. Mirollo and S.H. Strogatz. Synchronization of pulse-coupled biological oscillators. *SIAM J. Applied Mathematics*, 50(6):1645–1662, 1990.

[8] A. Nishwitz and H. Glunder. Local lateral inhibition - a key to spike synchronization. *Biological Cybernetics*, 73(5):389–400, 1995.

[9] A.R. Palmer and I.M. Winter. Cochlear nerve and cochlear nucleus responses to the fundamental frequency of voiced speech sounds and harmonic complex tones. *Advances in the Biosciences*, 83:231–239, 1992.

[10] J. O. Pickles. *An Introduction to the Physiology of Hearing*. Academic Press, 2nd edition, 1988.

[11] C.E. Schreiner and G. Langner. Periodicity coding in the inferior colliculus of the cat. ii. topographical organization. *Journal of Neurophysiology*, 60(6):1823–1840, 1988.

[12] L.S. Smith. Onset-based sound segmentation. In D.S. Touretzky, M.C. Mozer, and M.E. Hasselmo, editors, *Advances in Neural Information Processing Systems 8*, pages 729–735. MIT Press, 1996.

[13] A. van Schaik, E. Fragnière, and E. Vittoz. A silicon model of amplitude modulation detection in the auditory brainstem. In M.C. Mozer et al., editor, *Advances in Neural Information Processing Systems 9*, pages 741–747. MIT Press, 1997.

# The role of spatio-temporal neural response characteristics in the formation of synchrony

Ulrich Quill, Florentin Wörgötter, and Klaus Funke

Institute of Physiology, Department of Neurophysiology, Ruhr-University
Bochum, Germany

Anders Lansner

SANS - NADA - Kungl Tekniska Högskolan
Stockholm, Sweden

### Abstract

Although the functional role synchronous oscillations may play has been investigated in depth, the underlying processes and spatio-temporal aspects that establish the synchrony are still not thoroughly understood. Experimental studies suggest the existence of two kinds of oscillations: *stimulus-locked* and *stimulus-induced*. While stimulus-locked oscillations are systematically dependent on the stimulus, stimulus-induced oscillations (occurring in the $\gamma$ frequency range) show only little stimulus dependency. We propose a unifying approach which employs very generic connection structures. Different degrees of synchrony on different time scales are observed as an emergent feature of the network structure. Our model demonstrates that both, stimulus-locked and stimulus-induced oscillations are just two different states of the same system. A transition from one state to the other is observed, and the synchronous activity provides the basis for binding visual features.

## Introduction

Synchronization phenomena have long been an object of both, experimental and theoretical research in the neurosciences. Although synchronous oscillations at different frequencies have been reported by various groups ([7],[13]), Eckhorn et al. were the first to introduce the terminology of two different kinds of oscillations ([1],[3]): *stimulus-locked* and *stimulus-induced*. These oscillations occur at different stages of the visual pathway (retina, LGN, cortex), but were also observed in other parts of the brain, e.g. the auditory system.

*Stimulus-locked* synchronizations are directly driven by stimulus transients, to which they are temporally locked. Only strong transients, which e.g. occur during flashed stimuli, can elicit an oscillatory behaviour ([5],[6],[14],[15]).

*Stimulus-induced* synchronizations are supposed to be a results of self-organization processes among stimulus-driven "oscillators". Especially in the absence of strong stimulus transients, e.g. during slowly moving stimuli, these processes serve to generate synchronized oscillations in the $\gamma$-frequency range.

We propose a simple, generic network model which is able to elicit both types of synchronous activity depending on the stimulus situation. Furthermore it utilizes these oscillations in order to bind stimulus features in a temporal and spatial sense. Oscillating neurons synchronize spatially, depending on the spatial properties of the stimulus. The oscillations are also kept up for some time, allowing neurons from different areas to synchronize during the stimulus presentation.

## Model

The aim of our model is to show that the proposed transition from stimulus-locked to stimulus-induced oscillations is possible with a simple, generic, biologically plausible network. The model describes the information flow in the primary visual pathway. The afferent flow from LGN to cortical layers IV and VI is incorporated, as well as an excitatory feed-back loop from layer VI to the LGN and an inhibitory feed-back loop from the perigeniculate nucleus (PGN) back to the LGN.

The implementation is done using a simulator software developed in our group, which is built on top of the SPLIT library, which was developed in Anders Lansner's group at KTH, Stockholm ([9],[10]). This software package implements a realistic multi-compartmental neuron model with Hodgkin-Huxley type channel dynamics.

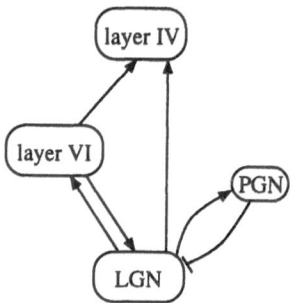

**Figure 1:** General connection structures between principal neuronal layers.

The network consists of 2 general types of neurons: thalamic and cortical. The primary connections between these functional layers are shown in figure 1. LGN cells are excitatorily connected with both types of cortical cells (layer IV and VI). These connections are topographically arranged with Hubel-Wiesel type receptive fields (RF) for the layer IV neurons. As our model is primarily aimed at demonstrating and examining synchronization phenomena, we confined ourselves to receptive fields comprised of simple precisely tuned feed-forward LGN-cortex structures, and did not incorporate more realistic models. The layer IV neurons are arranged in hyper-columns with preferred orientations varying from 0° to 180° in 10° steps. These afferent LGN-layer IV connections are fast and strong and act as the primary, stimulus-dependent driving force of the layer IV neurons.

Layer VI neurons also get input from the LGN, but their receptive fields are broader than those of layer IV neurons. There are two types of efferent connections from these layer VI neurons: firstly, divergent feed-forward projections onto layer IV with adjustable divergence parameters and secondly feed-back projections to the LGN with a point-spread-function (PSF) broader than the respective feed-forward RF (see figure 2, left). The layer VI neurons act as neural oscillators tuned to different eigen-frequencies. Thus they can perform two tasks. They can reflect tonic firing patterns from the LGN and maintain these patterns for some time even if LGN firing ceases. Furthermore the broader convergence and divergence of their connections helps to spread activity across neighbouring neurons (both in the LGN and cortical layer IV).

Finally, there is the LGN-PGN feed-back loop with excitatory feed-forward and inhibitory feed-back connections. This loop is an oscillator with alternating states and thus modulates the LGN activity with its resonance frequency (see figure 2, right).

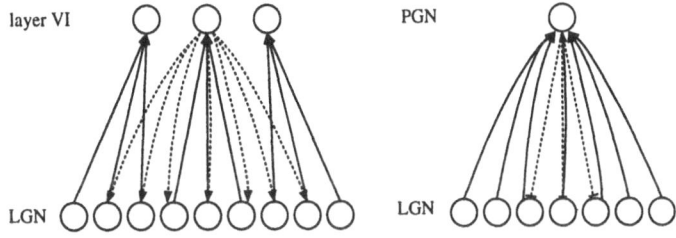

**Figure 2:** A detailed description of the information flow between LGN, layer VI, and PGN. Left: Layer VI neurons already have broad RFs (strong lines), but their feed-back connections reach even farther, thus projecting onto LGN neurons belonging to the RF of other layer VI neurons (dashed lines). Right: Both, the receptive fields and the PSFs of PGN neurons are not orientation tuned. The excitatory convergence to PGN is higher than the inhibitory divergence back to LGN.

# Results

*Flashed stimuli*

Our first experiments were carried out with flashed bar stimuli (light bar, switched on for about 800 ms). The final result of these simulations is that the cortical layer IV neurons exhibit an oscillatory activity with two main frequency components. First, there are alternating periods of activity and inactivity (corresponding to the $\alpha$-rhythm), the frequency of which is determined by the tuning of the LGN-PGN loop. Secondly, a higher frequency is observed during the active periods. Within these periods the layer IV neurons maintain a rather constant oscillatory activity as long as the stimulus lasts.

Different network effects must be examined to explain these different frequencies and the alternating periods of activity and inactivity. As long as the stimulus lasts, the LGN neurons receive a firing pattern composed of a strong and short phasic response followed by a longer lasting tonic response in the $\gamma$ range (as determined by the bandpass properties of the retina). The frequency of this tonic response depends on the stimulus properties (e.g. contrast) and decays slowly during a longer presentation due to adaptation processes in the retina.

The LGN response is inhibitorily modulated by the LGN-PGN loop. Tonic LGN activity stimulates the PGN neurons which in turn begin to fire, so that the LGN neurons are inhibited. With the decreasing LGN activity, the PGN neurons finally stop firing due to the missing input. So the inhibition stops, and the LGN neurons can again fire according to their retinal input. The LGN-PGN loop in this configuration acts as a neural oscillator modulating the thalamocortical spike trains. Without the influence of the layer VI neurons, the strong thalamocortical afferents are the driving force for the layer IV neurons, so that in this case, the layer IV neurons just follow the LGN activity pattern. The stimulus-locked oscillations observed in layer IV are therefore a feature of the strong thalamocortical connections.

The neural oscillators in layer VI are excited by the LGN neurons firing with their respective resonance frequencies. These oscillators only have to be started and can then maintain their activity for some time even with only little direct input from the LGN. Their broader RFs (compared to layer IV neurons) and their lateral connections to other oscillators with topographically adjacent RFs provide support for synchronization between these oscillators. As the driving input from the LGN is rather strong, the inactivity periods induced by the PGN inhibition are needed to establish and stabilize this synchronization. This synchronous activity (which is now stimulus-induced) is then transported on to layer IV, inducing synchronicity in that layer.

*Slowly moving stimuli*

With light bars moved slowly across the visual field (velocity about 1 degree/second), the layer IV neurons keep a rather steady firing pattern with a frequency which depends on the stimulus. This frequency is comparable to that of the "activity periods" observed during flashed stimuli, or can be a harmonics of it due to inhibition by PGN neurons ([5]). In this stimulus situation, the layer VI neurons are dominating the network behaviour. As strong transients are missing in the stimulus, the LGN-PGN loop can only contribute a tonic effect, which does not produce an alternating activity pattern, but instead keeps up a rather constant inhibition of medium strength. The LGN-layer IV afferents cannot evoke strong activity in layer IV. Instead, the layer VI oscillators are activated, which is possible because of the broader RFs in layer VI.

As already mentioned, the broader RFs of the layer VI neurons and their lateral interconnections allow synchronizations between different neural oscillators in layer VI, which in turn helps to synchronize layer IV neurons. The synchronization is based on the topographical structure of the stimulus. Different stimulus contrasts induce different LGN firing frequencies. And thus, because of sharply tuned bandpass characteristic of layer VI cells, different oscillators in layer VI are activated. These results provide a basic mechanism for binding similar stimuli which are topographically connected and separate dissimilar stimuli.

# Discussion

We could show that a very simple yet realistic model of the primary visual pathway is able to establish synchronous activity, which emerges in and propagates through the network spatially as well as temporally. First, there are fast, stimulus-forced oscillations which are directly driven by the strong LGN-CTX afferents. These are not generated in layer IV, though, but only mirror the retino-geniculate input. The slow decay in LGN activity, which develops during steady, flashed stimuli, is compensated for by local neural oscillators in cortical layer VI. These oscillators also drive the network in the presence of slowly moving stimuli. In this stimulus situation the LGN-PGN-layer IV loop is too weak to induce oscillations in layer IV. Instead, layer VI oscillators are activated and layer IV neurons couple to these oscillators.

This behaviour clearly suggests that there is a transition from stimulus-locked to stimulus-induced oscillations within the same network during the presentation of a flashed stimulus, as well as predominant stimulus-induced oscillations during slowly moving stimuli.

In our model, the intrinsic layer VI cell properties and the LGN-PGN loop are the exclusive sources of the observed rhythms. Neither the LGN nor the LGN-PGN loop generates a $\gamma$ rhythm. Further investigations have to be carried out to examine the stability of the observed oscillations. Until now we have not focused our attention on the fine-tuning of the model in order to bring oscillation frequencies in accordance with physiological data, with the fast oscillations being within the $\gamma$-range, the slow (LGN-PGN induced) oscillation being in the $\alpha$-range.

Due to the extended receptive fields of the layer VI neurons we could also observe a spread of activity across neighbouring neurons in layer IV. This suggests that the network is also capable of binding visual features on a low level of processing. We will further examine this effect to provide a profound explanation of the network features involved.

The authors acknowledge the support of the Deutsche Forschungsgemeinschaft (SFB 509) and the HFSP.

# References

[1] R. Eckhorn, R. Bauer, W. Jordan, M. Brosch, W. Kruse, M. Munk, and H. Reitboeck. Coherent oscillations: A mechanism of feature linking in the visual cortex? multiple electrode and correlation analysis in the cat. *Biological Cybernetics*, 60, 1988.

[2] R. Eckhorn and A. Frien. Neural signals as indicators of spatial and temporal segmentation coding in the visual system. In J. Mira-Mira, editor, *Proceedings of the International Conference on Brain Processes*. MIT-Press, 1995.

[3] R. Eckhorn, H. Reitboeck, M. Arndt, and P. Dicke. Feature linking via synchronization among distributed assemblies: Simulations of results from cat visual cortex. *Neural Computation*, 2:293–307, 1990.

[4] Ö. Ekeberg, P. Wallén, A. Lansner, H. Traven, L. Brodin, and S. Grillner. A computer based model for realistic simulations of neural networks. i. the single neuron and synaptic interaction. *Biological Cybernetics*, 65:81–90, 1991.

[5] K. Funke and F. Wörgötter. Temporal structure in the light response of relay cells in the dorsal lateral geniculate nucleus of the cat. *Journal of Physiology*, 485(3):715–737, 1995.

[6] K. Funke and F. Wörgötter. On the significance of temporally structured activity in the dorsal lateral geniculate nucleus (lgn). *Progress in Neurobiology*, 53:67–119, 1997.

[7] C. Gray. Synchronous oscillations in neuronal systems: Mechanisms and functions. *Journal of Computational Neuroscience*, 1:11–38, 1994.

[8] S. Grossberg and D. Somers. Synchronized oscillations during cooperative feature linking in a cortical model of visual perception. *Neural Networks*, 4:453–466, 1991.

[9] P. Hammarlund. *Techniques for efficient parallel scientific computing*. PhD thesis, Royal Institute of Technology, Stockholm, 1996.

[10] P. Hammarlund, Ö. Ekeberg, T. Wilhelmsson, and A. Lansner. Large neural network simulations on multiple hardware platforms. In *Proceedings CNS*96*, 1996.

[11] D. Kammen, P. Holmes, and C. Koch. Cortical architecture and oscillations in neuronal networks: feedback versus local coupling. In R. Cotterill, editor, *Models of Brain Function*, pages 273–284. Cambridge University Press, 1990.

[12] T. B. Schillen and P. König. Binding by temporal structure in multiple feature domains of an oscillatory neuronal network. *Biological Cybernetics*, 70:397–405, 1994.

[13] W. Singer and C. Gray. Visual feature integration and the temporal correlation hypotheses. *Annual Reviews of Neuroscience*, 18:555–586, 1995.

[14] F. Wörgötter and K. Funke. Fine structure analysis of temporal patterns in the light response of cells in the lateral geniculate nucleus of cat. *Visual Neuroscience*, 12:469–484, 1995.

[15] F. Wörgötter, R. Opara, K. Funke, and U. Eysel. Utilizing latency for object recognition in real and artificial neural networks. *NeuroReport*, 7:741–744, 1996.

# Three-Layered Neural Model between Cortical areas V1 and IT

Satoru Kato, Kunihito Yamamori and Susumu Horiguchi

Japan Advanced Institute of Science and Technology (JAIST)
Asahi-dai 1-1, Tatsunokuchi, Ishikawa, 923-1292, Japan
satoru@jaist.ac.jp , yamamori@jaist.ac.jp , hori@jaist.ac.jp

## Abstract

It is well known that orientation-selective cells in the striate cortex are organized as a spatial structure in the area V1 of the visual cortex, and stimulus-selective cells in the area IT only respond to simple geometrical patterns. However, the neural network structure and its learning principle between the area V1 and the area IT have not been studied sufficiently. This paper presents a hierarchical neural network model between the area V1 and the area IT as well as its learning principle based on Kohonen's self-organizing model. Experimental results show that the hierarchical neural network organizes orientation-selective cells in the area V1 and stimulus-selective cells responding to simple geometrical patterns in the area IT.

## 1 Introduction

Stimulus-selective cells in the visual cortex are classified according to functional matter. The orientation-selective cells known as typical stimulus-selective cells respond selectively to specific orientation of slit-patterns in the area V1. Recently, Tanaka, et al.[1] discovered stimulus-selective cells in the area IT of the macaque monkey which respond to specific geometrical patterns. Obermayer et al.[2] proposed a self-organizing neural network model which can organize cortical feature maps in the area V1 using Kohonen's self-organizing model.

In this paper, we propose a three-layered hierarchical neural network model and two-phase learning process applied to the proposed model. We suggest that the spatial map structures are organized on the second layer, and the stimulus-selective nodes are organized on the third layer, respectively.

## 2 The area V1 and the area IT

Cells in the cortical area V1 respond to optical stimuli such as orientation, color, and simple textures. Orientation selective cells respond selectively to slit-like patterns. These are organized as the spatial structure in the cortex, that is, columnar organization whose modules are perpendicular to the cortex [3]. A columnar module contains multiple cells which are overlapping but slightly different selectivity cluster together.

The stimulus-selective cells in the area IT respond to moderately complex optical features such as simple geometrical patterns( T-shaped pattern, stellar symbols, textures, and so on). These are not able to specify an object uniquely. It was pointed out that an activation of a few to several tens of cells with different critical feature is necessary to specify a particular natural object.

Regarding neural projection from V1 to IT, it is known that the ventral pathway is responsible for object vision and runs from the cortical area V1 to V2, thereafter to V4, and finally to the IT. The IT projects to various structures outside the visual cortex. There is a sequential cortical pathway from V1 to the anterior IT, and outputs from the pathway mainly originate in the anterior IT.

# 3 Three-Layered Network model for V1-IT

## 3.1 · Hierarchical three-layered network

Recent experimental discoveries about stimulus features in IT imply to us a hierarchical neural network between V1 and IT. The structure of the hierarchical neural network is shown in figure 1. The neural projection between retina and V1 is represented by one input layer and one competitive-layer having $30 \times 30$ and $20 \times 20$ nodes respectively, which is named as *V1-model* in figure 1(a). The input layer corresponds to the retina to express visual images consisting of $30 \times 30$ pixels. Output of each node ranges from 0.0 to 1.0.

The three-layered network model is adopted as a neural projection between V1 and IT. It has an additional competitive layer behind the first competitive layer. Figure 1(b) shows three layered network model, that is, *V1-IT model*. These layers correspond to retina, V1, and IT respectively. $W$ and $E$ denote the connective weight vectors and the input vectors for each competitive layer respectively. $E_{\text{layer1}}$ is equal to the input pattern $E_{\text{slit}}$ or $E_{\text{geom}}$. And $E_{\text{layer2}}$ is equal to an output vector $O_{\text{layer1}}$ from the first competitive layer.

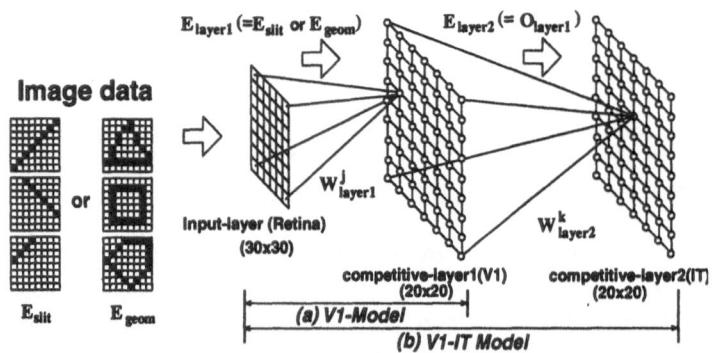

Figure 1: Hierarchical three-layered network

## 3.2 Basic self-organizing algorithm

Connective weight vectors $W^j_{\text{layer1}}$ and $W^k_{\text{layer2}}$ are organized by a self-organizing algorithm by Kohonen[4]. At first, a *winner-node* is chosen whose weight vector

$W^{\text{win}}_{\text{layer}N}$ is the closest to the input vector $E_{\text{layer}N}$.

Connective weight vector $W^M_{\text{layer}N}$ are corrected for the winner node and the neighboring nodes by input learning patterns. The correction of each element $w^{M,i}_{\text{layer}N}$ of connective weight vectors is given by

$$\Delta w^{M,i}_{\text{layer}N} = \begin{cases} \alpha_{\text{layer}N}(e^i_{\text{layer}N} - w^{M,i}_{\text{layer}N}) & \text{winner or neighboring nodes,} \\ 0 & \text{Other nodes,} \end{cases}$$

$$w^{M,i}_{\text{layer}N}(t+1) = w^{M,i}_{\text{layer}N}(t) + \Delta w^{M,i}_{\text{layer}N}, \tag{1}$$

where $M$ indicates $M$th node on each competitive layer, and $\alpha_{\text{layer}N}$ is a learning rate for each competitive layer. Another parameter $d_{\text{layer}N}$ is defined to determine neighboring nodes for the winner. It is a distance based on the *8-neighboring distance* from the winner to each node which will be corrected its weight vector. These parameters are linearly decreased with the iteration of learning.

## 3.3  Two phase learning

The learning process of the *V1-IT model* consists of two phases. Let $T$ is the total learning time which is divided to $T_1$ and $T_2$. Each $T_1$ and $T_2$ is the learning time of the first phase and the second phase, respectively.

When $0 \leq t \leq T_1$, the initial value of learning rates and the input vectors for each competitive layer are: $E_{\text{layer}1} = E_{\text{slit}}$, $E_{\text{layer}2} = O_{\text{layer}1}$, and $\alpha^{\text{Ini}}_{\text{layer}1} > 0$, $\alpha^{\text{Ini}}_{\text{layer}2} = 0$, thus the second competitive layer does not work in first phase. And when $T_1 < t \leq T_2$, each of them is: $E_{\text{layer}1} = E_{\text{geom}}$, $E_{\text{layer}2} = O_{\text{layer}1}$, and $\alpha^{\text{Ini}}_{\text{layer}1} = 0$, $\alpha^{\text{Ini}}_{\text{layer}2} > 0$, thus the first competitive layer does not learn anymore. Each element of the output vector $O_{\text{layer}N}$ is given by:

$$o^M_{\text{layer}N} = f(u^M_{\text{layer}N}) = \frac{1}{1 + \exp\{\beta(u^M_{\text{layer}N} - \theta)\}}, \tag{2}$$

$$u^M_{\text{layer}N} = \sum_i (w^{M,i}_{\text{layer}N} \cdot e^i_{\text{layer}N}), \tag{3}$$

where $f$ is a *sigmoid-function*, where $\beta$ determines a shape of sigmoid and $\theta$ is a threshold.

# 4  Spatial Organization in the V1-model

## 4.1  Self-organizing learning

In this section, we discuss the spatial organization in the V1-model generated by learning simulations in the first learning phase. Learning parameters $\alpha_{\text{layer}1}$ and $d_{\text{layer}1}$ are initialized at $\alpha^{\text{Ini}}_{\text{layer}1} = 0.2$ and $d^{\text{Ini}}_{\text{layer}1} = 10$, respectively. The initial state of connective weights are set random values from 0.0 to 1.0.

(a) Initial state         (b) After learning

Figure 2: Visualizations for a connective weight vector

Figure 2(a) shows the initial connective weights of the first competitive layer. The brightness of the dot is proportional to the value of the weight. Figure 2(b) shows the state of a connective weight vector after self-organized learning by applying 25000 slit patterns whose centers and orientations are arbitrary. Kohonen's self-organizing algorithm decreases the difference between input vectors and connective-weight vectors, thus a slit-like input pattern is reflected on the connective-weight space. Responses of each node in the competitive layer after self-organizing learning are discussed in following section.

## 4.2   Node responses in the V1-model after learning

To discuss the characteristics of the first competitive layer, we have used four slit-like patterns shown in the upper part of figure 3(a). The lower part of figure 3(a) shows response patterns for corresponding slit-like pattern. The activate level is visualized by the brightness. It is seen that a few nodes respond selectively to the specific orientation and center position. These simulation results imply to us that the V1-model achieves a topological mapping of the orientation and center position between the input layer and the first competitive layer by Kohonen's self-organizing learning algorithm. These phenomena correspond to the orientation selective cell in the cortical area V1 [5].

Each node can also detect a partial line segment of a simple geometrical pattern. In equation (2) and (3), connective weight vectors $W_{\text{layer} N}^{M}$ reflect input vectors. Therefore, when a simple geometrical pattern is applied to the input layer, each node will respond if the geometrical pattern includes a partial line segment which is similar to the node's connective weight vector.

Figure 3(b) and 3(c) explain this behavior. Figure 3(b) is a composition of the lower part of figure 3(a). Figure 3(c) shows the response pattern for a square which consists of four slits in figure 3(a). It is seen that figure 3(b) and 3(c) closely resemble each other.

# 5   Perception in the V1-IT model

## 5.1   Geometrical patterns

To investigate the V1-IT model and the two-phase learning, we have used some simple geometrical patterns as input patterns; diamonds, parallelograms, quadrilaterals, rectangles and triangles. The 288 test patterns are generated

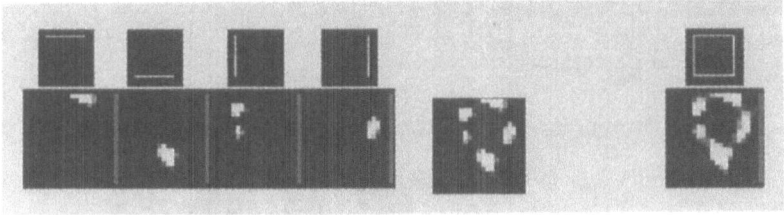

(a) Response patterns for each slit    (b) Composition of four response patterns    (c) Response pattern for square

Figure 3: Response pattern of nodes on the first competitive layer

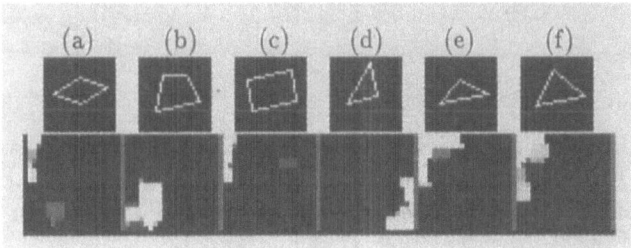

Figure 4: Response patterns for test patterns on the second competitive layer

by rotating these basic geometrical patterns, and the 144 patterns are used as an input pattern to the input layer for learning and the remained 144 patterns are used to investigate the activation of nodes in the second competitive layer.

## 5.2 Selectivity for geometrical patterns

Learning parameters $\alpha$ and $d$ in the second learning phase are chosen $\alpha_{layer2}^{Ini} = 0.2$, $d_{layer2}^{Ini} = 10$ respectively. To confirm the stimulus selectivity of the V1-IT model, the activations of nodes in the second competitive layer are investigated in detail. Figure 4 shows the test patterns and the node responses. It is seen that a cluster of nodes selectively respond to the simple geometrical patterns on the second competitive layer of the V1-IT model. However, it is observed that the activating regions are same for test patterns 4(a) and 4(c). This means that the activating regions for specific geometrical patterns do not occupy its own place. The proposed V1-IT model does not entirely simulate physiological facts by Fujita et al. [6].

## 5.3 Form perception

The form perception ability of the V1-IT model is discussed by applying patterns not including in the learning set to the V1-IT model.

Table 1 shows applied patterns and geometrical patterns which have the most similar response to the applied pattern in the learning set. The similarities between two responses are evaluated by $\cos \theta$ between these response vectors. This observation means the proposed three layered model can extract a property of shape from applied unknown geometrical patterns by using known geometrical patterns.

Figure 4: Form perception ability of the second competitive layer

| Input patterns | | | | |
|---|---|---|---|---|
| Learned patterns bringing most similar response | | | | |
| $\cos\theta$ | 0.99 | 0.51 | 0.84 | 0.87 |

# 6   Conclusions

The three-layered hierarchical network model and the two-phase learning algorithm are proposed for the cortical are V1 and IT. We have focused on the stimulus selective cells responding to the specific line segments or geometrical patterns. And we executed self-organizing simulations.

The simulations in the V1-model is qualitatively similar to the physiological knowledge for the area V1. Moreover, simulations in the V1-IT model show that the nodes responding only to a few specific geometrical patterns are organized on the second-competitive layer. The V1-IT model can extract a property of the shape within the applied geometrical patterns. However, activating regions for specific geometrical pattern don't occupy its own place on the second competitive layer. The spatial structure of cortical feature maps should be discussed more in the V1-IT model.

# References

[1] Tanaka,K. : "Neural Mechanisms of Object Recognition", *Science*, Vol.262, pp.685–688, 1993.

[2] K.Obermayer, H.Ritter, K.Schulten : "A principle for the formation of the spatial structure of cortical feature maps", *Proc. Natl. Acad. Sci. USA*, Vol.87, pp.8345–8349, 1990.

[3] Hubel,D.H. and Wiesel,T.N. : "Receptive fields and functional architecture of monkey striate cortex", *J.Physiol.,Lond.*, Vol.195, pp.215–243, 1968.

[4] Kohonen,T. : "Self-Organizing Maps", Springer-Verlag Berlin Heidelberg , 1995.

[5] Blasdel,G.,Salama,G. : "Voltage-sensitive dyes reveal a modular organization in monkey striate cortex", *Nature*, Vol.321, pp.579–585, 1986.

[6] Fujita,I.,Tanaka,K.,Ito,M. and Cheng,K : "Columns for visual features of objects in monkey inferotemporal cortex", *Nature*, Vol.360, pp.343–346, 1992.

# An Interruptible Connectionist Model for Real-Time Pattern Recognition

Jean-Denis Muller

CEA-DAM

B.P. 12, F-91680 Bruyères-le-Châtel, France

*muller@bruyeres.cea.fr*

### Abstract

We describe an approach used for the conception of neural real-time pattern recognition systems which are interruptible, that is, able to give answers before the computing completion. This method is based on dynamic coding of information in the neural network: the proposed neuron model performs a time integration of its inputs and emits binary spikes trains. We have designed a dynamic shared-weights multi-layer neural classifier with recognition rate close to a more classical pattern recognition network, suitable for real-time applications with early hypothesis production. Because of its low computing complexity, this model seems to be well suited to hardware implementations.

## 1    Introduction

Some real-time artificial vision applications need very fast operation. A possible solution is the realisation of specific electronic architectures using DSP, FPGA or ASIC, but this can be unsatisfactory without significant modifications in the software.

We study here the possibility of designing a fast connectionist system, able to give results at any time and compatible with the future conception of an integrated circuit. We are particularly interested in spiking neural networks [1], which explicitly model the dynamic of information exchange between neurons. We have already used such neuron models to design **SPIKE_4096** [2], a VLSI circuit performing very fast image segmentation using an original coding based on neural firing latency as described in [3, 4].

This coding has also been used for the conception of a connectionist model realising fairly advanced functions: fast extraction of salient characteristics in images using several filters with various spatial orientations and frequencies [5], or face localisation in visual scenes [6].

In this paper, we propose an interruptible multi-layer neural network for fast pattern recognition. This network is composed of neurons emitting binary spikes. We have experimentally checked that this dynamic network has a recognition rate close to results given by a more classical neural network. This model should open interesting perspectives for future real-time hardware realisations.

## 2 Model description

One of the current difficulties with multi-layer spiking neural networks is the lack of learning algorithms producing as good performances as static neural networks do, in spite of very interesting studies [7]. Thus, instead of searching for a learning method adapted to a given dynamic neural model, we have chosen to create a dynamic network inspired from a static neural network for which a learning algorithm exists already. We re-use the static network structure and synaptic weights: the only changes are the neuron model and the signals emitted by the neurons.

The chosen static model is *Le Net* [8], a shared-weights multi-layer perceptron with layers structured in maps. We have already used this type of neural network for classification of defects on Very Large Scale Integration masks [9] or texture analysis in satellite images [10]. In order to illustrate this communication, we have chosen a simple and didactic application: handwritten digit recognition.

In the network presented *Fig. 1*, the layer 1 (*resp.* layer 2, layer 3) is composed of maps Layer-1-$y_{(y=1\to5)}$ (*resp.* Layer-2-$y_{(y=1\to5)}$, Layer-3-$y_{(y=1\to5)}$). These layers are connected by 3×3 or 4×4 convolution masks excepting the last one, which is fully connected to Layer-3.

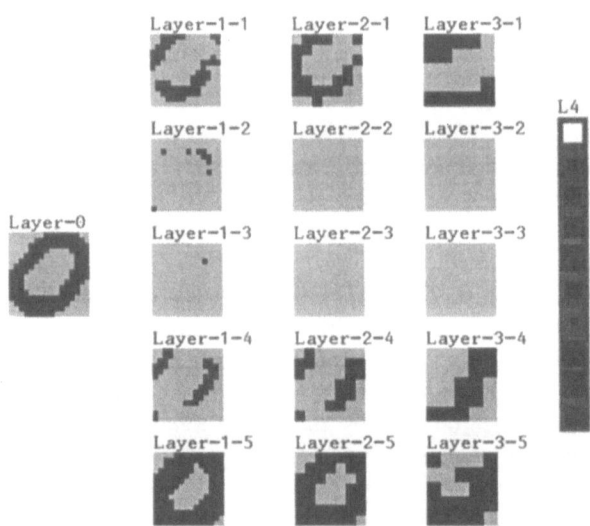

Figure 1: the working neural network before computing completion.
In layers 1, 2 and 3, one represents neuron binary outputs (black if active, grey if inactive).
In layer 4, one represents weighted sums (white if positive, black if negative).
Remark that the network has already given a response although four maps are still inactive.

The classical artificial neurons have continuous outputs. On the contrary, the neurons of our model have binary outputs. Thus, it is necessary to make up for the loss of information which was coded by the output level with the addition of a time integrating capacity. Since we only allow spiking communications between the neurons, this information has to be restored in the receiving neuron: this is done by a time integration of the spikes.

Let $i$ be an emitting neuron (layer $n$) and $j$ a receiving neuron (layer $n+1$). Let $\delta_i$ be the output of the neuron $i$, $w_{ji}$ the weight assigned to the connection neuron $i \rightarrow$ neuron $j$, $U_j$ the internal potential of the neuron $j$, $\theta_j$ its activation threshold and $\delta_j$ its output. Let us assume that:

$$U_j(t) = \min\left[\left(U_j(t-1) + \Delta\theta.\sum_i\left[w_{ji}.\delta_i(t)\right]\right), 1\right] \qquad \begin{array}{c}\textit{evolution law of the} \\ \textit{internal potential}\end{array} \qquad (Eq.\ 1)$$

$$\delta_j(t) = \begin{cases} 1 & \text{if } U_j(t) > \theta_j(t-1) \\ 0 & \text{otherwise} \end{cases} \qquad \begin{array}{c}\textit{condition of} \\ \textit{spike emission}\end{array} \qquad (Eq.\ 2)$$

$$\theta_j(t) = \theta_j(t-1) + \Delta\theta.\delta_j(t) \qquad \textit{desensitisation} \qquad (Eq.\ 3)$$

When neurons are activated, they emit spike trains. These spikes are modulated by the synaptic weights and time integrated by receiving neurons (*Eq. 1*). If the internal potential $U$ of a neuron exceeds a threshold $\theta$, this neuron emits a spike (*Eq. 2*) and its activation threshold is raised (*Eq. 3*). If the neuron remains inactive, the threshold is unchanged. In order to optimise computing time, neurons are authorised to emit when the first spike comes from the preceding layer: the initial threshold $\theta(t = 0)$ is set to zero. The initial potential $U(t = 0)$ is null.

When it is not saturated, the neuron emits a number of spikes proportional to its internal potential (*Eq. 1, 2, 3*). The associated static network must do the same: the output of the static neuron must be proportional to the weighted sum. An odd sigmoïdal activation function is not suitable: in this case, the neuron output would not be a linear function of the potential, and moreover this classical activation function has negative outputs which have no equivalent in the dynamic model, in which the considered output is a number of spikes. So we have chosen to change the activation function of the static model: we use a positive piecewise linear function (*Fig. 2*). Unfortunately, this activation function is not optimal and decreases the correct classification rate from 95 % to 87.5 %. We have experimentally noticed that this performance loss was mainly due to the difficulty for the back-propagation algorithm to converge with a non-odd activation function.

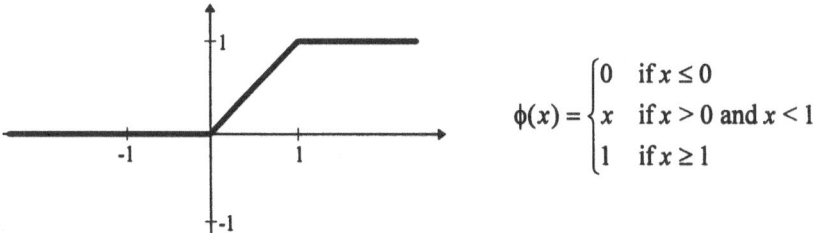

$$\phi(x) = \begin{cases} 0 & \text{if } x \leq 0 \\ x & \text{if } x > 0 \text{ and } x < 1 \\ 1 & \text{if } x \geq 1 \end{cases}$$

Figure 2: Activation function in the static network.

Since the dynamic neuron has no explicit activation function, its output non-linearity is performed by the potential saturation (*Eq. 1*).

# 3 Results

We have compared the two models average performance on classification of 16×16 handwritten digit images coded on 8 bits (*Fig. 3*). Because of the important influence of the computer implementation, this comparison is not absolutely fair: it does not depend only on the intrinsic qualities of the models. Particularly, in our current implementation the dynamic network is disadvantaged by management of complex structures like chained lists which slow down the simulations.

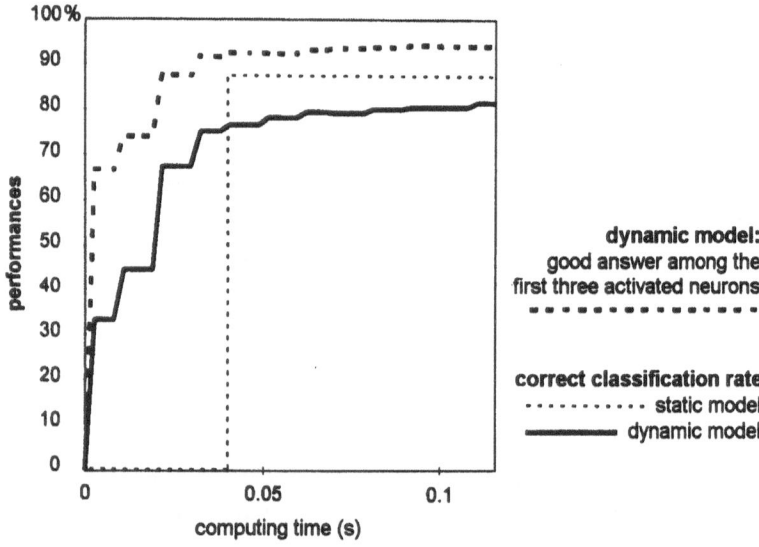

Figure 3: Average performance along time for the two models (Δθ = 0,05).

In our computer simulations, the time required by the static network to give its response is about 40 ms for a 87.5 % average classification rate. At the same time, the dynamic network has a 77 % average classification rate. It reaches static network performance in about 200 ms.

If one looks for the correct response among the first three activated neurons in the output layer of the dynamic network, one observes that the percentage reaches already 92 % after 40 ms. It would be possible to speed up the computing by increasing Δθ, but this leads to an important drop in performance for Δθ > 0,08, due to a too low computation precision.

One of the goals of this study is to obtain an interruptible system, suitable to a real-time context. The static model gives its response only after network computing completion, whereas the dynamic model can produce early hypothesis. From this point of view, the dynamic neural network is very attractive (*Fig. 3*): for example, after 10 ms only, it already reaches a correct classification rate which is the half of the static network performance and gives the correct class among the first three responses in 75 % of the cases.

# 4 Conclusions and perspectives

The spiking neural model proposed in this communication has proved its ability to get final performances close to the corresponding static model ones. Moreover, the spikes propagation mechanism allows the network to give continuously useful responses. Therefore, one can imagine to use this model for real-time applications requiring early hypothesis production (fast localisation of patterns or objects in large visual scenes, etc.). At the moment, we are applying this model to other pattern recognition applications. We have resolved the activation function problem using a bias shift. One of our future objectives is the design of an hardware implementation taking advantage of the low computing complexity of this spike coding.

*The author is grateful to Stéphanie Muller and François Durbin (CEA) for having read this article and proposed some relevant modifications.*

# References

[1] Eckhorn R., Reitboeck H. J., Arndt M., Dicke P. Feature linking via synchronization among distributed assemblies: Simulations of results from cat visual cortex. *Neural Computation*, 2, 293-307, 1990.

[2] Rebourg J. L., Muller J. D., Samuelides M. SPIKE_4096: A neural integrated circuit for image segmentation. Submitted to *ICES'98, International Conference on Evolvable Systems: From Biology to Hardware*, Lausanne, Switzerland, 23-26 sept. 1998.

[3] Thorpe S. J. Spike arrival times: A highly efficient coding scheme for neural networks. In Eckmiller, Hartman, and Hauske (Eds.), *Parallel Processing in Neural Systems*, North-Holland, Elsevier, 1990.

[4] Thorpe S. J., Fize D., Marlot C. Speed of processing in the human visual system. *Nature*, 381, 520-522, 1996.

[5] Thorpe S. J., Gautrais J. Rapid visual processing using spike asynchrony. NIPS'96, Denver, CO. In Jordan (Ed.), *Neural Information Processing Systems*, 9.

[6] VanRullen R., Delorme A., Gautrais J., Thorpe S. J. Face processing using one spike per neuron. *Biosystems*, in press.

[7] Samuelides M., Thorpe S. J., Veneau E. Implementing Hebbian Learning in a Rank-based Neural Network. *ICANN'97*, Lausanne, Switzerland, 8-10 oct. 1997.

[8] Le Cun Y., Jackel L., Bottou L., Brunot A., Cortes C., Denker J., Drucker H., Guyon I., Müller U., Säckinger E., Simard P., Vapnik V. Comparison of learning algorithms for handwritten digit recognition. *ICANN'95*, Paris, France, 9-13 oct. 1995.

[9] Muller J. D., Samuelides M. Integrated cooperation between unsupervised and supervised learning for a defect classification problem. *ICANN'92*, Brighton, UK. *Artificial Neural Networks*, Vol. 2, Aleksander, and Taylor (Eds.), Elsevier Science Publishers.

[10] Muller J. D., Cheynet P., Velazco R. Analysis and improvement of neural network robustness for on-board satellite image processing. *ICANN'97*, Lausanne, Switzerland, 8-10 oct. 1997.

# Implementation of Tunable Receptive Field (RF) Filters for Learning Retina Implants[*]

R. Hünermann and R. Eckmiller

Informatik VI (Neuroinformatik), Universität Bonn
D-53117 Bonn, F. R. Germany
Email: huenermann@nero.uni-bonn.de, URL: http://www.nero.uni-bonn.de

**Abstract**

Retina Implants for blind subjects with retinal degenerative disorders require retinal information processing in real time with many individually tunable spatiotemporal filters with antagonistic receptive field properties (RF filters). We describe a RF filter structure, function and tuning mechanism as well as the implementation of tunable RF filters in real time on a digital signal processor (DSP) for currently developed retina implants. Each tunable RF filter is capable of implementing a large variety of spatial and temporal mapping operations from spatiotemporal light patterns at the assigned array of photo sensors onto a corresponding single asynchronous stimulation pulse sequence at the output. For this purpose, our tunable RF filters incorporate a wide range of typical RF properties of primate retinal ganglion cells in their functional space. A number of spatial and temporal parameters in the RF filter algorithm is specified in order to assure smooth maneuvering within the functional space as well as tuning of a given RF filter to the desired RF function of a given contacted ganglion cell in a learning phase. Typical spatial and temporal information processing properties of tunable RF filters are presented.

## 1  Introduction

Retina implants, which are currently developed in Germany by a consortium of 14 expert groups, for blind subjects with retinal degenerative dysfunctions require the approximate technical simulation of retinal information processing in order to stimulate the output of retinal ganglion cells as closely as possible to the neural input, expected for each optic nerve fiber by the central visual system. To do so, a Retina Encoder (RE) has to map light patterns arriving at a photo sensor array onto asynchronous pulse trains, thus representing the individually expected antagonistic RF properties of the primate retina. These stimulation pulse trains have to be transmitted to implanted contacts in order to stimulate corresponding ganglion cells [2]. RE has to simulate various kinds of ganglion cells with their corresponding receptive field (RF) properties for each single implanted microcontact. In addition, RE has to offer the possibility for each simulated cell to adapt from one cell type to

[*] Supported by Federal Ministry for Education, Science, Research, and Technology (BMBF).

another during a dialog with the implant carrying patient [1][2][3].

We present a flexible, tunable model for RF filters with a sufficient number of adequate parameters for implementation of ganglion cell properties within their entire functional space. The whole RE currently consists of 200 RF filters each with individually tunable parameters, and is embedded in a simulation environment on a C80 DSP including utilization of its multiprocessing power in order to guarantee real time processing [5][6].

# 2 Tunable RF filters

RF filters have to model several spatiotemporal properties observed in various kinds of primate retinal ganglion cells. In the present implementation the model consists of two equally structured, independent pathways, an excitatory and an inhibitory one (see fig. 1). Each pathway has an receptive field as input at a suitable location on the photo sensor array. A spatial weighting function is used to calculate the dot product with the receptive field information on the photo sensor array. Usually, these weighting functions are two-dimensional gaussian functions, which can be tuned concerning their two spatial extensions, their position, and their volume.

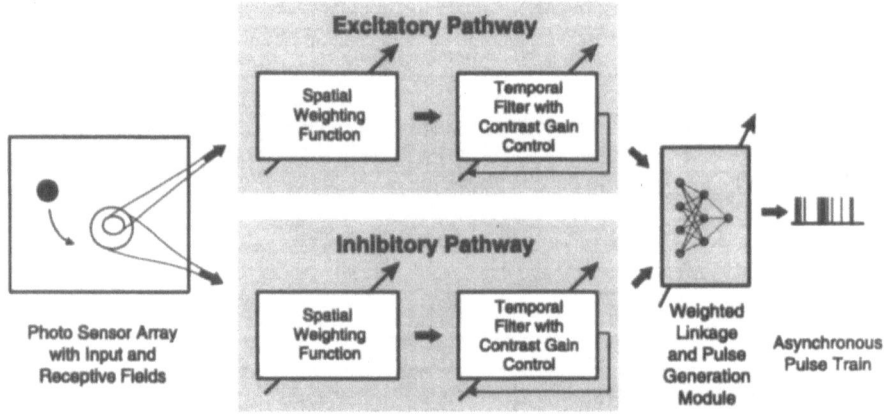

Figure 1: Scheme of tunable RF filter structure with receptive fields as input and asynchronous pulse trains as output.

The resulting time course is filtered afterwards by a temporal filter, which can be tuned in a continuos range from a transient band pass to a more sustained low pass. It is presently implemented by a cascade of a third order low pass and a first order high pass. In addition, the temporal filter has a contrast gain control, i.e. the temporal filter properties are dependent on presented contrast. The filter parameters are tuned towards a more transient filter with growing signal contrast. This phenomenon in retinal ganglion cells was observed and described by Victor [8]. The fundamental filter parameters can be tuned as well as their dependence on contrast. After both pathways were weighted and linked together, the result is transduced into pulse trains, which can be used to stimulate an appropriate ganglion cell in the retina.

This RF filter structure is capable to implement the main features of primate retinal ganglion cells as described in the literature. The antagonistic center-surround organization of RFs is taken into account by the two spatial weighting functions from which one has to be narrower than the other in order to describe On- or Off-center cells. The distinct temporal features of primate retinal M- and P-cells (see [7]) are modeled by those temporal filters, which can be tuned to get biologically plausible frequency spectra accordingly. The fact, that for example the temporal sensitivity of retinal ganglion cells depends on the spatial frequency of the presented stimulus indicates a spatiotemporal inseparability of the RF filter transfer function and can be described by two separate pathways [4]. This feature is also included in our model due to the excitatory and inhibitory pathways.

For RE implementation we currently use the digital signal multiprocessor TMS320C80 (C80) from Texas Instruments (TI). Its parallel processing power predestines the C80 for RE development, because besides its capability to ensure the adaptability and learning ability of 200 individual RF filters, which strictly need to be calculated in real time, the transformation of their membrane potentials into asynchronous pulse trains, and the generation of simulated input patterns for testing purposes also in real time can be guaranteed. The calculation frequency for the RE and the stimulus pattern generation is 100 Hz, i.e. all RF filters and the new input pattern have to be calculated in less than 10 ms. In addition to simulated input patterns, captured video frames from a standard CCD camera or an external photo sensor array can also be processed in real time (see also [6]).

# 3 Maneuvering Inside Functional Space of RF Filters

All possible combinations of RF filter parameter values define the RF filter state space. Its biologically plausible subspace shall be called the functional space of RF filters. Within this functional space RF filters need to be optimized with regard to the spatiotemporal function, which is expected by the central visual system from the randomly contacted ganglion cell in the retina. During this optimization process the parameters of each individual RF filter change from a predefined state to an optimal state concerning the perception of the implant carrying subject. Because this process is done by a dialog module [2][3], the individual RF filters are tuned only by their effect on perception without knowing the actual parameter values of RF filters. For that reason, mapping from RF filter parameter space to perception should be causal and continuos, i.e. small changes in parameters should result in small changes in perception. Outside the functional space of RF filters the condition of causality is violated, because the central visual system does not get biologically plausible input, which makes a smooth maneuvering impossible.

Fig. 2 exemplarily depicts subspaces of two different parameter spaces describing the same part of functional space. Points **A** and **B** represent two different states within the functional space. The direct connection between these two points as the intuitive way from one point to the other in parameter space is also shown. In representation a) this line leaves the functional space and therefore, RF filter would loose causality during tuning process. In that case, the patient would not be able to suggest

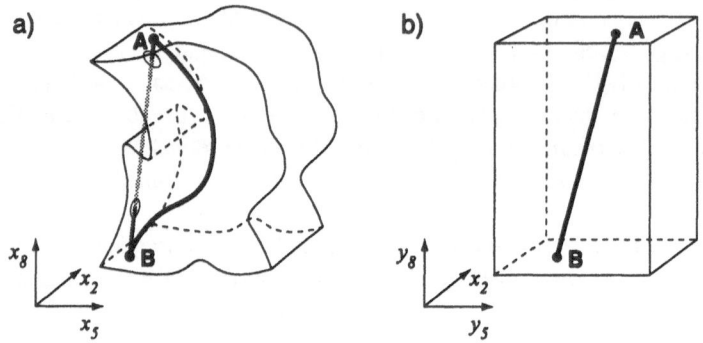

Figure 2: Subspaces of two different parameter representations
of an exemplary functional space

reasonable changes to the dialog module. Instead, tuning has to be done along any less intuitive curved line, which is not specified at all, inside functional space. The problem of representation a) is, that parameter restrictions of $x_5$ and $x_8$ concerning the functional space depend on each other. The restriction of parameter $x_2$, however, does not depend on $x_5$ and $x_8$ at all. On the other hand, representation b) supplies straight connections, which completely lie inside functional space, for any pair of allowed states, because the functional space got a convex representation by replacing the $(x_5, x_8)$-plane with the $(y_5, y_8)$-plane, which has independent restrictions.

Smooth maneuvering inside functional space of RF filters via a dialog module essentially requires independent parameter restrictions and subsequently a convex representation of functional space, because a dialog module needs to change parameters individually without additionally adjusting all the others. The described effects even multiply in high dimensional parameter space of RF filters.

# 4   Results

For the time filter subspace of RF filter functional space we already found a representation, which is convex and has independent parameter restrictions:

$$G(s) = \left(1 - \frac{a}{bs+1}\right) \cdot \left(\frac{1}{s+1}\right)^3$$

with $a \in [0;1]$ and $b \in [0;100]$

The parameters $a$ and $b$ of this representation have to be in the specified ranges, wherein all possible combinations result in biologically plausible transfer functions. Fig. 3 demonstrates

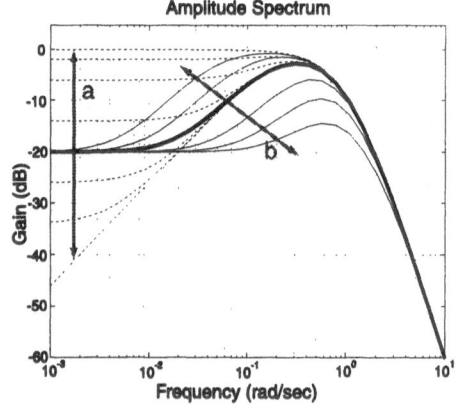

Figure 3: Time filter frequency spectra
dependent on special parameters (see text)

the variety of transfer functions available with these two parameters. Parameter $a$ controls the gain at low frequencies, whereas parameter $b$ controls the transient strength of the transfer function. The cut-off frequency of the low pass is set to one rad/sec. For illustration it was set to a fixed value, because it would just move the spectrum towards higher or lower frequencies.

If the transfer function was specified by a single fraction, the resulting term would have a lot more parameters to achieve similar tuning effects, which were additionally dependent on each other. The goal is to find a representation with as less independent parameters as possible for maximum flexibility, in order to describe the whole functional space. The given representation of our time filter structure as a first step allows smooth maneuvering within the whole biologically plausible subdomain without leaving it and without any singularities inside.

In order to analyze the spatiotemporal frequency spectrum of a specific RF-filter, we used a moving sinusoidal grating as input pattern, i.e. a light pattern which has a sinusoidal luminance distribution in one direction. The spatial frequency $f_S$ is the reciprocal value of the length in pixels for one period (cycles/receptor). The temporal frequency $f_T$ is the product of velocity $v$ of the moving grating and its spatial frequency, namely $f_T = v \cdot f_S$ (Hz). Because of a sinusoidal luminance distribution the stimulus generates only one distinct temporal frequency at a specific velocity with a fixed spatial period. For primate retinal ganglion cells the spatial weighting functions of RF-filters typically are rotationally symmetric, wherefore the gratings have to be moved only in one direction over the RF. Thus, a spatiotemporal frequency spectrum for a definite parameter set of a RF-filter can be determined and compared with physiologically measured data. We applied various different combinations of spatial and temporal frequencies, all with maximum contrast, and measured the amplitude of the simulated RF filter membrane potential. The corresponding impulse rate is used to create the spatiotemporal amplitude spectra as shown in figure 4.

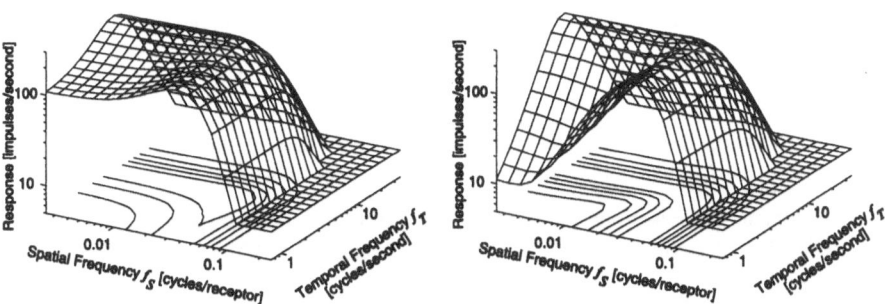

Figure 4: Typical spatiotemporal frequency spectra of P-type RF filter (left) and
M-type RF-filter (right)

The spatiotemporal spectra in figure 4 demonstrate the tunability of RF-filters. The left part of the figure shows a spectrum from an P-type filter, whereas the spectrum on the right side belongs to an M-type filter. The RF filter described by the left spectrum is sensible at low spatial frequencies, whereas the right one suppresses these frequencies and lets only higher ones pass. Clearly the spatiotemporal inseparability can be seen from the spectra in figure 4, i.e. the temporal frequency sensitivity is dependent from the spatial frequency and vice versa.

# 5 Conclusions

We described the present implementation of RF filters as they were used in our Retina Encoder with receptive fields as input and asynchronous pulse trains as output. It could be demonstrated that RF filters show similar spatiotemporal behavior as primate retinal ganglion cells. The functional space as a biologically plausible subspace of the representing parameter space for RF filters needs to be convex and the describing parameters need to be independent concerning their restriction for functional space. For a small parameter subspace, namely the linear part of the time filters, such a representation was found and presented. For a whole convex representation of the RF filter functional space further work has to be done. Especially, a flexible parameter representation has to be found in order to assure smooth maneuvering during a tuning procedure with suggested changes via a dialog module from an implant carrying patient.

# References

[1] Becker M, Eckmiller R. Spatio-temporal filter adjustment from evaluative feedback for a retina implant. In: Gerstner W et. al. (eds) Artificial Neural Networks - ICANN'97. Springer, 1997, pp 1181-1186 (Lecture notes in computer science no. 1327)

[2] Eckmiller R. Learning retina implants with epiretinal contacts. Ophthalmic Res. 1997; 29:281-289

[3] Eckmiller R, Hünermann R, Becker M. Dialog-based tuning of a Retina Encoder for Retina Implants. Inv. Ophthalmol. & Vis. Sci. 1998; (Suppl.) 39 (in press)

[4] Fleet D. J., Hallet P. E., Jepson A. D. Spatiotemporal Inseparability in Early Visual Processing. Biol. Cybern. 1985; 52:153-164

[5] Hünermann R., Becker M., Eckmiller R. Real Time Implementation of a Tunable Retina Encoder. Inv. Ophthalmol. & Vis. Sci.1997; (Suppl.) 38(4):S41

[6] Hünermann R., Eckmiller R. Verstellbare Spatiotemporale Rezeptive Feld Filter für Retina Encoder. In: Wiesböck J. (ed) DSP Deutschland '97. Design & Elektronik, München, 1997, pp 250-258

[7] Lee B. B., Pokorny J., Smith V. C., Kremers J. Responses to Pulses and Sinusoids in Macaque Ganglion Cells. Vision Research 1994, 34(23):3081-3096

[8] Victor J. D. The dynamics of the cat retinal X cell centre. J. Physiol. 1987; 386:219-246

# Rate and Temporal Coding with a Neural Oscillator*

Holger Bosch†, Ruggero Milanese†,
Abderrahim Labbi† and Jacques Demongeot‡

†Department of Computer Science, University of Geneva, Switzerland
{bosch,milanese,labbi}@cui.unige.ch
‡TIMC-IMAG, Faculté de Médecine, University of Grenoble, France

### Abstract

A neural oscillator capable of processing graded inputs is studied. The oscillator has two functional modes controlled by an external signal and codes information either by the amplitude of its oscillations or by the coordinates of its fixed point. Excitatory and inhibitory connections between coupled oscillators control their phase relations. Simulations and theoretical analyses show that any desired phase relation can be induced by an appropriate choice of connections. The capabilities of the oscillator model are demonstrated in an architecture for gray-level image segmentation.

## 1 Introduction

Computational models of information processing in the brain can be divided in two major classes, focusing respectively on the temporal pattern and on the average rate of each neuron's activity. One advantage of temporal coding is the possibility of dynamically encoding relationships between neurons, for instance by activity synchronization. This could provide a solution to the binding and the superposition catastrophy problems, as proposed in [7]. Experimental evidence for synchronized oscillations was provided by several researchers [5, 3]. Oscillatory activities in the visual cortex appear to be correlated if the corresponding units encode different parts or aspects of the same entity.

Several neuronal models have been proposed as a theoretical basis compatible with the experimental findings of oscillatory activity. Most of the proposed oscillators present limited capabilities to code sensory input, often restricted to binary coding and no graded input can be represented by a single oscillator, in contrast to the more classical rate coding scheme.

---

*The financial support of the Swiss National Science Foundation is gratefully acknowledged (grant no. 2100-045699.95).

Nevertheless, a few exceptions exist, among them the oscillator models proposed by Wang et al. [8] and by Niebur et al. [6] which are able to process graded inputs. The latter model is also able to switch between two different functional modes, i.e. fixed-point and oscillation mode.

In this paper, we propose a model based on the same principle, which will enable the use of both the classical rate coding scheme and the temporal coding scheme as well as the processing of graded inputs. Compared to the model by Niebur et al. [6], we rely on a more detailed and biologically plausible model, which is an extension of an oscillator model representing two interconnected populations of excitatory and inhibitory neurons originally derived by Atiya and Baldi [1]. Another property of our model is that it maximally simplifies the *read out problem*, i.e. the need to interpret the network state by an external observer. The network's output should directly be readable by an object recognition network without the need for complex, costly and time-consuming correlation analyses to distinguish between multiple objects.

The rest of this paper is organized as follows. The oscillator is introduced in Chapter 2, whereas Chapter 3 analyzes the phase relationships between a pair of coupled oscillators. Chapter 4 shows an application to object segmentation. A discussion concludes the paper.

## 2 The Two Neuron Oscillator

The model we propose was initially introduced and studied by Atiya and Baldi [1] without considering external inputs or networks of connected oscillators.

The basic model is a dynamical system of two variables $x$ and $y$:

$$\frac{dx}{dt} = -\frac{x}{\tau} + \tanh(\lambda x) - \tanh(\lambda y) - x_{in} \qquad (1a)$$

$$\frac{dy}{dt} = -\frac{y}{\tau} + \tanh(\lambda x) + \tanh(\lambda y) - y_{in} \qquad (1b)$$

Without external inputs, i.e. $x_{in} = y_{in} = 0$, the system oscillates under the condition $\lambda\tau > 1$ [1].

It can be shown that the vector field of the oscillator with no external inputs is invariant under rotations of $kT/4$ for $k \in N$: rotating the initial conditions by $kT/4$ is equivalent to rotating the trajectory of the original initial conditions by $kT/4$. For small input values $x_{in}$ and $y_{in}$, the system still oscillates, whereas it remains in a fixed point for large input values.

In order to achieve a flexible coding scheme, the inputs $x_{in}, y_{in}$ code two different types of signals. The value of $y_{in}$ represents the sensory input which modulates the amplitude of the oscillation and the coordinates of the fixed point. The value of $x_{in}$ represents a control signal which is meant to switch between oscillating and fixed point coding (Figure 1). The output of the oscillator is represented by the $x$ variable. Whereas the amplitude increases in function of the external input, the period remains almost constant, allowing stable phase relations between oscillators with varying inputs.

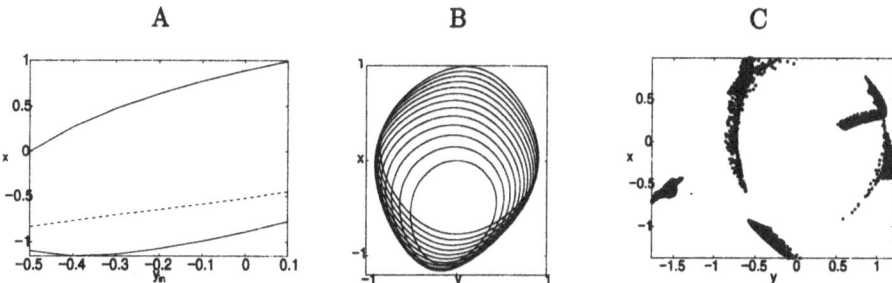

Figure 1: (A) Basic characteristics of the oscillator as a function of the external input. The dashed line shows the value of the $x$ unit in fixed-point mode whereas the others refer to units in oscillation mode. The dotted line represents the mean value and the two solid lines represent respectively the maximum and minimum value within one period. Both the fixed-point coordinate and the maximum value of the oscillation increase monotonically as a function of $y_{in}$. (B) Limit cycles of the model for different input values between $y_{in} = -0.5$ (smallest $x$ amplitude) and $y_{in} = 0.1$ (largest $x$ amplitude). (C) Four groups of oscillators in oscillation mode (on the right) and one group in fixed point mode (on the left).

## 3 Coupled Oscillators

In this section, we study a system of two coupled oscillators and investigate the influence of one and multiple connections on their phase lag. In general, a system of coupled oscillators is defined by

$$\frac{dx_i}{dt} = -\frac{x_i}{\tau_i} + f_{\lambda_i}(x_i) - f_{\lambda_i}(y_i) - x_{in} + \sum_{j \neq i}(c_{x_i x_j} f_{\lambda_i}(x_j) + c_{x_i y_j} f_{\lambda_i}(y_j))$$

$$\frac{dy_i}{dt} = -\frac{y_i}{\tau_i} + f_{\lambda_i}(x_i) + f_{\lambda_i}(y_i) - y_{in} + \sum_{j \neq i}(c_{y_i x_j} f_{\lambda_i}(x_j) + c_{y_i y_j} f_{\lambda_i}(y_j))$$

where $c_{z_i z_j}$ is the coupling strength from unit $z_j$ of oscillator $j$ to unit $z_i$ of oscillator $i$, with $z_i, z_j \in \{x, y\}$.

For oscillators without external inputs which are coupled with a simple unidirectional excitatory connection $x_1 \longrightarrow^+ x_2$ between the two excitatory units $x_1$ and $x_2$, the phase shift $p$ converges to a unique value $d \approx T/8$ independent of the initial conditions and the connection strength where $T$ is the period of the oscillators. Modifying the connection strength $c_{x_2 x_1}$ only influences the time needed to converge to $d$, which is reciprocal to the connection strength.

If the connection $x_1 \longrightarrow^+ x_2$ is replaced by another connection $z_1 \longrightarrow^\pm z_2$ or $z_2 \longrightarrow^\pm z_1$ (with $z \in \{x, y\}$, $z_i \longrightarrow^- z_j$ standing for an inhibitory connection) between the two oscillators, the induced phase difference can be analytically determined. Note that excitatory connections from inhibitory units and vice versa were also considered. These types of connections are not biological plausible but were added for completeness of the following

**Proposition:**
*Under the condition that the phase lag induced by $x_1 \longrightarrow^+ x_2$ converges to a constant value d, the phase lag $d^*$ induced by any other connection of type $z_1 \longrightarrow^\pm z_2$ or $z_2 \longrightarrow^\pm z_1$ with $z \in \{x,y\}$ exists and equals one of the values $d^* = \pm d + kT/8$ with $k \in \{1,3,5,7\}$ (cf. also Figure 2).*

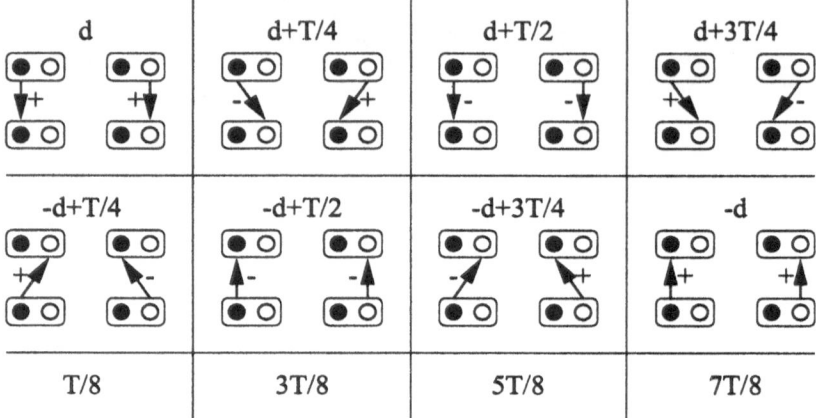

Figure 2: The induced phase differences for the different connections between the two oscillators. Filled (respectively, empty) circles indicate the $x$ (respectively, $y$) units. Since $d \approx T/8$ for a large parameter domain, all the phase differences equal in this case one of the four values $T/8$, $3T/8$, $5T/8$ and $7T/8$.

The proof relies on the symmetry of the vector field (see above). If two oscillators are coupled through more than one connection, simulations show that the resulting phase lag can be predicted by the following:

**Rule (Condensed Mean Value):** *Two equally weighted connections cancel each other out if their associated phase lags $d_{1,2}$ are opposite, i.e. $d_1 = d_2 + T/2 \bmod T$. If the weights $w_{1,2}$ are different, the weaker one is eliminated and the stronger one is adjusted to $w' = |w_1 - w_2|$. The phase lag between the two oscillators equals the mean of the phase lags of the remaining connections weighted by their adjusted weights.*

As a consequence, any desired phase lag between the oscillators can be realized by choosing a suitable connection scheme. If all connections cancel each other out, then the phase lag will not converge to a unique value but depends on the initial conditions of the two oscillators. In a similar context, Ermentrout et al. [4] approximate analytically the phase lag induced by multiple weak connections for any arbitrary oscillators. Using mean-field theory, they predict that the resulting phase lag is the mean value of the phase lags of each individual connection, weighted by the coupling strength.

For coupled oscillators with external inputs, the above results were verified with simulations showing that the phase lags depend only slightly on the inputs.

# 4 Application to object segmentation

In this section we demonstrate the capabilities of the proposed oscillator model by applying it to the problem of visual object segmentation. Temporal coding is only employed by units processing relevant, salient objects, whereas the other units adopt rate coding. This approach has the advantage that multiple salient objects can easily be segregated from each other in time through phase lags. Objects located in such an attention spot can be isolated by thresholding the oscillators activities (cf. Section 2).

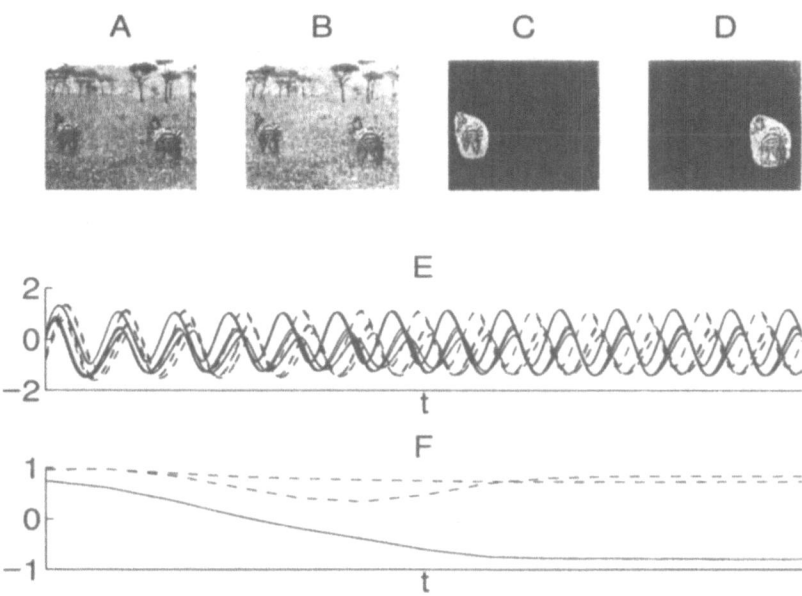

Figure 3: (A) Input image. (B) Oscillators in fixed-point mode. (C-D) Snapshots of the output map. (E) Output activities of units belonging to the right object (dashed lines) and left object (solid lines) as a function of time. Significant phase lags occur after $\approx 5T$. (F) Mean cross-correlation between oscillators belonging to the same object (dashed lines) and different objects (solid line).

The system consists of three layers: the oscillator map, the attention map and the output map. The oscillator map contains an array of oscillators which receive sensory input ($y_{in}$) from the input image and a control signal ($x_{in}$) from the attention map. The oscillators are mutually connected such that nearby units synchronize whereas remote units desynchronize. Suitable connection schemes can be determined through the results on coupled oscillators (cf. Section 3). A transfer function suppresses input from connections whose presynaptic activity is below a certain threshold to restrict mutual information exchange to units in oscillation mode. The attention map is a binary map representing regions of interest in the input image. A possible mechanism to extract such a map is described in [2]. Since its detailed description is beyond the scope of this article, it shall henceforth be regarded as provided by a mechanism external to the system. For the results presented below, the attention regions were hand-defined prior to the simulations. The output map is the $x$ variables of the oscillator map filtered by a threshold transfer function.

Simulation results are shown for an input image (Figure 3 A) containing two objects on a non-uniform background. Gray-level coding is absolutely necessary to recognize the objects, since the form alone would not provide enough information to identify the animal as a zebra.

## 5  Discussion

We presented a neural oscillator model able to process graded inputs, which we believe increases strongly the plausibility of the model as well as its suitability for real-world applications. The possibility to switch between oscillation and fixed point coding unifies the advantages of the classical rate coding scheme and the temporal coding scheme. The former is well suited to allow instant performance of basic operations in parallel, whereas the latter introduces the possibility to encode relations between units through phase lags.

## References

[1] A. Atiya and P. Baldi. Oscillations and synchronizations in neural networks: an exploration of the labeling hypothesis. *International Journal of Neural Systems*, 1(2):103–124, 1989.

[2] H. Bosch, R. Milanese, and A. Labbi. Object segmentation by attention-induced oscillations. In *IEEE International Joint Conference on Neural Networks*, Alaska, May 1998.

[3] R. Eckhorn, H. J. Reitboeck, M. Arndt, and P. Dicke. Feature linking via synchronization among distributed assemblies: Simulations of results from cat visual cortex. *Neural Computation*, 2(3):293–307, 1990.

[4] B. Ermentrout and N. Kopell. Learning of phase lags in coupled neural oscillators. *Neural Computation*, 6:225–241, 1994.

[5] C. Gray, P. Koenig, A. Engel, and W. Singer. Oscillatory responses in cat visual cortex exhibit inter-columnar synchronization which reflects global stimulus properties. *Nature*, 338:334–337, 1989.

[6] E. Niebur, C. Koch, and C. Rosin. An oscillation-based model for the neuronal basis of attention. *Vision Research*, 33(18):2789–2802, 1993.

[7] C. von der Malsburg and W. Schneider. A neural cocktail-party processor. *Biological Cybernetics*, 54:29–40, 1986.

[8] D. Wang and D. Terman. Image segmentation based on oscillatory correlation. *Neural Computation*, 9:805–836, 1997.

# Phase Transitions in Even Cyclic Inhibitory Networks

Valery D.Tsukerman

A.B.Kogan Research Institute for Neurocybernetics, Rostov State University,
Rostov-on-Don , Russia
e-mail: vdtc@krinc.rnd.runnet.ru

### Abstract

Foundations of a nonlinear dynamic theory of even cyclic inhibitory (ECI) networks are presented in this paper. Such networks are marked by a high degree of symmetry; however, their behavior may still be very complex, with quasiperiodic and chaotic dynamics due to the nonlinearity of the oscillatory units. Clearly nonlinear behavior generated by ECI-networks provides the system with particular features among which are the dynamic change of the effective degrees of freedom and controllable switches among attractors.

Two phase transitions were discovered in the hierarchical structures of ECI-networks. The first one was observed at low levels of the non-specific neuromodulatory inputs. In this case continuous activity broke down and large-size oscillatory clusters were formed. The second phase transition occurred at high potentials of the non-specific network control. Continuous oscillatory activity breaks down into separate pulses in this case.

## 1 Introduction

Oscillatory activity of the nervous system is a necessary component of various vegetative and locomotory functions in the organism [1,2]. It takes part in the implementation of other higher nervous functions, for example, speech production [3] , motion perception [4], etc. At last years we presented model of locomotor pattern activity in neural networks with cyclic inhibition [5,6]. Now a nonlinear dynamical theory of even cyclic inhibitory networks are presented in this paper. Our even cyclic inhibition theory contains the global feedback mechanism notwithstanding the local interactions of oscillatory units. Large structures of ECI-networks similar to spread two-dimension nets or crystal lattices are the convenient model for simulation of mechanisms of posture control, sensory-motor coordination, perception, etc. This paper presents our results of study of rhythmogenesis self-organization processes in ring structures.

The ECI-networks used are classified as follows:
(1) Homogeneous structures consisting of uniform oscillatory units.
(2) Structures with a pacemaker unit or leading semiring.
(3) Heterogeneous ring-like oscillatory structures.
Such functional classification is related to the difference in oscillators' generation rates which is of principal importance to the collective rhythmogenesis. The latter two structures are hierarchical ones with different degrees of oscillator domination.

## 2    A mathematical model of the ECI-network

Two levels of even cyclic inhibition were used in our simulation. At the low level, a strong ECI was applied inside a functional unit, the neuron module (Fig.1). Reciprocal inhibition of the conditional oscillatory neuron and the adaptive analog neuron of this module, due to the application of asymmetric parameters and a constant, but different excitatory synaptic drive, demonstrated a capability to switch between continuous activity and the burst one under the impact of the external inputs. This relaxation oscillator is a bistable unit, and a relatively more depolarized state of membrane potential corresponding to continuous oscillatory activity coexists with an oscillatory mode corresponding to spindle-like activity.

The dynamics of the relaxation oscillator may be described by the following system (1) of differential equations:

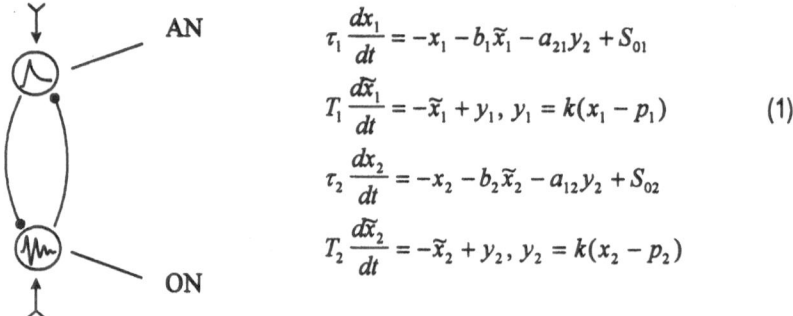

$$\tau_1 \frac{dx_1}{dt} = -x_1 - b_1 \tilde{x}_1 - a_{21} y_2 + S_{01}$$

$$T_1 \frac{d\tilde{x}_1}{dt} = -\tilde{x}_1 + y_1, \ y_1 = k(x_1 - p_1) \qquad (1)$$

$$\tau_2 \frac{dx_2}{dt} = -x_2 - b_2 \tilde{x}_2 - a_{12} y_2 + S_{02}$$

$$T_2 \frac{d\tilde{x}_2}{dt} = -\tilde{x}_2 + y_2, \ y_2 = k(x_2 - p_2)$$

Figure 1: Reciprocal inhibition of the AN-neuron and ON-neuron constituting the relaxation oscillator.

Here $x_1$ , $x_2$ are the membrane potentials of the AN-neuron and the ON-neuron, respectively; $x_1$, $x_2$ are the degrees of adaptation to the constant level input; $\tau_1, \tau_2$, $T_1$, $T_2$ are time constants; $b_1$, $b_2$ are relaxation parameters; $a_{12}$, $a_{21}$ are inhibitory connection weights; $p_1, p_2$ are the thresholds of the neurons; $S_{01}$, $S_{02}$ are the external potential neuron's inputs; $y_1$, $y_2$ are the output neuron's values, defined as:   $y_{1,2} = (x_{1,2} - p_{1,2})$, at $x_{1,2} > p_{1,2}$;   $y_{1,2} = 0$, at $x_{1,2} \leq p_{1,2}$. Hereafter the parameter $k$ is equal to 1.

Computational studies of the module indicate that two distinct modes of activity may coexist under a given set of parameters and this bistability can be modulated by the potential of the external neuron's inputs:

$$\tau_1 = 0,01; \quad 4 \leq T_1 \leq 6; \quad b_1 = 4; \qquad 1,9 \leq a_{12} = a_{21} \leq 2,3;$$

$$\tau_2 = 0,5; \quad 0,4 \leq T_2 \leq 0,8; \quad 20 < b_2 \leq 42; \quad 0,0742 \leq S_{01} \leq 2,167; \quad S_{02} = 1.$$

At the high level, a weak ECI was used among the neural modules in a rhythmogenic structure as in an example in Fig.2. The inhibitory connection weights $a_{ij}$ among modules were constant and did not exceed 0.3. Numerical simulation of the system (2) was carried out in accordance with the network classification presented above. In this case the main network tuning parameters were the relaxation ones $b_i$ of the ON-neurons and/or the potentials of the neuromodulatory inputs of the AN-neurons.

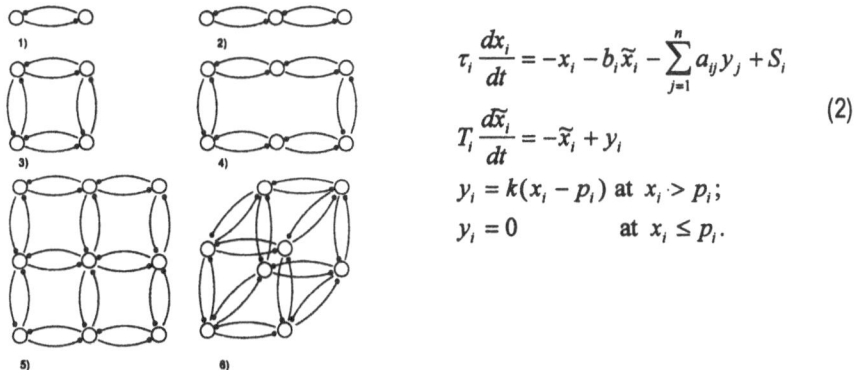

$$\tau_i \frac{dx_i}{dt} = -x_i - b_i \tilde{x}_i - \sum_{j=1}^{n} a_{ij} y_j + S_i$$

$$T_i \frac{d\tilde{x}_i}{dt} = -\tilde{x}_i + y_i$$

$$y_i = k(x_i - p_i) \text{ at } x_i > p_i;$$

$$y_i = 0 \qquad \text{at } x_i \leq p_i.$$

(2)

Figure 2: Examples 1) - 6) of ECI-networks and the corresponding system of differential equations

# 3 Nonlinear dynamical modes of the homogeneous structures

The numerical simulation of the homogeneous system of oscillators shows the presence of two generation modes: (1) In the upper part of the mentioned AN-inputs interval the network demonstrates continuous oscillatory activity of all units. The main phenomenon is the semi-periodic synchronization of the nearest neighbour oscillators and coherent rhythms of the far, viz., separated by another one, oscillators. (2) In the lower part of this interval the network demonstrates spindle-like activity. There exist two possibilities for the implementation of multistable rhythmic patterns under a certain set of the network parameters. The first one is to apply very small increments of the potential inputs; in the second one, short-time and very-small-amplitude impact (of the order of $10^{-15} - 10^{-17}$) on some of the network neurons. An increase in the number of the oscillatory units in the network induces new attractors and corresponding patterns.

# 4 Nonlinear dynamical modes of the heterogeneous structures

This series of numerical simulation experiments explored the following modes:
(1) Oscillatory rings with strong unilateral domination.
(2) Oscillatory rings with hierarchical structure.
    The difference in the values of the parameters given was as follows: in the first case one half of the ring differed markedly from the other one, while within each semiring the difference in the values of these parameters was two orders lower. In the second case all units of the ring were ranked by the relaxation values, significantly different from each other. Dynamical modes connected, as above, with some critical value of the neuromodulatory input were obtained in both cases. If the input is greater, the network oscillates continuously, with the semi-periodic synchronization of the unilateral oscillatory units and a constant phase shift among the units of different semirings. When the modulatory inputs are close to the critical value, increasing phase shifts in bilateral units were observed. The self-organization processes in this mode lead to weak phase tuning and synchronization of high-frequency

oscillations, which is then followed by the reiteration of the desynchronization process.

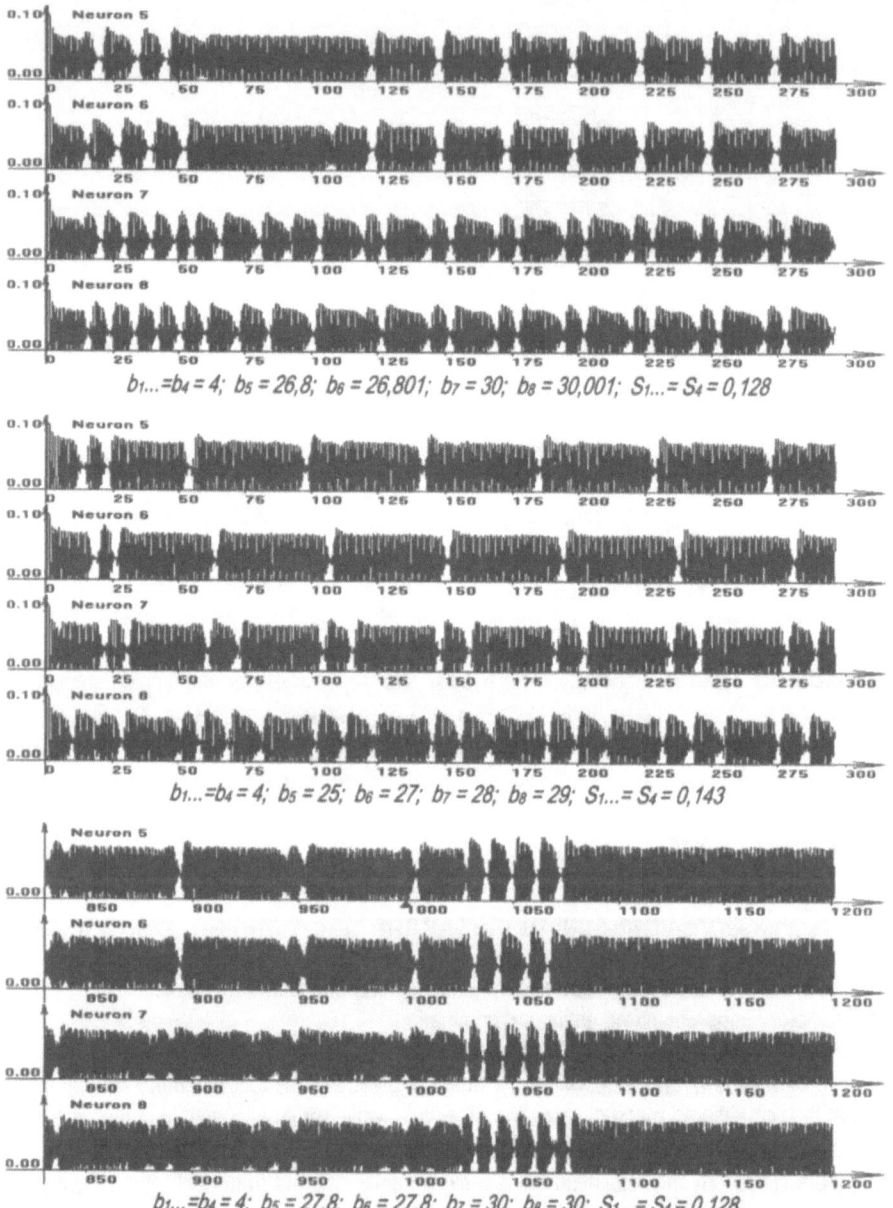

Figure 3: Two different cluster orderings (top) and the phase transition from cluster activity to continuous one under the impact of the single pulse (bottom) with an amplitude of $1*10^{-15}$ and duration of $1*10^{-3}$.

Below the critical value an oscillatory mode with rhythm amplitude reset and strong phase tuning appears. Large-size cluster formation, due to the split of continuous high-frequency oscillations and singularities of the system, is observed. A new rhythm is born in this mode, viz., a new macrotemporal order takes place. Further decrease of the potential of the control inputs leads to two consecutive phenomena: the simultaneous diminution of the same-size clusters and the split of clusters into smaller fragments. The stable rhythms are represented by clusters of several constant sizes. As a rule, rhythms represented by clusters of various sizes are unstable, thus contributing to transition from quasiperiodic to chaotic dynamics.

Macrotemporal order reflecting the self-organization processes in transition from multicluster chaos to synchronized activity may be obtained by a weak potential shift of the neuromodulatory inputs or by a single pulse. At even lower values of the neuromodulatory potential the network generates spindles.

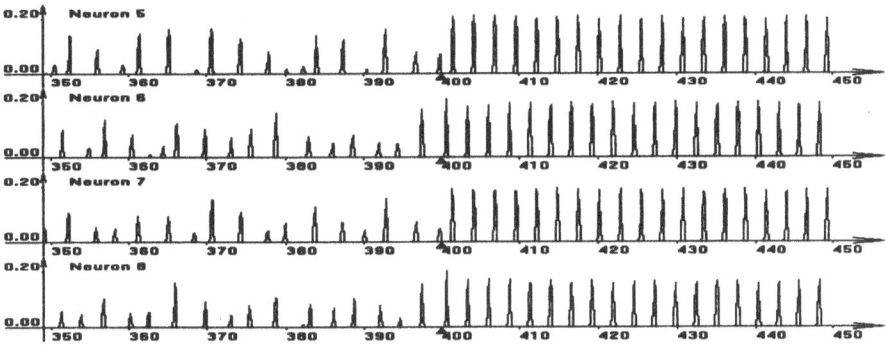

$$b_1...=b_4 = 4; \ b_5 = 26,8; \ b_6 = 26,801; \ b_7 = 30; \ b_8 = 30,001; \ S_1...= S_4 = 2,3$$

Figure 4. Chaotic pulse mode regularized by the impact of the single pulse with an amplitude of 0.16 at the time moment 400.

The behaviour of the hierarchical rings above the critical level is of special interest. As mentioned above, high-frequency oscillations are generated in this mode. A further increase of modulatory potential reveals another, greater, critical value when continuous oscillations split again, but this time into trains of short single spikes (positive half-waves). It should also be noted here that in certain intervals of the modulatory values such sequences either become chaotic, or assume a regularized, ordered pattern form. Therefore, given large-scale initial data, the behaviour of the system (2) for the hierarchical structures is similar to the low-potential mode with spindle-like activity.

Also, modes of so-called floating synchronization (intermittency modes) were observed in the hierarchical structures between ordered oscillatory modes and chaotic ones. Periodic rhythm capture and oscillatory units synchronization (with the subsequent desynchronization) occurred in this mode.

# 5 Discussion

Several dynamical modes take place in the hierarchical ring-like structures of ECI-networks. A conjecture may be made as to the relationship between those modes and two qualitatively important events, phase transitions. One of them is realized in the lower part of the non-specific control interval, where the breakdown of continuous activity leads to the formation of large clusters of rhythmic activity. The other phase transition is observed at a high level of the control and leads to the split of the continuous oscillations to the series (sequences) of single pulses.

Switching between regularized dynamics (order) and chaotic one is possible for all types of oscillatory activity, viz., spindles, clusters, and spice sequences. This is accomplished by parametrization (in our case, by the relaxation parameters $b_j$). These most important parameters of order [7], which in our model determine the extent of adaptation of the oscillatory neurons, simultaneously change the strength of the interoscillatory connections in the ECI-network. This allow synchronize the oscillator units activity in the listed regimes, i.e. switching from the chaotic to the regular dynamics type. Therefore, input signals, by changing the network's connections, are capable of switching dynamical modes by virtue of the internal self-organization processes.

Let us finally consider how these results may be construed. The neurobiological meaning of phase transitions may be related to some qualitative changes in the state of the organism, e.g., for the 'vigilance-sleep' transition, which is the specific case when spindles are observed. Another feasible example of qualitative differences in the state of sensorimotor cortex is the regulation of posture at rest and locomotion (run).

# References

[1] Tosini G., Menaker M. Multioscillatory circadian organization in a vertebrate, *Iguana iguana*. J. Neuroscience, 1998, 1, 18(3): 1105-1114.

[2] Selverston A.I. Modulation of circuits underlying rhythmic behaviors. J.Comp.Physiol. 1995, A 176 : 139-147.

[3] Saltsman E.L., Munhall K.G. A dynamical approach to gestural patterning in speech production. Ecological Psychology, 1989; 1: 333-382.

[4] Giese M.A, Schöner G., Hock H.S.. Neural field dynamics for motion perception. In: Proceedings of the ICANN 1996, Bochum, Germany, 1996.

[5] Tsukerman V.D., Makarova L.S. Local mechanisms underlying coordination of movements. Proceedings in nonlinear science. Neurocomputers and attention., vol 1,. Manchester Univ. Press. Manchester, New York. 1991, pp. 117 - 127.

[6] Tsukerman V.D. Coordination of rhythmic motor pattern activity in neural networks with cyclic inhibition. In: Proceedings of KRINC/LACOS Workshop on Robot Vision, Rostov-on-Don, 1996, pp.90-104.

[7] Haken H. Synergetics. An introduction. Springer-Verlag, Berlin, Heidelberg, New York, 1978.

# The Basal Ganglia viewed as an Action Selection Device

Kevin N. Gurney, Tony J. Prescott, and Peter Redgrave

Department of Psychology, University of Sheffield,
Western Bank, Sheffield S10 2TP, UK.
Email: k.gurney, t.j.prescott, p.redgrave·@sheffield.ac.uk

## Abstract

The *action selection* problem describes the task of resolving conflicts between the different functional systems that can control behavior. This paper reviews the role of the *basal ganglia* (BG) summarising evidence that they function within the vertebrate brain architecture as a specialized action selection device. There is a rich connectivity within the BG whose function is not well understood. We outline a new computational model of BG intrinsic pathways which demonstrates that these circuits could allow the BG to implement clean switching between competing functional systems.

## 1 The role of the basal ganglia in action selection

An important task for the vertebrate nervous system is the resolution of conflicts between functional units that are physically separated within the brain but are in competition for common resources. For instance, the neural systems involved in tasks such as feeding, drinking, and escape, are located at widely distributed sites and at multiple levels of the neuraxis, yet are in competition for the use of the same effector mechanisms. The task of resolving such conflicts has been the subject of much research in ethology and artificial intelligence (see [1,2]) where it is termed the *action selection* problem. We have argued in [2] that the requirement for effective action selection favors the evolution of centralised switching devices, and that in the vertebrate brain, the *basal ganglia,* a group of functionally-related, central brain structures, have evolved to fill this role. This paper briefly reviews neuroscientific evidence for the involvement of the BG in action selection and outlines a new computational model of BG intrinsic circuitry viewed as implementing a switching device.

The principal components of the primate basal ganglia are the *striatum,* the *globus pallidus (GP),* and the *subthalamic nucleus (STN)* in the forebrain, and the *substantia nigra (SN)* in the midbrain. The globus pallidus contains two separate areas which are termed the internal and external segments *(GPi and GPe).* Homologous structures (though often with different names) are found in the nervous systems of other vertebrate classes [3]. The BG are illustrated in a schematic drawing of a generalized mammalian brain in figure 1. The proposal that the BG performs action selection in the vertebrate brain is not a radical perspective on BG function but rather derives from a growing consensus that a key function of these structures is to enable desired actions and to inhibit undesired, potentially competing, actions (see [2,4] for review). This literature suggests the following view of the functional architecture of the BG. Activity relating to 'bids' for access to common resources (e.g. muscle groups) appears to be continuously projected to the input side of the BG from relevant functional sub-systems in the midbrain and forebrain of the animal. This activity may form the 'common currency' in which the relative salience of competing requests can be effectively compared. Internal circuitry within the BG then determines a 'winner' whose contact with the output mechanisms is specifically disinhibited. The following briefly summarizes some of the key findings in support of this proposal, further details and supporting evidence are described in [2].

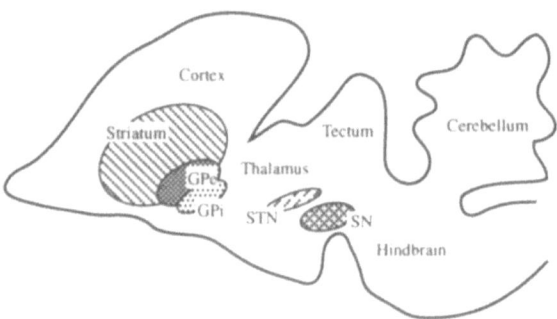

Figure 1: Schematic diagram of a saggital section through a generalized mammalian brain showing the principal BG structures, adapted from [3].

Anatomical evidence shows that cortical and midbrain sensorimotor systems, plus several of the forebrain limbic structures, communicate directly with motor and pre-motor mechanisms in the brainstem and spinal cord. However, these systems also project, usually via a collateral (split) pathway, to the striatum, the main input center of the BG (see figure 2), this branch could allow them to enter into a competition for control of motor outputs hosted within the BG. Afferents from a wide range of sensory and motivational systems also arrive at BG input neurons. These connections could allow both extrinsic and intrinsic factors to enter into a "vast machinery" of context-specific filters in the striatum [4], influencing the strength of rival bids, and hence the currently preferred course of action. The input connectivity of the BG therefore indicates that it is well placed to resolve the problem of selecting an appropriate action for a given circumstance.

The principal output structures of the BG are the SN and GPi. Neurons in both these structures are tonically active and inhibitory and project to all the different sensorimotor systems that are in contact with the striatum. This tonic inhibition acts as a brake on the target systems thereby denying them access motor circuitry. Signals emanating from the striatum inhibit the inhibitory BG output centers so *disinhibiting* selected systems (see [2, 4, 5] for review, and see figure 2 for an illustration of this double-inhibitory pathway). In the absence of such signals there can be no voluntary movement. The BG thus seems to hold a 'veto' over midbrain and forebrain systems that seek access to the motor outputs which is relinquished, for a selected action, through the mechanism of *disinhibition*.

Projection lines through the various sub-components of the basal ganglia appear to be largely organized into segregated parallel 'channels'. This segregation is maintained in the disinhibitory output projections. Behavioral studies indicate that although the architecture of these channels is similar throughout most of the BG, different areas are functionally heterogeneous. For instance, restricted lesions at different locations in the striatum effect different actions such as forelimb manipulation, biting and gait. This would suggest that the circuitry in these local areas in the striatum may primarily be used to resolve conflicts between competitors bidding for incompatible uses of specific groups of muscles. More generally, each local group of parallel circuits may be competing for a single output mechanism thereby forming a single, multi-way 'switch'. If this interpretation is correct then the BG may provide an array of similar switching devices.

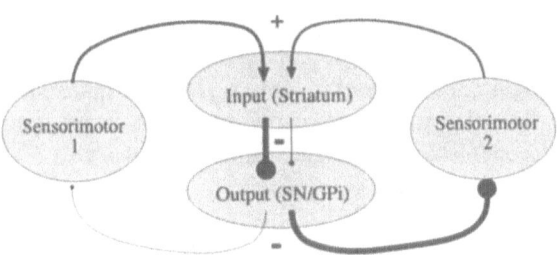

Figure 2: Functional diagram of the principal hypothesized selection mechanism. Sensorimotor systems project to the striatum, the main BG input structure. Intrinsic striatal circuitry resolves the selection competition in favor of the strongest competitors (here 1) and selectively inhibits neurons in SN/GPi switching off their tonic inhibitory control of winning sensorimotor systems whilst maintaining or increasing inhibition on losers (here 2). (Thicker lines indicate stronger excitatory or inhibitory links.)

In [2] we summarize evidence that various aspects of behavior selection and switching are effected by neurochemical or neurophysiological interventions in BG structures. The BG are also implicated in a number of human brain disorders including Parkinson's disease, Huntingdon's Disease, and Tourette's syndrome, whose symptoms may be interpretable as resulting from the failure, or inappropriate operation, of selection mechanisms. BG structures are important in instrumental conditioning and in various forms of sequential learning suggesting that they are appropriately designed for adaptive tuning of selection mechanisms. It has recently been suggested that the theory of temporal difference learning in actor-critic mechanisms could be used to understand the learning architectures embedded in the BG (see [6]). This suggests the prospect of a fruitful interaction between work on artificial reinforcement learning systems and the understanding of adaptive BG processes.

The BG have been implicated in a wide range of processes that includes aspects of motor control, perception, learning, and memory (see [2, 4, 6]). BG involvement in so many diverse functions suggests to us that it may play a similar function in multiple domains—that is, selecting between competitors that require access to some limited resource be it motor, cognitive, or memorial.

# 2 A new model of intrinsic basal ganglia function

A number of computational models of BG function, at both the cellular and circuit level, have been investigated (see [6]), however, there are as yet few models that capture the distinctive neurodynamics of BG circuits while mimicking their behavioral functions [5]. Our current research is directed at developing models of exactly this sort, and, as a first step, we have constructed a system-level simulation of the mechanisms operating within a single BG selection circuit.

There is a rich connectivity within the BG whose function is not clear. Our initial work has resulted in a simulation in which different functional components of the intrinsic BG circuit are modeled as leaky integrator units. Our investigations of the behavior and dynamical properties of this model are beginning to provided valuable insights into the possible role of each of the component pathways. A full quantitative description of the model and simulation results will be published elsewhere [7], here we briefly outline some of our principal findings.

As shown in figure 3, our model is composed of two functional subsystems, one which performs the selection process *per se* (the *selector* sub-system*)*, and another which adaptively controls the former (the *adaptive controller)*.

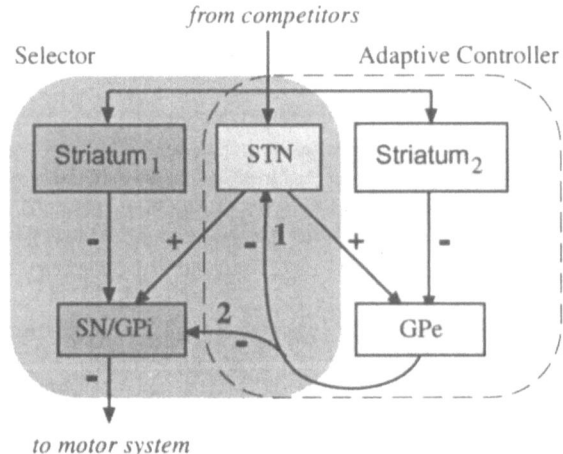

Figure 3: Functional model of the intrinsic circuitry of a single basal ganglia 'switch'. The gray area encloses the components of the *selector* subsystem, while the dotted line encloses the components of the *adaptive controller* subsystem (note the subthalamic nucleus belongs to both).

The selector sub-system resolves the competition for specific output mechanisms by providing off-centre, on-surround activation of SN/GPi (the BG output structures). Excitatory input to a population of neighbouring neurochemically defined striatal neurons (*striatum1* in the diagram) encodes the salience of competing bids. Activated striatal neurons directly inhibit neurons in SN/GPi thereby providing the off-centre effect. Through a second input pathway, the subthalamic nucleus (STN) provides diffuse excitation to SN/GPi. This pathway acts as the on-surround, ensuring the inhibition of losing and inactive competitors. Within the striatum the contrast between stronger and weaker competitors is further enhanced through recurrent reciprocal inhibition.

The *adaptive controller* sub-system is structurally similar to the *selector* but is based around a different neurochemically-defined cell population in the striatum (*striatum2*), and provides an indirect link to BG output structures via the globus pallidus external segment (*GPe*). Two likely functions of this subsystem are, via pathway *1* in figure 3, to make the selection circuit robust to variation in the number of competitors and their relative levels of support; and, via pathway 2, to act as a gain control on the output signal strength of competitors. A further function of the adaptive controller could be to enhance the high frequency response of the system thereby allowing faster switching.

Simulation results illustrating the operation of the switching mechanism and the effect of 'lesioning' pathway 1 are illustrated in figure 4. The top row of graphs (1-3) illustrates the normal operation of the BG switching mechanism (figure 3) for three model competitors with different 'salience' strengths and onset times. In each graph the solid line indicates the activity of the excitatory input to the striatum and STN (the salience signal), and the dotted line the activity of the inhibitory BG output from SN/GPi to the motor system. Competitors with output close to zero are selected, those with high output are suppressed. Competitor 1 (salience=0.6, onset=1) is activated first and is selected, 3 (salience=0.5, onset=2) fires next but has lower salience than 1 and is therefore not selected, 2 (salience=0.8, onset=3) is activated last but has the highest salience and is therefore selected while 1 and 3 are suppressed. The BG output signals indicate reasonably clean switching between competitors, in other words, the selection competition is resolved rapidly and decisively in favor of the strongest competitor. This simulation run is repeated in the bottom row of graphs (1*–3*) with the inhibitory GPe–STN link removed (pathway 1 in figure 3). Clean switching is compromised (winners are not effectively disinhibited) as the 'lesioned' network is inappropriately sensitive to the number of active competitors and their relative levels of support (see, for instance, the increase in inhibitory output to competitor 1 at t=2 when 3 becomes active).

Aside from input driven effects, the influence of the adaptive control pathway can be modified in the biological setting by changing levels of the neuromodulator dopamine. Dopamine enhances the response of striatal cells projecting directly to BG output structures (*striatum1*) while suppressing those that project indirectly via GPe (*striatum2*), it therefore appears to alter the balance between the different BG intrinsic pathways. This effect has been incorporated into the model and it appears that increased dopamine has the potential to make the current selection more vulnerable to alternative competitors; in effect dopamine is capable of dynamically modulating the sensitivity of the switch. Parkinson's disease is associated with abnormally low levels of striatal dopamine, simulating this deficit within a BG model such as our own could therefore improve our understanding this disorder.

The simulation has been the subject of mathematical analyses which make explicit its functional and parametric dependencies, and will facilitate comparisons with other models. The analytic approaches that we are currently using rely on approximating the nonlinear output characteristics of each subsystem in a piecewise-linear scheme. In the spirit of our system-level investigation, our focus is not on the details of neural non-linearities but rather, in describing enough of the system's gross properties to capture its main consequences. This should allow an examination of the dynamics under small signal changes which, although of interest in its own right has a further significance within our general framework. Specifically, the transient characteristics of the signals generated by the model (e.g. the output traces in figure 4) can be thought of as a 'fingerprint' for the underlying architecture. Comparing this fingerprint with that indicated by the neurophysiological data should provide a further means for evaluating and refining our model.

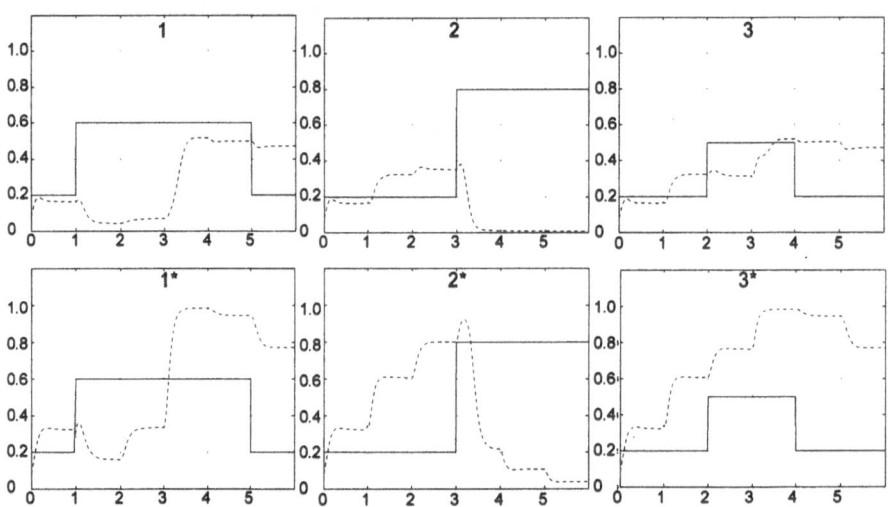

Figure 4: Simulation results. The solid line indicates excitatory input to the BG (salience) and the dotted line inhibitory BG output to the motor system. Neural response is constrained to lie between 0 and 1 on the y-axis. The time-scale on the x-axis is notional.

# References

1.      Maes, P., *Modelling adaptive autonomous agents*, in *Artificial Life: An Overview*, C.G. Langton, Editor. 1995, MIT Press: Cambridge, MA. p. 135-162.

2.      Prescott, T.J., P. Redgrave, and Gurney, K.N., *Layered control architectures in robots and vertebrates*. Adaptive Behavior, In Press.

3.      Medina, L. and A. Reiner, *Neurotransmitter organization and connectivity of the Basal Ganglia in vetebrates: implications for the evolution of the basal ganglia*. Brain Behavior and Evolution, 1995. **46**(4-5): p. 235-258.

4.      Mink, J.W., *The basal ganglia: focused selection and inhibition of competing motor programs*. Progress In Neurobiology, 1996. **50**(4): p. 381-425.

5.      Alexander, G.E., *Basal ganglia*, in *The Handbook of Brain Theory and Neural Networks*, M.A. Arbib, Editor. 1995, MIT Press: Cambridge, MA.

6.      Houk, J.C., J.L. Davis, and D.G. Beiser, *Models of Information Processing in the Basal Ganglia*. 1995, Cambridge, MA: MIT Press.

7.      Gurney, K.N., T.J. Prescott, and P. Redgrave, *A model of intrinsic processing in the basal ganglia*. In Preparation.

# Parameters Estimation in the Diffusion Model for Multidimensional Neural Data

Olivier François, Maryse Béguin, Lamine Mohamed Abdallahi

LMC/IMAG, BP 53 38041 Grenoble cedex 9

France

## Abstract

The aim of this study is to propose a new model to analyze multisite spatiotemporal neuronal data, and to provide a new method to estimate the parameters of the model from real data. This model is based on the principle of dynamical diffusion. The parameters of the model can be understood in terms of frequencies. These parameters can be estimated from experimental data thanks to a mean field approximation of the model. The results are compared to standard statistics and Markov random field and give further insights into the understanding of the data.

## 1 Introduction

The analysis of neural network activity is a crucial step for understanding the brain's behavior. In practice, optical recordings or multielectrode recordings are made on some specific areas of an animal cortex, thus providing a huge amount of spatiotemporal data. The interpretation of these data may be difficult and many paradigms and models have been proposed to get useful insights into the structure of these data. In any case, these models rely on some parameters for which the problems are twofold. On one hand the parameters should be easy to interpret. On the other hand one should be able to provide some statistical procedures to estimate these parameters from the real data. The models previously used [1, 2] are inspired from the statistical physics formalism and the interpretation of their parameters in the biological context might be unadapted. In any case, the estimation of the parameters may present some difficulties. This study introduces a new model to proceed with multidimensional neural signals and describes a new statistical method to compute the parameters of this model from real data. The paper is organized as follows. In section 2 the *diffusion model* (DM), is described. The theoretical important steps to deal with DM are given and the statistical procedures to estimate the parameters are detailed. A link and a comparison with other methods studied in the literature is given. The method has been experimented on real data gathered from the auditory cortex of a guinea-pig for which spatiotemporal patterns have been observed [3]. The results are presented in the section 3 and are compared to the results obtained with classical statistics.

## 2  Method

### 2.1  Description

In this section the diffusion model (DM) is described and the method used for the statistical estimation of the parameters is given. It is assumed that binary data have been obtained from optical diodes measurements made with a grid. This grid can be represented as a lattice endowed with a graph structure. The vertices of the graph, denoted by $S$, are called the sites and the edges of the grid may be seen as the possible connection between the sites. Two connected sites are said to be neighbors. For all site $i$, the subset of its neighbors is denoted by $N_i$. It is assumed that the cardinality of $N_i$ and $S$ are constants equal to $m$ and $n$ respectively. The data are supposed to translate the overall activity of the sites. The value 1 is coding for an active state whereas the value 0 is coding for a passive state. Details about the way by which the data are obtained will be given in section 3. Formally the data consist of a collection of $n$-dimensional binary vectors $x = (x_1, \ldots, x_n)$ translating the activities of the sites at each instant. The data are supposed to be the observation of a spatiotemporal stochastic process. The space state of this process is $\{0,1\}^n$. It is assumed that in a very small interval of time $dt$, only one site can change its value. Then, a vector $x = (x_1, \ldots, x_i, \ldots, x_n)$ is changed to the vector $x^i$ defined by $x^i = (x_1, \ldots, 1 - x_i, \ldots, x_n)$. The evolution of the activity of the whole network is supposed to be Markovian. The process studied is defined by

$$P(X(t + dt) = x^i \,|\, X(t) = x) = \lambda(x, x^i)dt + o(dt) \quad , \tag{1}$$

where $X(t)$ denotes the observed vector at instant $t$. The transition rates are given by

$$\lambda(x, x^i) = \begin{cases} \delta & \text{if } x_i = 1 \\ \lambda + \frac{\mu}{m} \sum_{j \in N_i} x_j & \text{if } x_i = 0 \end{cases} \tag{2}$$

These rates are supposed to be nonnegative. In this model, the global activity observed on the sites is separated in two components. The first component integrates the activity due to an external signal, and the second one integrates the internal activity due to the interactions between sites. The parameter $\lambda$ can be thought as coding for the rate of arrival of the external signal. The parameter $\mu$ may be interpreted as a diffusion rate and codes for the internal part. The last parameter $1/\delta$ can be interpreted as the mean duration of the activity on each site. This parameter can be easily computed from the real data. Provided they are nonnegative the parameters have the dimension of a frequency and can be measured from the real data in inverse milliseconds.

This paper focuses on the estimation of these parameters from binary real data. To achieve this goal two kinds of approximation are needed. The first one is a mean field approximation. It enables to obtain a statistic for the parameter $\mu$, provided that the other two are known

$$\mu \approx \frac{(\delta + \lambda)}{\hat{u}(1 - \hat{u})} \left( \hat{u} - \frac{\lambda}{\lambda + \delta} \right) . \tag{3}$$

where $\hat{u} = \frac{1}{n}\sum_{i=1}^{n} x_i$. Equation (3) will be justified in section 2.2. A second approximation is needed to estimate $\lambda$. This approximation fits the diffusion model with a Markov random field (MRF) model. The likelihood function of the MRF is defined as

$$L(\alpha, \beta, x) = \exp(H(x))/Z(\alpha, \beta), \quad x \in \{0,1\}^S \tag{4}$$

where $H(x) = \sum_{i \in S} \alpha x_i + \frac{1}{2}\sum_{i \in S}\sum_{j \in N_i} \beta x_i x_j$, $Z(\alpha, \beta)$ is the normalization constant, and $\alpha, \beta$ the parameters of the model. The parameters are estimated according to the pseudo likelihood method. Accordingly, an approximation of $\lambda$ can be deduced,

$$\lambda \approx \frac{\delta\hat{u}}{n} \sum_{i \in S} \frac{e^{\alpha+\beta\sum_{j\in N_i} x_j} - \frac{\sum_{j\in N_i} x_j}{m(1-\hat{u})}}{\hat{u} - \frac{1}{m}\sum_{j\in N_i} x_j}. \tag{5}$$

The point of view of the DM is dynamical and is close to the approach by point processes [4]. Other models proposed to deal with binary data are based on Markov random fields, and among them the most used in the last years is the Ising model [1, 2] defined by equation (4). Traditionally, $\alpha$ is called the singleton potential and it codes for the external signal which arrives on each site, whereas $\beta$ is called the pair potential and codes for the interaction between sites. The advantage of using MRF model is that robust estimations of the parameters have been proposed and tested. MRF models are very useful and allows to understand the interactions between sites. However the drawback of MRF models is that the parameters have no dimension. On the opposite the parameters of the diffusion model have the dimension of a frequency and may be fruitfully used for interpretation in the context of multidimensional data.

## 2.2 Mean field analysis

This section details the mean field approximation and justifies equation (3). The mean field approximation presumes that all the sites are connected. The fully connected DM can be solved using the aggregation method. Let $N(t)$ be the process defined by $N(t) = \sum_{j \in S} X_j(t)$, where $X_j(t)$ denotes the activity of the site $j$ at instant $t$. The process $N(t)$ is a birth and death process with values in $\{0, \ldots, n\}$. The birth rate (from $k$ to $k+1$ for $k < n$) is equal to $(n-k)(\lambda + \frac{\mu k}{n})$ and the death rate (from $k$ to $k-1$ for $k > 0$) is equal $k\delta$. Let $(p_j)_{0 \le j \le n}$ denotes the stationary distribution of $N(t)$, and let $u_n$ denotes the probability for each site to be active in the stationary regime. According to a standard argument, the following equation is verified:

$$p_k = \frac{\lambda}{\delta^k}\binom{n}{k}(\lambda + \frac{\mu}{n})\ldots(\lambda + \frac{\mu(k-1)}{n})p_0. \tag{6}$$

We study the behavior of this quantity as $n$ goes to infinity. Let $Z_n$ be equal to $1/p_0$. Then $Z_n$ appears as a sum of exponential quantities and can be written as follows:

$$Z_n = 1 + \sum_{k=1}^{n} \exp\left[ n \left( \frac{1}{n}\log\binom{n}{k} + \frac{1}{n}\sum_{i=0}^{k-1}\log(\lambda + \frac{\mu i}{n}) - \frac{k}{n}\log\delta \right) \right] . \quad (7)$$

For large $n$, $Z_n$ is equivalent to the dominant term of the exponentials, thus leading to

$$\frac{1}{n}\log Z_n \approx \max_{k \in \{1...n\}} \left\{ \frac{1}{n}\log\binom{n}{k} + \frac{1}{n}\sum_{i=0}^{k-1}\log(\lambda + \frac{\mu i}{n}) - \frac{k}{n}\log\delta \right\} . \quad (8)$$

Using Stirling's formula one gets (for large $n$ and $k$ fixed), $\frac{1}{n}\log\binom{n}{k} = -I(\frac{k}{n})(1 + \epsilon(\frac{k}{n}))$, where $I(u) = -u\log u - (1-u)\log(1-u)$, and $\lim_{x\to 0}\epsilon(x) = 0$. As well, using Riemann's integration, one gets

$$\frac{1}{n}\sum_{i=0}^{k-1}\log(\lambda + \frac{\mu i}{n}) = \left( \int_0^{\frac{k}{n}}\log(\lambda + \mu x)dx \right)\left( 1 + \epsilon'(\frac{k}{n}) \right) , \quad (9)$$

with $\lim_{x\to 0}\epsilon'(x) = 0$. Then,

$$\lim_{n\to +\infty}\frac{1}{n}\log Z_n = \max_{u \in [0..1]} \left\{ I(u) - u\log(\delta) + \int_0^u \log(\lambda + \mu x)dx \right\} . \quad (10)$$

The maximum is obtained for the value $u* = \lim_{n\to +\infty} u_n$ in $[0,1]$ solution of the following quadratic equation,

$$\lambda + (\mu - \lambda - \delta)u - \mu u^2 = 0 , \quad (11)$$

which is equivalent to equation (3).

## 3  Data Analysis

The method described above has been tested on real data gathered from the auditory cortex of a guinea-pig (layer II and III) [3]. Optical recordings using a $12 \times 12$ photodiode array were monitored after a contralateral ear stimulation of 7 KHz. The experimental protocol is fully described in [3]. For each site, two variables have been extracted from the recorded continuous signals. The first variable gives the instant at which the activity begins at the site. This variable ranges between 20-35 milliseconds. The second variable is the duration of this activity (at the same site). The mean duration (averaged over all sites) is equal to $1/\delta = 4.63$ milliseconds. From these variables a collection of 40 binary vectors representing the activity of the sites each half millisecond has been constructed. The size of each vector is equal to $n = 144$. These vectors correspond to the history of the process in the period 20-40 milliseconds. Each half millisecond, different statistics have been computed from the binary data. These statistics are the mean activity: $\hat{u} = \sum_{i=1}^{n} x_i/n$, the covariance between neighboring sites: $\hat{c} = \frac{1}{nm} \sum_{(i,j); j \in N_i} x_i x_j - \hat{u}^2$, the parameters $\alpha$ and $\beta$ of the MRF model, and the parameters $\lambda$ and $\mu$ of the DM. The mean activity sums up the overall activity of the network. The covariance gives some information about the spatial structure of the data. Indeed, it quantifies the interaction between neighboring sites. A positive value of $\hat{c}$ indicates that the interaction is excitatory whereas a negative one means that the interaction is inhibitory. The parameters of the MRF have been used to compute $\lambda$ and $\mu$. Besides, they are given in figure 2 for sake of comparison. The parameter $\lambda$ and $\mu$ may be interpreted as frequencies. For instance, $\mu = 0.2$ means that the action of a site on its neighbors is excitatory and that the frequency of excitatory events is 1 each 5 $ms$ (an excitatory event may be viewed as the emission of a spike or group of spikes that activates a neighbor). As well, the absolute value of negative rates can be interpreted as frequencies of inhibitory events (an inhibitory event is viewed as the emission of a spike or a group of spikes that decreases the probability for a site to be active).

The results of the computations are displayed in Figures 1, 2 and 3. The mean activity increases from 0.0 to 37%, reaches a peak after 28 milliseconds and decreases slowly downto 0.0. From the shape of the covariance, one can observe that the interaction is excitatory between 20 and 32 $ms$. Then it becomes inhibitory between 32 and 38 $ms$. The estimation of the parameters according to MRF or DM yield qualitative differences. The curves of $\beta$ and $\mu$ confirm the analysis of the interaction made with the covariance. However, we emphasize that the shape of $\mu$ matches accurately with the covariance (actually, there is an evidence that a linear relation between $\hat{c}$ and $\mu$ exists). On the opposite the relationship between $\beta$ and $\hat{c}$ is less clear. A second observation is that it seems difficult to retrieve $\hat{u}$ from $\alpha$ and $\beta$. In contrast, it is possible to reconstruct $\hat{u}$ from $\lambda$ and $\mu$ according to the mean field equation (3). The curve of $\lambda$ presents two maxima, the first one is after 23 $ms$ and the second one after 32 $ms$. Meanwhile the value of $\mu$ reaches a peak after 25 $ms$. The first maximum of $\lambda$ is followed by an increase of the diffusion rate $\mu$. During 23 and 30 $ms$, diffusion contributes significantly to the overall activity. Then after 30 $ms$, the interaction becomes inhibitory and the remaining activity is mainly explained by the external signal.

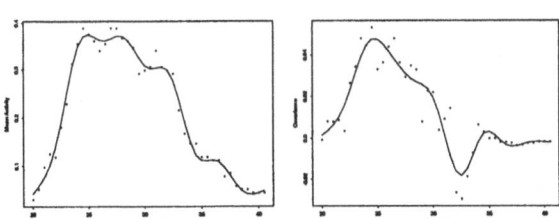

Figure 1: Mean activity value and empirical covariance.

A new statistical method has been proposed to process multidimensional binary data. This method relies on the principle of diffusion of activity between

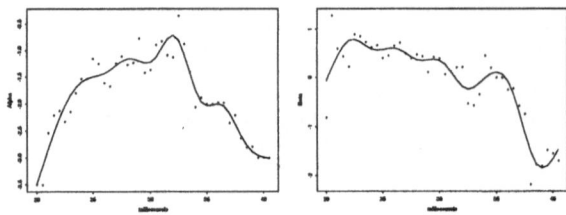

Figure 2: Parameters of the MRF model: $\alpha$ and $\beta$.

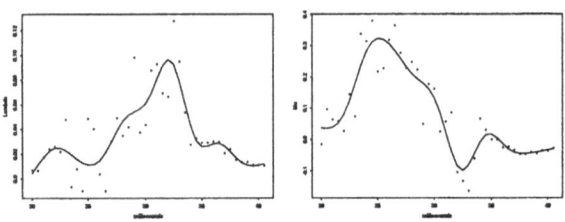

Figure 3: Parameters of the DM: $\lambda$ and $\mu$.

neighboring sites instead of that of pair interaction used in MRF models. The estimation of the parameters of the model is achieved thanks to a mean field approximation of the model. This method has been successfully applied to real data collected from the auditory cortex of a guinea-pig. The results confirm the conclusions obtained from the standard statistics and enables further interpretations on frequencies of either inhibitory or excitatory events.

# References

[1] Hervé T, Dolmazon JM, Demongeot J. Random field and neural information. Proc Natl Acad Sci USA 1990; 87:806-810.

[2] Makarenko VI, Welsh JP, Lang EJ, Llinas R. A new approach to the analysis of multidimensional neuronal activity: Markov random fields. Neural Networks 1997; 10:785-789.

[3] Horikawa J, Hosokawa Y, Kubota M, Nasu M, Taniguchi I. Optical imaging of spatiotemporal patterns of glutamatergic excitation and GABAergic inhibition in the guinea-pig auditory cortex. J Physiol 1996; 497:629-638.

[4] Perkel DH, Gerstein GL and Moore GP. Neuronal spike trains and stochastic point processes. II. Simultaneous spike trains. Biophys J 1967; 7:419-440.

# Asynchronous simulation of large networks of spiking neurons and dynamical synapses

Maurizio Mattia, Paolo Del Giudice

Physics Laboratory, Istituto Superiore di Sanità – Roma, Italy

`mattia@jupiter.roma1.infn.it, paolo@ibmteo.iss.infn.it`

D. J. Amit

Physics Dept., Università di Roma "La Sapienza", Rome, Italy

and Racah Inst. of Physics, The Hebrew University, Jerusalem, Israel

`damita@ilios.fiz.huji.ac.il`

## 1   Simulating the coupled dynamics of IF neurons and plastic synapses

Numerical simulations of networks of 'integrate and fire' (IF) neurons are typically 'synchronous': for a given time resolution $\Delta t$ the differential equations expressing the neural dynamics are numerically solved, with step $\Delta t$. In practical terms, the simplest integration methods can be used, and simulations of reasonably large networks (of order $10^4$ neurons) are within the capabilities of a fast workstation. The need to introduce synaptic dynamics coupled to the evolution of neural states, makes large scale simulations unfeasible. There are two main reasons: 1) the number of synapses grows with the square of the number of neurons, at fixed connectivity. With synapses as dynamical variables, the computational load becomes excessive. 2) the typical time scale for synaptic modifications is generally assumed to be much longer than that of neural dynamics. Thus, to observe effects of synaptic dynamics on neural activities, much longer simulations (in 'biological times') are required.

A way out is in the observation that in biologically plausible regimes neurons emit spikes (which are very localized in time) at low rates: from a few Hz in spontaneous activity, to tens of Hz when excited by a stimulus. Besides, the fraction of neurons in a population which are excited by a stimulus is very low. In the interval between two incoming spikes, the depolarization of an IF neuron varies deterministically, so it is pointless to update it in a sequence of little steps $\Delta t$. On the other hand, $\Delta t$ cannot be chosen to be too large, because neural dynamics undergoes fast transients, which can produce significant effects, at the level of both neural (e.g. cross correlations) and synaptic dynamics (see below).

A possible alternative is to let an incoming spike act asynchronously as a trigger signal for updating of the target neuron's state, and to interpolate the deterministic evolution across the interval between the last spike received and the present one. This would make the computational load roughly proportional to the total number of spikes emitted in the network, a favorable situation for low rates. Furthermore, this procedure eliminates an intrinsic temporal cutoff, and the numerical experiment could capture transients at arbitrarily short time scales.

What's more important, if the synaptic dynamics is spike-driven, the asynchronous approach provides a major advantage: the synaptic dynamics does not increase the computational complexity. In a network of $N$ neurons, with average connectivity $c$, each spike affects $cN$ neurons, and hence only $cN$ synapses on average. If a neuron spikes at $\nu$ Hz, the average rate of synaptic updates provoked by the activity of that neuron is $\nu cN$. By contrast, in a synchronous approach with step $\Delta t$ the number of synaptic updates per second would be $cN/\Delta t$; $\Delta t$ has to be chosen much smaller than the average interval between two successive afferent spikes, which implies $\Delta t < 1/(\nu N)$, giving an average of synaptic updates equal to $c\nu N^2$. Therefore, spike-driven synaptic dynamics leads to a linear scaling with $N$ of the computational complexity per neuron, compared with the $N^2$ dependance in the 'synchronous' case. In fact it would roughly double the load compared to simulation with fixed synapses.

Such an approach is necessary and promising; however, its implementation is not simple. It has been tried in simplified conditions of a uniform, deterministic, synchronizing network with fixed synapses ([4]). Sources of difficulty are random spikes from outside the network, random connectivity, random delays in spikes propagation, which, together with a spike-driven synaptic dynamics, entail a complex management of temporal hierarchy of events in the simulated network.

Here we report an implementation of a simulation of this type, focusing on the strategy followed to solve the critical problems, and provide an example.

## 2 Modus operandi

We highlight the main features of the algorithm, which are independent of details of the neural and synaptic dynamics (as long as the first is IF and the second is driven by spikes). The core of the algorithm is the management of the temporal hierarchy of the spikes which are generated in the network.

A data structure is prepared, in which the synapses are organized in matrix-structured layers; each column $j$ of each layer $l$ contains the set of synapses situated on the axon of neuron $j$, by which a post-synaptic neuron receives spikes emitted by neuron $j$ with discrete delay $d_l$.

Suppose a spike is emitted by neuron $j$ at time $T$. The pair of labels $(j, T)$ (an 'event') enters a first queue $Q_0$ associated with the first synaptic layer (with minimal delay $d_0$), and the time label is updated to $T + d_0$, which is when the spike will affect its target neurons with delay $d_0$. The 'event' waits to be handled until its temporal label becomes the oldest in the queue (it is in the first position in the queue). When its time comes, it is transferred to column $j$ of the first synaptic layer, whose elements identify the post-synaptic partners of neuron $j$ with delay $d_0$. Target neurons for the spike are sequentially addressed, and their depolarization is updated: 1) knowing the arrival time of the last spike that reached the target neuron, the value of its depolarization is calculated, based on the deterministic decay, 2) the post-synaptic contribution due to the spike is added, 3) the new value of the depolarization is compared with the threshold for emission of a spike, 4) if spike emission occurs, the new event enters the queue $Q_0$ with time label $T + d_0$.

At this point, also the synaptic efficacies are updated by the combination of their deterministic decay and the spike triggered change. For the particular choice of synaptic dynamics, see below.

Once all the post-synaptic updates associated with delay $d_0$ are completed, the event is appended to the next queue $Q_1$, attached to the layer with delay $d_1$, and it is handled as above.

Queues are filled in such a way that the time labels are automatically ordered inside each queue (the first element is garanteed to be the oldest of its queue), without need of further sorting. However, it is necessary to sort the first events in all queues, in order to choose the oldest among all events, to be processed. With a small number of allowed discrete values for the delays, this is a negligible additional load.

The assumption of monotonic decay of the depolarization between two incoming spikes, ensures that a spike can only be generated at the time a spike arrives.[1]

In addition to recurrent spikes, neurons in the local network also receive spikes from outside: low frequency external spikes implement the background activity in'the area surrounding the local module; high rate external spikes code for afferent stimuli. The resulting external current is assumed to be random in both cases (a superposition of poissonian trains of spikes with appropriate rates). In the simulation, trains of external spikes with assigned frequencies are produced by a single poissonian engine, and then randomly targeted to neurons in the network, in such a way that each neuron sees the desired statistics of incoming spikes. Random selection of the receiver ensures that trains of external spikes afferent on different neurons are statistically independent. Each new external spike is stored in a 'register', and its time label participates in the sorting process together with the top elements of the queues; as soon as the event in the register has been processed, it is replaced by a new one.

For a network composed of several populations of neurons (e.g. a population of inhibitory neurons and several populations of excitatory neurons, activated by different external stimuli), one spike generator with appropriate frequency, and one register are required for each population (details will be described elsewhere).

---

[1] Situations, such as dynamical conductances examined in [3] or, more in general, EPSPs with a finite rise time, can also be handled with a slight modification of the sorting strategy.

# 3 Neural and synaptic dynamics

We adopt, as an example, a linear IF neuron. Such neuron is suitable for VLSI implementation, and retains most of the collective features of the leaky IF neuron [1]. The choice of the neuron is not critical for the algorithm. The recurrent network includes $N_E$ excitatory neurons (possibly including subpopulations activated by different external stimuli) and $N_I$ inhibitory neurons. Each neuron's depolarization integrates linearly the afferent stream of spikes, with a constant decay between two subsequent incoming spikes. Each spike from neuron $j$ contributes a jump in the depolarization $V_i$ of receiving neuron $i$, equal to the synaptic efficacy $J_{ij}$. If $V_i$ crosses a threshold $\theta$, a spike is emitted, $V_i$ is reset to 0, where it is forced to remain for a time $\tau_{arp}$, the absolute refractory period. A 'reflecting barrier' at 0 forces $V_i$ to stay 0 whenever it is driven towards negative values, either by the constant decay, or by inhibitory spikes. The first and third boxes from the top in part A of in figure 1 show sample time evolutions of the depolarization.

As an example of spike-driven synaptic dynamics we adopt the one defined in [2], where the motivations and an analytical treatment of the model are provided, together with results from a VLSI implementation (for another model of a spike-driven synaptic dynamics, see [5]). It is a specific implementation of a stochastic, hebbian learning dynamics. The synapses connecting pairs of excitatory neurons are plastic, all the others are fixed. The synaptic efficacy $J$ takes one of two values, $J_0$–depressed and $J_1$–potentiated, and the learning dynamics evolves as a sequence of random transitions between $J_0$ and $J_1$, driven by the pre- and post-synaptic activities. An internal synaptic variable ('synaptic potential', $V_J$) describes the short time response of the synapse to pre- and post-synaptic events, and determines the transitions between the states $J_0$ and $J_1$, as follows (see the second box in part A of figure 1): $V_J$ is confined to vary in the interval $[0, 1]$. Upon arrival of a pre-synaptic spike, $V_J$ undergoes a positive (negative) jump of size $a$ ($b$) if the post-synaptic depolarization is found to be above (below) a threshold $\theta_V$ (not the spike emission threshold, $\theta > \theta_V$). The accessible interval for $V_J$ is split in two parts by a synaptic threshold $\theta_J$. The synaptic efficacy $J = J_0$ if $V_J < \theta_J$ and $J = J_1$ if $V_J > \theta_J$; whenever a jump brings $V_J$ above (below) $\theta_J$, $J$ undergoes a transition: $J_0 \rightarrow J_1$ ($J_1 \rightarrow J_0$). In the time between two successive pre-synaptic spikes, $V_J$ is linearly driven towards 0 (1), if the last spike left it below (above) $\theta_J$. The extreme values 0 and 1 act as reflecting barriers for $V_J$. Regardless neuronal and synaptic dynamics, the event-driven algorithm is suitable for any IF-type neuron and spike-driven synaptic dynamics. In particular, leaky IF neurons have also been implemented, results will be shown elsewhere.

# 4 An illustrative example

An example of the time evolution of a recurrent network of linear IF neurons, subjected to the above synaptic dynamics is a network of 1500 neurons

(1200 excitatory, 300 inhibitory) connected by synapses with 10% connectivity ($\sim 144,000$ plastic synapses). In the initial state, 10% of the synapses are potentiated; the neural parameters and the synaptic efficacies are chosen in order to have mean spontaneous activity of about 8 Hz.[2] Box B1 in figure 1 shows a snapshot of the network in its initial state: the big square sprayed with black and white dots provides a representation of the excitatory-excitatory part of the synaptic matrix - white (black) dots stand for potentiated (depressed) synapses, and their coordinates inside the square identify pre- and post-synaptic neuron. To help inspection of the synaptic matrix, the average rates of excitatory neurons labelling each axis of the synaptic matrix are shown in the (identical) horizontal and vertical rectangular boxes in B1...B4 (rates are averaged in 500 ms before the acquisition of the synapses).

The learning protocol is as follows (see boxes B2...B4 in figure 1): there are two stimuli ($S_1, S_2$), which are directed to neurons $1 \ldots 120$ and $121 \ldots 240$ respectively. After 1 s of spontaneous activity stimulus $S_1$ is presented for 1 second (B2), then, after one further second in absence of stimuli, $S_2$ is also presented for 1 second (B3); the network is then left in absence of stimuli for 4 seconds (B4).

In qualitative agreement with expectations (see [2]) on the synaptic dynamics, it is seen from the figures that: potentiation occurs for high pre- and post-synaptic frequencies (during stimulation); depression occurs for high pre- and low post-synaptic frequencies; for low pre-synaptic frequencies synapses remain essentially unaffected.

# References

[1] S. Fusi, M. Mattia (1998) Collective behavior of networks with linear (VLSI) Integrate and Fire Neurons, submitted to *Neural Computation*; M. Mattia (1997) Thesis, Rome University "La Sapienza" (unpublished)

[2] M. Annunziato, D. Badoni, S. Fusi, A. Salamon (1998) Analog VLSI implementation of a stochastic dynamical synapse, *these proceedings*; M. Annunziato and S. Fusi (1998) Queue theory spike-driven synaptic dynamics, *these proceedings*

M. Annunziato (1995) Thesis, Rome University "La Sapienza" (unpublished)

[3] N. Brunel and S. Sergi (1998), Firing frequency of leaky integrate-and-fire neurons with synaptic currents dynamics, submitted

[4] Tsodyks, M., I.Mit'kov, H.Sompolinsky (1993): Pattern of synchrony in inhomogeneous networks of oscillators with pulse interactions. *Phys. Rev. Lett.*, **71**, 1280

[5] P. Hafliger, M. Mahowald, and L. Watts (1996) A spike based leaning neuron in analog VLSI. *Advances in neural information processing systems*, **9** 692

---

[2]For these parameters, the simulation runs on a Pentium II PC with a ratio $\sim 6$ between machine time and 'biological time'.

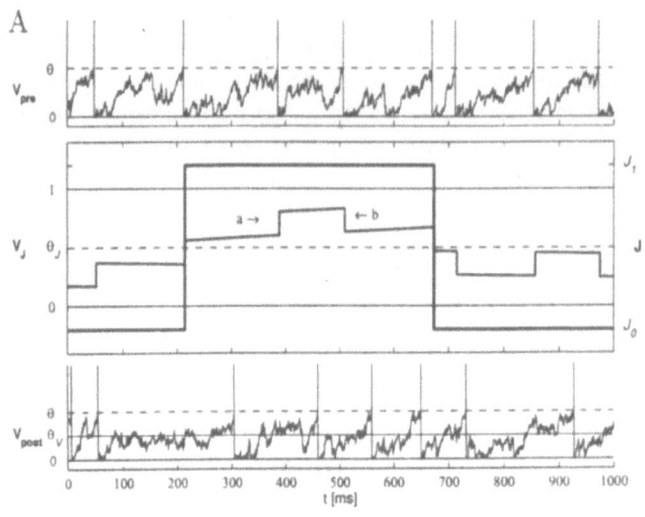

A

**Fig. 1.** A: Illustration of the synaptic dynamics. For one synapse, a sample time evolution of the potential $V_J$ and the efficacy $J$ is shown (A2), together with the depolarization of its pre- and post-synaptic neurons (A1 and A3, respectively). B1...B4: Successive snapshots of the excitatory-excitatory part of the synaptic matrix, during a learning session. White (black) pixels stand for potentiated (depressed) synapses (see text for explanation).

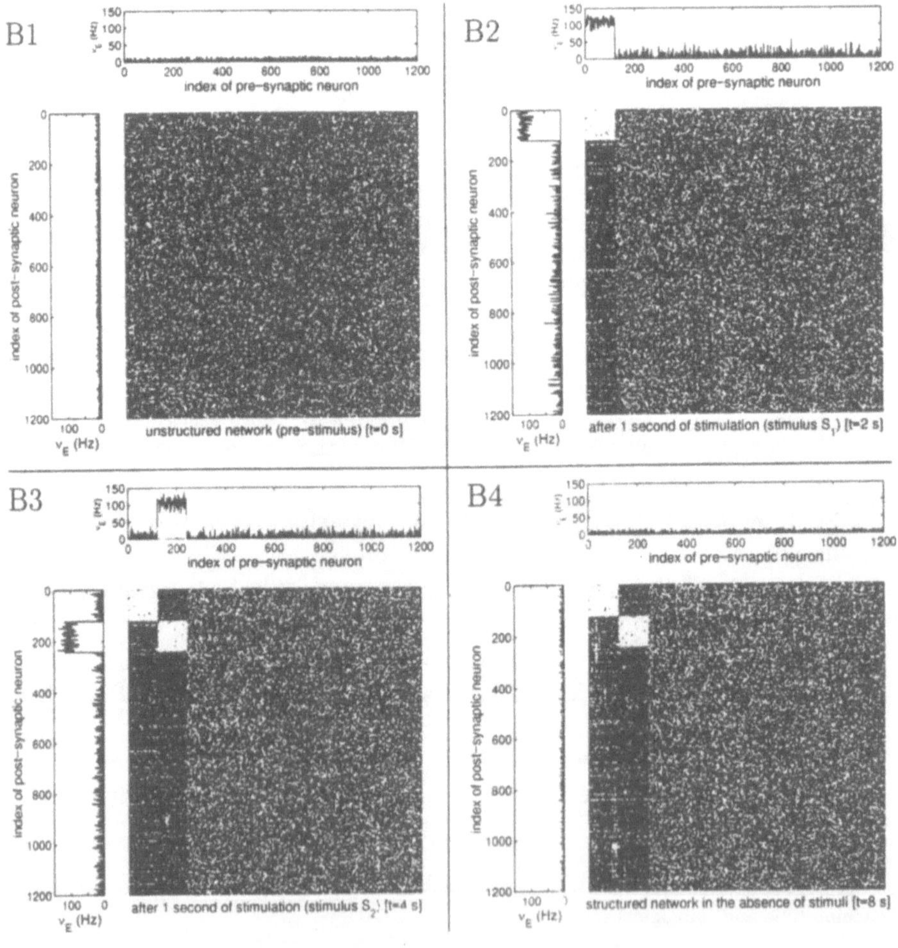

B1

unstructured network (pre–stimulus) [t=0 s]

B2

after 1 second of stimulation (stimulus $S_1$) [t=2 s]

B3

after 1 second of stimulation (stimulus $S_2$) [t=4 s]

B4

structured network in the absence of stimuli [t=8 s]

# Novelty Learning in a Discrete Time Chaotic Network

Emmanuel Daucé
ONERA-CERT/DTIM
2, avenue E.Belin
31055 Toulouse cedex
France
e-mail: dauce@cert.fr

Dr. Bernard Doyon
Unité INSERM 455
Service de Neurologie - CHU PURPAN
31059 Toulouse cedex
France
e-mail: doyon@purpan.inserm.fr

### Abstract

Although extraordinarily complexes, the mental processes can be regarded as products of the neuronal dynamical system. In this context, biological observations make it possible to emit the conjecture that recognition of a form or a stimulus leads to a reduction of neuronal dynamics. This paper proposes a generic model for the study of such dynamics by learning random stimuli. We implement a Hebb-like learning rule, which reinforces the innovation in a network stimulated by a random input. The network learns to react specifically to one or more learned inputs. An estimation of the networks reactivity after learning brings encouraging results in terms of capacity. Then the question of dynamical coding is evoked in terms of limit cycles associated to specific patterns.

## 1 Introduction

Neurobiology shows every day the extreme richness of the processes developped by the brain for data processing. The modeling of neural networks as non-linear dynamical systems can clarify some aspects of neuronal computation[4][5]. The model we present here is an asymmetrical discrete-time recurrent network, well designed for the development of complex dynamics. A statistical analysis of this model has been previously published [1].

The way of processing information with chaos is our main concern. Freeman's paradigm will be our biological background. On the basis of recordings on the olfactory bulb of rabbits and cats, Freeman associates the recognition of a known odor to a specific limit cycle attractor, the waking rest activity being chaotic[7].

In section 2 we present the basic properties of the model. In section 3 we point out the non-homogeneity of neuronal activity. In section 4 we precise the characteristics of the pattern-forced dynamics. In section 5 we present the learning process and measure its efficacy. We then discuss the problems of capacity and dynamical coding. We conclude in section 6.

## 2   A discrete-time dynamical system

Our model comprises $N$ neurons, connected by synaptic weights $J_{ij}$ (weigh of the signal of neuron $j$ towards neuron $i$). The network is fully connected ($J_{ij} \neq 0$ for $i \neq j$) and has no memory of its former state ($J_{ii}=0$). The weights $J_{ij}$ are independent samples of a centered normal law: $\mathcal{P}_{Jij} = \mathcal{N}(0, \frac{1}{N-1})$, $i \neq j$. Thanks to the scaling law, the variance of the sum of the input weights of a neuron, is constant $\forall N$.

The dynamics equation is the following one:

$$(1) \begin{cases} u_i(t+1) = \sum_{j=1}^{N} J_{ij} x_j(t) - \theta \\ x_i(t+1) = f_g(u_i(t+1)) \end{cases}$$

The transfer function $f_g$ is a sigmoid whose values are on $[0,1]$ and whose gain is $g$ - $f_g(u)=(1+\tanh(g.u))/2$ -. Each neuron calculates its output $x_i(t)$ from its local field $u_i(t)$. The threshold $\theta \geq 0$ is equal for every neuron. The slope of the transfer function $g/2$ represents the sensibility to local field variations.

The system develops a rich dynamical behaviour by tuning $g$. For low values of $g$, the system converges to a fixed point. A continuous increase of $g$ leads the system from fixed point to chaos following a quasi-periodicity route [3]. The evolution of the stationary dynamics attractors with $g$ are presented on figure 1.

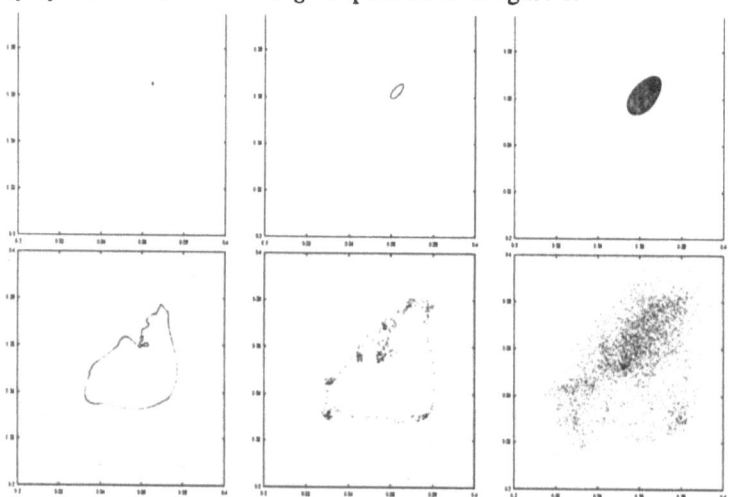

Figure 1: Representation in the $(t,t+1)$ space of the mean output signal with -first line- $g=4$ (fixed point), $g=4.42$ (cycle), $g=4.5$ ($T2$ torus), -second line- $g=5$ (frequency locking), $g=5.5$ (frontier of chaos) and $g=6$ (chaos). The chaoticity of a dynamics is determined with the largest Lyapunov exponent. $N=100$, $\theta=0.3$.

An analytical approach has been developped using Mean Field Theory (MFT), which statistically describes the network behavior at the thermodynamic limit ($N\rightarrow\infty$). A small number of parameters ($\theta$ and $g$ for the model presented here) are enough to give the evolution laws of $u_i(t)$ and $x_i(t)$. Depending on $\theta$ and $g$, the stationary dynamical mode is either a fixed point, or the Gaussian process described below [1]:

$$(2)\begin{cases} U(0) = U \\ U(t) = U + B(t) \end{cases}$$

where $U$ is a Gaussian random variable and $B(t)$ a centered white noise. The MFT description of a network's dynamic is very accurate, but ignores some finite size particularities, like quasi-periodicity routes, seen as a succession of dynamical states between fixed point and stochastic dynamical activity.

## 3 Neuronal activity

The vector of mean local fields $U=(\overline{u_i(t)})_{i=1..N}$ has approximatively a Gaussian distribution. The transfer function being a sigmoid with values on [0,1], the neurons whose mean local field is very negative have an output close to zero. Their local field signal $u_i(t)$ is flattened through the sigmoïd. Those neurons are almost silent. In the same way, the neurons with very positive mean local field saturate and behave like additional thresholds. In both cases, silent and saturated neurons do not favour the propagation of dynamics through the network. Dynamics are carried by the other ones that we will call dynamic neurons. Those neurons amplify the signal $u_i(t)$ and propagate it through the network.

We define the *amplification rate* of a neuron $i$ as the ratio var$(x_i(t))$/var$(u_i(t))$, roughly equal to $f_g'(U_i)^2$. If the amplification is >1, the neuron is said dynamic. Elsewhere, if the amplification is <1, the neuron is either silent or saturated, depending on its mean local field. In a typical network with size $N=900$, sigmoid gain $g=6$, threshold $\theta=0.3$, the proportions of silent, dynamic and saturated neurons are respectively 51%, 40% and 9% (mean on 5 networks). Figure 2, whose samples are taken out of a network with chaotic dynamics, presents these three types of neuronal activity. Note there is no correlation between the three signals.

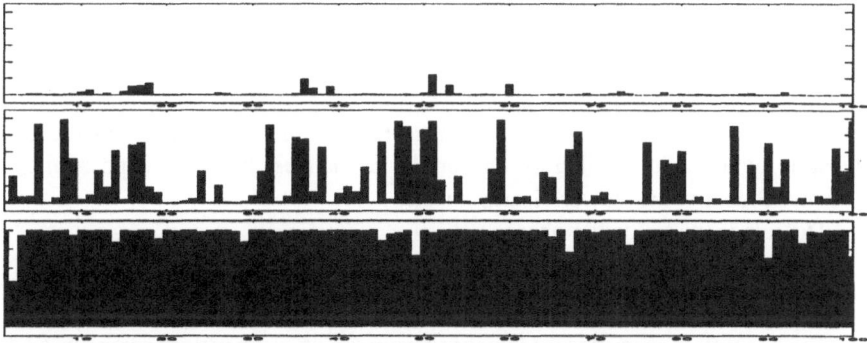

Figure 2: Output of 3 neurons out of the same network, on 100 time steps, $N=100$, $g=6$, $\theta=0.3$. 1st: almost silent activity. 2nd: dynamic activity. 3rd: saturation.

# 4 Reactivity to stimulations

We present to the network a spatial binary signal. We are interested in the dynamical response of the network to this information. The input set (pattern) $I=(I_i)_{i=1..N}$ is a vector of $N$ binary values. A subset of size $n<<N$, is equal to $a$, while the remainder is null. In our simulations, we take $a=1$.

$$(3)\begin{cases} u_i(t+1) = \sum_{j=1}^{N} J_{ij}x_j(t) - \theta + I_i \\ x_i(t+1) = f_g(u_i(t+1)) \end{cases}$$

When the pattern is present, the characteristics of the dynamical system are modified. We call it forced dynamics (3). The mean output of the directly excited neurons increases significantly, and, by reaction, dynamics as a whole are reorganized.

Starting from chaotic spontaneous dynamics, we call *reactivity* the proportion of patterns which lead the system towards non-chaotic dynamics. This reactivity has to be evaluated before any learning. For instance, with $N=400$, $g=6$ and $\theta=0.3$, on the basis of 10 network simulations with chaotic spontaneous dynamics, the reactivity (measured on samples of 100 random patterns) is 3%. A low reactivity means that chaos is stable in the spontaneous dynamics, so that the forced dynamics may remain chaotic for almost every pattern presentation.

Hebbian learning will be applied on the forced dynamics. The aim is to induce a specific reactivity to several random patterns arbitrarily chosen.

# 5 Learning

Let us specify the context of learning. In our simulations, on the basis of a chaotic spontaneous dynamics, one will say there is recognition if the presentation of a learned pattern leads to non-chaotic dynamics, of type $T2$ torus, limit cycle or fixed point (the dynamics are non-chaotic if largest Lyapunov exponent is $< 0$). This may be viewed as an interpretation of Freeman's paradigm, in which chaos would be related to an "I don't know" state, and cycle dynamics to a recognition state [7].

Our learning rule is derived from Hebb's rule. In a classical Hebbian process, the synaptic weights are reinforced when the pre and post-synaptic signals are correlated. Here, our learning process takes place when something new happens in the local fields of the neurons. When the neuronal activities·change after pattern presentation, the synaptic modifications will rely on correlated changes in pre- and post-synaptic activity. Globally, the network learns novelty.

The rule is the following one:

$$(4) \quad J_{ij}(T+1) = J_{ij}(T) + \frac{\alpha}{N}\Delta_i(T) \cdot (X_j(T) - s) \cdot \Theta(\Delta_j(T) - s)$$

The quantity $\Delta_i$ measures the change in neuron $i$, after the pattern presentation, by making the difference of the mean outputs after and before presentation. Parameter $\alpha$ gives the intensity of learning. $X_j$ is the mean output of input neuron $j$. $\Theta()$ is the unit step function which selects the input neurons, according to the value $s$. We have two time scales during the learning process : a fast one (given by $t$) for the iteration of the network dynamics (3) and a slow one (given by $T$) for the iteration of learning (4). In practice, one learning step is processed every 200 time steps.

The selected weights move almost continuously according to the sign of $\Delta_i$, which tends to shift the neuron mean local field towards extreme (positive or negative) values. The initial change in mean output is amplified, and every neuron tends towards silence or saturation, which leads in both cases to a fall of the global dynamics. If the process is not stopped, it leads to a fixed point. In our experiments, we stop the process as soon as a limit cycle is reached. When the learning process ends, the network is able to *recognizes* the pattern, i.e. every new presentation of the pattern will lead to the same limit cycle reached at the end of the learning process (while the spontaneous dynamics remain chaotic).

Up to now, no analytical theory is available for the study of learning. The results below consequently come from simulated data, with networks of size $N=400$, $g=6$ and $\theta=0.3$. Our learning parameters are $\alpha=1$, $s=0.5$, $n=0.05N$. To learn a pool of $P$ patterns, a cross training is carried out. One learning step is processed for each pattern of the pool, and this operation is reiterated until the network is reactive to every pattern of the pool. At each learning step, because of the selection on the input weights, only 5% of the weights are actually modified, of about $10^{-1}/N$ by weight. We verified that random modifications of the same order of magnitude do not lead to any change in the dynamics.

In order to have an insight into specificity of learning, we add a gaussian noise to previously learned patterns, and measure the specific reactivity to different classes of patterns made with a noise of 5%, 10%, 15% and 20% (% based on signal/noise ratio). The results are in Table 1.

Table 1: Specific reactivity to learned patterns with an additive gaussian noise, on one network, $N=400$, $g=6$ and $\theta=0.3$ (measured on 100 noisy versions on the basis of a pool of 5 learned patterns)

| noise | 5% | 10% | 15% | 20% |
|---|---|---|---|---|
| reactivity | 88% | 87% | 78% | 76% |

We can see that the specific reactivity remains high even for relatively loud noise (20%), which shows the robustness of our learning scheme.

The evolution of (non-specific) reactivity with $P$ can help for an estimation of the effective capacity. We made a measure on random networks with $N=400$, $g=6$ and $\theta=0.3$. Ten networks were to learn 1 random pattern, ten others were to learn 2 random patterns, etc... until $P=10$. For each value of $P$, on each network, we measured reactivity after learning by presenting random non-learned patterns. We observed that reactivity does not increase significantly until $P=8$ (it remains lower than 5%), and then strongly increases for $P=9$ (reactivity=30%) and $P=10$ (reactivity=70%). So the capacity of our network seems to be overrun for $P=9$. However, a complete measure of the capacity should take into account others parameters, especially $g$ (which determines the degree of chaos of the dynamics) and $n$ (number of neurons excited by the pattern). Anyway, these measures give clues for a general trend. The ability to discriminate between learned and non-learned patterns disappears while the system tries to learn too many things.

The last point is about dynamical coding. After learning, each learned pattern can be associated to a pattern of activity, made of $N$ local limit cycles. For characterizing the attractor, we look at the mean output signal $m_{net}(t)=<x_i(t)>$. The question comes wether this limit cycle is specific of the pattern. It is difficult to give a formal answer to that question. We have observed that noisy versions of a learned pattern can induce strong changes in the characteristics of the attractor. This neighbour attractor can as well be a fixed point or a torus or a even strange attractor. Nevertheless, some general characteristics like gravity center or winding number remain in the same range than the original ones[2]. Moreover, the spatial repartition of neuronal activity remains very close from the original one. We compared on 5 network ($N=400$, $g=6$ and $\theta=0.3$) the vectors of mean outputs $(X_i)_{i=1..N}$ for regular and noisy versions of 1 learned pattern. The mean correlation between the two vectors is found to be 0,93 (noise=10%) and 0,87 (noise=20%). So, the coding of the input may both be seen in the topological characteristics of the attractor and in the spatial repartition of neuronal activity.

# 6   Conclusion

The learning scheme described in this article illustrates the rich dynamical behaviour of random recurrent neural networks. We have seen that the simulation of a learning rule, which uses the properties of individual neuronal dynamics, gives a good insight into the mechanism of learning and recognizing spatial patterns. Our network is moreover robust to noise addition. Some complementary simulations have to be made out in order to explore the question of network capacity. At last, the role of oscillations in neuronal computation could be explored in a more complex architecture including several clusters, and be relied to neurophysiological works[6].

# References

[1] Cessac B: Increase in complexity in random neural networks. J de Phys I 1995; I.5: 409-432.

[2] Daucé E, Quoy M, Cessac B, Doyon B, Samuelides M: Self-Organization and Dynamics reduction in in recurrent networks: stimulus presentation and learning. Neural Networks. In press.

[3] Doyon B, Cessac B, Quoy M, Samuelides M: Chaos in neural networks with random connectivity. Int. J. of Bifurcation and Chaos 1993; 3-2: 279-291.

[4] Ginzburg I, Sompolinsky H: Theory of correlation in stochastic neural networks. Phys Rev E 1994; 50: 3171-3191.

[5] Herrmann M, Hertz J, Prugel-Bennett A: Analysis of synfire chains. Network: Computation in neural systems 1995; 6: 403-414.

[6] Gray, C M: Synchronous oscillations in neural systems: mechanisms an functions. J. Comput. Neurosciences 1994, 1: 11-38.

[7] Skarda C A, Freeman W J: How brain makes chaos in order to make sense of the world. Behav & Brain Sci 1987; 10 161-195.

# Spike-Based Hebbian Learning for Stimulus Discrimination

Jan Storck*

Fakultät für Informatik, Technische Universität München

80290 München, Germany

Gustavo Deco†

Siemens AG, Corporate Technology, ZT IK 4,

81739 München, Germany

### Abstract

We analyse the impact of Hebbian Learning on a network of spiking neurons. The network consists of pyramidal cells each of which is connected to other pyramidal cells and also locally to an inhibitory stellate cell. The neurons are described by the spike response model (SRM). The network can be driven by different classes of sensorial stimulus. The learning results in stimulus dependent spatio-temporal spike patterns, which are given by clusters of synchronously firing neurons, such that the stimuli can be easily discriminated. A two-picture experiment serves as illustration.

## 1  Introduction

*How* information is carried in the brain and *why* it is carried in this way and not in others requires analysing the question which *first principles* guide the neural processing of information and knowledge. There seems to be a consensus that the neural system encodes information by action potentials or "spikes" which characterize neural firing events [1]. In the framework of *timing coding*, spatio-temporal firing patterns therefore encode information about sensorial stimuli. In other words, different classes of stimuli can be discriminated by different kinds of spatio-temporal firing patterns. Recently, maximization of mutual information – as means to describe discriminability – was proposed by Deco and Schürmann [2] for achieving this goal. The aim of this work is to investigate if Hebbian Learning can be a first principle leading also to this goal. This would mean that in contrast to a biologically unplausible global optimization of a MI-based cost function, the herein introduced Hebbian-based approach would use only local information for that purpose and hence can be viewed as fully unsupervised. Our implementation of Hebbian Learning in the context of pulse coded neurons is biologically motivated by the experimental work of Markram et al. [3] and the theoretical work of Senn et al. [4] and is also in some

---

*Jan.Storck@mchp.siemens.de and storck@informatik.tu-muenchen.de

†Gustavo.Deco@mchp.siemens.de

aspects similar to one suggested by Gerstner et al. [5]. Simply put, synaptic efficacies are increased if presynaptic firing precedes postsynaptic and decreased if it's the other way round, where the amount of change in each case depends on a function of the time difference between the two spikes. The network consists of pyramidal cells each of which is connected to other pyramidal cells and locally to an inhibitory stellate partner cell. The neurons are described by the spike response model (SRM) of Gerstner et al. [6]. The network can be driven by different classes of sensorial stimulus. The learning results in stimulus dependent spatio-temporal spike patterns, which are given by clusters of synchronously firing neurons. These can be interpreted as Hebbian cell assemblies [7]. It will be seen that the resulting spike patterns exhibit cluster synchronization which makes a clear discrimination possible. Even more, this discrimination can be done already after a very short time, namely as soon as the first wave of the cluster of synchronized spikes appears.

# 2 Hebbian Learning for Spiking Neurons

## 2.1 Architecture and Neuron Model

The architecture is schematically presented in Fig. 1 (left) and was introduced by Gerstner et al. [6] in the context of associative memories. The connection topology takes into account that pyramidal cells establish both long-ranged and short-ranged connections whereas inhibitory stellate cells are primarily local. The network includes these neurophysiological facts, and therefore we consider a structure of connected pyramidal cells and associating to each one an inhibitory stellate local partner. We implement this constraint by considering in this case only negative synapses. When we talk about the network structure, by the terminus neuron from now on we actually refer to this pair of cells. The neurons are modelled according to the spike response model (SRM) of Gerstner et al. [6]. For introducing the SRM we closely follow the presentation of Ritz et al. [8] and adapt it or extend it for our purpose where necessary. In the SRM all details of processes on the biochemical level such as ion channels or neuro transmitters as well as the branching of axons and dentritic trees are neglected. The phenomenological model of neuron $i$ is based on two experimentally measurable quantities: its threshold $\theta$ and its absolute refractory time $\tau_{ref}$. Neuronal activity is modelled by a binary variable $S_i(t) \in 0, 1$. A spike lasts for one elementary time step $\Delta t$. The dynamic evolution of neuron $i$ is determined by the probability to fire in the next time step given its local field $h_i(t)$

$$P_F[S_i(t + \Delta t) = \frac{1}{2}\{1 + \tanh[\beta(h_i(t) - \theta)]\} \tag{1}$$

where the noise parameter $\beta$ models the uncertainty of the neuronal threshold process due to synaptic noise. The local field $h_i(t)$ consists of four parts

$$h_i(t) = h_i^{syn}(t) + h_i^{inh}(t) + h_i^{ext}(t) + h_i^{ref}(t) \tag{2}$$

where $h_i^{syn}(t)$ equals the sum of the synaptic inputs of all the other pyramidal cells to which neuron $i$ is connected via synaptic efficacies. The synaptic response is modelled by

$$h_i^{syn}(t) = \sum_{j=1}^{N} J_{ij} \sum_{\tau=0}^{\infty} \epsilon(\tau) S_j(t - \tau - \Delta_i^{ax}) \quad \text{where } \epsilon(t) = \frac{t}{\tau_\epsilon^2} \exp(-t/\tau_\epsilon). \quad (3)$$

Each pyramidal cell is in addition connected to its inhibitory partner cell: $h_i^{inh}(t)$ describes the input from this local inhibition. This means that after each firing the pyramidal cell will be reached by a strong IPSP from its inhibitory partner cell after a certain delay time, decreasing strongly the probability of additional firing of this cell in the nearer future. Technically the inhibitory cell is modelled simply by the input $h_i^{inh}(t)$, where

$$h_i^{inh}(t) = \sum_{\tau=0}^{\tau_{max}} \eta(\tau) S_i(t - \tau - \Delta_i^{inh}) \quad \text{and} \quad \eta(t) = J^{inh} \exp(-t/\tau_\eta). \quad (4)$$

The summation limit $\tau_{max}$ is assumed to be variable. The summation is stopped at the first non-vanishing term. The external stimulus is nominated by $h_i^{ext}(t)$. Finally, $h_i^{ref}(t)$ models the cell's refractory behaviour with

$$h_i^{ref}(t) = \begin{cases} -R & : \quad \text{for } t_F \leq t \leq t_F + \tau_{ref} \quad \text{with } R \gg 1. \\ 0 & : \quad \text{else} \end{cases} \quad (5)$$

This excludes any firing for time $\tau_{ref}$ after one spike has been emitted.

The neurons are coupled by directed links such that if neuron $i$ is connected to neuron $j$ through synaptic efficacy $J_{ji}$ then there exists also connection $J_{ij}$. Thus, the neurons build a recurrent dynamic network. Each synaptic efficacy has a corresponding delay time $\Delta_i^{ax}$ modelling the time interval the spike needs to travel along the axon. Similarly, for the local inhibition there is also a predefined delay $\Delta_i^{inh}$ for each neuron. To inhibit the network from full synchronization due to the Hebbian Learning, synapses are decreased by a small percentage decay $\nu$ in each learn iteration. This simulates natural neural networks' important characteristic of oblivion and introduces the necessary competition to achieve discrimination of different stimulus patterns.

## 2.2 Learning Rule

For adjusting the synaptic efficacies $J_{ij}$ in order to learn we use a spike-based version of Hebbian Learning similar to that proposed by Gerstner et al. [5]. In Hebbian Learning, a synaptic efficacy is changed by a small amount $\Delta J_{ij}$, if presynaptic spike arrival and postsynaptic firing coincide. This simultaneity constraint is implemented by a learning window $W(s)$ where $s$ is the time difference between the postsynaptic firing and the arrival time of a presynaptic spike. For the biological motivation of $W(s)$ and its actual form see also Senn et al. [4]. In our model $W(s)$ has two regimes (Fig. 1 right). For $s > 0$, $W(s)$ is positive. Thus, the efficacy of synapses which are repeatedly active shortly

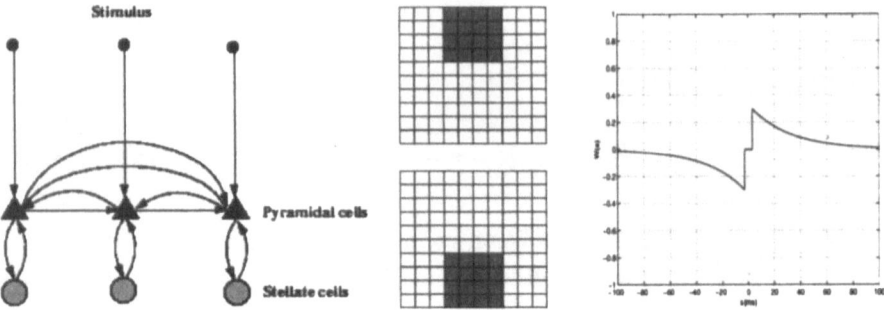

Figure 1: Left: The Neural Network structure. Middle: Stimulus 1 (top) and Stimulus 2 (bottom). Right: The learning window $W(s)$.

before a postsynaptic spike occurs is increased. The efficacy of synapses which are active shortly after the postsynaptic spike is decreased. More concretely, during learning, synaptic efficacies are changed according to

$$\Delta J_{ij} = \varepsilon \sum_f \sum_n W(t^n - t_j^f) \tag{6}$$

with learn factor $\varepsilon = 0.0002$. The sum runs over sufficiently many spike arrival times $t_j^f < t$ and postsynaptic firings $t^n < t$. We take

$$W(s) = \begin{cases} -0.3\exp((s+d)/30) & \text{for} \quad s < -d \\ 0.3\exp(-(s-d)/30) & \text{for} \quad s > d \\ 0 & \text{for} \quad s \in [-d; d] \end{cases} \tag{7}$$

with $d = 3$ms. Hence, if pre- and postsynaptic spikes are sufficiently close in time (i.e. $s \in [-d; d]$) $\Delta J_{ij}$ stays unchanged.

## 3 Experiment

We consider a visual stimulus distributed in a $10 \times 10$ matrix pixel-grid (see Fig. 1 middle) with one neuron per pixel. This gives a network of 100 pyramidal and 100 stellate cells. Each pixel has a unique direct and fixed connection with each pyramidal cell. We use direct-neighbour-connectivity which means that each pyramidal cell is connected to its four direct neighbours in its adjacent lines and columns respectively (with marginal neurons accordingly having less connections). However, a fully connected network is also thinkable. The neurons are numbered columnwise starting from 0. The axonal transmission delay is chosen randomly in the range between 0 and 2 ms for the connections between the pyramidal cells and between 3 and 6 ms for the local inhibitory synapses (Ritz et al. [8]). Synaptic efficacies are decreased by $\nu = 0.01\%$ in each learn iteration. The synaptic strength saturates at a minimum of $-0.01$ and at a maximum of 2 with 'bounce-back-possibility' so that synapses can recover from hitting both of the extremes. Synapses are initialized randomly between $[-0.0001; 0.0001]$. Local inhibition $J^{inh}$ is constantly set to $-2$ for each neuron. Neuron parameter values are $\Delta t = 1$ms, $\theta = 0.12$, $\tau_{ref} = 1$ms. The latter limits the maximal firing rate to 500Hz. However, with local inhibition it ranges significantly below this maximum anyway. The window size for Hebbian Learning is set to 100 which means that at time $t$ only pre- and postsynaptic spikes that appeared after time $t - 100$ms are included in the calculation. Weight update happens every 100th iteration of neuron update, therefore ruling out that a spike is considered more than one time in the Hebbian Learning rule. Hence, the number of learn iterations is one hundredth of the total number of iterations. The two different stimulus patterns from Fig. 1 (middle) are

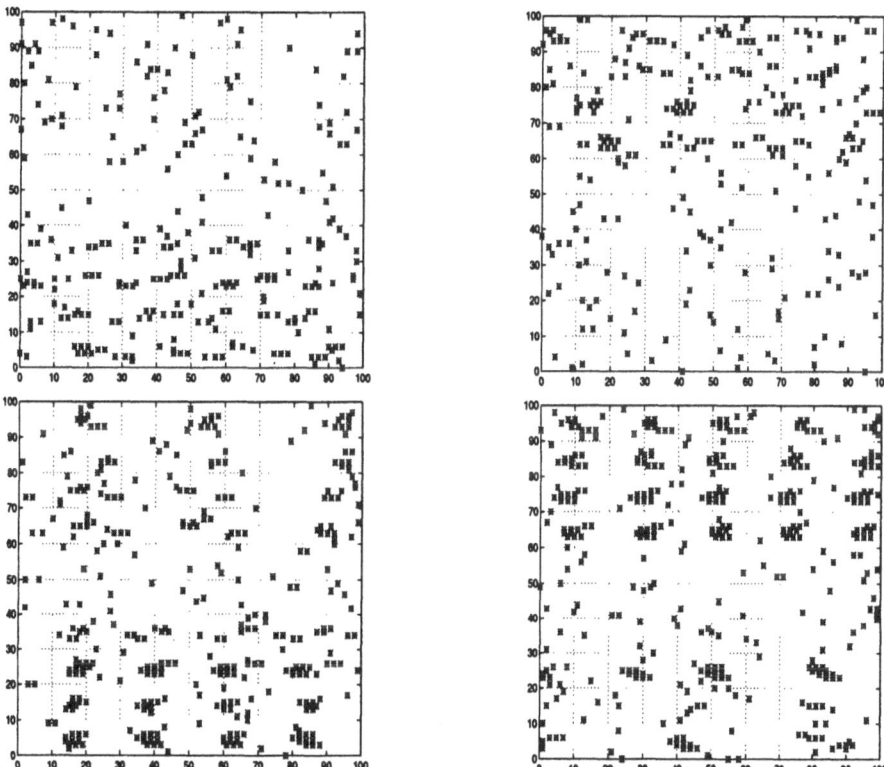

Figure 2: Top left: Spike pattern of stimulus 1 with random initialization of the synaptic efficacies (before learning). Top right: Same for stimulus 2. Bottom left: Spike pattern of stimulus 1 simulated for 100ms after $1.4 \times 10^6$ iterations where synaptic adaptation happened every 100th iteration (horiz. axis: time in ms; vert. axis: index of neuron). A star signals the presence of a spike. Bottom right: Same for stimulus 2. After learning, waves of regular clusters appear for stimulus 1 at the bottom half and for stimulus 2 at the top half of the diagram.

alternately presented to the net in intervals of 500ms. The external stimulus

$h_i^{ext}(t)$ of neuron $i$ is given by the product of the corresponding bit map pixel (either 0 or 1) and the stimulus level which is 0.2 in our case. The learning task consists of finding an internal representation of the information given in the two presented stimuli which leads to a high discriminability (see Fig. 2). It should be stressed that discriminability is not achieved by a complex spatio-temporal pattern, but by clusters of synchronized spikes. Hence, the pattern structure is so clear that a decision can already be made after the first spike wave of the corresponding cell assembly. For this reason, it seems highly probable that the discriminability is achieved in minimal time as suggested by the MTMR(Min.-Time-Max.-Reliability)-principle of Deco and Schürmann [2].

# 4   Conclusions

To summarize, we have shown that Spike-Based Hebbian Learning is a promising candidate for a first principle which guides the network to build stimulus dependent clusters of synchronously firing neurons in a local and unsupervised way. These clusters can also be interpreted as stimulus dependent cell assemblies. One achieves a high discriminability of different sensorial stimuli. In a future work we are planning to use our learn paradigm also for the classification of temporal sequences.

# References

[1] Rieke F, Warland D, de Ruyter van Steveninck R, and Bialek W (1997) Spikes: Exploring the neural code. Cambridge: The MIT Press.

[2] Deco G and Schürmann B (1998) Spatio-temporal coding in the cortex: information flow based learning in spiking neural networks, submitted.

[3] Markram H, Lübke J, Frotscher M, and Sakmann B. (1997) Regulation of synaptic efficacy by coincidence of postsynaptic APs and EPSPs. Science, 275: 213-215.

[4] Senn W, Tsodyks M, and Markram H (1997) An algorithm for synaptic modification based on exact timing of pre- and post-synaptic action potentials. In Gerstner W, Germond A, Hasler M, and Nicoud J-D (eds.) Artificial Neural Networks - ICANN '97, Lausanne, Springer-Verlag, Heidelberg, pp 121-126 (Lecture Notes in Computer Science no. 1327).

[5] Gerstner W, Kempter R, van Hemmen JL, and Wagner H (1996) A neuronal learning rule for sub-millisecond temporal coding. Nature, 383: 76-78.

[6] Gerstner W, Ritz R and Van Hemmen JL (1993) A biological motivated and analytically soluble model of collective oscillations in the cortex. Biological Cybernetics, 68: 363-374

[7] Hebb, DO (1949) The organization of behavior - a neurophysiological theory. New York: John Wiley.

[8] Ritz R, Gerstner W and Van Hemmen JL (1994) In: Models of neural networks II, edited by Domany E, Van Hemmen JL and Schulten K. Berlin: Springer Verlag.

# Poster Presentations:
# Connectionist Cognitive Science and Artificial Intelligence

# What Type of Finite Computations Do Recurrent Neural Networks Perform?

Roman Pozarlik

Software Engineering Group, Institute of Engineering Cybernetics
Janiszewskiego Str. 11-17, PL 50-372 Wroclaw, Poland

http://www.pwr.wroc.pl/~rpoz/

**Abstract**

A recurrent neural network working with finite precision is formally equivalent to some Finite–State Automaton (FSA). However, since this equivalence is true for any finite model of computations, including Turing Machine with a finite tape, it does not actually explain the mechanisms of such a network's operation. So what is the type of finite computations that recurrent nets perform?

In this paper a formal model of finite computations in Simple Recurrent Networks with binary hidden neurons is presented. The model is an automaton whose states are binary vectors; transitions on such states are formalised using the idea of causal computations. It is postulated that although computationally equivalent to an FSA, the model is not limited in performing its cognitive computations. This proposition is posed because the model can modify its states in a massively parallel manner and in doing that it is even more powerful than the above mentioned Turing Machine. It is suggested that the classical notion of computational power should be reformulated in favour of finite computations.

## 1 Introduction

A finite–size Recurrent Neural Network (RNN) working with finite precision is finite and thus formally equivalent to some Finite–State Automaton (FSA). Much work has been devoted to extract FSA from a trained RNN (e.g. see [3]). Does it mean that RNNs implement FSAs? Siegelmann & Sontag [6] have shown that it is possible to construct a RNN with Turing–equivalent computational power. However, a finite–dimensional RNN can robustly – in the presence of noise – perform only finite state computations [1]. It seems that either RNNs perform some finite computations and we must reformulate our understanding of the notion of computational power, or nature really prefers infinite computations and we must strive to cope with them [6, 7]. In this paper, we will argue for the first solution.

Note that an equivalence of RNN and FSA does not explain how the network works because it holds true for any finite model of computations. For instance, a Bounded Turing Machine (BTM), whose tape is finite, is a model of finite computations. Yet, it seems to be somehow more powerful than the equivalent FSA. It means that doing infinite computations is not deciding in solving most practical problems. In our approach, we postulate that computational power should be understood as the model's ability to modify its own states. Finite RNNs are equivalent to FSAs because they are finite, like BTM. However, there is no limitation, as in the case of BTM, on their ability to modify their own states in massively parallel computational steps [4]. Thus, although finite, they have the same or higher computational power than the BTM does (the power being associated with the ability to use the storage which is a multi–dimensional state space).

In this paper, we will try to show that using the idea of *causal computations* [4] it is possible to find a common explanation of operation for RNNs, both finite and infinite. However, our considerations here are limited to finite models. In particular we will study the Elman–Style Simple Recurrent Network (SRN) [2] with binary hidden neurons. We will consider the SRN states as macro–states of an automaton and define a transition function for them. A macro–state may represent a BTM state [4]. We call the resulting model as *Neural Finite–State Transducer* (NFST).

The paper is organised as follows. Section 2 contains an analysis of SRN performance leading to the definition of NFST. Section 3 cursorily discusses the fundamental paradigm of SRN computations called *causal computations* and concludes that the model is general enough to encompass other models of RNNs. The Conclusions are the last part and close the paper.

# 2 Finite computations of discrete-time/discrete-state RNNs

It is useful to start with a definition of computations. Here, as usual, we will define them as a mapping $C$ from a set of input strings over an alphabet $\Sigma_1$ to a set of output strings over an alphabet $\Sigma_2$. For the remainder of the paper $C$ is an FSA computation which means that there exists an FSA performing a mapping equivalent to $C$.

In the analysis that follows we treat RNNs as dynamical Boolean systems whose behaviour can be described by a set of Boolean equations. To make our analysis more specific we will consider an Elman–Style SRN [2] with threshold (step–function) elements in the hidden layer.

## 2.1 An Elman–Style SRN

An Elman–Style SRN, shown in Figure 1, contains four layers of neurons: input, output, hidden layer, $x_m$, and context layer, denoted $x_{n-1}$. $n$ indicates a time step.

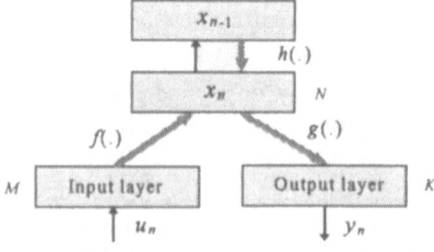

Figure 1: An Elman–Style Simple Recurrent Network (description in text).

Connections from $x_n$ to $x_{n-1}$ are „one-to-one". Arrows in bold indicate layers of trainable, full connections. In general, these should be multilayer connections because $x_n$ may not be linearly separable. We use functions $f(.)$, $g(.)$ and $h(.)$ to describe their mappings. The hidden layer is composed of threshold elements, that is $x_n \in \{0, 1\}^N$. The input and output neurons are also of this type.

In general, we can study the model's performance from the perspective of dynamical Boolean system described by two following equations

$$x_n = F(u_n, x_{n-1}): \{0, 1\}^M \times \{0, 1\}^N \to \{0, 1\}^N, \tag{1}$$

$$y_n = g(x_n): \{0, 1\}^N \to \{0, 1\}^K. \tag{2}$$

$F(.)$ has the following form for the SRN

$$x_n = \theta(f(u_n) + h(x_{n-1}) - T), \tag{3}$$

where $\theta(x)$ is a threshold function ($\theta(x) = 1$ if $x > 0$ and $\theta(x) = 0$ for $x \leq 0$) and $T$ is an $N$-dimensional vector of thresholds. Moreover $f: \{0, 1\}^M \to R^N$ and $h: \{0, 1\}^N \to R^N$ are linear functions. The network computes a Boolean function given by equation (1) and using equation (2) implements some FSA. This FSA can be extracted from the network. However, in the presented approach $x_n$ is treated as a macro–state of an automaton, for which $F(.)$ is a transition function. Note that $g(x_n)$ acts as an inference engine and requires the 'appropriate' vectors $x_n$. $F(.)$ must provide such vectors, which implies what the SRN should compute. But let us first consider what the SRN actually computes.

### 2.1.1   What does the SRN compute?

Any Boolean function can be expressed in the canonical sum form. Let us now prove the following lemma.

**Lemma 1.** Any function $F: \{0, 1\}^M \times \{0, 1\}^N \to \{0, 1\}^N$: $F(u_n, x_{n-1}) \mapsto x_n$, can be expressed by the following formula

$$x_n = \sum_k \alpha_k(u_n) \vee \beta_k(u_n) \wedge \gamma_k(x_{n-1}), \tag{4}$$

where $\Sigma$ denotes the Boolean OR operation (applied componentwise), $\alpha_k: \{0,1\}^M \to \{0,1\}^N$, $\beta_k: \{0,1\}^M \to \{0,1\}^N$ and $\gamma_k: \{0,1\}^N \to \{0,1\}^N$ are some functions and $k \leq M$.

*Proof.* Let us consider $i$-th element of $x_n$, denoted $x_{n,i}$ and $k$-th input vector, denoted $u_{n,k}$. For every $u_{n,k}$ we can define $\beta_k$ such that $\beta_{k,i}(u_{n,k}) \equiv 1$ and $\beta_{j,i}(u_{n,k}) \equiv 0$ when $j \neq k$ (for any $n$). Now, if $x_{n,i} = 1$ whenever $u_{n,k}$ is the input vector (for any $n$), then we should take $\alpha_{k,i}(u_{n,k}) \equiv 1$. Otherwise we take $\alpha_{k,i}(u_{n,k}) \equiv 0$ and $\gamma_{k,i}(x_{n-1}) = 1$ for every $x_{n-1}$, for which $F_i(u_{n,k}, x_{n-1}) = 1$; $\gamma_{k,i}(x_{n-1}) = 0$ otherwise. The number of $x_{n-1}$ is finite. $\square$

In many cases the number of terms in equation (4) can be smaller than $M$. But what type of mapping does this equation reveal?

### 2.1.2  What should the SRN compute?

Any inferences should be permitted. Thus if $g(x_n) \neq g(x_k)$ then it must be $x_n \neq x_k$. In general, $x_n$ should differ from all $x_k$, if they are proceeded by sequences of states in $C$, different than $x_n$. The following definition expresses it in a more formal way.

**Definition 1.** Let $x_1 x_2 \ldots x_{n-1} x_n$ be computations in $C$. $x_n$ is *unique* in $C$ if for any other sequence $x_1 x_2 \ldots x_{k-1} x_k$ in $C$ if $x_1 x_2 \ldots x_{n-1} \neq x_1 x_2 \ldots x_{k-1}$ then $x_n \neq x_k$.  □

In general, a model of computations should be powerful enough to provide a unique representations for every $x_n$ in $C$. However, no finite model of computations fulfils this requirement if $C$ is associated with a regular language and iteration is to be dealt with as well. Thus we consider $x_1 \ldots x_i^* \ldots x_n$ to be the same as $x_1 \ldots x_i \ldots x_n$, and say that $x_n$ should be unique in $C$ exact to iteration.

What is the fundamental principle which different RNN models could use to construct unique representations for $x_n$? In general, $x_n$ is an effect of subsequent causes, i.e. all states in the state sequence $x_1 x_2 \ldots x_{n-1}$. Thus, all the models share the principle of physical causality. We say that $C$ is *causal* if $x_n$ depends causally on $x_k$ for all $k < n$. This definition needs some explanation. First, the term *causal* refers to such a relation where a finite set of causes implies a finite set of effects while *deterministic* refers to a situation in which a finite set of causes implies one effect. Second, all current models of computations are causal. Yet, besides being causal they are also structural [4]. There is no reason to insist that causality can be achieved *only* by manipulating structured representations. Thus we can formulate the following.

**Corollary 1.** Causal computations are performed on causal representations which do not have to be (directly or indirectly) symbolic structured representations.

We cannot discuss the issue here but we will go back to it in Section 3 (see also [4, 5]). Now we will present the equation for causal construction of states.

## 2.2  Neural Finite–State Transducers

Note that equation (4) can be rewritten as $x_n = \alpha(u_n) \vee \beta(u_n) \wedge \gamma(x_{n-1})$ for $x_n \in \{0, 1\}$ $^{N \times M}$. Thus this conjunction can provide any representation for the first $N$ elements of $x_n$ and can be an equation for causal construction of states (although it will not provide some representations for the whole $x_n$). We can redefine this equation in the form shown in equation (5), which results in a more compact definition for $\gamma(.)$. For this equation the following theorem can be proved.

**Theorem 1.** It is possible to define functions $\alpha: \Sigma \rightarrow \{0, 1\}^N$, $\beta: \Sigma \rightarrow \{0, 1\}^N$, and $\gamma: \{0, 1\}^N \rightarrow \{0, 1\}^N$ in such a way that $x_n \in \{0, 1\}^N$, where $x_m$ defined as follows:

$$x_n = \alpha(u_n) \vee \beta(u_n) \wedge (x_{n-1} \vee \gamma(x_{n-1})), \tag{5}$$

will be unique in $C$.

This theorem follows from Lemma 1 (a constructive proof can be found in [5]). Now that we have the equation for causal construction of states, we can define the model of causal computations, called *Neural Finite–State Transducer* (NFST).

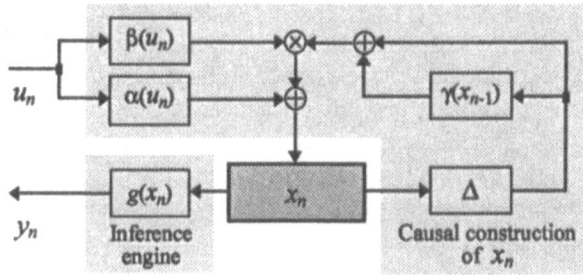

Figure 2: The schema of NFST ($\otimes$ and $\oplus$ denote logical sum and conjunction, respectively; $\Delta$ is a unit delay element).

The model is composed of two parts: one that forms states causally and one which is an inference engine. RNNs will differ mostly in the first part. Whilst $u_n$'s arrive on input a representation is formed in $x_n$ on which $g(x_n)$ makes an inference.[1] Assuming that $g(x_n)$ can realize any Boolean function and $x_n$ is unique, any finite computations can be carried out by the NFST.

### 2.2.1 Properties of NFST

NFST is a model of finite computations but it is not an ordinary FSA because its states, described by $x_m$, are not atomic. Therefore it has the following properties.

Firstly, the transition graph, which is normally drawn for an FSA, does not define how NFST works because transitions depend on the content of states. Such states can be compared with each other to explain such features of computations as systematicity and generalisation. Note that this property is much closer to dynamical systems than to symbolic models.

Secondly, NFST implements limited recursion. It does not have an external storage but inside its states it can accumulate information about executed processes. If the representations used are distributed then free space for keeping this information can be very limited, even if states have a high dimension.

Finally, NFST is an example of nonstructural approach to information processing. It means that $x_n$ are not required to encode structured representations. They must be unique. Although a systematic application of the state construction equation may impose some structure on $x_m$, it is not the symbolic structure to which structure–sensitive procedures are applied and by which other structures emerge.

## 3 Causal computations of discrete/continuous RNNs

From the perspective of causal computations it is possible to explain such dichotomies like parallel/serial and discrete/continuous computations. Serial computations of TMs require structured representations [4]. Massively parallel computations natu-

---

[1] Note that this NFST need not function as an SRN of the same dimension.

rally imply the nonstructural (in the sense described above) processing. Discrete RNNs implement, in our view, this style of processing.

Continuous RNNs and other continuous dynamical systems seem to be suited for causal computations better than discrete models. An equation of causal construction can also be proposed for such models and their dynamics may even enrich computations with such a new quality like infinite number of states. Note that such a quality is not, in our view, crucial for performing intelligent computations. What is important is how powerful the model is in modifying its states, regardless on whether the number of states is finite. It can clearly be seen by comparing FSA and BTM, either of which is of the same computational power.

# 4 Conclusions

The model of finite computations introduced in this paper raises three issues. First, it answers the question what is the type of finite computations performed by a Boolean SRN. Further, it shows that these finite computations, though equivalent to the FSA ones, have properties typical for automata with higher computational power. Finally, it induces a fundamental paradigm of causal computations, which applies to other models of RNNs as well as continuous dynamical systems.

# References

[1] Casey, M. *The Dynamics of Discrete-Time Computations, with Application to Recurrent Neural Networks and Finite State Machine Extraction*. Neural Computation, 8, 1996, pp. 1135--1178.

[2] Elman, J. L. *Distributed Representations, Simple Recurrent Networks, and Grammatical Structure*. Machine Learning, 1991, 7, pp. 195--225.

[3] Omlin, C. W. & Giles, C. L. *Extraction of Rules from Discrete-Time Recurrent Neural Networks*. Neural Networks, 9:1, 1996, pp. 41–52.

[4] Pozarlik, R. *Connectionist Symbol Processing with Causal Representations*. In: Proc. of the 4th Neural Computation and Psychology Workshop: Connectionist Representations, Bullinaria, J. A., Glasspool, G. W. & Houghton, D. (eds.), Springer Verlag, London, 1998, pp. 331--342.

[5] Pozarlik, R. *Integrating Parallel Computations and Sequential Symbol Processing in Neural Networks*. In: Proc. of the IEEE World Congress on Computational Intelligence, 4-9 May, Anchorage, IEEE, 1998, pp. 1444--1449.

[6] Siegelmann, H. T. & Sontag, E. D. *On the computational power of neural nets*. J. Comp. Syst. Sci., 50, 1995, pp. 132–150.

[7] Zeng, Z., Goodman, R. M., & Smyth, P. *Discrete recurrent neural networks for grammatical inference*. IEEE Trans. on Neural Networks, 1994, 5:2, pp. 320--330.

# Statistical Estimation in Conceptual Spaces

Aapo Hyvärinen

Helsinki University of Technology
Laboratory of Computer and Information Science
Rakentajanaukio 2C, 02150 Espoo, Finland
Email: aapo.hyvarinen@hut.fi

## Abstract

Concepts are often considered to be represented by prototypes in conceptual spaces. In this paper, we consider how such a representation could be utilized from a statistical viewpoint. Modelling the probability distribution of objects by a gaussian mixture model, we show how such a representation helps to cope with missing and noisy data. In particular, maximum likelihood estimation gives simple and intuitively appealing methods for predicting missing attributes and reducing observational noise.

## 1 Introduction

In the prototype theory of concept representation [7, 3], it is assumed that a concept (e.g. 'cat', 'chair') is represented as a point in multidimensional space, sometimes called a conceptual space. The point corresponds to the propotype of the concept, which is either a point with especially characteristic attributes or the best example(s) of the concept. The prototype theory is closely connected to the theory of conceptual spaces [4], in which concepts correspond to convex regions in a space defined by some features.

In this paper, we propose to model the data (objects) in conceptual spaces using statistical models. Statistical models can provide principled methods for such problems as categorization, prediction of missing values, and noise reduction. We show the implications of a specific data model (gaussian mixture model) for these problems. Some existing empirical evidence seems to be in line with these implications, supporting the validity of this approach.

## 2 Modelling distributions in conceptual spaces

In this section, we show how to model the distribution of observations (i.e. objects) in the conceptual space by a gaussian mixture model.

In many cases, the objects can be considered to be clustered (concentrated) around the prototypes. It can be assumed that only a few objects have equal distances to two or more prototypes, because such objects would be difficult to represent with the prototypes, which would mean that the prototypes are not

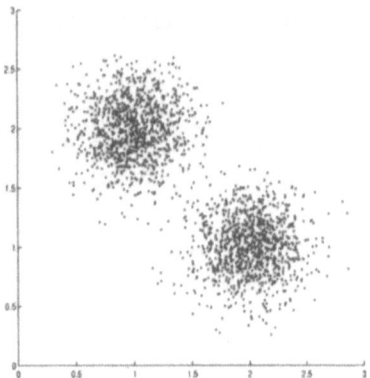

Figure 1: Illustration of gaussian clusters in the conceptual space. For simplicity, the conceptual space is depicted as a two-dimensional plane. Points represent objects, or their representations.

well chosen, or that the object set cannot be adequately represented by prototypes at all. Therefore, it seems reasonable to model the overall distribution of objects (or, rather, their representations) in the conceptual space by a cluster model. The classical statistical cluster model is the gaussian mixture model, see e.g. [2, 6].

Formally, denote by $S$ the conceptual space, assumed here to be equivalent to an $n$-dimensional euclidean space, and by $\mathbf{p}_i \in S, i = 1, ..., m$ the prototypes for a given domain of concepts. Denote by $\mathbf{x} \in S$ a random vector that gives the observed data, i.e. the observations of (the representations of) objects in the conceptual space $S$. We approximate the probability distribution $p$ of $\mathbf{x}$ as a gaussian mixture model

$$p(\mathbf{x}) = \sum_{i=1}^{m} \frac{\pi_i}{\sigma_i^m} \varphi(\|\frac{1}{\sigma_i}(\mathbf{x} - \mathbf{p}_i)\|^2) \tag{1}$$

where the $\sigma_i$ are parameters controlling the width of the clusters around the prototypes, $\varphi(u) = \exp(-u^2/2)/\sqrt{2\pi}$ is the standardized gaussian density, and the $\pi_i$ are the weights (probabilities) of the clusters. For simplicity, we assume here that the covariance matrices of the clusters equal identity.

Figure 1 illustrates such mixture densities. In the figure there are two clusters (concepts), whose prototypes are in the points (1,2) and (2,1). The observed objects are depicted as points on the plane, and they are clearly clustered around the prototypes.

This statistical formalization allows us to use the conventional methods of statistical estimation in conceptual spaces. This gives simple and principled answers to the following questions:

1. Given an observed data point (e.g. perception of an object), what is the probability that is belongs to a given concept?

2. If some of the values of **x** are missing, how should they be predicted?

3. If a noisy version of **x** is observed (as is, in practice, always the case), how could noise be reduced?

The first point is answered by classical decision theory [2, 6], and we shall therefore not treat it in this paper. In the next sections, we propose answers to the two latter questions, using maximum likelihood estimation theory.

# 3 Maximum likelihood estimation of missing data

First, assume that we observe only some of the attributes (components) of **x**. Collect these attributes in the vector $\mathbf{x}_o$, and the missing attributes in $\mathbf{x}_-$. The maximum likelihood principle states that we should predict the $\mathbf{x}_-$ so as to maximize the conditional probability:

$$\max p(\mathbf{x}_-|\mathbf{x}_o) = \max p(\mathbf{x}_-, \mathbf{x}_o)/p(\mathbf{x}_o) \tag{2}$$

Since the term $p(\mathbf{x}_o)$ is not a function of the predictor, the predictor $\hat{\mathbf{x}}_-$ is obtained by maximizing the joint probability density $p(\mathbf{x}_-, \mathbf{x}_o) = p(\mathbf{x})$, as given by Eq. (1).

Assume that the clusters have little overlap, i.e. most of the observations can be reasonably well categorized to one of the concepts. Assume further that attributes of $\mathbf{x}_o$ determine the concept with sufficient certitude. Then straightforward calculations show (see Appendix) that the maximum likelihood prediction $\hat{\mathbf{x}}_-$ is obtained by the following procedure:

1. Classify the object using the available attributes $\mathbf{x}_o$ by associating it with one of the prototypes.

2. Take as $\hat{\mathbf{x}}_-$ the prototype's values of the missing attributes.

Classification in the first step can be performed by classical Bayes classifiers [2, 6].

This prediction is clearly very similar to the intuitive method of prediction often used in schema/script/frame theory [9, 5, 8], which consists of filling in the missing 'slots' with their default values. The above discussion can be considered a statistical justification of such a procedure.

This procedure is illustrated in Fig. 2, for the data depicted in Fig. 1. The cluster centers are depicted as circles. Only one of the attributes of the object, i.e. the value 2.5 of the horizontal axis is observed. The object is associated with the most probable prototype; since the clusters have here equal weight, this is the nearest prototype. Then the value of the vertical axis that corresponds to the nearest prototype, i.e. 1.0, is predicted for the missing attribute. The point marked with asterisk is thus the predicted point in the conceptual space.

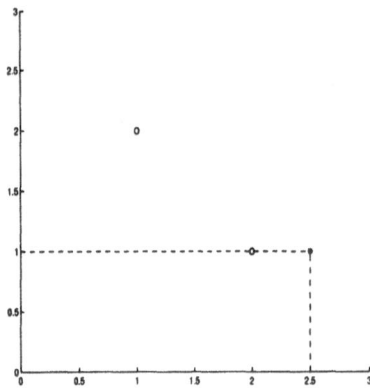

Figure 2: Illustration of prediction (regression) in conceptual spaces. Only one of the attributes of the object, i.e. the value 2.5 of the horizontal axis is observed. The value of the vertical axis that corresponds to the nearest prototype, i.e. 1.0, is predicted for the missing attribute. Asterisk shows the predicted point in the conceptual space, and circles show cluster centers.

# 4 Denoising by maximum likelihood estimation

A second application of the statistical framework is in noise reduction. Assume that the observed data $\mathbf{x}$ contains noise, i.e. it is actually a sum of the real data $\mathbf{s}$ and a gaussian noise vector $\mathbf{n}$:

$$\mathbf{x} = \mathbf{s} + \mathbf{n}. \tag{3}$$

The original data $\mathbf{s}$ is assumed to follow the mixture model in Eq. (1).

The noise $\mathbf{n}$ could correspond to, e.g., imprecision in observations due to short exposure times or distracting objects, or the effects of deteriorating memory traces due to forgetting.

In the Appendix it is shown that the maximum likelihood estimate of $\mathbf{s}$ for given $\mathbf{x}$ can be (approximately) obtained by the following procedure:

1. Classify the observation $\mathbf{x}$ to concept $I$, i.e. associate it to the most probable cluster (prototype), $\mathbf{p}_I$.

2. Translate the observation towards the prototype to obtain the denoised estimate:

$$\hat{\mathbf{s}} = \frac{1}{\sigma_I^2 + \sigma_n^2}[\sigma_I^2 \mathbf{x} + \sigma_n^2 \mathbf{p}_I] \tag{4}$$

where $\sigma_I$ is the width of the cluster corresponding to the concept $I$, and $\sigma_n$ is the standard deviation of the noise (i.e. noise has covariance matrix $\sigma_n^2 \mathbf{I}$).

The denoising estimator given above could be interpreted as a conservative estimator. The observation is translated towards its most probable prototype,

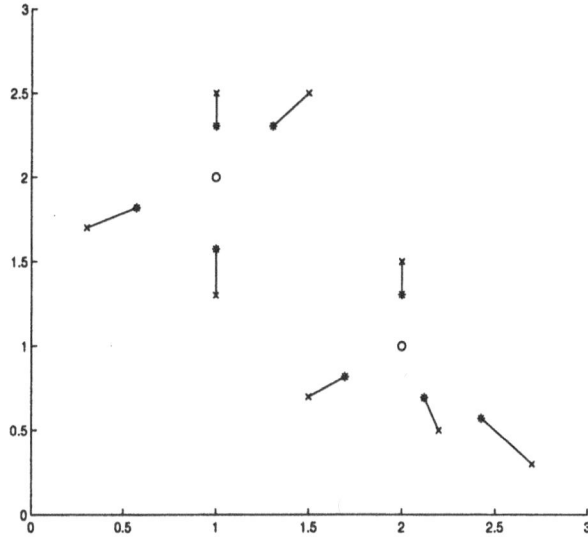

Figure 3: Illustration of denoising in conceptual spaces. Data points marked with x's were observed. The denoised versions are given with asterisks, connected with a line to the corresponding data points. The effect of denoising is to move the point towards the nearest cluster centers (depicted as circles).

'to be on the safe side'. This estimation method is illustrated in Fig. 3 for the data shown in Fig. 1. Observed data point are marked with x's, and the corresponding predictions are marked with asterisks. The figure shows clearly how the denoising moves the data points towards the nearest (or, in general, most probable) cluster centers.

In schema theory [1, 3], there seems to be experimental results that show exactly this phenomenon. For example, in Bartlett's experiments [1], the subjects recalled a story as being much closer to their schema-based expectations than it actually was. This may be in part due to a conservative denoising estimator.

## 5  Conclusion

We presented a statistical framework for object representation in conceptual spaces. Considering the objects to be clustered near the prototypes, we proposed to model their distribution by a gaussian mixture model. This enabled us to use statistical estimation theory to propose methods for predicting missing values, and to reduce noise in the observations. Missing values are predicted in this simplified framework by taking the default values of the most probable prototype. Denoising is performed by a conservative estimation in which the observed point is translated towards the associated prototype. Some existing experimental evidence seems to support our modelling approach. The approach also implies testable hypotheses for future research.

# A   Mathematical appendix

First we prove the result in Section 3. The assumptions imply that the clusters are well separated in the conceptual space as well as in the subspace of $\mathbf{x}_o$. This means that the likelihood can be well approximated by taking just the most probable cluster in the sum in Eq. (1). Then we maximize $p$ in the vicinity of the prototype that is nearest to $\mathbf{x}_o$ in the metric of that subspace; the maximum is clearly obtained when the missing values are given the prototypical values.

To prove the results in Section 4, assume again that the clusters are sufficiently well separated in space. Then the likelihood of $\mathbf{s}$ can be approximated by using only the likelihood of the most probable cluster, and the likelihood of the noise (assumed to have covariance $\sigma_n^2 \mathbf{I}$):

$$\log L(\mathbf{s}) = -\frac{1}{2\sigma_I^2}\|\mathbf{p}_I - \mathbf{s}\|^2 - \frac{1}{2\sigma_n^2}\|\mathbf{x} - \mathbf{s}\|^2 \tag{5}$$

To solve for the maximum likelihood estimator, find a point where the gradient with respect to $\mathbf{s}$ vanishes:

$$\frac{1}{\sigma_I^2}(\mathbf{p}_I - \mathbf{s}) - \frac{1}{\sigma_n^2}(\mathbf{x} - \mathbf{s}) = 0 \tag{6}$$

which gives the maximum likelihood estimator in Section 4.

# References

[1] F. C. Bartlett. *Remembering: A study in experimental and social psychology.* Cambridge University Press, 1932.

[2] C. M. Bishop. *Neural Networks for Pattern Recognition.* Clarendon Press, 1995.

[3] M. W. Eysenck and M. T. Keane. *Cognitive Psychology: A Student's Handbook.* Lawrence Erlbaum, 1990.

[4] P. Gärdenfors. Conceptual spaces as a basis for cognitive semantics. In A. Clark et al, editor, *Philosophy and Cognitive Science*, pages 159–180. Kluwer, 1996.

[5] M. Minsky. A framework for representing knowledge. In P. H. Winston, editor, *The Psychology of Computer Vision.* 1975.

[6] Brian D. Ripley. *Pattern recognition and Neural networks.* Cambridge University Press, Cambridge, UK, 1996.

[7] E. Rosch. Principles of categorisation. In E. Rosch and B. B. Lloyd, editors, *Cognition and Categorisation.* Lawrence Erlbaum, 1978.

[8] D. E. Rumelhart. The representation of knowledge in memory. In R. C. Anderson, R. J. Spiro, and W. E. Montague, editors, *Schooling and the acquisition of knowledge.* 1977.

[9] R. C. Schank and R. P. Abelson. *Scripts, Plans, Goals, and Understanding.* Lawrence Erlbaum, 1977.

# A Connectionist account of Spanish determiner production

Andrew Nix, Neil Davey, David Messer, Pamela Smith

University of Hertfordshire

Hatfield, UK

### Abstract

A Connectionist Network that models the production of simple phonologically coded Spanish Noun Phrases is described. The training data uses type/token frequencies taken directly from a Spanish child's linguistic environment. The training set increases in size in a manner which mirrors the increasing complexity of the real linguistic environment. The results show that the model can learn the task and generalise to unseen Noun Phrase combinations. Moreover the generalisation performance is of a similar nature to that of Spanish children.

## 1 Introduction

Research into the acquisition of Spanish gender has revealed that masculine expressions are acquired with equal if not greater ease than feminine expressions despite the fact that masculine articles have irregular morphology. Evidence in support of this interpretation is present in previous research which has shown that, Spanish children pay more attention to morphophonological cues present in nouns than to natural semantics when assigning gender. They are better at producing the correct determiner when given a noun with masculine cues and are more likely to assign masculine gender to nouns with ambiguous cues (Pérez-Pereira, 1991).

The research described here is an attempt to model Spanish determiner production using a connectionist network and is characterised by the following key points:

- The training data uses type/token frequencies taken directly from a Spanish child's linguistic environment.

- The training set increases in size in a manner which mirrors the increasing complexity of the real linguistic environment.

- The results show concordance with the results of Pérez Pereira (1991) - the model produces similar behaviour, to a child, with respect to gender assignment generalisation.

# 2 Spanish Gender Harmony

Spanish nouns which end with **-a** are generally feminine, while those ending with **-o** are generally masculine. However, there are many exceptions to this pattern.

Spanish determiners have to agree in gender with the noun. Feminine determiners are regular in that they are all marked with the -a suffix. Masculine determiners, however, exhibit varying degrees of irregularity in the singular but are regular in the plural form.

| Determiner | Feminine | | Masculine | |
|---|---|---|---|---|
| English equivalent | *Singular* | *Plural* | *Singular* | *Plural* |
| The | la | las | el | los |
| a/some | una | unas | un | unos |
| This/these | esa | esas | ese | esos |
| That/those | esta | estas | este | estos |
| (an)other/other | otra | otras | otro | otros |

These five different determiner types were used in this study. While the task of gender agreement might seem a trivial task with regular nouns, problems occur when children are confronted with nouns with ambiguous cues such as mano, día, calle, coche, etc. In a study of Spanish children Pérez Pereira (1991) found that they are better at producing the correct determiner when given a noun with masculine cues and more likely to assign masculine gender to nouns with ambiguous cues. These findings are hard to explain, given the greater complexity of the masculine determiner system.

# 3 Network Architecture

A feedforward network with 59 input units, 20 hidden units and 35 output units was presented with phonological representations of Spanish nouns together with a 3-bit code to distinguish which determiner (DET) type was to be produced (See Figure 1). Phonemes were encoded using a 7-bit phonological representation (Nix, 1997, Plunkett & Marchman, 1993). The task for the network was to produce the correct phonological representation of the determiner, from a specification of the type of determiner required, e.g.:

> Input = *<definite article>* + /gato/
> Output = /el/
>
> Input = *<indefinite article>* + /pupa/
> Output = /una/

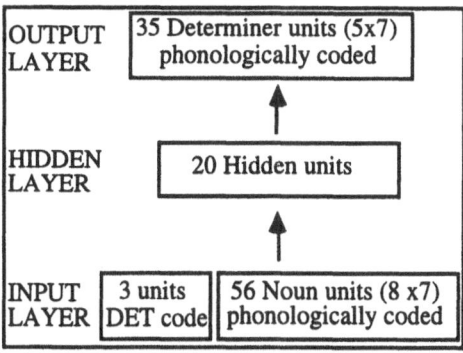

Figure 1: The feedforward network trained to produce phonologically coded determiners

## 4    Training

A longitudinal study of a child, María, conducted by Susana López-Ornat in Madrid was used as the basis of the simulation (López-Ornat, 1994). María was recorded in conversation with her parents and members of her family a total of 568 times between the ages of 1;7 and 2;11. Transcriptions of the *parental productions*, that were made available in a machine-readable form, were organised into three-monthly chunks which would form the basis of an incremental training regime for the network (Plunkett & Marchman, 1993). The database consisted of 6 three-monthly transcription files which reflected the items that María had been exposed to at 1;7, 1;10, 2;1, 2;4, 2;7 and 2;11:

### Incremental training data - NP's per training file

| María's age | 1;7 | 1;10 | 2;1 | 2;4 | 2;7 | 2;11 |
|---|---|---|---|---|---|---|
| Incremental lexicon size | 351 | 922 | 1398 | 1803 | 2048 | 2285 |

Training took place using backpropagation with a learning rate of 0.1 and a momentum of 0.5 with weights being updated after each pattern. The network was trained for 100 epochs on the training set at 1;7 and the weights were saved. The lexicon was then increased to the 1;10 stage and was trained for a further 100 epochs and the weights were saved. This process was repeated until the network was being trained on the full lexicon of 2,285 patterns

## 5    Results

### 5.1  Test Set

To discover whether the network was able to generalise to novel DET+NOUN combinations, a test set was constructed. The 16 most frequent nouns were extracted from the lexicon and presented to the network in DET+NOUN combinations not present in the training set:

### 16 most frequent nouns used in the testing

| Feminine | | Masculine | |
|---|---|---|---|
| **Regular** | **Irregular** | **Regular** | **Irregular** |
| cosa | mano | cuento | nene |
| niña | calle | perro | pie |
| vaca | vez | beso | día |
| caca | leche | culo | coche |

This resulted in a 98 pattern test set consisting of the various unseen DET+NOUN combinations. The network's performance on the test patterns was calculated at the six points along the incremental training regime where the weights had been saved.

## 5.2 Overall Results

The results show that the network can learn to produce the correct determiner in unseen, novel, combinations. Although initially, feminine DETs are produced with a slightly higher degree of success, performance on masculine DETs soon overtakes:

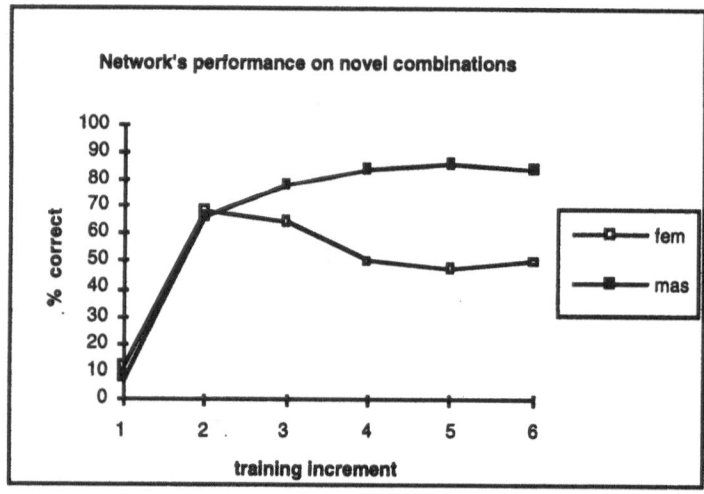

After the 2nd increment, performance on the feminine nouns falls before settling at a level of about 55% at increments 4, 5 and 6.

## 5.3 Regular vs. Irregular

The next graph shows the same information broken down into regular and irregular items.

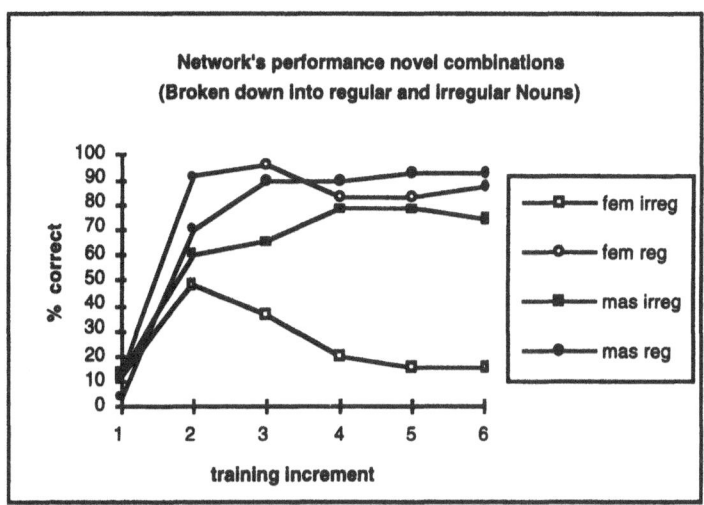

Performance on regular feminine determiners peaks at almost 100% at increment 3 only to be overtaken by masculine regulars at increment 4. Despite this, performance is still remarkably strong - never falling below 80% after its peak. It is clear from the graph that the poor overall performance on feminine determiners is largely due to the irregular items. Many of the irregular feminine DET+NOUN combinations presented to the trained network resulted in masculine DETs being produced at the output layer.

## 6 Discussion

The results of this experiment show that a connectionist network can be trained using child directed speech, to produce Spanish determiners. The successful performance on masculine noun-determiner pairs led us to analyse the type/token frequencies in the training set (see Figure 2).

It seems that the higher proportion of irregular masculine types and tokens is responsible for this pattern of results. Irregular nouns are more frequently presented to the child with a masculine determiner. Thus, when presented with the task of producing a determiner to accompany an ambiguous noun the child is more likely to produce a masculine one. Given that this training data are taken from speech directed to a young child, it suggests that the different frequencies in child-directed speech in Spanish, hitherto unnoticed, may account for the way in which the child learns the masculine forms at the same age as the feminine forms despite the differences in regularity.

This supports a theory that the acquisition of noun phrase morphology in Spanish may be a largely data-driven process owing much to the type and token frequencies of child-directed speech.

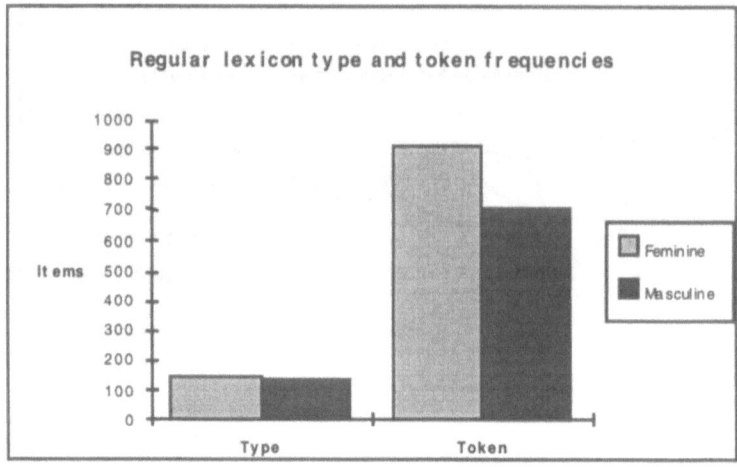

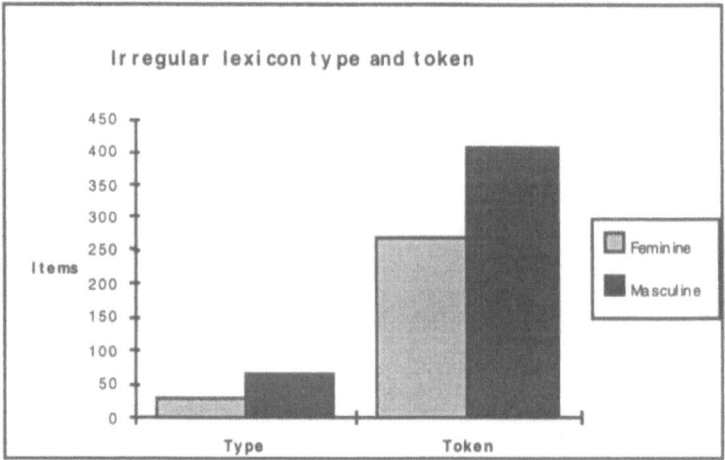

Figure 2: Type and Token Frequencies for both regular and irregular items in the training set

# References

[1] López-Ornat, S. (1994). *La adquisición de la lengua española.* Madrid: Siglo XXI.

[2] Nix, A. J., (1997). A connectionist enquiry into the production of Spanish noun phrases. Ph.D. thesis. University of Hertfordshire.

[3] Pérez-Pereira, M. (1991). The acquisition of gender: What Spanish children tell us. *Journal of Child Language,* 18(3): 571-590.

[4] Plunkett, K., & Marchman, V. (1993). From rote learning to system building: acquiring verb morphology in children and connectionist nets. *Cognition,* 48: 21-69.

# Learning Decompositional Structures in a Network of Max-Π Units with Exponents as Connection Strengths

Harald Hüning*

Neural Systems, Electrical Engineering, Imperial College

London SW7 2BT, U.K.

Email h.hueningⓒic.ac.uk

## Abstract

A catalytic reaction network model is used for learning combinations of symbols. By variation of connection parameters, which are exponents here, the network architecture is modified such that it resembles the structure of the data. The advantage of learning the structure is the generalisation to unseen data, in an example of word recognition the unseen data is a word in a new font or case. The network consists of max-Π units whose output is integrated in a competitive dynamics. Such networks exhibit autocatalytic sets, which have been regarded as connectionist by Farmer [1]. Here learning is introduced into the catalytic network model and a general module is defined that can learn AND/OR combinations. Intermediate state values are employed when modules change their function. The resulting network structure and internal representations encourage applying this method to inference problems.

## 1 Introduction

Sigma-Pi units [2] formally occur in population dynamic models of chemical reactions. A sum of products of state variables corresponds to several different reactions giving the same chemical product. A product term corresponds to a reaction requiring several partners, where a metabolic reaction involves several dynamic equations, but *catalysis* affects only a single equation.

Networks of catalytic couplings exhibit *autocatalytic sets*, where several state variables are cooperatively high while other variables are forced towards zero by competition. The dynamics of autocatalytic sets can be considered as a many-take-all mechanism similar to a winner-take-all mechanism.

Farmer [1] has proposed that autocatalytic sets belong to a generalised class of connectionist systems. While Farmer considers the state variables of catalysts appearing in product terms as the connection parameters, here every factor of the products is weighted by an exponent as parameter. These connection strength parameters avoid the need for adding and removing equations

---

*The work is carried out at Imperial College London as part of a European Union training project financed by the European Commission (programme: Training and Mobility of Researchers).

like in the evolution of an autocatalytic set [3], and allow the change of the system by a learning algorithm.

## 2 Modules for autocatalytic sets

The formation of cooperating groups is studied with the following dynamical system, where $n$ modules compute scores or growth terms $g_i$ which are integrated under competition to give the states $x_i$:

$$\tau \dot{x}_i = g_i(x) - x_i f\left(g_1(x), g_2(x), \ldots, g_n(x)\right) \tag{1}$$

with $f\left(g_1(x), g_2(x), \ldots g_n(x)\right)$ taking the maximum of all growth terms $g_j(x)$ divided by the target value for the maximum state (here 0.5), and

$$
g_i(x) = \max(b_{i1}\, x_1^{w_{i11}}\, x_2^{w_{i21}} \cdots x_n^{w_{in1}}, b_{i2}\, x_1^{w_{i12}}\, x_2^{w_{i22}} \cdots x_n^{w_{in2}}, \ldots
$$
$$
\ldots, b_{im}\, x_1^{w_{i1m}}\, x_2^{w_{i2m}} \cdots x_n^{w_{inm}})
$$

where $b_{i1} \ldots b_{im}$ are the biases of different product terms and $w_{ijk}$ with $1 \leq k \leq m$ are *weighting exponents* for each factor in the products. An exponent of 0 makes a factor neutral to 1. Eventually a normalised version of the weighting function is used $\frac{x^w}{x^x} = x^{w-x}$ with an upper bound of 1 and a minimum state value of $x \geq 0.001$.

The reason to take the maximum within a growth term and not the sum like in models of reactions is that when using a sum a large number of low inputs can cause a state to assume a too high value in a transitory period.

The second maximum decision in function $f$ in (1) achieves a controlled level of the state values necessary for threshold decisions. An analysis for this general system is not known [4], so one has to rely on simulations and the analysis of very restricted cases [5] [6].

A module is defined as the implementation of one equation $i$ in the dynamical system, and self-feedback is disallowed. Input is presented to the system by supplying external states, which could be the output of other pattern classifiers.

## 3 Many-take-all learning rule

From preliminary experiments follows that adapting the weighting exponents towards the mean of the input or feedback states (cf. competitive learning [7]) results in more non-zero exponents than the size of the autocatalytic sets. Thus a threshold $T_{q,ik}$ is applied for setting non-zero exponents based on the conditional frequencies $q_{ijk}$ according to the following batch learning scheme:

1. reset frequency counters $q_{ijk}$ and minimum records $m_{ijk}$
2. apply input as external states (to some of the $x_j$)
3. run until a stable state is reached (all $|\Delta x_i|$ small)
4. in all modules $i$ whose state $x_i$ is above $T_{\text{learn}}$ update $q_{ijk}$ and $m_{ijk}$, given that the product term $k$ has the highest score:
   - increment the frequency counters $q_{ijk}$ where $x_j > T_{\text{learn}}$
   - $m_{ijk}$ records the minimum of $x_j$ that crossed $T_{\text{learn}}$

5. repeat 2. - 4. until the end of the training set is reached

6. set the weighting exponents for each module $i$, product term $k$:

$$w_{ijk} = \begin{cases} m_{ijk} & \text{for } q_{ijk} \geq T_{q,ik} \\ 0 & \text{otherwise} \end{cases}$$

with $T_{q,ik} = \max_j q_{ijk} - tol$ where $tol$ is a tolerance parameter

7. commit a new module or split modules and repeat from 1.

Similar to the maximum response selection in competitive learning, the decision which modules are allowed to learn is made depending on their activity (state) by the learning threshold $T_{\text{learn}}$. The weighting exponents are set to the minimum of the states which are desired to lead to a high recognition score, so with the normalised weighting function a factor of 1 is obtained for all higher state values (above $w_{ijk}$).

# 4 Learning of a decompositional structure

Figure 1 shows two input channels that carry letters represented in a binary 1-of-5 code. When input 1 and input 2 are used together, a constraint is imposed such that several symbols in one input do not occur together with certain symbols of the other input according to the Figure. In the initial configuration a dedicated module is employed for each input letter, see Fig. 2 epoch 1.

| Letter pair | Input 1 | Input 2 |
|:---:|:---:|:---:|
| af | 10000 | 10000 |
| ag | 10000 | 01000 |
| ah | 10000 | 00100 |
| bf | 01000 | 10000 |
| bg | 01000 | 01000 |
| bh | 01000 | 00100 |
| cf | 00100 | 10000 |
| cg | 00100 | 01000 |
| ch | 00100 | 00100 |
| di | 00010 | 00010 |
| dj | 00010 | 00001 |
| ei | 00001 | 00010 |
| ej | 00001 | 00001 |

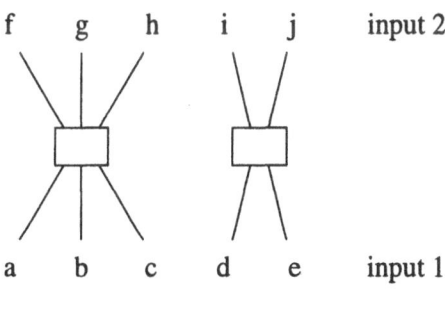

Figure 1: Training data with a decompositional structure. **Left:** Input 1 carries a localist code for the letters a to e and input 2 codes for f to j. **Right:** Structure of the data. The lines represent that a/b/c only occur together with f/g/h, and d/e with i/j.

**Epoch 1, 2: Commitment of modules with function OR** Inputs 1 and 2 are at first provided separately in time, in the first epoch a, b, c, d, e and then f, g, h, i, j. For each letter a single module in the initial configuration responds. An additional uncommitted module performs frequency

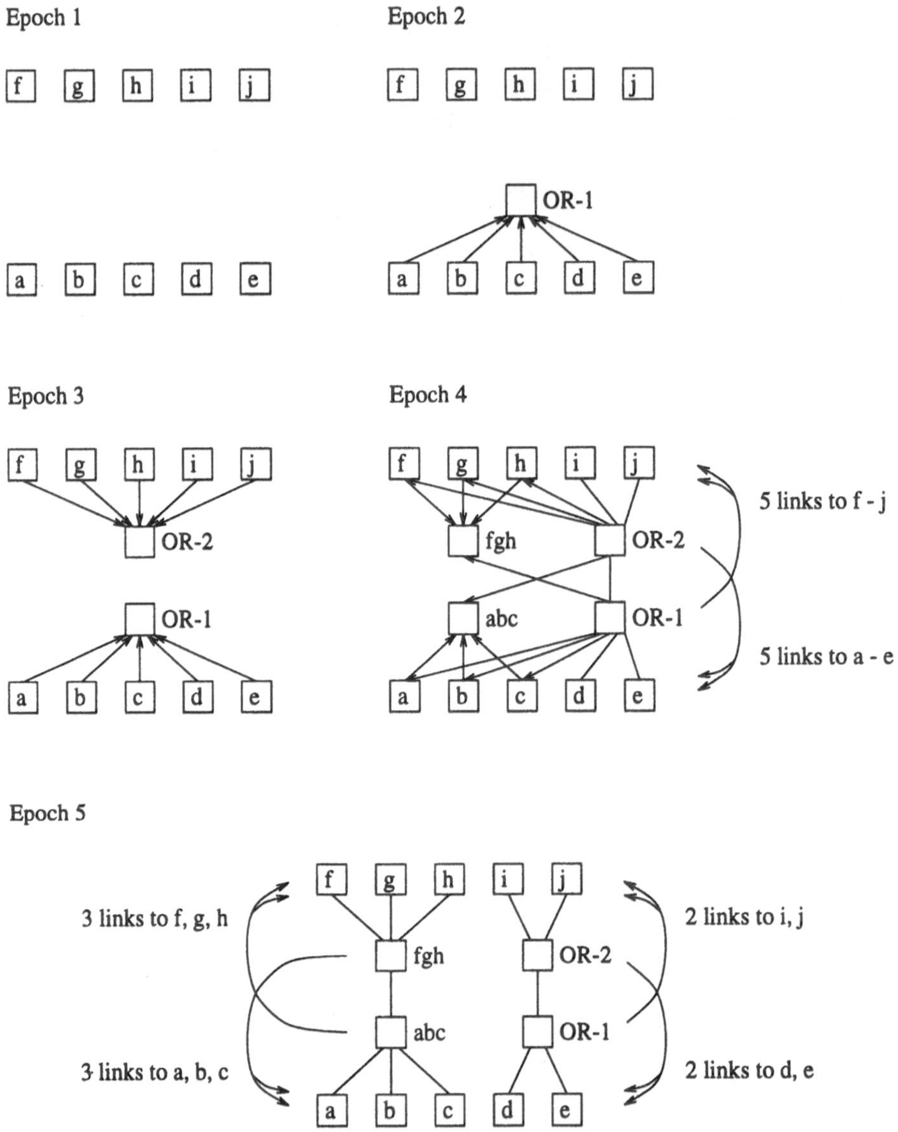

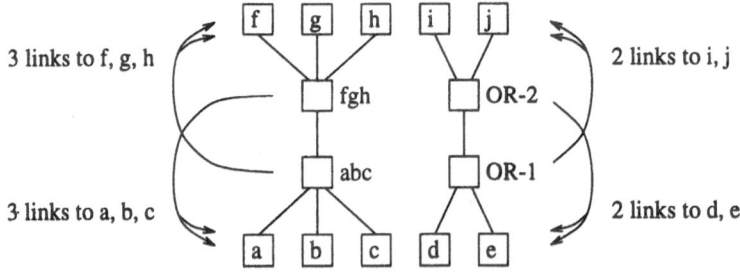

Figure 2: Network architectures for the simulation epochs shown. Lines denote a bidirectional coupling and arrows a unidirectional coupling, i.e. a single non-zero weighting exponent. **Epoch 1, 2:** Input 1 and 2 are presented separately. **Epoch 3:** The complete data is presented from now on. **Epoch 4:** Network architecture after splitting. The arcs on the right denote unidirectional links, i.e. the letter modules have a non-zero exponent for the state of OR-1/2, but not vice versa. **Epoch 5:** The network architecture is now complete, but the biases and weighting exponents are still increasing in further epochs.

measurements without the condition $x_i > T_{\text{learn}}$, because its state is always zero. When the module becomes committed it assumes the function OR with as many product terms as there are non-zero frequencies $q_{ijk}$, so five product terms are generated in this example. Fig. 2 shows the resulting module 'OR-1' after epoch 1 and the second new module 'OR-2' after epoch 2.

**Epoch 3: Splitting of modules** From epoch 3 on the training data from Figure 1 is always applied completely, not separately. This allows the modules OR-1/2 to measure the frequencies in the other input channel as shown in table 1, for example at the product terms responding to a, b, c in module OR-1 the frequency 1 (of total 3) will be measured for f, g, h, but 0 for i, j.

| module | \multicolumn{12}{c}{connected module} | total |
|---|---|---|---|---|---|---|---|---|---|---|---|---|---|

| module | a | b | c | d | e | f | g | h | i | j | OR-1 | OR-2 | total |
|---|---|---|---|---|---|---|---|---|---|---|---|---|---|
| | 3 | 0 | 0 | 0 | 0 | 1 | 1 | 1 | 0 | 0 | - | 3 | 3 |
| | 0 | 3 | 0 | 0 | 0 | 1 | 1 | 1 | 0 | 0 | - | 3 | 3 |
| OR-1 | 0 | 0 | 3 | 0 | 0 | 1 | 1 | 1 | 0 | 0 | - | 3 | 3 |
| | 0 | 0 | 0 | 2 | 0 | 0 | 0 | 0 | 1 | 1 | - | 2 | 2 |
| | 0 | 0 | 0 | 0 | 2 | 0 | 0 | 0 | 1 | 1 | - | 2 | 2 |
| | 1 | 1 | 1 | 0 | 0 | 3 | 0 | 0 | 0 | 0 | 3 | - | 3 |
| | 1 | 1 | 1 | 0 | 0 | 0 | 3 | 0 | 0 | 0 | 3 | - | 3 |
| OR-2 | 1 | 1 | 1 | 0 | 0 | 0 | 0 | 3 | 0 | 0 | 3 | - | 3 |
| | 0 | 0 | 0 | 1 | 1 | 0 | 0 | 0 | 2 | 0 | 2 | - | 2 |
| | 0 | 0 | 0 | 1 | 1 | 0 | 0 | 0 | 0 | 2 | 2 | - | 2 |

Table 1: Conditional frequencies $q_{ijk}$ of above-threshold inputs from connected modules as measured in modules OR-1 and OR-2. Each row of this table corresponds to one product term, whose exponents are non-zero for those factors where the frequency equals the total (last column).

The block structure of the frequencies (here with value 1) is taken as the criterion to split the modules such that each resulting module has a single block. The module OR-1 splits off three product terms, giving the new module 'abc' as shown in Fig. 2 epoch 4, and similarly OR-2 splits off 'fgh'.

**Epoch 4 onwards: Iterations** After the splitting the modules OR-1 and OR-2 no longer respond to three of the five inputs they responded to earlier, see Fig. 2. However, the other modules have already learned (and thus 'expect') the response of the OR-modules. This problem is solved by slowly increasing the bias from zero for all new modules, which makes the response of new modules so low in the first few epochs that other modules can start learning their states, but do not depend on them with respect to crossing the learning threshold. After epoch 4 the new responses are learned, see Fig. 2. Iterations of training epochs are required until the bias of the new modules reaches the level of 1.

# 5 Discussion

The dynamics of autocatalytic sets are used to determine the states of all modules, similar to the pattern completion by Hopfield networks. However, due to the max-Π units the attractor in most cases does not depend on the initial state, but only on the external input states.

The learning example is a building block of learning part-whole relationships with exchangeable parts, like the exchangeable d/e here for one of the parts and i/j for the whole. Considering alternative representations of words, for example when d, e correspond to different fonts for one letter, then the learning of a word can immediately be generalised to all fonts that have been learned on the level of letters. As has been discussed by Hinton [8], a replicated structure for each part has the advantage that top-down feedback can be used for the disambiguation of the parts.

Further work will deal with the comparison of the method arising from this work to a connectionist inference architecture [8]. The internal representations learned in the hidden modules (OR-1/2, abc, fgh) may be regarded as partitions of the input space like the distributed representations discussed by Hinton. Using the current method the conditions for correct inference may be better defined.

This method of learning the structure of data does not require an exponential increase of memory with the size of the problem. Weaknesses of this first approach are 1) it is not defined when no more modules need to become committed, 2) unidirectional links are remaining in the architecture.

# References

[1] Farmer JD, A rosetta stone for connectionism. *Physica D* 1990; 42: 153-187.

[2] Williams RJ, The logic of activation functions. In DE Rumelhart, JL McClelland (Eds.) *Parallel Distributed Processing*, vol. 1, pp. 423-443, Cambridge, MA: MIT Press, 1986.

[3] Bagley RJ, Farmer JD, Fontana W, Evolution of a Metabolism. In CG Langton, C Taylor, JD Farmer, S Rasmussen (Eds.) *Artificial Life II*. Santa Fe Institute Proceedings Vol. X, (pp. 141-158). Redwood City, CA: Addison-Wesley, 1992.

[4] Stadler PF, Fontana W, Miller JH, Random catalytic reaction networks. *Physica D* 1993; 63: 378-392.

[5] Eigen M, Schuster P, The Hypercycle. A Principle of Natural Self-Organization. Part B: The Abstract Hypercycle. *Die Naturwissenschaften* 1978; 65: 7-41.

[6] Hofbauer J, Sigmund K, *The Theory of Evolution and Dynamical Systems*. Cambridge University Press, 1988.

[7] Hertz J, Krogh A, Palmer RG, *Introduction to the theory of neural computation*. Redwood City, Addison-Wesley, 1991.

[8] Hinton, GE, Mapping part-whole hierarchies into connectionist networks. *Artificial Intelligence* 1990; 46: 47-75.

# Fuzzy Heterogeneous Neural Networks for Signal Forecasting

Lluís Belanche, Julio J. Valdés and René Alquézar

Dept. Llenguatges i Sistemes Informàtics, Universitat Politècnica de Catalunya

{belanche, valdes, alquezar}@lsi.upc.es

Barcelona, Spain

### Abstract

Fuzzy heterogeneous neural networks are recently introduced models based on neurons accepting heterogeneous inputs (i.e. mixtures of numerical and non-numerical information possibly with missing data) with either crisp or imprecise character, which can be coupled with classical neurons. This paper compares the effectiveness of this kind of networks with time-delay and recurrent architectures that use classical neuron models and training algorithms in a signal forecasting problem, in the context of finding models of the central nervous system controllers.

## 1 Introduction

A fuzzy heterogeneous neuron is defined as a mapping $h : \hat{\mathcal{H}}^n \to \mathcal{R}_{out} \subseteq \mathbb{R}$, satisfying $h(\phi) = 0$ ($\phi$ is the empty set). Here $\mathbb{R}$ denotes the reals and $\hat{\mathcal{H}}^n$ is a cartesian product of an arbitrary number of *source sets*. Source sets may be families of extended reals $\hat{\mathcal{R}} = \mathbb{R} \cup \{\mathcal{X}\}$, extended fuzzy sets $\hat{\mathcal{F}}_i = \mathcal{F}_i \cup \{\mathcal{X}\}$, and extended finite sets of the form $\hat{\mathcal{O}}_i = \mathcal{O}_i \cup \{\mathcal{X}\}$, $\hat{\mathcal{M}}_i = \mathcal{M}_i \cup \{\mathcal{X}\}$, where each of the $\mathcal{O}_i$ has a full order relation, while the $\mathcal{M}_i$ have not. In all cases, the special symbol $\mathcal{X}$ denotes the unknown element (missing information) and it behaves as an incomparable element w.r.t. any ordering relation. According to this definition, neuron inputs are possibly empty arbitrary tuples, composed by $n$ elements among which there might be reals, fuzzy sets, ordinals, nominals and missing data [1], [2]. Heterogeneous neurons are classified according to the nature of their image set (which need not be restricted to a subset of the reals). In the present study, since the image set is given by $\mathcal{R}_{out}$ the model is of the *real kind*, which is easily coupled with other classical neuron models (i.e. accepting only real inputs), thus leading to hybrid networks in a straightforward way. These networks have been used successfully in classification problems reported elsewhere [1], but their potential of application in other fields was not yet assessed experimentally. The purpose of this paper is to explore further the performance of fuzzy heterogeneous networks (in hybrid architectures) in a signal forecasting task concerning the central nervous system control.

The paper is organized as follows. Section 2 reviews the concept of fuzzy heterogeneous neurons and their use in configuring hybrid networks, while section 3 describes the problem at hand, covering also the different neural paradigms compared to the one presented, the experiment setup and the obtained results. Finally, section 4 presents the conclusions.

## 2 Heterogeneous Neural Networks

A particular class of heterogeneous networks (HNNs) is constructed by considering $h$ as the composition of two mappings, that is, $h = f \circ s$ , such that $s : \hat{\mathcal{H}}^n \to \mathcal{R}' \subseteq \mathbb{R}$ and $f : \mathcal{R}' \to \mathcal{R}_{out} \subseteq \mathbb{R}$. The mapping $h$ can be considered as a $n$-ary function parameterized by a $n$-ary tuple $\tilde{w} \in \hat{\mathcal{H}}^n$ representing neuron's weights, i.e. $h(\hat{\tilde{x}}, \tilde{w}) = f(s(\hat{\tilde{x}}, \tilde{w}))$. In particular, function $s$ represents a *similarity* and $f$ a squashing non-linear function with its image in $[0, 1]$. Accordingly, the neuron is sensitive to the degree of similarity between its inputs –composed in general by a mixture of continuous and discrete quantities possibly with missing data– and its weights. More precisely, $s$ is understood as a *similarity index*, or proximity relation (transitivity considerations are put aside). That is, a binary, reflexive and symmetric function $s(x, y)$ with image on $[0, 1]$ such that $s(x, x) = 1$ (strong reflexivity). The concrete instance of the model under study in the present paper uses as aggregation function a *Gower-like* similarity index in which the computation for heterogeneous entities is constructed as a weighted combination of partial similarities over subsets of variables. This coefficient has its values in the real interval $[0, 1]$ and for any two objects $i$, $j$ given by tuples of cardinality $n$, is given by $s_{ij} = \frac{\sum_{k=1}^{n} g_{ijk}\, \delta_{ijk}}{\sum_{k=1}^{n} \delta_{ijk}}$ where $g_{ijk}$ is a similarity *score* for objects $i$, $j$ according to their value for variable $k$. These scores are in the interval $[0, 1]$ and are computed according to different schemes for numeric and qualitative variables. The factor $\delta_{ijk}$ is a binary function expressing whether objects $i$, $j$ are comparable or not according to their values w.r.t. variable $k$. Gower's original definitions [3] for real-valued and discrete variables are kept, although other similarity functions are possible. For variables representing fuzzy sets, similarity relations from the point of view of fuzzy theory have been defined elsewhere [4] and different choices are possible. In our case, if $\mathcal{F}_i$ is an arbitrary family of fuzzy sets from the source set, and $\tilde{A}, \tilde{B}$ are two fuzzy sets such that $\tilde{A}, \tilde{B} \in \mathcal{F}_i$, the following similarity relation is used:

$$g(\tilde{A}, \tilde{B}) = \sup_{x} (\mu_{\tilde{A} \cap \tilde{B}}(x)) \text{ where } \mu_{\tilde{A} \cap \tilde{B}}(x) = min(\mu_{\tilde{A}}(x), \mu_{\tilde{B}}(x)).$$

For the activation function, a modified version of the classical logistic is used, which is an automorphism of the real interval $[0, 1]$.

$$f(x, p) = \begin{cases} \frac{-p}{(x-0.5)-a(p)} - a(p) & \text{if } x \leq 0.5 \\ \frac{-p}{(x-0.5)+a(p)} + a(p) + 1 & \text{otherwise} \end{cases}$$

where $a(p)$ is an auxiliary function given by $a(p) = \frac{-0.5 + \sqrt{0.5^2 + 4*p}}{2}$ and $p$ is a real-valued parameter controlling the curvature, set in the experiments to 0.1. The general training procedure for the HNN is based on genetic algorithms, since the heterogeneity of the variables involved and the non-differentiability of the similarity function prevent the use of gradient-based techniques [1].

# 3   A case study in signal forecasting

## 3.1   Problem description

The problem studied consists of forecasting the output signals of the Central Nervous System (CNS) controllers of the hemodynamical system. This system, together with the CNS control, form the cardiovascular system. The CNS generates the regulating signals for the blood vessels and the heart, and it is composed of five controllers: *heart rate, peripheral resistance, myocardial contractility, venous tone* and *coronary resistance*. All of these controllers are single-input/single-output (SISO) systems driven by the same input variable, namely the *carotid sinus pressure*. Whereas the structure and functioning of the hemodynamical system are well known and a number of quantitative models, mostly based on differential equations, have been developed, the functioning of the CNS control is of high complexity and still not completely understood. Although some differential equation models for the CNS have been postulated, these models are not accurate enough, and therefore, the use of other modeling approaches like neural networks or qualitative methodologies may offer an interesting alternative for capturing the behaviour of the CNS control [5].

## 3.2   Neural approaches used in the experiments

Two types of neural network architectures can be used for learning tasks involving a dynamic input/output relation, such as prediction and temporal association: *time-delay neural networks* (TDNNs) and *recurrent neural networks* (RNNs) [6]. The HNN model is to be compared to a RNN and two different TDNN models described below.

### 3.2.1   Time-delay neural networks

If some fixed-length segment of the most recent input values is considered enough to perform the task successfully, then a temporal sequence can be turned into a set of spatial patterns on the input layer of a multi-layer feedforward net trained with an appropriate algorithm such as backpropagation. These architectures are called TDNNs, since several values from an external signal are presented simultaneously at the network input using a moving window (shift register or tapped delay line) [6]. A main advantage of TDNNs in front of RNNs is their lower cost of training, which is very important in case of long training sequences. TDNNs have been applied extensively in recent years to different tasks, in particular to prediction and system modeling [7]. In the case of learning a SISO controller, with an input real-valued variable $x(t)$ and an output real-valued variable $y(t)$, the output layer of a TDNN consists of a single output unit that will provide the predicted value for $y(t)$, whereas the input layer holds some previous values $y(t-1), \ldots, y(t-m)$ and some recent values of the input variable $x(t), x(t-1), \ldots, x(t-p)$, from which the value $y(t)$ could be estimated (i.e. a total number of $m+p+1$ input units). Additionally, a hidden layer of $N$ units (to be determined) is required. In the present study, two different TDNN approaches that differ in the training method have been tested: a standard backpropagation algorithm (TDNN-BP) using sinusoidal units, and a hybrid procedure composed of repeated cycles of simulated annealing coupled with a conjugate gradient algorithm (TDNN-AC) [8]. For the latter, hyperbolic tangent units form the hidden layer whereas the output layer is composed by a linear neuron. It should be noted that the HNN model as used here (TD-HNN) can be viewed as a TDNN that incorporates heterogeneous neurons and is trained by means of genetic algorithms.

### 3.2.2 Recurrent neural networks

In recent years, several RNN architectures including feedback connections, together with their associated training algorithms, have been devised to cope naturally with the learning and computation of tasks involving sequences and time series [6]. A type of RNN that has been proven useful in grammatical inference through next-symbol prediction is the first-order augmented single-layer RNN (or ASLRNN) [9], which is similar to Elman's SRN [10] except that is trained by a true gradient-descent method, using backpropagation for the feed-forward output layer and Schmidhuber's RTRL algorithm [11] for the fully-connected recurrent hidden layer. Although the use of sigmoidal activation functions has been common in RNNs, a better learning performance can be achieved using other activation functions such as the sine function [9]. Such networks with sinusoidal units can be seen as generalized discrete Fourier series with adjustable frequencies [7]. Hence, the ASLRNN model used here was built up with sinusoidal units.

## 3.3   Experiment setup

The data used in the training and test phases of the experiments came from a single subject. Five CNS control models, namely, *heart rate, peripheral resistance, myocardial contractility, venous tone* and *coronary resistance*, were inferred for this subject by means of the neural approaches aforementioned. The input and output signals of the CNS controllers were recorded with a sampling rate of 0.12 seconds from simulations of a purely differential equation model. This model had been tuned to represent a specific patient suffering from coronary arterial obstruction, by making the four *different* physiological variables (right auricular pressure, aortic pressure, coronary blood flow, and heart rate) of the simulation model agree with the measurement data taken from the patient. The training set was composed of 1,500 data points for each controller, whereas six data sets not used in the training process (600 points each) were used as forecasting targets, containing signals that represent specific morphologies. The HNN and the TDNN architectures were fixed to include 1 output unit, 8 hidden units, and 7 input units, corresponding to the values $x(t)$, $x(t-1)$, $x(t-2)$, $x(t-3)$, $y(t-1)$, $y(t-2)$ and $y(t-3)$, where $x(t)$ denotes the current value of the input variable and $y(t-1)$ denotes the value of the controller output in the previous time step. All inputs to the HNN were treated as fuzzy sets and the similarity relation given in Section 2 was used. The first-order ASLRNN architecture also included 1 output and 8 hidden units, but just

2 input units, corresponding to the values $x(t)$ and $y(t-1)$, though in this case the hidden layer incorporated additional weights for the feed-back connections.

In the testing process, the normalized mean square error (in percentage) between the predicted output value, $\hat{y}(t)$, and the controller output, $y(t)$, was used to determine the quality of each of the inferred models. This error is given by $MSE = \frac{E[(y(t)-\hat{y}(t))^2]}{y_{var}} \cdot 100\%$ where $y_{var}$ denotes the variance of $y(t)$.

For each CNS controller and neural approach three different training trials were run using a different random weight initialization. The HNN was trained using a standard genetic algorithm with the following characteristics: binary-coded values, probability of crossover: 0.6, probability of mutation: 0.01, number of individuals: 100, linear scaling with factor: 1.5, selection mechanism: tournament. The algorithm stopped when no improvement was found for the last 1,000 generations (typical values were about 5,000). On the other hand, the TDNN-BP and ASLRNN nets were allotted 3,000 epochs using a small learning rate of $\alpha = 0.025$ to allow a smooth minimization trajectory. These parameters were tuned after some preliminary tests. For each run, the network yielding the smallest MSE error on the training set during learning was taken as the controller model. The TDNN-AC was trained in only one run and the process was stopped when a reasonable error was attained.

## 3.4 Results

The nets resulting from the training phase were applied to the training set and to the six test data sets associated with each controller. The normalized MSE errors for these sets were calculated, together with their averages for the different training runs and test sets. The summary of the errors obtained by the different neural approaches is displayed in Table 1.

|  | TD-HNN | | TDNN-BP | | TDNN-AC | | ASLRNN | |
|---|---|---|---|---|---|---|---|---|
|  | Train. | Test | Train. | Test | Train. | Test | Train. | Test |
| HRC | 0.11% | 0.18% | 1.15% | 1.52% | 0.15% | 0.13% | 1.63% | 1.91% |
| PRC | 0.09% | 0.12% | 0.94% | 1.27% | 0.26% | 0.14% | 0.84% | 1.10% |
| MCC | 0.03% | 0.06% | 0.81% | 1.33% | 0.09% | 0.08% | 0.71% | 1.18% |
| VTC | 0.03% | 0.06% | 0.81% | 1.33% | 0.09% | 0.08% | 0.71% | 1.18% |
| CRC | 0.10% | 0.11% | 0.47% | 0.66% | 0.03% | 0.04% | 0.41% | 0.53% |
| mean | 0.07% | 0.11% | 0.84% | 1.22% | 0.12% | 0.09% | 0.86% | 1.18% |

Table 1: Average normalized MSE errors for the training sets (left) and test sets (right) of the CNS controller models inferred by each neural approach.

It is interesting to observe the excellent results yielded by the models inferred by both the HNN and the TDNN-AC, especially as compared to the TDNN-BP and ASLRNN, which showed an almost identical prediction performance, possibly caused by a short depth of temporal dependencies in the modeled system (i.e. all relevant past information could be included in the moving window that selects the inputs of a TDNN).

# 4 Conclusions

Heterogeneous neural networks have been successfully tested in a signal forecasting task (learning central nervous system controllers). The learning and generalization performance of HNNs are comparable to that of TDNNs trained with sophisticated optimization algorithms and better than that of TDNNs trained with backpropagation and RNNs trained with a true gradient-descent algorithm. However, further experiments addressing this kind of problems should be carried out towards a better understanding of their capabilities.

# References

[1] Valdés, J.J., García, R., A model for heterogeneous neurons and its use in configuring neural networks for classification problems, *Procs. of IWANN'97, Intl. World Conf. on Artificial and Natural Neural Networks.* Lecture Notes in Computer Science 1240, Springer-Verlag, pp. 237-246.

[2] Belanche, Ll., Valdés, J.J., Using Fuzzy Heterogeneous Neural Networks to Learn a Model of the Central Nervous System Control. Procs. of *EUFIT'98, 6th European Congress on Intelligent Techniques and Soft Computing.* Aachen, Germany.

[3] Gower, J.C., A general coefficient of similarity and some of its properties, *Biometrics*, 27, pp. 857-871, 1971.

[4] Dubois D., Prade H., Esteva F., García P., Godo L., López de Mántaras, R., Fuzzy set modeling in case-based reasoning. Intl. Journal of Intelligent Systems, 1997 (to appear).

[5] J. Cueva, R. Alquézar and A. Nebot, "Experimental comparison of fuzzy and neural network techniques in learning models of the central nervous system control", *Proc. of EUFIT'97, 5th European Congress on Intell. Tech. and Soft Comput.*, Aachen, Germany, pp.1014-1018, September 1997.

[6] J. Hertz, A. Krogh and R.G. Palmer, *Introduction to the Theory of Neural Computation*, Addison-Wesley, Redwood City, 1991.

[7] A. Lapedes and R. Farber, *Nonlinear signal processing using neural networks: prediction and system modeling*, Tech. Rep. LA-UR-87-2662, Los Alamos National Laboratory, Los Alamos NM, 1987.

[8] Ackley, D. *A connectionist machine for genetic hillclimbing* Kluwer Acad. Press, 1987.

[9] J.M. Sopena and R. Alquézar, Improvement of learning in recurrent networks by substituting the sigmoid activation function. *ICANN'94, Proc. Int. Conf. Artif. Neural Networks*, Sorrento, Italy, Springer-Verlag, Vol.1, pp.417-420, 1994.

[10] J.L. Elman, Finding structure in time. *Cogn. Sci.* Vol. 14, pp.179-211, 1990.

[11] J. Schmidhuber, A fixed size storage $O(n^3)$ time complexity learning algorithm for fully recurrent continually running networks, *Neural Computation* Vol. 4, pp.243-248, 1992.

# Poster Presentations:
# Autonomous Robotics and Adaptive Behavior

# Behavioural Coordination in Acoustically Coupled Agents

Ezequiel A. Di Paolo

School of Cognitive and Computing Sciences, University of Sussex,
Brighton, BN1 9QH, U.K., Email: ezequiel@cogs.susx.ac.uk

### Abstract

Approaching behaviour is studied in simulated agents interacting acoustically. A genetic algorithm is used to evolve a fully recurrent, continuous neural network for controlling the agents. Evolved agents actively discriminate the location of external sources of sound. Their own signalling behaviour is integrated with their search behaviour and sensor gain regulation through *self-hearing*. Coupled agents show signs of structural congruence as they perform dancing patterns in space, while the same agents behave very differently when acting on their own or in the presence of a source of sound that imitates their signal patterns.

## 1 Introduction

From the days when W. Grey Walter [2] wired a light bulb into the steering-motor circuit of his phototactic *M. Speculatrix* and let two of them dance around the lab floor until relatively recently, attempts to study the behavioural basis of social coordination have been few and far between. Much of the work in biology concerning social behaviour has concentrated in understanding the functional aspects of coordination between organisms and their stability under evolutionary pressures, despite many well-known cases where the strictly *behavioural* aspects are the most intriguing[1].

In this paper we explore some of the issues that arise when such coordination is achieved through acoustic interaction. The motivation behind this choice lies in the non-trivial constraints posed by the use of a sound channel which affects all participants, the sound producer included and how these constraints can be met by the adaptive behaviour of *simulated* embodied agents controlled by time-continuous, recurrent neural networks, [1]. Agents are evolved using a genetic algorithm as a search technique so that when placed on a flat arena they will approach one another and try to remain proximate for the longest time possible by means of noisy acoustic interaction. We will show how the resulting behaviour makes sense only from the perspective of the whole embodied agent, and how such behaviour is so integrated that perturbation of any component (like self-hearing) is enough for the whole behaviour to collapse. Finally, we will show how coordination is achieved by means of a dynamic and structural congruence between the interacting agents by contrasting how agents behave individually and in pairs.

---

[1] E.g., sustained duetting in *Laniarius aethiopicus* and other East African birds, [3]

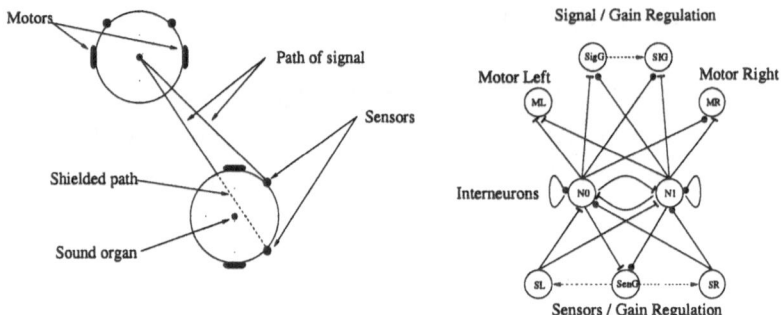

Figure 1: Agent's body, and neural network (only two inter-neurons shown).

## 2   Model

Each agent is modelled as a circular body with two diametrically opposed motors and two sound sensors symmetrically placed at 45 degrees to the motors (see Figure 1). Motors can drive the agent backwards and forwards in a 2-D unstructured and unlimited arena where they move freely except when they collide with each other[2]. A sound organ is located at the center of the body. Sound is *modelled* as an instantaneous, additive field of single frequency and time-varying intensity which decreases with the square of the distance from the source. The sound organ regulates the intensity of the sound produced. Since the task the agents must perform involves some sort of spatial discrimination this must be provided by the relative activity of the sensors. These are physically separated so that in general they will be influenced by different external intensities, however this difference provides poor discrimination especially if we add background noise. In many natural cases spatial discrimination involves the attenuation of intensity as sound travels *through the body* which is linked to the angular movement of the agent, except in the case of sound produced by itself[3]. This "self-shielding" mechanism is modelled as a linear attenuation without diffusion proportional to the distance travelled by the signal within the body. If there is a direct line between sensor and source the perceived intensity is 100 % of the intensity (at that time and position), otherwise the intensity is reduced up to a minimum of 10 % in the case when the sensor is directly opposed to the external sound source (see Figure 1).

A 4-node, fully recurrent, continuous inter-neuron network was used as the agent's controller[4]. Dynamical neural networks have proven to be a powerful tool for studying adaptive behaviour, [1], especially when dealing with time

---

[2] Collisions are modelled as point elastic: no energy loss and no angular effects.

[3] Other natural mechanisms involve differences in time of arrival to the ears, (effective at low frequencies, $< 1400 Hz$ in humans) or specific combinations of frequency filtering and delaying. At this early stage these remain beyond the scope of our model.

[4] For reasons of space, full details on parameter ranges will be available, along with the simulation code, from the following webpage: http://www.cogs.susx.ac.uk/users/ezequiel/coord/coord.html.

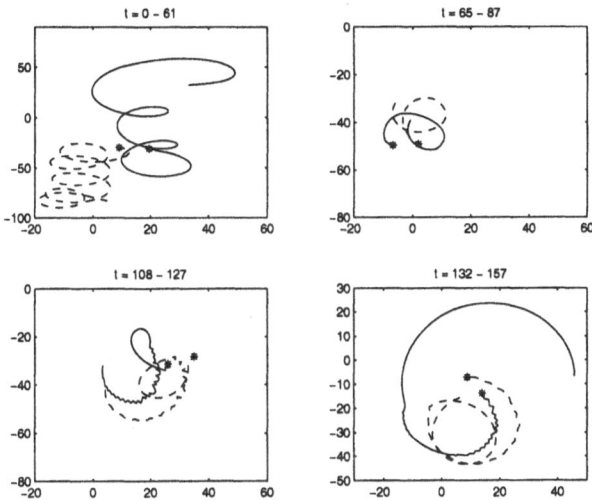

Figure 2: Spatial trajectories of interacting agents (bodies not shown).

constraints becomes an essential part of adaptation. Sensors and effectors are also constituted by neurons which connect with all 4 nodes in the inter-neuron network. Inter-neurons and effectors neurons obey the following law:

$$\tau_i \dot{y}_i = -y_i + \sum_j w_{ji} z_j; \quad z_j = logistic(y_j + b_j).$$

while sensory neurons obey: $\tau_i \dot{y}_i = -y_i + I_i$, where $y_i$ represents the cell potential, $\tau_i$ the decay constant, $b_i$ the activation threshold, $z_i$ the firing frequency, $w_{ij}$ the strength of synaptic connection from node $i$ to node $j$ and $I_i$ the incoming current into sensory neuron. In some cases sensors are directly regulated by their participation in the network dynamics (i.e. by incoming synapses). We chose not to model direct synapses from the inter-neuron network into the sensory networks, and instead we added an effector that directly regulates the sensory gain. This is done in the hope that it will facilitate analysis. The gain of effectors can be regulated as well. In all cases presented here we have used only two regulatory neurons, one for regulating the gain of both sensors, and another one for regulating the gain of the sound organ. Gain regulation gives a minimal bodily plasticity. There is no other form of structural change.

In order to constraint sound production to realistic behaviours we allow sensors to "burn up" if the cell potential of the sensory neurons exceeds certain limits due to intense sounds. A burnt sensor results in zero fitness for the agent in that run. Resulting structures can be viewed as approaching natural cases where viable behavioural (and evolutionary) trajectories are characterized by a certain equilibrium between the autonomy of the nervous system and the autonomy of the individual cells.

A rank based selection genetic algorithm was used as a search technique

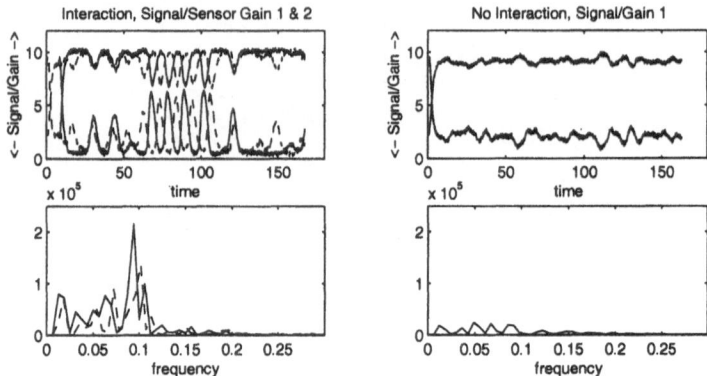

Figure 3: Signals, sensors gains (top) and frequencies spectra (bottom).

with a fixed population of 90 agents evolving for up to 1000 generations. Each agent was selected an average of ten times (five guaranteed) to play with a randomly chosen different agent in the population which was introduced in the arena at a random time after the first one. Fitness values were averaged over all the trials. Fitness was allocated in terms of how much the agents approached each other at the end of the run, and how much time they spent within a distance of 4 body radii of each other.

All parameters (weights, gains and biases) were encoded in a real-valued vector of fixed dimension. Each component specified a parameter by a value in the interval [0,1], (later scaled appropriately). An agent with $N$ inter-neurons and $N_{SE}$ sensors/effectors would have a genome size of $(N + N_{SE})(N + 2)$. Symmetry between left and right hemispheres was enforced only for biases and gains but not for weights and some of the gain parameters were directly regulated by the agent (see above), so the resulting genome size was usually less than the above quantity. No crossover operator was used, and mutation consisted in perturbing the genome vector in a random direction with probability $\mu = 0.005$ by a distance chosen uniformly in the interval [0,1].

## 3 Integrated behaviour and coordination

Different approaching behaviours evolved successfully after a few hundred generations in separate runs; we will report on the most common of them. This behaviour involves strong angular movement during the initial phase (see Figure 2, top left), which is an effective way of actively using the mechanism of self-shielding to produce differences of integrated activity between the two sensors. Less "dynamic" approaching strategies would, after initial orientation, fail to discriminate further because attenuation would be similar for both sensors. Signals are produced in an oscillatory pattern, which is coordinated with the sensor regulation so that sensory gain is reduced for high intensities (Figure 3, top left). Signalling behavior does not only perform the function of a beacon, but it is fully integrated into the search behaviour of the signal producer as can be seen in the fact that if we perturb the ability of *self*-hearing while leaving the rest intact, search behaviour collapses and agents rotate on the spot[5].

If we extend the interaction time, agents perform a behaviour similar to a dance alternating between "leader" and "follower" (Figure 2). Figure 3, top left, shows periods of entrainment between the signals in an anti-phase mode resembling turn-taking, and the Fourier transform of the signals (bottom). Coordination patterns are not permanent, but can be lost and regained. Although both agents were taken from a same evolved population their structures are not identical. If we place each agent on its own in the arena they will wander trying to pick up some external signal, however, we observe (Figure 3 right) that the "natural" signalling behavior is both very different from the interactive case and also between the agents (not shown). The result of ongoing mutual interaction is the triggering of internal structural and dynamic changes (particularly during transients) that drive the agents to possess congruent characteristics manifested in the entrainment of signals. As a consequence of these changes agents become "tuned" to one another thus allowing coordination to be sustained for long periods, and to be regained if it is briefly lost.

In order to verify that a period of un-coordinated and transient coupling is required before the achievement of coordination, an immobile sound generator (beacon) was placed in the arena. This beacon imitates the long term signalling behaviour of the coordinating agents. If we place an agent in the arena and activate the beacon after a while, we would expect the agent to coordinate its signals with the beacon's. It does not happen. The agent signals sporadically with long periods of "silence" (although it approaches the beacon). Since the beacon is non-plastic, it is clear that it cannot actually *interact* with the agent during the transients when attunement occurs, (the different long-term dynamics are shown in Figure 4 for two inter-neurons for one agent).

# 4 Discussion

These experiments help us to explore the biological significance of social coordination through acoustic interaction. There is no trivial functional decomposition of behaviour into *social* and *non-social* categories: emitted sound is used both to regulate movement and to signal spatial position and movement is used both to discriminate internal signals from external sources and to approach the latter. Nor is there any trivial way of separating one agent's behaviour from the other's once certain degree of structural congruence has been obtained through sustained interaction. After an initial transient, the whole agent-agent sys-

---

[5]Note that we are concerned with acoustic interactions in which self-stimulation is *unavoidable*. It is possible to evolve structures that will perform the approaching behaviour without self-hearing but this is not the question we ask here. In general, if we perturb self-hearing overall behaviour will change, it is not clear that it will change so dramatically even for small perturbations (like a 5 % reduction in self-hearing), thus proving the difficulty of trying to functionally decompose behaviour into purely social/non-social categories. Signalling behaviour plays a role in both.

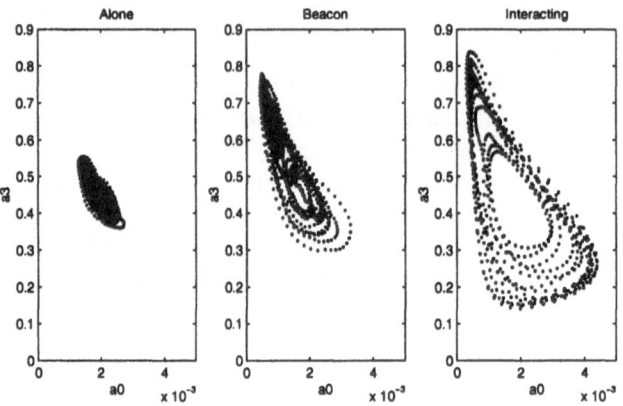

Figure 4: Attractor dynamics for a given agent when acting alone, in the presence of a periodic beacon, and interacting.

tem attains a state of coherence which is manifested in the fact that different "natural" patterns in the isolated agents are replaced by a single coordinated behaviour. Without such transients agents produce different behaviours even in the presence of a beacon that imitates their signalling patterns.

Since behavioural coordination lies at the basis of all animal communication (human language included) exploring its mechanisms becomes a significant task if we are interested in understanding social behaviour. Questions like what causes a pattern of sustained interaction to break off, how a history of interaction affects future encounters, what counts as an invariant during interaction, etc., need further exploration. We think that the use of simple models such as this one is a promising route.

## Acknowledgements

Thanks to Inman Harvey and Phil Husbands for valuable comments. The author acknowledges the support received from the *Consejo de Investigaciones Científicas y Técnicas de la República Argentina* and an Overseas Research Students Award.

## References

[1] R. D. Beer and J. C. Gallagher. Evolving dynamical neural networks for adaptive behavior. *Adaptive Behavior*, 1 - 1:91 – 122, 1992.

[2] W. Grey Walter. An imitation of life. *Scientific American*, pages 42 – 45, May 1950.

[3] W. H. Thorpe. *Duetting and antiphonal song in birds: its extent and significance*. E. J. Brill, Leiden, 1972.

# Self-Localization by Hidden Representations

Michael Herrmann, Klaus Pawelzik, Theo Geisel
Max-Planck-Institut für Strömungsforschung
Bunsenstraße 10, 37073 Göttingen, Germany
email: {michael,klaus,geisel}@chaos.gwdg.de

### Abstract

We present a framework for generating representations of space in an autonomous agent which does not obtain any direct information about its location. Instead the algorithm relies exclusively on sensory input and internal estimations of actions. The activations within a neural network are propagated in time depending on internal estimations of actions. Sensory input connections are adapted according to a Hebbian learning rules derived from the prediction error on sensory inputs one step ahead. During exploration of the environment the respective cells develop location and direction selectivity even when relying on highly ambiguous stimuli.

## 1   Introduction

Placed in an unknown environment humans, animals, and robots receive only incomplete information about their current location via their senses. Nevertheless, we and many other animals very quickly develop a sufficiently precise knowledge of where we are and at which direction we are heading. In rat hippocampus this localization ability depends on visual input [1], and has been shown to persist also in dark [2]. Hence, one should assume some kind of representation of the environment rather independent from direct visual input. For an autonomous robot that uses infrared or touch sensors, the localization task is conceptually more difficult: Local, qualitative and highly ambiguous external inputs are to be combined with the internal motion information to form a representation of the environment.

The problem of determining location from motion can be solved in principle by path integration. Due systematic errors and cumulating noise, however, calibration procedures and exploitation of information provided by landmarks and compasses is needed in order to obtain meaningful results. In particular in situations as described above path integration algorithms shall provide short-term accuracy only and are to be supported by correction mechanisms that evolve in a self-organizing fashion.

A previous attempt to approach these problems has been taken up in Ref. [3]. There the basic idea consisted in applying Bayes' formula to arrive at predictions of the current position conditioned on sensory inputs from conditioned predictions of inputs. In our approach no direct information whatsoever about the actual position and allocentric orientation of the agent is taken into account. In contrast to [3], where specific information was obtained from long range sensors, our agent is equipped only with sensors that merely indicate the existence of an obstacle immediately in front, left or right of the robot. Further, our agent is not equipped with a compass which had to be used in [3]. Due to the latter fact, the model in [3] cannot be considered as a hidden Markov model, whereas this theoretical framework can be directly applied in the present more general setting.

Our approach assumes a neural activity dynamics which depends on internal estimations of movements and which is modulated by external sensory input. The parameters are adapted in a self-supervised manner such that expected observations maximally match the sensory input. At no point the "objective" location is mapped into the neuronal activities. Instead, an observer can deduce that the agent appears to "know" its location because after adaptation particular neurons are active only when the agent is at certain locations.

The scheme will be detailed in the following section. The third section of the present paper is devoted to two experiments with a robot simulator [4]. The discussion briefly addresses applications of the present approach in robotics and neural modeling.

## 2 Hidden Markov model of spatial localization

The internal states of the robot are represented by discrete vectors $\xi$ corresponding to a $d$ dimensional grid of nodes. The robot receives $S$ different kinds of inputs from the environment and has internal sensations of its motor actions. The probability $P(\xi)$ of an internal state will be referred to also as the activity of unit $\xi$, i.e., if we assume some normalization scheme for the total activity the following probabilistic scheme can be interpreted as a description of the dynamics of a simple neural network.

Past knowledge about the conditioned mean frequency $P_t(s_t|\xi)$ of the present stimulus $s_t$ is used to improve the prior distribution $P_t(\xi)$ via Bayes' formula

$$P_t(\xi|s_t) = \frac{P_t(s_t|\xi) P_t(\xi)}{\sum_{\xi'} P_t(s_t|\xi') P_t(\xi')}. \tag{1}$$

In this way the activity of units that are unlikely to correspond to the present sensation is suppressed in favor of units that predict $s_t$. In order to postulate an adaptation rule for $P_t(s_t|\xi)$ we consider the likelyhood the network assigns to $s_t$, which is given by $\sum_\xi P_t(s_t|\xi) P_t(\xi)$. The cost function

$$E = \frac{1}{2} \sum_s \left( \delta(s_t, s) - \sum_\xi P_t(s_t|\xi) P_t(\xi) \right)^2 \tag{2}$$

compares this likelyhood with the presence of a certain input. Here, the presence of an input is formally expressed by the probability density $\delta(s_t, s)$, with

$\delta(s, s') = 1$ if $s = s'$, $\delta(s, s') = 0$ otherwise. Gradient descent on $E$ results in

$$\Delta P_{t+1}(s|\xi) = \left(\delta(s_t, s) - \sum_{\xi} P_t(s|\xi)P_t(\xi)\right) P_t(\xi), \quad s = 1, \ldots, S, \qquad (3)$$

which can be interpreted as a Hebbian correlation rule between input and internal activity supplemented by subtractive normalization.

For correlating the time course of the robot's position to its internal state the latter is subject to a temporal evolution as well. If the path integration errors are assumed to be independent at each time step, this evolution can be modeled as a Markov process. A transition matrix $T$ propagates the activity distribution towards a new distribution.

$$P_{t+1}(\xi) = \sum_{\xi'} T(\xi; \xi') P_t(\xi') \qquad (4)$$

Although it is in principle possible to estimate the transition matrix from the sequence of observation $s_t$, this is unfeasible if the number of internal states is large. The complexity can be reduced by additional assumptions. We have chosen to restrict the robot's actions to either turns or straight movements of unit step size. Further, we assume the internal state to be decomposed into two components that potentially represent, respectively, position and direction estimations: $\xi = (\mathbf{i}, \mathbf{k})$, where $\mathbf{i} = (i_1, i_2)$, $i_1, i_2 \in \{1, \ldots, N\}$ is a two dimensional integer vector and $\mathbf{k}$ is a two dimensional vector of unit length (internal and external *units* can be different), whose argument represents a direction $\phi = 2\pi n/K$ with $n \in \{0, \ldots, K-1\}$. $\mathbf{k}$ serves to simplify the formal notation, whereas numerically $\phi$ is used instead, such that the effective dimension of $\xi$ is $d = 3$. The straight motion of the robot from the objective position $x'$ to $x$ ideally corresponds to a shift of single-peaked activity distribution with the (continuous) mean vector $\overline{\mathbf{i'}}$ to the new center $\overline{\mathbf{i}}$, which is calculated relying on the current direction estimate represented by a k-distribution around $\overline{\mathbf{k'}}$, i.e. $\overline{\mathbf{i}} = \overline{\mathbf{i'}} + \overline{\mathbf{k'}}$. On the other hand, a turn of the robot leaves the center of the i-distribution unchanged, whereas the k-distribution is rotated by an angle $\Delta\phi_t$, which is a function of the motor actions. $\mathbf{k}_t$ denotes the unit vector with argument $\Delta\phi_t$. For straight motion we set $\|\mathbf{k}_t\| = \Delta\phi_t = 0$. After learning i is supposed to be related to the position of the robot by a fixed affine transformation and $\mathbf{k}$ codes the robot's heading direction up to a fixed rotation.

In this way $T(\xi, \xi') = T(\mathbf{i}, \mathbf{k}; \mathbf{i'}, \mathbf{k'})$ (cf. Eq. (4)) can be parametrized by the variances of positional and directional distributions, $\sigma_p$ and $\sigma_d$, resp.

$$T(\mathbf{i}, \mathbf{k}; \mathbf{i'}, \mathbf{k'}) = \exp\left(\frac{\langle \mathbf{k}, \mathbf{k'} + \mathbf{k}_t \rangle - 1}{\sigma_d}\right) \exp\left(-\frac{\|\mathbf{i} - (\mathbf{i'} + \mathbf{k'})\|^2}{2\sigma_p^2}\right), \qquad (5)$$

where $\langle \cdot, \cdot \rangle$ is the scalar product. Denoting the first term in (5) by $T_{\mathbf{k}_t}^{(d)}$ and the second one by $T_{\mathbf{k'}}^{(p)}$ the sum in (4) can be decomposed in order to obtain a numerically treatable probabilistic model of the robot's behavior to be referred to in the following section.

$$P_{t+1}(\mathbf{i}, \mathbf{k}) = \sum_{\mathbf{k'}} T_{\mathbf{k}_t}^{(d)}(\mathbf{k}; \mathbf{k'}) \sum_{\mathbf{i'}} T_{\mathbf{k'}}^{(p)}(\mathbf{i}; \mathbf{i'}) P_t(\mathbf{i'}, \mathbf{k'}) \qquad (6)$$

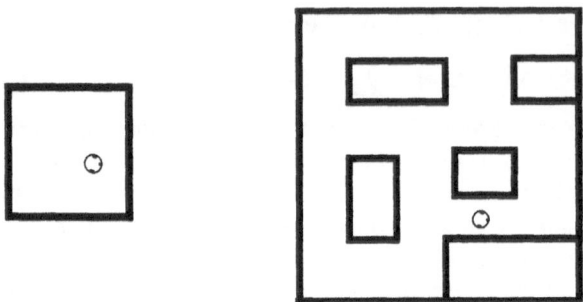

Figure 1: Screenshots from the Khepera Simulator [4] displaying *(left)* the square and *(right)* maze environments used in the simulations.

# 3   Robot experiments

We tested our model by simulating a mobile robot in the Khepera simulator [4]. Two stages of environment complexity were chosen: a simple square course and a maze with several obstacles, cf. Fig. 1. The challenge in the first environment consists in the highly ambiguous sensory input which is not specific to particular locations, whereas the second one requires larger computational resources while providing more specific information (which is still ambiguous).

The robot is controlled by a Braitenberg algorithm (cf. [5]; a version of rule "3a" is used, that is built-in in the simulator), which makes the robot turn left if an obstacle is recorded by sensors on the right and vice versa. In order to prevent the robot from getting trapped in a corner of the maze this behavior is supplemented by turns of random angle occurring on average every 100 steps. The algorithm for spatial localization turned out to be rather insensitive on the behavior of the robot provided that it guarantees a sufficient exploration of the environment. External sensations are made up from three groups of infrared sensors, i.e. $S = 8$ referring to the eight possible combinations of either one out of the two front, left or right sensors being activated. In the square (maze) problem $20 \times 20$ ($32 \times 32$) spatial units have been provided though only a fraction of these becomes activated in the stationary phase. In both cases 48 directional units have been used.

For evaluating the robot's performance the baseline $\phi = 0$ of the internal direction estimation was compared to the value with the direction of the physical $x$-axis, cf. Fig. 2. The robot is able to achieve an approximately consistent orientation in the environment after 1000 to 10000 time steps (traversing the maze a few hundred times) in dependence on task complexity and network size. Consistent orientation indicates that the ambiguities occurring in the square task are resolved into different internal activity patterns. Concerning the representation of the global spatial structure of the environment we found it useful to plot (Fig. 3) for each unit the mean objective location of the robot weighted by the activity of the respective unit. The results for the maze task (Fig. 3, *right*) suggest that a one-to-one correspondence between unit activities and the spatial structure of the environment is not necessary for consistent orientation.

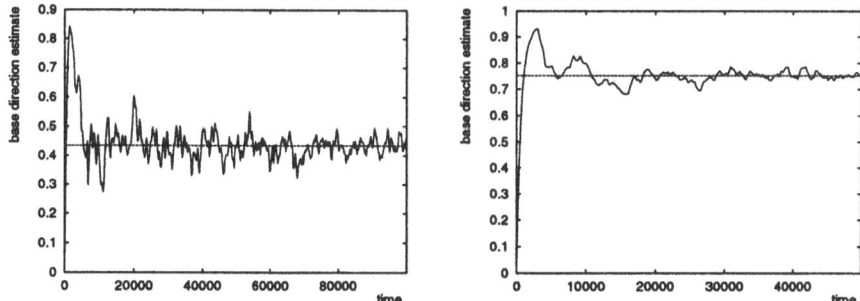

Figure 2: Time evolution of the direction estimate achieved by the robot. Displayed is the scalar product of the internal $k = (1, 0)^T$ direction and the heading direction of the robot. The solid lines are fits to the curves. Since the baseline of the internal direction estimate is arbitrarily fixed, only the consistent relation between internal estimate and actual heading direction is of behavioral importance. Larger fluctuations in the square *(left)* are due to less specific information available compared to the maze *(right)*.

# 4  Discussion

Robotics applications will involve more complex environments and behaviors as well as larger numbers of environmental parameters and sensory modalities than considered in the present experiments. In order to restrict the computatiohal requirements additional modules need to be included. This concerns firstly the selection of input states. Here, schemes for performance-related vector quantization are available [6]. Secondly, the occurrence of spatially homogeneous regions allows to contract several internal states into a single one. A hierarchical organization may allow to represent similar regions in the environment by identical regions in the cortex.

Whether or not the present approach bears some explanatory power for spatial localization of animals depends on the plausibility of its neural implementation rather than merely on a successful performance in this task. Eqs. (3) and (5) can be understood, resp., as a Hebbian learning rule and a linearized neural dynamics. A major challenge arises from the multiplicative interaction required by Eq. (1), which is according to present experimental evidence unlikely to occur in single neurons. However, preliminary results [7] exist that demonstrate how multiplicative responses can arise through population effects.

# 5  Conclusion

Understanding intelligent systems — both natural and artificial — does not consist in merely presenting the map from sensory inputs to behavioral outputs, but should include the quest for the internal "representation" of the agent's environment. Certain aspects of the representation may have come about by evolution or basic design (such as the dimensional structure of the states in our robot) or they refer to features which are acquired by life-long learning

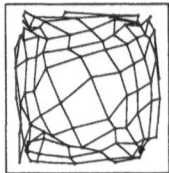
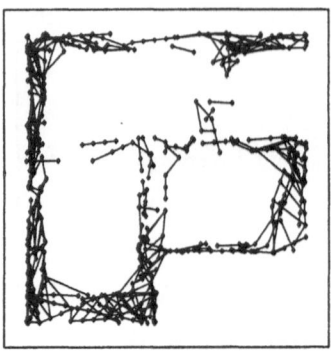

Figure 3: Place field centers defined as the true location of the robot averaged over the unit's activity. Lines between points indicate that the corresponding units are nearest neighbors with respect to the cortical connectivity, allowing thus to visualize the reconstructed topology of the environment, cf. Fig. 1. *(left)* The wide spacing in the center of the square course reflects higher positional uncertainty due to the lack of stimuli in this region, whereas twists in the corners are caused by different stimuli received in small nearby regions. *(right)* In the maze task the topology of the internal states is somewhat disrupted due to the complexity of the environment. This does not affect the orientation capabilities of the robot as can be seen in Fig. 2.

(such as the variances in Eq. 5), whereas other ones directly relate to events in the current environment. The present approach emphasizes the fact that such representations cannot be considered as an *image* of the environment, but are defined by their action-dependent predictive power with respect to future sensory inputs. Only upon inspection by an external observer (such as in Fig. 3) the representations of a behaving being may be understandable in terms of a physical reality.

# References

[1] Wilson MA, McNaughton BL. Evolution and dynamics of the hippocampal ensemble code for space in a novel environment. Science 1993, 261:1055-1058.

[2] Quirk GJ, Muller RU, Kubie JL. The firing of hippocampal place cells in the dark depends on the rat's recent experience. J. Neurosci. 1998, 10:2008-2017.

[3] Oore S, Hinton GE, Dudek G. A mobile robot that learns its place. Neur. Comp. 1997, 9:683-699.

[4] Michel O. Khepera Simulator version 2.0: Freeware mobile robot simulator. University of Nice Sophia–Antipolis, 1996. Downloadable at http://wwwi3s.unice.fr/~om/khep-sim.html.

[5] Braitenberg V. Vehicles: Experiments in synthetic psychology. MIT Press, Cambridge MA, 1986.

[6] Herrmann M, Der R. Efficient $Q$-learning by division of labor. In: Proc. ICANN'95, vol. 2. EC2 & Cie, Paris, 1995, pp 129-134.

[7] Salinas E, Abbott LF. A model of multiplicative neural responses in parietal cortex. Proc. Natl. Acad. Sci. USA 1996, 93:11956-11961.

# Reinforcement Learning of Collision-free Motions for a Robot Arm with a Sensing Skin

Pedro Martín[1]

Dept. of Computer Science, University of Jaume I

12071 Castellón, Spain

martin@inf.uji.es

José del R. Millán

ISIS, Joint Research Centre, European Commission

21020 Ispra (VA), Italy

jose.millan@jrc.it

## 1 Introduction

Sensory information is fundamental for autonomous robots that face unknown environments. On-line sensing allows a robot arm to modify its motion in real time to cope better with the environment. Reactive systems (e.g., [1]) are appropriate to generate on-line motions from local sensory data. A reactive controller can be implemented automatically by using artificial neural networks and reinforcement learning (RL) [2,3,4]. RL allows a neural network to acquire reaction rules while the robot arm interacts with its environment. We have previously demonstrated the feasibility of RL to acquire sensor-based reaching strategies for simulated multi-link planar manipulators [5]. In this paper, we extend this work to a real manipulator, namely a Zebra ZERO, that has a whole-arm sensing skin with sonar proximity sensors (see Fig. 1a). We describe a neural reactive controller that learns goal-oriented obstacle-avoiding motion strategies for such a manipulator in unknown 3D environments. The controller is made up of two main modules: a reinforcement-based action generator (AG) and a goal vector generator (GG). The AG uses local sensory data and position information to determine an appropriate deviation from the goal vector given by the GG. The task of collision-free reaching can be decomposed into two sequential subtasks: Negotiate Obstacles (NO subtask) and Move to Goal position (MG subtask). When the robot arm is not near the goal position and detects an obstacle in its way to the goal, the best strategy is to focus on negotiating the obstacle—moving along an efficient trajectory is not so important. The robot arm deals with the MG subtask later on, when the goal

[1] Pedro Martín was at the Joint Research Centre of the European Commission in Ispra (Italy) while this research was done. His stay was supported by a grant from Fundació Caixa-Castelló (Spain). He also acknowledges partial support from CICYT (grant TAP 95-0710-C-01).

position is nearby. As each subtask may require different motion strategies, the AG has two different reinforcement-based modules. When the robot is dealing with the NO subtask, the GG provides a fixed goal vector in the joint space that allows the AG to learn goal-independent motion strategies. For the MG subtask, the GG efficiently computes a goal vector in the configuration space by using a differential inverse kinematics (DIV) neural network.

The Zebra ZERO (see Fig. 1a) is a 6 dof manipulator with a gripper and a force sensor attached to the wrist. All its six joints are revolute: three joints are associated with the arm (first three links) and the other three with the wrist. The controller generates angular motions for the three joints of the arm $(q_1, q_2, q_3)$. In the experiment, the angular movements of the joints are constrained in degrees as follows: $q_{1min} \leq q_1 \leq q_{1max}$, $q_{2min} \leq q_2 \leq q_{2max}$, and $q_{3min} \leq q_3 \leq q_{3max}$; where $(q_{1min}, q_{1max}) = (-180, 180)$, $(q_{2min}, q_{2max}) = (10, 65)$ and $(q_{3min}, q_{3max}) = (-50, 0)$. The sensing skin has 13 ultrasonic smartsensor modules. The sensors provide a value in the range [1..255] that corresponds to the proximity of an object between 0.1 and 25.5 inches approximately. While performing the action given by the neural controller, the robot stops and retracts if any of the sensors detects an obstacle closer than a prespecified safety distance $d_s = 2$ inches.

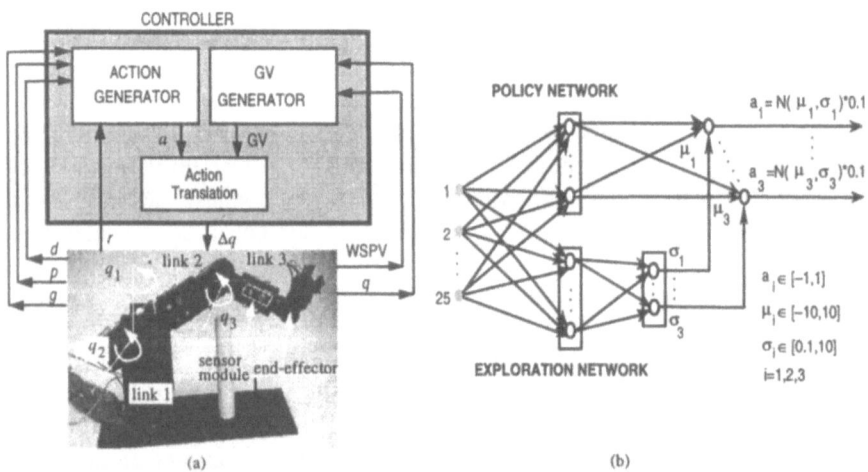

(a)                                    (b)

Figure 1: (a) The robot arm and the controller architecture. (b) Actor

## 2 Controller architecture

Fig. 1a shows how the controller interacts with the robot arm environment. The AG receives 25 real-valued inputs that range in the interval [0..1]. These inputs are divided into three groups: goal, position and sensory inputs. Goal inputs $g$ provide information about how close the robot arm is getting to the goal position. They correspond to the components of the workspace shortest path vector (WSPV) normalized between 0 and 1. Given the robot end-effector location $(\phi_1, \phi_2, \rho)$ and the goal location $(\theta_1, \theta_2, P)$, both in spherical coordinates, the WSPV is given by

$WSPV=(\theta_1-\phi_1,\theta_2-\phi_2,P-\rho)$. The NO module of the AG does not make use of the goal inputs, as we want the NO module to learn goal-independent motion strategies. Position inputs $p$ give information about the current joint configuration of the robot arm: $p_1=|\theta_1-q_1|/(q_{1max}-q_{1min})$, $p_i=|q_i|/(q_{imax}-q_{imin})$, $i=2,3$. The input $p_1$ is not supplied to the NO module since the way in which the robot can negotiate a certain obstacle while getting closer to the goal is independent of the angular value of the first joint. Three kinds of sensory inputs $d$ can be distinguished: *front*, *back* and *joint-limit*. The front and back components correspond to the sonar readings from the sensing skin. They are computed as follows: $d_i=1-(v_i/255)$, $i=1,13$, where $v_i$ is the range value given by the corresponding sonar sensor. The front inputs come from the sensors that lie in the direction of rotation that the first joint must follow to approach the goal. The joint-limit components are associated with virtual sensors. They give information about the proximity of the joints to their physical limits. There are two joint-limit inputs for each joint: $d_{imin}=1-(q_i-q_{imin})/17.2$, $d_{imax}=1-(q_{imax}-q_i)/17.2$, $i=1,2,3$. The joint-limit inputs are forced to be zero when their computation gives a negative value. A sensory input has a value close to 1 when its associated sensor is detecting a very near obstacle.

Finally, the AG gives an action vector $a=(a_1,a_2,a_3)$ that specifies a movement in relative spherical coordinates with regard to the goal vector (GV) given by the GG. All the components are real values in the interval $[-1,1]$. $a_1$ and $a_2$ represent the horizontal and the vertical angular deviation from the GV, respectively. $a_3$ corresponds to the length of the action vector as a proportion of the length of the GV. The actual angular increments $\Delta q=(\Delta q_1,\Delta q_2,\Delta q_3)$ supplied to the robot joints are calculated from the action vector.

The reinforcement signal $r$ is used to update the neural modules of the AG. It is computed in the same manner for both the NO and MG modules. The AG is severely punished when it runs into a *virtual collision* (i.e., if any sensor detects an obstacle closer than a safety distance): $r=-20$. For collision-free situations, $r$ pays attention to approach the goal properly: $r=-15e^{-4pg}$, where $pg = (1-|a_1||a_2|)|\Delta q|/|GV|$ measures the progress of the robot to the goal. Note that the reinforcement signal is only based on local information.

## 3  Action generator

The decomposition of the collision-free reaching problem in the NO and MG subtasks is achieved by defining a suitable region of reachability around the goal position. Given the end-effector location $(\phi_1,\phi_2,\rho)$ and the goal location $(\theta_1,\theta_2,P)$, both in spherical coordinates, the robot arm is considered to be inside the region of reachability when $|\theta_1-\phi_1|<20$ degrees. As each subtask may require different motion strategies, the action generator has two connectionist reinforcement-based modules, the NO module and the MG module. In addition, the AG has another simple reactive module that allows the robot to follow the GV (FG module). The FG module is used when the robot's sensors do not detect any obstacle along the direction to the goal. The NO module is in operation when an obstacle is detected along the direction to the goal and the robot arm is outside the region of reachability. The MG module is switched on when the end-effector enters the region of reachability. These last two modules have an actor-critic architecture [2]. The actor of each module aims at learning to perform those actions that optimize the cumulative reinforcement along the trajectory to the goal. Each actor is made up of two artificial neural networks: the policy network, which codifies the situation-action rules, and the exploration network (see Fig. 1b), which allows to explore the continuous action space. Both networks have one hidden layer of 50 logistic units. The policy network has Gaussian units that generate the three action components by using random Gaussian distributions $N(\mu_i, \sigma_i)$, $i=1,2,3$. The mean value $\mu_i$ for the Gaussian unit $i$ is the net input coming from the hidden layer to that unit. The standard deviation $\sigma_i$ of each Gaussian unit is provided by the corresponding output unit of the exploration network, which is linear. Figure 1b shows how the action components are finally calculated. The structural assignment problem is tackled by means of a REINFORCE algorithm to adjust the weights of the policy and exploration networks [4]:

$$\Delta w_{ij} = \alpha(r-b)e_{ij} = \alpha(r-b)\frac{\partial \ln N}{\partial w_{ij}}$$  (1)

where $\alpha$ is the learning rate, $b$ is the reinforcement baseline, and $e_{ij}$ is the characteristic eligibility of $w_{ij}$. In order to deal with the temporal credit assignment problem, the critic of each module uses *temporal differences (TD) methods* [3] to learn the cumulative reinforcement, TD(0) in particular. Each critic is also implemented as an artificial neural network with a linear output unit and one hidden layer of 50 logistic units.

## 4 Goal vector generator

The GG provides suitable goal vectors for the different modules of the AG. Whenever the NO module of the AG is in operation the following goal vector in the joint space in degrees is provided: GV=(20,0,0), if the angle of the first joint must be increased to approach the goal location and GV=(−20,0,0), in the other way around. For the MG module, the GG produces suitable goal vectors with a differential inverse kinematics (DIV) neural network. Such DIV network is also useful for redundant manipulators. This network is inspired by the *distal learning* approach [6] to the inverse kinematics of robot manipulators. Let $q$ and $p$ be the current configuration and the current end-effector location of the robot arm in spherical coordinates. Let $p^*$ be the goal location in spherical coordinates where the end-effector must be finally placed. The robot's task can be stated as a differential inverse kinematics problem: "Given the difference $\Delta p = p^* - p = WSPV$ in the workspace between the current location of the end-effector $p = F(q)$ and a goal location $p^*$, obtain a goal vector $GV$ that makes the robot approach the closest goal configuration $q^*$." Function $F$ stands for the forward kinematics. We have applied a backpropagation inversion algorithm [7] to a feedforward neural network that has been previously trained to approximate the robot forward kinematics in spherical coordinates. The network has one hidden layer of 50 units with logistic units and has been trained through backpropagation with momentum. A gradient search in input space is conducted by this algorithm through several iterations. The search generates a sequence of input vectors $q=q^0,q^1,\ldots,q^n$ such that

$$E = \frac{1}{2}\left(F\left(q^n\right)-p^*\right)^T\left(F\left(q^n\right)-p^*\right) \qquad (2)$$

is minimized. That error function is the same used to train the neural network with the forward kinematics. The inversion procedure finishes when $E<\xi$, where $\xi$ is the error tolerance to consider that a goal configuration is achieved. From the last configuration of the sequence, a GV= $q^n-q$ is obtained. Suitable goal vectors for both the MG and FG modules of the AG are obtained in just one iteration of the neural inversion. See [5] for the benefits of the DIV neural network over other approaches to compute goal vectors.

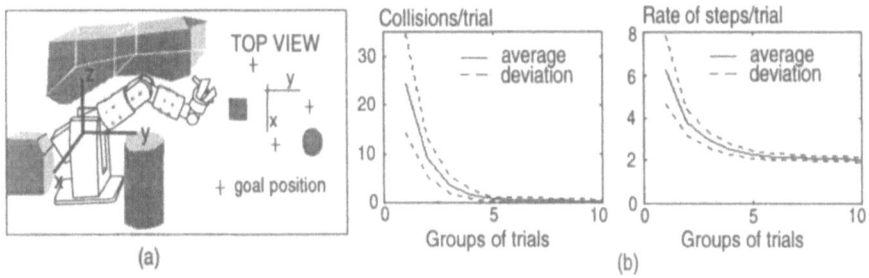

Fig. 2: (a) Environment used in the experiments (profile and top view). (b) Results.

## 5  Experimental setup and results

A high level task for a manipulator generally requires the end-effector of the manipulator to be placed sequentially at different goal locations. A suitable learning method for acquiring collision-free motions must allow the robot to learn appropriate motions while trying to accomplish high level tasks. Our controller copes with such a realistic demand. An experiment has been carried out in the environment shown in Fig. 2a to test the performance of the controller. Three different goal locations were considered. The experiment consists of 20 runs, differing in the initial seed of the random number generator. All runs are made of 300 trials. At the beginning of a trial, the robot is commanded to reach one of the goal locations. Once the arm arrives there, a new trial starts whose target is one of the two remaining goal locations (selected at random). Thus, the starting and target locations change from trial to trial. For each run, the performance of the controller is measured every group of 30 trials according to two parameters: the number of collisions and the rate of steps (see Fig. 2b). Collisions fall down below 1 per trial after only 150 trials (around 5 hours of robot operation) and the number of steps is also reduced to 2.3 times the number of steps needed for the manipulator to follow the shortest path in the configuration space (i.e., straight line between the starting and goal configurations). The performance of the controller keeps on improving as learning proceeds. It is worth noting that since the learning task is defined by the goal location and the obstacles as perceived by the sensors, in this experiment the robot arm faces a different environment at each trial. However, the controller succeeds in learning environment-independent motion strategies. This is mostly due to two factors, namely, the NO module is supplied with local goal-independent information and the reinforcement signal. The input codification, which exploits the symmetry of the sonar information (front and back sensory inputs), has also been essential to achieve such a performance of the controller.

# 6 Conclusions

We have proposed a controller to learn obstacle-avoiding reaching strategies for sensor-based multilink robot arms. This approach has been tested on a real robot arm, namely a Zebra ZERO, that has a sonar sensing skin mounted along its links. The controller consists of two main modules: a reinforcement-based action generator (AG) and a goal vector generator (GG). The collision-free reaching task is decomposed into two sequential subtasks: negotiate obstacles (NO) and move to goal (MG). The AG is implemented with two connectionist actor-critic modules, one for each subtask. The GG makes use of a differential inverse kinematics neural module to obtain suitable goal vectors for the MG module. The NO module is supplied with a fixed goal vector that is independent of the goal location. The use of this goal vector, as well as the environment-independent input codification and reinforcement schedule for the controller, facilitates the learning of the motion strategies in realistic situations with different goal locations at each trial.

# References

[1] Brooks R. A. A robust layered control system for a mobile robot. IEEE Journal of Robotics and Automation 1986; 2:14-23

[2] Barto A. G., Sutton R. S., Anderson C. W. Neuronlike adaptive elements that can solve difficult learning control problems. IEEE Trans. on Systems, Man, and Cybernetics 1983; 13: 834-846.

[3] Sutton, R. S. Learning to predict by the methods of temporal differences. Machine Learning 1988; 3:9-44.

[4] Williams R. J. Simple statistical gradient-following algorithms for connectionist reinforcement learning. Machine Learning 1992; 8:229-256.

[5] Martín P., Millán J. del R. Learning reaching strategies through reinforcement for a sensor-based manipulator. Neural Networks 1998; 11:359-376.

[6] Jordan M. I., Rumelhart, D. E. Forward models: Supervised learning with a distal teacher. Cognitive Science 1992; 16:307-354.

[7] Kindermann J., Linden A. Inversion of neural networks by gradient descent. Journal of Parallel Computing 1992; 14:277-286.

# Multitask Pattern Recognition for Vision-Based Autonomous Robots

Rich Caruana
JustResearch, 4616 Henry St.
Pittsburgh, PA 15213
caruana@cs.cmu.edu

Joseph O'Sullivan
Carnegie Mellon University
Pittsburgh, PA 15213
josullvn@cs.cmu.edu

## Abstract

This paper uses Multitask Pattern Recognition (MTPR) to improve the accuracy and robustness of neural net based vision systems. MTPR trains neural nets on a set of auxiliary recognition problems at the same time the net is trained on the main recognition task. The predictions made for the auxiliary tasks are not used, but the internal features learned by the net for them improve performance on the main recognition task. The auxiliary tasks allow us to focus attention towards features that learning would otherwise ignore. MTPR is broadly applicable. It improves performance on a simulated ALVINN domain 10%-30%.

## 1 Multitask Pattern Recognition

Figure 1 shows a neural net trained to do road following[8]. The net inputs are the retina of a camera. The main pattern recognition task is training the net to steer the vehicle given images of the road on the retina. But the net also has additional outputs for extra pattern recognition tasks. The extra outputs are fully connected to a hidden layer shared with the main steering direction task. Backpropagation is done in parallel on all outputs in this multitask net.

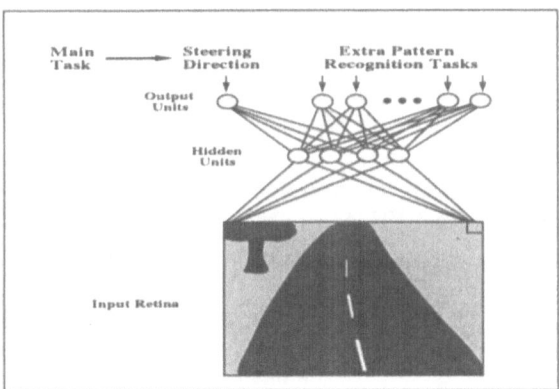

Figure 1: MTPR for an autonomous vehicle learning to steer from road images.

Because all outputs share a hidden layer, internal representations learned in the hidden layer for one task can be used by other tasks. Moreover, some features may develop to support several tasks that would not have developed in nets trained on single tasks. Sharing what is learned by different tasks while tasks are trained in parallel is the central idea in multitask learning[1, 3, 4, 5, 9].

MTPR uses the information in the training signals of the extra tasks to bias learning towards a better solution for the main task. MTPR has been shown to be effective in learning systems based on decision trees[4], nearest neighbor[3, 4, 10], and artificial neural nets[6]. Here, we apply MTPR to pattern recognition for autonomous vehicles (ALVINN)[8]. We discuss how neural net based MTPR boosts the performance of learning robots and show that adjusting the learning rates of extra tasks improves the performance of MTPR.

## 2   The SYN-ALVINN Domain

The data in SYN-ALVINN is generated with a road image simulator developed by Pomerleau to permit rapid testing of learning methods for road-following domains[8]. Figure 2 shows several of the generated road images.

Figure 2: One and two-lane roads generated with Pomerleau's road simulator.

Road images are inputs to the net in Figure 1. This simple fully-connected feedforward net is not told how input pixels are spatially related. The vehicle is randomly positioned on the road, to the left or right of where good driving would keep it. This promotes robustness by insuring the net learns to recover from a broad range of situations, but makes learning harder. The images contain one or two-lane roads. This also makes learning harder; steering direction depends on whether one is driving on a single lane road, or on a two-lane road where the vehicle should be centered in one lane instead of at the road center.

The principal task in SYN-ALVINN is to predict steering direction. We used eight additional pattern recognition tasks for MTPR:

- whether the road is one or two lanes
- location of centerline (2-lane roads only)
- location of left edge of road
- location of right edge of road
- location of road center
- intensity of road surface
- intensity of region bordering road
- intensity of centerline (2-lane roads only)

These additional tasks are computed from internal variables in the simulator. We modified the simulator so that training signals for these extra tasks are added to the synthetic data with the training signal for the main steering task.[1]

---

[1]Extra tasks such as centerline location can be used for real autonomous vehicles. While

# 3 Results

Table 1 shows the average performance of ten runs of single and multitask pattern recognition on SYN-ALVINN using nets with one hidden layer. We tried STPR nets with 2, 4, 8 or 16 hidden units. Each single task net has one output. The MTPR net has 16 hidden units, and 9 outputs. We did not optimize the size of the MTPR net and believe it would perform better with a larger hidden layer. Because this experiment compares the best results from multiple runs of STPR nets with varying net size with single runs using unoptimized MTPR net sizes, it is biased in favor of the STPR nets.[2]

Table 1: Performance of single and multitask nets on the SYN-ALVINN domain. Underlined entries in the STPR columns are the single task runs that performed best. The single MTPR run outperforms the best of the four STPR runs on 6 of the 9 tasks. (Differences significant at $p = .05$ are marked with *.)

| | ROOT-MEAN SQUARED ERROR ON TEST SET | | | | | |
|---|---|---|---|---|---|---|
| **TASK** | **Single Task (STPR)** | | | | **MTPR** | Change MTPR |
| | 2HU | 4HU | 8HU | 16HU | 16HU | to Best STPR |
| 1 or 2 Lanes | .201 | .209 | .207 | <u>.178</u> | <u>.156</u> | -12.4% * |
| Left Edge | <u>.069</u> | .071 | .073 | .073 | <u>.062</u> | -10.1% * |
| Right Edge | .076 | .062 | .058 | <u>.056</u> | <u>.051</u> | -8.9% * |
| Line Center | .153 | <u>.152</u> | .152 | .152 | <u>.151</u> | -0.7% |
| Road Center | .038 | <u>.037</u> | .039 | .042 | <u>.034</u> | -8.1% * |
| Road Greylevel | <u>.054</u> | .055 | .055 | .054 | <u>.038</u> | -29.6% * |
| Edge Greylevel | <u>.037</u> | .038 | .039 | .038 | <u>.038</u> | 2.7% |
| Line Greylevel | .054 | .054 | <u>.054</u> | .054 | <u>.054</u> | 0.0% |
| Steering | .093 | <u>.069</u> | .087 | .072 | <u>.058</u> | -15.9% * |

Entries under the STPR and MTPR headings are the test set error.[3] The last column is the percent reduction in error of MTPR over the best STPR run. Negative percentages indicate MTPR performs better. On the main steering task, MTPR outperforms STPR 15%. It does this without any extra training patterns: exactly the same training patterns are used for both STPR and MTPR. The only difference is that multitask training patterns have training signals for all nine tasks, whereas single task training patterns have training signals for only one task at a time. (Experiments show that the benefit from MTPR is not due to regularization caused by the extra tasks injecting noise into the net[4].)

---

it is hard to build vision systems to accurately locate centerlines in forward looking images, it is easy to accurately locate centerlines along the side of the vehicle. Dead reckoning can compute where the centerlines would have been in the images collected from a forward-looking camera, and this allows those images to be augmented with the extra task training signals.

[2]Nets using 2 hidden layers yielded similar results.

[3]Most differences between STPR with different size nets are random variation. However, the right edge task may be learned better, and prefer more hidden units, than the left edge task because images of 2 lane roads are biased by driving on the right.

Figure 3 shows the performance on the main steering task as the number of examples increases. Neither method performs well with little data. As the number of examples increase, MTPR outperforms STPR. When MTPR sees 25–65 examples, it performs as well as STPR with 45–120 examples. When there is enough data for STPR to perform well, there is no benefit to MTPR.

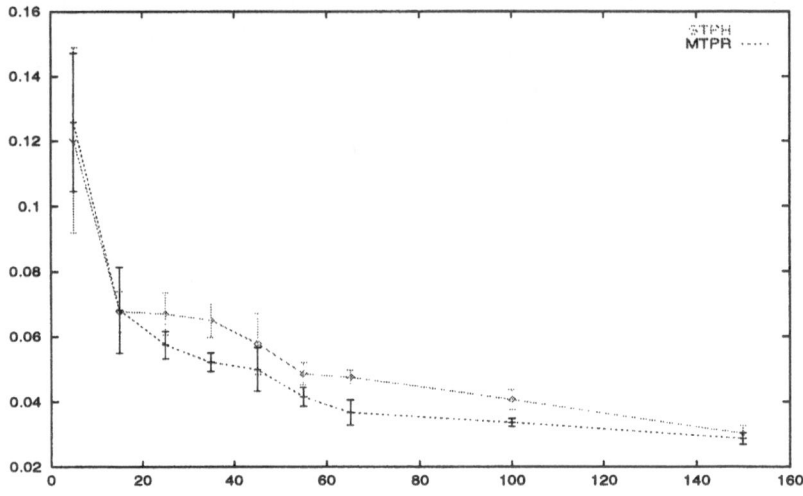

Figure 3: Test set RMSE for steering direction as training set size increases.

# 4  Using Extra Tasks to Focus Attention

Pattern recognition systems often learn to use large or ubiquitous patterns. MTPR can coerce learning to attend to small or less common patterns it might otherwise ignore by forcing the net to learn internal representations for tasks that depend on the small or less frequent patterns. Nets trained for road following have difficulty learning to use lane markings because lane markings are small and move a lot. The extra tasks in Section 2 require the MTPR net to learn internal representations for road stripes. Since the MTPR net learns to steer with this hidden layer, the steering task uses some of the hidden representation developed for the stripe tasks. (See [2, 7, 8] for alternate approaches.)

We did an experiment where nets were *trained* on road images containing center-stripes, but nets were *tested* on images with the center-stripes removed. The STPR net error increased by a factor of 2.0 when center-stripes were removed. The MTPR net error increased by a factor of 3.1. We conclude that MTPR's steering predictions are, in fact, more sensitive to center-stripes.

# 5  Learning Rates and Early Stopping

The MTPR net for SYN-ALVINN has nine outputs. Figure 4 shows learning curves for four of these tasks. Each plots the root-mean-squared-error of one output on a test set as the net trains.

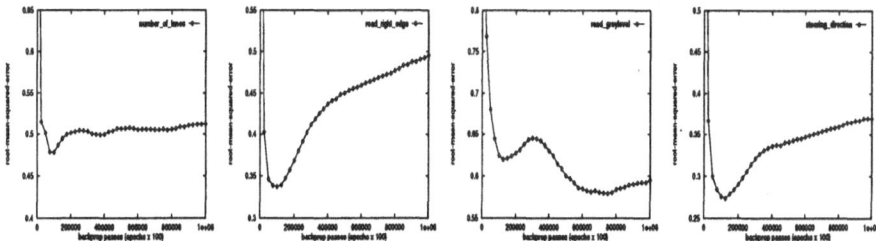

Figure 4: Test-Set Learning Curves for Tasks 1, 3, 6, and 9 (task 9 is steering direction) from MTPR Net Trained on Nine ALVINN Tasks.

From the graphs it is clear that the best place to halt training differs for each task. Task 6 reaches peak performance at 775,000 backprop passes, but the main steering task performs best at 125,000 passes. There is no one epoch where training can be stopped to achieve maximum performance on all tasks.

We applied gradient descent to the learning rates of the extra tasks to find the learning rate for each extra task that maximized performance on the main steering task. Table 2 shows the change in RMS Error for the nine tasks after learning rate optimization for the main task. Optimizing the learning rates for the extra MTPR tasks improved performance on the main steering task an additional 11.1%. This improvement is over and above the original improvement of 15% for MTPR over STPR observed in Section 3.[4]

Table 2: Change in performance of MTPR on the nine tasks after learning rates are optimized to improve performance on the main steering direction task.

| TASK | % Change |
|---|---|
| 1: 1 or 2 Lanes | -2.8% |
| 2: Left Edge | -3.5% |
| 3: Right Edge | -19.4% |
| 4: Line Center | -17.0% |
| 5: Road Center | -24.1% |
| 6: Road Greylevel | +10.4% |
| 7: Edge Greylevel | -15.0% |
| 8: Line Greylevel | -8.6% |
| 9: Steering Direction | -11.1% |
| Average | -9.0% |

Performance on 7 of the 8 extra tasks also improves. This is surprising because the criterion optimized is performance on the main task, not the performance on any of these extra tasks. What is good for the goose appears to be

---

[4]Applying gradient descent to the learning rates is computationally expensive, and is not currently feasible for more than a few dozen tasks.

good for the gander. Interestingly, optimizing the learning rates on the extra tasks has a significant effect on the rate at which the main task is learned. This is interesting because we do not adjust the learning rate of the main steering task. Any change in how fast the main task is learned is due to changes in how other tasks it shares the hidden layer with are learned.

# 6 Summary

In this paper we demonstrate the benefit of MTPR in a synthetic ALVINN domain. Although the MTPR net has more free parameters, the extra information in the training signals for the extra tasks allows MTPR to improve steering direction prediction accuracy up to 15% in this domain. For moderate sample sizes, MTPR is able to achieve similar accuracy to STPR, but using only half as much training data. We also show how optimizing the learning rates for the extra tasks yields an additional 11% improvement. MTPR is a general method that can be applied to many problems in vision-based robotic systems. See [4] for a review of other machine learning domains to which MTPR is applicable.

# References

[1] Y.S. Abu-Mostafa, "Learning From Hints in Neural Networks," *Journal of Complexity* 6:2, pp. 192–198, 1989.

[2] S. Baluja, and D.A. Pomerleau, "Using the Representation in a Neural Network's Hidden Layer for Task-Specific Focus of Attention". In C. Mellish (ed.) IJCAI-95: Montreal, Canada. pp 133-139, 1995.

[3] J. Baxter, "Learning Internal Representations," *8th ACM Conference on Computational Learning Theory*, (COLT-95), Santa Cruz, CA, 1995.

[4] R. Caruana, "Multitask Learning," Doctoral Thesis, Carnegie Mellon University: *CMU-CS-97-203*, 1997.

[5] R. Caruana, "Multitask Learning: A Knowledge-Based Source of Inductive Bias," *Proceedings of ICML-93*, Amherst, 1993, pp. 41-48.

[6] R. Caruana, "Learning Many Related Tasks at the Same Time With Backpropagation," *Advances in Neural Information Processing Systems 7*, 1995.

[7] Davis, I. and Stentz, A., "Sensor Fusion for Autonomous Outdoor Navigation Using Neural Networks," *IEEE Intelligent Robots and Systems Conference*, 1995.

[8] Pomerleau, D. A., "Neural Network Perception for Mobile Robot Guidance," Doctoral Thesis, Carnegie Mellon University: *CMU-CS-92-115*, 1992.

[9] S. C. Suddarth and Y. L. Kergosien, "Rule-injection Hints as a Means of Improving Network Performance and Learning Time," *Proceedings of the 1990 EURASIP Workshop on Neural Networks*, 1990, pp. 120-129.

[10] S. Thrun and J. O'Sullivan, "Discovering Structure in Multiple Learning Tasks: The TC Algorithm", Proceedings of ICML-96, Torino, Italy, 1996.

# Three Principles of Hierarchical Task Composition in Reinforcement Learning

H. Vollbrecht

Neural Information Processing Department
University of Ulm
hans@neuro.informatik.uni-ulm.de

**Abstract:**

We present three principles of hierarchical task composition within a single agent using reinforcement learning to solve continuous control problems. We consider complex tasks having goals defined as conjunctions of subgoals, each learned by a separate task with Q-learning. However, subgoals may depend on each other, requiring particular task composition principles. In the first principle, the Q-function of some task gets underlaid with the Q-function of an avoidance task, resulting in a composition in which the latter may put a veto on an action of the former. The second principle uses explicit task activation as a hierarchical relation between two tasks. Subtask activation lasts just one time-step the length of which is adapted to the particular subtask's state-space discretization. In the third principle, two tasks are related to each other such that the hierarchically higher one perturbs the goal state of the lower one in the direction of its own goal. These principles define interaction in a multi-layer architecture, with sequential task composition within each layer, and with each maintaining the system in an equilibrium condition. The approach is demonstrated with the task in which a truck navigates backwards to a docking point.

## 1 Introduction

Model-free reinforcement learning for control problems has become increasingly popular in the last years. The perspective to learn control tasks from scratch, without supplying a model of the process dynamics, is clearly attractive at first sight. However, for practically relevant tasks having complex state spaces it is unrealistic to be learnt from scratch, without any predefined structure. Several approaches for structuring the learning problem and the agent's internal representations have been proposed. For an overview see [1]. One such approach tries to break down a task into subtasks, each responsible for a goal condition simpler than that of the whole task. The objective is to reduce the complexity of the state space and the depth of temporal propagation of the experienced reward. We present here our work on this approach. The target application domain consists in continuous, deterministic control problems. Our example task domain is navigating a trailer truck backwards [2]. Tasks are defined by a reward function $r$ based on predicates on the state variables. Complex tasks can be defined by conjunction (or disjunction) of these predicates. Composition structure for defining task goals with reward functions thus reduces to logical operators, without any hierarchical structure.

However, defining goal structure is much easier than defining structure in learning. In our approach, the agent learns different tasks for different goal predicates independently, each with the same learning algorithm (Q-learning, [3]) but with a different state space. How can independently learned tasks be combined? Since the solution paths to different goals in continuous control problems are notoriously dependent on each other, composition of elementary behaviors to solve a complex task may require a learning process by itself. We found three principles having basic relevance for predefining composition structure in this learning process ([5,6] present other approaches). In the first principle, the *veto principle*, the Q-function of some task gets underlaid with the Q-function of a reactive avoidance task, resulting in a composition in which the latter may put a veto on an action of the former, but leaving it otherwise completely uninfluenced. The second principle, the *subtask principle*, uses explicit task activation as a hierarchical relation between two tasks. Unlike the common approach in which the activated (sub)task is active until subgoal achievement (useful for disjunction of subgoals), we let it perform just one time-step whose length is adapted to the particular subtask's state-space discretization. In the third principle, the *perturbation principle*, two tasks are related to each other such that the hierarchically higher one perturbs the goal state of the lower one in the direction of its own goal. These principles define a multi-layer architecture, with sequential task composition on each layer, and each keeping the system in an equilibrium condition which can be perturbed by the next higher one.

In section 2, we overview the system characteristics and the basic learning theory. In section 3, we present the example domain. Details of the three composition principles are given in section 4. Results are presented and discussed in section 5.

## 2   Q-learning for continuous control processes

A process $\varphi_{x,a(.)}$ is described by a state space $S \subset \mathbb{R}^n$, a control space $Act \subset \mathbb{R}^m$, and a differential equation

$$\dot{\varphi}(t) = f(\varphi(t), a(t)), \qquad \varphi(0) = x, \qquad (\varphi(t) \in S, a(t) \in Act), \text{ with solution } \varphi_{x,a(.)}.$$

The control problem is defined as finding an optimal control law $\pi^* : S \to Act$ such that for any $x \in S$ the reward functional $J_\infty$ with

$$J_t(x, a(.)) := \int_0^t \gamma^\tau \cdot r(\varphi_{x,a(.)}(\tau)) d\tau$$

is maximized at $a(.)$ with $a(\tau) = \pi^*(\varphi_{x,a(.)}(\tau))$. $r$ specifies the reward, $\gamma \in (0,1)$ is a discount factor for future reward, and $J_t(x,a(.))$ defines the accumulated, discounted reward encountered by the process $\varphi_{x,a(.)}$ during the time interval $[0,t]$. Our control agent uses a basic time step $h$, and a finite action set $A \subset \{a:[0,k\cdot h] \to \mathbb{R}^m \mid k \in \mathbb{N}\}$ of constant functions, with $k$ the action duration $(a_{dur})$, and $a(t) \equiv a_{int}$ the action "intensity". It is well known from dynamic programming [4], that the equation

$$Q_h(s,a) = r_h(s,a) + \gamma^{a_{dur}\cdot h} \cdot \max_{a' \in A} \{Q_h(\varphi_{s,a}(a_{dur} \cdot h), a')\}$$

with $r_h(s,a) = J_{a_{dur}\cdot h}(s,a)$ , has a unique fixpoint $Q^*_h{:}S{\times}A{\to}\mathbb{R}$ (when $\gamma{<}1$ and $r$ bounded). $Q^*_h$ is the optimal Q-function which defines through $\pi^*(s) :=$ argmax$_{a\in A}\{Q^*_h(s,a)\}$ an optimal control law as defined previously.

Our agent adapts $a_{dur}$ to the discretization of state space, such that autotransitions in cells of the state space are avoided. The agent learns the discretization itself using a kd-tree structure (fig. 1)[1]. A Q-learning on this state space is then defined by[2]

$$Q_h^{(n+1)}(i_n,a_n)=(1-\alpha_n)\cdot Q_h^{(n)}(i_n,a_n)+ \alpha_n\cdot(r_h(s_n,a_n)+\gamma^{(a_n)_{dur}\cdot h}\cdot\max_{a'\in A}\{Q_h^{(n)}(\hat{i}_n,a')\}) \quad (*)$$

with $s_n\in S_{i_n}$, $\varphi_{s_n,a_n}((a_n)_{dur}\cdot h)\in S_{\hat{i}_n}$, and $S=\bigcup_{i=1}^N S_i$, the partitioning of state space

# 3  The TBU example

In the Truck-Backer-Upper example [2], the control agent has to navigate a trailer truck to a docking point (fig. 2). The truck can move only backwards, and it has to avoid an inner blocking between cab and trailer, too large a steering angle, and hitting the wall. The dynamics of this system are highly nonlinear, and the final task $T_{final}(\dot{\Phi}_{int}=0 \wedge \Phi_{ext}=0 \wedge x=0 \wedge y=0)$[3] has a complex state space $(\dot{\Phi}_{int},\Phi_{int},\Phi_{ext},x,y)$. The following list shows some examples of simpler tasks:

*avoidance tasks:*
$T°_1(\Phi_{int}\geq\Phi_{intern\_max} \mid (\dot{\Phi}_{int},\Phi_{int}))$ — avoid blocking between cab and trailer
$T°_2(\text{"cab or trailer hit the wall"} \mid (\dot{\Phi}_{int},\Phi_{int},\Phi_{ext},y))$ — avoid wall collision
*goal seeking tasks:*
$T_3(\Phi_{goal}=0 \mid (\dot{\Phi}_{int},\Phi_{int},\Phi_{goal}))$ — line up the trailer with the goal point
*state maintaining tasks:*
$T_4(\dot{\Phi}_{int}=0 \mid (\dot{\Phi}_{int},\Phi_{int}))$ — run on a straight line
$T_5(\Phi_{goal}=0 \wedge \Phi_{int}=0 \mid (\dot{\Phi}_{int},\Phi_{int},\Phi_{goal}))$ — line up cab & trailer with goal point

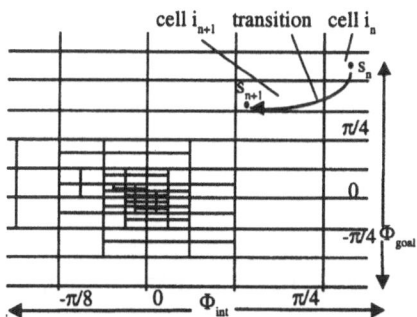
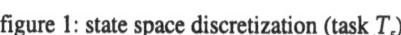
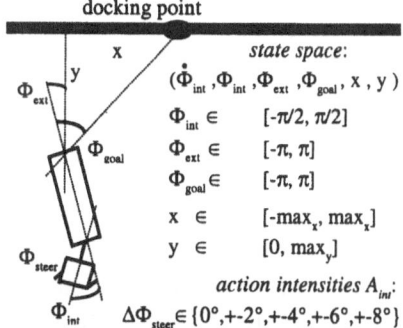

figure 1: state space discretization (task $T_5$)      figure 2: the TBU-example

---

[1] goal seeking tasks: highest discretization resolution is in (positive) reward region; avoidance tasks: highest resolution at the border of "collision avoidable"- and "collision inevitable"-regions, much like in the Party-Game algorithm of Moore [7].

[2] note that $(a_n)_{dur}$ is a function of state $i_n$, and $\alpha_n$ is a function of $i_n$ and action $a_n$.

[3] we use the notation $T(p)$ or $T(p\mid S)$ for a task $T$ with goal predicate $p$ and state space $S$. $T°(p\mid S)$ denotes an avoidance task with $p$ defining the collision region.

# 4 Hierarchical task composition

A strategy for dealing with large state spaces and deep temporal credit assignment is to structure the learning problem into a hierarchy of simpler learning problems. This hierarchy can reflect hierarchical state space discretization/aggregation [5], action hierarchies, or hierarchical goal composition [6]. In our approach, we developed hierarchical composition principles based on goal conjunction, with goals whose solution paths may interact, state spaces not necessarily representing physical space (see figure 1), and single task learning in a hierarchy of tasks, in an incremental bottom-up manner, where a task learns under the influence of already learned tasks at lower hierarchical levels.[4] In the next three subsections, we present composition principles in which tasks $T_1, T_2$ ... with goals $p_1, p_2$ ... are combined to carry out a combined task with goal $p_1 \wedge p_2 \wedge$ ... .

*The veto principle*
In this principle, an avoidance task $T_1$ may veto the action selected by some task $T_2$ (for example, take tasks $T^\circ_1$ and $T_3$ from section 3). First, task $T_1$ is learned in isolation. Then $T_2$ learns its Q-function. During this learning and later also during agent controlling, the veto principle is applied to the *action selection* for $T_2$ following the policy $\pi(s) := \text{argmax}_{a \in A}\{Q_{combined}(s,a)\}$ where

$$Q_{combined}(s,a) = Q^N(s,a) + c_1 \cdot \frac{1}{1 + e^{c_2 \cdot (Q_{avoid}(s,a) - c_3)}}$$

with $Q^N$ the Q-function of $T_2$ (normalized to [0,1]), $Q_{avoid}$ the Q-function of $T_1$, and $c_3$ a critical value (we use half of the negative reward for collision). The effect of this combination using a squashing function is a policy that uses $Q_{avoid}$ in an ε-margin around the collision part of state space (future collision inevitable), but that remains mostly unaffected by $Q_{avoid}$ in the rest of the non-collision part, thus resulting in a pure veto (0-1) interaction. Note that task $T_2$ does *not learn* the combined Q-function $Q_{combined}$. This is advantageous because, depending on the granularity of state-space discretization, $T_2$ shall eventually evaluate an action as good in cell "$i$" even when in a certain percentage of the cases this action will be vetoed[5], depending on the nondeterministic position within the cell.

*The subtask principle*
A task $T$ activates another task $T_i \in \{T_1, T_2 .. T_m\}$, and this activation represents an action $a_j$ of $T$ evaluated by its Q-function. Thus, for subtask $T$ , we have $A_{int} = \{\text{activate}(T_j) \mid 1 \leq j \leq m\}$. Unlike the common approach in which an activated (sub)task $T_j$ is active until achievement of its goal $p_j$ (useful for finding the shortest path for a predicate disjunction), we let it perform just one time-step of length $a_{dur} \cdot h$ with action $a = \pi_j(i)$, where $\pi_j$ is $T_j$'s policy and $i$ the actual state cell in $T_j$'s state-space discretization. Note that $a_{dur}$ is adapted to $T_j$'s state-space discretization, and activate$(T_j)_{dur}$ which defines how many times $T_j$ will be activated in sequence, is

---

[4] because of frequently encountered instabilities when learning tasks simultaneously.
[5] in this case, Q-updating as defined by (*) will simply not be executed

adapted to the specific state-space discretization of $T$. In this context, task $T$ will learn the goal composition $\wedge_{1 \leq j \leq m} p_j$ with already learned subtasks $T_j$ by finding the correct sequence of subtask activations. As a TBU example, take $T = T_5$ , with subtasks $T_3$ and $T_4$ learnt in isolation without knowing that their solution paths interact. This is what the combined task $T$ learns about. Because of this interaction, it must have control on single time-steps of each subtask and not merely on achievement of the (sub)goals $p_j$ . Our discretization strategy which applies maximum resolution in the goal region, leads to consistent integration of action durations of the supertask and of its subtasks.

*The perturbation principle*

Here, two tasks $T$ and $T'$ are related to each other such that the hierarchically higher one, say $T$, perturbs with his actions the goal state of the lower one, $T'$, in the direction of its own goal (rewarded state region) . This is accomplished by activating $T'$ after execution of any action $a$ (with duration $a_{dur}$) of $T$ and, unlike the subtask principle, leaving $T'$ active, and $T$ interrupted, until $T'$ reaches again its goal region. $T'$ then returns control to $T$, and only at this point the (perturbation-) action $a$ of $T$ is concluded and thus evaluated. See figure 3 for the general principle and figure 4 for a TBU example. This approach results in the *reduction of state-space complexity* for the perturbing task $T$ in the way that its state space $S_T$ reduces to $S_T|_{p'} := \{ s \in S_T \mid p'(s) \text{ is true} \}$ with $p'$ the conjunction of all goals of lower levels.

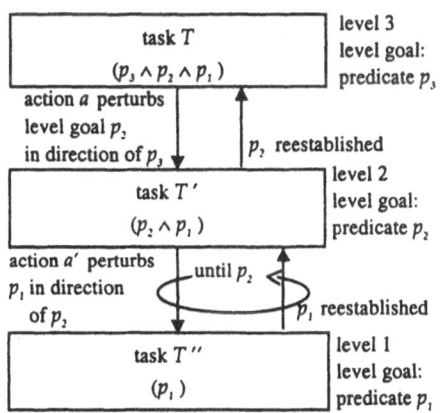

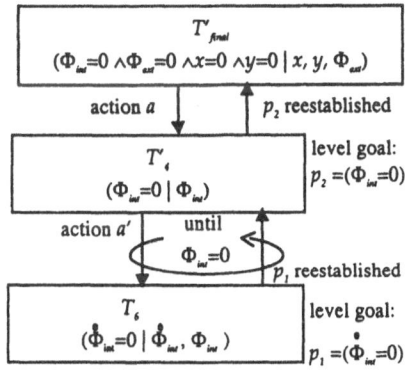

figure 3: perturbation interaction at three levels, for one action $a$ of task $T$ at level 3

figure 4: composition for $T'_{final}$ in a perturbation hierarchy. Note the simplification in state space (see $T_{final}$ and $T_4$ from chapter 3)

# 5 Results

The principles described in section 4 have been tested in the TBU example. It consists of some 9 tasks on 3 levels in the perturbation hierarchy. State space complexity ranges from 16 states (cells) in $T_4'(\Phi_{int}=0 \mid \Phi_{int})$ to 770 states in $T°_2$("hit the wall" $\mid (\Phi_{int}, \Phi_{ext}, y)$. The two most complex tasks are the final task of docking at the goal point with 550 states ($T'_{final}$ ,see fig. 6) and the wall avoidance task ($T°_2$ , see fig. 5). $T°_2$ has been learned after 5000 trials, and the final task has been learned after 10000 trials, starting from uniformly distributed initial positions on state space.

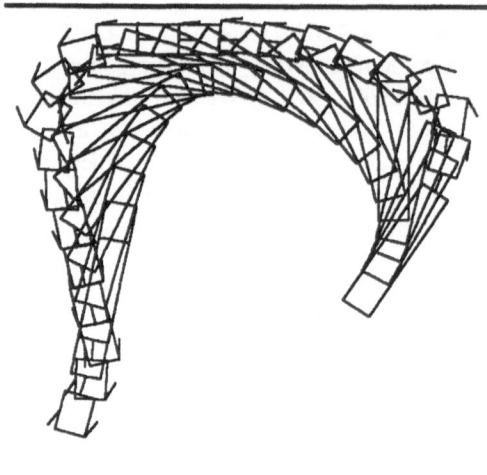

figure 5: wall avoidance task $T°_2$

figure 6: final docking task

# 6 Conclusions

A hierarchical multi-task architecture has been presented that addresses the concern of reducing state space and temporal propagation complexity in reinforcement learning for continuous control problems. Three principles of hierarchical task composition have been defined which we believe are substantial in this concern, and of general relevance. Future research will investigate the opposite problem: learning to decompose tasks within the framework of our task composition principles. This includes a more formal approach to finding conditions of convergence to (at least sub-optimal) solution paths.

# References

[1] L.P. Kaelbling, M.L. Littman, A.W. Moore, (1996): Reinforcement Learning: a Survey. In *Journal of Artificial Intelligence*, pp. 237-295

[2] D. Nguyen, B. Widrow, (1991): The Truck Backer-Upper: An Example of Self-Learning in Neural Networks. In Miller et al: Neural Networks for Control, MIT Press, Cambr.

[3] C. Watkins, (1989): Learning from Delayed Rewards. PhD thesis, Cambridge, UK

[4] D.O. Bertsekas, (1987): Dynamic Programming. Prentice-Hall, N.J.

[5] P. Dayan, G.E. Hinton, (1993): Feudal Reinforcement Learning. In *Advances in Neural Information Processing Systems 5*, San Mateo, CA

[6] S.P. Singh, (1993): Learning to Solve Markovian Decision Processes. PhD thesis, University of Massachusetts. (CMPSCI Tech. Report 93-77)

[7] A.Moore, (1994): The parti-game algorithm for variable resolution RL in multidimensional state spaces. In: *Advances in Neural Information Processing Systems 6*, San Mateo

# On-line EM Algorithm and Reinforcement Learning

Shin Ishii †‡

†Nara Institute of Science and Technology
Ikoma, Nara 630-0101, Japan
E-mail: ishii@is.aist-nara.ac.jp

Masa-aki Sato ‡

‡ATR Human Information Processing Research Laboratories
Seika, Kyoto 619-0288, Japan

### Abstract

We previously proposed an on-line EM algorithm for Normalized Gaussian Network (NGnet), which is a network of local linear regression units. In this article, we will apply our approach based on the on-line EM algorithm to reinforcement learning problems. We will examine a task for swinging-up and stabilizing a single pendulum with a limited torque, and a task for stabilizing a double pendulum. As a result, our approach is much more efficient than that based on the gradient descent algorithm.

## 1 Introduction

Normalized Gaussian Network (NGnet) [4] is a network of local linear regression units. This model softly partitions the input space by using normalized Gaussian functions and each local unit linearly approximates the output within its partition. The NGnet can be interpreted as an output of a stochastic model with hidden variables. We previously proposed an on-line EM algorithm for the NGnet [5]. The algorithm works well even in dynamic environments, e.g., when the input-output distribution changes in time.

In this article, we apply the NGnet trained by the on-line EM algorithm to reinforcement learning problems. In the conventional actor-critic architecture [1], the actor produces a control signal according to its policy, and the critic estimates the discounted future return, which is called the value function. The value function depends on the actor's policy. In the learning process, the actor is changed so that the value function increases, while the critic is updated so as to approximate the value function for the current policy. The critic learning can be regarded as a function approximation problem in a dynamic environment, because the policy changes as the learning proceeds.

We consider optimal control problems for deterministic nonlinear dynamical systems having continuous state/action spaces. As the first experiment, we examine a task for swinging-up and stabilizing a single pendulum with a limited torque [3]. As the second experiment, we examine a task for balancing a double pendulum where a torque is applied only to the first pendulum. Our approach based on the on-line EM algorithm shows good performances in these experiments.

## 2 NGnet and on-line EM algorithm

The NGnet [4], which transforms an $N$-dimensional input vector $x$ to a $D$-dimensional output vector $y$, is defined by the following equations.

$$y = \sum_{i=1}^{M} \frac{G_i(x)(W_i x + b_i)}{\sum_{j=1}^{M} G_j(x)} \tag{1a}$$

$$G_i(x) \equiv (2\pi)^{-N/2} |\Sigma_i|^{-1/2} \exp\left[-\frac{1}{2}(x - \mu_i)' \Sigma_i^{-1} (x - \mu_i)\right]. \tag{1b}$$

$M$ denotes the number of units, and the prime ($'$) denotes a transpose. $G_i(x)$ is an $N$-dimensional Gaussian function, which has an $N$-dimensional center $\mu_i$ and an $(N \times N)$-dimensional covariance matrix $\Sigma_i$. $W_i$ and $b_i$ are a $(D \times N)$-dimensional linear regression matrix and a $D$-dimensional bias vector, respectively. Subsequently, we use notations $\tilde{W}_i \equiv (W_i, b_i)$ and $\tilde{x}' \equiv (x', 1)$.

The NGnet can be interpreted as a stochastic model, in which a pair of an input and an output, $(x, y)$, is a stochastic event. For each event, a unit index $i \in \{1, ..., M\}$ is assumed to be selected, which is regarded as a hidden variable. The stochastic model is defined by the probability distribution for a triplet $(x, y, i)$, which is called a complete event:

$$P(x, y, i|\theta) = (2\pi)^{-(D+N)/2} \sigma_i^{-D} |\Sigma_i|^{-1/2} M^{-1} \tag{2}$$

$$\times \exp\left[-\frac{1}{2}(x - \mu_i)' \Sigma_i^{-1} (x - \mu_i) - \frac{1}{2\sigma_i^2}(y - \tilde{W}_i \tilde{x})^2\right].$$

Here, $\theta \equiv \{\mu_i, \Sigma_i, \sigma_i^2, \tilde{W}_i \mid i = 1, ..., M\}$ is a set of model parameters.

We can easily prove that the expectation value of the output $y$ for a given input $x$, i.e., $E[y|x] \equiv \int y P(y|x, \theta) dy$, is identical to equation (1). Namely, the probability distribution (2) provides a stochastic model for the NGnet.

From a set of $T$ events (observed data) $(X, Y) \equiv \{(x(t), y(t)) \mid t = 1, ..., T\}$, the model parameter $\theta$ of the stochastic model (2) can be determined by the EM algorithm [2]. The EM algorithm repeats the following E- and M-steps.

E (Estimation) step:    Let $\bar{\theta}$ be the present estimator. By using $\bar{\theta}$, the posterior probability that the $i$-th unit is selected for $(x(t), y(t))$ is given as

$$P(i|x(t), y(t), \bar{\theta}) = P(x(t), y(t), i|\bar{\theta}) / \sum_{j=1}^{M} P(x(t), y(t), j|\bar{\theta}). \tag{3}$$

M (Maximization) step:    Using the posterior probability (3), the expected log-likelihood $L(\theta|\bar{\theta}, X, Y)$ for the complete events is defined by

$$L(\theta|\bar{\theta}, X, Y) = \sum_{t=1}^{T} \sum_{i=1}^{M} P(i|x(t), y(t), \bar{\theta}) \log P(x(t), y(t), i|\theta). \tag{4}$$

$L(\theta|\bar{\theta}, X, Y)$ is maximized with respect to $\theta$. A solution is given [7] by

$$\mu_i = \langle x \rangle_i(T) / \langle 1 \rangle_i(T) \tag{5a}$$

$$\Sigma_i^{-1} = [\langle xx' \rangle_i(T)/\langle 1 \rangle_i(T) - \mu_i(T)\mu_i'(T)]^{-1} \tag{5b}$$

$$\tilde{W}_i = \langle y\tilde{x}' \rangle_i(T)[\langle \tilde{x}\tilde{x}' \rangle_i(T)]^{-1} \tag{5c}$$

$$\sigma_i^2 = \frac{1}{D}\left[\langle |y^2| \rangle_i(T) - \mathrm{Tr}\left(\tilde{W}_i\langle \tilde{x}y' \rangle_i(T)\right)\right]/\langle 1 \rangle_i(T), \tag{5d}$$

where the symbol $\langle \cdot \rangle_i$ is defined by

$$\langle f(x,y) \rangle_i(T) \equiv \frac{1}{T}\sum_{t=1}^{T} f(x(t), y(t))P(i|x(t), y(t), \bar{\theta}). \tag{6}$$

The EM algorithm introduced above is based on batch learning [7], namely, the parameters are updated after seeing all of the observed data. We introduce here an on-line version [5] of the EM algorithm, where the weighted mean (6) is replaced by

$$\ll f(x,y) \gg_i (T) \equiv \eta(T)\sum_{t=1}^{T}(\prod_{s=t+1}^{T}\lambda(s))f(x(t),y(t))P(i|x(t),y(t),\theta(t-1)).\tag{7}$$

Here, $\theta(t-1)$ is the estimator after the $(t-1)$-th observed data $(x(t-1), y(t-1))$. The discount factor $\lambda(t) \in [0,1]$ is introduced for forgetting the effect of earlier inaccurate estimator. $\eta(T) \equiv (\sum_{t=1}^{T}(\prod_{s=t+1}^{T}\lambda(s)))^{-1}$ is the normalization coefficient, and it is iteratively calculated by $\eta(t) = (1 + \lambda(t)/\eta(t-1))^{-1}$. The modified weighted mean $\ll \cdot \gg_i$ can be step-wisely obtained by

$$\ll f(x,y) \gg_i (t) = \ll f(x,y) \gg_i (t-1) \tag{8}$$
$$+\eta(t)[f(x(t),y(t))P_i(t) - \ll f(x,y) \gg_i (t-1)],$$

where $P_i(t) \equiv P(i|x(t), y(t), \theta(t-1))$. Using the modified weighted mean, the new parameters are obtained by the following equations.

$$\mu_i(t) = \ll x \gg_i (t)/ \ll 1 \gg_i (t) \tag{9a}$$

$$\tilde{\Lambda}_i(t) = \frac{1}{1-\eta(t)}\left[\tilde{\Lambda}_i(t-1) - \frac{P_i(t)\tilde{\Lambda}_i(t-1)\tilde{x}(t)\tilde{x}'(t)\tilde{\Lambda}_i(t-1)}{(1/\eta(t)-1) + P_i(t)\tilde{x}'(t)\tilde{\Lambda}_i(t-1)\tilde{x}(t)}\right] \tag{9b}$$

$$\tilde{W}_i(t) = \tilde{W}_i(t-1) + \eta(t)P_i(t)(y(t) - \tilde{W}_i(t-1)\tilde{x}(t))\tilde{x}'(t)\tilde{\Lambda}_i(t) \tag{9c}$$

$$\sigma_i^2(t) = \frac{1}{D}\left[\ll |y|^2 \gg_i (t) - \mathrm{Tr}\left(\tilde{W}_i(t) \ll \tilde{x}y' \gg_i (t)\right)\right]/ \ll 1 \gg_i (t), \tag{9d}$$

where $\tilde{\Lambda}_i(t) \equiv [\ll \tilde{x}\tilde{x}' \gg_i]^{-1}$. $\Sigma_i^{-1}(t)$ can be obtained from the following relation with $\tilde{\Lambda}_i(t)$.

$$\tilde{\Lambda}_i(t) \ll 1 \gg_i (t) = \begin{pmatrix} \Sigma_i^{-1}(t) & -\Sigma_i^{-1}(t)\mu_i(t) \\ -\mu_i'(t)\Sigma_i^{-1}(t) & 1 + \mu_i'(t)\Sigma_i^{-1}(t)\mu_i(t) \end{pmatrix}. \tag{10}$$

In this algorithm, there is no need to calculate the matrix inverse, while it is needed in the batch EM algorithm.

In the on-line EM algorithm, we also employ dynamic unit manipulation mechanisms in order to efficiently allocate the units [5]. They are unit production, unit deletion and unit division, and they are done according to a probabilistic interpretation. In addition, a regularization method for the co-variance matrices $\Sigma_i$ $(i = 1, ..., M)$ is employed in order to deal with a singular input distribution [5]. We have found that the regularization is important for attaining a good performance in reinforcement learning applications.

# 3 Reinforcement learning applications

We apply our approach to reinforcement learning problems. We consider op-timal control problems for deterministic nonlinear dynamical systems having continuous state/action spaces. An actor-critic architecture [1] is used for the learning system.

For the current state, $x_c(t)$, of the controlled system, the actor outputs a control signal (action) $u(t)$, which is given by the policy function $\Omega(\cdot)$, i.e., $u(t) = \Omega(x_c(t))$. The controlled system changes its state to $x_c(t+1)$ after receiving the control signal $u(t)$. Following that, a reward $r(x_c(t), u(t))$ is given to the learning system.

The objective of the learning system is to find the optimal policy function that maximizes the discounted future return defined by

$$V(x_c) \equiv \sum_{t=0}^{\infty} \gamma^t r(x_c(t), \Omega(x_c(t)))\big|_{x_c(0)=x_c} , \tag{11}$$

where $0 < \gamma < 1$ is a discount factor. $V(x_c)$, which is called the value function, depends on the current policy function $\Omega(\cdot)$ employed by the actor.

The Q-function is defined by

$$Q(x_c, u) = \gamma V(x_c(t+1)) + r(x_c, u), \tag{12}$$

where $x_c(t) = x_c$ and $u(t) = u$ are assumed. The value function can be obtained from the Q-function:

$$V(x_c) = Q(x_c, \Omega(x_c)). \tag{13}$$

The Q-function should satisfy the consistency condition:

$$Q(x_c(t), u(t)) = \gamma Q(x_c(t+1), \Omega(x_c(t+1))) + r(x_c(t), u(t)). \tag{14}$$

In our approach, the policy function and the Q-function are approximated by the NGnet's, which are called the actor-network and the critic-network, re-spectively. The learning process proceeds as follows. For the current state $x_c(t)$, an action $u(t)$ is generated by the current actor-network as shown later. At the next time step, the learning system gets the next state $x_c(t+1)$ and the re-ward $r(x_c(t), u(t))$. The critic-network is trained by the on-line EM algorithm. The input to the critic-network is $(x_c(t), u(t))$. The target output is given by the right hand side of (14), where the Q-function and the policy function are calculated using the current critic- and actor- networks, respectively.

The actor-network is defined as a variation of the NGnet:

$$u = \Omega(x_c) = u_{max} \cdot \tanh\left(\sum_{i=1}^{M_0} \omega_i \mathcal{N}_i(x_c) + \epsilon\right), \tag{15}$$

where a random noise $\epsilon$ is added in the training phase in order to explore the state space. The maximum value of the torque is assumed to be fixed at $u_{max}$. Then, the output of the actor-network is filtered through the sigmoidal function, $\tanh(\cdot)$. The actor-network is trained by the gradient ascent method so that the Q-function value increases [6].

$$\Delta\omega(t) \propto \frac{\partial\Omega}{\partial\omega}(x_c(t)) \cdot \frac{\partial Q}{\partial u}(x_c(t), u(t)). \tag{16}$$

We do not change the centers and the covariance matrices of the actor-network.

The first experiment is a task for swinging up and stabilizing a single pendulum with a limited torque controller [3]. The state of the pendulum is represented by $x_c \equiv (\phi, \dot{\phi})$, where $\phi$ and $\dot{\phi}$ denote the angle from the upright position and the angular velocity of the pendulum, respectively. The reward $r(x_c(t), u(t))$ is assumed to be given by $\tilde{r}(x_c(t+1))$, where

$$\tilde{r}(x_c) = \exp(-(\dot{\phi})^2/(2\nu_1^2) - \phi^2/(2\nu_2^2)). \tag{17}$$

$\nu_1$ and $\nu_2$ are constants. This reward encourages the pendulum to stay high.

After releasing the pendulum from a vicinity of the upright position, the control and the learning process of the actor- and critic- networks is conducted for 7 seconds. This is a single episode. The reinforcement learning is done by repeating these episodes.

After 40 episodes, the system is able to make the pendulum achieve an upright position from almost every initial state. Figure 1 shows a control process using the trained actor-network. The pendulum is initially set at the lowest position with a zero velocity, i.e., $\dot{\phi} = 0$ and $\phi = -\pi$. The dotted, dashed, and solid lines denote $\phi$, $\dot{\phi}$, and the control signal $u$ produced by the actor-network, respectively. Since the maximum torque generated by the controller is limited, the system inverts the pendulum after swinging it several times. Figure 2 shows stroboscopic time-series of the pendulum's state in the same control sequence.

According to our former experiment, in which the critic-network was a center-fixed NGnet trained by the gradient descent algorithm, a good control was obtained after about 2,000 episodes. This former approach used 18,081 ($= 21^2 \times 41$) center-fixed critic units, while our new approach based on the on-line EM algorithm uses 114 center-adapted critic units. Both model use 441 ($= 21^2$) center-fixed actor units. The computation times (user times) on DEC Alpha Station 600 5/266 for the former and new approaches are 1153 sec. and 61 sec., respectively. Therefore, our new approach is able to obtain a good control much faster than the former approach.

Next, we apply our approach to a task for balancing a double pendulum where a torque is applied only to the first pendulum. The state of the pendulum is represented by $x_c \equiv (\phi_1, \phi_2, \dot{\phi}_1, \dot{\phi}_2)$, where $\phi_1$ and $\phi_2$ are the first pendulum's angle from the upright direction and the second pendulum's angle from the first pendulum's direction, respectively. The reward is given by the height of the second pendulum's end from the lowest position.

After 50 episodes, the system is able to make the double pendulum straight up, when the initial state of the pendulum is fairly close to the upright position. Figure 3 shows a control sequence, where the dotted, dashed, and solid lines denote $\phi_1$, $\phi_2$, and the produced control signal $u$, respectively.

In the double pendulum task, our new approach uses 10,000 center-fixed actor units, and 93 center-adapted critic units. Our former approach did not work even when 10,000 actor units and 100,000 critic units were prepared.

# 4 Conclusion

In this article, we applied the NGnet trained by the on-line EM algorithm to reinforcement learning problems. In particular, we examined a task for swinging-up and stabilizing a single pendulum with a limited torque, and a task for balancing a double pendulum. The experimental results show that our approach can be applied to reinforcement learing problems for continuous state/action spaces. Moreover, our approach is much more efficient than that based on the gradient descent algorithm.

# References

[1] Barto, A. G., et al. (1983). *IEEE Transactions on SMC.*, **13**.

[2] Dempster, A. P., et al. (1977). *Journal of Royal Statistical Society B*, **39**.

[3] Doya, K. (1996). In *NIPS 8* (pp. 1073-1079), MIT Press.

[4] Moody, J., & Darken, C. J. (1989). *Neural Computation*, **1**.

[5] Sato, M., & Ishii, S. (1998). *ATR Technical Report*, **TR-H-243**, ATR.

[6] Sofge, D. A., & White, D. A. (1992). In *Handbook of Intelligent Control* (pp. 259-282), Van Nostrand Reinhold.

[7] Xu, L., et al. (1995). In *NIPS 7* (pp. 633-640), MIT Press.

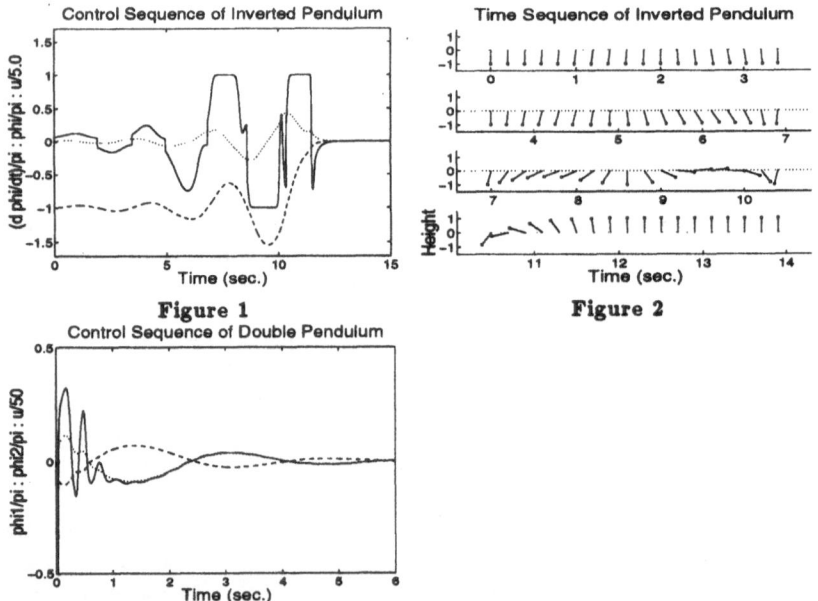

**Figure 1**

**Figure 2**

**Figure 3**

# Embedding Knowledge in Reinforcement Learning

G. Hailu, G. Sommer

Christian Albrechts University, Department of Cognitive Systems
Preusserstrasse 1-9, D-24105 Kiel, Germany
gha@informatik.uni-kiel.de

**Abstract.** In almost all real systems where reinforcement learning is applied, it is found that a knowledge free approach doesn't work. The basic RL algorithms must sufficiently be biased to achieve a satisfactory performance within a bounded time. This bias takes different forms. In this paper, in addition to reflex rules [6], environment (domain) knowledge is embedded into the learner. Environment knowledge gives leverage to the adaptive state space construction algorithm by splitting *key* states quickly. The learner is tested on a B21 robot for a goal reaching task. Experimental results show that after few trials the robot has indeed learned the right situation action rules that unfold its path.

## 1 Introduction

For more than a decade reinforcement learning (RL) has been studied extensively and its properties are well understood. One of its nice property is that it allows agents to be programmed by reward and punishment, without the need to specify how the task is achieved. Unfortunately, it has an inherent problem - its learning time increases exponentially with the size of the state space. Consequently, RL has remained difficult to implement in realistic domains that are characterized by large state and action spaces - typically robot domain. Yet, despite this inherent problem, there is still a surge of interest in putting RL onto a real robot.

Researchers have tried to overcome the inability of RL to scale well to learning tasks with large state and action spaces. Mahadevan *et al.* [4] have decomposed the task into sets of simple sub-tasks each with its own prewired applicability predicate. Matarić [5] has minimized the state space by transforming state-action pairs to condition-behavior pairs and maximized learning by designing reward rich heterogeneous reinforcement. Recently, Millán [6] has tremendously accelerated RL by integrating it with reflex rules that focus exploration where it is mostly needed. The common characteristic of the above examples is that the basic RL algorithm has been endowed with some *built-in knowledge*. In each case, however, the built-in knowledge has different forms and is used for different purposes : in [4] to break down and to arbiter tasks, in [5] to design rich reward and, in [6] to focus exploration.

This paper is concerned with using environment knowledge to pre-structure the state space. There is a conflict between the required number of states and the actual states constructed by the *adaptive state space construction* algorithm

[6]. This is not because of the algorithm, but because of the particular platform we are working with. To resolve this conflict, the controller is shaped to accommodate implicit environment knowledge that enables the algorithm to construct appropriate state space during the course of learning.

## 2   The robot task

The B21 robot from RWI has been used as our experimental platform. The robot is a four-wheeled cylindrical synchro-drive with two parts : a base and an enclosure. The base carries 32 infra-red (IR) and 32 tactile sensors. Whereas the enclosure carries 24 tactile, 24 IR and 24 sonar sensors.

The task of the robot is to reach a specified goal $p_g$ through a (sub)-optimal path. Optimality is defined on a certain payoff function. For every action the robot has chosen, it receives an immediate reinforcement $r_t$ that has two components. The first component penalizes the robot when it either collides with or approaches an obstacle. Whereas the second component penalizes the robot in proportion to the angle between the robot heading and the vector connecting the current robot and goal locations. The immediate reinforcement value is the sum of these two components and the payoff function is defined as the sum of immediate reinforcements the robot receives until it reaches the goal, i.e., $R = \sum_t r_t$.

In any mobile robot control, determining the robot position is one of the crucial issue. In the presented learning system the robot position has been used in two ways. First to decode the relative distance between the robot and the goal (section 3) and second to provide a part of the reinforce function from which the robot learns. From the two, the latter one is more sensitive to the inaccuracy of robot position. Because it leads to inconsistent reinforce function that makes learning difficult or even impossible[†]. In this work, *dead reckoning* method has been used to obtain the robot position $p_r(t)$. However, to get a satisfactory reading, we have exploited the crucial property of the robots' dead reckoning system. Dead reckoning system performs satisfactory provided that the robot does not move for a prolonged periods of time with out reaching the goal. This characteristic puts directly a limit on how far and how hidden the goal should be placed from the robot.

## 3   Embedding

The input $x=[s,d]$ to the controller is a vector of 32 elements, each between $[0, 1]$. The first 24 elements are normalized depth readings of the sonar sensors[‡], while the remaining eight inputs are codified distance between the robot and

---

[†]Noting this, Millán [7] has eliminated the dependence of the reinforcement value on the odometry reading by building other types of sensors that are capable of detecting the goal *directly*.

[‡]Since IRs are short range ($\approx 0.3m$) proximity sensors, they are used here in emergency routine only.

the goal. In the work of [6], where sensor values are made independent of the robot heading, the input to the controller turns out to be a function of the robot position (if sensors noise is neglected), i.e., $x = [s(p_r(t)), d(p_r(t), p_g)]$. In this case, key states that require different actions are easily split.

For most platforms, however, the sensors can not be aligned independently of the base. Consequently, the perceived sensory data would be different every time the robot visits a given location at different headings, i.e., $x = [s(p_r(t), \theta_r(t)), d(p_r(t), p_g)]$. This results in huge states which the adaptive state space constructor could not cope with identifying and splitting key states quickly. In order to overcome this problem, we have *embedded* environment knowledge [3] [8] into the learning architecture. The environment in which the robot operates is first partitioned into four regions that are considered to be the same for the purpose of learning and generating actions. These are: a concave region that misleads and fold the robot path, a door region through which the robot has to carefully pass, a corridor where the goal is located, and a vast space inside the room from where the robot starts off. These partitions together with their corresponding metric data are supplied to the controller as built-in knowledge. From the metric data and the robot position $p_r(t)$, disjointed rules have been written to single out a particular partition where the robot is in. This early splitting of the state space based on prior environment knowledge can be viewed as one way of giving leverage to the adaptive state space constructor so that during the course of learning it can construct appropriate states for each partitions.

Apart from environment knowledge, two fuzzy behaviors [9] *obstacle avoidance* and *goal following* are used as reflex that enable the controller to act initially in some reasonable way. The reflex delivers the next robot heading $\alpha$ whenever it is requested. The fuzzy reflex works as follows. First, the range of possible robot heading has been fuzzified into three fuzzy sets: left, straight and, right. The obstacle avoidance behavior receives the range data of the sonars and outputs a vector $\alpha_a$ - whose elements indicate the activation levels of the above fuzzy sets. Likewise, the goal following behavior inputs the acute angle $\theta$ between the robot heading and the vector connecting the current robot and goal locations and outputs a similar vector $\alpha_g$. A simple behavior blender with constant *desirability functions* $d_a$ and $d_g$ ($d_g \leq d_a$) is used to combine the output of the two behaviors, i.e., $\alpha_f = d_g \alpha_g + d_a \alpha_a$. Subsequently, a defuzzifier decodes the fused vector $\alpha_f$ to a crisp value $\alpha$ using centroid technique.

# 4 Controller

The architecture of the controller is an actor critic type that is proposed by [6]. In most actor critic systems, two networks are adapted over time - an action and a critic network. In the proposed architecture, however, these networks are integrated into one network. Besides, unlike the former ones where the training rules adjusts certain weights of the action or critic networks, the latter one adapts directly the critic and action values. The controller consists of a gradually growing RBF neurons in the input layer and a stochastic neuron in the output layer. Whenever a new situation is perceived, the controller uses the built in knowledge to associate the situation to one of the partitions discussed in section 3. Within the partition existing neurons (if any) compete to win the situation. If a winning neuron exists, it will be connected to the output layer to generate action. Action is generated by exploring a restricted area around a prototypical action. To enforce exploration a *Gaussian stochastic unit* with parameters $(\mu, \sigma)$ is introduced at the output layer [2]. The extent of the exploration is determined by the critic (utility) value $u_j$ and the temperature factor $T(n)$, i.e., $\sigma = T(n)f(u)$. At the end of every trial the temperature is cooled down so that the stochastic unit produces a progressively deterministic output [1].

The adaptive state construction algorithm introduces a new neuron into the selected subspace when existing neurons can not generalize the current situation or if a selected neuron has performed poorly for the previous situation. When a new neuron is created four learning parameters $(p_j, u_j, w_j$ and, $c_j)$ are attached to it [6]. Each of the parameters are adapted by different adaptation algorithm and error sources. The utility value of the winning neuron $u_j(t)$ is updated by *temporal difference* (TD) method [10]. Williams' REINFORCE algorithm [11] is employed to adapt the weight $w_j$. Depending on the performance of the winning neuron, its center position $c_j$ is either shifted toward the previous sensation or left untouched. The prototypical action $p_j$ is overridden by a more accurate learned action when the robot reaches the goal through a trajectory whose total payoff is greater than the maximum payoff so far obtained.

# 5 Experimental results

Figure 1 depicts the trajectories of the robot in the first and the last trials and figure 2 shows the learning curves of the controller against the number of trials. Ten sets of experiments, each consisting of 20 trials were carried out. The vertical error bars (fig.2) indicate the variations of the learning curves. During the first few trials, the robot has taken many steps (fig.2(b)) to reach the goal, thereby incurring a high payoff (fig.2(c)), and the number of neurons added to the network has grown sharply (fig.2(a)). As trials goes on, however, the robot has started to unfold its path and neurons are added to the network at a reduced slope than earlier trials. On the sixth trial and afterwards the robot has straighten its path, except at the eighth trial where the robot left the optimum path in search for a better one. In subsequent trials, however, the robot has returned to its previous performance and followed the same path with out significant divergence through out the remaining trial. A similar phenomena is also observed in the work of [6].

Comparing the final network performance Table 1 with that of [6] the following observation can be made. First, since neurons are not shared across partitions, the total number of neurons in this small environment is almost equal to that obtained in the large environment of [6]. Second, due to sensory alignment scheme of [6], the network size has already ceased to grow during the last few trials and hence, the variances of the final network performance are smaller than the one reported here. To obtain a similar performance on B21, we are emulating the turret motor in our subsequent work. Instead of using the sensory sequence that point in the direction of the robot heading, the controller can mentally rotate (in the reverse direction of the base rotation) the sensors in such a way that the new sensory sequence points always towards the goal.

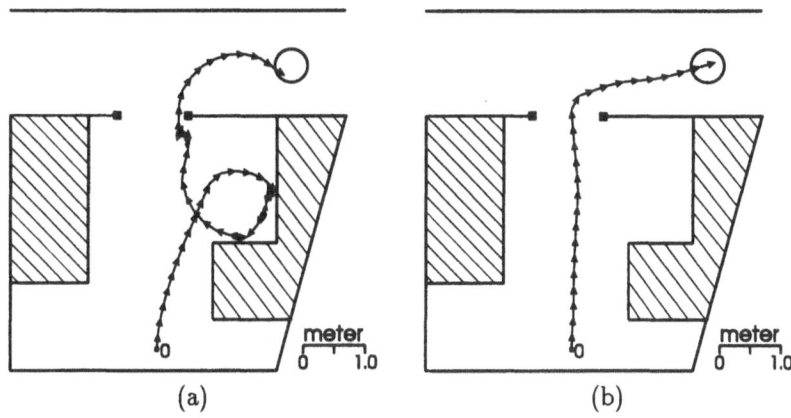

Figure 1: Trajectories a) first trial, b) last trial

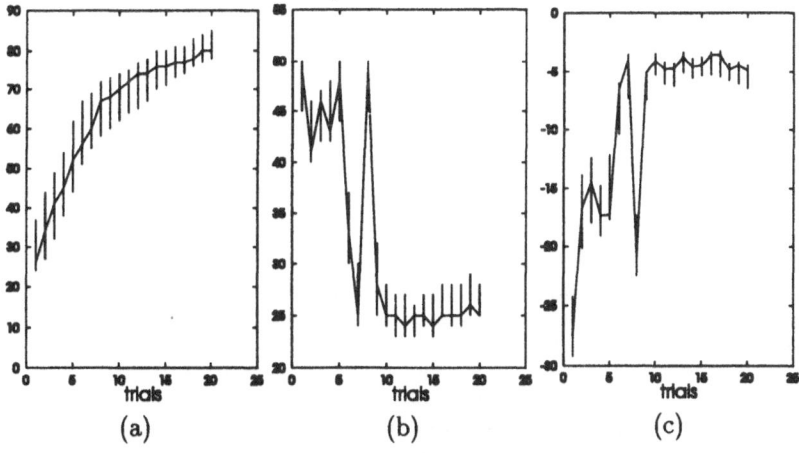

Figure 2: a) Number of neurons, b) Number of steps, c) Total reinforcements

Table 1: Final Network Performance

| Quantities | Mean | Variance |
|---|---|---|
| Number of neurons | 82.5 | 4.7286 |
| Number of steps | 27.7 | 1.9000 |
| Total reinforcement | -6.24 | 0.8752 |

# 6 Conclusion

Two kinds of built-in knowledge have been used to support RL on B21 robot. The first one is *a priori* environment knowledge to pre-structure the state space rapidly. Whereas the second one is two fuzzy behaviors combined with fixed desirability values to focus exploration. Experimental results have shown that the robot has indeed learned to unfold its path and to consistently follow a trajectory that has a minimum payoff value.

# 7 Acknowledgment

We would like to extend our thanks to J. R. Millán for a useful discussion on TESEO's architecture that helped us greatly in this work. The support given to the first author by DAAD under grant code 413/ETH-4-BOA is also acknowledged.

# References

[1] A. G. Barto, S. J. Bradtke, and S. P. Singh. Learning to act using real time dynamic programming. Technical report, University of Massachusetts, Amherst MA 01003, 1993.

[2] V. Gullapalli. A stochastic reinforcement learning algorithm for learning real valued function. *Neural Networks*, 3:671–692, 1990.

[3] L. P. Kaelbling. *Learning in Embedded Systems*. MIT Press, 1993.

[4] S. Mahadevan and J. Connell. Automatic programming of behavior-based robots using reinforcement learning. *Artificial Intelligence*, 55:311–365, 1992.

[5] M. J. Matáric. Reward functions for accelerated learning. In *Proceedings of the Eleventh International Conference on Machine Learining*. Morgan Kaufmann, 1994.

[6] J. R. Millán. Rapid, safe and incremental learning of navigation stratagies. *IEEE Transactions on Systems, Man, and Cybernetics*, 26(3):408–420, June 1996. Special Issue on Learning Autonomous Robots.

[7] J. R. Millán. Incremental acquisition of local networks for the control of autonomous robots. In *7th International Conference on Artificial Neural Networks*, pages 739–744, Lausanne, Switzerland, 1997.

[8] U. Nehmzow, T. Smithers, and J. Hallam. Steps towards intelligent robots. Technical Report 502, Universty of Edinburgh, 1990.

[9] D. W. Payton, J. K. Rosenblatt, and D. M. Keirsey. Plan guided reaction. *IEEE Transaction on Systems, Man, and Cybernetics*, 20(6):1370–1382, November 1990.

[10] R. S. Sutton. Learning to predict by the methods of temporal differences. *Machine Learning*, 3(1):9–44, 1988.

[11] R. J. Williams. Simple statistical gradient-following algorithms for connectionist reinforcement learning. *Machine Learning*, 8:229–256, 1992.

# Multistage STM in a Multilayer Hebbian Learning Architecture for Local Navigation[*]

Andreas Bühlmeier[1,2], Markus Rossmann[1],
Karl Goser[1], Gerhard Manteuffel[3]

[1]LS Bauelemente der Elektrotechnik, Universität Dortmund, Germany,

[2]FB-3 Informatik, Universität Bremen, Germany,

[3]FBN, Dummerstorf, Germany

Email: buehlmei@luzi.e-technik.uni-dortmund.de

**Abstract**

In this paper we motivate and present a novel neural network architecture that includes multi-stage short-term memory (STM) and multilayer Hebbian learning. We apply this network as an adaptive steering assistant for an electrically driven wheelchair, which is equipped with tactile, sonar and other sensors. The influence of the adaptive controller increases with the probability that user commands result in collisions.

## 1 Motivation

Building an adaptive system based on Hebbian learning offers new insights in three principal directions. Firstly, results contribute to our understanding of how living systems may work since Hebbian mechanisms are still the most likely physiological explanation for their learning ability. Secondly, transferring knowledge from biology into a technical system shows, through success or failure, where a neuromorphic approach [3] is viable or not. Thirdly, Hebbian learning is especially interesting for hardware implementations and therefore offers new concepts of how to build advanced microelectronic circuits, e.g. in the nanometer scale[4].

Up to now, not too many complete systems based on Hebbian learning have been shown that go beyond one-layer feedforward networks. In this paper we show that for the adaptation of local navigation of a wheelchair, a multi-stage short-term memory and multi-layer learning is needed in general. To show this, the consideration of four prototypical situations is sufficient, although the system also copes with other configurations. We show how biologically plausible neuromimetic circuits are arranged in a network to meet the needs stated above. Based on earlier work [1][2], we present a novel Hebbian learning architecture that includes classical and operant conditioning, a multi-stage short-term memory and a neural layer for adaptive clustering. We demonstrate the application of the neural architecture to the adaptive steering control of an electrically driven wheelchair that is equipped with tactile, sonar and other sensors.

---

[*] Supported through the Deutsche Forschungsgemeinschaft (grant Go379/12-3).

## 2 The Neuromimetic Circuit

Our approach is to employ a biologically plausible neuronal model that allows adaptation and short-term dynamics. The main features of the neuromimetic circuits can be summarized as follows:

- Each synapse includes first order low pass filters with different time constants for onset and offset to provide short-term dynamics.
- Excitatory, inhibitory and plastic inputs are provided, each of which can be shunted by additional shunting inputs, which is used to approximate multiplication of signals.
- Hebbian heterosynaptic long-term potentiation with active decay is employed as a hardware friendly and robust learning rule.
- Weight sum limitation reduces the risk of weight saturation in the neurons.
- Output is clipped to non-negative values that are lower than an upper bound.

Heterosynaptic Hebbian learning as implemented here, requires that an input $S_i$ is above a threshold $\theta_1$ and that the sum of all other inputs, $r_i = \Sigma_{j \neq i} w_j S_j$, is larger than a second threshold $\theta_2$. As a result, a single synapse can not cause its own potentiation. Weight saturation effects are reflected by the term $(w_{max} - w_i)$, where $w_{max}$ is the upper bound of all weights $w_i$. To provide further stability to the adaptation process, weight increment is not possible when the sum of weights exceeds a threshold $w_{sm}$. For the $i$-th input $S_i$ change of the modifiable weight $w_i$ is:

$$\frac{dw_i}{dt} = \begin{cases} K_1(w_{max} - w_i) & \text{if } S_i \geq \theta_1 \wedge r_i \geq \theta_2 \wedge \sum_j w_j < w_{sm} \\ -K_2 w_i & \text{if } S_i \geq \theta_1 \wedge r_i < \theta_2 \\ -K_3 w_i & \text{else} \end{cases} \qquad (1)$$

Threshold $\theta_r$ and the limited weight sum $w_{sm}$ control secondary conditioning effects, i.e. whether it is possible that conditioned stimuli act like a reinforcer. If all $S_i \leq 1$ and $\theta_r > w_{sm}$, then no secondary conditioning occurs [1].

## 3 The Adaptive Steering Assistant Application

Figure 1A shows the electrically driven wheelchair (Meyra™ model 3.422), which is used as an experimental platform and prototype application [1]. Steering the wheelchair is not simple for users who are not experienced whereas an experienced user may perform better than any automatic control. An experienced user does not appreciate if any of his commands are altered. Since a user's ability changes with time and can not be predicted effectively, we suggest an *adaptive* steering assistant that interferes with the user only when steering mistakes become probable. We base the adaptation on reinforcement signals, i.e. we define tactile signals as negative reinforcement signals. Tactile signals trigger retreat movements, too. We also define that moving backwards due to the steering assistant is a global cost that is to be avoided through appropriate forward steering commands. Signals of sonar sensors are associated by the steering assistant for steering control.

## 3.1 The Need for Multistage STM and Multilayer NN

Figure 1B shows four different situations that typically occur when the naive driver of the wheelchair interacts with obstacles. In situations 1a and 1b steering to the right hand side is not appropriate and should therefore be suppressed under the assumption that all destinations can be reached by going forward. Situations 2a and 2b show the analog for steering to the left hand side. The problem in the situations depicted is that if only the current sonar reading is taken into account, then, situations 1a and 2b can not be separated. Providing the controller with a memory overcomes this problem because it can represent that the obstacle is on one side at first and is perceived later on the other side.

Adding the steering angle to the input pattern instead of providing memory can be used to distinguish between the situations, but does not help to learn the task at hand, i.e. it can not be used to learn, which action is appropriate. The system would learn that there are two patterns, 1a and 2b that caused costs, but without considering past signals the controller can only say that when an obstacle is detected on the right hand side and any steering is performed, then, this steering angle is not suitable. Following this idea, the trained controller would switch to steering to the right hand side in situation 2b, but it also learnt that this is inappropriate as experienced in setup 1a. Hence, we decided to add a multi-stage STM to the system.

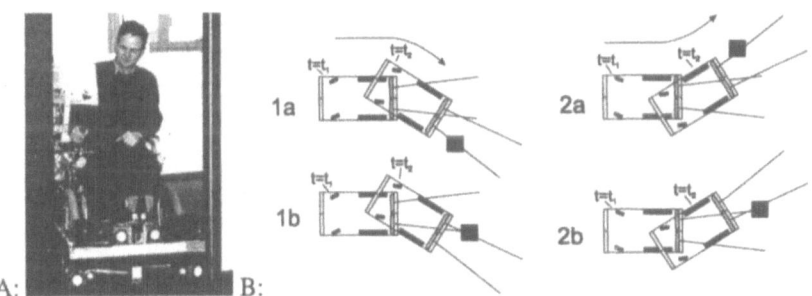

Figure 1. **A:** A wheelchair equipped with sonar, tactile and infra-red sensors serves as an prototype application. **B:** Typical situations that force the wheelchair to retreat.

Since we do not assume to have precise knowledge which distances should be considered, we define that a "1" is delivered by the sensor when an obstacle is sensed in a predefined interval of distance reading, and a "0" stands for no obstacle detected. The "0s" and "1s" provide an input vector to the associative part of the steering assistant. For the four situations, the two sensors and two points in time, these vectors are listed in the following table:

| | Before reinforcement ($t_1$) | | At reinforcement ($t_2$) | |
|---|---|---|---|---|
| Situation | Left sensor | Right sensor | Left sensor | Right sensor |
| 1a | 0 | 0 | 0 | 1 |
| 1b | 0 | 1 | 1 | 0 |
| 2a | 0 | 0 | 1 | 0 |
| 2b | 1 | 0 | 0 | 1 |

Table 1: Typical situations with resulting input vectors to the associative element.

The input vectors in Table 1 show that when computing activation using the dot product, then, an element that adapted situation 1a will also be activated by input vector 2b. Complement coding, which provides normalized input, is not a solution to this problem. For example, a linear associative element that adapted equally to the inputs 1a and 1b will provide the same activation when input 2a is attached.

Based on the considerations above we suggest the architecture shown in Figure 2.

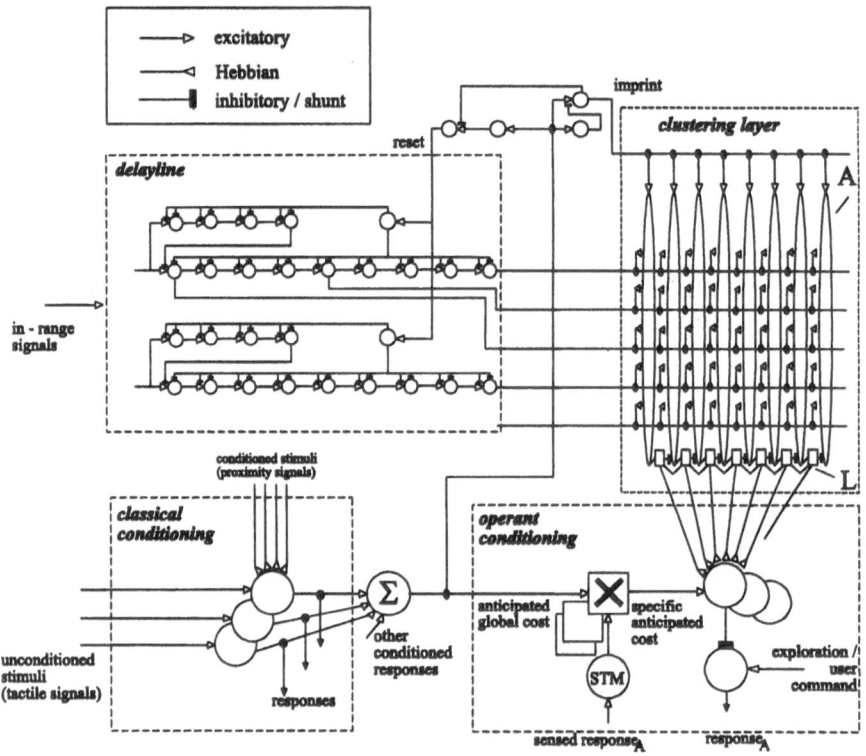

Figure 2: The novel neural network architecture.

The architecture comprises four main parts:

1. The **Classical Conditioning Module** includes reflexes as a basic set of knowledge to handle critical events, such as performing a retreat movement after touching an obstacle. Through adaptation, the actions are anticipated. The sum of all responses serves as an global reinforcement signal.

2. **Delay Lines** provide a STM that stores past signals, which are reset with the offset of a reinforcing event.

3. The **Clustering Layer** is used to provide non-linear separation of input data. "A" neurons adapt to specific input patterns that occur simultaneously with a reinforcer. To prevent that all "A"-neurons adapt to the same input pattern, lateral interaction is supplied through a number of neurons depicted by "L"

units. The activation of an "A"-neuron acts as output only if its current activation is near its maximum activation in the past.

4. The **Operant Conditioning Module** maps the output of the clustering layer to steering angles.

Earlier versions of each module were described in [1][2], partly for different purposes. The complete architecture, however, was not shown before.

# 4   Experimental Results

We follow two different approaches to investigate the network's behavior. Offline learning is effective to exactly replicate a situation, whereas on-line learning is performed to assess the system's performance in a realistic setup.

At first, we recorded sensor readings that include all four situations depicted in Figure 1B. The activation that occurs in the delay lines when the robot encounters situation 1b is shown in Figure 3 for the moment of reinforcement. An obstacle is sensed on the right hand side at first as indicated by the high activation at tap 9 of the right hand side sensor. Due to the movement, the obstacle is sensed on the left hand side later, which is represented in the high activation.

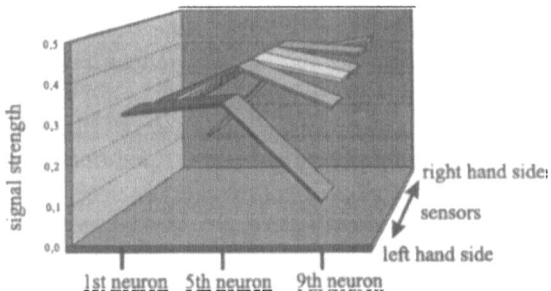

Figure 3:  Activation in the delay lines for situation 1b in Figure 1.

Figure 4 displays the activation of "A" units of the clustering layer. Sensor data of all four situations shown in Figure 1B were recorded and processed offline by the network four times in succession. The diagram on the left hand side shows the data for the "A" unit of the first group of neurons. Arrows indicate the reinforcing events. As expected, the first group associates the first reinforcing event and all other situations cause a smaller activation. The same data is shown for the fourth group's "A" neuron on the right hand side. This unit is suppressed by other groups for all situations except the fourth one and hence is activated higher by the fourth situation after two runs. Responses in the second and third group are analogous.

On-line learning was performed steering the wheelchair manually. All four situations occurred again, however at different locations, such that more variations were included in the sensor data. After about three minutes, the controller effectively  interfered with user commands and prevented collisions. The system's performance does not depend on the exact shape of obstacles. Of course, it is necessary that obstacles can be detected by the sensors.

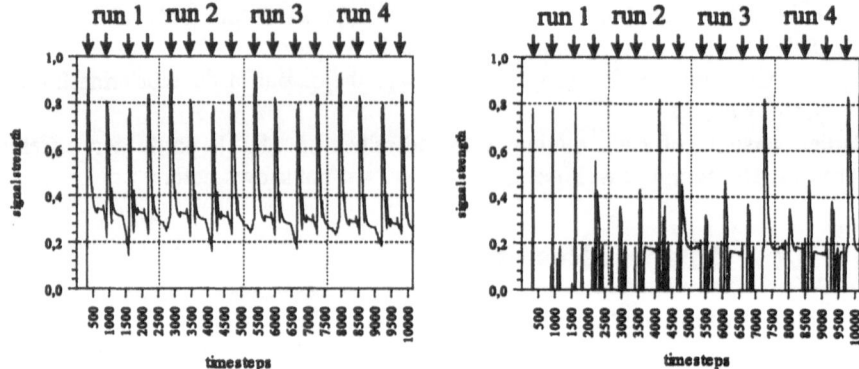

Figure 4: Output of "A" units of the first "A" neuron on the left hand side and the fourth unit on the right hand side. Output signals of "A" neurons that are close to their maximum activation are forwarded to the "operant conditioning module" for motor control.

# 5    Conclusion

The main contribution described in this paper is the Hebbian learning architecture that uses only one type of processing unit: the circuit described in section 2, which is implemented in hardware with minor changes [5].

Important details of the architecture are the reset of the STM, such that each episode is separated and the clustering layer that enables non-linear separation of input vectors. In addition, we presented a useful application of learning with a novel approach of human-machine interaction based on reinforcement learning as an alternative to supervised learning.

# References

[1] Bühlmeier A. Analog Neural Networks in Autonomous Robots. Fachbuchreihe Wissenschaft - Erfurt: AWOS Publishing. ISBN 3-932649-11-7, 1997

[2] Bühlmeier A, Steiner P, Rossmann M, Goser K, Manteuffel, G. Hebbian Multilayer Network in a Wheelchair Robot. In: Gerstner W, Germond A., Hasler M., Nicoud J-D. (eds) Artificial Neural Networks - *ICANN 97*. pp 727 - 732. Springer Verlag. 1997

[3] Douglas R, Mahowald M, Mead, C (1995) Neuromorphic Analogue VLSI. *Ann. Rev. Neuroscience*, 18: 255-281.

[4] Goser K, Pacha C, Kanstein A, Rossmann M. L. (1997) Aspects of Systems and Circuits for Nanoelectronics. *Proceedings of the IEEE*, Vol.84, No.4, April, Special Issue on Nanometer-Science and Technology, pp. 558-573

[5] Rossmann M, Burwick C, Bühlmeier A, Manteuffel G, Goser K. Neural Dynamics in Real-Time for Large Scale Biomorphic Neural Networks. ICANN 98. Springer Verlag

# Diploid Robots Adapting to Fast Changing Environments

Raffaele Calabretta, Stefano Nolfi, Domenico Parisi

Institute of Psychology, National Research Council, Rome - Italy

raffaele@caio.irmkant.rm.cnr.it

Riccardo Galbiati

Department of Biology, University Tor Vergata,

Rome - Italy

## Abstract

In most work applying genetic algorithms to populations of neural networks there is no real distinction between genotype and phenotype. In nature both the information contained in the genotype and the mapping of the genetic information into the phenotype are usually much more complex. Moreover, the genotypes of many organisms exhibit diploidy, i.e., they include two copies of each gene whose expression is governed by some dominance rules. We briefly review the literature on diploidy and we present our own model which in the present paper is applied to populations of organisms living in a fast changing environment. Our results show that diploidy produce better performance than haploidy in this type of environments.

# 1   Introduction

In most work on genetic algorithms there is no real distinction between genotype and phenotype ([1]) since genetic traits tend to map one-to-one into phenotypic traits. In real organisms the genes-to-phenotype mapping is much more complex. One dimension of this complexity is that in most organisms the genome exhibits diploidy, i.e., it contains two copies of each gene. Several theoretical analyses have been proposed to explain the importance of diploidy and how it could have evolved. Diploids are believed to adapt better and faster than haploids for several reasons: (a) diploids can mask the effect of deleterious mutations which usually affect the recessive features of a trait; (b) overdominance (i.e. a positive interaction between different alleles in the expression of a trait) may improve the adaptability of evolving individuals; (c) there may be a larger occurrence of favorable and initially partially dominant mutations ([2]; [3]). On the other hand, because diploids are subjected to a larger number of mutations with respect to haploids, a long-term reduction in fitness should be expected unless dominance is complete or strong epistatic effects are present (i.e. decoupling of phenotypic expression relative to its genetic background; [3]) ([4]; [5]).

In a preceding paper ([6]) we presented an Artificial Life model for investigating these issues and compared the performance of haploid and diploid populations of ecological neural networks living in both fixed and changing environments. In this paper we add some new interesting results we have obtained with our simulations in an environment that changes each generation and we discuss the insights this approach can provide for understanding the conditions in which diploidy can enhance the adaptation of asexual organisms.

## 2   Simulations

We ran a set of simulations in which the task is to explore an environment and to return, once in a while, to a "food" area where individuals can reintegrate the energy consumed during the exploration (for more details see [6]). The organism was a computer simulation of a miniature mobile robot (Khepera) whose features are described in a previous paper ([7]). The environment was a rectangular box of 60x35 cm and contained a circular food area of 20 mm diameter.

The behavior of each individual robot was controlled by a feed-forward neural network. The network included 10 input units (8 units encoding the activation level of the 8 infrared sensors, 1 unit encoding the current energy level of the robot, and 1 unit encoding whether the robot is inside or outside the food area) and 2 output units encoding the speed of the robot's two wheels. Note that the robot can sense the food area only when it is in the food area itself. Therefore, when the robot must reintegrate its reserve of energy it must find the food area by exploring the environment.

A genetic algorithm ([8]) was used to evolve the connection weights of the neural networks. An initial population of 100 neural networks was generated by assigning random values to the 22 weights of each network (20 weights connecting the 10 input units to the 2 output units and 2 bias weights for the 2 output units). The 100 individuals were tested to determine their fitness by placing each of them in a separate copy of the environment. Each individual was placed in the box with a randomly selected orientation and it was allowed to move for 2,000 cycles each corresponding to 100 ms of real time. This process was repeated three times (epochs) for a total of 6,000 cycles. The environment was ideally divided up into cells of 2x2 cm and individuals were scored for the total number of cells visited in each epoch when the robot's energy was above 0. (Cells visited when the energy level was 0 did not produce an increase in fitness). The 20 individuals that obtained the highest fitness score were allowed to reproduce by generating five copies of their genotype with the addition of random mutations. The 20x5 new individuals constituted the next generation that was tested exactly like the first one. The process was continued for 300 generations.

We used two different types of genetic coding for our neural networks and we ran two different sets of simulations. The first type of genetic coding was haploid, the second diploid.

The haploid genotype included 22 chromosomes ($n$), one for each of the 22 connection weights of the neural network. Each chromosome is a sequence of 8 bits (0 or 1) which coded for a specific value of the corresponding connection weight.

Normal binary coding was used to translate the 8 bits into a weight value between -10.0 and +10.0.

The diploid genotype included 22 pairs of chromosomes (*2n*) (see [6], Figure 1). Each pair of genes coded for two possibly different values of the corresponding connection weight. In the case of diploid genotypes, each of the two homologous chromosomes consisted of a sequence of 10 bits: 8 coding for the corresponding weight value (*structural genes*) and the remaining 2 coding for the dominance/recessivity mechanism (*dominance modifier genes*) that decides which of the two copies of the encoded weight value is phenotypically realized. (Cf. [6] for details.)

# 3 Experiments and Results

In Calabretta *et al.* ([6]) we described a number of simulations in which the environment was either stable or changing during the evolutionary process. In the stable environment the position of the food area was 17.5 on the x coordinate and 17.5 on the y coordinate, and it remained the same for the entire course of evolution (300 generations). In the changing environment the position of the food area remained the same (17.5/17.5) for the first 59 generations, and then it started to alternate between two different positions (17.5/17.5 and 50/10) every 25 generations.

For each of the two conditions we investigated the performance of haploids and diploids with different mutation rates: 1%, 2%, or 3% of the bits of the genotype randomly selected were replaced with a new randomly selected value. (Notice that diploids have twice the structural bits of the haploids plus the modifier bits.) In particular, we were interested in examining the consequences of different mutation rates for haploid and diploid individuals with respect to the distribution of fitness values among the 100 individuals of each generation.

When the environment was stable we observed that diploids with 1% mutation rate outperformed both haploids with 1%, 2% and 3% mutation rate, and diploids with 2% and 3% mutation rate. The picture was reversed when we compared the average performance of each generation, with haploids with 1% mutation rate scoring the highest values. This result can be explained by considering that random changes produced by mutations have negative consequences most of the time and only occasionally produce an improvement. Therefore, lower mutation rates result in better performance on average while higher mutation rate may produce better performance if we consider the best individual of each generation. In addition, mutations tend to have larger effects on diploids than on haploids because, unlike haploids, diploids include modifier genes and mutations in modifier genes have in general much larger consequences than mutations in structural genes. Therefore, average performance is lower in diploids. However, diploids occasionally are affected by adaptive mutations which have a large effect and therefore may show better performance if we look at the best individual of each generation (see [6], Figure 4).

When the environment changed every 25 generations (after an initial period in which it remained fixed) the results were about the same as those already described. However in the changing environment diploid populations exhibit another feature: the capacity to keep a sort of genetic «memory» of the past recorded in their non-expressed genes (the shielding effect); in fact they recovered faster and with a lesser decrease in fitness than haploids after environmental change.

In the present paper we report the results of a new set of experiments in which the environment changes every other generation instead of every 25 generations. In this case we observed a remarkable difference between haploid and diploid individuals (see Figure 1). Haploids with all mutation rates (1%, 2% and 3%) are unable to adapt to the highly frequent changes in the environment and they show a significant drop in both average and peak fitness every other generation. On the contrary, diploids with a mutation rate of 1% are able to tolerate the rapid changes in the environment without displaying a significant loss in both their average and peak fitness.

# 4 Discussion

To our knowledge Ng and Wong's work ([9]) is the only one in which performances of haploid and diploid populations evolving in changing environments are compared using a dominance mechanism similar to our own. Ng and Wong challenged the classical Smith and Goldberg's ([10]) triallelic scheme opposing a new diploid scheme and dominance change mechanism with which they demonstrated the best performance of diploids in tracking environmental changes. However, our approach is different in two main aspects. First, our dominance decision mechanism is somewhat different in that it uses two modifier biallelic genes for every structural gene that allow dominance shift when they are mutated. Second, we applied this mechanism to a genetic algorithm encoding neural networks weights: the nonlinear interaction between weights results in strong epistatic effects that utterly shield the effect of deleterious (and favorable) mutations in both haploids and diploids.

The higher performance of diploids with a 1% mutation rate can be explained with the effect of dominance shift and epistasis which allow, through mutations, the adaptively neutral change of non-expressed genetic information and the occasional emergence of new adaptive complexes that may later be useful especially after environmental change. On the other hand, in haploid organisms only epistatic effects are present and a greater deal of old and new (mutated) information is directly expressed in the phenotype and therefore subject to selection. Nevertheless, our previous results are similar to those of Ng and Wong ([9]) especially with respect to the role of mutation rate and the buffering capacity of diploids in response to environmental fluctuations. However, when the environment changes more frequently (i.e., every other generation) diploids with a mutation rate of 1% show dramatically their ability to find very quickly a solution which is good for both environments (for both average and best fitness values).

Despite the enormous simplification of the model, our simulations confirm some of the results obtained with real organisms. In particular our results

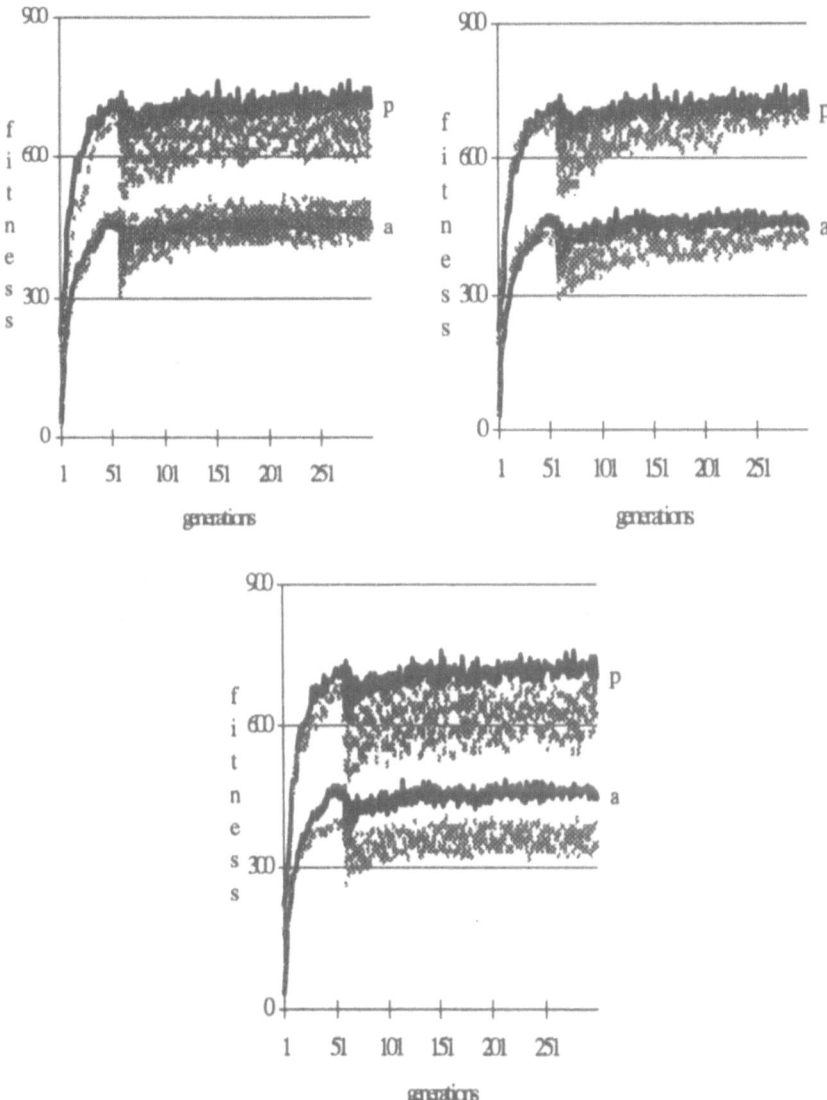

Figure 1: Average and peak fitness in a diploid population with mutation rate of 1% (black curve in the left and right graph on the top and in the bottom graph) and in a haploid population (gray curve) with mutation rates of 1% (left graph on the top), 2% (right graph on the top), and 3% (bottom graph). Individuals were placed in an environment that after 59 generations started to change every other generation.

may support Paquin and Adams' ([2]) claims concerning the role of diploidy in short-term adaptation to new environments in asexual organisms lacking the mechanism of recombination as a source of genetic variability.

# 5 Conclusions

In this paper we have examined the consequences of a more realistic genetic coding for neural networks in which the genotype and the phenotypical neural network are distinct entities and in which the genotype may exhibit diploidy, i.e., it may include two copies of each gene whose expression is governed by some dominance rules. Although diploids tend to have lower performance than haploids on average, they may produce better performances from the point of view of the best individual of each generation. Moreover, we observed that diploidy is useful to produce individuals able to adapt to nonstationary environments, especially in the case of fast changing environments. In these environments diploid individuals succeed in adapting to the quick environmental changes while haploid individuals tend to fail.

# References

[1] Nolfi S, Parisi D. Genotypes for neural networks. In: Arbib MA (ed) The handbook of brain theory and neural networks. MIT Press, Cambridge, 1995, pp 431-434

[2] Paquin C, Adams J. Frequency of adaptive mutations is higher in evolving diploid than haploid yeast population. Nature 1983; 230:495-500

[3] Kondrashov AS, Crow JF. Haploidy or diploidy: which is better? Nature 1994; 351:314-31

[4] Goldstein DB. Heterozygote advantage and the evolution of a dominant diploid phase. Genetics 1992; 132:1195-1198

[5] Orr HA, Otto SP. Does diploidy increase the rate of adaptation? Genetics 1994; 136:1475-1480

[6] Calabretta R, Galbiati R, Nolfi S, Parisi D. Two is better than one: a diploid genotype for neural networks. Neural Processing Letters 1996; 4:1-7

[7] Nolfi S, Floreano D, Miglino O, Mondada F. How to evolve autonomous robots: different approaches in evolutionary robotics. In: Brooks RA, Maes P (eds) Artificial Life IV: Proceedings of the Fourth International Workshop on the Synthesis and Simulation of Living Systems. MIT Press, Cambridge, 1994, pp 190-197

[8] Holland JJ. Adaptation in natural and artificial systems. University of Michigan Press, Ann Arbor, 1975

[9] Ng KP, Wong KC. A new diploid scheme and dominance change mechanism for non-stationary function optimization. In: Eshelman L (ed) Proceedings of the Sixth International Conference on Genetic Algorithms. Morgan Kaufmann, San Mateo, 1995, pp 159-166

[10] Smith RE, Goldberg DE. Diploidy and dominance in artificial genetic search. Complex Systems 1992; 6:251-285

# Dynamical Adaptation of a Neural-Net Based Agent

Masato Ito

Department of Computer Science, Keio University

Yokohama, JAPAN

masato@mt.cs.keio.ac.jp

Jun Tani

Sony Computer Science Laboratory Inc.

Tokyo, JAPAN

tani@csl.sony.co.jp

## Abstract

This paper shows a study of the dynamical adaptation of a neural-net based agent. We use a recurrent neural net (RNN) scheme which combines prediction and reinforcement learning. We investigated how the internal dynamics of the RNN and the resultant goal-directed behaviors are self-organized while the internal rewarding system of the agent is dynamically changed according to the agent's own goal-achievements. Our simulation results showed that limit cycling dynamics appear as a solution for the goal-achievement. The whole history of the adaptation process fluctuated by repeating intermittent phase transitions of the internal RNN dynamics from one limit cycling to another with generating diverse goal-directed behaviors.

## 1   Introduction

The dynamical systems approach using a recurrent neural network (RNN) has been actively studied in the domain of adaptive behavior. Pollack and Tani showed that the RNN can learn the grammatical structure from the string sequences[3] or the sensory-motor flow[5] on the static environments. Beer[1] showed that the agent using the RNN was adapted to not only the static environments but also the dynamic environments through its reactive behavior. However, these works are not adequate to the point of view how the dynamical structure attained in the RNN adapts to the dynamic environments.

Lin[2] showed that the reinforcement learning model using the RNN succeeded in the simple static environments, which include "hidden state problem". In this paper, we apply Lin's model to learning on the dynamic environments. We conduct the experiment of the learning on the simple discrete world expressed as a finite state machine. This world is dynamic in the sense that the reward value located in some states is changed dynamically due to the goal achievement of a learning agent. Therefore, the agent needs to learn incrementally and to rebuild internal representation of the world. In this paper, we observe how structure in the RNN dynamically adapts through the goal-achievement interactions.

# 2 Task and Models

In this section, we describe the task in the experiment and the model accomplishing it.

## 2.1 Task

We assume that the agent travels the world, which simplifies the workspace adopted for the physical robot navigation experiments[5], expressed as a finite state machine shown in Fig.1. The agent in certain state maneuvers by selecting the binary action $\{0, 1\}$ into the next state. Then the agent can receive one of the sensory inputs $\{A, B, C, D, E\}$. We assume the non-markovian environment where the agent cannot identify each state only by current sensory inputs. In this world, there are three states $\{4, 7, 9\}$ in which the "food" is located, and the agent can gain the reward there. The reward value of "food" in a initial condition is 0.5. While the reward value of "food" which the agent has eaten decreases by 0.01, others increase by 0.003. Although the finite state machine structure in this world is quite static, the agent constitutes dynamic environments in the sense that the goal achievement behavior of the agent changes the reward value of "food". The task of the agent is to learn an adequate action leading to the "food" and an analogical model of the world through its travel.

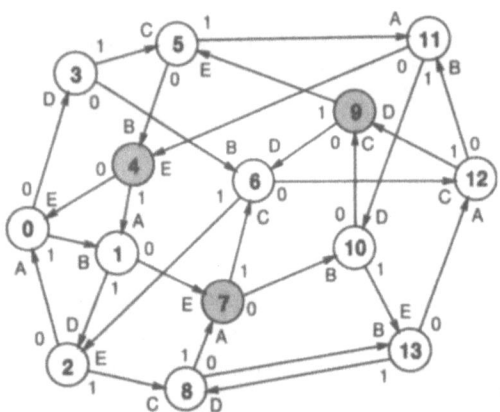

Figure 1: Finite State Machine World

## 2.2 Models

We use the model shown in Fig.2, which is the *recurrent-model* proposed by Lin, to investigate the learning in the dynamic environment.

This model consists of two modules. The model learning module implemented as the RNN learns to predict both the next sensory inputs and the reward value from the current sensory inputs and the binary action. The RNN

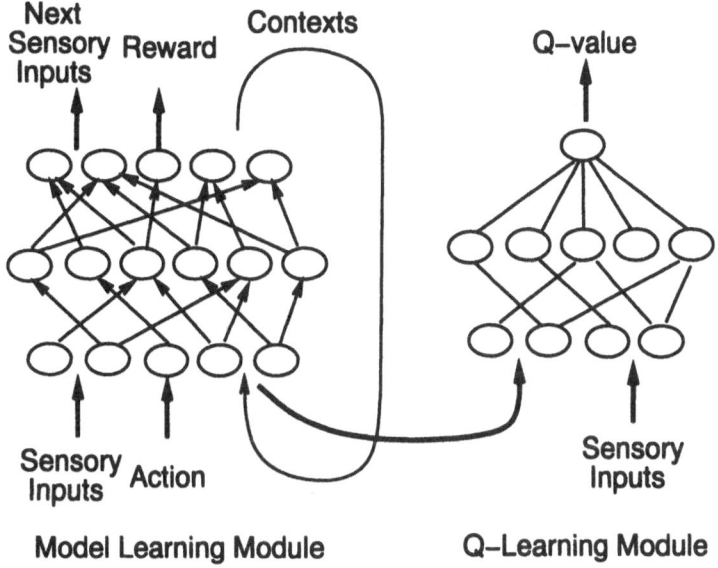

Figure 2: Model

used in the experiment has 7 input units ( 5 current sensory inputs, 2 current action inputs ), 6 output units ( 5 next sensory inputs, 1 reward value ), 8 context units and 14 hidden units. The Q-learning module implemented as the feed-forward neural networks learns to predict the Q-value, which represents the quality of action in a state, from the sensory inputs and the context given by the model learning module. We use two networks to assign to each binary action to estimate each Q-value independently. The binary action is selected according to probabilities promotional to each Q-value. A Boltzmann distribution is used to determine this probability $p$, as follows:

$$p(action_i) = \frac{\exp(Q_i/T)}{\sum_{j \in \{0,1\}} \exp(Q_j/T)} \tag{1}$$

where $Q_i$ is the Q-value for the binary action $i$. The temperature value $T$, which represents the degree of "adventurousness" of the agent, is changed as follows to adapt to the dynamic environment.

The trial of the agent in the experiment consists of two phases, the behavior phase and the learning phase. In the case that the behavior phase where the agent can eat "food" twice within regular steps succeeds, the temperature value goes down and the agent turns to the learning phase. Conversely, in case of failure, the temperature value goes up and the agent starts next behavior phase from start state (0) without any learning. In the learning phase, two modules are trained using the experience ( sensory inputs, binary action and reward ) during the travel. The training of the RNN is conducted using the back-propagation through time (BPTT) algorithm[4] for 200 steps and the training of the Q-nets are conducted using back-propagation algorithm. After the training, the agent restarts to travel again from the state where the behavior phase was ended continuously. This trial is repeated 500 times, while the RNN incrementally learns the experiences. The agent living in this world must not aim at one goal but continue to travel from current goal to next goal, because the attraction of the goal is dynamically changed by the behavior of the agent.

# 3 Experiment

In this section, we investigate how the agent of the above model is adapted to the dynamic environment.

In the early stage of the trial, the action is determined nearly at random, because the temperature value is high, the Q-nets are not trained enough and the RNN does not supply reliable contexts to the Q-learning module. While the agent explores the world and eats the "food" several times, the agent learns an adequate state-action map and constructs its internal state to some degree. Although the agent cannot learn all the path in the world, in the state near the "food" the agent can predict next sensory inputs accurately and take a correct action toward the "food". In the trial till 209 steps, the trajectory, through which the agent goes, has no special regularity.

In the trial from 210 to 260 steps, the agent continues to go through the same trajectory, where the agent goes the round of three states periodically $(9 \rightarrow 6 \rightarrow 12)$, shown in Fig.3 (a). After that the agent goes out this trajectory, but the agent finds two different trajectories in due order. In the trial from 270 to 331 steps, the trajectory, where the agent goes the round of six states periodically $(4 \rightarrow 0 \rightarrow 3 \rightarrow 6 \rightarrow 12 \rightarrow 11)$, is generated shown in Fig.3 (b). In the same way, Fig.3 (c) is the trajectory $(7 \rightarrow 6 \rightarrow 2 \rightarrow 8)$ in the trial from 443 to 455 steps. We observed these three different sequential patterns in the trial.

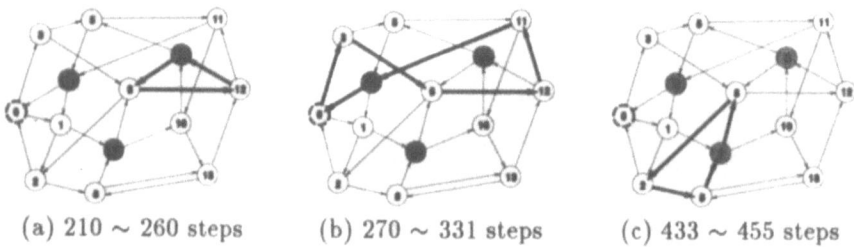

(a) $210 \sim 260$ steps     (b) $270 \sim 331$ steps     (c) $433 \sim 455$ steps

Figure 3: Patterns of remarkable trajectory in the FSM world

We consider the internal structure of the agent during taking these trajectories. We examined the dynamical structure self-organized in the RNN by conducting a phase-space analysis. The recursively activated for 5,000 steps with establishing the self-feedback from the sensory prediction outputs to the inputs and using the action selection by the Q-learning module without the Boltzmann Distribution. Fig.4 (a) shows the phase space of the RNN in the trial 230 steps. We took $c_1$ as average activation over a half of context units and

$c_2$ as that over the other half. We see that an attractor of the limit cycle with the periodicity three is generated. This periodicity three is consistent with the number of the state on the trajectories shown in Fig.3 (a). In the same way, the periodicity six (Fig.4 (b)) and the periodicity four (Fig.4 (c)) correspond to the trajectory shown in Fig.3 (b) and (c), respectively. The internal state suited to the behavior of the agent is generated.

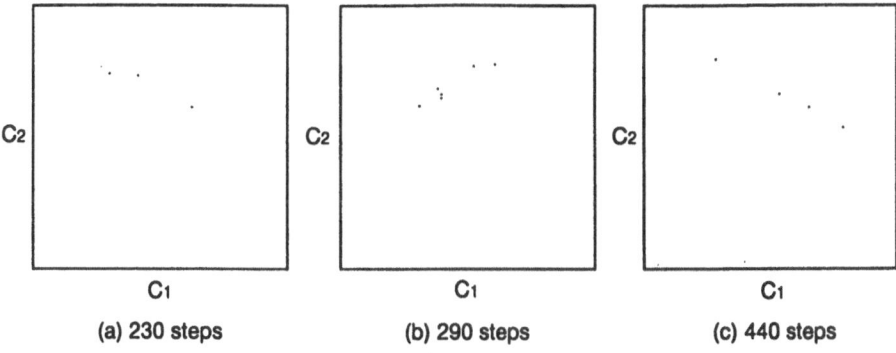

Figure 4: Phase space analysis of the structure self-organized in the RNN

The same trajectory does not continue all through a trial. While the agent keeps eating the same "food", the reward value of the "food" decreases gradually. Then the Q-value of the action toward the "food" decreases. The agent goes out the trajectory, because the action selection in the trajectory becomes unstable. The agent restarts to explore the world, and then finds different trajectories where the agent will get more reward. The internal structure of the agent is rebuilt according to such dynamical change of the interactions.

## 4    Conclusion

In this paper, we conducted the experiment of the learning in the dynamic environment using the model which combines the RNN with the reinforcement learning. We observed the structure emerged through the interaction between the reward-based action and the environment from the dynamical systems view. We showed that the dynamical structure with a limit cycle was self-organized corresponding to the behavior of the agent going round of several states periodically. The agent can learn the subset of the environment taking the form of *dynamical closure*. It was further observed as different sequential patters in the trial. In the adaptation process the internal dynamics repeated the phase transition from one limit cycling to another, by which diverse goal-directed behaviors were generated. In the future research, we examines this agent adaptation scheme in more complex real world environment.

# References

[1] Randall D. Beer. A dynamical systems perspective on agent-environment interaction. *Artificial Intelligence*, 1995.

[2] Long Ji Lin and Tom M. Mitchell. Reinforcement learning with hidden states. In *Proceedings of the Second International Conference on Simulation of Adaptive Behavior*, 1992.

[3] Jordan B. Pollack. The induction of dynamical recognizers. *Machine Learning*, 1991.

[4] David E. Rumelhart, James L. McClelland, and the PDP Research Group. *Parallel Distributed Processing: Explorations in the Microstructure of Cognition*. MIT Press, 1986.

[5] Jun Tani. Model-based learning for mobile robot navigation from the dynamical systems perspective. *IEEE TRANSACTIONS ON SYSTEMS, MAN, AND CYBERNETICS*, 1996.

# Poster Presentations:
# Hardware and Implementations

# A Reconfigurable Neuroprocessor with On-chip Pruning

Jean-Luc Beuchat and Eduardo Sanchez

Logic Systems Laboratory, Swiss Federal Institute of Technology
CH – 1015 Lausanne, Switzerland ·
E-mail: {name.surname}@di.epfl.ch

### Abstract

The appearance of fast reconfigurable FPGA circuits brings about new paths for the design of neuroprocessors. A learning algorithm is divided into different steps that are associated with specific FPGA configurations. The training process then consists of alternating computing and reconfiguration stages. Such a method leads to an optimal use of hardware resources. This new method is applied to the design of a neuroprocessor implementing multilayer perceptrons with on-chip training and pruning. All arithmetic operations are carried out with fixed-point numbers. The first step of our work is the simulation of limited precision training and pruning algorithms. Our experiments demonstrate that this representation is well suited for this task. This paper also presents the principles of our hardware implementation, focusing in particular on the pruning mechanisms.

## 1 Introduction

Artificial neural networks can solve complex problems such as handwritten pattern recognition and time series prediction. Though software simulations are essential when one sets about to study a new algorithm, they can not always fulfill real-time criteria required by some practical applications. In order to exploit the inherent parallelism of artificial neural networks, hardware implementations are essential.

Among the many neuroprocessors described in the literature, we distinguish between two main design philosophies. The first approach involves the design of a highly parallel computer and a programming language dedicated to neural networks. Many algorithms can be implemented on the same system. Nevertheless, programming such a machine is often arduous. The second approach involves the design of a specialized chip for a given algorithm, thus avoiding the tedious programming task. The main drawback lies in the need for a different chip for each algorithm.

Fast reconfigurable FPGA circuits offer new paths for neuroprocessor implementations. A learning algorithm is split into several sequentially executed steps, each of which is associated with a specific FPGA configuration. Such an approach leads to an optimal use of hardware resources. The reconfiguration paradigm also allows the implementation of multiple algorithms on the same hardware.

We apply these principles to the design of a reconfigurable neuroprocessor able to run different learning rules for multilayer perceptrons. The performance of such systems depends on the network topology (number of neurons and interconnection scheme). Pruning or growing algorithms allow to find an optimal topology during training. However, we did not find any hardware implementation of such algorithms in the literature. Thus, preliminary experiments (with a software tool) must be carried out to determine a topology, before the beginning of the training on the neuroprocessor. In order to solve this problem, we study the hardware implementation of pruning algorithms. Section 2 briefly describes two pruning algorithms considered in our work. As FPGAs are not well suited for floating-point computation [1], we use fixed-point numbers to carry out all arithmetic operations. Section 3 presents results of simulations which prove the efficiency of such a representation. Then, we discuss several problems related to the hardware implementation of neural networks (section 4). Finally, section 5 presents our concluding remarks and some future extensions of our work.

## 2 Pruning Algorithms

When we train a system by examples, it is generally illusory to provide every possible input pattern. Therefore, an important issue of training is the capability of the network to generalize, that is, cope with previously unseen patterns. However, generalization depends on the network topology. A rule of thumb for obtaining a good generalization is to use the smallest network able to learn the training data [2]. Training successively smaller networks is a time-consuming approach. Among the efficient processes to determine a good topology, one can cite genetic algorithms, growing methods, and pruning algorithms, the latter of which are used herein.

With pruning algorithms one trains a network that is larger than necessary and deletes superfluous elements (units or connections). These algorithms can be classified into two general categories: sensitivity estimation and penalty-term methods. Algorithms within the first category measure the sensibility of the error to the removal of a connection or a unit. Then, elements with the smallest sensibilities are pruned. Methods belonging to the second category suggest new error functions which drive weights to zero during training.

Pruning connections sometimes leads to a situation where some neurons have no more inputs or outputs. Such neurons, called *dead units*, can be deleted.

Let us introduce some notations before discussing two pruning algorithms. $E^\rho$ denotes the mean square error in the backpropagation algorithm (computed for an input pattern $\rho$). $w_{n_i m_j}$ is the weight between neuron $i$ in layer $n$ and neuron $j$ in layer $m$. The net input for neuron $j$ in layer $m$ is denoted by $h^\rho_{m,j}$. $a^\rho_{m_j}$ is the activity (or activation value) of neuron $j$ in layer $m$.

## 2.1 A Sensibility Estimation Algorithm

Autoprune [3] is an efficient sensitivity estimation algorithm. It computes for each weight a statistic test $T(w_{m-1_i m_j})$, assuming that $w_{m-1_i m_j}$ becomes zero during training:

$$T(w_{m-1_i m_j}) = \log \frac{\sqrt{n} \cdot \left| n \cdot w_{m-1_i m_j} - \eta \cdot \sum_{\rho=1}^{n} \frac{\partial E^\rho}{\partial w_{m-1_i m_j}} \right|}{\eta \cdot \sqrt{n \cdot \sum_{\rho=1}^{n} \left( \frac{\partial E^\rho}{\partial w_{m-1_i m_j}} \right)^2 - \left( \sum_{\rho=1}^{n} \frac{\partial E^\rho}{\partial w_{m-1_i m_j}} \right)^2}}, \quad (1)$$

where $\eta$ is the learning rate and $n$ denotes the number of training patterns.

## 2.2 A Penalty-Term Method

We will now focus on a penalty-term algorithm suggested by Ishikawa [4]. Let us consider the slightly modified error function:

$$E_\lambda^\rho = E^\rho + \lambda \cdot \sum_{m,i,j} |w_{m-1_i m_j}|. \quad (2)$$

Differentiating equation (2) with respect to the synaptic coefficient $w_{m-1_i m_j}$ leads to a new update rule :

$$\Delta_\lambda w_{m-1_i m_j} = -\eta \cdot \frac{\partial E^\rho}{\partial w_{m-1_i m_j}} - \lambda \cdot \text{sgn}(\text{w}_{m-1_i m_j}). \quad (3)$$

Equation 3 drives synaptic coefficients to zero. Weights are removed when they decrease below a given threshold.

# 3 Experiments

As mentioned in the introduction, floating-point numbers are not well suited for our hardware implementation. We use fixed-point numbers (two's complement numbers) to carry out additions, subtractions, and multiplications.

We have performed a series of experiments to evaluate the efficiency of such a limited-precision system. Four learning rules (Backpropagation, Non-Linear Backpropagation [5], Resilient Backpropagation and Weighted Error Function [6]) and the two pruning algorithms previously discussed have been implemented in SNNS [7], a neural network simulator. A first version of these algorithms performs all arithmetic operations with floating-point numbers, while a second one exploits fixed-point numbers.

The training and pruning algorithms have been applied to eight problems from the Proben1 [8] data set and generalization capabilities have been studied. Our experiments have demonstrated that limited-precision (generally 3 bits for the integral part and 12 bits for the fractional part) algorithms have the same performance as floating-point algorithms. (Note that weights sometimes become huge when the network is trained with Resilient Backpropagation. More bits are then needed for the integral part of numbers.)

We found that Ishikawa's algorithm is especially suitable for our hardware implementation. Equation 3 prevents an important growth of synaptic coefficients, thus bounding the number of required bits for the integral part of the numbers.

# 4  Hardware Description

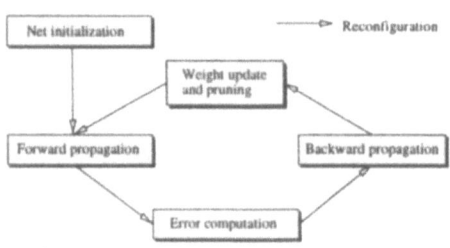

Figure 1: *A possible decomposition of the backpropagation algorithm.*

A learning algorithm consists of several steps, including: network initialization, forward propagation, error computation, backward propagation, weight update, and pruning. Each step requires specific hardware resources, e.g., network initialization makes use of a random number generator, which is unused in the following steps.

Fast reconfigurable FPGA circuits [9] offer new possibilities for designing hardware neural networks. The learning algorithm is divided into several sequentially executed stages, each of which is associated with an FPGA configuration. A possible decomposition scheme is depicted in Fig. 1. Note that the reconfiguration time is of crucial import—if this process needs more time than computation, such an approach is not appropriate. In order to realize an efficient system one must carefully choose the FPGA family.

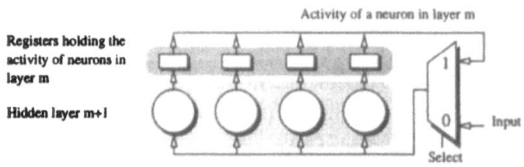

Figure 2:  *The time-multiplexed interconnection scheme.*

When designing the hardware architecture of our neural network we first observed that a time-multiplexed interconnection scheme provides a good trade-off between speed and scalability (Fig. 2). The main idea is to connect all outputs of hidden layer $m$ and inputs of hidden (or output) layer $m + 1$ to a common bus; the same hardware is reused for all layers of the network. The multiplexor allows to provide the network with an input signal or an activation value of a hidden unit. All synaptic weights are stored in a memory associated with the FPGA(s). We will focus herein on forward propagation of a signal (the backward propagation obeys the same principles). The first neuron in layer $m$ places its activity $a^{\rho}_{m,1}$ on the bus. All neurons in layer $m + 1$ read and multiply it by the appropriate synaptic weight $w_{m_1 m+1_j}$

and finally store the result. Simultaneously, we load the weights from the next layer-$m$ neuron to layer $m + 1$. This process is sequentially repeated for every neuron in layer $m$. Each processing element in layer $m + 1$ accumulates the results of the successive multiplications.

Due to this interconnection scheme, each neuron is a very simple processing element. Figure 3a depicts the architecture used during the forward propagation step. A register stores the weight value involved in the next multiplication. We have associated with each weight a special bit, called $\overline{\text{Pruning}}$, which indicates whether a connection has been pruned

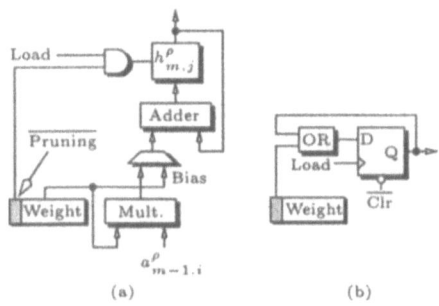

Figure 3: *(a) Architecure of a neuron. (b) Dead unit detection mechanism.*

(in which case, this bit is set to 0) or not. Combined with a load signal, it enables the accumulation of a multiplication result. We now describe the implementation of the two pruning algorithms.

1. **Autoprune.** The sums $\sum_{\rho=1}^{n} \frac{\partial E^{\rho}}{\partial w_{m-1;m_j}}$ and $\sum_{\rho=1}^{n} \left( \frac{\partial E^{\rho}}{\partial w_{m-1;m_j}} \right)^2$ are calculated during training. However, the statistical test $T(w_{m-1;m_j})$ involves the computation of a logarithm and a squared root. We investigated two different approaches to solve this problem. The first one suggests the exploitation of logarithm properties to convert equation (1) into a sum of logarithms. The logarithm is then estimated by a piecewise linear function. The second approach consists of computing the logarithms and squared roots on a microcontroller.

2. **Penalty-term method.** The implementation of Ishikawa's algorithm is straightforward. Remember that the backpropagation update rule involves an adder. The addition (or subtraction) of $\lambda$ is carried out by the same adder. If the absolute value of the result is below a given threshold, the weight is removed (i.e., its $\overline{\text{Pruning}}$ bit is set to zero).

We now have to provide our neuroprocessor with a means for detecting dead units. The mechanism illustrated in Fig. 3b solves this problem. Assume that a neuron $j$ in layer $m$ has no more inputs. All $w_{m-1;m_j}$ coefficients are loaded when a signal is forward-propagated through the network. As the $\overline{\text{Pruning}}$ bits associated with the $w_{m-1;m_j}$ are set to zero, the flip-flop output remains zero as well. Consider a neuron with no outputs. The backward-propagation process involves all weights $w_{m_j m+1_k}$ whose $\overline{\text{Pruning}}$ bits are equal to zero. Consequently, the detection of such dead units occurs during this step. Once a dead unit has been detected, a signal is sent to a global controller which manages the network topology. As the activities of neurons are sequentially placed on the bus, the deletion of dead units increases the learning speed.

# 5 Conclusions

This paper has presented two pruning algorithms and some results of their simulation with fixed-point numbers. Our experiments proved that this number representation scheme is well suited for our application. We then discussed some hardware implementation details. An interesting issue is the pruning system. Dead unit removal and training are executed concurrently. Our solution is more efficient than the one used in some software simulators, where the removal of dead units and the training stages are done sequentially.

We now have to complete the hardware implementation and to evaluate the performance of our neuroprocessor (we have actually been working with Altera Flex 10K130 FPGAs). Further improvements will require new hardware. One of the suggested implementations of Autoprune involves a commercial microprocessor in order to carry out the logarithm computations. The Core+ chip announced by Motorola merges a ColdFire processor with FPGAs. If the multilayer perceptron is distributed in $n$ FPGAs, a Core+ based system allows to compute $n$ statistical tests concurrently. Another possible enhancement lies in the partial reconfiguration paradigm.

Finally, reconfigurable systems offer some other interesting prospects. The architecture depicted in Fig. 2 is well suited for the learning process. However, it should be interesting to increase the parallelism when training is over. Therefore, we plan to design special FPGA configurations for the recall process.

# References

[1] Villasenor J, Mangione-Smith WH. Configurable Computing. Scientific American, June 1997, pp. 54-59.

[2] Reed R. Pruning Algorithms – A Survey. IEEE Transactions on Neural Networks 1993; 4(5):740-747

[3] Finnoff W, Hergert F, Zimmermann HG. Improving Model Selection by Nonconvergent Methods. Neural Networks 1993; 6:771-783

[4] Ishikawa M. Structural Learning with Forgetting. Neural Networks 1996; 9:509-521

[5] Hertz J, Krogh A, Lautrup B, Lehmann T. Non-Linear Back-Propagation: Doing Back-Propagation without Derivatives of the Activation Function. IEEE Transactions on Neural Networks 1997; 8(6):1321-1327

[6] Sakaue S, Kodha T, Yamamoto H, Maruno S, Shimeki Y. Reduction of Required Precision Bits for Back-Propagation Applied to Pattern Recognition. IEEE Transactions on Neural Networks 1993; 4(2):270-275

[7] Zell A, Mamier G, Vogtet M et al. SNNS, User Manual, Version 4.1. Technical Report 6/95, Institute for Parallel and Distributed High Performance Systems, University of Stuttgart, 1995

[8] Prechelt L. Proben1 - A Set of Neural Network Benchmark Problems and Benchmarking Rules. Technical Report 21/94, Fakultät für Informatik, University of Karlsruhe, 1994

[9] Sanchez E. Field Programmable Gate Array (FPGA) Circuits. In: Sanchez E, Tomassini M (eds) Towards Evolvable Hardware : The Evolutionary Engineering Approach. Springer-Verlag, 1996, pp 1-18 (Lecture Notes in Computer Science no. 1062)

# Laser Neural Network Demonstrates Data Switching Functions

E.C. Mos and H. de Waardt

Department of Electrotechnical Engineering, Eindhoven University of Technology
P.O. Box 513, 5600 MB Eindhoven, The Netherlands
mosec@natlab.research.philips.com

J.J.H.B. Schleipen

Philips Research Laboratories,
Prof. Holstlaan 4, 5656 AA Eindhoven, The Netherlands

### Abstract

By use of an extended-cavity laser diode setup, all-optical thresholding and weighting is combined to obtain an all-optical neural network. The operation of the neural threshold function is realised by controlling the optical feedback in the extended cavity. An optical vector matrix multiplier provides weighting of inputs, the outputs are in the spectral distribution of the laser output power. We demonstrate up to 32 neurons and 12 inputs in our experimental network by use of a fast liquid crystal display in the optical vector matrix multiplier. A $\delta$-rule type algorithm is adapted to train our winner-take-all neural network. Measured characteristics of this algorithm and results of training data switch functions are presented.

## 1 Introduction

Neural networks can play a role in all-optical routing devices in telecommunication systems. For this application area, the neural threshold function should preferably operate in the optical domain. In this way, the speed of the telecommunication system will not be hindered by the conversions between the optical and the electrical domain.

In previous work [1,2], we presented our laser neural network (LNN). We used the sensitivity of a laser diode to external optical feedback to implement an all-optical, single layer winner-take-all neural network. With an experimental LNN we demonstrated training of some small-sized functions (4 inputs and 5 outputs), including the XOR function, to prove the operation principle of the LNN concept.

In this study, we examine the capabilities of our LNN towards the application area of optical telecommunications. For this, the number of neurons and the number of inputs of our experimental LNN are increased. The stochastic learning algorithm used in our previous experiments is replaced by a $\delta$-rule type learning algorithm, also known as Widrow-Hoff learning [3]. It is adapted for our neural network by implementing the winner-take-all nature of the LNN.

# 2 Laser neural network operation principle

In the following, the operation principle of our neural network is briefly presented in a rather simplified manner. For a detailed description of the operation principle, including numerical simulations, we refer to [1].

## 2.1 Laser diodes, optical feedback, and threshold function

A laser diode consist of an optical gain medium inside an optical resonator. The resonator is formed by the two cleaved facets of the laser chip acting as mirrors. When an anti-reflection coating is applied to one facet of such a laser diode, laser operation is lost. It can be selectively restored by adding external optical feedback by means of a mirror and some optics (See Figure 1).

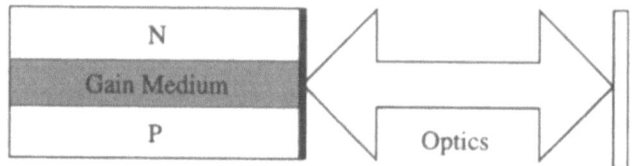

Figure 1 Schematic drawing of an anti-reflection coated laser diode in an external cavity

As a function of the amount of feedback, the output power of the laser diode can be observed to behave just like a neural threshold function. The origin of this threshold lies entirely in the optical domain. If the amount of optical feedback reaches some threshold value, the gain of the laser diode will compensate all optical losses. At this level the laser starts to lase, causing a sharp increase in the output power. At still higher levels of feedback, the output power of the laser will saturate to a maximum value.

For this effect, no change in current through the laser diode is necessary. As a result, the operation speed of the threshold function is only limited by the length of the external cavity, and the time constants associated with the gain medium. The ultimate operation speed is estimated to be compatible to that of high speed telecommunication networks.

## 2.2 Winner-take-all neural network

The laser cavity can support different wavelengths, matching the length of the laser cavity. The threshold function outlined above, applies for all of these longitudinal cavity modes. If, by means of an optical vector matrix multiplier, the amount of optical feedback is made proportional to a weighted sum of inputs for each longitudinal mode individually, its output power will exhibit a threshold response to a weighted sum of inputs. Thus, each longitudinal mode will behave like an artificial neuron and a single layer neural network is formed.

Due to a phenomenon known as mode competition, not all longitudinal cavity modes can lase at the same time. This, in turn, implies that not all neurons of our one layer neural network can be active simultaneously. It turns out in practice that this effect is strong enough to turn the neural network into a winner-take-all type neural network, in which only one neuron can be active in each instance.

# 3 Experiment

## 3.1 Optical setup

To accommodate the LNN with as many inputs and outputs as possible, we used a fast, deformed helix ferroelectric (DHF) [4] LCD with 32×32 pixels. Compared to standard liquid crystal materials, the DHF material has a much shorter response time. As a result, a learning supervisor PC can update the weight matrix about 125 times per second in our current setup.

To optimise the feedback efficiency of the setup, the linear external cavity of Figure 1 in which the laser light passes all optical components twice, is replaced by a loop mirror external cavity.

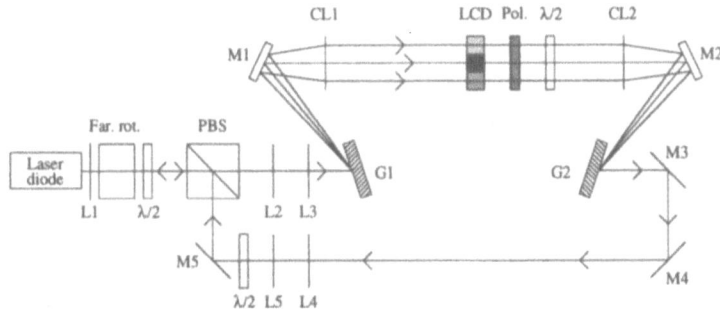

Figure 2 Optical setup of the laser neural network

Figure 2 shows a laser diode that is coupled to a loop mirror formed by a polarising beam splitter (PBS) and a number of mirrors (M1-M5). In the loop, an optical vector matrix multiplier is placed, consisting of the LCD, a number of lenses (L2-L5), cylindrical lenses (CL1 and CL2) and two gratings (G1-G2). With this optical vector matrix multiplier the amount of feedback through the loop is made proportional to a weighted sum of inputs for each longitudinal mode of the laser diode individually. In the figure, the direction of the optical signal is indicated with arrows. This direction is maintained via polarisation control by use of the Faraday rotator (Far.Rot), the PBS and a number of $\lambda/2$-plates ($\lambda/2$).

## 3.2 Learning algorithm

To train our experimental LNN, a supervisor PC is used that can receive measured spectra from the LNN and send weight matrices and input vectors to the LNN. On the PC we implemented a modified version of the $\delta$-rule algorithm [3].

Because we are mainly interested in training digital functions to our network, that has analogue outputs, digital decision levels have to be defined. These levels correspond to the 'on' and 'off' values of a neuron. Instead of using fixed values for these decision levels, a target extinction ratio between these levels is used in the learning algorithm. In this way we do not need to determine the 'on' and 'off' levels for each neuron. As an additional advantage, a noise margin can be trained in a rather straightforward manner by simply training the network with a larger extinction ratio than needed.

The learning algorithm is presented below. It starts by calculating an initial guess for the weight matrix based on the winner-take-all inequalities [2] of the function to be trained. Then, iteratively, the weight matrix is updated using a delta rule with learning parameter $\eta$ (in step 4a) until the error measure equals zero. When an appropriate weight matrix is found, and this solution is tested a number (N) of times to verify the stability of the solution, the algorithm stops.

**Basics of our modified $\delta$-rule algorithm**

```
Initialize:
    Get desired 'on'/'off' extinction ratio Er,
    Get learning rate η
    Calculate starting matrix W̿₀
Repeat
1: Show all input patterns Ī_p to the network with W̿
    Measure all corresponding output patterns Ō_p,meas
2: Calculate new target output values for each neuron j:
        'On' value:   O_j,des-on = Er × maximum of
            O_p,meas,j'    all p for which j is losing neuron,
            O_p,meas,k'    all p for which j is winning neuron, all k≠j
        'Off' value:  O_j,des-off = (1 / Er) × minimum of
            O_p,meas,j'    all p for which j is winning neuron
            O_p,meas,k'    all p for which j is losing neuron, all k≠j
3: Calculate error measure, Error=ΣpΣj |O_j,des - O_j,meas|
4a:If (Error > 0) then for all inputs, I, and all neurons, j:
        Calculate W_i,j = W_i,j + ΣpΣiΣj I_p,i × η × (O_p,j,des - O_p,j,meas)
4b:If (Error = 0) then check solution W̿ N times:
        repeat step 1 and 3, accumulate error measure
Until (Error =0)
```

In step 2, the digital decision levels are recalculated for each neuron using the target extinction ratio, $Er$, and the measured output values of the neuron in the 'on' and 'off' states. To account for the winner-take-all nature of our network, also the measured output values for all other neurons are included in this calculation.

# 4  Results

With the experimental LNN described in section 3 we were able to define up to 32 neurons and up to 12 inputs per neuron. With a selection of these neurons, we demonstrated some training examples towards the application area of optical telecommunications using the algorithm of section 3.2. We also verified the use of the extinction ratio $Er$.

## 4.1 Training examples

In the 1:8 data switch function, the idea is to switch a data bit to one out of eight channels, selected by a 3 bit address. We trained this function to the LNN by use of a learning set resulting from the truth table of Table 1. In this table, the data bit is the second bit of the input vector, whereas the address is represented by the last three bits and the first bit is always on. The weights corresponding to this input element can control the threshold level of the neurons. The output vector has 8 destination output channels and one dump channel.

### Table 1 Truth table of 1:8 data switch

| Input | Output | Input | Output |
|-------|-----------|-------|-----------|
| 10000 | 100000000 | 11000 | 000000001 |
| 10001 | 010000000 | 11001 | 000000001 |
| 10010 | 001000000 | 11010 | 000000001 |
| 10011 | 000100000 | 11011 | 000000001 |
| 10100 | 000010000 | 11100 | 000000001 |
| 10101 | 000001000 | 11101 | 000000001 |
| 10110 | 000000100 | 11110 | 000000001 |
| 10111 | 000000010 | 11111 | 000000001 |

When the data bit is off (first column of Table 1), one of the 8 output channels should be active, depending on the address represented by the 3 bit address. If the data bit is on (last column of Table 1), the dump output should be active. In this way, the inverse of the data bit is routed to one of the 8 output channels.

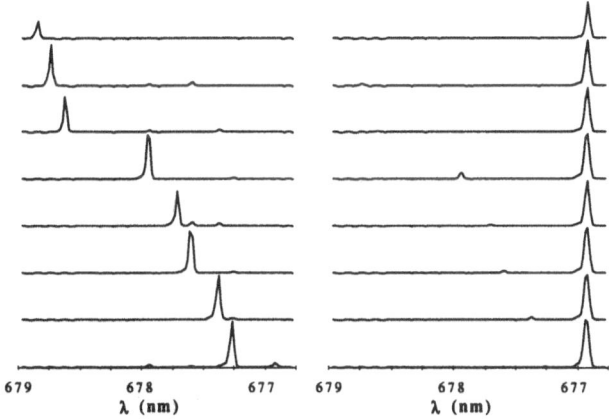

Figure 3 Output spectra, i.e. neural outputs, of a trained LNN for the 1:8 data switch

After training, we measured the spectral output of the LNN for all input vectors. In Figure 3 the resulting spectra are plotted in the same order as used in Table 1. The Figure shows a single active output wavelength for each input vector. Some wavelengths were not used in this experiment. Using a training set similar to that of Table 1, we also successfully trained a 1:16 data switch function to our network.

## 4.2 Varying the extinction ratio Er

We trained the 1:8 data switch function with a number of target extinction ratios. Aiming at a minimum extinction ratio of 4, we trained the network with $Er = 4, 5, 6$ and 7. After training we tested the behaviour of the LNN by showing 1000 input vectors to it. We repeated this for 10 learning trials for each value of $Er$. Since noise is obviously present in our setup, the fraction of these tests for which the measured extinction ratio was below 4 can be used as an indication of the noise margin. The higher this fraction, the smaller the noise margin. With $Er$ set to 4, this fraction was 4.5%, for $Er = 5$ it was 1.0%, for $Er = 6$ about 0.3%, and with $Er$ set to 7 it was below 0.1%.

# 5 Conclusions

We demonstrated an all-optical neural network with up to 32 neurons and 12 inputs. The resulting network is a winner-take-all type neural network. With it, we demonstrated training of a 1:8 and a 1:16 data switch function.

These functions could be used in telecommunication systems where data is to be routed to one of 8 (16) destination channels. For this application, high speed optically addressed optical modulators will be required to replace the input mask of the vector matrix multiplier.

By training the LNN with a higher extinction ratio than needed, the noise margin of the trained network can be enhanced.

Future work will focus on experimentally verifying the operation speed of the LNN, transferring the optical inputs from the transmission domain to the optical power domain and miniaturisation of the optical setup.

# References

[1] Colak SB, Schleipen JJHB, Liedenbaum CTH. Neural network using longitudinal modes of an injection laser with external feedback. IEEE Transaction on Neural Networks 1996; 7; 6: 1389-1400

[2] Mos EC, Schleipen JJHB, de Waardt H. Optical mode neural network by use of the nonlinear response of a laser diode to optical feedback. Appl. Opt. 1997; 36: 6654-6663

[3] Widrow B, Hoff ME. Adaptive switching circuit. in: Western Electronic Show and Confention, Confention Record. Institute of Radio Enigineers, 1960, Part 4, pp 96-104 Reprinted in: Anderson j, Rosenfeld E. (eds) Neurocomputing. MIT Press, Cambridge, 1989, pp 126 134

[4] Verhulst AGH, Cnossen G, Fünfschilling F, Schadt M. A wide-viewing-angle video display based on the deformed-helix ferroelectric liquid-crystal effect and a diode active matrix. Journal of the SID 1995; 3/3: 133-138

# A New Stochastic Learning Algorithm for Analog Hardware Implementation

Massimo Conti, Simone Orcioni, Claudio Turchetti

Department of Electronics, University of Ancona, Ancona, Italy

e_mail: max@eealab.unian.it

## Abstract

This paper presents a new stochastic learning algorithm suitable for analog implementation. The Neural Network is partitioned into subnetworks and learning is applied to each subnet in turn. Numerical simulations show an improvement in learning accuracy and a less critical dependence on noise amplitude and annealing parameters. The capability of the algorithm to reduce the sensitivity of the network to weight variation is investigated. The hardware implementation of the algorithm in an analog neural network shows a reduction of 75% in the area occupied by the learning circuitry with respect to a possible implementation without partition in subnetwoks.

## 1  Introduction

Learning algorithm is critical in digital or analog hardware implementation of Neural Networks due to the complex operations required. Some algorithms such as Backpropagation, in which the weight changes depend on gradient of the error, fails when the error objective function has a great number of local minima, as in general occur in practical applications. Many proposed solutions use a perturbation inserted in the network in order to get out from local minima. The introduction of noise during learning allows the solution to reach the global optimum (some examples of stochastic learning are: simulated annealing, introduction of weight noise during training[1], Random Weight Change (or Brownian weight movement) [2]). Furthermore it has been shown that training with noise may improve the generalization capability of the network and reduce the negative effects of the weight variations after learning [1].

Usually all the weights of a neural network are adjusted at the same time during learning. In this work the network is subdivided in a fixed number of neurons to form some clusters of neurons and learning is applied to one cluster at a time. The topology of the network is fixed and does not change during learning.

## 2  Random Weight Change

In the following we will refer to a learning algorithm suitable for analog implementation, called Random Weight Change (RWC) [2], based on the well known Brownian motion equation, defined by a couple of first order differential equations:

$$\begin{cases} \dfrac{dw(t)}{dt} = \mu v(t) \\[2mm] \dfrac{dv(t)}{dt} = \begin{cases} \eta[n(t)-v(t)] & c(t) > 0 \\ 0 & c(t) \le 0 \end{cases} \quad with \quad c(t) = u\left( \dfrac{dE(w(t))}{dt} \right) \\[4mm] v(t) = 0 \quad for \; |E(t)| < \varepsilon \end{cases} \qquad (1)$$

where $w$ is the weight vector, and $n(t)$ is a random process vector, whose components are statistically uncorrelated with zero mean and standard deviation $\sigma$ vanishing as $t \to \infty$, $u(.\,)$ is the step function, $E(t)$ is the error, $\eta$ and $\mu$ are learning parameters. When the trajectory of $w(t)$ proceeds in a region where the energy increases, the equation becomes the Langevin's equation describing Brownian motion. Conversely, when the error decreases, that is the trajectory is moving on the right direction, the same direction is maintained.

The advantages of a stochastic learning algorithm (such as Simulated Annealing or RWC) with respect to gradient based algorithms (BackPropagation) are the ability to avoid local minima of the error function and the low computational complexity, which leads to a simplification in hardware implementation. On the other hand convergence speed is usually lower.

Although the convergence property of stochastic algorithms such as Simulated Annealing has been mathematically proved, an accurate tuning of learning parameters, for example the annealing function, is in practice necessary to reach a good solution in a short time. Annealing function must be chosen in order to reduce noise variance in dependence of time and error, that is

$$\sigma^2_n(t) \to 0, \qquad as \; m \to \infty \quad or \quad E(m) \to 0;$$

where $E$ is the error and $m$ is the iteration number. On the basis of many simulations performed we concluded that noise variance must have an upper and an lower bound in order to ensure a good stability of the learning algorithm. In fact, if the noise amplitude is too high the algorithm will diverge, on the other hand it is convenient that noise is zero only at the end of learning. Hence the noise standard deviation chosen has the following dependence on iterations and error:

$$\sigma_n(E,m) = \frac{a}{1 + h_E \exp(\alpha_E(E - E_0)) + h_m \exp(\alpha_m(1/m - 1/m_0))} \qquad (2)$$

where $h_E, \alpha_E, h_m, \alpha_m, m_0$ are parameters which control the annealing speed.

## 3   Cluster Random Weight Change

We expect that the probability of finding the correct search direction decreases as the number of weights increases. Furthermore, the complexity of the optimization grows exponentially with the dimension of the search space.

In this work we propose a cyclic learning of fixed weight clusters, called Cluster Random Weight Change (CRWC). The network is subdivided in a fixed number of neurons to form some groups or clusters of neurons, as shown in Fig. 1. Learning is applied to one cluster at a time for a fixed number of steps, keeping the other weights fixed to the value they reached at the end of their learning time. This cyclic procedure is repeated until the error is acceptable. In this way in each step the dimension of the search space is not too big, hence the annealing parameters is not so critic. The error is calculated as the difference between the desired output and the output of the complete network.

In the following a brief description of the algorithm used is reported:

```
m=0
while ( E < E_desired and  m < itermax)
    for j=1 to k   /* for each cluster */
        i=0
        E_average=0
        while ( i< iter )
            calculate error E
            E_average=E+E_average
            update j-th cluster of weights  using RWC
            m=m+1
            i=i+1
        end while
        E_average=E_average/iter
        update noise variance σ_n²(E_average,m)
    end for
end while
```

Where k is the number of clusters, M the number of weights for each cluster, N=k*M the total number of weights, iter is the number of iteration for each cluster.

The following considerations must be taken into account in the choice of the dimension of each cluster (M) and the number of iterations for each cluster (iter):

1) learning complexity for each cluster increases exponentially with M;
2) many iterations may be wasted for a cluster for which the optimum value has been already reached (this may happen when M is too low and iter is too high);
3) the hardware complexity of the learning circuit increases by increasing M, while the complexity of the multiplexer which control the distribution of the learning current to each cluster increases with the number of clusters k.

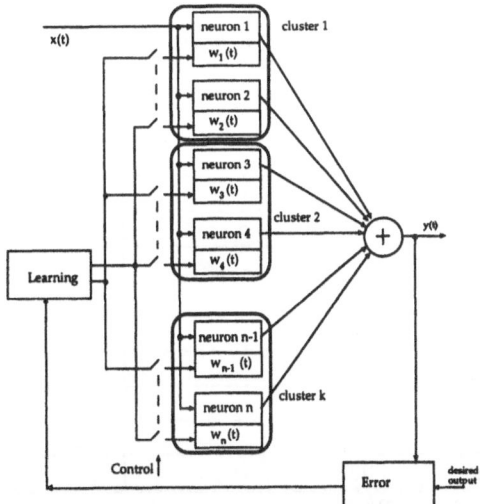

Fig.1 Example of Cluster Network partition

# 4 Results: Numerical Simulations

The proposed algorithm is general and can be applied to many optimization problems such as learning of different types of Neural Networks. In the following example it has been applied to a class of Neural Networks called Approximate Identity Neural Networks (AINN) [3] whose input-output relationship is:

$$\Lambda(x,w) = \sum_{j=1}^{N} c_j \sum_{i=1}^{P} \left\{ \tanh\left( \frac{n_{ij}(x_i - t_i) + \sigma_{ij}}{2} \right) - \tanh\left( \frac{n_{ij}(x_i - t_i) - \sigma_{ij}}{2} \right) \right\}$$

where $w=\{n,t,\sigma\} \in \Re^r$ is the weight vector, $x \in \Re^P$ the input vector.

The example is the training of an AINN with a one dimensional input. The simulations have been performed with Matlab. The desired functions are:

(1) $f = 0.6\exp(-10x^2)$; (2) $f = 0.6\sin(\pi x)$; (3) $f = 0.6\sin(\pi x/2)\cos(\pi x/2)$

Two examples of learning are reported in Fig. 2 for the two algorithms. It can be seen that in general CRWC reaches low values of the error after an higher number of iterations, but does not escape far from the good results obtained like RWC does. The same consideration can be drawn from the results shown in Table 1 which reports the mean value (obtained over 15 simulations with different random sequences) of the error reached by the two algorithms.

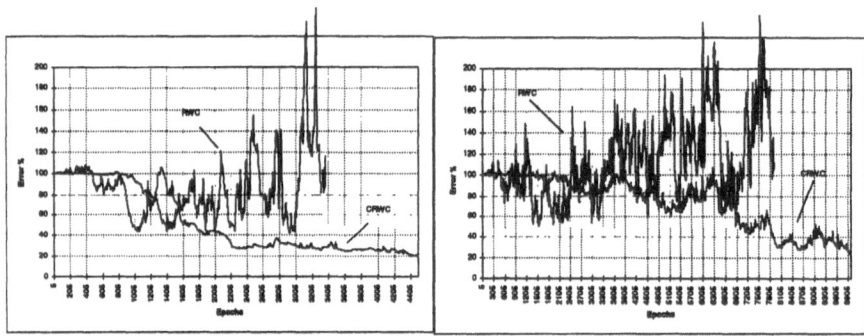

Fig.2 Error during learning with CRWC or RWC for the desired function (2) and (3).

| desired function | Algorithm | Number of neurons | Number of clusters | mean value of Error % | mean value of the iterations |
|---|---|---|---|---|---|
| 1 | CRWC | 2 | 2 | 5.0 | 3717 |
| 1 | RWC | 2 | - | 10.5 | 6731 |
| 1 | CRWC | 6 | 2 | 5.5 | 6453 |
| 1 | RWC | 6 | - | 6 | 6643 |
| 1 | CRWC | 8 | 4 | 5.2 | 6508 |
| 1 | RWC | 8 | - | 12 | 3788 |
| 2 | CRWC | 12 | 3 | 11.8 | 8150 |
| 2 | RWC | 12 | - | 13.4 | 7100 |
| 3 | CRWC | 12 | 3 | 15.85 | 3792 |
| 3 | RWC | 12 | - | 21.3 | 9105 |

Table 1 (iter=100, $\eta$=1, $\mu$=0.5, a=1, $h_g$=5, $h_n$=5, $\alpha_g$=10, $\alpha_n$=10).

# 5 Sensitivity to weight variations

The behavior of a Neural Network implemented in hardware differs from that predicted by mathematical modeling due to circuit nonidealities and technological random tolerances of parameters. In particular these random variations can affect the expected performances of the networks especially when they are implemented with analog circuits and/or off chip learning is adopted so that the settled weights differ from those numerically computed. Some works have been recently presented [1] showing that the introduction of noise during learning may reduce the sensitivity of the output of the neural network to weight variation at the end of learning and it may increase the generalization capability of the network.

In this Section the capability of CRWC to reduce the sensitivity to weight variation is investigated. Two networks with 16 neurons have been trained with CRWC and RWC to approximate a sinusoid. The algorithms have been stopped when they reached approximately the same value of the error $E_0$, in order to make a comparison between the sensitivity of the two networks. At the end of learning perturbations are added to the value $w_0$ of the weights following a random sequence with uniform distribution in the range $[-\delta_n, +\delta_n]$. The mean value $E_m$ and the standard deviation $\sigma_E$ of the error has been calculated and reported in Table 2 for different values of $\delta_n$. The results show a lower sensitivity to weight variations for the network trained with CRWC.

| $\delta_n$ | $E_0(w_0)$ % | | $E_m$ % | | $\sigma_E$ | |
|---|---|---|---|---|---|---|
| | CRWC | RWC | CRWC | RWC | CRWC | RWC |
| 0.003 | 17 | 17.4 | 16.8 | 21 | 0.02 | 0.44 |
| 0.005 | 17 | 17.4 | 17.4 | 23.4 | 0.07 | 1.02 |
| 0.01 | 17 | 17.4 | 19.2 | 29.5 | 0.27 | 3.85 |

Table 2

# 6 Experimental Results: On-Chip Learning

In previous Sections we noticed that clustering may increase or at least maintain the same performances of learning. An evident advantage of this technique is the reduction of the chip area occupancy in the case of on-chip learning.

We implemented cluster learning on a chip, named AINN296 [4], fabricated in the ES2 1.0μm CMOS technology. The network implemented has a 2-dimension input vector, 16 neurons, 112 weights, each one with a storage circuit, 28 internal uncorrelated noise sources.

Since the CRWC is a stochastic algorithm, a noise source is required by each couple of OTAs, and all sources should be uncorrelated, to make the random component added to a weight independent from all the others. Therefore, we made use of a shift register generating a maximum-length Pseudo Random Bit Sequence, and XOR gates to produce 28 versions of the original sequence, properly shifted to make them practically uncorrelated. By adding gain controlled low pass filters to each digital output, 28 uncorrelated analog noise generators have been obtained. Noise amplitude is externally controlled in order to reduce noise variance as required by the learning algorithm to reach convergence.

The internal structure of the chip is evidenced in the photo of Fig.3. The various parts have been arranged in a modular way in order to obtain a 4x4 array of neurons, whose weights, 7 for each neuron, are controlled during the training phase by the learning circuitry in the top. This circuitry implements the CRWC algorithm, and contains 28 couples of OTA-based integrators, each couple being fed by a noise source and providing a weight-change line for the array. An externally piloted row selection circuit ("decoder" block) enables for learning only one row at a time, i.e. it connects the weight nodes of that row to the corresponding learning lines, while values of the other weights are kept stored in a multistable (5-value) circuit. The multistable system is obtained by connecting in parallel a negative resistance circuit with a 5 level current quantiser. In this way, 28 lines can be shared by all the rows in the array, thus avoiding the need of 112 learning blocks, one for each weight. In this example (4 clusters with 4 neurons each) a reduction of 75% in the area occupied is obtained for the learning part. Preliminary measurements show acceptable performances for each single block of the chip implemented.

## 7   Conclusions

An improvement in convergence stability is reached by partitioning a Neural Network and changing only one cluster at a time during learning.

A reduction in the area required in an IC hardware implementation is an important feature of the learning algorithm proposed.

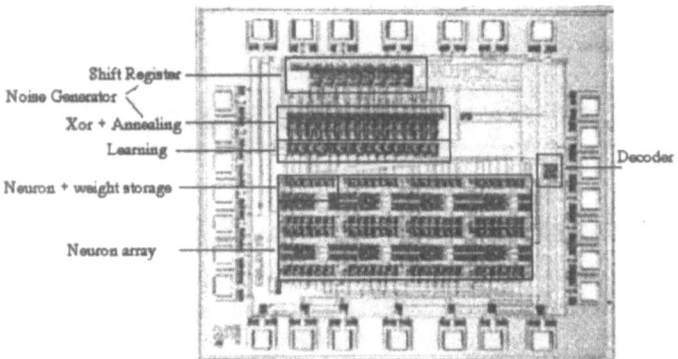

Fig. 3 Photo and internal structure of the chip AINN296.

## References

[1] A.F.Murray,P.J.Edwards, "Enanched MLP Performance and Fault Tolerance Resulting from Synaptic Weigth Noise During Training", IEEE Transactions on Neural Networks, Vol.5,No.5, pp.792-802, September 1994.

[2] M.R.Belli, M.Conti, C.Turchetti,"Analog Brownian Weight Movement for Learning of Artificial Neural Networks", Proc. of ESANN'95, Brussels, p.75-80, April 19-21 1995.

[3] M.Conti, S.Orcioni, C.Turchetti, "A class of neural networks based on approximate identity for analog IC's hardware implementation", IEICE Trans. on Fundamentals, Japan, Vol.E77-A, n.6, pp. 1069-1079, June 1994.

[4] M.Conti, G.Guaitini, C.Turchetti, "A CMOS analog Neuro-chip with stochastic learning and multilevel weight storage", Proc. of ISCAS97, June 1997, Hong Kong.

# An Analog Neural Signal Processor for Embedded Applications*

Mario Costa, Davide Palmisano and Eros Pasero[t]

Dipartimento di Elettronica, Politecnico di Torino

corso Duca degli Abruzzi, 24 10129 TORINO, ITALY

### Abstract

An analog signal processor is developed and implemented using on-chip non-volatile analog weight storage. Both synapses and neurons are small size devices (about 6500 $\mu m^2$ and 18500 $\mu m^2$, respectively) and can easily be arranged to create large networks on chip. An extremely modular system can be obtained to build multi layer networks of various size and topology suitable to be employed in several applications as non-linear control system. A special programming and erasing procedure based on Fowler-Nordhein tunneling mechanism is developed in order to get the required resolutions on synaptical weights.

## 1 Introduction

The use of neural networks as an engineering tool is widely accepted in many fields such as non linear processing and control systems. Though analog hardware implementations of artificial neural networks are really attractive because they can be easily interfaced to physical systems with no A/D, D/A and sampling devices, only a few analog neural integrated circuits are by now available [1], [2]. This is because the requirements of efficient analog memories, linearity for synapses and large number of computing units are hard to be satisfied with today technology.

Analog weights storage is performed by means of floating gate transistors. The high amount of time required to program these devices makes an on-chip training procedure quite unefficient. Training is performed off-chip based on I/O tables describing the behavior of real devices [3]; weight values are then downloaded to chip according to the training results. The neural chip becomes then an analog processor whose weights play a role similar to that of software programs for a DSP. To overcome mismatches and degradations for stored quantities a fine weight tuning with chip-in-the-loop must be done after training [3].

We will first show how flash transistors can be used and programmed as analog memories and then we will describe basic circuits for synapses and neurons.

---

*This work was partially funded by MURST Progetto 40% "Microarchitetture Neurali"

[t]email costa@polito.it, palmy@neuronica.polito.it, pasero@polito.it

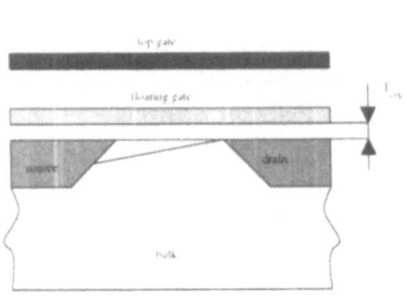

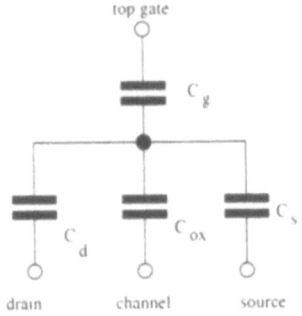

Figure 1: Structure of a flash transistor

Figure 2: Capacitive equivalent model of a flash cell

## 2   Flash transistors as analog memories

Flash transistors are double gate fully overlapped transistors as shown in figure 1 [4]. An equivalent model based on the capacitive coupling between all electrodes can be deduced, as shown in figure 2 [5]. A common way to analyze flash transistors is to consider the floating gate as the gate of a classical MOS transistor. Its saturation current is

$$I_D = \mu C_{OX} \frac{W}{L} \left[ \frac{C_G V_{GS} + C_D V_{DS} - V_{TH} (C_{OX} + C_D + C_G + C_S)}{0.5 C_{OX} + C_D + C_G + C_S} \right]^2 \quad (1)$$

where all capacitors are indicated in figure 2. Threshold voltage $V_{TH}$ can be modified by injecting carriers from channel to floating gate by Fowler-Nordheim tunneling or by channel hot electrons injection. Apart from charge leakage phenomena, the floating gate is isolated from any other electrode. Thus the imposed threshold voltage will stay unchanged unless a reverse process occurs. A resulting precision for threshold voltage of about 50 mV was measured for a period of several months up to two years depending on the technology used [6]. We make use of double slope ramps [4] to force a tunneling current described by the Fowler-Nordheim expression [7] both for writing and erasing the cell. To write the cell a 7 V ÷ 12 V 10 ms programming ramp is applied to the top gate while both drain and source are grounded. To erase the cell a 12.5 V 100 ms ramp is applied to the drain while both source and gate are grounded.

## 3   The chip architecture

We built an analog asynchronous neural chip using flash EEPROM devices for non-volatile analog weight storage with a 0.7 $\mu$m double polysilicon process. The microphotography of a 4 mm$^2$ reduced test version of the chip is shown in figure 3 and a conceptual block diagram is shown in figure 4. Sixteen neurons with sixteen synapses each are present on the chip. Each synapsis provides an output current as a function of the corresponding input and weight values. A training algorithm based on RPROP [8] was developed to work with both measured or simulated I/O data for such synapses [3] giving good results in spite

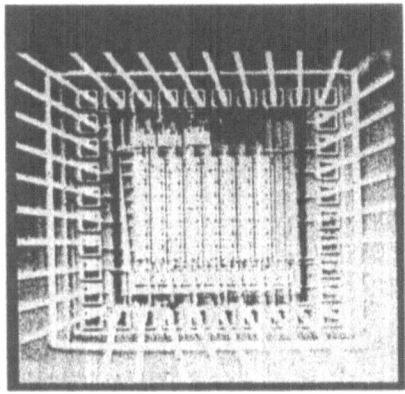

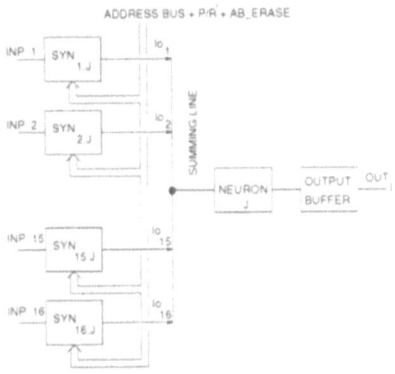

Figure 3: Test chip microphotography

Figure 4: Block diagram of a neuron with synaptical connections

of their inherent non-linearity. Weights are related to the threshold voltage of a pair of flash transistors and can be modified by injecting or extracting electrons from the floating gate [6].

All the synaptical currents are summed into the neuron input node. Neurons give out a voltage in response to the input current according to the transfer function (the activation function). Synapses are arranged as an array. An address bus with a dedicated static decoder is used to address each synaptical cell. It is possible to build a large variety of feed-forward networks by enabling or disabling each synaptical connection as required by the chosen topology and by suitably connecting some input and output pins together to create multi-layer architectures. Larger networks can be obtained by cascading more chips.

Output signals coming from neurons have a dynamic range compatible with the input range of synapses to achieve a modular structure. So, input and output voltages range from 0 V up to 5 V, whereas threshold voltage swing can be increased by injecting electrons from floating gate in order to induce a channel, as in a depletion device [9]. A rail-to-rail output buffer is connected to each neuron to properly drive the load . Buffer features, such as slew rate, settling time and gain-bandwidth product, come from a trade-off between performance and power consumption. If greater load driving capabilities are required, off-chip buffers can be used as well.

The circuit has two different operating modes, named respectively programming and reading mode. In the former flash transistors threshold voltages are imposed to configure the desired neural architecture whereas in the latter the circuit processes the inputs and gives the neuron outputs. By using two control signals named $P/\bar{R}$ and $AB\_ERASE$ and by imposing the proper input voltages on the address bus and input lines it is possible to select the operation mode.

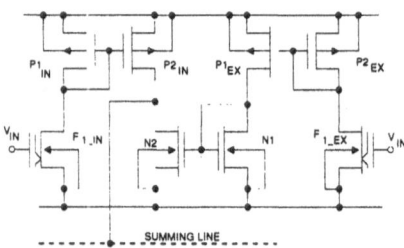

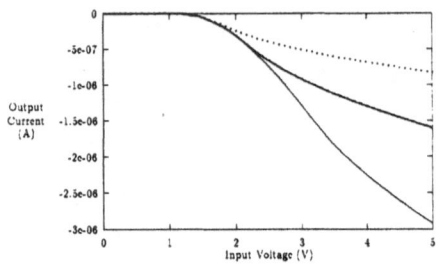

Figure 5: Simplified schematic of the synapse

Figure 6: Synaptical current as function of input and weight

# 4 The synapses

Synapses give an output current function both of the input voltage and of the weight. As it is shown in figure 5, every synapsis is made of two parts named respectively inhibitory side ($F_{1_{IN}}$ and $P_{1_{IN}} - P_{2_{IN}}$) and excitatory side ($F_{1_{BX}}$ and $P_{1_{BX}} - P_{2_{BX}}$). The output contribution of each side are named inhibitory and excitatory currents. The resulting current of the synapsis is the sum of both two contributions with a mirroring of the excitatory term performed by the unitary NMOS current mirror ($N_1$ and $N_2$) in order to make a real subtraction. The output current of each synapsis is then given, according to the flash cell model [10] and [9], by

$$I_{OUT} = \mu C_{OX} \frac{W}{L} A k_1 \left( \frac{C_G V_{IN} + C_D V_{DS} - V_{TH_1} (C_{OX} + C_D + C_G)}{0.5 C_{OX} + C_D + C_G} \right)^2$$

$$- \mu C_{OX} \frac{W}{L} A k_2 \left( \frac{C_G V_{IN} + C_D V_{DS} - V_{TH_2} (C_{OX} + C_D + C_G)}{0.5 C_{OX} + C_D + C_G} \right)^2$$

where A is the mirroring ratio and $k_1$ and $k_2$ are two quantities used to take into account the channel length modulation effect for each side. The output current of the synapsis is shown in figure 6.

By holding the summing line voltage at a constant value no mutual interactions between any pair of synapses exists [9]. To minimize fluctuations of summing line a reduced output current range is necessary and therefore a mirroring ratio of 24:1 is imposed for the PMOS current mirrors. A drain and gate enable transistor able to manage high voltages are added to each flash transistor in the synapsis to allow the writing and erasing operations.

# 5 Neuron Circuit

Neuron performs a twofold function. Firstly converts the sum of the synaptical currents into a voltage by applying a non-linear function (close to the classical sigmoidal one). Second it must hold the summing line voltage as close as possible to $\frac{V_{DD}}{2}$ in order to minimize mutual interactions between different synapses.

The neuron is shown in figure 7. It is a transresistance amplifier made by a CMOS inverter with a non-linear resistive feedback. As long as the neuron

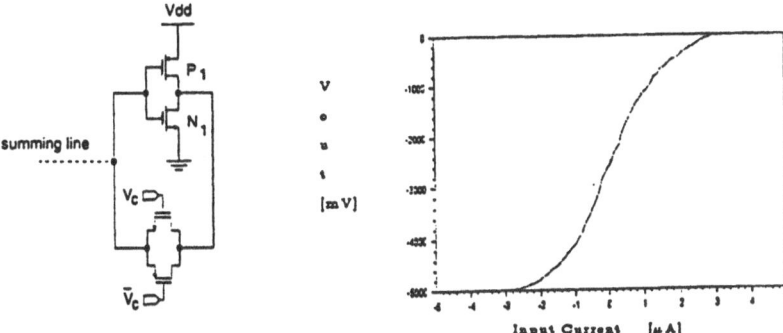

Figure 7: Simplified schematic of the neuron

Figure 8: Measured Neuron Transfer Function

works into its linearity range, the input node can be considered a virtual ground point in small signal analysis and therefore fluctuations are function of output voltage together open loop gain: the larger is the gain and the smaller will be the resulting fluctuations. When input current drives neuron in saturation it is no longer necessary to hold input line voltage because mutual interactions can now be neglected. Resistive feedback is designed so that only one synapsis can drive the neuron over its full dynamic range and is made by a couple of voltage controlled MOS transistors. Particular care is spent in drawing the layout to minimize offset and other parasitic effects for this device. The measured neuron transfer function is shown in figure 8 and main characteristics are reported in table 1.

Table 1: Measured Results ($C_L$=10 pF)

| | |
|---|---|
| power supply | 5 V |
| quiescent current | 7 mA |
| Common mode Input Range | $0.8 \div 5$ V |
| Output Voltage Swing | $10$ mV $\div 5$V |
| Neuron PSRR @ 250 kHz | 34 dB |
| Offset Voltage | 20 mV |
| Synaptical Max Output Current | 7 $\mu$A |
| Synapse area | 6524 $\mu m^2$ |
| Neuron area | 18520 $\mu m^2$ |

# 6    Conclusions

Neural signal processors are an interesting solution in many fields where off-chip training is possible. Analog implementations offer greater interfacing capabilities with respect to digital ones. By eliminating the assumption of linear working for synapses and by using floating gate devices for weight storage very compact devices and therefore high integration levels are allowed. Weights are mapped onto the threshold voltage of a flash transistor pair. Double slope ramps are used to write and erase synapses and experimental results showed how sufficient resolution on weights can be obtained without any feedback-based programming scheme. Weights can then be tuned by using local weight perturbation.

# References

[1] A. J. Montalvo, R. S. Gyurcsik, and J. J. Paulos, "An Analog VLSI Neural Network with On-Chip Perturbation Learning," *IEEE Journal of Solid-State Circuits*, vol. 32, no. 4, pp. 535–543, April 1997.

[2] G. Cauwenberghs, "An Analog VLSI Recurrent Neural Network Learning a Continuos-time Trajectory," *IEEE Transanctions on Neural Networks*, vol. 7, no. 2, pp. 346–361, July 1996.

[3] M. Costa, D. Palmisano, and E. Pasero, "An Analog Neuronless Reconfigurable Neural Network," in *ECS97 Electronics Circuits and Systems Conference*, V. Stopjanková, I. Mucha, and N. Frištacký, Eds., Bratislava (SK), September 1997, pp. 315–321.

[4] M. Costa, D. Palmisano, and E. Pasero, "NESP: a NEural Signal Processor," in *ISCAS95 1995 International Symposium on Circuits and Applied Systems*, IEEE, Ed., April 1995, vol. 3, pp. 2189–2193.

[5] A. Kolodny, S. T. K. Nieh, B. Eitan, and J. Shappir, "Analysis and Modelling of Floating-Gate EEPROM Cells," *IEEE Transanctions on Electron Devices*, vol. ED-33, no. 6, pp. 835–842, June 1986.

[6] C. K. Sin, A. Kramer, V. Hu, R. R. Chu, and P. K. Ko, "EEPROM as Analog Storage Devices with Particular Applications in Neural Networks," *IEEE Transanctions on Electron Devices*, vol. 39, no. 6, pp. 1410–1419, June 1992.

[7] C. Y. Wu and C. F. Chen, "Physical Model for Characterizing and Simulating a FLOTOX EEPROM Device," *Solid-State Electronics*, vol. 35, no. 5, pp. 705–716, May 1992.

[8] M. Riedmiller and H. Braun, "A Direct Adaptive Method for Faster Backpropagation Learning:the RPROP Algorithm.," in *Proc. ICNN93 International Conference on Neural Networks*, San Francisco, 1993, IEEE.

[9] D. Palmisano, *Progetto di Circuiti Neuronali Analogici*, Ph.D. thesis, Politecnico di Torino, Torino, (I), February 1998.

[10] A. Kramer, C. K. Sin, V. Hu, R. R. Chu, and P. K. Ko, "Compact EEPROM-based Weight Functions," in *Advances in Neural Information Processing Systems 3*, pp. 1001–1007. Morgan-Kaufmann, San Mateo, CA, 1991.

# The NeuroAccess System

Erich Schikuta, Christian Brunner and Christian Schultes

Institute for Applied Computer Science and Information Systems,
Department of Data Engineering, University of Vienna,
Rathausstr. 19/4, A-1010 Vienna, Austria

### Abstract

In this paper the *NeuroAccess* system is presented, an artificial neural
network simulator based on a relational database system. It provides a
novel approach for the conceptual and physical integration of neural networks into the relational model. An object oriented approach is followed
for description of the paradigm hierarchies, and it is shown, how it can
be mapped to the relational model. A handling scheme for the application data set, as input, output and training information, stored in the
relational database together with the neural networks is depicted.

## 1 Introduction

Many different systems artificial neural networks simulators (ANNS), tools for
the easy and software supported creation and administration of neural networks were developed in the last few years. System can be found both for
special types of neural networks only, e.g. Aspirin/MIGRAINES [Lei91], and
as a comprehensive tool for a variety of known network paradigms, as SNNS
[Z+92] to name only a few. Basically all these systems, reaching from highly
sophisticated interactive systems to programming language extensions, share
the same common problems,

- a proprietary software environment, which faces the user with the problem
  to cope with a new and/or complex tool,

- most of these systems present a stand-alone environment, which is not
  capable to interconnect to other software systems, and

- all these systems lack a generalized framework for handling data sets
  and neural networks homogeneously. During the training phase and the
  evaluation phase of a neural net the user has to feed the net with large
  amounts of data. Conventionally data sets are mostly supported via
  sequential files only and the definition of the input stream, output or
  target stream into a neural net is often extremely clumsy.

We see a solution to these problems by the embedding of neural networks as
basic elements into the well-known and standardized environment of database
systems (see [Sch96a] for object-oriented systems).

In this paper we present the *NeuroAccess system*, a relational framework for embedding of neural network into an off-the-shelf relational database management system (DBMS). It is our objective to consider neural networks as conventional data in the database system. From the logical point of view a neural network is a complex data value and can be stored as a normal data object.

The usage of a database system as an environment for neural networks provides both *qualitative* and *quantitative* advantages. The user has powerful tools and models at hand, like data definition and manipulation languages, report generators or transaction processing. These tools provide a unified framework for both handling neural networks and the input/output data streams of these networks. A homogeneous and comprehensive user interface is provided to the user. Modern database systems allow the administration of objects efficiently. This is provided by a 'smart' internal level of the system, which exploits well studied and well known data structures, access paths, etc. A whole bunch of further concepts is inherent to these systems, like models for transaction handling, recovery, multi-user capability, concurrent access etc. This places an unchallenged platform in speed and security for the definition and manipulation of large data sets at users disposal.

Our approach of *embedding* of neural networks into database systems follows an opposite direction compared to conventional approaches. We move the neural networks to the database systems, and not the data to the neural network simulators. This situation is depicted by figure 1.

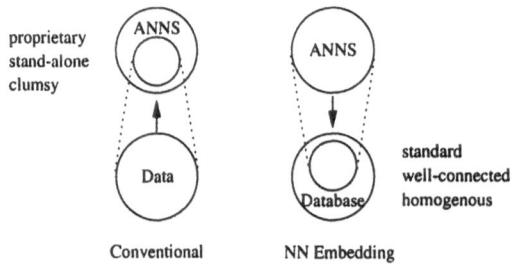

Figure 1: Conventional versus NN embedding approach

In the remainder of this paper we present our NeuroAccess System, describe the underlying relational data model, the mapping of the network paradigms to an object-oriented type hierarchy, the extensibility of the system and the easy handling of large data sets by the novel datastream concept .

# 2 The NeuroAccess System

The NeuroAccess system was realized using the MS Access database system running on Windows95/NT[1]. The relations of the database system are used to model the static components and the Visual Basic application language for the dynamic components. These application code is stored in the database together with the static information, grouped accordingly to the specific network paradigm. Until now NeuroAccess supports 10 network paradigms, which are Backpropagation (3 different variations), Counterpropagation, Hopfield, Boltzmann, Art1, ART2, Jordan, and Elman networks. In figure 2 a screen shot of the NeuroAccess system is given. The main window for the choice of the paradigm in focus, the object window of a backpropagation network and the respective structure and error curve window of a training phase can be identified.

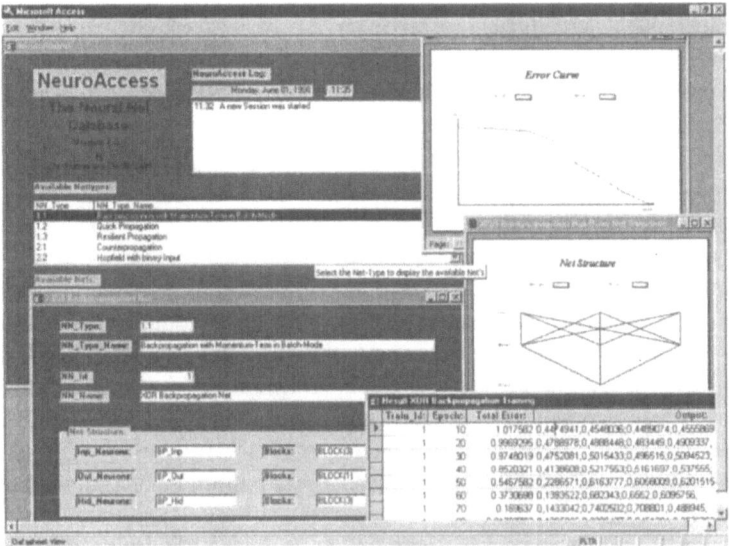

Figure 2: NeuroAccess screens

## 2.1 The Database Model

A neural network object is defined by its static and dynamic components. These components are dependent on the appropriate network paradigm. Different paradigms show different static and dynamic properties.

In the following we use the object-oriented modeling approach due to its expressive power. We show in section 2.2 how the object-oriented structures

---

[1] MS Access and Windows95/NT are trademarks of the Microsoft Cooperation

are mapped to the relational model, which is the model in focus.

The static neural network components comprise all information stored in relations, as neural network specific parameters, links, training objects, evaluation objects, etc.

The 'NeuralNet' type is a sub-type of the general object type of the database system. Sub-types of this 'NeuralNet' type can be classified into specialized neural network types according to their network paradigm. The network paradigm is defined by a specialization, a sub-type of 'NeuralNet'. This sub-type (which inherits all characteristics of its super-type) provides the specific and necessary attributes dependent on the network paradigm. Combined with the definition of the paradigm are the dynamics (the dynamic behavior) of the network.

The dynamic components of the neural network object are the typical operations on neural network, the training and evaluation phase. The algorithms for these phases are realized by routines coded in the internal database application code (this is in our NeuroAccess implementation Visual Basic). Additionally these routines keep certain consistency assertions on the static components after execution of specific phases, as insertion of link weights after a training, results after an evaluation phase, and so on. Training and evaluation sessions are represented by training and evaluation objects according to the proposed NeuroAccess data base framework.

## 2.2 Type Hierarchy Mapping

To provide the expressiveness and flexibility of the object oriented framework in a relational system a mapping from the object-oriented neural network specification to the relational data modeling environment has to be defined. In the relational model all specific neural network types along a type hierarchy are represented by single relations. The relationship between these types is expressed in our approach by a hierarchical numbering scheme stored together with the relation. Different types on the same hierarchy are mapped to different numbers. Subtypes are expressed by the number of their supertype, a dot, and the specific number of the respective subtype, similar to a categorization scheme. This approach is depicted by figure 3. The hierarchy paths are attributed with the respective number codes.

## 2.3 Extensibility

An important aspect is the extensibility of the system. This is reached by a modularized paradigm approach. New modules have to follow a specified programming style to make it possible to integrate them easily into the NeuroAccess environment. Thus users have the possibility to shape the system to their needs by changing existing or adding new paradigms easily. All these implementations can be done without leaving the comfortable database environment.

According to the new paradigm the user has to provide specific database tables and forms for the neural network, the training, and the evaluation objects.

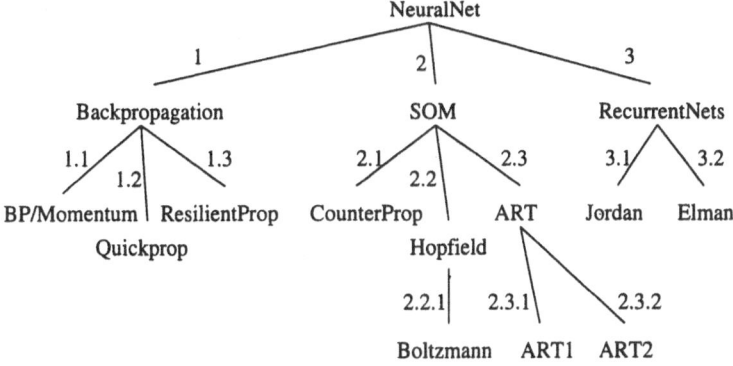

Figure 3: Type hierarchy mapping

Further the paradigm has to be established as a new module in the database realized in the specific database language (a form of Visual Basic). Thus a standardized interface was defined which eases this process for the user. This modularization allows also different versions of the NeuroAccess supporting different network paradigms.

## 2.4 Datastream Concept

One powerfull component of the NeuroAccess system is its flexible datastream concept, a novel approach to handle large data sets in the context of an neural netwrok simulator.

There are three ways to specify data sets (e.g. input streams) in the NeuroAccess system. These possibilities are

- taxative data specification,

- internal data stream declaration, and

- external data stream declaration.

The first possibility, the taxative specification, is to specify the data values explicitly. This is very useful to create small data sets for simple neural network applications. But it is not comfortable to enter large data streams to solve more complex problems.

So the well-known apparatus of the SQL database manipulation language is used to create suitable input streams. It is possible to describe the desired data stream by SQL statements which will be evaluated to real data values. So it is easily possible to use real data sets, which are stored in data base tables, as training set for neural networks. For example, the statement `select * from TrainData` can provide training data to a training object. This method can be applied for the specification of the input and target patterns during the training phase and equally well for the input patterns during the evaluation phase. This method is not only applicable to internal data stream declarations, which means accesses to locally stored data only. Even more this approach allows to input data of external data sources into the NeuroAccess system, which can be specified by the ODBC interface of the Access system. We call this approach the external data stream specification. These pattern sets are described by their location on the net and can be accessed by specification of the description of the SQL statement and the necessary access privileges. This provides an unrestricted access and integration of data sources available via the network.

Thus the same tool can both be used for administration and analysis of the stored information. So it is easily possible to use 'real world' data sets as training set for neural networks and to analyze other (or the same) data with trained networks.

# 3  Conclusions and Future Research

We presented in this paper the NeuroAccess system, a relational model for the embedding of neural networks into data base systems. It delivers an extensible, comfortable, and powerful neural network tool embedded into the relational MS Access database system. This approach provides a homogeneous and natural environment for the administration and handling of neural networks to the user.

In the near future work on a Java frontend will commence to access the stored information via the World-Wide-Web. Further parallelization approaches similar to [Sch96b] for inter- and intra-operation parallelization of the neural network dynamics (training and evaluation) using the Java RMI interface are planned.

# References

[Lei91]   R.R. Leighton. The aspirin/migraines software tools user's manual. Technical Report MP-91W00050, The MITRE Corporation, McLean, Virginia, 1991.

[Sch96a]  E. Schikuta. Neudb'95: An sql based neural network environment. In Shun-ichimeri et al., editors, *Progress in Neural Information Processing, Proc. Int. Conf. on Neural Information Processing, ICONIP'96*, pages 1033–1038, Hong Kong, September 1996. Springer-Verlag, Singapore.

[Sch96b]  E. Schikuta. Parallelism in the neudb system. In *Proc. 2nd Int. Conf. on Massively Parallel Computing Systems*, pages 122–129, Ischia, Italy, May 1996. IEEE Computer Society Press.

[Z+92]    A. Zell et al. Snns, stuttgart neural network simulator, user manual. Technical Report 3/92, Univ. Stuttgart, 1992.

# Recognizing Handwritten Digits with a Dedicated Analog VLSI Feature Extractor

G.M. Bo, D.D. Caviglia, and M. Valle

Department of Biophysical and Electronic Engineering
University of Genoa, Via all'Opera Pia 11/A, 16145 Genova, Italy
ph:+39 10 3532287, fax: +39 10 3532777, e-mail: gian_bo@dibe.unige.it

### Abstract

The classification of handwritten digits through an analog feature extractor chip and a neural classifier is discussed in this paper. The chip implements a feature extraction algorithm onto analog circuits; it extracts a set of 112 features from the input character (32×24 binary pixel matrix). The features, coded by current signals, are given in input to a neural classifier which performs the recognition task. The chip validation results are reported: a set of handwritten digits have been classified by a neural network implemented by a software simulator. The resulting classification error rate has been successfully compared with the ones obtained by a high level model of the chip and to those obtained with other techniques reported in the literature.

## 1 Introduction

Among the different techniques for handwritten character recognition reported in the literature (see among others [1]), we consider in this paper those based on the following processing steps:

- the characters are pre-processed to extract a set of features;

- the features feed a classifier that performs the recognition task.

The reduction of information redundancy during the pre-processing step is the main advantage of this approach. An effective classification and reduced error rates can be achieved.

Since software implementations of feature extraction algorithms may require great computational power, dedicated hardware implementations have to be taken into account for real time target applications (i.e. post automation, check reading, etc.). In particular, analog VLSI implementations will be valuable if high processing speed and low power consumption are pursued.

In this paper the classification of handwritten digits through an analog VLSI feature extractor chip based in the algorithm discussed in [2] is presented. The chip extracts from handwritten characters, represented by 32×24 binary pixels matrices, a set of 112 features coded by current signals. Such features are given in input to a software implemented neural classifier (a Multi Layer

Perceptron network, MLP [6]) performing the recognition task. The chip has been fabricated and succesfully tested.

This paper is organized as follows. In Section 2 a brief description of the feature extraction algorithm is given. In Section 3 the analog VLSI implementation of the feature extractor is presented. In Section 4 the classification results obtained from the recognition of handwritten digits from NIST 19 database are presented and discussed. The conclusion are drawn in Section 5.

## 2  The Feature Extraction Algorithm

The algorithm rationale [2] consists of extracting local information: the character matrix (i.e. $32 \times 24$ binary pixels) is partitioned into 16 sub-matrices of $9 \times 7$ pixels each, partially overlapped by one or two columns and/or rows. On each sub-matrix seven pattern recognition operators are applied to obtain a set of seven features: 112 features per character are thus obtained.

Let us consider a generic sub-matrix: the first feature, called smoothing (SM), evaluates the density of active pixels. It is defined as the weighted sum of pixel values in the sub-matrix (see [2] for details).

The other features evaluate to which extent the sub-matrix shapes match six major directions. In particular the horizontal, vertical, $\pm 45$ deg, and $\pm 60$ deg directions have been taken into account (HR, VR, P45, M45, P60, and M60 respectively). For each direction, a set of pixel paths have been defined: the sums of active pixels in each path are computed and an exponential function is applied to the sums. The exponential function is able to emphasize the most distinctive shape of the character. Finally, all the contributions are summed up to get the feature value [2].

## 3  Analog VLSI Implementation of the Feature Extractor

The feature extractor algorithm has been directly mapped onto analog circuits. The pixel information is coded by current signals, then the sum operator is easily implemented. A proper circuit based on a MOS transistor biased in the weak inversion region implements the exponential operator. More details about the analog VLSI implementation of the feature extraction algorithm can be found in [4].

The chip photograph and characteristics are shown in Fig. 1. One can see the modules corresponding to the 16 sub-matrices placed on a $4 \times 4$ array. The circuit has been fabricated through the EUROCHIP Service by using the ATMEL ES2 ECPD10 technology (a CMOS digital process with 1.0 $\mu$m channel length). The chip has 156 pads, and occupies an area of $7.4 \times 9$ mm$^2$. The measured average power consumption is 60 mW, and the character recognition throughput is of about 140 KChar/s.

In Fig. 2 the measured feature values obtained from the processing of three synthetic plus one real characters are reported: in each figure the character and

| Technology process | ES2 CMOS 1.0 $\mu$m |
|---|---|
| Character size | 32×24 binary pixels |
| Input data size | 32 bits |
| Throughput | 140 KChar/s |
| Chip size | 7.4×9 mm$^2$ |
| Transistor count | 28000 |
| Power consumption | 60 mW |

Figure 1: The chip photograph and characteristics.

the results of the processing are drawn. The feature values are coded by grey levels: the minimum and maximum values are coded by black and white levels respectively.

The character of Fig. 2.a has a horizontal shape on the top (sub-matrices 1 to 8) and a vertical shape on the bottom (sub-matrices 9 to 16): the SM, P45, M45, P60, and M60 have consequently small values. The HR feature has high values at the top, and the VR feature has high values at the bottom, according to the character shape.

Similar considerations can be pointed out regarding Fig. 2.b and Fig. 2.c. In Fig. 2.b the character has a +45 deg shape at the top, and a +60 deg shape at the bottom. In Fig. 2.c the character has a -45 deg shape on the top, and a -60 deg shape on the bottom. The results obtained from the feature extraction process are in accordance with the character shapes.

In Fig 2.d the features obtained from a real character are reported: the character is a "3" (NIST 19 database [3]). Again, the feature values are in accordance with the shape of the input character.

# 4 Validation of the Optical Recognition System

## 4.1 Validation Set-Up

The NIST 19 database [3] has been used to validate the feature extractor chip. It contains 402953 samples of segmented handwritten digits, each one represented by a matrix of 128×128 binary pixels. From this database we extracted 3000 characters divided into two subsets:

- 2000 characters (200 per class) from the Handwriting Sample Forms (HSF) HSF_1, used as training set;

- 1000 characters (100 per class) from HSF_4, used as test set [5].

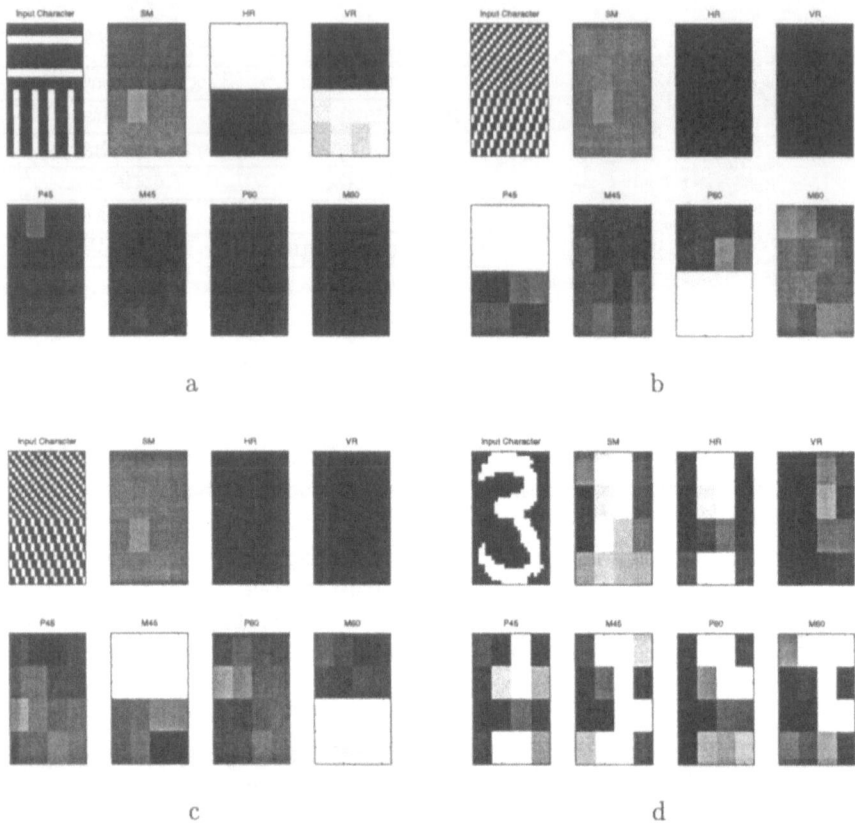

Figure 2: Chip measurement representations.

The 3000 characters have been normalized as 32×24 binary pixel matrices and given in input to the chip. An 8-bits ADC converts the output analog features (coded by currents) into digital values that vary in the range [-1,+1].

## 4.2 Simulation Results

The MLP network used for the classification task has 112 inputs (the number of features per character), 48 hidden neurons, and 10 output neurons (one per class).

The network has been trained by using the standard *by pattern* Back Propagation (BP) algorithm [6]. The classification task has been performed in two ways.

- With a rejection criterion: an input pattern will be classified if the difference between the maximum and second maximum of the network output values is greater than a predefined threshold, otherwise it will be rejected.

- Without a rejection criterion: all the patterns are classified.

In both cases, the classification is based on the highest output value.

In Fig. 3 the accuracy rate on the test set versus the rejection rate is shown. The MLP was trained with the whole training set (2000 patterns). The accuracy rate is defined as the ratio between the number of correct patterns and the number of accepted patterns (i.e. not rejected, classified anyway).

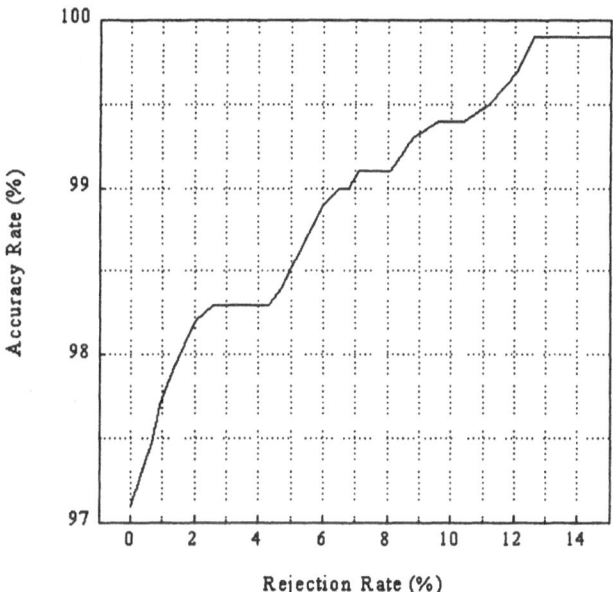

Figure 3: Accuracy rate vs rejection rate.

It is worth noting that an accuracy rate of 99% is obtained with a rejection rate of about 6.5% (93.5% of acceptance). An accuracy rate of 99.9% is obtained with a rejection rate of 13%.

These figures are satisfactory, taking into account even the small size of the training set and compare favourably to those discussed in [7].

## 5  Conclusion

The aim of our research is to develop an Optical Character Recognition system based on neural models and implemented by analog computational modules. In this paper the results obtained from the classification of handwritten digits through analog VLSI feature extractor chip have been presented and discussed. The chip extracts a set of 112 features based on local information from pixel matrices coding handwritten characters. The features are given in input to a software implemented neural classifier (a MLP network) which performs the recognition task.

The chip implementing the feature extractor module has been designed and fabricated: experimental results validate the chip functionality.

## Acknowledgements

The authors wish to thank F. Orengo and G. Curatolo for their help in performing the chip measurements.

# References

[1] S.N. Shrihari. *Recent Advances in Off-line Handwriting Recognition at CEDAR.* In Proc. of the Fifth Int. Workshop on Frontiers in Handwriting Recognition, IWFHR'96, pp. 1-15, 1996.

[2] D.D. Caviglia, M. Valle, A. Rossi, M. Vincentelli, G.M. Bo, P. Colangelo, P. Pedrazzi, and A.M. Colla. *Feature Extractor Circuit for Optical Character Recognition.* Electronics Letters, Vol. 30, No. 10, pp. 769-760, 1994.

[3] P.J. Grother. *NIST Special Database 19.* National Institute of Standards and Technology, 1995.

[4] G.M. Bo, D.D. Caviglia, and M. Valle. *An Analog VLSI Implementation of a Feature Extractor for Real Time Optical Character Reacognition .* IEEE Journal of Solid State Circuits, Vol. 33, No. 4, pp. 556-564, 1998.

[5] P.J. Grother. *Cross Validation Comparison of NIST OCR Databases.* In D.P. D'Amata Editor, volume 1906, SPIE, San Jose, 1993.

[6] J. Herzt, A. Krogh, and J.L. Palmer. *Introduction to the Theory of Neural Computation.* Addison-Wesley, 1989.

[7] S.B. Cho. *Neural-Netwok Classifiers for Recognizing Totally Unconstrained Handwritten Numerals.* IEEE Transacions on Neural Networks, Vol. 8, No. 1, pp. 43-53, 1997.

# Author Index